Stochastic Modeling and Mathematical Statistics

A Text for Statisticians and Quantitative Scientists

Logistic Regression Models
J.M. Hilbe

Richly Parameterized Linear Models: Additive, Time Series, and Spatial Models Using Random Effects
J.S. Hodges

Statistics for Epidemiology
N.P. Jewell

Stochastic Processes: An Introduction, Second Edition
P.W. Jones and P. Smith

The Theory of Linear Models
B. Jørgensen

Principles of Uncertainty
J.B. Kadane

Graphics for Statistics and Data Analysis with R
K.J. Keen

Mathematical Statistics
K. Knight

Nonparametric Methods in Statistics with SAS Applications
O. Korosteleva

Modeling and Analysis of Stochastic Systems, Second Edition
V.G. Kulkarni

Exercises and Solutions in Biostatistical Theory
L.L. Kupper, B.H. Neelon, and S.M. O'Brien

Exercises and Solutions in Statistical Theory
L.L. Kupper, B.H. Neelon, and S.M. O'Brien

Design and Analysis of Experiments with SAS
J. Lawson

A Course in Categorical Data Analysis
T. Leonard

Statistics for Accountants
S. Letchford

Introduction to the Theory of Statistical Inference
H. Liero and S. Zwanzig

Statistical Theory, Fourth Edition
B.W. Lindgren

Stationary Stochastic Processes: Theory and Applications
G. Lindgren

The BUGS Book: A Practical Introduction to Bayesian Analysis
D. Lunn, C. Jackson, N. Best, A. Thomas, and D. Spiegelhalter

Introduction to General and Generalized Linear Models
H. Madsen and P. Thyregod

Time Series Analysis
H. Madsen

Pólya Urn Models
H. Mahmoud

Randomization, Bootstrap and Monte Carlo Methods in Biology, Third Edition
B.F.J. Manly

Introduction to Randomized Controlled Clinical Trials, Second Edition
J.N.S. Matthews

Statistical Methods in Agriculture and Experimental Biology, Second Edition
R. Mead, R.N. Curnow, and A.M. Hasted

Statistics in Engineering: A Practical Approach
A.V. Metcalfe

Beyond ANOVA: Basics of Applied Statistics
R.G. Miller, Jr.

A Primer on Linear Models
J.F. Monahan

Applied Stochastic Modelling, Second Edition
B.J.T. Morgan

Elements of Simulation
B.J.T. Morgan

Probability: Methods and Measurement
A. O'Hagan

Introduction to Statistical Limit Theory
A.M. Polansky

Applied Bayesian Forecasting and Time Series Analysis
A. Pole, M. West, and J. Harrison

Statistics in Research and Development, Time Series: Modeling, Computation, and Inference
R. Prado and M. West

Introduction to Statistical Process Control
P. Qiu

Sampling Methodologies with Applications
P.S.R.S. Rao

A First Course in Linear Model Theory
N. Ravishanker and D.K. Dey

Essential Statistics, Fourth Edition
D.A.G. Rees

Stochastic Modeling and Mathematical Statistics: A Text for Statisticians and Quantitative Scientists
F.J. Samaniego

Statistical Methods for Spatial Data Analysis
O. Schabenberger and C.A. Gotway

Large Sample Methods in Statistics
P.K. Sen and J. da Motta Singer

Decision Analysis: A Bayesian Approach
J.Q. Smith

Analysis of Failure and Survival Data
P. J. Smith

Applied Statistics: Handbook of GENSTAT Analyses
E.J. Snell and H. Simpson

Applied Nonparametric Statistical Methods, Fourth Edition
P. Sprent and N.C. Smeeton

Data Driven Statistical Methods
P. Sprent

Generalized Linear Mixed Models: Modern Concepts, Methods and Applications
W. W. Stroup

Survival Analysis Using S: Analysis of Time-to-Event Data
M. Tableman and J.S. Kim

Applied Categorical and Count Data Analysis
W. Tang, H. He, and X.M. Tu

Elementary Applications of Probability Theory, Second Edition
H.C. Tuckwell

Introduction to Statistical Inference and Its Applications with R
M.W. Trosset

Understanding Advanced Statistical Methods
P.H. Westfall and K.S.S. Henning

Statistical Process Control: Theory and Practice, Third Edition
G.B. Wetherill and D.W. Brown

Generalized Additive Models: An Introduction with R
S. Wood

Epidemiology: Study Design and Data Analysis, Third Edition
M. Woodward

Experiments
B.S. Yandell

Texts in Statistical Science

Stochastic Modeling and Mathematical Statistics

A Text for Statisticians and Quantitative Scientists

Francisco J. Samaniego

University of California

Davis, USA

CRC Press
Taylor & Francis Group
Boca Raton London New York

CRC Press is an imprint of the
Taylor & Francis Group an **informa** business
A CHAPMAN & HALL BOOK

CRC Press
Taylor & Francis Group
6000 Broken Sound Parkway NW, Suite 300
Boca Raton, FL 33487-2742

Printed on acid-free paper
Version Date: 20150325

International Standard Book Number-13: 978-1-4665-6046-8 (Hardback)

Visit the Taylor & Francis Web site at
http://www.taylorandfrancis.com

and the CRC Press Web site at
http://www.crcpress.com

Dedication and Acknowledgments

It seems fitting that I dedicate this book to my students. I've had thousands of them over my long career, and I've learned a lot from them, from the genuine curiosity from which many questions came and from their growth as they confronted the intellectual challenges I set before them. The challenge of challenging them to think hard about new ideas has shaped who I am as a teacher. In particular, the many students I have taught in my probability and mathematical statistics courses over the years contributed mightily to the development of this book. I am grateful for their trust as I guided them through what occasionally looked to them like a minefield. Teaching these students has been a joy for me, and I will always appreciate the gracious reception they have given me as their teacher. The students whom I taught in the fall and winter quarters of this academic year were especially helpful to me as I "classroom tested" the penultimate version of the book. Readers have been spared hundreds of typos that were fixed due to their watchful eyes.

I would like to thank a number of individuals who provided significant assistance to me in the writing of this text. I consider the problems to be the most important part of the book, and I am grateful to Apratim Ganguly, Kimi Noguchi, Anzhi Gu, and Zhijie Zheng for patiently working through hundreds of problems and providing solutions that could be made available to students and/or instructors who use the text. The steady and sage advice of my editor John Kimmel is much appreciated. A special benefit of John's stewardship was the high quality of the reviewers that he recruited to comment on early versions of the text. They all contributed to making this a better book. My sincere thanks to Adam Bowers (UC San Diego), James Gentle (George Mason University), Solomon Harrar (University of Montana), Wesley Johnson (UC Irvine), Lawrence Leemis (William & Mary University), Elena Rantou (George Mason University), Ralph P. Russo (U. of Iowa), and Gang Wang (DePaul University). I am also most grateful to Gail Gong for her helpful advice on Chapter 12 and to Ethan Anderes for his help with the graphics on the book's cover. Christopher Aden took my somewhat primitive version of the text and made it sing on Chapman and Hall's LaTeX template. Thanks, Chris, for your timely and high-quality work! Finally, I thank my wife, Mary O'Meara Samaniego, for her patience with this project. I am especially grateful for her plentiful corporal and moral support. And a special thanks to Elena, Moni, Keb, Jack, and Will. It's hard for me to imagine a more supportive and loving family.

F.J. Samaniego
June 2013

Contents

Preface for Students

Let me begin with a sincere welcome. This book was written with you in mind! As you probably know, new textbook projects are generally reviewed pretty carefully. The reviewers either tell the publisher to tank the project or they give the author their best advice about possible improvements. I'm indebted to the reviewers of this book for providing me with (a) much constructive criticism and (b) a good deal of encouragement. I especially appreciated one particular encouraging word. Early on, a reviewer commented about the style of the book, saying that he liked its conversational tone. He/she said that it read as if I was just talking to some students sitting around my desk during office hours. I liked this comment because it sort of validated what I had set out to do. Reading a book that uses mathematical tools and reasoning doesn't have to be a painful experience. It can be, instead, stimulating and enjoyable; discovering a new insight or a deeper understanding of something can be immensely satisfying. Of course, it will take some work on your part. But you know that already. Just like acquiring anything of value, learning about the mathematical foundations of probability and statistics will require the usual ingredients needed for success: commitment, practice, and persistence. Talent doesn't hurt either, but you wouldn't be where you are today if you didn't have that. If you concentrate on the first three attributes, things should fall into place for you. In this brief preface, my aim is to give you some advice about how to approach this textbook and a course in which it is used. First, I'd recommend that you review your old calculus book. It's not that calculus permeates every topic taken up in this book, but the tools of differential and integral calculus are directly relevant to many of the ideas and methods we will study: differentiating a moment-generating function, integrating a density function, minimizing a variance, maximizing a likelihood. But calculus is not even mentioned until the last couple of sections of Chapter 2, so you have time for a leisurely yet careful review. That review is an investment you won't regret.

Most students will take a traditional-style course in this subject, that is, you will attend a series of lectures on the subject, will have the benefit of some direct interaction with the instructor and with graduate teaching assistants, and will work on assigned problem sets or on problems just for practice. While there is no unique strategy that guarantees success in this course, my prescription for success would certainly include the following: (1) Read ahead so that you place yourself in the position of knowing what you don't understand yet when you attend a lecture on a given topic. If you do, you'll be in a good position of focus on the particular elements of the day's topic that you need more information on, and you'll be prepared to ask questions that should clarify whatever seems fuzzy upon first reading. (2) Work as many problems as you have time for. "Practice makes perfect" is more than a worn out platitude. It's the truth! That's what distinguishes the platitudes that stick around from those that disappear. (3) Try to do problems by yourself first. The skill you are hoping to develop has to do with using the ideas and tools you are learning to solve new problems. You learn something from attempts that didn't work. But you learn the most from attempts that do work. Too much discussion or collaboration (where the answers are revealed to you before you've given a problem your best effort) can interfere with your learning. (4) While it's true that mastering a new skill generally involves some suffering, you should not hesitate to seek help after giving a problem or a topic your honest effort. Receiving helpful hints from an instructor, TA, or tutor is generally more beneficial than just having the solution explained to you. It's also better than total frustration. So put in a decent effort and, if a problem seems resistant to being solved, go have a chat with your instructor or his/her surrogate. (5) Mathematical ideas and tools do not lend themselves to quick digestion. So give yourself some time to absorb the material in this course. Spreading out a homework assignment over several days, and studying for a test well before the eve of the exam, are both time-honored study habits that do help. To help make learning this subject less painful for you, I've included many reader-friendly explanations, hints, tutorials, discussion, and occasional revelations of the "tricks of the trade" in the text.

I'd like to give you a few tips about how to "attack" this book. First, the book has lots of "problems" to be solved. I've placed exercises at the end of every section. I encourage you to work all of them after you've read a section, as they represent an immediate opportunity to test your understanding of the material. For your convenience, and because I am, regardless of what you may have heard, a compassionate person who wants to be helpful, I've included, in an Appendix, the answers (or helpful comments) for all the exercises. So you can check your answer to confirm whether or not you've nailed the exercise. Some of these exercises may be assigned for homework by your instructor. It's OK, while you are first learning the subject, to be working toward a particular answer, although you will usually get the most benefit from looking up the answer only after you've solved, or at least seriously attempted to solve, the exercise. I should add that not all the exercises are simple applications of the textual material. Some exercises address rather subtle notions within a given topic. If you are able to do all the exercises, you can be confident that you've understood the material at a reasonably high level.

Now, let me give you a heads up about the sections in the book that are the most challenging. These sections will require special concentration on your part and may benefit from some collateral reading. I recommend that you spend more time than average reading and digesting Section 1.8 on combinatorics, Section 2.8 on moment-generating functions, Section 3.6 on "other" continuous distributions (since you may need to study and learn about these on your own), Section 4.6 on transformation theory, Sections 5.3 and 5.4 on the Central Limit Theorem and on the delta method, Section 6.3 on Fisher information and the Cramér-Rao inequality, Section 6.4 of sufficiency, completeness, and minimum variance unbiased estimators, Section 10.4 on optimality in hypothesis testing, Section 11.3 on properties of estimators in regression, Section 12.1 on nonparametric estimation, and Section 12.2 on the nonparametric bootstrap. I mention these sections specifically because they will require your careful attention and, perhaps, more "practice" than usual before you feel you have a good grasp of that material. You may wish to read more, and work additional problems, on these topics. Supporting material can be found in books that I've marked with an asterisk (*) in the bibliography.

I wish you the best as you begin this exploration into stochastic modeling and mathematical statistics. I'll be very interested in hearing about your experience and also in having your feedback on the book. Feel free to contact me with your comments.

Francisco J. Samaniego
University of California, Davis
fjsamaniego@ucdavis.edu

Preface for Instructors

There are quite a few textbooks that treat probability and mathematical statistics at the advanced undergraduate level. The textbooks used in courses on these topics tend to fall into one of two categories. Some of these texts cover the subject matter with the mathematical rigor that a graduate school–bound mathematics or statistics major should see, while the remaining texts cover the same topics with much less emphasis on mathematical developments and with more attention to applications of the models and statistical ideas they present. But isn't it desirable for students in a "theoretical" course to be exposed to serious statistical applications and for students in an "applications-oriented" course to be exposed to at least some of the mathematics that justifies the application of statistical modeling and inference in practice? This book offers instructors the flexibility to control the mathematical level of the course they teach by determining the mathematical content they choose to cover. It contains the mathematical detail that is expected in a course for "majors," but it is written in a way that facilitates its use in teaching a course that emphasizes the intuitive content in statistical theory and the way theoretical results are used in practice.

This book is based on notes that I have used to teach both types of courses over the years. From this experience, I've reached the following conclusions: (1) the core material for both courses is essentially the same, (2) the ideas and methods used in mathematical proofs of propositions of interest and importance in the field are useful to both audiences, being essential for the first and being helpful to the second, (3) both audiences need to understand what the main theorems of the field say, and they especially need to know how these theorems are applied in practice, (4) it is possible, and even healthy, to have theory and application intertwined in one text. An appealing byproduct of this comingling of mathematical and applied thinking is that through assigned, recommended, or even optional reading of sections of the text not formally covered, an instructor can effectively facilitate the desired "exposure" of students to additional theoretical and applied aspects of the subject.

Having often been disappointed with the quantity and range of the problems offered in textbooks I've used in the past, I embarked on the writing of this book with the goal of including tons of good problems from which instructors could choose. That is not to say that you won't have the inclination to add problems of your own in the course that you teach. What I'm really saying is that you may not have to work as hard as usual in supplementing this book with additional problems. Every section ends with a small collection of "exercises" meant to enable the student to test his/her own understanding of a section immediately after reading it. Answers to (or helpful comments on) all the exercises are given at the end of the book. A sizable collection of problems is gathered at the end of each chapter. For instructors who adopt this book as a text for a course, a Solutions Manual containing detailed solutions to all the even-numbered problems in the text is available from Chapman and Hall.

This book is intended as a text for a first course in probability and statistics. Some students will have had a previous "pre-calculus" introduction to statistics, and while that can be helpful in various ways, it is by no means assumed in this text. Every new idea in the text is treated from scratch. What I expect is that a course from this book would be a student's first calculus-based statistics course and their first course emphasizing WHY (rather than HOW) probability and statistics work. The mathematical prerequisite for this text is a course on differential and integral calculus. While the stronger the mathematical background of the student, the better, students who have taken a calculus sequence for majors in the sciences (i.e., non-math majors) will do just fine in the course. Since it's not uncommon for a student's calculus skills to get rusty, an early review of one's old calculus text is recommended. Occasional calculus tutorials in the text (e.g., on integration by parts, on changing variables of integration, and on setting limits of double integrals) are aimed at assisting students in their ongoing review.

In the Statistics Department at the University of California, Davis, separate courses are offered on probability and mathematical statistics at the upper division level. The year-long, more mathematical course is intended for Statistics and Mathematics majors, but is also taken by a fair number of students in computer science and engineering and by other "non-majors" with strong mathematical backgrounds and interests. The alternative course is a two-quarter sequence (known as the "brief course") which is taken by applied statistics and applied mathematics majors, by students working on a minor in statistics, by graduate students in quantitative disciplines ranging from engineering to genetics to quantitative social science, and by a few ambitious undergraduates. The first group is, typically, already familiar with mathematical argumentation, and although the second group is capable of digesting a logical mathematical argument, they will need some careful guidance and encouragement before they get comfortable.

If your course is mostly taken by stat and math majors, it can be thought of as the first course above. If such students are in the minority, your course may be thought of as the second. Both groups will get a solid grounding in the core ideas in probability and statistics. If this book is used in the course for majors, then most of the theorems treated in the book can be proven in the classroom or assigned as homework when not given in the text. The notions of combinatorial proofs and mathematical induction, which would typically be skipped in the brief course, can be treated and applied as in the text. When they are skipped, it is useful to state certain results proven by these methods that arise later in the text. In the first course, the instructor may wish to include problems involving proofs in both homework and exams. In the second course, I generally assign some "doable" proofs for homework, but exams don't ask for proofs. In both courses, I like to give open-book, problem-solving exams. After all, life itself is an open-book problem-solving exam. The present book retains the main topics, tools, and rigor of traditional math-stat books, and is thus suitable for a course for majors. But the book also contains careful intuitive explanations of theoretical results that are intended to provide students with the ability to apply these results with confidence, even when they have not studied or fully digested their proofs. I have found that this latter goal, while ambitious, is achievable in the classroom. This text is aimed at replicating successful classroom strategies in a text having the academic goals described above.

Several sections of the book are labeled as "optional." I believe that the entire book is appropriate for the audience in the first course mentioned above, the course for "majors," and a two-semester or three-quarter course can very comfortably accommodate all twelve chapters of the text and still leave room for additional topics favored by the instructor. For the course aimed at non-majors, I include below a chapter-by-chapter discussion of how the text might be used. I have always contended that our main mission as teachers of a mathematical topic is (1) to make sure that the topic (the idea, the method, or the theorem statement) is clear and well understood and (2) to make sure that students understand why, when, and how the result may be applied. These goals can often be accomplished without a formal proof, although a proof does have a nice way of convincing a reader that a proposition in unquestionably true. What are this text's "special features"? Here is a "top-10 list," in the order in which various topics arise.

Some of the topics mentioned would be needed in a course for majors but can be trimmed or skipped in a brief course. The text (1) emphasizes "probability models" rather than "probability theory" per se, (2) presents the "key" stochastic tools: a careful treatment of moment-generating functions, bivariate models, conditioning, transformation theory, computer simulation of random outcomes, and the limit theorems that statisticians need (various modes of convergence, the central limit theorem, the delta method), (3) presents a full treatment of optimality theory for unbiased estimators, (4) presents the asymptotic theory for method of moments estimators (with proof) and for maximum likelihood estimators (without proof, but with numerous examples and a formal treatment of the Newton-Raphson and EM algorithms), (5) devotes a full chapter to the Bayesian approach to estimation, including a section on comparative statistical inference, (6) provides a careful treatment of the theory and applications of hypothesis testing, including the Neyman-Pearson Lemma and Likelihood Ratio Tests, (7) covers the special features of regression analysis and analysis of variance which utilize the theory developed in the core chapters, (8) devotes a separate chapter to nonparametric estimation and testing which includes an introduction to the bootstrap, (9) features serious scientific applications of the theory presented (including, for example, problems from fields such as conservation, engineering reliability, epidemiology, genetics, medicine, and wild life biology), and (10) includes well over 1000 exercises and problems at varying levels of difficulty and with a broad range of topical focus. When used in ways that soften the mathematical level of the text (as I have done in teaching the brief course some 20 times in my career), it provides students in the quantitative sciences with a useful overview of the mathematical ideas and developments that justify the use of many applied statistical techniques.

What advice do I have for instructors who use this book? My answer for instructors teaching the first course described above, the course for "majors," is fairly straightforward. I believe that the text could be used pretty much as is. If an instructor wishes to enhance the mathematical level of the course to include topics like characteristic functions, a broader array of limit theorems, and statistical topics like robustness, such topics could be logically introduced in the context of Chapters 2, 5, 6, and 7. It seems likely that a year-long course will allow time for such augmentations. Regarding the augmentation of topics covered in the text, I believe that the most obvious and beneficial addition would be a broader discussion of linear model theory. The goal of Chapter 11 is to illustrate certain ideas and methods arising in earlier chapters (such as best linear unbiased estimators and likelihood ratio tests), and this goal is accomplished within the framework of simple linear regression and one-way analysis of variance. An expansion of my treatment of nonparametric testing in Chapter 12 is another reasonable possibility. My own choices for additional topics would be the Wilcoxon signed-rank test and the Kolmogorov-Smirnov tests for goodness of fit. Another topic that is often touched on in a course for majors is "decision theory." This is briefly introduced in Chapter 9 in the context of Bayesian inference, but the topic could easily have constituted a chapter of its own, and a broader treatment of decision theory could reasonably be added to a course based on the present text.

My advice to instructors who use this text in teaching something resembling the brief course described above is necessarily more detailed. It largely consists of comments regarding what material might be trimmed or skipped without sacrificing the overall aims of the course. My advice takes the form of a discussion, chapter by chapter, of what I suggest as essential, optional, or somewhere in between.

Chapter 1. Cover most of this chapter, both because it is foundational, but also because the mathematical proofs to which the students are introduced here are relatively easy to grasp. They provide a good training ground for learning how proofs are constructed. The first five theorems in Section 1.3 are particularly suitable for this purpose. Section 1.4 is essential. Encourage students to draw probability trees whenever feasible. Teach Bayes' Theorem as a simple application of the notion of conditional probability. The independence section is straightforward. Students find the subject of combinatorics the most difficult of the chapter. I recommend doing the poker examples, as students like thinking through these problems. I recommend skipping the final two topics of Section 1.8 . The material on multinomial coefficients and combinatorial proofs can be recommended as optional reading.

Chapter 2. I recommend teaching all of Chapter 2, with the exception of Section 2.4 (on mathematical induction) even though Professor Beckenbach's joke (Theorem 2.4.3) offers some welcome comic relief for those who read the section. If you skip Section 2.4, I suggest you state Theorem 2.4.1, Theorem 2.4.2 and the result in Exercise 2.4.2 without proof, as these three facts are used later. Students find Section 2.8 on moment-generating functions to be the most challenging section in this chapter. It's true that mgfs have no inherent meaning or interpretation. The long section on them is necessary, I think, to get students to appreciate mgfs on the basis of the host of applications in which they can serve as useful tools.

Chapter 3. The first two sections are fundamental. Section 3.3 can be treated very lightly, perhaps with just a definition and an example. In Section 3.4, the material on the Poisson process and the gamma distribution (following Theorem 3.4.4) may be omitted without great loss. It is an interesting connection which provides, as a byproduct, that the distribution function of a gamma model whose shape parameter α is an integer may be computed in closed form. But the time it takes to establish this may be better spent on other matters. (Some assigned reading here might be appropriate.) Section 3.5 is essential. Section 3.6 can be left as required reading. The models in this latter section will occur in subsequent examples, exercises, and problems (and probably also exams), so students would be well advised to look at these models carefully and make note of their basic properties.

Chapter 4. This chapter contains a treasure trove of results that statisticians need to know, and know well. I try to teach the entire chapter, with the exception of Section 4.4 on the multinomial model, which might be considered as optional. Included are definitions and examples of bivariate (joint) densities (or pmfs), marginal and conditional densities, expectations in a bivariate and multivariate setting, covariance, correlation, the mean and variance of a linear combination of random variables, results on iterated expectations (results I refer to as Adam's rule and Eve's rule), the ubiquitous bivariate normal model, and a comprehensive treatment of methods for obtaining the distribution of a transformed random variable $Y = g(X)$ when the distribution of X is known. Section 4.7 can be covered briefly. Students need to see the basic formulae here, as many examples in the sequel employ order statistics.

Chapter 5. A light treatment is possible. In such a treatment, I recommend establishing Chebyshev's inequality and the weak law of large numbers, the definition of convergence in distribution, a statement of the Central Limit Theorem, and a full coverage of the delta method theorem.

Chapter 6. I recommend treating the first two sections of this chapter in detail. In Section 6.3, I suggest omitting discussion of the Cauchy-Schwarz inequality and stating the Cramér-Rao Theorem without proof. From Section 6.4, I would suggest covering "sufficiency" and the Rao-Blackwell theorem and omitting the rest (or relegating it to assigned or optional reading). Section 6.5 on BLUEs is short and useful. Section 6.6 has important ideas and some lessons worth learning, and it should be covered in detail.

Chapter 7. Section 7.1 may be covered lightly. The discussion up to and including Example 7.1.1 is useful. The "big o" and the "small o" notation are used sparingly in the text and can be skipped when they are encountered. Sections 7.2 and 7.3 are the heart of this chapter and should be done in detail. Section 7.4 treats an important problem arising in statistical studies in epidemiology and provides an excellent example of the skillful use of the delta method. I recommend doing this section if time permits. Otherwise, it should be required reading. Section 7.5 should be done in some form, as students need to be familiar with at least one numerical method which can approximate optimum solutions when they can't be obtained analytically. Section 7.6 on the EM algorithm covers a technique that is widely used in applied work. It should be either covered formally or assigned as required reading,

Chapter 8. I suggest covering Sections 8.1 and 8.2, as they contain the core ideas. Sample size calculations, covered in Section 8.3, rank highly among the applied statistician's "most frequently asked questions," and is a "must do." Section 8.4 on tolerance intervals is optional and may be skipped.

Chapter 9. This chapter presents the Bayesian approach to estimation, pointing out potential gains in estimation efficiency afforded by the approach. The potential risks involved in Bayesian inference are also treated seriously. The basic idea of the approach is covered in Section 9.1 and the mechanics of Bayesian estimation are treated in Section 9.2. Sections 9.3 and 9.4 provide fairly compelling evidence, both empirical and theoretical, that the Bayesian approach can be quite effective, even under seemingly poor prior assumptions. These two sections represent an uncommon entry in courses at this level, a treatment of "comparative inference" where the Bayesian and classical approaches to estimation are compared side by side. Section 5 treats Bayesian interval estimation. If an instructor is primarily interested in acquainting students with the Bayesian approach, a trimmed down coverage of this chapter that would accomplish this would restrict attention to Sections 9.2 and 9.5.

Chapter 10. This is the bread and butter chapter on hypothesis testing. Section 10.1 contains the needed concepts and definitions as well as much of the intuition. Section 10.2 treats the standard tests for means and proportions. The section is useful in tying the general framework of the previous section to problems that many students have seen in a previous course. Section 10.3 on sample size calculations for obtaining the desired power at a fixed alternative in an important notion in applied work and should be covered, even if only briefly. Section 9.4 presents the Neyman-Pearson Lemma, with proof. The proof can be skipped and the intuition of the lemma, found in the paragraphs that follow the proof, can be emphasized instead. The examples that complete the section are sufficient to make students comfortable with how the lemma may be used. Sections 10.5 and 10.6 cover important special topics that students need to see.

Chapter 11. Sections 11.1, 11.3, and 11.4 carry the main messages of the chapter. Section 11.2, which presents the standard tests and confidence intervals of interest in simple linear regression, can be skipped if time is precious.

Chapter 12. Section 12.1 treats nonparametric estimation of an underlying distribution F for either complete or censored data. In a brief presentation of this material, one might present expressions for the two resulting nonparametric MLEs and an example of each. The nonparametric bootstrap is described in Section 9.2. The widespread use of the bootstrap in applied work suggests that this section should be covered. Of the three remaining sections (which are each free standing units), the most important topic is the Wilcoxon Rank-Sum Test treated in Section 12.5. Sections 12.3 and 12.4 may be treated as assigned or recommended reading.

Instructors who have comments, questions, or suggestions about the discussion above, or about the text in general, should feel free to contact me. Your feedback would be most welcome.

Francisco J. Samaniego
University of California, Davis
fjsamaniego@ucdavis.edu

1

The Calculus of Probability

1.1 A Bit of Background

Most scientific theories evolve from attempts to unify and explain a large collection of individual problems or observations. So it is with the theory of probability, which has been developing at a lively pace over the past 350 years. In virtually every field of inquiry, researchers have encountered the need to understand the nature of variability and to quantify their uncertainty about the processes they study. From fields as varied as astronomy, biology, and economics, individual questions were asked and answered regarding the chances of observing particular kinds of experimental outcomes. Following the axiomatic treatment of probability provided by A. N. Kolmogorov in the 1930s, the theory has blossomed into a separate branch of mathematics, and it is still under vigorous development today.

Many of the early problems in mathematical probability dealt with the determination of odds in various games of chance. A rather famous example of this sort is the problem posed by the French nobleman, Antoine Gombaud, Chevalier de Méré, to one of the outstanding mathematicians of his day, Blaise Pascal: which is the more likely outcome, obtaining at least one six in four rolls of a single die or obtaining at least one double six in twenty-four rolls of a pair of dice? The question itself, dating back to the mid-seventeenth century, seems unimportant today, but it mattered greatly to Gombaud, whose successful wagering depended on having a reliable answer. Through the analysis of questions such as this (which, at this point in time, aptly constitutes a rather straightforward problem at the end of this chapter), the tools of probability computation were discovered. The body of knowledge we refer to as the probability calculus covers the general rules and methods we employ in calculating probabilities of interest.

I am going to assume that you have no previous knowledge of probability and statistics. If you've had the benefit of an earlier introduction, then you're in the enviable position of being able to draw on that background for some occasionally helpful intuition. In our efficiency-minded society, redundancy has come to be thought of as a four-letter word. This is unfortunate, because its important role as a tool for learning has been simultaneously devalued. If some of the ideas we cover are familiar to you, I invite you to treat it as an opportunity to rethink them, perhaps gaining a deeper understanding of them.

I propose that we begin by giving some thought here to a particularly simple game of chance. I want you to think about the problem of betting on the outcome of tossing two newly minted coins. In this particular game, it is clear that only three things can happen: the number of heads obtained will be zero, one, or two. Could I interest you in a wager in which you pay me three dollars if the outcome consists of exactly one head, and I pay you two dollars if either no heads or two heads occur? Before investing too much of your fortune on this gamble, it would behoove you to determine just how likely each of the three outcomes is. If, for example, the three outcomes are equally likely, this bet is a real bargain, since the 2-to-1 odds in favor of your winning will outweigh the 2-to-3 differential in the payoffs. (This follows from one of the "rules" we'll justify later.) If you're not sure about the relative likelihood of the three outcomes, you might pause here to flip two coins thirty times and record the proportion of times you got 0, 1, or 2 heads. That mini-experiment should be enough to convince you that the wager I've proposed is not an altogether altruistic venture.

The calculus of probability can be thought of as the methodology which leads to the assignment of numerical "likelihoods" to sets of potential outcomes of an experiment. In our study of probability, we will find it convenient to use the ideas and notation of elementary set theory. The clarity gained by employing these simple tools amply justifies the following short digression.

In the mathematical treatment of set theory, the term "set" is undefined. It is assumed that we all have an intuitive appreciation for the word and what it might mean. Informally, we will have in the back of our minds the idea that there are some objects we are interested in, and that we may group these objects together in various ways to form sets. We will use capital letters A, B, C to represent sets and lower case letters a, b, c to represent the objects, or "elements," they may contain. The various ways in which the notion of "containment" arises in set theory is the first issue we face. We have the following possibilities to consider:

Definition 1.1.1. (Containment)

(i) $a \in A$: a is an element in the set A; $a \notin A$ is taken to mean that a is not a member of the set A.

(ii) $A \subseteq B$: The set A is contained in the set B, that is, every element of A is also an element of B. A is then said to be a "subset" of B.

(iii) $A \subset B$: A is a "proper subset" of B, that is, A is contained in B ($A \subseteq B$) but B contains at least one element that is not in A.

(iv) $A = B : A \subseteq B$ and $B \subseteq A$, that is, A and B contain precisely the same elements.

In typical applications of set theory, one encounters the need to define the boundaries of the problem of interest. It is customary to specify a "universal set" U, a set which represents the collection of all elements that are relevant to the problem at hand. It is then understood that any set that subsequently comes under discussion is a subset of U. For logical reasons, it is also necessary to acknowledge the existence of the "empty set" $\varnothing$, the set with no elements. This set constitutes the logical complement of the notion of "everything" embodied in U, but also plays the important role of zero in the set arithmetic we are about to discuss. We use arithmetic operations like addition and multiplication to create new numbers from numbers we have in hand. Similarly, the arithmetic of sets centers on some natural ways of creating new sets. The set operations we will use in this text appear in the listing below:

Definition 1.1.2. (Set Operations)

(i) A^c: "the complement of A," that is, the set of elements of U that are not in A.

(ii) $A \cup B$: "A union B," the set of all elements of U that are in A or in B or in both A and B.

(iii) $A \cap B$: "A intersect B," the set of all elements of U that are in both A and B.

(iv) $A - B$: "A minus B," the set of all elements of U that are in A but are not in B.

We often use pictorial displays called Venn diagrams to get an idea of what a set looks like. The Venn diagrams for the sets in Definition 1.1.2 are shown below:

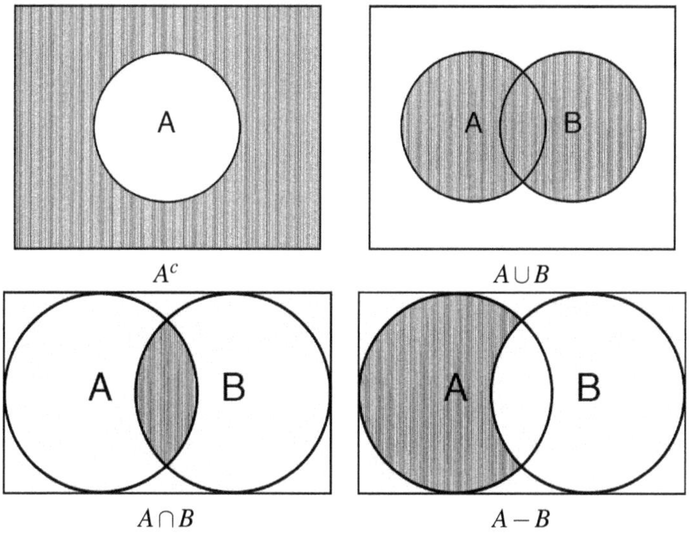

Figure 1.1.1. Venn diagrams of A^c, $A \cup B$, $A \cap B$, and $A - B$.

Venn diagrams are often helpful in determining whether or not two sets are equal. You might think "I know how to check set equality; just see if the sets contain the same elements. So what's the big deal?" This reaction is entirely valid in a particular application where the elements of the two sets can actually be listed and compared. On the other hand, we often need to know whether two different methods of creating a new set amount to the same thing. To derive a general truth that holds in all potential applications, we need tools geared toward handling abstract representations of the sets involved. Consider, for example, the two sets: $A \cap (B \cup C)$ and $(A \cap B) \cup (A \cap C)$. The equality of these two sets (a fact which is called the distributive property of intersection with respect to union) can be easily gleaned from a comparison of the Venn diagrams of each.

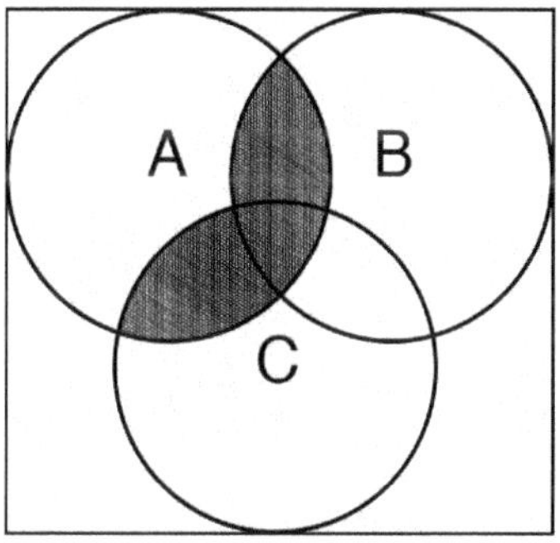

Figure 1.1.2. The Venn Diagram of either $A \cap (B \cup C)$ or $(A \cap B) \cup (A \cap C)$.

Often, general properties of set operations can be discerned from Venn diagrams.

Among such properties, certain "decompositions" are especially important in our later work. First, it is clear from the Venn diagram of $A \cup B$ that the set can be decomposed into three non-overlapping sets. Specifically, we may write

$$A \cup B = (A - B) \cup (A \cap B) \cup (B - A). \tag{1.1}$$

Another important decomposition provides an alternative representation of a given set A. For any two sets A and B, we may always represent A in terms of what it has (or hasn't) in common with B:

$$A = (A \cap B) \cup (A \cap B^c). \tag{1.2}$$

In a Venn diagram showing two overlapping sets A and B, combining the sets $(A \cap B)$ and $(A \cap B^c)$ clearly accounts for all of the set A, confirming the validity of the identity of (1.2). The identity also holds, of course, if A and B do not overlap. Why?

There is something special about the sets involved in the unions which appear in (1.1) and (1.2). Sets D and E are said to be "disjoint" if $D \cap E = \varnothing$, that is, if they have no elements in common. Both of the identities above represent a given set as the union of disjoint sets. As we will see, such representations are often useful in calculating the probability of a given set of possible experimental outcomes. As helpful as Venn diagrams are in visualizing certain sets, or in making the equality of two sets believable, they lack something that mathematicians consider to be of paramount importance: rigor. A rigorous mathematical argument consists of a logical sequence of steps which lead you from a given statement or starting point to the desired conclusion. Each step in the argument must be fully defensible, being based solely on basic definitions or axioms and their known consequences. As an example of a "set-theoretic proof," we will provide a rigorous argument establishing a very important identity known as De Morgan's Law.

Theorem 1.1.1. (De Morgan's Law for two sets)
For two arbitrary sets A and B,

$$(A \cup B)^c = A^c \cap B^c. \tag{1.3}$$

Proof. The two sets in (1.3) can be shown to be equal by showing that each contains the other. Let's first show that $(A \cup B)^c \subseteq A^c \cap B^c$. Assume that x is an arbitrary element of the set $(A \cup B)^c$ that is, assume that

$$x \in (A \cup B)^c.$$

This implies that

$$x \notin (A \cup B),$$

which implies that

$$x \notin A \text{ and } x \notin B,$$

which implies that

$$x \in A^c \text{ and } x \in B^c,$$

which implies that

$$x \in A^c \cap B^c.$$

The same sequence of steps, in reverse, proves that $A^c \cap B^c \subseteq (A \cup B)^c$. The most efficient proof of (1.3) results from replacing the phrase "implies that" with the phrase "is equivalent to." This takes care of both directions at once. ■

It is easy to sketch the Venn diagrams for the two sets in De Morgan's Law. You might decide that the two ways of convincing yourself of the law's validity (namely, the proof and the picture) are of roughly the same complexity, and that the visual method is preferable because of its concreteness. You should keep in mind, however, that convincing yourself that something is true isn't the same as formulating an air-tight argument that it's true. But there's an additional virtue in developing a facility with the kind of mathematical reasoning employed in the proof above. The bigger the problem gets, the harder it is to use visual aids and other informal tools. A case in point is the general version of De Morgan's Law, stating that, for n arbitrary sets $A_1, A_2, \ldots, A_n$, the compliment of the union of the sets is equal to the intersection of their compliments.

Theorem 1.1.2. (General De Morgan's Law) For n arbitrary sets $A_1, A_2, \ldots, A_n$,

$$\left(\bigcup_{i=1}^{n} A_i\right)^c = \bigcap_{i=1}^{n} (A_i^c). \tag{1.4}$$

The general statement in (1.4) cannot be investigated via Venn diagrams, but it is easy to prove it rigorously using the same logic that we used in proving De Morgan's Law for two sets. Give it a try and you'll see!

Exercises 1.1.

1. Suppose $U = \{1,2,3,4,5,6,7,8,9,10\}$ and the sets A, B, and C are given by $A = \{2,4,6,8,10\}$, $B = \{2,5,6,7,10\}$, and $C = \{1,6,9\}$. Identify each of the following sets:
   $(a) A \cup B$, $(b) A \cap B$, $(c) A - B$
   $(d) A \cup B^c$, $(e) A \cap B \cap C$, $(f) B \cap (A \cup C)^c$
   $(g) (A \cap C) \cup (B \cap C)$, $(h) (A - C) \cup (C - A)$, $(i) A^c \cap B \cap C^c$.

2. Let A, B, and C be three overlapping sets. Draw the Venn diagram for $A \cup B \cup C$, and note that this union can alternatively be viewed as the union of seven disjoint sets. Using the set operations defined in Section 1.1, give each of these seven sets a name. (And I don't mean Jane, Harry, Gustavo, ...)

3. Construct two Venn diagrams which show that the set $(A \cap B) - C$ is equal to the set $(A - C) \cap (B - C)$.

4. For arbitrary sets A and B, give a set theoretic proof that $A \cap B^c = A - B$.

5. For arbitrary sets A and B, prove that $A \cup B = A \cup (B - A)$.

6. Let $A_1, \ldots, A_n$ be a "partition" of the universal set U, i.e., suppose that $A_i \cap A_j = \varnothing$ for $i \neq j$ (or, simply, the As are disjoint sets), and that $\bigcup_{i=1}^{n} A_i = U$. Let B be an arbitrary subset of U. Prove that

$$B = \bigcup_{i=1}^{n} (B \cap A_i). \tag{1.5}$$

   (Note that the identity in (1.5) is still true if the As are not disjoint, as long as the condition $\bigcup_{i=1}^{n} A_i = U$ holds.)

1.2 Approaches to Modeling Randomness

Randomness is a slippery idea. Even the randomness involved in a simple phenomenon like flipping a coin is difficult to pin down. We would like to think of a typical coin toss as yielding one of two equally likely outcomes; heads or tails. We must recognize, however, that if we were able to toss a coin repeatedly under precisely the same physical conditions, we would get the same result each time. (Compare it, for example, to physical processes that you can come close to mastering, like that of throwing a crumpled piece of paper into a waste basket two feet away from you.) So why do we get about as many heads as tails when we toss a coin repeatedly? It's because the physical processes involved are well beyond our control. Indeed, they are even beyond our ability to understand fully.

Consecutive coin tosses are generally different enough from each other that it is reasonable to assume that they are unrelated. Moreover, the physical factors (like force, point of contact, direction of toss, and atmospheric conditions, as when someone right in front of you is breathing heavily) are sufficiently complex and uncontrollable that it is reasonable to assume that the outcome in a given toss is unpredictable, with heads neither more nor less likely than tails. Thus, in modeling the outcomes of successive coin tosses, a full understanding of the mechanics of coin tossing is generally impossible, but also quite unnecessary. Our models should reflect, of course, the information we happen to have about the phenomenon we are observing. If we were able, for example, to ensure that our coin tosses were so precise that the coin made exactly eight full revolutions in the air at least 70 percent of the time, then we would wish our model for the outcomes of such tosses to reflect the fact that there was at least a 70 percent chance that the initial and final states of the coin agreed.

What is important about the models we will employ to describe what might happen in particular experiments is that they constitute reasonable approximations of reality. Under normal circumstances, we'll find that the coin tossing we do is quite well described by the assumption that heads and tails are equally likely. Our criterion for the validity of a model will be the level of closeness achieved between real experimental outcomes and the array of outcomes that our model would have predicted. Our discussion of randomness will require a bit of jargon. We collect some key phrases in the following:

Definition 1.2.1. (Random Experiments) A *random experiment* is an experiment whose outcome cannot be predicted with certainty. All other experiments are said to be *deterministic*. The *sample space* of a random experiment is the set of all possible outcomes. The sample space plays the role of the universal set in problems involving the corresponding experiment, and it will be denoted by S. A single outcome of the experiment, that is, a single element of S, is called a *simple event*. A *compound event* is simply a subset of S. While simple events can be viewed as compound events of size one, we will typically reserve the phrase "compound event" for subsets of S with more than one element.

Developing a precise description of the sample space of a random experiment is always the first step in formulating a probability model for that experiment. In this chapter and the next, we will deal exclusively with "discrete problems," that is, with problems in which the sample space is finite or, at most, countably infinite. (A countably infinite set is one that is infinite, but can be put into one-to-one correspondence with the set of positive integers. For example, the set $\{2^n : n = 1, 2, 3, \dots\}$ is countably infinite.) We postpone our discussion of "continuous" random experiments until Chapter 3. We turn now to an examination of four random experiments from which, in spite of their simplicity, much can be learned.

Example 1.2.1. Suppose that you toss a single coin once. Since you catch it in the palm of one hand, and quickly turn it over onto the back of your other hand, we can discount the possibility of the coin landing and remaining on its edge. The sample space thus consists of the two events "heads" and "tails" and may be represented as

$$S_1 = \{H, T\}.$$

■

Example 1.2.2. Suppose that you toss a penny and then a nickel, using the same routine as in Example 1.2.1. The sample space here is

$$S_2 = \{HH, HT, TH, TT\}.$$

■

You will recall that in the preceding section, I mentioned that only three things could happen in this experiment, namely, you could get either 0, 1, or 2 heads. While this is true, it is important to recognize that this summary does not correspond to the most elementary description of the experiment. In experiments like this, which involve more than one stage, it is often helpful to draw a "tree" picturing the ways in which the experiment might play out. Here, the appropriate tree is shown below.

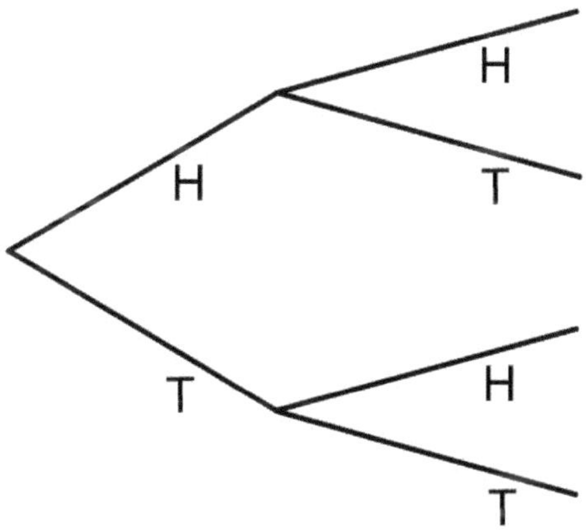

Figure 1.2.1. A tree displaying the sample space for tossing two coins.

Example 1.2.3. A single die is rolled. The sample space describing the potential values of the number facing upwards is

$$S_3 = \{1, 2, 3, 4, 5, 6\}.$$

■

Example 1.2.4. A pair of dice, one red, one green, are rolled. The sample space for this experiment may be represented as a tree with six branches at the first stage and six branches at the second stage, so that the two-stage experiment is represented by a tree with 36 paths or possible outcomes. The rectangular array below is an equivalent representation in which the first digit represents the outcome of the red die and the second digit corresponds to the outcome for the green die. The sample space S_4 is shown below.

$$S_4 = \left\{\begin{matrix} 1,1 & 1,2 & 1,3 & 1,4 & 1,5 & 1,6 \\ 2,1 & 2,2 & 2,3 & 2,4 & 2,5 & 2,6 \\ 3,1 & 3,2 & 3,3 & 3,4 & 3,5 & 3,6 \\ 4,1 & 4,2 & 4,3 & 4,4 & 4,5 & 4,6 \\ 5,1 & 5,2 & 5,3 & 5,4 & 5,5 & 5,6 \\ 6,1 & 6,2 & 6,3 & 6,4 & 6,5 & 6,6 \end{matrix}\right\}$$

Figure 1.2.2. The sample space S for rolling two dice.

■

We now turn our attention to the problem of assigning "probabilities" to the elements of S. We will use the word "stochastic" as the adjective form of the noun "probability." A stochastic model is a model for a random experiment. It provides the basis for attaching numerical probabilities or likelihoods to sets of outcomes in which we might be interested. There are various schools of thought concerning the construction of stochastic models. Here, we will follow a route that is generally called the "classical" or the "frequentist" approach. Prominent among alternatives is the "subjective" school of probability, an approach that we will explore a bit in Section 1.5 and take up more seriously in Chapter 9. The classical approach to probability is based on the notion that the likelihood of an event should reveal itself in many repetitions of the underlying experiment. Thus, if we were able to repeat a random experiment indefinitely, the relative frequency of occurrence of the event A would converge to a number, that number being the true probability of A.

It is helpful to begin with simpler, more concrete ideas. After all, that is how the field itself began. When the sample space is finite, as it is in the examples considered above, the specification of a stochastic model consists of assigning a number, or probability, to each outcome, that number representing, in the modeler's mind, the chance that this outcome will occur in a given trial of the experiment. Let's reexamine the four examples with which we began. If you had to specify stochastic models for these experiments, what would you do? It's not uncommon in these circumstances to encounter an irresistible urge to assign equal probabilities to all the simple events in each experiment. Pierre Simon Laplace (1749–1827), a brilliant mathematician who was eulogized as the Isaac Newton of France, had enough respect for this urge to have elevated it to the lofty status of a principle. Roughly speaking, his Principle of Insufficient Reason says that if you don't have a good reason to do otherwise, a uniform assignment of probabilities to the sample space is appropriate. In simple examples such as the ones before us, the principle does indeed seem compelling. Thus, in situations in which we are reasonably sure we are dealing with "fair" coins and "balanced" dice, we would consider the uniform stochastic model to be appropriate. More generally, we will call this first approach to the assignment of probabilities to simple events the "intuitive" approach, since it generally relies on intuitive judgments about symmetry or similarity.

Suppose we assign each of the two simple events in S_1 the probability 1/2, recognizing that, under normal circumstances, we believe that they will occur with equal likelihood. Similarly, suppose each simple event in S_2 is assigned probability 1/4, each simple event in S_3 is assigned probability 1/6, and each simple event in S_4 is assigned probability 1/36. We then have, at least implicitly, a basis for thinking about the probability of any compound event that may interest us. To make the transition to this more complex problem, we need the link provided by the natural and intuitive rule below.

Computation Rule. For any compound event A in the finite sample space S, the probability of A is given by

$$P(A) = \sum_{a \in A} P(a), \tag{1.6}$$

where $\sum$ is the standard symbol for addition.

The number $P(A)$ should be viewed as an indicator of the chances that the event A will occur in a single trial of the experiment. The intuition behind this computation rule is fairly basic. If the two simple events a_1 and a_2 have probabilities p_1 and p_2 respectively, then, in many trials, the proportion of trials in which a_1 occurs should be very close to p_1, and the proportion of trials in which a_2 occurs should be very close to p_2. It follows that the proportion of trials in which the outcome is either a_1 or a_2 should be close to $p_1 + p_2$. Since the probability we assign to each simple event represents our best guess at its true probability, we apply to them the same intuition regarding additivity. As we shall see, the assumption that probabilities behave this way is a version of one of the basic axioms upon which the theory of probability is based.

Let's apply our computation rule to the experiment of rolling a pair of balanced dice. One of the outcomes of special interest in the game of Craps (in which two dice, assumed to be balanced, are rolled in each play) is the event "the sum of the digits (or dots) facing up is seven." When this happens on a player's first roll, the player is an "instant winner." The computation rule enables us to obtain the probability of this event:

$$P(\text{sum is seven}) = P(16) + P(25) + P(34) + P(43) + P(52) + P(61) = 6/36 = 1/6.$$

The probabilities of other compound events can be computed similarly:

$$P(\text{At Least One Digit is a 4}) = 11/36$$

$$P(\text{First Digit} = \text{Second Digit}) = 1/6$$

$$P(\text{First Digit} > \text{Second Digit}) = 5/12.$$

The latter computation can be done easily enough by identifying the 15 simple events, among the 36, which satisfy the stated condition. It is useful, however, to get accustomed to looking for shortcuts. Could you have obtained the answer 5/12 just by reflecting on the symmetry of the situation? Another type of shortcut reveals itself when we try to compute the probability of an event like "the sum of the digits is at least five." Again, we could collect the simple events for which this happens, and add their probabilities together. Even in the simple problem we are dealing with here, this seems like too much work. It's appealing to exploit the fact that the complementary event "the sum is less than five" is a lot smaller. Indeed, there are just six simple events in the latter compound event, leading to the computation

$$P(\text{Sum is at Least 5}) = 1 - P(\text{Sum is Less Than 5}) = 1 - 1/6 = 5/6.$$

The alternative computation is based on the fact that the probabilities assigned to the 36 sample events in this experiment add up to 1. You will have the opportunity to gather additional experience with the computation rule in the exercises at the end of this section.

So far, we have dealt only with experiments in which the assignment of probabilities to simple events can be done intuitively from our assumptions of symmetry, balance, or similarity. We believe that the chances that a card drawn randomly from a standard 52-card deck will be a spade is 1/4 simply because we see no reason that it should be more or less likely than drawing a heart. You will notice, however, that the applicability of the computation rule is not limited to uniform cases; it simply says that once the probabilities of simple events are specified, we can obtain probabilities for compound events by appropriate addition.

The theory of probability, just like that of physics, economics, or biology, had better give answers that are in general agreement with what we actually observe. If the foundations of these subjects were radically inconsistent with what we see in the world around us, we would discard them and keep looking for a good explanation of our individual and collective experience. Your "intuition" about a random experiment is only trustworthy if it is compatible with your past experience and if it helps you predict the outcomes of future replications of the same kind of experiment. The real basis for modeling a random experiment is past experience. Your intuition represents a potential shortcut, and whenever you use it, you need to entertain, at least momentarily, the idea that your intuition may be off the mark. If you saw someone roll five consecutive sevens in the game of Craps, wouldn't the suspicion that the dice are unbalanced creep into your consciousness? But how do we approach the assignment of probabilities to simple events when we are unable to rely upon our intuition? Since most of the situations in which we wish to model randomness are of this sort, it is essential that we have a trustworthy mechanism for treating them. Fortunately, we do, and we now turn our attention to its consideration. For reasons that will be readily apparent, we refer to this alternative approach to stochastic modeling as the "empirical" approach.

Consider now how you might approach stochastic modeling in an experiment in which a thumb tack, instead of a fair coin, was going to be flipped repeatedly. Suppose a "friend" offers you the opportunity to bet on the event that the tack lands with the point facing downwards (D), and proposes that he pay you \$3 each time that happens, with you paying him \$2 whenever the point faces upwards (U). In order to approach this bet intelligently, you need a stochastic model for the two possible outcomes D and U, down and up. But where does such a model come from? Would you trust your intuition in this instance? Since your intuition about this experiment is probably a bit hazy, and since your financial resources are not unlimited, that seems pretty risky. What is your best course of action? "I'll let you know tomorrow" seems like a prudent response. What you need is some experience in flipping thumb tacks. In a twenty-four hour period, you could gather a great deal of experience. Suppose you find that in one thousand flips, your thumb tack faced downwards only 391 times. What you have actually done is manufactured a stochastic model for flipping a thumb tack. Based on your experience, it is reasonable for you to assign the probabilities $P(D) = .391$ and $P(U) = .609$ to the simple events in your experiment. Is this model a good one? It is probably quite good in terms of predicting how many times you would observe the event "D" in a few future flips. It is a model that is undoubtedly better than the one you would come up with based on raw intuition, but it is probably not as reliable as the model you could construct if you'd had the patience to flip a thumb tack ten thousand times. On the basis of your model for this experiment, you should turn down the wager you have been offered. In Exercise 2.2.5, you will be asked to confirm, by making a suitable computation, that this bet is stacked against you.

Most instances in which we confront the need for stochastic modeling don't actually involve wagering, at least not explicitly. Probability models are used in predicting the chances that it will rain tomorrow, in determining the appropriate premium for your automobile insurance (a touchy subject, I know), and in predicting how well an undergraduate student might do in law school. These are problems in which intuition is hard to come by, but they are also problems that arise in real life and require your, or at least someone's, occasional attention. How does an insurance company go about determining their costs in offering life insurance to your Aunt Mable? Suppose, for simplicity, that Aunt Mable wants a straightforward policy that will pay her heirs $1,000,000 if she dies before age sixty-five, and pays them nothing otherwise. An insurance company must consider a number of relevant issues, among them the chances that a thirty-two-year-old woman in good health will live to age sixty-five. You might think that, since there is only one Aunt Mable, one can't experiment around with her as easily as one can flip a thumb tack. Indeed, the thought of flipping Aunt Mable even a couple of times is enough to give one pause. But insurance companies do just that; they treat Aunt Mable as if she were a typical member of the population of all thirty-two-year-old women in good health. Under that assumption, they determine, from an appropriate volume of the mortality tables they maintain very diligently, what proportion of that population survives to age sixty-five. In fact, in order to set an annual premium, a company will need to have a stochastic model for Aunt Mable surviving one year, two years, ..., thirty-three years. The number of insurance companies seeking Aunt Mable's business, and the size of the buildings in which they operate, attest to the success this industry tends to have in constructing stochastic models as needed. Actuarial science, as the mathematical theory of insurance is called, is an area of inquiry that is centuries old. Halley's life tables, published in 1693, estimated the probability that a thirty-two-year-old female would survive to age sixty-five to be 177/403.

The art world has been rocked by recent reassessments of the authenticity of certain paintings that have traditionally been attributed to Rembrandt. The "connoisseurship" movement has, as you might imagine, made more than one museum curator sweat. "Authenticity studies" often involve the development of stochastic models in an attempt to determine the chances that a particular body of work could have been done by a given individual. A statistically based comparison of work that may or may not have been performed by the same individual has become a standard tool in such studies. One of the best-known examples involves the use of statistical techniques in settling a long-standing historical debate involving the authorship of the *Federalist Papers*. This rather famous piece of detective work is described below.

In 1787–88, a series of anonymous essays were published and distributed in the state of New York. Their clear purpose was to persuade New Yorkers to ratify the Constitution. In total, eighty-five essays were produced, but serious interest in precisely who wrote which ones arose only after Alexander Hamilton's death in a duel in 1804. By a variety of means, definitive attribution appeared to be possible for seventy of the papers, with Hamilton identified as the author of forty-one, and James Madison being credited with writing fourteen of the others. Of the fifteen papers whose authorship was uncertain, three were determined to have been written by a third party, and each of a group of twelve was variously attributed to Hamilton or to Madison, with scholars of high repute stacking up on both sides in a vigorous game of academic tug of war. Enter Mosteller and Wallace (1964), two statisticians who were convinced that the appropriate classification could be made through a careful analysis of word usage. In essence, it could be determined from writings whose authorship was certain that there were words that Hamilton used a lot more than Madison, and vice versa. For example, Hamilton used the words "on" and "upon" interchangeably, while Madison used "on" almost exclusively. The table below, drawn from the Mosteller and Wallace study, can be viewed as a specification of a stochastic model for the frequency of occurrence of the word "upon" in arbitrary essays written by each of these two authors. The collection of written works examined consisted of forty-eight works known to have been authored by Hamilton, fifty works known to have been authored by Madison, and twelve *Federalist Papers* whose authorship was in dispute. By itself, this analysis presents a strong argument for classifying at least eleven of the twelve disputed papers as Madisonian. In combination with a similar treatment of other "non-contextual" words in these writings, this approach provided strong evidence that Madison was the author of all twelve of the disputed papers, essentially settling the authorship debate.

Rate/1000 Words	Authored by Hamilton	Authored by Madison	12 Disputed Papers
Exactly 0	0	41	11
(0.0, 0.4)	0	2	0
[0.4, 0.8)	0	4	0
[0.8, 1.2)	2	1	1
[1.2, 1.6)	3	2	0
[1.6, 2.0)	6	0	0
[2.0, 3.0)	11	0	0
[3.0, 4.0)	11	0	0
[4.0, 5.0)	10	0	0
[5.0, 6.0)	3	0	0
[6.0, 7.0)	1	0	0
[7.0, 8.0)	1	0	0
Totals:	48	50	12

Table 1.2.1. Frequency distribution of the word "upon" in 110 essays.

Exercises 1.2.

1. Specify the sample space for the experiment consisting of three consecutive tosses of a fair coin, and specify a stochastic model for this experiment. Using that model, compute the probability that you (a) obtain exactly one head, (b) obtain more heads than tails, (c) obtain the same outcome each time.

2. Suppose a pair of balanced dice is rolled and the number of dots that are facing upwards are noted. Compute the probability that (a) both numbers are odd, (b) the sum is odd, (c) one number is twice as large as the other number, (d) the larger number exceeds the smaller number by 1, and (e) the outcome is in the "field" (a term used in the game of Craps), that is, the sum is among the numbers 5, 6, 7, and 8.

3. A standard deck of playing cards consists of 52 cards, with each of the 13 "values" ace, 2, 3, ..., 10, jack, queen, king appearing in four different suits—spades, hearts, diamonds, and clubs. Suppose you draw a single card from a well-shuffled deck. What is the probability that you (a) draw a spade, (b) draw a face card, that is, a jack, queen, or king, and (c) draw a card that is either a face card or a spade.

4. Refer to the frequency distributions in Table 1.2.1. Suppose you discover a dusty, tattered manuscript in an old house in Williamsburg, Virginia, and note that the house displays two plaques, one asserting that Hamilton slept there and the other asserting that Madison slept there. Your attention immediately turns to the frequency with which the author of this manuscript used the word "upon." Suppose that you find that the rate of usage of the word "upon" in this manuscript was 2.99 times per 1000 words. If Hamilton wrote the manuscript, what probability would you give to the event that the manuscript contains fewer than 3 uses of the word "upon" per 1000 words? If Madison wrote the manuscript, what probability would you give to the event that the manuscript contains more than 2 uses of the word "upon" per 1000 words? What's your guess about who wrote the manuscript?

1.3 The Axioms of Probability

Mathematical theories always begin with a set of assumptions. Euclid built his science of plane geometry by assuming, among other things, the "parallel postulate," that is, the assertion that through any point outside line A, there exists exactly one line parallel to line A. The mathematical theory that follows from a set of assumptions will have meaning and utility in real-world situations only if the assumptions made are themselves realistic. That doesn't mean that each axiom system we entertain must describe our immediate surroundings. For example, Einstein's Theory of Relativity is based on a type of non-Euclidean geometry which discards the parallel postulate, thereby allowing for the development of the richer theory required to explain the geometry of time and space. Since the purpose of any mathematical theory is to yield useful insights into real-world applications, we want the axioms used as the theory's starting point to agree with our intuition regarding these applications. Modern probability theory is based on three fundamental assumptions. In this section, we will present these assumptions, argue that they constitute a reasonable foundation for the calculus of probability, and derive a number of their important implications.

When you go about the business of assigning probabilities to the simple events in a particular random experiment, you have a considerable amount of freedom. Your stochastic model for flipping a thumb tack is not likely to be the same as someone else's. Your model will be judged to be better than another if it turns out to be more successful in predicting the frequency of the possible outcomes in future trials of the experiment. It still may be true that two different models can justifiably be considered "good enough for government work," as the old saying goes. We will not ask the impossible—that our model describe reality exactly—but rather seek to specify models that are good enough to provide reasonable approximations in future applications. In the words of statistician George E. P. Box, "all models are wrong, but some are useful."

While two models may assign quite different probabilities to simple events, they must have at least a few key features in common. Because of the way we interpret probability, for example, we would not want to assign a negative number to represent the chances that a particular event will occur. We thus would want to place some intuitively reasonable constraints on our probability assignments. The following axiom system is generally accepted as the natural and appropriate foundation for the theory of probability. Let S be the sample space of a random experiment. Then the probabilities assigned to events in S must satisfy

Axiom 1. For any event $A \subseteq S, P(A) \geq 0$.

Axiom 2. $P(S) = 1$.

Axiom 3. For any collection of events $A_1, A_2, \ldots, A_n, \ldots$ satisfying the conditions

$$A_i \cap A_j = \varnothing, \text{ for all } i \neq j, \tag{1.7}$$

the probability that at least one of the events among the collection $\{A_i, i = 1, 2, 3, \ldots\}$ occurs may be computed as

$$P(\bigcup_{i=1}^{\infty} A_i) = \sum_{i=1}^{\infty} P(A_i). \tag{1.8}$$

The first two of these axioms should cause no anxiety, since what they say is so simple and intuitive. Because we generally think of probabilities as long-run relative frequencies of occurrence, we would not want them to be negative. Further, we know when we define the sample space of a random experiment that it encompasses everything that can happen. It thus necessarily has probability one. Axiom 3 is the new and somewhat exotic assumption that has been made here. Let's carefully examine what it says.

In Section 1.1, we referred to non-overlapping sets as "disjoint." In probability theory, it is customary to use the alternative phrase "mutually exclusive" to describe events which do not overlap. The message of Axiom 3 may be restated as: For any collection of mutually exclusive events in a discrete random experiment, the probability of their union is equal to the sum of their individual probabilities. The assertion made in Axiom 3 is intended to hold for a collection of arbitrary size, including collections that are countably infinite. When Axiom 3 is restricted to apply only to finite collections of mutually exclusive events, it is called "the axiom of finite additivity." Otherwise, we call it "the axiom of countable additivity." The argument over which of the two is more appropriate as part of the axiomatic foundation of probability theory is not fully settled, though the proponents of the more general form of the axiom far outnumber the champions of finite additivity. The latter group argues that our intuition regarding additivity is based on our experience, which can only involve finite collections of events. Thus, we should not assert, as a fundamental truth, a rule for combining infinitely many probabilities. The other side argues that countable additivity does agree with our intuition and experience, and, in fact, enables us to obtain intuitively correct answers to problems that simply can't be treated otherwise. The following example illustrates this point.

Example 1.3.1. Suppose you toss a fair coin until you get a head. Let X represent the number of tosses it takes you. What probability would you attach to the event that X will be an odd number? It is clear that $P(X = 1) = 1/2$, since the very first toss will yield a head with probability $1/2$. If a fair coin is tossed twice, the probability that you get the outcome TH, that is, a tail followed by a head, is $1/4$. In your sequence of tosses, this is the one and only way in which the event $\{X = 2\}$ can happen. It follows that $P(X = 2) = 1/4$. This same logic extends to arbitrary values of X, yielding, for $n = 1, 2, 3, 4, \ldots$, the result that

$$P(X = n) = 1/2^n. \tag{1.9}$$

With the axiom of countable additivity, we can represent the desired probability as

$$\begin{aligned} P(\text{X is odd}) &= P(X = 1) + P(X = 3) + P(X = 5) + \ldots \\ &= 1/2 + 1/8 + 1/32 + \ldots. \end{aligned} \tag{1.10}$$

If you are acquainted with geometric series, you might recognize the infinite sum in (1.10) as something you can evaluate using an old familiar formula. If not, you needn't worry, since the appropriate formula, which will turn out to be useful to us in a number of different problems, will be introduced from scratch in Section 2.6. For now, we will take a handy shortcut in evaluating this sum. Since the sample space in this experiment can be thought of as the set of positive integers $\{1, 2, 3, \ldots\}$, the sum of all their probabilities is 1. Think of each of the terms you wish to add together as being paired with the probability of the even number that follows it. Thus, 1/2 goes with 1/4, 1/8 goes with 1/16, etc. For each term that is included in your series, a term with half its value is left out. If we denote the sum of the series in (1.10) by p, you can reason that, since the sum of all the terms $P(X = k)$ for $k \geq 1$ is 1, you must have

$$p + (1/2)p = 1.$$

From this, you may conclude that $p = P(X \text{ is odd}) = 2/3$. ■

What does your own intuition have to say about the probability we have just calculated? Perhaps not very much. The two-to-one ratio between the probabilities of consecutive integer outcomes for X is itself an intuitively accessible idea, but it is not an idea that you would be expected to come up with yourself at this early stage in your study of the subject. What should you do when your intuition needs a jump-start? You proceed empirically. You will find that it's not that difficult to convince yourself of the validity of the answer obtained above. Take a coin out of your pocket and perform this experiment a few times. You will see that the event $\{X \text{ is odd}\}$ does indeed occur more often that the event $\{X \text{ is even}\}$. If you repeat the experiment enough, you will end up believing that the fraction $2/3$ provides an excellent forecast for the relative frequency of occurrence of an odd value of X. This, and many examples like it, give us confidence in assuming and using the axiom of countable additivity.

Let us now suppose that we have specified a stochastic model for the sample space S of a random experiment, and that the probabilities assigned by our model obey the three axioms above. From this rather modest beginning, we can derive other rules which apply to and facilitate the calculation of probabilities. Some of these "derived" results will seem so obvious that you will want to react to them with the question "What's to prove here?" This reaction is especially understandable when the claim we wish to prove seems to be nothing more than an alternative way to state the contents of one or more axioms. Here is the attitude I suggest that you adopt in our initial mathematical developments. Erase everything from your mind except the three axioms. Suppose that this constitutes the totality of your knowledge. You now want to identify some of the logical implications of what you know.

If some of the early results we discuss seem obvious or trivial, keep in mind that one must learn to walk before one learns to run.

Each of the conclusions discussed below will be stated as a theorem to be proven. Our first result is an idea that we used, on intuitive grounds, in some of our probability computations in the last section. We now consider its formal justification.

Theorem 1.3.1. For any event $A \subseteq S$, the probability of A^c, the complement of A, may be calculated as

$$P(A^c) = 1 - P(A). \tag{1.11}$$

Proof. Note that the sample space S may be represented as the union of two mutually exclusive events. Specifically, $S = A \cup A^c$, where $A \cap A^c = \varnothing$. It follows from Axioms 2 and 3 that

$$1 = P(S) = P(A \cup A^c) = P(A) + P(A^c),$$

an equation that immediately implies (1.11). ■

Theorem 1.3.2. The empty event $\varnothing$ has probability zero.

Proof. Since $\varnothing = S^c$, we have, by Theorem 1.3.1 and Axiom 2,

$$P(\varnothing) = 1 - P(S) = 0.$$

■

One simple consequence of Theorem 1.3.2 is the fact that it makes it apparent that the finite additivity of the probabilities assigned to the simple events of a given sample space is a special case of the countable additivity property stated in Axiom 3. To apply Axiom 3 to a finite collection of mutually exclusive events $A_1, A_2, \ldots, A_n$, one simply needs to define $A_i = \varnothing$ for $i > n$. With this stipulation, we have that

$$\bigcup_{i=1}^{\infty} A_i = \bigcup_{i=1}^{n} A_i,$$

so that

$$P(\bigcup_{i=1}^{n} A_i) = P(\bigcup_{i=1}^{\infty} A_i) = \sum_{i=1}^{n} P(A_i) + \sum_{i=n+1}^{\infty} P(\varnothing) = \sum_{i=1}^{n} P(A_i).$$

Another consequence of Axiom 3 is the important

Theorem 1.3.3. (The Monotonicity Property) If $A \subseteq B$, then $P(A) \leq P(B)$.

Proof. Since $A \subseteq B$, we may write

$$B = A \cup (B - A).$$

Moreover, the events A and $B - A$ are mutually exclusive. Thus, we have, by Axiom 3,

$$P(B) = P(A) + P(B - A). \tag{1.12}$$

Since, by Axiom 1, $P(B - A) \geq 0$, we see from (1.12) that $P(B) \geq P(A)$, as claimed. ■

You will notice that the first axiom states only that probabilities must be positive; while it seems obvious that probabilities should not exceed one, this additional fact remains unstated, since it is implicit in the axioms, being an immediate consequence of the monotonicity property above. As is typical of axiom systems in mathematics, Axioms 1–3 above represent a "lean" collection of axioms containing no redundancy. In the present context, we derive from our axioms, and their three known consequences proven above, that no event A can have a probability greater than 1.

Theorem 1.3.4. For any event $A \subseteq S, P(A) \leq 1$.

Proof. By Theorem 1.3.3 and Axiom 2, we have that $P(A) \leq P(S) = 1$. ■

The axiom of countable additivity gives instructions for computing the probability of a union of events only when these events are mutually exclusive. How might this axiom give us some leverage when we are dealing with two events A and B that have a non-empty intersection? One possibility: we could capitalize on the identity (1.1.1) in Section 1.1, which expresses $A \cup B$ as a union of disjoint sets. By Axiom 3, we could then conclude that

$$P(A \cup B) = P(A - B) + P(A \cap B) + P(B - A). \tag{1.13}$$

This formula, while perfectly correct, is not as convenient as the "addition rule" to which we now turn. In most applications in which the probability of a union of events is required, the probabilities on the right-hand side of 1.13 are not given or known and must themselves be derived. A more convenient formula for $P(A \cup B)$ would give this probability in terms of the probabilities of simpler, more basic events. Such a formula is provided in

Theorem 1.3.5. (The Addition Rule) For arbitrary events A and B,

$$P(A \cup B) = P(A) + P(B) - P(A \cap B). \tag{1.14}$$

Remark 1.3.1. If you consider the Venn diagram for $A \cup B$, you will see immediately why merely adding the probabilities P(A) and P(B) together tends to give the wrong answer. The problem is that this sum will count the event $A \cap B$ twice. The obvious fix is to subtract $A \cap B$ once. We now show that this is precisely the right thing to do.

Proof. Note that $A \cup B$ may be written as $A \cup (B - A)$; since $A \cap (B - A) = \varnothing$, Axiom 3 implies that

$$P(A \cup B) = P(A) + P(B - A). \tag{1.15}$$

Similarly, since $B = (A \cap B) \cup (B - A)$, and $(A \cap B) \cap (B - A) = \varnothing$, we also have that

$$P(B) = P(A \cap B) + P(B - A). \tag{1.16}$$

We may rewrite (1.16) as

$$P(B - A) = P(B) - P(A \cap B). \tag{1.17}$$

Substituting (1.17) into (1.15) yields the addition rule in (1.14). ■

Proving the extension of the addition rule below is left as an exercise.

Theorem 1.3.6. For any events A, B, and C,

$$\begin{aligned} P(A \cup B \cup C) = P(A) + P(B) + P(C) - P(A \cap B) \\ - P(A \cap C) - P(B \cap C) + P(A \cap B \cap C). \end{aligned} \tag{1.18}$$

A full generalization of the addition rule will be a useful formula to have in your toolbox. The formula is known as the Inclusion-Exclusion Rule and is stated below. The theorem is true for any collection of events in an arbitrary sample space. A proof for the discrete case, that is, for the case in which the sample space is finite or countably infinite, uses Theorem 2.4.2 (the binomial theorem), which we have yet to discuss, so we will state the result, at this point without proof. Notice that when $n = 3$, the result reduces to Theorem 1.3.6. (See Problem 1.8.15 for a proof when S is finite and $n = 4$.)

Theorem 1.3.7. (The Inclusion-Exclusion Rule) Consider the events $A_1, A_2, \ldots, A_n$ in a given random experiment. Then

$$\begin{aligned} P\left(\bigcup_{i=1}^{n} A_i\right) = \ & \textstyle\sum_{i=1}^{n} P(A_i) - \sum_{1\leq i<j\leq n} P(A_i \cap A_j) \\ & + \textstyle\sum_{1\leq i<j<k\leq n} P(A_i \cap A_j \cap A_k) \\ & - \cdots + (-1)^{n+1} P\left(\bigcap_{i=1}^{n} A_i\right). \end{aligned} \tag{1.19}$$

We now come to a somewhat more complex claim, a property of probabilities called "countable subadditivity." It is a result which is useful in providing a lower bound for probabilities arising in certain statistical applications (for example, when you want to be at least 95% sure of something). In addition to its utility, it provides us with a vehicle for garnering a little more experience in "arguing from first principles." Before you read the proof (or any proof, for that matter), it is a good idea to ask yourself these questions: i) Does the theorem statement make sense intuitively? ii) Do I know it to be true in any special cases? and iii) Can I see how to get from what I know (that is, definitions, axioms, proven theorems) to what I want to prove? The last of these questions is the toughest. Unless you have a mathematical background that includes a good deal of practice in proving things, you will tend to have problems getting started on a new proof. At the beginning, your primary role is that of an observer, learning how to prove things by watching how others do it. In the words of the great American philosopher Yogi Berra, you can observe a lot by watching. Eventually, you learn some of the standard approaches, just like one can learn some of the standard openings in the game of chess by watching the game played, and you find yourself able to construct some proofs on your own. You should be able to do well in this course without perfecting the skill of mathematical argumentation, but you will have difficulty understanding the logic of probability and statistics without gaining some facility with it. The proof of countable subadditivity is our first serious opportunity to exercise our mathematical reasoning skills. The inequality below is generally attributed to the English mathematician George Boole.

Theorem 1.3.8 (Countable subadditivity). For any collection of events $A_1, A_2, \ldots, A_n, \ldots,$

$$P\left(\bigcup_{i=1}^{\infty} A_i\right) \leq \sum_{i=1}^{\infty} P(A_i). \tag{1.20}$$

Remark 1.3.2. While the inequality in (1.20) is written in a way that stresses its applicability to countably infinite collections of events, it applies as well to the special case of the finite collections of events. To obtain inequality for k events, say, simply define $A_i = \varnothing$ for $i > k$. As we proceed to the proof, our first order of business will be to put ourselves in a position to use "the hammer," that is, countable additivity, which is the only thing we really know about handling probabilities for infinite collections of events. The challenge in doing this is that Axiom 3 only applies to non-overlapping events, while the events here are arbitrary and may, in fact, overlap in weird and exotic ways. We begin the proof of (1.20) by trying to express unions of the A's as unions of mutually exclusive events.

Proof. Let's define a new collection of events, a collection that is mutually exclusive yet accounts for all the elements contained in the events $A_1, A_2, \ldots$. Let

$$B_1 = A_1,$$

and let

$$B_2 = A_2 - A_1.$$

We may describe the event B_2 as "the new part of A_2," that is, the part of A_2 that has not been accounted for in A_1. Clearly, $B_1 \cup B_2 = A_1 \cup A_2$ and, in addition, $B_1 \cap B_2 = \varnothing$. Now, let

$$B_k = A_k - (\bigcup_{i=1}^{k-1} A_i) \text{ for all integers } k \geq 3.$$

Each set B can be thought of as the "new part" of the corresponding A. In general, B_k is the set of all elements of A_k which are not contained in any of the preceding $(k-1)$ A's. Note that any two of the B events are mutually exclusive, that is, for any $i < j$,

$$B_i \cap B_j = \varnothing. \tag{1.21}$$

This follows from the fact that, for arbitrary $i < j, B_i \subseteq A_i$ while $B_j \subseteq A_i^c$, since B_j only contains elements that were not members of any preceding event A, including of course, the event A_i. Moreover, for every positive k, a simple set theoretic argument shows that

$$\bigcup_{i=1}^{k} B_i = \bigcup_{i=1}^{k} A_i. \tag{1.22}$$

(See Exercise 1.3.3. Actually, don't just see it, do it!) Using the same basic argument, we may conclude that

$$\bigcup_{i=1}^{\infty} B_i = \bigcup_{i=1}^{\infty} A_i. \tag{1.23}$$

Now we are in a position to apply Axiom 3, which together with Theorem 1.3.3, yields

$$\begin{aligned} P(\bigcup_{i=1}^{\infty} A_i) &= P(\bigcup_{i=1}^{\infty} B_i) \\ &= \sum_{i=1}^{\infty} P(B_i) \\ &\leq \sum_{i=1}^{\infty} P(A_i) \end{aligned}$$

and completes the proof. ■

Over the years, I have seen the mathematical foundations of probability engender a variety of reactions in my students, reactions ranging from serenity to gasping for air. If you aren't feeling entirely comfortable with what we have done so far, it's important for you to separate your reactions into two parts—the "I would never have thought of this myself" part and the "it doesn't seem to make sense" part. The first part is nothing to worry about. Kolomogorov was a pretty smart guy, and the idea that the axioms we have examined are precisely what is needed for a coherent theory of probability is far from obvious. If it was, someone would probably have said so before 1933. It is important, however, for you to recognize that the developments above make logical sense. Understanding the mathematical reasoning connecting them will give you a certain confidence in using them to solve problems. It might be worth your while to copy the statements of the theorems we have proven thus far onto a clean sheet of paper, close the book, and see if you can provide the proofs yourself. If you can't, read them again, and then try once more to reproduce them.

We close this section by considering some problems that call for the application of some form of the addition rules discussed above.

Example 1.3.2. The most basic approach to a probability computation is to identify all the simple events contained in the event of interest, and to add up their probabilities. Since simple events are, of necessity, mutually exclusive, the approach is justified by Axiom 3; for finite sample spaces, it was stated explicitly in (1.2.1), and was used in some of our early computations. Let's use it now in some applications with a little more complexity. Suppose first that a fair coin is tossed four times. What is the probability that you obtain exactly two heads? In identifying the simple events that correspond to this outcome, it is helpful to draw, or at least picture mentally, the corresponding tree:

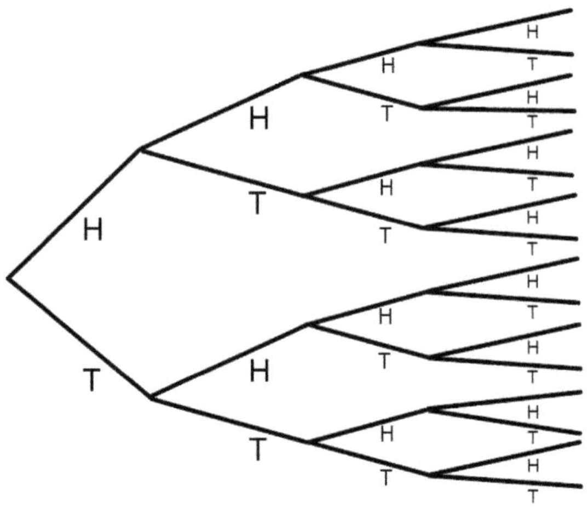

Figure 1.3.1. The sample space for the experiment of tossing a coin four times.

From this tree, it is easy to spot the simple events we are interested in. Since the sixteen simple events can be regarded as equally likely, it is clear that

$$P(\text{Exactly 2 heads in 4 tosses}) = 6/16 = 3/8.$$

Coin tossing can provide a host of similar questions; their conceptual difficulty gets no greater than that of the problem we have just solved, but the computation gets increasingly awkward as the number of tosses grow. Consider, for example, the probability of getting exactly five heads in ten tosses of a fair coin. If you are willing to draw the tree corresponding to this experiment, it's possible that you have too much time on your hands. You could, instead, concentrate on identifying only those paths with exactly five heads, but even that chore is quite formidable. If you had the patience to do it, you would find that there are 252 ways to get a string of five heads and five tails. Since the tree we have alluded to contains a total of 1024 branches, we may conclude that

$$P(\text{Exactly 5 heads in 10 tosses}) = 252/1024 \approx .2461.$$

Clearly, we will want to have better tools for attacking problems like this. ■

Example 1.3.3. The same approach we have used on coin tossing will apply to repeated draws, without replacement, from a finite population. Let's suppose, for example, that two cards are drawn at random from a standard 52-card deck. The tree representing this experiment is too large to draw in complete detail, but it looks something like this:

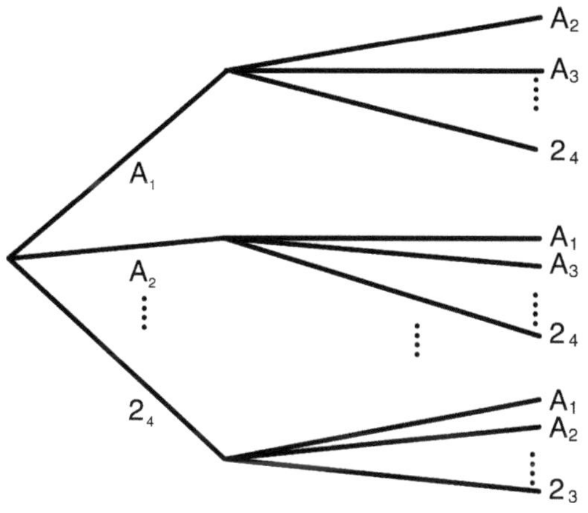

Figure 1.3.2. Sample space for drawing 2 cards from a standard deck w/o replacement.

From the figure above, it is clear that there are a total of $52 \times 51 = 2652$ different paths in this tree. Assuming that each of them is equally probable, we can again use the basic computation rule to obtain probabilities of interest. For instance, using the tree above as a crutch, we could compute probabilities like the following:

$$P(\text{Two Aces}) = \frac{4 \cdot 3}{52 \cdot 51} = \frac{1}{221},$$

and

$$P(\text{No Aces}) = \frac{48 \cdot 47}{52 \cdot 51} = \frac{188}{221}.$$

The remaining possibility, that of getting exactly one ace, can then be easily treated using Theorem 1.3.1:

$$P(\text{Exactly One Ace}) = 1 - 189/221 = 32/221.$$

■

Example 1.3.4. The U.S. National Center for Health Statistics reported that an estimated 36.7% of the adult population in 1970 were self-proclaimed "smokers." A national Health Interview Survey showed that this figure had fallen to 29.8% by 1985. The fact that this percentage was as large as it was in 1985 bears witness to the tremendous power and influence of advertising. By 1970, the connection between smoking and lung cancer was not only scientifically substantiated, but it had also been widely publicized and was almost universally known. This particular fifteen-year period was, however, the age of the Marlboro Man and of the catchy phrase "you've come a long way, baby." Thus, while many smokers became "former smokers" during this period, the percentage of adults that indicated that they had never smoked actually decreased from 1970 to 1985. To complete the 1985 profile, we note that the adult population targeted by the aforementioned survey was 53% female, and that female smokers constituted 14.5% of the total population. Let's suppose that a randomly chosen individual from this population was subjected to some follow-up questioning. What are the chances that this individual is (a) a smoking male, (b) a non-smoking male, or (c) either a female or a smoker or both? You might give these questions a little thought before reading on.

Let's introduce some simplifying notation, say M and F, for the individual's gender and S or N to denote whether the individual smokes or not. From the above, we take the following probabilities as known:

$$P(S) = .298, P(F) = .530, P(S \cap F) = .145.$$

The answer to question (a) may be obtained using Axiom 3. Since

$$P(S) = P(S \cap M) + P(S \cap F),$$

we have that

$$.298 = P(S \cap M) + .145,$$

so that

$$P(S \cap M) = .153.$$

Part (b) may be answered similarly; since

$$P(M) = P(S \cap M) + P(N \cap M),$$

we may utilize the answer obtained in part (a) to get

$$.470 = .153 + P(N \cap M),$$

yielding

$$P(N \cap M) = .317.$$

The answer to part (c) may be obtained from the addition rule in Theorem 1.3.5:

$$P(S \cup F) = P(S) + P(F) - P(S \cap F) = .298 + .530 - .145 = .683.$$

In examples such as this one, where the events we are interested in can be classified by reference to a couple of factors (here, gender and smoking status), I generally recommend an organizational tool which gives you a quick global picture of the population you are dealing with. I call the tool a "box description of the population." This box is simply a two-dimensional array which helps record the given information. In the present problem, this box, with what we know duly incorporated, is shown below.

	Male	Female	Totals
Smoker		0.145	0.298
Non-smoker			
Totals		0.530	

For this particular box, any three pieces of information allow us to determine quite readily all the remaining entries. This process will seem so easy that you may not even be aware that you are using our axioms and theorems as you go. One does, however, use the fact that the interior boxes are mutually exclusive events, so that the probability of any union of them is appropriately computed by applying Axiom 3. From the completed box, which is given below, the three probabilities we computed above seem to have been given to you for free.

	Male	Female	Totals
Smoker	0.153	0.145	0.298
Non-smoker	0.317	0.385	0.702
Totals	0.470	0.530	1.00

Table 1.3.1. A box description of the population of smokers in 1985.

Once a box like this has been filled in, you have a complete description of the sample space for the experiment of drawing a person at random from the population of interest. So any probability of interest could be obtained from the probabilities displayed. For example, the probability that the individual drawn is either a female or a non-smoker, but not both, may be obtained as $P(F \cap N^c) + P(F^c \cap N) = .145 + .317 = .462$. ■

Exercises 1.3.

1. Let A and B be two events. Identify the conditions on A and B that must hold in order to guarantee the validity of the identity: $P(B - A) = P(B) - P(A)$. Assuming that these conditions hold, prove this identity.

2. Prove Theorem 1.3.6 (Hint: First apply Theorem 1.3.5 to the pairwise union of the events $(A \cup B) \cup C$; then use it to re-express the probability $P(A \cup B)$; finally, note that the event $(A \cup B) \cap C$ may be written as $(A \cap C) \cup (B \cap C)$, so that Theorem 1.3.5 may be applied again.)

3. Provide a set-theoretic proof of the identities (1.22) and (1.23).

4. Consider further the experiment of tossing a fair coin until you obtain a head. Find the probability that the number of tosses required to do this (a) is no greater than three, (b) is greater than four, and (c) is a multiple of three. (See Example 1.3.1.)

5. Suppose two cards are drawn at random (without replacement) from a standard deck. What is the probability that (a) the cards have the same value (e.g., two jacks) and (b) the first

card has a larger value than the second? For concreteness, assume in part (b) that an ace has the lowest value, i.e., the value one.

6. Psychology is a highly diverse academic discipline. The psychology majors at Freudian Slip U. are required to take two particular upper division courses: Psychology 139 (Dreams, Pipe-Dreams, and Outright Fantasies) and Psychology 153 (Numbers, Numeracy, and Numerology). It is a rare student indeed who does outstanding work in both courses. It is known that the chances of getting an A in PSY 139 is .4 and the chances of getting an A in PSY 153 is .3, while the chances of getting an A in both courses are .05. What are the chances that a randomly selected student will get exactly one A in the two courses?

1.4 Conditional Probability

Lloyd's of London built a reputation in the insurance industry by their willingness to insure anything, at any time, in any place. Their approach seems remarkably bold; it suggests that Lloyd's is prepared to construct and trust a stochastic model for the occurrence of any future event. In a sense they are, but success in the insurance business has as much to do with writing good contracts as it does with building good models. Good modeling relies on one's ability to use the information available to characterize the inherent uncertainty in a situation. A good contract will provide for the payment of a premium that comfortably protects the insurer from a reasonable amount of inaccuracy in the model. I am not able to teach you a great deal about contract writing—you might have to go to law school to get good guidance on that. On the other hand, using available information in your modeling is a topic that we will go into in considerable detail. We will see that the stochastic model one would want to use in a given situation is highly dependent on what one happens to know. The probability of living for another year is quite different for you, for someone twice your age, and for someone who has just undergone cancer surgery. The influence that a known condition might have on the specification of probability reveals itself quite clearly in the following simple example.

Suppose we have a box containing ten marbles, some large, some small, some red, some white; the exact contents of the box is represented below:

	Large	Small	Totals
Red	3	2	5
White	4	1	5
Totals	7	3	10

Table 1.4.1. A population of marbles.

Assume that the marbles are well mixed, and that a single marble is drawn from the box at random. Since it is reasonable to think of each of the ten marbles as having the same chance of selection, you would undoubtedly compute the probability that a red marble is drawn as

$$P(R) = \frac{5}{10} = \frac{1}{2}. \tag{1.24}$$

Suppose, however, that I was the one to draw the marble, and without showing it to you, I proclaimed that the marble drawn was large. You would be forced to rethink your probability assignment, and you would undoubtedly be led to the alternative computation

$$P(R|L) = \frac{3}{7}. \tag{1.25}$$

The new symbol introduced in (1.25) is pronounced "the probability of red, given large"; that is, the vertical slash in (1.25) is simply read as "given." Shortly, a formula for computing such a "conditional probability" will be introduced. Before proceeding to a more formal treatment of the topic, let's pause to reflect on the intuition behind the computation in (1.25). The reasoning behind this computation is simple. The known condition leads us to focus on a new sample space. Within that new sample space, we attach equal probability to each marble; after all, the marbles were equally likely to begin with, and it thus seems reasonable to regard all large marbles as equally likely among themselves. When we restrict our attention to the sample space consisting of seven large marbles, and recognize the uniform probability assignment as reasonable, the appropriate probability associated with drawing a red marble is intuitively clear. The basic intuition behind conditional probability computation is well exemplified by what we have done here. When the outcome of an experiment is known to belong to a particular subset of the sample space (the event B, say), one can compute conditional probabilities, given B, by treating B itself as the new sample space, and focusing on the relative likelihoods of occurrence of the simple events within B. This is especially easy to do when our original stochastic model is uniform, but also holds true for more general problems. This background intuition can serve as a useful way for you to check an answer derived by the formulaic approach to conditional probability based on the following:

Definition 1.4.1. (Conditional Probability) Consider a pair of events A and B associated with a random experiment, and suppose that $P(B) > 0$. Then the conditional probability of A, given B, is defined as

$$P(A|B) = \frac{P(A \cap B)}{P(B)}. \tag{1.26}$$

Remark 1.4.1. Note that the conditional probability $P(A|B)$ remains undefined when the event B has probability zero. In discrete problems, one does not encounter the need to condition on an event that has no a priori chance of happening. This need does arise in the continuous case, however. Our treatment of conditional probability will be suitably generalized in Chapter 3.

In our experiment with ten marbles, the formal definition of conditional probability in (1.26) yields

$$P(R|L) = \frac{P(R \cap L)}{P(L)} = \frac{3/10}{7/10} = \frac{3}{7}. \tag{1.27}$$

This computation agrees with our intuitive derivation. I invite you to confirm that these two approaches give the same numerical answers for other conditional probabilities (like $P(R|S)$, $P(S|W)$, and $P(L|R)$) in this problem.

As the discussion above suggests, the stipulation that a particular event B is known to have occurred defines for us a new sample space and a new probability assignment for the simple events in that sample space. We can think of the conditional probabilities that result from (1.26) as defining a new stochastic model. As such, it should obey the constraints we place on all stochastic models, that is, it should obey the three basic axioms of probability theory. If your original stochastic model obeys these axioms, and if conditional probabilities are defined according to (1.26), it can be proven rather easily that these new probabilities will obey the axioms as well. When $P(B) > 0$, it is obvious from (1.26) that we will have $P(A|B) \geq 0$ and $P(S|B) = 1$. Proving that conditional probabilities, given a fixed event B, must be countably additive, provided the original probabilities are, is left as a problem at the end of the chapter. (See Problem 1.7.)

Because conditional probabilities "inherit" the three characteristics assumed to hold for the original probability function P, they also inherit all properties derived from these assumptions. Thus, for example, we don't need to prove from scratch that the conditional version of Theorem 1.3.1 holds. We have already proven that this property holds for any set function satisfying the three fundamental axioms. We therefore may assert that for any fixed event B with $P(B) > 0$, the identity

$$P(A^c|B) = 1 - P(A|B) \tag{1.28}$$

holds for all events $A \subseteq S$. It will not be uncommon for us to solve a problem in more than one way; in that spirit, you are asked in Exercise 1.4.1 to provide a direct argument proving (1.28). Please understand that I do this only to help you gain experience in, and confidence with, the tools of probability theory. Proving a theorem isn't like killing a cockroach, where stepping on it several extra times seems advisable. A theorem will not be more true after you have proven it a second time (using an alternative approach). On the other hand, you will have a better understanding for having done it! Trust me.

To a poker player, conditional probabilities are everything. During a poker game, you need to assess the chances, given the cards you have seen dealt, that one of the other players will beat your hand. While conditional probabilities are often of interest in their own right, their greatest utility occurs in the computation of certain "unconditional" probabilities. We will be particularly interested in computing the probability of the simultaneous occurrence (that is, the intersection) of two or more events. The role that conditional probability can play in such a computation becomes clear if we think of an experiment as evolving in stages. Suppose we are interested in the event $A \cap B$. We may represent the possible outcomes of the experiment (to the extent that they relate to A and B) in the following tree:

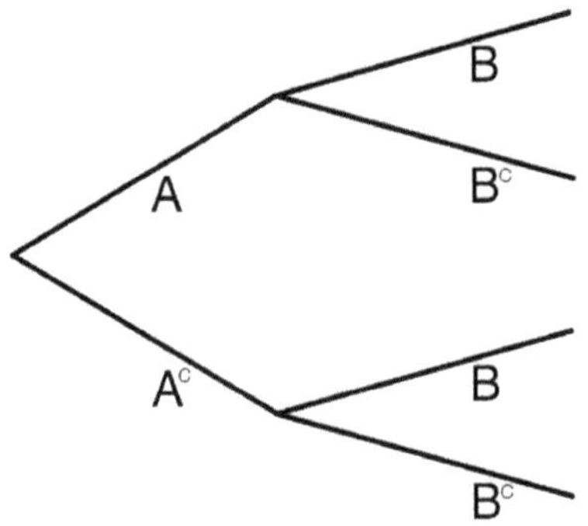

Figure 1.4.1. A tree for a small two-stage experiment.

The event $A \cap B$ is the upper path of this tree and stands for the outcome "A and B both happen." It is clear that $A \cap B$ occurs if, and only if, A happens in the first stage, and then, given that A has happened, B happens in the second stage. The identity linking these two ways of thinking about the event $A \cap B$ is stated in the important

Theorem 1.4.1. (The Multiplication Rule) For any events A and B such that $P(B) > 0$,

$$P(A \cap B) = P(A) \cdot P(B|A). \tag{1.29}$$

The identity in (1.29) follows from the definition of $P(B|A)$ upon multiplying $P(B|A)$ by $P(A)$. It is important to note that, in calculating the probability of the simultaneous occurrence of two events A and B, the event which one chooses to condition on is optional. In other words, one may calculate $P(A \cap B)$ as in (1.29), or one may calculate it as

$$P(A \cap B) = P(B) \cdot P(A|B). \tag{1.30}$$

Both formulas will result in the same answer. The particulars of the problem of interest will usually dictate which of the two orders is best or is more natural. For example, if we are calculating the probability of drawing two aces at random, without replacement, from a standard deck, one would typically condition on the outcome of the first draw to obtain

$$P(A_1 \cap A_2) = P(A_1) \cdot P(A_2|A_1) = \frac{4}{52} \cdot \frac{3}{51} = \frac{1}{221}. \tag{1.31}$$

Remark 1.4.2. It is perhaps worth mentioning that the situations for which the multiplication rule doesn't provide an answer (for example, when $P(B) = 0$) can be handled quite easily, as follows. Since $A \cap B$ is a smaller event than B itself, all of our stochastic models will assign probability zero to $A \cap B$ when the event B has probability zero.

Finally, we note the fact that there is nothing sacred about the number two. The multiplication rule may be extended to apply to the intersection of an arbitrary collection of events. For example, the multiplication rule for three events A, B, and C (for which $P(A \cap B) > 0$) says that

$$P(A \cap B \cap C) = P(A) \cdot P(B|A) \cdot P(C|A \cap B). \tag{1.32}$$

We can and will use the multiplication rule in computing probabilities for multi-stage (say n-stage) experiments. The general multiplication rule by which the probability of the joint occurrence of n events may be computed is given by

$$P(A_1 \cap A_2 \cap \cdots \cap A_n) = P(A_1) \cdot P(A_2|A_1) \cdots P(A_n|\bigcap_{i=1}^{n-1} A_i). \tag{1.33}$$

Let's now turn our attention to a series of examples in which the power and utility of the multiplication rule become apparent. As we have seen, it is convenient to think of experiments with several aspects as evolving in stages. If I hand you two cards from a well-shuffled deck, there is no intrinsic necessity to think of them as a first card and a second card, but it is very convenient to do so. The following examples will help you to understand why.

Example 1.4.1. Suppose two cards are drawn without replacement from a 52-card deck. What is the probability that both are spades? This question can be answered without drawing a "probability tree," but I will draw one anyway. I consider the probability tree to be the single most useful tool in dealing with probability computations for multistage experiments, and I'd like you to get accustomed to drawing them. Let's denote the event "spade" by the letter S and the event "non-spade" by the letter N. The tree for this particular problem is shown below.

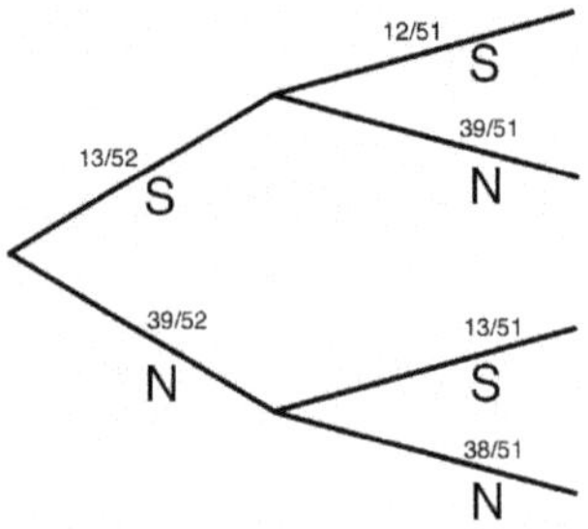

Figure 1.4.2. A tree for drawing spades in two random draws from a standard deck.

The language I will use when referring to trees is the following: each individual segment of the tree (there are four of them in the tree above) will be called a branch; a set of branches that form a continuous pathway from left to right will be called a path. For example, the upper path in the tree above represents the event $S_1 \cap S_2$, a spade in each draw, and consists of the sequence of individual branches S_1 and S_2. You will notice that I have attached a probability to each branch in the tree. The probabilities of the branches S_1 and N_1 are unconditional, because the experiment is just beginning. The probability of an outcome of any subsequent draw is a conditional probability; its numerical value will depend on what happened in earlier draws. For the upper path, we have entered the probability $P(S_1) = 13/52$ for the first branch and $P(S_2|S_1) = 12/51$ for the second branch. In the problem at hand, and in most multistage experiments, we are interested in the probability of one or more paths of the tree. You can think of each path as a simple event; the paths of the tree are precisely the mutually exclusive ways in which the experiment can turn out. Path probabilities are thus the fundamental building blocks for probability computations in this type of experiment. Where do path probabilities come from? From the multiplication rule. That is why the tree is so useful. It helps you line up all the components of a potentially complex probability computation, and it makes the computation look easy. In the present problem, we compute the probability of the upper path as

$$P(S_1 \cap S_2) = P(S_1) \cdot P(S_2|S_1) = \frac{13}{52} \cdot \frac{12}{51} = \frac{1}{17}. \tag{1.34}$$

The probabilities of the three other paths of the tree in Figure 1.4.2 are obtained similarly. It is clear that the computation in (1.34) could have been done without drawing the tree. The primary value of the tree lies not in the fact that it helps record the elements from which an individual path probability is obtained. Helpful as that may be, the main value of the tree is that it gives you an overview of the experiment as a whole. The single most common mistake made by nouveau-probabilists is the failure to account for all the ways that a particular event can happen. A probability tree will help you avoid this mistake. It also puts you in a position to obtain the probability of complex events, since any event can be thought of as a certain collection of paths. The experiment of drawing two cards from a deck is too simple to provide a good example of this latter point (the probability of getting exactly one spade best exemplifies it), but the point will become entirely believable as we proceed. ■

Example 1.4.2. Playing poker is fun. By the time you have finished this chapter, you may have formed the impression that I have a perverse fascination with the game. In the words of the elder, late Mayor Daley of Chicago, I want you to know that I resent this *insinuendo*. My fascination with poker is not perverse! Be that as it may, we will be considering a good many questions that arise in poker, primarily because answering these questions will involve us in a deep and instructive excursion into the probability calculus, but also because I want you to be well prepared for the class field trip to Las Vegas. (Later in the book, I'll also briefly discuss blackjack, although this is a topic that merits—and has been the focus of—some nice books dedicated to it.) The question I wish to discuss here is a simple extension of the example above. Suppose five cards are drawn without replacement from a 52-card deck, as is the case, for example, in the poker game called "5-card stud." Your hand is called a "flush" if all five cards are of the same suit. Drawing five spades is one of four ways of getting a flush. Your intuition should tell you that drawing five spades is no more or no less likely to occur than drawing five hearts, say, so that we can represent the probability of a flush as

$$P(\text{flush}) = 4 \cdot P(\text{five spades}).$$

The problem is thus reduced to drawing spades or non-spades in each of five draws. The probability tree for this experiment has thirty-two paths, and it isn't worth drawing as preparation for computing the single path probability we are interested in. It is nonetheless useful to reflect upon what the tree for this experiment looks like. It will be our practice to draw trees when they are not too large, and to imagine what they look like when they are too bulky to draw. The particular path of interest here has probability

$$\begin{aligned} P(\text{five spades}) &= P(S_1 \cap S_2 \cap S_3 \cap S_4 \cap S_5) \\ &= P(S_1) \cdot P(S_2|S_1) \cdot P(S_3|S_1 \cap S_2) \cdot P(S_4|S_1 \cap S_2 \cap S_3) \cdot P(S_5|S_1 \cap S_2 \cap S_3 \cap S_4) \\ &= \frac{13}{52} \cdot \frac{12}{51} \cdot \frac{11}{50} \cdot \frac{10}{49} \cdot \frac{9}{48} \\ &\cong .0005. \end{aligned}$$

We thus have that

$$P(\text{flush}) \cong 4(.0005) = .002.$$

In five-card stud, a flush is a fairly rare event. It tends to happen once in every five hundred hands. No wonder it ranks quite highly in the poker hierarchy. ■

Example 1.4.3. Picture a box containing five marbles—three red and two white. Suppose you randomly draw two marbles from the box without replacement. (Perhaps it goes without saying, but the draws are also made without peeking.) This two-stage experiment may be represented by the following probability tree:

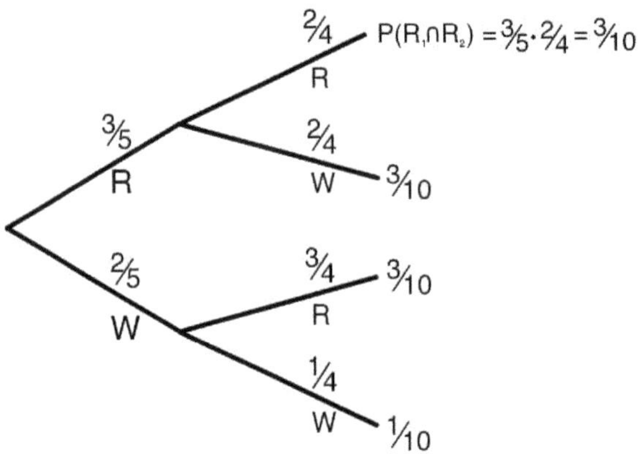

Figure 1.4.3. A tree for drawing two marbles w/o replacement from a box.

Having the four path probabilities in hand, we are now prepared to answer any question about this experiment. What about the probability of getting at least one red marble?

$$P(\text{at least one red}) = 1 - P(WW) = 9/10.$$

What is the probability that the marbles are the same color? Just add up the path probabilities corresponding to that outcome:

$$P(\text{same color}) = P(RR) + P(WW) = 4/10.$$

Given that the marbles selected are the same color (SC), what is the probability that they are both red (RR)? Applying the definition of conditional probability in (1.26), we get

$$P(RR|SC) = \frac{P(RR \cap SC)}{P(SC)} = \frac{P(RR)}{P(SC)} = \frac{3/10}{4/10} = \frac{3}{4}. \tag{1.35}$$

The middle step in (1.35) is a bit subtle. You will run into this little maneuver again, so it is worth pausing at this point to make sure you've got it. It's not often the case that the set A is the same as the set $A \cap B$. Indeed, it is only true when $A \subseteq B$. But that is precisely what is happening here. The event SC consists of the two simple events RR and WW. Since $RR \subseteq SC$, it follows that $RR \cap SC = RR$. One final question: if I told you that the second marble drawn was red, what probability would you ascribe to the event that the first marble was white? This question may seem backwards, since we are accustomed for the outcome of the first stage to be revealed before the outcome of the second stage. Still, it is a perfectly legitimate question concerning a particular conditional probability, and can be answered without great difficulty:

$$P(W_1|R_2) = \frac{P(W_1 \cap R_2)}{P(R_2)} = \frac{3/10}{6/10} = \frac{1}{2}. \tag{1.36}$$

The answer in (1.36) should appeal to your intuition. After all, if you know a red marble was selected on the second draw, then the marble selected on the first draw must have come from a population of four marbles, two of which were white. It is worth making special note of the computation in (1.36). In Section 1.5, we'll introduce a special formula for computing a "backwards" probability, that is, the probability of the outcome of an early stage of an experiment given the outcome of a later stage. It should be clear from the computation in (1.36) that we can do this simply by availing ourselves of the definition of conditional probability. ■

Example 1.4.4. I need to wean you away from the idea that probability trees are necessarily nice and symmetric. They are nice all right, but if they are going to help us in real-life problems, they better be able to incorporate a little more complexity. Let's take a look at a non-symmetric situation. Suppose you draw two marbles at random, without replacement, from a box containing four red marbles and one white one. The probability tree for this experiment is pictured below:

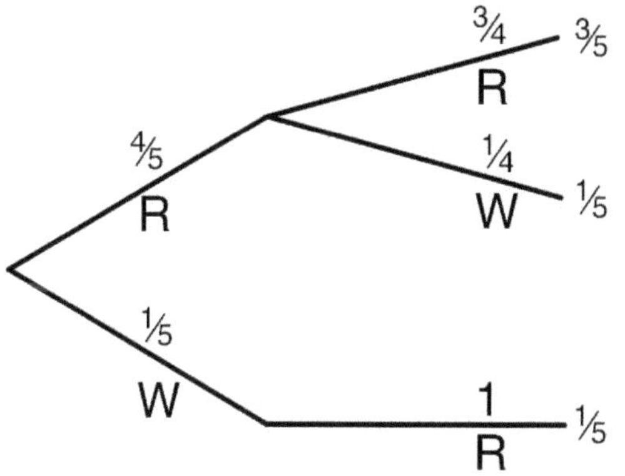

Figure 1.4.4. Another tree for drawing two marbles w/o replacement from a box.

You will note that the lower portion of the tree above has narrowed, because there aren't as many options left when a white marble is drawn at stage one as there would be otherwise. Also, I have assigned probability one to the branch R_2, since that event will occur with certainty, given W_1. Some people like to handle this situation by drawing a symmetric tree, and attaching the probability zero to branches that can't occur. Since probability trees grow as we proceed from left to right, and they can get bulky rather quickly, I recommend that you drop branches you don't need. Now, as both a logical and a computational exercise, let's compare the probabilities $P(W_2)$ and $P(W_2|R_1)$. The first is obtained by identifying paths resulting in the event W_2, and the second may be recognized as a branch probability in the second stage of the tree above. Thus,

$$P(W_2) = P(R_1 \cap W_2) = 1/5 \text{ while } P(W_2|R_1) = 1/4.$$

These computations are simple enough, but doesn't it seem a bit paradoxical that these two probabilities are different? After all, W_2 can only occur if R_1 occurs, so the occurrence of the event R_1 seems implicit in the computation of $P(W_2)$. Exercise 1.4.3 asks you to reconcile this apparent paradox. ■

Example 1.4.5. When is the last time you were the life of the party? Well, my friend, knowing some probability theory can change your life. Here is some ammo for seizing center stage at the next beer bust you attend (provided, of course, that it is not a beer bust thrown for the people taking this particular class with you). I'm going to sketch for you the solution of a problem that is known, in the esoteric literature of the stochastic sciences, as "the birthday problem." Don't let this highly technical moniker dissuade you from reading on. The problem is this: What is the probability that at least two people, in a room with n people, have matching birthdays? To solve this problem, we need to make some assumptions. For the sake of simplicity, let's assume that there are 365 days every year. If you had 35 people in attendance, and asked any of them to guess at the chances of a matching pair of birthdays in the room, many of them are quite likely to say $1/10$, which is much lower than the true probability. Their mistake is that they are providing the (approximately) right answer to the wrong question. One tenth is approximately the probability that your birthday will match someone else's, but that is only one of the many ways in which there could be two matching birthdays among 35 people. We will obtain the probability of at least one match by computing the probability of the complimentary event "no matches," which is much simpler. It is helpful to view the birthday problem as a multistage experiment, with the n people revealing their birthdays in some specific order. Let D_i represent the event that the i'th person's birthday does not match the birthday of any predecessor. The probability that the n birthdays in the room are all different and distinct may be computed as

$$P(\text{no match}) = P(D_2) \cdot P(D_3|D_2) \cdots P(D_n|\bigcap_{i=2}^{n-1} D_i)$$

$$= \frac{364}{365} \cdot \frac{363}{365} \cdots \frac{365-n+1}{365},$$

or, in the product notation we will employ from time to time,

$$P(\text{No Match}) = \prod_{i=2}^{n} \frac{365-i+1}{365}. \tag{1.37}$$

The larger n is, the smaller is the probability of finding no matching birthdays. What few expect, though, is that this probability declines so rapidly as a function of n. It first dips below .5 when $n = 23$, is first less than .3 when $n = 30$, and it is approximately equal to .035 when $n = 50$. So the next time you are at a party with a sizable crowd, and you sense that the time is right for your fifteen minutes of fame, position yourself to be overheard casually musing that the chances are pretty good that there are two people at the party with the same birthday. Oh yes, and watch out for the one thing that can go wrong—a pie in the face! ■

Exercises 1.4.

1. Use the definition of conditional probability to construct a direct proof of the identity in (1.28). Specifically, show that, if $P(B) > 0$, then $P(A|B) + P(A^c|B) = 1$.

2. Give a direct proof of the addition rule for conditional probabilities: if $P(C) > 0$, than for any events A and B, $P(A \cup B|C) = P(A|C) + P(B|C) - P(A \cap B|C)$.

3. Explain how the probabilities $P(W_2)$ and $P(W_2|R_1)$ in Example 1.4.4 are related. (Hint: It is helpful to think of the event W_2 as a union of two mutually exclusive events: $W_1 \cap W_2$ and $R_1 \cap W_2$.)

4. Sampling with, rather than without, replacement is an alternative way to draw items from a finite population. For obvious reasons, it is not as widely applicable as sampling without replacement. If you were selecting people to interview, for example, you would ordinarily want to exclude the possibility of interviewing the same person twice. There are interesting differences (and also some interesting similarities) between the probabilities attached to an event when one samples one way or the other. Consider, for example, drawing two cards at random with replacement from a 52-card deck. (This means that the first card is replaced, and the deck is suitably shuffled before the second card is drawn.) (a) Draw the probability tree for this experiment (assuming that we are only interested in the occurrence or non-occurrence of spades). Compare these probabilities with the answers when sampling is without replacement. (b) What is the probability of drawing a spade on the second draw when sampling is done with replacement? Without replacement? Explain intuitively why these answers are the same.

5. Two marbles are randomly drawn without replacement from a box containing three red marbles, two white marbles, and four blue marbles. Find the probability that the marbles selected (a) are the same color, (b) include at least one blue marble. Also, find the probability (c) that both marbles are red, given that neither of the marbles are blue.

6. Six angry students enter an elevator on their way to complain to their instructor about a tricky question on an exam. What are the chances that at least two of these students have birthdays in the same month?

7. Your company purchases computer chips from companies A, B, and C with respective probabilities $1/3$, $1/2$, and $1/6$. The probabilities of a chip being defective are $p_A = 1/10$, $p_B = 1/20$, and $p_C = 1/30$, respectively. What is the probability that a randomly chosen chip will be defective? If a chip is chosen at random and is found defective, what's the probability that it was purchased from company A?

1.5 Bayes' Theorem

The notion of a repeatable experiment has been central to our approach to stochastic modeling. We wish now to consider the possibility that one might apply the language and calculus of probability more broadly than this. Does it make sense, for example, to speak of the chances that a particular team will win the Super Bowl? Proponents of what is called "subjective probability" hold that using the ideas of probability in such situations is perfectly reasonable. Indeed, it often seems imperative that we be able to assess the chances that a particular event will occur in a unique, one-time experiment. For example, a patient considering a new, radical surgical procedure is entitled to have a rough idea of the chances for success. To the subjectivist school, probability is precisely the right way to quantify uncertainty. One way to identify the probability that should be assigned to a given event is to determine the odds that would make a disinterested person indifferent to betting on its occurrence or nonoccurrence. If, for example, I believe that a flipped thumb tack will point downwards with probability 1/3, then I would be indifferent to an offer of two-to-one odds in favor of its landing upwards. But this approach applies to once-in-a-lifetime events, like the outcome of this year's Super Bowl, as well as it does to experiments on which we can gather some experience. In the subjectivist view, you always have some initial opinion about the experiment you are interested in. From this intuition, sketchy as it may be, you have the means to formulate an initial stochastic model for the experiment. In situations in which you can gather experience, you will need to have a mechanism for updating your opinion. To see how this might work, consider the following:

Example 1.5.1. Suppose that an acquaintance of yours owns a two-headed quarter. He doesn't always use it in coin tossing, but you know that he uses it occasionally. If you are willing to believe, at least initially, that he chooses randomly between a fair coin and the two-headed coin, you might assign probability 1/2 to the event that he uses the two-headed coin on the next coin toss you witness. The probability tree appropriate to this experiment is shown in the figure below. The initial "F" is used to indicate that a fair coin was used in the toss.

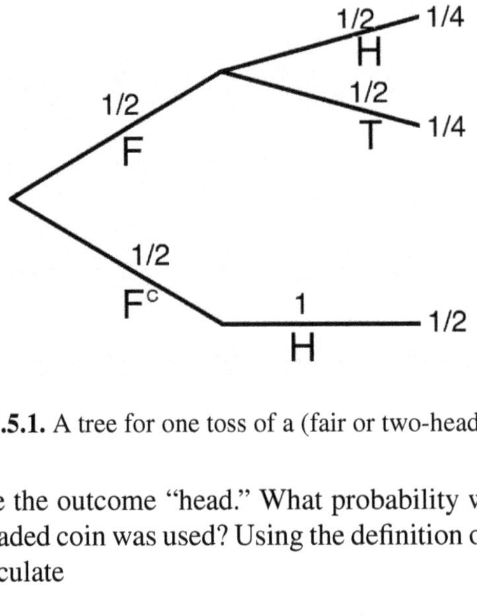

Figure 1.5.1. A tree for one toss of a (fair or two-headed) coin.

Suppose you observe the outcome "head." What probability would you now assign to the event that the two-headed coin was used? Using the definition of conditional probability in (1.30), you would calculate

$$P(F^c|H) = \frac{P(F^c \cap H)}{P(H)} = \frac{1/2}{3/4} = \frac{2}{3}.$$

It seems reasonable, doesn't it, that you would be a little more suspicious of the coin after observing a head than you were before making that observation? ■

The example above is a typical application of Bayes' Theorem, a result that is at the heart of a two-hundred-year-old debate in probability and statistics. As you will see, the theorem itself is perfectly innocuous. In its simplest form, it can be viewed as giving a formula for computing certain conditional probabilities in two-stage experiments. While the formula is indisputably correct, the way it is interpreted and used tends to separate people into two camps. The "Bayesian camp," which espouses the utility and validity of subjective probability, uses Bayes' Theorem to update the subjective opinion held prior to an experiment, thereby obtaining a new (posterior) opinion in the light of new evidence. The "frequentist camp" holds that stochastic modeling only has meaning when the model can be interpreted in terms of relative frequencies in repeated trials of an experiment. "Frequentists" might also use Bayes' Theorem in probability calculations; they would require, however, that each individual probability used in modeling an experiment have an interpretation as a long-run relative frequency.

The famous essay by Thomas Bayes which presented the theorem that bears his name was published in 1764, three years after his death. It was brought to the attention of the Royal Society (in London) by Bayes' friend, Richard Price, who was, like Bayes, a minister in the Church of England. Not much is known about how Bayes came to the result, but some have speculated that his interest in contributing to the philosophical dialogue of the day concerning formal proofs of the existence of God may have led him to his inquiry into how initial beliefs might be adjusted on the basis of new information. The setting of Bayes' Theorem can be pictured in terms of the two-stage experiment below:

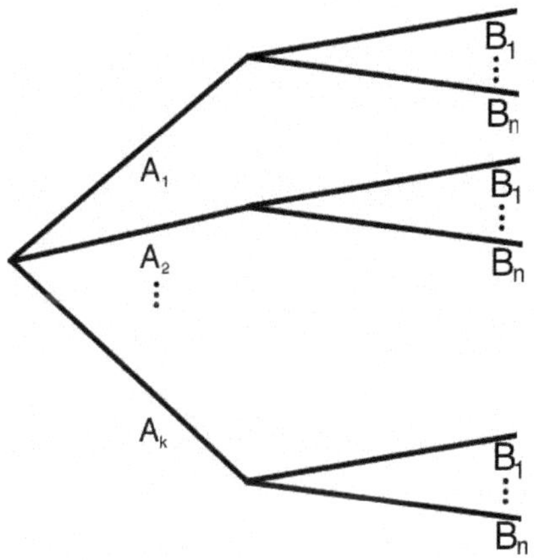

Figure 1.5.2. A two-stage probability tree with $k \times n$ paths.

In this context, Bayes' Theorem simply provides a way of computing the probability of the first stage outcome A_i given that the outcome B_j of the second stage.

Theorem 1.5.1. Let $A_1, A_2, \ldots, A_k$ be the possible outcomes of the first stage of a two-stage experiment, and let $B_1, B_2, \ldots, B_n$ be the possible outcomes of the second stage. Then

$$P(A_i|B_j) = \frac{P(A_i)P(B_j|A_i)}{\sum_{r=1}^{k} P(A_r)P(B_j|A_r)}. \tag{1.38}$$

Proof. From the definition of conditional probability, we know that

$$P(A_i|B_j) = \frac{P(A_i \cap B_j)}{P(B_j)}. \tag{1.39}$$

The numerator in (1.39) is just the probability of a single path in the probability tree in Figure 1.5.1 corresponding to the outcome "A_i followed by B_j," while the denominator may be computed by summing the probabilities of all the mutually exclusive ways in which the second stage outcome B_j can happen. By the multiplication rule, we have

$$P(A_i \cap B_j) = P(A_i) \cdot P(B_j|A_i). \tag{1.40}$$

Recall from Exercise 1.1.6 that the event B may be decomposed into a union of mutually exclusive events, that is,

$$B_j = \bigcup_{r=1}^{k} (A_r \cap B_j).$$

Thus, finite additivity and the multiplication rule justify the expression

$$P(B_j) = \sum_{r=1}^{k} P(A_r) \cdot P(B_j|A_r). \tag{1.41}$$

Substituting (1.40) and (1.41) into (1.39) yields the expression in (1.38). ■

Remark 1.5.1. The formula in (1.38) looks more imposing than it really is. If you keep in mind that the ratio in Bayes' formula is simply the ratio of one path probability to the sum of an appropriate collection of path probabilities (viz., the probabilities of those paths yielding the outcome B_j), then the computation becomes both intuitive and quite automatic. The fact that we can solve Bayes-type problems (see Example 1.4.3 and Exercise 1.4.7) without formally introducing Bayes' Theorem should convince you that this is so.

You are probably aware of some of the issues that arose in the public debate about the wisdom of mandatory AIDS testing. Those who expressed reservations about such testing often cited the likelihood of a false positive reaction as their reason. Understanding the issue of "false positives" is difficult without doing a Bayes-type calculation. The following example is typical of the calculation involved.

Example 1.5.2. Suppose that 5% of the population have a certain disease, and that the test administered for detecting the disease is reasonably accurate. Specifically, suppose that a subject who has the disease will test positive with probability .98. As is the case with most diagnostic procedures, however, there is a chance that some people without the disease will also test positive. Let's assume that the probability of a false positive in this instance is .1. When a randomly chosen subject is tested, the full testing procedure may be represented as in the tree below, where "*D*" stands for disease and "+" stands for a positive test result.

The probability of interest in this experiment is $P(D|+)$, that is, the chance that an individual who tests positive will actually have the disease. Bayes' Theorem (applied either formally or informally) yields the answer:

$$P(D|+) = \frac{P(D \cap +)}{P(+)} = \frac{.049}{.049 + .095} = .34.$$

This probability may be quite a bit smaller than you expected. The test itself can hardly be blamed, since its 98% accuracy on infected subjects, and its relatively low error rate on the rest, together suggest that the test should be a pretty good diagnostic tool. It is the rareness of the disease that conspires against the procedure's efficacy. In mass testing, the preponderance of individuals without the disease will tend to produce a fair number of false positives. Because of the possible adverse effects of notifying someone of a positive

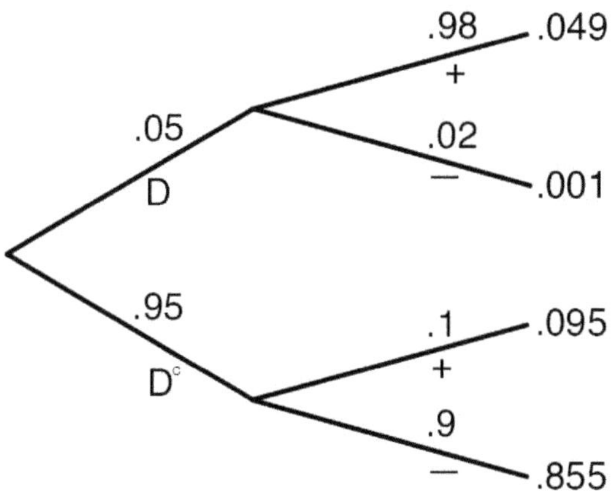

Figure 1.5.3. A probability tree for a medical diagnosis.

test result when the chance of error is high, proponents of mass testing have experienced some difficulty in convincing policy makers to accept their position. ■

Exercises 1.5.

1. You give a friend a letter to mail. He forgets to mail it with probability .2. Given that he mails it, the Post Office delivers it with probability .9. Given that the letter was not delivered, what's the probability that it was not mailed?

2. The Rhetoric Department offers the popular course "Statistically Speaking" two times a year. Students seem to like the course, both because it teaches them to integrate quantitative analyses into their argumentation, and because it is an easy A. The final (and only) exam in the course is a fifty-question multiple-choice test. If rhetoric major Rhett Butler knew the answer to forty of the questions, and selected an answer at random (from among five possible answers) for the remaining ten questions, what is the probability that he actually knew the answer to a particular question that he got right?

3. Tim, Jim, and Slim have been detained as suspects in a one-man bank robbery. They are all reputed to have the know-how and the inclination for such work. Tim has been quoted as saying that he likes the hours; Jim claims that some of his best friends are bankers; and Slim is known to have a Robin Hood complex. A priori, the police regard each as equally likely to have pulled the current heist. It is also known, however, that the likelihood of using an assault rifle during the robbery varies, being 80% for Tim, 40% for Jim, and only 10% for Slim (who has a strong preference for a bow and arrow). The physical evidence in the present case includes the fact that an assault rifle was used. Given this fact, what are the posterior probabilities that Tim, Jim, or Slim was the perpetrator?

4. Mr. Phelps plays poker on Wednesday nights. About 70% of the time, he will return home inebriated. Mrs. Phelps will retire before he comes home, but she leaves the porch light on for him. When Mr. Phelps comes home inebriated, he forgets to turn off the porch light with probability .6; otherwise, he forgets with probability .2. On those Thursday mornings

on which Mrs. Phelps finds the porch light on, what probability should she ascribe to her husband having been inebriated the night before?

5. A Chicago radio station is giving away pairs of concert tickets to the next listener who calls in. It seems that three singing groups, of varying popularity, are all appearing in the Chicago area next month. The best known is the Irish rock group U2. The others are the farm-friendly band "Ewe Two" and the madrigal ensemble "You Too (Brutus)." If the caller correctly answers a U2 trivia question, he/she gets to pick randomly from among 5 pairs of U2 tickets and 3 pairs of Ewe Two tickets while, if the trivia question is answered incorrectly, he/she will pick among 2 pairs of You Too tickets and 3 pairs of tickets to see Ewe Two. By sheer luck, you are selected to participate. The station announces that a U2 buff of long standing, like you, should have a 75% chance of getting the trivia question right. What might a friend who sees you at the Ewe Two concert conclude about the probability that you answered the question correctly? What might you conclude about her taste in music? (Don't forget you are there too! But, of course, she *paid* for her ticket.)

6. It is known that 60% of suspected criminals tell the truth on a given incriminating question in a lie detector test. Assume that when a suspect tells the truth, the test classifies the answer as a lie with probability .05. If the suspect does not tell the truth, the answer is classified as a lie with probability .9. Restaurateur Moshe Porque's answer to a question about tax evasion was classified as a lie. Find the probability he was telling the truth. (It's helpful to use the notation T & T^c with regard to whether a suspect is telling the truth and L & L^c for classifying the result of the test.)

1.6 Independence

If you are able to analyze a situation, carefully scrutinizing the facts on both sides of the relevant issues, and come to a decision or a conclusion without pausing to wonder what your friends, teachers, parents, or the CIA might think, you probably fit the standard description of an "independent thinker." Our independence is usually judged in terms of the extent to which external forces influence us. Our intuitive notions about independence are not too far from the stochastic version we will introduce in this section. We will encounter the need to talk about the independence of two events in a random experiment. Independence can be defined in terms of the influence that the occurrence of one event has on the chances that the other event will occur. Intuitively, an event A would be considered as independent of the event B if the knowledge that the event B has occurred has no effect on the probability that A occurs. Indeed, we will take this property as our working definition of independence. Given an event B for which $P(B) > 0$, we will say that the event A is independent of B if and only if

$$P(A|B) = P(A). \tag{1.42}$$

Similarly, when $P(A) > 0$, we say that B is independent of A if and only if

$$P(B|A) = P(B). \tag{1.43}$$

By the multiplication rule in Theorem 1.4.1, we have that if $P(B) > 0$ and (1.42) holds, then

$$P(A \cap B) = P(A) \cdot P(B|A) = P(A) \cdot P(B). \quad (1.44)$$

Similarly, if $P(A) > 0$ and (1.43) holds, then

$$P(A \cap B) = P(B) \cdot P(A|B) = P(A) \cdot P(B). \quad (1.45)$$

Note that the identity $P(A \cap B) = P(A) \cdot P(B)$ holds as well when either $P(A) = 0$ or $P(B) = 0$, making this latter identity slightly more general than those in (1.42) and (1.43). Because of its generality, the identity is often taken as the formal definition of independence. We will record this fact as follows.

Definition 1.6.1. In any given random experiment, the events A and B are said to be independent if, and only if,

$$P(A \cap B) = P(A) \cdot P(B). \quad (1.46)$$

Remark 1.6.1. It is clear from the discussion above that whenever the appropriate conditional probabilities are well defined, the conditions in (1.42) and (1.43) imply, and are implied by, the identity in (1.46). The definition above has the slight advantage of being valid and applicable whether or not the probabilities $P(A)$ and/or $P(B)$ are positive. If a given event has probability zero, it is automatically independent of every other event.

Recognizing that two events are independent is not always a simple matter. The independence of two events can only be confirmed by calculating the relevant probabilities, that is, by checking that (1.42) or (1.43) hold. The answer you derive will usually agree with your intuition, although we will encounter a situation or two where the answer is surprising. The following examples will help give you a better intuitive grasp of the idea.

Example 1.6.1. Suppose a single marble is to be drawn at random from the population of fifteen marbles pictured below.

	Large	Small	Totals
Red	3	2	5
White	6	4	10
Totals	9	6	15

Table 1.6.1. Population of 15 marbles of varying size and color.

Is the event "Red" independent of the event "Large"? We need to check whether or not the occurrence of one affects the chances that the other will happen. We thus compute and compare the two probabilities

$$P(R) = 5/15 = 1/3$$

and

$$P(R|L) = 3/9 = 1/3.$$

We see that these two events are indeed independent. Knowing that the marble drawn was large has no affect whatever on the chances that the marble drawn was red. This example carries another lesson: it shows that independence is a pretty fragile property. If you remove any of the fifteen marbles from this population, and then draw your one marble, it is no longer true that color is independent of size. It is clear that the independence encountered in this example is easily disturbed. ■

Example 1.6.2. You know those national promotions that try to sell magazine subscriptions by enticing you with the message that you may already have won ten million dollars? Well this year, rumor has it that because the magazine industry is having a tough time, they will be cheaping out. They will be trying to entice you with the chance to win a car. If you are one of the lucky twenty-six grand prize winners, you will be given a car drawn randomly from the population of cars shown below.

Color→ Model↓	Metallic Beige	Standard Beige	Metallic Blue	Standard Blue	Totals
Convertible	2	4	3	1	10
Sedan	3	5	5	3	16
Totals	5	9	8	4	26

Table 1.6.2. Population of 26 prize automobiles.

Is the model you draw independent of the car's color? Definitely not. The conditional probability of drawing a convertible varies with the color of the car drawn, from a low of 1/4 if the car is Standard Blue to a high of 4/9 if the car is Standard Beige. Is the model of your car independent of the fact that the paint treatment on the car you draw was metallic? It certainly is. Notice that

$$P(\text{convertible}) = 10/26$$

and

$$P(\text{convertible} \mid \text{metallic paint}) = 5/13.$$

On the other hand, the event "convertible" is not independent of either of the events "Metallic Beige" or "Metallic Blue." ■

Let us consider an example in which the independence of two events may not be intuitively obvious.

Example 1.6.3. Suppose that a fair coin is tossed independently three times. Consider the following events:

A = exactly one H occurs in the first two tosses and

B = exactly one H occurs in the last two tosses.

Are the events A and B independent? To answer this question, we must compute three probabilities. It is clear that $P(A) = 1/2 = P(B)$, as both events are based on two independent coin tosses, and the chances of obtaining a single head in each of these experiments is clearly 1/2. How about $P(A \cap B)$ in the general experiment? Well, if you examine the eight possible outcomes of this experiment, you will find that there are exactly two ways in which A and B could both happen. We may thus compute the probability of this latter event as

$$P(A \cap B) = P(HTH) + P(THT) = 1/8 + 1/8 = 1/4.$$

Since $P(A \cap B) = P(A) \cdot P(B)$, we conclude that A and B are indeed independent. ■

Independence occurs most often in random experiments when we repeat the same identical experiment a number of times. The two most common circumstances in which we encounter such repetitions are well exemplified by consecutive coin tosses and by randomly drawing cards with replacement from a standard deck. Suppose we toss a fair coin four times. The natural extension of the multiplication rule (1.29) to this situation yields

$$\begin{aligned} P(\text{four heads}) &= P(H_1 \cap H_2 \cap H_3 \cap H_4) \\ &= P(H_1) \cdot P(H_2) \cdot P(H_3) \cdot P(H_4) \\ &= 1/2 \cdot 1/2 \cdot 1/2 \cdot 1/2 \\ &= 1/16. \end{aligned}$$

Since there are four mutually exclusive ways to get exactly three heads in four tosses, and each of these ways has probability 1/16 of occurrence, we have

$$P(\text{exactly three heads}) = 1/4.$$

Suppose five cards are drawn with replacement from a 52-card deck. What is the probability of getting five cards of the same suit (the "with replacement" version of a flush)? Again, by the multiplication rule for independent events, we have

$$\begin{aligned} P(\text{flush}) &= 4 \cdot P(5 \text{ spades}) \\ &= 4 \cdot (1/4)^5 \\ &= 1/256. \end{aligned}$$

Referring back to Example 1.4.2, you can see that you're about twice as likely to draw a flush if cards are drawn with replacement than you are if they are drawn without replacement. This fact is only of academic interest, of course; there is no known card game that calls for drawing cards with replacement. It would simply be too much trouble to play.

I have found over the years that students often get independent events confused with mutually exclusive events. This appears to be due to the words involved, which seem somewhat similar, rather than to the closeness of the ideas themselves. Actually, mutually exclusive events are about as dependent as events can get. Recall the events A and B are mutually exclusive if $A \cap B = \emptyset$. This means that these events cannot happen simultaneously. If A occurs (that is, if the outcome of the experiment is a member of A), then B cannot possibly occur. It follows that for two mutually exclusive events A and B for which $P(A) > 0$ and $P(B) > 0$, we have $P(B|A) = 0 \neq P(B)$, that is, A and B are not independent. It should be intuitively clear that the occurrence of one of two mutually exclusive events carries a lot of information about the chance that the other event will occur, while the occurrence of one of two independent events carries none.

Exercises 1.6.

1. Suppose that the events A and B satisfy (1.42), where $P(B) > 0$. If $P(A) > 0$, show that A and B also satisfy (1.43).

2. Assume that, when a thumb tack is flipped, it will land with its point facing downwards with probability 1/3. Find the probability that it will land pointing downwards exactly twice in four independent flips.

3. Sally will go to the homecoming dance with probability .5, while John will go with probability .2. If the probability that at least one of them goes is .6, can Sally's decision on whether or not to go be considered independent of John's?

4. Three people work independently on deciphering a coded message. Their probabilities of success are 1/2, 1/4, and 1/8, respectively. What is the probability that the message will be decoded?

5. At an upcoming holiday gathering, cousins William and Brenda will play a game, repeatedly and independently, until one of them wins. A given game ends in a tie with probability p_1. The probability that William wins an individual game is p_2, while the probability that Brenda wins an individual game is p_3. Show that the probability that William is the eventual winner is $p_2 / (p_2 + p_3)$.

1.7 Counting

Don't be offended. I know you have been counting for years. But this section isn't about counting, it's about **COUNTING**. I mean counting in really big problems, where even taking off one's shoes and socks doesn't help. A state lottery is a good example. In many states, whether or not you win is determined by the closeness of the match between the numbers you choose and the numbers chosen randomly by the state from among some finite set of integers. For concreteness, let's suppose that you are required to select six numbers, and that the state also selects six numbers from among the integers 1 through 53. Unfortunately, you have to pick first. You can win some cash if at least three of your numbers match the state's, and you will share in the grand prize, always several million dollars, if you happen to match all six. The question naturally arises: how many ways can six numbers be chosen from fifty three? What are your chances of winning something? Anything? The whole enchilada? Since there are millions of different ways in which a group of six numbers can be selected from a collection of 53 numbers, the answers involve **COUNTING**. We will return to these questions once we have developed some industrial-strength tools in this area.

Many counting problems can be attacked pretty effectively by thinking of them as multi-stage experiments. In such problems, we are often able to obtain the count we need via the following fundamental rule. We state it without proof, as the validity of the rule is transparently valid in view of the tree in Figure 1.7.1.

Theorem 1.7.1. (The Basic Rule of Counting) If the first stage of an experiment can occur in n ways, and the second stage can occur in k ways, then the two-stage experiment has $n \times k$ possible outcomes.

The tree corresponding to the experiment covered by this rule is pictured below:

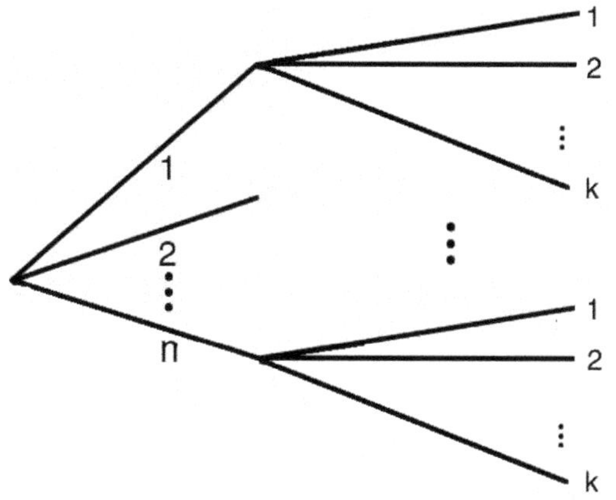

Figure 1.7.1. A tree for a two-stage experiment.

Remark 1.7.1. The basic rule of counting presumes that the number of possibilities at the second stage does not depend on the outcome of the first stage. This, of course, is the reason that the tree shown above is symmetric. While this does constitute a limitation on the rule's applicability, it is a relatively minor one, since a great many of the counting problems encountered in practice are of the symmetric type. Another point worth interjecting here is that the basic rule extends in the obvious way to experiments with any finite number of stages. As long as the number of possible outcomes at each stage does not depend on what happened at earlier stages, the number of outcomes of the experiment as a whole is obtained by multiplication.

Example 1.7.1. There is a little French restaurant in Davis named "Chez Eats." They have a different, hand-written menu every day. They always offer three appetizers, five entrees, and four desserts. People tend to think of this as a small menu. But consider, s'il vous plait, that it is possible for you to have quite a few different meals on any given day, $3 \times 5 \times 4 = 60$ to be exact. By the way, I recommend, when it's available, their Boeuf a la Provençalé, which they serve beautifully with Pommes de Terre Duchesse and Legumes Verié au Beurre. I'm assuming of course, that you like roast beef, mashed potatoes, and mixed vegetables seared in butter. ■

Example 1.7.2. It's about a two-hour drive from my home to North Lake Tahoe. On some of our occasional day trips to Tahoe, my family would humor me by agreeing to stop at a very uncool restaurant that happens to serve dynamite omelets. (See, my "fixation" on poker has been exaggerated—I do have other interests!) What originally caught my eye about the place was a billboard which states that they can serve you over 1,000 different omelets. My family likes to remind me of things like "you can only eat one at a time" and "they may all be terrible." I try to take their well-intentioned advice in stride, partly because I like to remind them of stuff too. After our first visit to the place, at which time I rendered both of those reminders null and void, my family seemed to withdraw into a quiet acceptance of the fact that this restaurant would henceforth be a part of our lives. What about their "1,000 omelet" claim? It's true! And it doesn't take a fifty-page menu to describe your choices. Basically, they've got ten ingredients, and you get to choose which ones you would like to have tossed into your omelet. There are indeed over 1,000 (actually, 2^{10}) possibilities, from the plain omelet with nothing in it to the everything omelet with all ten things in it. The first few branches of the tree corresponding to building an omelet are shown below.

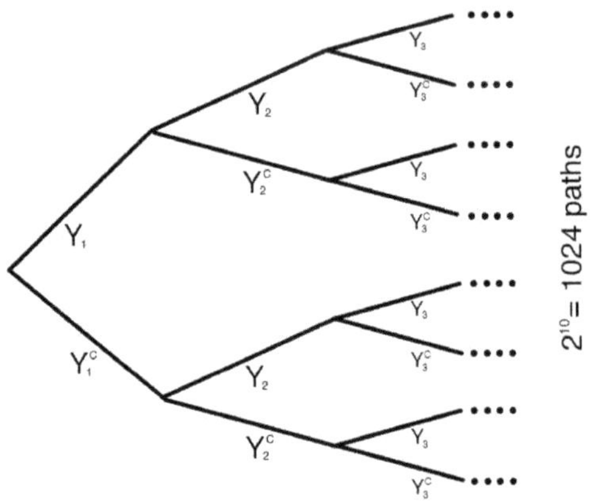

Figure 1.7.2. A tree for choosing which of 10 ingredients you want in your omelet. (The indicator "Y_i" means "yes" for the $i^{\underline{th}}$ ingredient.)

In the tree above, each ingredient is a stage. One has the choice of "yes" or "no" when deciding on a given ingredient. By the way, their okra, anchovy, mince, chorizo, and gouda cheese omelet is to die for! They call it their "Very Gouda Omelet." The full tree describes the $2^{10} = 1024$ different omelets they offer. It is surprising that more businesses have not picked up on the advertising potential of the basic rule of counting. ■

Example 1.7.3. I would like to tell you about Idaho license plates, but I don't happen to know anything about them. So, with apologies, here is my second consecutive California example. A California license plate is a sequence of numbers and letters arranged in the order NLLLNNN. How many possible plates are there? The basic rule of counting yields the answer: $10 \times 26 \times 26 \times 26 \times 10 \times 10 \times 10$ or $175,760,000$. There are, of course, a few three-letter words that aren't allowed, but these absent license plates are more than compensated for by the option people have to order personalized plates. Clearly, it should be a while before the state runs out of plates of the current design. ■

Enough frivolity. We need to develop some useful counting formulas. The formulas we will use most often involve the mathematical symbol "$n!$". When you first encounter this symbol, you might be tempted to pronounce it like you would if you were in the cheering section at a football game: "Gimme an N." "N!" (This joke works better if you pronounce the second N really loud.) The name we use for the symbol $n!$ is "n factorial". Its technical definition is given below.

Definition 1.7.1. Let n be a positive integer. Then $n!$ represents the product of all the integers from n on down to one, that is,

$$n! = n \cdot (n-1) \cdot (n-2) \cdot \cdots \cdot 3 \cdot 2 \cdot 1. \tag{1.47}$$

The definition above is easy to work with. For example, you can always get the next factorial from the last one, since $n! = n(n-1)!$. It's clear that $1! = 1$, $2! = 2 \cdot 1 = 2$, $3! = 3 \cdot 2 = 6$, $4! = 4 \cdot 6 = 24$, etc. Since $n!$ gets large quite quickly, you will tend to use your calculator to compute it when n is large. There is one important case not covered by the definition above. We will have occasion to use the number $0!$ in our counting, and we thus need to provide the following addendum to our definition of $n!$:

Definition 1.7.2. For $n = 0$, $n!$ is defined as

$$0! = 1. \tag{1.48}$$

This latter definition is a mathematical convention, much like a convention introduced in algebra courses: for any $x \neq 0$, $x^0 = 1$. When we define $0!$ as 1, the counting formulas we are about to introduce will make sense and give correct answers in all cases in which $0!$ pops up. (They are also correct, of course, when $0!$ doesn't pop up.)

The two types of counting problems we will treat here in detail differ from each other in one essential respect: in one type of problem, you are concerned with the order in which items occur, while in the other type of problem, you are not. We have already encountered both problem types. Two license plates are not the same just because they contain the same numbers and letters. The order in which the numbers and letters are arranged is important. On the other hand, two ham and cheese omelets are certainly the same whether the ham was added before the cheese or after it.

In many of the situations we will face, we will speak of forming groups of objects selected from a population of n "distinguishable" objects. The group of balls, numbered 1 to 15, used in the game of pool is an example of a population of distinguishable objects. A collection of fifteen red marbles is not. When the items in your population can be separately identified, each subgroup of a given size can be distinguished from every other. We use the word *permutation* in reference to an ordered collection of objects, and the word *combination* for collections in which order is irrelevant. One way to remember which is which is to note that "o" (for "order") and "p" (for "permutation") are adjacent letters in the alphabet. We will deal with the idea of permutation first.

Definition 1.7.3. A *permutation* of a collection of n objects is a fixed ordering of these objects.

The six permutations of the numbers 1, 2, and 3 are:

$$(1,2,3),\ (1,3,2),\ (2,1,3),\ (2,3,1),\ (3,1,2) \text{ and } (3,2,1). \tag{1.49}$$

Why would you want to consider these six orderings to be different from one another? Sometimes it will be because each ordering represents a label, like a license plate or a social security number. Other times, it will be because the order of the items selected determines some subsequent action. When the items on a list will be treated differently, you will need to keep track of the order in which they occur. In choosing club officers from among three volunteers, for example, it might be agreed in advance that the first person selected will serve as club president, the second as vice president, and the third as secretary treasurer. In that case, the six permutations in (1.49) could be thought of as the six ways in which candidates 1, 2, and 3 could fill the three jobs.

Consider the process of drawing k objects from a population of n distinguishable objects, keeping track of order. The number of different ways this can be done, first when the draws are made with replacement and secondly when the draws are made without replacement, is identified in the two fundamental counting formulas below:

Theorem 1.7.2. The number of ways of obtaining a permutation of k objects drawn *with replacement* from a population of n objects is denoted by $P^*(n,k)$, and is given by

$$P^*(n,k) = n^k. \tag{1.50}$$

The number of ways of obtaining a permutation of k objects drawn *without replacement* from a population of n objects is denoted by $P(n,k)$, and is given by

$$P(n,k) = \frac{n!}{(n-k)!} \tag{1.51}$$

Proof. Think of the process of drawing k items from a population of n items as a k-stage experiment. When sampling is done with replacement, the number of possible outcomes at each stage is the same, namely, n. By the basic rule of counting, the number of ways this can be done is $n \times n \times n \times \cdots \times n = n^k$. This proves (1.50). When sampling is done without replacement, the number of possible outcomes decreases by one from one stage to the next. Thus, the multiplication rule in this case yields $n \times (n-1) \times (n-2) \times \cdots \times (n-k+1)$. This product is most conveniently written as the ratio in (1.51). ■

To make the distinction between these two formulas concrete, let's find the number of permutations, with and without replacement, when we draw two items from a population of size four. Suppose we wish to choose two letters from among the letters {A,B,C,D}, keeping track of order. With replacement, there are $P^*(r,2) = 4^2 = 16$ ways to do this. The tree representing the possible selections is displayed below:

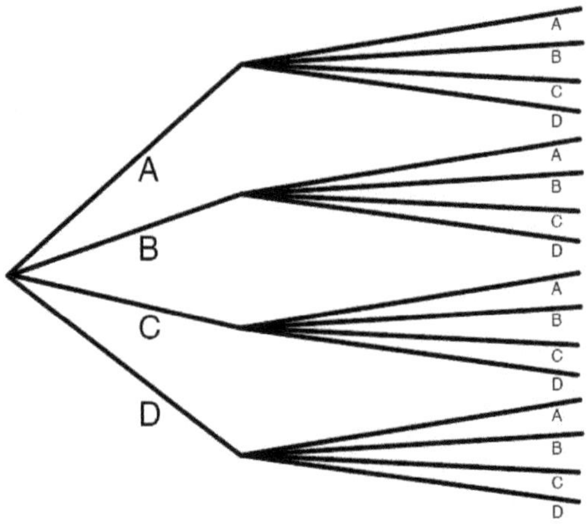

Figure 1.7.3. A tree for permutations with replacement.

This tree shows the sixteen different two-letter "words" you can form from your four-letter alphabet. If we were drawing letters without replacement, so that no letter could be used twice, there would be $P(4,2) = 4 \cdot 3 = 12$ possible words. These are displayed in the smaller tree:

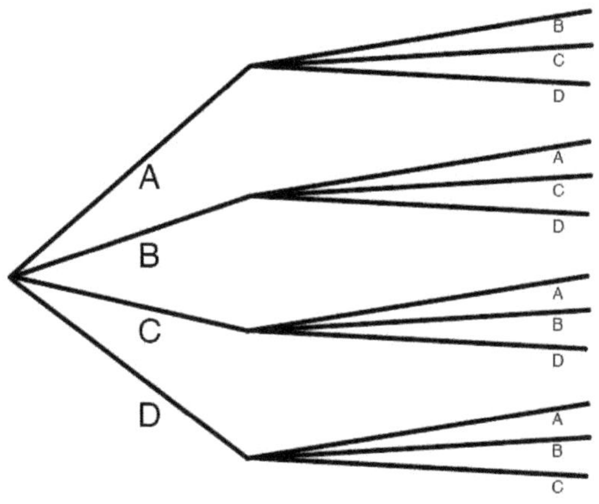

Figure 1.7.4. A tree for permutations without replacement.

Sampling with replacement from a finite population is not very common. Most applications in which permutations arise will involve sampling without replacement, and will therefore call for use of the counting formula for $P(n,k)$ rather than the formula for $P^*(n,k)$. The count, and idea, embodied in $P(n,k)$ occurs often enough that it has been given a name: we refer to $P(n,k)$ as "the number of permutations of n things taken k at a time." The formula $P^*(n,k)$ has no name; when you need it, try something like "hey you." A special case of $P(n,k)$ that occurs with some frequency is

$$P(n,n) = n!. \tag{1.52}$$

This formula simply counts the number of ways that you can order n objects.

There is one important class of problems that involves both of these counting formulas, that is, both P and P^*. The birthday problem is an example. In "matching problems" like this, we are generally interested in the probability that all of the items drawn from a given population are different. When you draw items without replacement from a finite population of distinguishable items, matches are impossible. Matches are possible, on the other hand, when draws are made with replacement. There is a natural containment of the experiment of sampling without replacement within the larger experiment of sampling with replacement. Check it out for yourself. Aren't all the paths in the tree in Figure 1.7.4 contained among the paths of the tree in Figure 1.7.3?

The question we are concerned with in matching problems is this: when you are sampling with replacement, what are the chances that there are no matches among your first k draws? We interpret "sampling with replacement" here to mean that, at every stage, the set of possible outcomes is the same, and each of them has an equal chance of selection. If you randomly draw two letters with replacement from the population $\{A,B,C,D\}$, it is evident from the tree in Figure 1.7.3 that the probability they don't match is $12/16 = 3/4$. The same logic extends to more imposing questions like "what's the probability that no two people among n have matching birthdays?" If you draw repeatedly with replacement from the population of 365 possible birthdays, then the number of possible ordered lists of n birthdays is $P^*(365,n)$. On the other hand, the number of ordered lists with no repetitions is $P(365,n)$. We can thus identify the desired probability of no matching birthdays as

$$P(\text{no match}) = \frac{P(365,n)}{P^*(365,n)}. \tag{1.53}$$

Compare this with the answer derived in Example 1.4.5.

Example 1.7.4. Queen Elizabeth I of England (1533–1603) was fond of picnics. It is said that William Shakespeare attended several and that Sir Walter Raleigh attended most of them. An Elizabethan picnic consisted of the Queen and seven guests. The eight of them would sit on the ground in a circle, each buoyed by an overstuffed goose down pillow. The picnics were legendary for the sparkling conversation, but they were also famous as a culinary experience. Because the Queen always served herself first and then passed food to the right, sitting at the Queen's right was considered the greatest honor, and sitting as close to her right as possible was considered desirable. The phrases "right-hand man" and "left-handed compliment" might have originated at these picnics. How many different seating orders were possible at these picnics? Should Ben Jonson have been miffed because he never got closer than fourth place at the three picnics he attended? The first question is easily answered since, after the queen has been seated, one merely needs to count the number of permutations possible for the seven places to her right. Thus, there are $7! = 5,040$ possible seating orders. If Ben Jonson is placed in any one of the last four places, there are $6! = 720$ possible orderings of the other guests; there are thus $4 \cdot 720 = 2,880$ such seatings. If seating order had been determined at random, the chances that Jonson would get an unfavorable seating on a given occasion is $2880/5040 = 4/7$ (which will agree with your intuition), and the chances it would happen on three independent seatings is $(4/7)^3 = .187$. That likelihood is not so small that Jonson could justifiably conclude that his seating was discriminatory. ■

Let us now give some thought to situations in which order is unimportant or irrelevant. Such problems arise when our interest centers on the *set* of items selected; the set will be the same, regardless of the order in which the items were selected. For example, the value of your poker hand depends only on the cards you are holding, and not on the order in which they were dealt.

Suppose you have a population of n items, and you wish to assemble a sub-population of k distinct items from that population. In sampling from the population, you obviously would choose k items *without replacement*, as you do not wish to allow for possible duplications in separate draws. We will therefore restrict our attention to sampling without replacement, which is the only case of any practical interest. We use the term *combination* when referring to a set of objects for which order is irrelevant. How many possible combinations are there of n things taken k at a time? The answer is contained in the following:

Theorem 1.7.3. The number of ways of drawing a combination of k objects without replacement from a population of n objects is denoted by $C(n,k)$, and is given by

$$C(n,k) = \frac{n!}{k!(n-k)!}. \tag{1.54}$$

Proof. The number of permutations of n things taken k at a time is given by the ratio

$$P(n,k) = \frac{n!}{(n-k)!}$$

It is clear that this formula over-counts the number of sets of size k that can be formed. The over-counting is systematic however, and can be fully characterized as follows. For any fixed set of size k drawn without replacement from this population, there are $k!$ distinct permutations that can be formed from them. Thus, if we had in hand the number $C(n,k)$ of sets of size k, we could obtain the number of possible permutations by the necessary relationship

$$P(n,k) = k!C(n,k);$$

the latter equation may be rewritten as (1.54). ■

Remark 1.7.2. Some authors use the notation C_k^n for what I have called $C(n,k)$. The most commonly used notation for combinations is the symbol

$$\binom{n}{k} = \frac{n!}{k!(n-k)!}. \tag{1.55}$$

This new symbol is pronounced "n choose k," and this will be our standard notation for combinations as we proceed. You will notice some inherent symmetry in the combinations formula; it is clear, for example, that

$$\binom{n}{k} = \binom{n}{n-k}. \tag{1.56}$$

This identity is obvious algebraically when one replaces the two symbols by what they stand for, but it also follows from the fact that choosing k items among n to be in a set is equivalent to choosing the $(n-k)$ items to be left out.

Counting with the aid of formula (1.55) looks imposing, but isn't really that bad. Even though the factorials themselves are big numbers, the computation of a combination is often greatly simplified by the cancellation that occurs in the ratio of factorials. For example, the number of sets of size 15 that can be formed from a population of size 20 is

$$\begin{aligned}\binom{20}{15} &= \frac{20!}{15!5!}\\ &= \frac{20\cdot 19\cdot 18\cdot 17\cdot 16\cdot 15!}{15!\cdot 5\cdot 4\cdot 3\cdot 2\cdot 1}\\ &= \frac{20\cdot 19\cdot 18\cdot 17\cdot 16}{5\cdot 4\cdot 3\cdot 2\cdot 1}\\ &= 19\cdot 3\cdot 17\cdot 16\\ &= 15,504.\end{aligned}$$

If your calculator does factorials, you are on easy street. If it does combinations, you might as well start planning what you are going to be doing with all your spare time.

Earlier in this section, we treated the problem of drawing two letters from among the population {A,B,C,D}. We noted that there are 16 permutations possible if repetitions are allowed, and twelve possible permutations if they are not. The $\binom{4}{2} = 6$ combinations (i.e., sets) of four letters taken two at a time are shown below:

$$\{A, B\}, \{A, C\}, \{A, D\}, \{B, C\}, \{B, D\} \text{ and } \{C, D\}.$$

Let's consider some more colorful examples.

Example 1.7.5. Kindergarten teacher Monica Speller gave each of her students a box of twenty crayons on the first day of class. Miss Speller (an unfortunate moniker for a teacher) has told her students that she would like them to draw a new picture every day using exactly three different colors. She also wants the three colors chosen to be a different combination each day. Given that her kindergarten class will meet 150 times during the course of the year, is she asking the impossible? Anyone who knows a five-year-old will know that she is! But suppose she is willing to replace crayons as needed, and will personally keep track of the color combinations used so far. Are there at least 150 combinations possible? Sure. There are $\binom{20}{3} = 1,140$ possible combinations, enough to last these kids through the middle of seventh grade. ■

Example 1.7.6. Let's take another look at the lottery (in an imaginary principality referred to as "the State"). After the public has selected its numbers, the State, using a time-honored procedure involving air-blown ping pong balls, will identify the six numbers against which a player's chosen numbers must be matched. The State's numbers may be thought of as having been chosen at random from the integers $1,2,\cdots,52,53$. The number of possible sixsomes that can be chosen is

$$\binom{53}{6} = \frac{53!}{6!47!} = 22,957,480.$$

A "Lotto" ticket costs one dollar. You win five dollars when exactly three of your choices match up with the State's. How many possible ways are there of obtaining exactly three matches? Think of Lotto as a two stage experiment. The State will pick six numbers at random. Stage one consists of picking three numbers to match three of your six. This can be done in $\binom{6}{3} = 20$ ways. Stage two consists of picking three numbers that are different than any of yours. This can be done in $\binom{47}{3} = 16,215$ ways. Thus, the chances that exactly three of your numbers match numbers selected by the State is given by the ratio

$$P(\text{three numbers match}) = \frac{\binom{6}{3}\binom{47}{3}}{\binom{53}{6}} = \frac{20 \times 16215}{22957480} = .01413.$$

As small as this probability is, it is a lot bigger than your chances of winning more. Verify, in Exercise 1.7.4, that the probability that four or more of your numbers match up with the State's is .00072. It certainly seems that you are throwing money away when you play the lottery. What is clear, at least, is that you will rarely win anything, and if you do, it's likely to be a measly five dollars. Actually, many states return about half of their lottery income to the players, so, if you played the lottery a lot, you might figure that you stand to get 50 cents back on every dollar you spend. That doesn't sound like the kind of investment one should mortgage one's house for. In spite of the fact that Lotto is stacked against you, there is a characteristic of the way it's played (having to do with what happens when no one has a perfect match in several consecutive lotteries) that can change the game into something favorable. The subject of the lottery is taken up again in Exercise 2.3.3. ■

Put on your poker hat, my friend. The time has come to compute some poker probabilities. Actually, there are many different kinds of poker, and a goodly number of games which masquerade as poker. We will restrict our attention here to a game called "five-card stud poker," a game with no wild cards and a game in which you are dealt five cards at random and must play these precise cards at their face value. The cards numbered 2 through 10 have the value of their number, while a jack is worth 11, a queen 12, and a king 13. An ace is assigned the highest value, 14, although it can also be used as a 1 when forming the straight A, 2, 3, 4, 5. A standard 52-card deck consists of four cards of each of the thirteen "numbers," where cards with a given number come in one of the four "suits" — spade, heart, diamond, or club. The hierarchy among poker hands is based directly on the likelihood of their occurrence. From best to worst, the possible types of hands you can have are straight flush, four of a kind, full house (e.g., three 10's and two 8's), flush, straight, three of a kind, two pair, one pair, and nothing (five cards of different values, but neither a flush nor a straight).

Example 1.7.7. Let's begin by retracing some familiar ground. We have already computed the probability of getting a flush. We did not, however, apply any high-powered counting techniques to the problem. Our counting formulas apply to this problem as follows. Putting a poker hand together is nothing other than forming a set of size five from a population of 52 distinguishable objects. The number of possible poker hands is thus equal to

$$\binom{52}{5} = \frac{52!}{5!47!} = 2,598,960. \tag{1.57}$$

Since the number of ways you can assemble a set of 5 spades, chosen from among the population of 13 spades, is $\binom{13}{5}$, we have

$$P(\text{five spades}) = \frac{\binom{13}{5}}{\binom{52}{5}} = .000495,$$

so that the probability of a flush is obtained as

$$P(\text{flush}) = 4 \cdot P(\text{five spades}) = .00198,$$

where the 4 above accounts for the fact that a flush can be obtained in any of four suits. ■

The calculation of probabilities in poker will involve all of the counting ideas we have seen so far. The basic rule of counting plays an essential role. We will also need to take order into account for some calculations, and ignore order in others. The three poker hands whose probabilities we calculate below are meant to help you distinguish between situations which require keeping track of order from those which don't. Let's start with the probability of getting a full house.

Example 1.7.8. The basic rule of counting will come into play in this example because we will take the view that putting together a full house is a four-stage experiment: pick a value you will be taking three cards from, take three cards of this value, pick a value you will be taking two cards of, take two cards of this value. Notice that the two values that are picked play different roles; if you pick a king in stage one and a four in stage three, you get a different full house than if the order was reversed. Since we are assuming that cards are dealt randomly, so that all the possible poker hands are equally likely, we can represent the probability of a full house as the ratio of the number of ways of getting a full house to the total number of poker hands possible. Visualizing the situation as a four-stage experiment, we thus obtain

$$P(\text{full house}) = \frac{13 \cdot \binom{4}{3} \cdot 12 \cdot \binom{4}{2}}{\binom{52}{5}} = .00144. \tag{1.58}$$

To emphasize the fact that we are keeping track of the order of the two values we pick, let's do the computation again, but in a different way. Think of the process of forming a full house as a three-stage experiment: pick two ordered values, take three cards of the first value, take two cards of the second value. Visualizing the process this way leads to

$$P(\text{full house}) = \frac{P(13,2)\binom{4}{3}\binom{4}{2}}{\binom{52}{5}}, \tag{1.59}$$

which, of course, is the same answer as obtained above. The most common mistake made when computing probabilities in this type of problem is the mistake of ignoring order. If you were to replace P(13,2) by $\binom{13}{2}$ in (1.59), you would be off by a factor of two. In not accounting for order, you would be implicitly equating full houses having three kings and two fours with full houses having three fours and two kings. If you make that mistake while playing poker, it could very well cost you some money. ■

Example 1.7.9. By way of contrast, let's now compute the probability of getting five cards having different values. Note that this is not necessarily a bad poker hand. It might, in fact, be a straight or a flush, or even a royal flush (10, J, Q, K, A of the same suit), the best possible hand in poker games without wild cards. Usually, however, a hand with five different values is a loser, as is reflected in the name "a toilet flush" that is often used to describe it. Assembling a hand with five different values in it can be thought of as a ten-stage experiment in which five values are chosen and then one card is taken from each of the chosen values. The difference between this and the previous example is that each of the five values here will play the same role after it has been identified. There is no need to know the order in which the values were chosen, since we will simply proceed to do the same thing with each, namely, pick a single card of that value. This leads to the use of combinations rather than permutations in choosing the five values involved. We therefore obtain

$$P(\text{five different values}) = \frac{\binom{13}{5}\binom{4}{1}^5}{\binom{52}{5}} = .50708. \tag{1.60}$$

■

Example 1.7.10. Let us now consider a hybrid example which utilizes ordering on a certain part of the problem and ignores ordering in the remaining part. Let's calculate the probability that you draw the hand "three of a kind" when dealt a five card poker hand at random. Such a hand will contain three cards having the same number and two cards with two different numbers. A hand with 3 jacks, a 7, and a 4 would be an example. It is helpful to think of assembling the hand as a five-stage experiment. We will need to choose a value (among 2, 3, . . . , J, Q, K, A) to get three cards from, then choose three cards of that value, then choose two additional values from which a single card will be drawn and then choose one of the four cards from each of these latter two values. I will need to keep track of order, at least partially. Simply choosing the three numbers that will be in my hand will not do, as I also need to know precisely which of them corresponds to my threesome. So I will proceed as follows. I will first pick a number that I will be drawing three cards from. I can choose that number in 13 ways. I will then choose those three cards, which I can do in $\binom{4}{3}$ ways. Then I will choose the two numbers (from the remaining twelve) that I will get singletons from. I don't need to keep track of the order of the latter two numbers because these numbers will not be treated differently. I can choose these two numbers in $\binom{12}{2}$ ways. Then I will choose one card with each of these numbers. Taking these actions in sequence, I get

$$P(\text{three of a kind}) = \frac{13\binom{4}{3}\binom{12}{2}\binom{4}{1}\binom{4}{1}}{\binom{52}{5}} = .02113. \tag{1.61}$$

■

Finding the probabilities for the complete collection of standard poker hands is the subject of Problem 66 at the end of the chapter.

The final topic we will treat in this section is an important extension of the formula for combinations to problems dealing with "partitions." We can think of the formula $\binom{n}{k}$ itself as solving a partition problem. The process of selecting a set of size k from a population of size n is equivalent to creating a partition; we have, in effect, partitioned the population into two groups, the k items in the set of interest and the $n-k$ items outside of it. Since there are $\binom{n}{k}$ different sets of size k, there are precisely that many possible partitions of the population into groups of size k and $n-k$. Now consider the following more general question: for any fixed group sizes $\{k_1, k_2, \ldots, k_r\}$ for which

$$\sum_{i=1}^{r} k_i = n. \tag{1.62}$$

In how many different ways can a population of n items be partitioned into groups of these sizes? The answer is contained in the result below.

Theorem 1.7.4. Let n be a positive integer and $k_1, k_2, \ldots, k_r$ be nonnegative integers such that (1.62) holds. The number of partitions of n objects into groups of sizes $k_1, k_2, \ldots, k_r$ is given by

$$\binom{n}{k_1, k_2, \ldots, k_r} = \frac{n!}{k_1!k_2!\ldots k_r!}. \tag{1.63}$$

Proof. Think of the development of a partition of n objects into r groups as an r-stage experiment. In the first stage, we simply wish to form a group of size k_1 from population of size n. There are $\binom{n}{k_1}$ ways of doing this. Given that the first group has been formed, we are left with a population of size $n-k_1$. In stage two, we will form a group of size k_2 from these remaining $n-k_1$ objects. We can do that in $\binom{n-k_1}{k_2}$ ways. Proceeding in this manner, we can identify the number of ways in which the ith stage can be executed as $\binom{n-\sum_{j=1}^{i-1} k_j}{k_i}$. Applying the basic rule of counting, we thus evaluate the desired number of partitions to be the product

$$\binom{n}{k_1} \cdot \binom{n-k_1}{k_2} \cdots \binom{n-\sum_{j=1}^{r-2} k_j}{k_{r-1}} \binom{n-\sum_{j=1}^{r-1} k_j}{k_r}. \tag{1.64}$$

which, when written as

$$\frac{n!}{k_1!(n-k_1)!}\cdot\frac{(n-k_1)!}{k_2!(n-k_1-k_2)!}\cdots\frac{(n-k_1-\cdots-k_{r-2})!}{k_{r-1}!(n-k_1-\cdots-k_{r-1})!}\cdot\frac{(n-\sum_{j=1}^{r-1}k_j)!}{k_r!0!},\tag{1.65}$$

is an expression which reduces to (1.63) after successive cancelations. ■

The counting formula in (1.63) applies to so-called multinomial problems, that is, to multistage experiments which have multiple (say r) possible outcomes at each stage. If there are n stages in the experiment, then the basic rule of counting tells us that, overall, there are r^n ways in which the experiment can turn out. You can visualize this experiment by picturing a tree in which each path expands into r new branches at each stage. What does the multinomial counting formula keep track of? For any fixed set of r nonnegative integers $k_1,\ldots,k_r$ satisfying (1.62), this formula gives the number of paths in the tree that contain exactly k_1 outcomes of the first type, k_2 outcomes of the second type, ..., k_r outcomes of the rth type. For concreteness, suppose we draw four times with replacement from the population of three letters {A,B,C}. There are $3^4 = 81$ possible outcomes of this experiment. How many of them consist of two A's, one B and one C? The multinomial counting formula gives us

$$\binom{4}{2,1,1}=\frac{4!}{2!1!1!}=12.$$

The twelve paths fitting this description are listed below:

AABC	AACB	ABAC	ACAB
ABCA	ACBA	BAAC	BACA
BCAA	CAAB	CABA	CBAA

Example 1.7.11. Picture yourself as a teacher. Suppose that you have given a multiple-choice test with 50 five-answer questions and that you suspect that a certain very chummy trio of students in your class cheated. How would you go about analyzing the test results in seeking to confirm your suspicions? Here's one possibility. You might first recognize that wrong answers are more revealing than right answers. This is because a right answer has several plausible explanations, including the possibility, which is every teacher's hope, that the students do know something. When students get a question wrong, it is reasonable to conclude that they don't know that particular part of the subject matter fully. From there, one might leap, not entirely capriciously, to the notion that a student will be selecting an answer at random. Suppose, on this particular exam, there were twelve questions that all three students got wrong. Suppose you decide to keep track of the number of different answers these students selected (either 1, 2, or 3) on each of these questions, and you determine that the students chose the same wrong answer on nine of them, once choosing two different answers, and twice selecting three different answers. To attack the hypothesis that they were cheating, you need to know how many ways what you have observed can happen, and, ultimately, how rare this event is. The multinomial counting formula will provide us with an answer to the first question, and to several closely related questions; we will return to the question of just how incriminating the observed agreement among these students is after we have introduced the multinomial distribution in Chapter 4. To put yourself in a position to count, you need to think of this situation as a twelve-stage

experiment with three possible outcomes at each stage, these being 1, 2, or 3 different answers on the question involved. It then becomes clear that what you have observed is the event consisting of 9 occurrences of the first outcome, 1 of the second, and 2 of the third. For short, call this event simply (9,1,2). The number of ways this event can happen is

$$\binom{12}{9, 2, 1} = \frac{12!}{9!2!1!} = 660. \qquad (1.66)$$

In trying to determine how rare you should consider the event "nine common wrong answers," you would need to know how large the probability is of getting less than this under random guessing. Later, we will compute the (simpler) complementary probability of seeing nine or more common wrong answers. That computation will require a count of the number of ways each version of "nine or more" can happen. There are actually ten different versions of the event "nine or more," among them, for example, the event (10, 0, 2). The number of ways this latter possibility can happen is

$$\binom{12}{10, 0, 2} = \frac{12!}{10!0!2!} = 66.$$

Answers are requested for the remaining eight possibilities in Exercise 1.7.5. ■

Example 1.7.12. Deoxyribonucleic acid, better known as DNA, has been described as the basic building block of life. DNA is composed of two intertwined strands, each consisting of an ordered sequence of constituents called nucleotides. A gene is nothing more than a sequence of nucleotides (often numbering in the hundreds of thousands) along a molecule of DNA. Nucleotides are usually coded by letters, since DNA strands for a given organism will involve only a limited number of different ones. A DNA sequence found in the X chromosome of the fly *Drosophilamelanogaster* is shown below:

$$GATAGATCAACC \qquad (1.67)$$

While slight variations in a DNA sequence will sometimes have no discernible genetic significance, two organisms can be considered to be genetically identical only if their various DNA sequences match exactly. Even though there are only four nucleotides involved in the sequence (1.67), you can see that there is substantial potential for genetic variation. How many different DNA sequences of length twelve can one form from five A's, three C's, two G's and two T's? If their order didn't matter, the answer would be one. Since the order is precisely what matters here, we will take the following tack. Think of the problem as a 4-stage experiment. Suppose we have twelve empty slots in which to place our twelve letters. First, we pick the five slots to put A's in. There are $\binom{12}{5}$ ways to do this. From the remaining seven slots, we pick three slots to put C's in. This can be done in $\binom{7}{3}$ ways. You should recognize this construction as the one we used to develop the multinomial formula. The number of possible sequences thus turns out to be

$$\binom{12}{5, 3, 2, 2} = \frac{12!}{5!3!2!2!} = 166,320$$

The fact that the multinomial formula in (1.63) applies in this example is no accident.

The number of arrangements of our twelve letters can indeed be thought of as being in one-to-one correspondence with the number of partitions of twelve slots into groups of sizes 5, 3, 2, and 2. When a partition is identified, you know exactly how many of each letter will be used, and when an ordering is identified, you can determine where each of the letters go. This logic naturally extends to the problem of determining the number of distinguishable orderings of n objects when there are k_1 of type 1, k_2 of type 2, ..., and k_r of type r. The multinomial formula provides the answer. ■

An Aside: The Idea of a "Combinatorial Proof" (Optional)

Mathematicians take pleasure in proving things. Developing an airtight proof for something that you think is true is very satisfying. As mathematical frameworks get more complex, there are often several different approaches one can take to proving a particular claim, and all might be perfectly valid. Mathematicians reserve the term "elegant" for a proof that is clear, correct, and succinct. An elegant proof is of course the ideal, but we all know that any valid proof will do. Still, a one-line proof of a seemingly complex statement is not only satisfying, but it can appear to those who appreciate such things as truly beautiful, and the person who thought of it first can justifiably take pride in the achievement. There is a special style of proof that goes by the name "combinatorial," and the approach generally leads to an argument that is quick, simple, and once you've thought about it for a moment, totally obvious. In other words, combinatorial proofs are often elegant, and they are sometimes far simpler than any alternative approach you can think of. As a mathematics major in college, I enjoyed proving things as much as the next person (er, I mean, as much as the next math major), but when I first encountered the field of combinatorics, I was bowled over by how incredibly neat it was. It was the "combinatorial proof" that impressed me. To me, it was like a "proof by conversation." You could prove something simply by telling a story (the "right" story), and once the story was told, the result you wanted to prove was seen to be clearly true. Since some of the problems at the end of this chapter make reference to combinatorial arguments, I end the chapter with an example of such an argument. I believe that you, too, will be impressed with the approach.

Example 1.7.13. Consider the following combinatorial identity for positive integers k, m and n for which $k \leq min\{m,n\}$,

$$\binom{m+n}{k} = \binom{m}{0}\binom{n}{k} + \binom{m}{1}\binom{n}{k-1} + \cdots + \binom{m}{k}\binom{n}{0}. \tag{1.68}$$

You might well ask: why should we care about this identity? I could use a parent's favorite reply: Because I said so!!! But, actually, there are some good reasons for wanting to know this. Notice, for example, that the combinatorial formulas on the right-hand side (RHS) of (1.68) are of lower order (that is, involve smaller numbers) than the formula on the left-hand side (LHS). This means that you can calculate the value on the left from smaller numbers that you may already have in hand. This is how tables are often built. The identity in (1.68) is what we call a "recursive relationship," and such relationships are often useful in computation (either directly or with computer assistance).

Let's consider how we might prove the identity in (1.68). Since k, m and n are arbitrary positive integers, subject only to the constraint that $k \leq min\{m,n\}$, an algebraic proof (which would replace each combination symbol by a fraction of factorials) is not really workable as m and n grow—there would be more fractions than you could write down. Getting a bit ahead of ourselves, we could develop a proof of this identity using an inductive argument (as explained in Section 2.4), but that's not entirely simple either. A combinatorial argument for (1.68) would simply tell a story, with a slightly different but equivalent version applied to each side of the identity. The story for the LHS of (1.68) is trivial. If you want to choose a group of k objects from the collection of $m+n$ objects, there are, clearly, $\binom{m+n}{k}$ ways of choosing such a group. Now, suppose that m of those objects are RED and n of the objects are BLUE. Let's examine the RHS of (1.68) . The first term is the number of ways in which you could choose k objects such that all k of them were blue. The second term is the number of ways for which one was red and the rest were blue. Proceeding in this fashion, you will account for all the possible sets with i red objects and $(k-i)$ blue ones, where $i=0,1,2,\ldots,k$. Since these $k+1$ possibilities (all possible because $k \leq m$ and $k \leq n$) account for all of the ways in which a set of k objects can be assembled, the RHS must in fact be equal to the LHS. QED, as mathematicians are fond of saying. That acronym stands for the Latin phrase "quod erat demonstrandum," which translates to "that which was to be demonstrated." In this book, we use the less ostentatious little black square instead. ■

Exercises 1.7.

1. Morse code is made up of dots and dashes. A given sequence of dots and dashes stands for a letter. For example, $-\cdot-$ might be one letter, and $\cdot\cdot-\cdot$ might be another. Suppose we are not interested in our own alphabet, but in a more general alphabet with more letters, and suppose we use a Morse code with at least one, and at most n, dots and dashes. How many different letters could be represented by such a code?

2. Certain members of the eight-member city council are feuding at present and absolutely refuse to work together on council projects. Specifically, Mr. T refuses to work with Ms. B, and Mr. U refuses to work with Dr. P. How many three-person committees can be formed (to serve as the city's Public Relations Task Force) that involve only council members willing to work together amicably?

3. Refer to Example 1.7.6. Suppose you have selected the six numbers 1, 14, 25, 31, 37 and 41 to play in the next lottery game. If you had thought of these numbers earlier, on March 7, 2010, for example, you would have won a share in that day's thirty million dollar jackpot. C'est la vie. Anyway, these numbers are as good as any others in future Lotto games, so you are going with them. What are the chances that, among the State's six randomly chosen numbers, the number of matches with your choices is (a) four, (b) five, or (c) six?

4. The game of seven-card stud is played by dealing each player a random seven-card hand. A player will then discard two of the cards in the original hand and play his or her best five-card hand as in a game of five-card stud. Suppose you are dealt a seven-card hand. What is the probability that, after discarding two of your cards, you will play a full house. (Hint: This can happen only if your hand contains one of the following value configurations: $(3,3,1)$, $(3,2,2)$, or $(3,2,1,1)$. An example of a $(3,3,1)$ configuration would be 3 kings, 3 tens, and 1 three. Note that there is no possibility that such a hand will contain four of a kind or a straight flush, hands that beat a full house.)

5. Refer to Example 1.7.11. Let the vector (x,y,z) represent the event that three students got the same wrong answer x times, two different answers y times and three different answers

z times among the twelve questions that they all got wrong. As mentioned in the example above, there are a total of ten vectors for which x is at least 9. We know that $(9,2,1)$ can occur in 660 ways, and that $(10,0,2)$ can occur in 66 ways. Identify the remaining relevant vectors, and count the number of ways each of them can occur.

1.8 Chapter Problems

1. Let A, B, and C be arbitrary sets contained in the universal set U. Provide a set-theoretic proof of each of the following laws:

   (a) A Commutative Law: $A \cup B = B \cup A$

   (b) An Associative Law: $(A \cup B) \cup C = A \cup (B \cup C)$

   (c) A Distributive Law: $(A \cup B) \cap C = (A \cap C) \cup (B \cap C)$

2. Prove the general version of DeMorgan's Law stated in Equation (1.4). Show that (1.4) implies the following alternative version of DeMorgan's Law: $(\bigcap_{i=1}^{n} A_i)^c = \bigcup_{i=1}^{n} A_i^c$.

3. Let A, B, and C be three events in a random experiment with sample space S. Write expressions for each of the following sets in terms of the set operations "union," "intersection", "complement," and "difference":

   (a) only A occurs

   (b) A and B occur but C does not occur

   (c) exactly one of the events occurs

   (d) at least one of the events occurs

   (e) at most one of the events occurs

   (f) exactly two of the events occur

   (g) at least two of the events occur

   (h) at most two of the events occur

   (i) all three events occur

   (j) none of the events occur

4. In the game of roulette, each of the numbers 1 to 36 have an equal chance of occurrence, along with the special outcomes 0 and 00. (a) Suppose you bet on the number 12 on the first spin and on the number 18 on the second spin. What are that chances that at least one of your numbers wins? (b) Suppose you bet on the two numbers 12 and 18 on the first spin. What are the chances that one of your numbers wins? (c) Explain precisely why the answers to (a) and (b) are different.

5. A college student taking three courses this term is virtually certain that she will get an A or a B in each. (a) Draw a tree representing the sample space for this "experiment." (b) If her chances for an A are 30%, 60%, and 80% in the three courses, respectively, what's the probability that she gets more As than Bs?

6. Three cards are drawn at random from a 52-card deck. (a) What is the probability that exactly two aces are drawn? (b) Assuming that you draw at least one ace and at least one face card, what's the probability that you draw an ace before you draw a face card?

7. Let $A_1, A_2, \ldots, A_n, \ldots$ and B be events in the sample space S. Assume that the events $\{A_i \cap B\}$ form a partition of B, that is, assume that for $i \neq j, (A_i \cap B) \cap (A_j \cap B) = \varnothing$ and that $\bigcup_{i=1}^{\infty}(A_i \cap B) = B$. Prove that

$$P(\bigcup_{i=1}^{\infty} A_i|B) = \sum_{i=1}^{\infty} P(A_i|B).$$

(Note that if $\{A_i\}$ is a partition of S, then $\{A_i \cap B\}$ is necessarily a partition of B.)

8. Prove the Bonferroni inequality: for any collection of n sets $\{A_i : i = 1, \ldots, n\}$,

$$P(\bigcap_{i=1}^{n} A_i^c) \geq 1 - \sum_{i=1}^{n} P(A_i). \tag{1.69}$$

(Hint: Use a little set theory along with the "countable subadditivity" property.)

9. The workers in a particular factory are 65% male, 70% married, and 45% married male. If a worker is selected at random from this factory, find the probability that the worker is (a) a married female, (b) a single female, (c) married or male or both.

10. An urn contains six chips numbered 1 through 6. Three chips are drawn out at random. What is the probability that the largest number drawn is 4?

11. Consider an infinite collection of events $\{A_i, i = 1, 2, 3, \ldots\}$, each with the property that $P(A_i) = 1$. Prove that $P(\bigcap_{i=1}^{\infty} A_i) = 1$.

12. There seems to be some nasty bug making its way around school these days. Professor I. C. Buggs is a bit paranoid about getting a cold or flu from a student, so he brings certain instruments to his class every day. During his lecture today, his Sneezometer tells him that 30% of his students sneezed during his lecture. His Coughometer says that 40% of them coughed during the lecture. His CS-meter determined that 18% of his students did both. If one student was chosen at random, what is the probability that he or she either sneezed or coughed during the lecture, but didn't do both?

13. Let A and B be two events in a random experiment. Prove that the probability that exactly one of the events occurs (that is, the event $(A \cap B^c) \cup (A^c \cap B)$ occurs) in a given trial of this experiment is equal to $P(A) + P(B) - 2P(A \cap B)$.

14. Which is the more likely outcome, obtaining at least one six in four rolls of a single balanced die or obtaining at least one double six in twenty-four rolls of a pair of balanced dice?

15. Prove the Inclusion-Exclusion formula (see Theorem 1.3.7) for $n = 4$ under the assumption that the sample space is finite or countably infinite. (Hint: Let x be an arbitrary simple event. Suppose that x is contained in exactly k of the sets A_1, A_2, A_3, A_4, where $k = 1, 2, 3,$ or 4. Show that the probability of x is counted exactly once on the right-hand side of (1.14).

16. Four marbles are chosen from a bowl containing 4 red, 3 white, and 2 blue marbles. Find the probability that at least one marble of each color is selected.

17. At Betty Boop's birthday party, each of her ten guests received a party favor. The nicest two were wrapped in white paper, the next three nicest party favors were wrapped in blue, and the remaining five were wrapped in red. As it happens, the Bobsy twins (Bobo and Mimi) attended the party, and they tend to have huge fights when they don't get exactly the same thing. Mrs. Boop plans to distribute the party favors completely at random. (a) What's the probability that neither of the Bobsy twins got favors wrapped in white paper? (b) Given that neither of the Bobsy twins got a white favor, what's the probability that their party favors were the same color?

18. Four male members of the Davis Opera Society went to see La Boheme at the Mondavi Center last night. They turned in their top hats at the hat check stand and they were placed by themselves in a small box. When they left, the hat check clerk was nowhere to be seen. Since all four hats looked the same, they each took one of the hats at random. What's the probability that none of them got their own hat? (Hint: draw a tree.)

19. Urn #1 contains 4 red balls and 2 white balls. Urn # 2 contains 3 red balls and 4 white balls. An urn is selected at random (by tossing a fair coin), and then a ball is drawn randomly from the selected urn. (a) Find the probability that a red ball is selected. (b) Given that a red ball was selected, what is the probability that it came from urn #1?

20. Let A and B be two events in a finite or countably infinite sample space S of a random experiment. Suppose that A and B are mutually exclusive. Prove that

$$P(A|A\cup B)=\frac{P(A)}{P(A)+P(B)}. \tag{1.70}$$

21. A statistical colleague of mine went to a research conference last month. On the second day, he had the choice of attending one of three sessions scheduled simultaneously, one on logistics (L), one on cybernetics (C), and one on nanostatistics (N), and he chose one of them *at random*. Now it's well known among the eggheads who attend these conferences that the probabilities of each of these sessions being stimulating, boring, or insufferable (S, B or I) are, respectively, (3/6, 2/6, 1/6) for session L, (1/6, 3/6, 2/6) for session C, and (1/6, 1/6, 4/6) for session N. What is the probability that the session my colleague chose to attend was insufferable? Given that the session my colleague attended was insufferable, what are the chances that the session was on cybernetics? (Hint: draw a tree.)

22. Given: $P(A)=1/2, P(B|A)=1/3$, $P(B|A^c)=2/3$, $P(C|A\cap B)=1/4$, $P(C|A\cap B^c)=1/2$, $P(C|A^c\cap B)=2/3$, $P(C|A^c\cap B^c)=1/6$, calculate the probabilities: $P(B)$, $P(C)$, $P(A\cap B\cap C)$, $P(A|B\cap C)$, $P(A\cap B|C)$ and $P(A|C)$.

23. Sacramento High School has fielded a 3-student team known as the ABCs. The three team members now go by the nicknames "A," "B," and "C." Based on past performance, it is known that A will spell a given word correctly with probability 9/10, B with probability 7/10, and C with probability 6/10. Each new word is presented to a team member chosen at random. (a) What is the probability that the ABCs will produce the right spelling for a given word? (b) Given that a word was spelled right, what's the probability that the word was spelled by B? (c) This week, the opposing team spelled 8 words right out of the 10 they were given to spell. What's the probability that the ABCs won the spelling bee outright, that is, spelled at least 9 of their 10 words right?

24. Pakaf Airlines services the cities of Islamabad and Kabul. On any given Tuesday, a random passenger is traveling on business (B) with probability .6, is traveling for leisure (L)

with probability .3, and is traveling for trouble-making (T) with probability .1. The probability that Airport Security detains (D) a traveler varies: $P(D|B) = .2, P(D|L) = .3$ and $P(D|T) = .9$. (a) What's the probability that a random traveler is detained? (b) Given that a particular traveler is detained, what the probability he/she is traveling on business? (c) Suppose that 5 travelers on a given Tuesday flight are traveling for trouble-making. What's the probability that at least 4 of them are detained?

25. The final episode of a new reality TV show ends with the starring bachelorette standing before three closed doors. Behind each door is a homely man (one fat, one bald, one ugly) and a handsome man. (Three different handsome men were used.) In each case, the homely man is holding one million dollars. She will open a door at random and choose one of the men behind it. She has agreed to marry the man she chooses through this process. Suppose that the bachelorette will choose the homely man with probability 3/4, 2/3, and1/2 (depending on whether he's fat, bald, or ugly). (a) What is the probability that she will marry a homely man. (b) This show was filmed several months ago. Given that this former bachelorette's husband is handsome, what's the probability that he was paired with the ugly man?

26. There is no grey area for college student Will B. Stubboure. Will knows what he knows, and unabashedly guesses at everything else. On the multiple choice final exam in his Introductory Psychology class, he will know the correct answer to a random question with probability .7. When he doesn't know the correct answer to a question, he chooses one of the *five* possible answers at random. (a) What's the probability that Will gets a randomly chosen question right? (b) Given that he got a particular question right, what's the probability he actually knew the answer? (c) Assuming that his performance on each question is independent of his performance on any other, what is the probability that he gets exactly two of the first three questions right?

27. This problem is a simple abstraction of how contagion works (for example, how colds spread). Consider three urns. Urn 1 has 1 red and 2 white marbles. Urns 2 and 3 both have, initially, 1 red and 1 white marble. A marble is chosen at random from urn 1 and is placed into urn 2. Then a marble is drawn at random from urn 2 (which now contains 3 marbles) and is placed into urn 3. Finally, a marble is drawn from urn 3 (which also now contains 3 marbles). Let R_i and W_i represent the events of drawing a red or a white marble from urn i. Show that $P(R_3|W_1) < P(R_3) < P(R_3|R_1)$. (Hint: Everything works out nicely by analyzing a probability tree with 8 paths corresponding to three draws with two possible outcomes each.)

28. A widget dealer purchases 50% of his widgets from factory A, 30% of his widgets from factory B, and 20% of his widgets from factory C. Suppose that the proportion of defective widgets among widgets purchased from factories A, B, and C are .02, .03, and .05, respectively. (a) What is the probability that a randomly chosen widget from this dealer's warehouse is defective? (b) Given that a randomly chosen widget is defective, what are the probabilities that it came from factory A, factory B, or factory C?

29. One of the games at an elementary school carnival involves drawing a marble at random from one of three urns. Each urn contains five marbles. Urn 1 contains 2 red marbles, Urn 2 contains 3 red marbles, and Urn 3 has 4 red marbles, with all remaining marbles being blue. You pay to play, and you lose your money if you draw a red marble. Suppose the urn you draw from is selected by tossing a fair coin twice, with Urn k selected if you get $k-1$ heads (for $k = 1, 2$, and 3). (a) What's the probability that you draw a red marble? (b) Given that you drew a red marble, what's the probability that you drew from Urn 3?

30. A traveling salesman asked his travel agent to book him a room at one of two Santa Fe motels, the Laid Back Inn (LBI) and the Laid Way Back Inn (LWBI). One-fifth the rooms at the first have Jacuzzis, while one-half of the rooms at the second do. Assume that the probability that his agent booked him at the LWBI is 2/3 (and 1/3 at the LBI). (a) What's the probability that the salesman gets a room with a Jacuzzi? (b) Given that he didn't get a Jacuzzi, what's the probability that he was booked at the Laid Back Inn? (c) If the salesman stays at one of these two Santa Fe motels on each of four consecutive trips, what is the probability that he gets his first Jacuzzi-equipped room on his fourth visit?

31. Michael Macho is a bit of a tyrant. This morning, he barked at his wife: "Doris, darn my socks!" Doris replied: "Michael, damn your socks." Michael was forced to operate as usual, randomly drawing one sock at a time from his sock drawer, without replacement, until he finds two good ones. Today, the drawer contained 4 good socks and 7 socks with holes in them. (a) What's the probability that he gets exactly one good sock in his first two draws from the drawer? (b) Given that the first two socks drawn are of the same type, what's the probability they are both good? (c) What's the probability that he draws the second good sock on the fifth draw from the drawer?

32. Tom, Dick, and Harry play a round-robin tennis match every Saturday. Each of them plays one set with each of the other two players (so that three sets are played in all). The outcome of each set is independent of the outcome of any other. Tom beats Dick two thirds of the time, Tom beats Harry half of the time, and Dick beats Harry three fourths of the time. Calculate the probability that on a given Saturday, each player wins one set and loses one set.

33. Prove that, for arbitrary events A, B, C in a random experiment, for which $P(B \cap C) \times P(B \cap C^c) > 0, P(A|B) = P(A|B \cap C)P(C|B) + P(A|B \cap C^c)P(C^c|B)$.

34. (The Prisoner's dilemma) Jim, Joe, and Jed are in prison. Jim learns, via the prison's underground grapevine, that two of the three of them are going to be released. He has befriended the jailer, and believes that he would get a reliable answer to the request: "Please tell me the name of a jailmate who is going to be released." After further thought, Jim decides not to ask the question after all. He figures that his chances of being released are now 2/3, but that if he knows the identity of the fellow prisoner who will be released, then he will be one of the two remaining prisoners, reducing his chances of release to 1/2. Is Jim right? Explain.

35. In the game of craps, played with a pair of balanced dice, the shooter rolls the two dice, and wins his bet immediately if he rolls a 7 or an 11. A first roll of 2, 3 or 12 results in an immediate loss. If the shooter rolls any other number (i.e., 4, 5, 6, 8, 9, 10), that number is designated as the "point." The shooter then rolls the dice repeatedly until he/she obtains the point, resulting in a win, or obtains a 7, resulting in a loss. Calculate the probability of winning in the game of craps. (Hint: The probability that your point is 4, for example, is 3/36. The probability that you roll a 4 before a 7, given that your point is 4, is $\sum_{i=0}^{\infty} P(\{\text{next } i \text{ rolls are neither 4 nor 7}\} \cap \{\text{a 4 on roll } i+1\})$. It follows that

$$\begin{aligned} P(\text{win}|\text{point} = 4) &= \sum_{i=0}^{\infty} (27/36)^i (3/36) \\ &= (3/9) \sum_{i=0}^{\infty} (27/36)^i (9/36) \\ &= (3/9) = 1/3. \end{aligned}$$

Note that for $0 < a < 1$, $\sum_{i=0}^{\infty} a^i(1-a) = 1$. Now, compute and add together the probabilities of winning on the initial roll and winning by rolling one of the six points above and then rolling that point again before rolling a 7. (E.g., P(point = 4 and you win) $= P(4) \times P(\text{Win}|4) = 1/36$).

36. Suppose n people are in a room, where $n \leq 12$. Find an expression for the probability that at least two of these people have birthdays in the same month. Assume that the birthday of a random individual is equally likely to be any one of the 12 months of the year.

37. Prove, for arbitrary events A and B, that $P(A|A \cup B) \geq P(A|B)$.

38. An urn contains 6 blue balls, 4 red balls, and 2 white balls. Balls are randomly selected from the urn, without replacement, until all have been drawn. What is the probability that a white ball is drawn before the first red ball is drawn.

39. A loaded die is rolled. For i = 1, 2, 3, 4, 5, and 6, the probability that the digit i faces up is $P(i) = i/21$. If the roll of the die yields the outcome i, a fair coin is tossed i times. Find the probability that, in the coin-tossing portion of the experiment, at least one heads is obtained.

40. Thirty percent of professional baseball players chew tobacco. Forty percent have bad breath. Twenty-five percent have both properties. Suppose that a professional baseball player is selected at random. (a) What is the probability that this player chews tobacco but doesn't have bad breadth? (b) Given that this player has bad breadth, what's the chance that he doesn't chew tobacco?

41. From DNA evidence, the police know that one of the Didit brothers pulled off last weekend's solo jewelry theft in Beverly Hills. The lead detective authoritatively asserts that I. Didit did it with probability .6 and that U. Didit did it with probability .4. Also, police records show that I. Didit wears black on 30% of the jewel thefts attributed to him, while U. Didit wears black 60% of the time for such thefts. (a) Assuming that one of the Didit brothers really did last weekend's jewelry theft, what's the probability that the thief wore black? (b) An eyewitness could not identify the thief but swears that the thief was not wearing black. Given this fact, what's the probability that U. Didit didn't do it (i.e., I. Didit did it)?

42. Restaurateur S. Yogi Wilkinson believes that a diner's choice of first course has a strong relationship to that diner's choice of second course. Yogi points out that the chances that a random diner has soup as a first course is .4 (with salad chosen 60% of the time). Further, he says that those who order soup will have a meatless second course with probability .2, while those who order salad follow up with a meatless dish with probability .4. Given that a random diner had a meat dish as a second course, what's the probability that this dish was preceded by a salad?

43. Every Saturday morning, I glance out my window and look over my lawn. With probability 2/3, I decide to mow it. (Actually, my wife claims this probability is somewhat lower.) With probability 1/3, I decide to flop down on the couch and watch cartoons. What is the probability that on the next five Saturdays, I will mow the lawn at least 4 times?

44. Papa, a doting grandfather, likes to spoil his grandson J.T. So does J.T.'s doting auntie Bee. At their request, J.T. revealed his three "dream Christmas presents." Both Papa and Bee

plan to give J.T. one. The choices are A) a trip to Disneyland, B) the latest mega-play station, and C) a fully functional "junior" golf cart. Papa and Bee choose gifts independently. Papa chooses A, B, and C with probabilities 3/6, 2/6, and 1/6. Bee chooses A, B, and C with probabilities 1/8, 5/8, and 2/8. (a) What's the probability that Papa and Auntie Bee give J.T. two different presents? (b) Given that Papa and Bee give J.T. different presents, what's the probability that one of them is a trip to Disneyland?

45. Just before going on stage, a magician crams 3 bunnies and 2 squirrels into his top hat. At the beginning of his act, he takes off his hat, nervously wipes his brow, and shows the audience what appears to be an empty hat. Then he sets the top hat on a table and proceeds to draw animals from the hat. (a) What is the probability that one of the first two draws is a squirrel and the other is a bunny? (b) Given that the same kind of animal was drawn in the first two draws, what is the probability that they were both bunnies? (c) What is the probability that the second bunny was drawn on the fourth draw?

46. The Pennsylvania School of Psychic Arts and Sciences has 1000 students. The admission criteria at PSPAS include the requirement that students have vivid dreams every night. It is known that the chance that a given dream will be predictive (that is, entirely come true within 30 days) is 1 in 10,000. What is the probability that at least one student at the school will have a predictive dream in the next seven days?

47. (The Monte Hall Problem) A 1960s TV game show hosted by Monty Hall featured a contestant trying to guess which of 3 doors he/she should open. Behind one was a fabulous prize, like the keys to a new sports car. There was nothing behind the other two doors. The contestant would pick a door (presumably at random), and then Monty would ask "Are you suuuuuure?" Before allowing the contestant to answer, Monty would open one of the doors not chosen and show that there was nothing behind it. (Monty could always do this because he knew where the prize was.) The contestant was then allowed to stick with the first choice or to switch to the other unopened door. Thus, the contestant had to choose between the two strategies: (a) switch doors or (b) stick with his/her initial choice. Calculate the probability of winning the prize when strategy (a) is used. (Hint: Draw a tree in which stage 1 represents placing the prize at random behind doors 1, 2, and 3, and stage 2 represents the contestant choosing a door at random. Each path of this tree has probability 1/9. Apply the switching strategy and determine how many of the paths result in winning the prize.)

48. This intriguing problem was posed by the legendary mathematician Pierre Simon Laplace in 1812. There are $k + 1$ boxes, numbered $i = 0, 1, 2, \ldots, k$. The box numbered i has i white marbles and $(k - i)$ red marbles. You select a box at random, and draw n marbles in a row from that same box, replacing the chosen marble after each draw. Given that you draw a white marble every time (i.e., you get n white marbles), find *an expression* for the probability that the next (the $(n + 1)$st) draw from this box will be white. (Hint: Draw a probability tree. By the way, the expression you are after is a bit messy [being a ratio of two sums].)

49. An urn contains 5 red marbles, 3 blue ones, and 2 white ones. Three marbles are drawn at random, without replacement. You got to the party late, but you overheard someone say that the third marble drawn was not white. Given this fact, compute the conditional probability that the first two marbles were of the same color.

50. Suppose two cards are to be drawn randomly from a standard 52-card deck. Compute the

probability that the second card is a spade when the draws are made a) with replacement and b) without replacement. Compare these two answers and reflect on whether the result you got works more generally (that is, with more than 2 draws). Now, use your intuition to obtain the probability that the 52nd draw, without replacement, from this 52-card deck will be a spade.

51. Suppose that two evenly matched teams (say team A and team B) make it to the baseball World Series. The series ends as soon as one of the teams has won four games. Thus, it can end as early as the 4th game (a "sweep") or as late as the 7th game, with one team winning its fourth game compared to the other team's three wins. What's the probability that the series ends in 4 games, 5 games, 6 games, 7 games?

52. A hunter in an African jungle notices a menacing, seemingly hungry tiger directly ahead of him. His rifle has six bullets in it. The tiger rushes toward him, and he fires six shots in rapid succession. If the shots are considered independent events and the chances he will hit the tiger with any given shot is .6, what are the chances that he lives to tell about the adventure? (Assume that a single hit is sufficient to stop the tiger.)

53. Three bums catch a late night bus to Davis Community Park. The bus makes three stops at the park. Each bum chooses to exit at a stop at random, independently of the choices made by others. Find the probability that the bums exit at three different stops.

54. Two events A and B are "pairwise independent" if $P(A \cap B) = P(A) \times P(B)$. Three events A, B, and C are "mutually independent" if each pair of events are independent and if, in addition, $P(A \cap B \cap C) = P(A) \times P(B) \times P(C)$. Now, suppose that a bowl contains four balls numbered 1, 2, 3, and 4. One ball is drawn at random from the bowl. Let A = {1, 2}, B = {1, 3}, and C = {1,4}. Consider the three events defined as: The number on the ball that is drawn is in the set A, in the set B, or in the set C. Show that the events are pairwise independent but not mutually independent.

55. The subject of Reliability Theory lies on the interface of Statistics and Engineering and concerns the performance characteristics of engineered systems. Two of the best known system designs are the "series system," which works as long as all of its components are working, and the "parallel system," which works as long as at least one of its components is working. Suppose such systems have components that work independently, each with probability p. (a) Calculate the probability that a series system with *n* components will work. (b) Calculate the probability that a parallel system with *n* components will work.

56. A bridge system in 5 components has the structure pictured below.

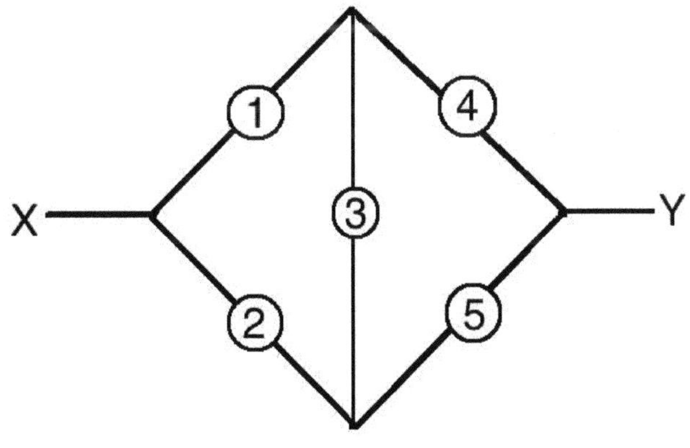

The system works if there is a path of working components that connects the X on the left side to the Y on the right side. The minimal "path sets," whose working guarantees that the system will work, are {1, 4}, {2,5}, {2, 3, 4}, and {1, 3, 5}. If all five components operate independently and each has probability p of working, what is the probability that the bridge system works? (Hint: a system works if and only if at least one of its minimal path sets is working. Apply Theorem 1.3.7.)

57. Suppose that k and n are integers such that $0 \leq k \leq n$. (a) Replace each of the combinatorial symbols in the identity below by the appropriate fraction and prove the identity by an algebraic argument:

$$\binom{n-1}{k} + \binom{n-1}{k-1} = \binom{n}{k} \tag{1.71}$$

(b) Give a "combinatorial argument" showing that this identity holds. (Hint: Think about choosing a set of k people from a group of n people, one of whom is your friend Sally.)

58. Give a combinatorial argument for the following identity: For any integers k and n such that $1 \leq k \leq n$,

$$\binom{n}{2} = \binom{k}{2} + k(n-k) + \binom{n-k}{2}.$$

Confirm the validity of this identity with an algebraic proof.

59. Suppose you have k indistinguishable white balls and $(n-k)$ indistinguishable green balls. If you line up the n balls in a row, how many distinctly different orderings are possible?

60. The Casino game "Keno" is notorious for yielding one of a gambler's poorest possible expected payoffs. The game is typically played as follows. A player will select n integers between 1 and 80, inclusive, where $2 \leq n \leq 10$. Then the house will choose 20 integers between 1 and 80. The player receives a payoff which depends on how many of the house's numbers he/she has matched. The payoff table used by the house looks attractive but, in reality, substantially favors the house. (a) Obtain a general formula for the probability that the player matches k of the house's numbers, for $k = 0, 1, \ldots, n$. (b) What is the probability that the player will match $k = 5$ house numbers when he selected $n = 10$ on his/her Keno card?

61. A University President's Student Advisory Committee consists of 8 students, two from each class (Sr., Jr., So., and Fr.). A subcommittee of 4 students is to be chosen to provide student input on the question of whether the campus should create a formal undergraduate degree program in Facebook Studies. (a) How many different subcommittees can be formed? (b) How many different committees can be formed if the advisory committee must have a representative from each class? (c) Suppose the subcommittee is formed completely at random. What is the probability that two of the four classes are excluded from the committee?

62. You may have heard of the famous French mathematician Pierre de Fermat who, in 1637, noted, without giving a formal proof, that a particular mathematical claim (namely, that the equation $x^n + y^n = z^n$ has no integer solutions (x, y, z) if n was an integer greater than 2) was true. The claim seemed like quite a bold leap, given that such equations have infinitely many integer solutions when $n = 2$, a fact proven by Euclid about so-called Pythagorean triples. Fermat claimed he had a "marvelous" proof of his claim, but that there wasn't enough room in the margin of the book in which he had written the claim to fit in the proof.

The result became known as "Fermat's last theorem," and it intrigued mathematicians for over 350 years. It was finally proven by Princeton University's Andrew Wiles in 1994 in a manuscript summarizing years of work in which major new mathematical tools had been invented for application to the problem. (You can read about this stunning mathematical accomplishment in the Simon Singh's (1997) engaging book *Fermat's Enigma*). Here's another problem due to Fermat. It should take you significantly less than 350 years to prove it, especially if you follow the lines of the combinatorial argument suggested in the hint. Prove Fermat's combinatorial identity: for any positive integers $k \leq n$,

$$\binom{n}{k} = \sum_{i=k}^{n} \binom{i-1}{k-1}. \tag{1.72}$$

(Hint: For $i \geq k$, let i be the largest integer in a given set of k integers chosen from the collection $\{1, 2, \ldots, n\}$. For a fixed such integer i, how many different sets of size k can be formed?)

63. A set of dominos can be described as a collection of chips, each with two integers on it drawn from a set of n integers. Order plays no roll, as the chip [2, 3] occurs only once and can be played as either [2, 3] or [3, 2]. The standard set of dominos uses $n = 7$ integers and contains 28 pieces. (a) Find the general formula for the number of dominos in a set based on n integers. (Hint: You can do this by drawing triangular array that starts with the domino [1, 1] in the first row, dominos [2, 1] and [2, 2] in the second row ... and dominos $[n, 1]$, $[n, 2]$,..., $[n, n]$ in the nth row. Then just add the row totals.) (b) Notice that your answer to part (a) is actually a combinations symbol. Give a combinatorial argument which obtains this same answer. (Hint: Imagine $n+1$ numbered red marbles in a row. Suppose you pick two of them and mark them with an X. Argue that the number of ways you can do that is in one-to-one correspondence with the process of constructing a set of dominos from n integers. To get this correspondence, it's helpful to give a special interpretation to the cases in which the integer $(n+1)$ is chosen.)

64. There are 5 males and 5 females taking Statistics 177, "Little Known and Rarely Used Statistical Methods." A team of three students is needed for a panel that will discuss the instructor's research achievements. (a) How many different teams can be formed? (b) How many teams are there in which both genders are represented? (c) If John will only participate if Wendy participates and Tom will participate only if his team is all male, how many different teams are possible?

65. What is the probability that all four suits are present in a random five-card poker hand?

66. A poker hand is obtained by drawing five cards at random without replacement from a standard 52-card deck. Find the probability of the following poker hands: (a) a straight flush (5 cards in a row of the same suit; ace can be high or low) (b) four of a kind (4 cards of one value, one of another) (c) a straight, but not a straight flush, (d) two pairs (2 cards of one value, 2 of another value, 1 of a third value) (e) one pair (2 cards of one value, 1 each from three other values) (f) a "Toilet Flush" (5 different values, but not all in a row (a straight) and not all the same suit (a flush) nor both (a straight flush).

67. Since your statistics class this term has been such fun, the class has decided to have a potluck dinner after finals. The instructor has graciously offered his home for the occasion. A committee of 8 students (3 males and 5 females) has been selected to organize the dinner. Several subcommittees will be formed to coordinate particular tasks. (a) If three committee

members are selected at random to coordinate main dishes, what are the chances that this subcommittee has both male and female members? (b) If each subcommittee is formed completely at random, what are the chances that two 3-person subcommittees will overlap, that is, have at least one member in common. (c) If Miller and Madison are on the eight-person committee but refuse to serve together on the 3-person dessert committee (having, as they do, very different opinions about what makes an appropriate dessert), how many different 3-person dessert committees can be formed?

68. In the poker game of seven-card stud, players are dealt seven cards at random from a standard deck, and they toss out two and play their best five-card hand. Suppose you are dealt seven random cards. Find the probability that your best hand is "two pair," that is, among your seven cards, you got two values twice and three different single values (e.g., two Ks, two 5s, one Q, one 8, one 2). You may assume that the dealer has imposed the "house rule" that *straights and flushes have no value.* (Without that rule, you'd have to consider that a hand with exactly two pairs and three singletons might actually contain a flush or a straight and thus would not really be played as "two pair." In this problem, you can ignore that possibility.)

69. This year's Insurance Sales Person (ISP) Banquet in Sacramento was attended by n ISPs. (a) It is a tradition at this banquet for each ISP to shake hands with every other ISP exactly once (so as to save their handshaking energy for all the other hands they have to shake on a daily basis). How many handshakes occurred at the banquet? Give a formula and explain your logic. (b) Another tradition at this banquet is that all ISPs who attend are seated at one huge round table. (Clearly, an RSVP is required for anyone who wishes to attend.) How many different seating arrangements are possible? Give a formula and explain your logic. (Hint: Start by seating the "host." Note that a particular seating arrangement is identical to another if the same ordered sequence of $(n-1)$ people sit to the immediate left of the ISP who is hosting the banquet, regardless of where the host is seated. Take "mirror image" seating orders as identical; for example, if $n = 4$, H, 1, 2, 3 is considered the same arrangement as H, 3, 2, 1.)

70. In a cereal promotion during the "Harry Potter era," every box of Cocoa Puffs cereal contained a Harry Potter collector's card. There are 20 different cards in all, and the one in any given box was chosen completely at random. Three big Harry Potter fans each bought several boxes. In the end, Wolfgang had 5 cards, Jessica had 4, and Frank had 6. What are the chances that all fifteen of these cards are different? (Hint: Conceptually, this problem bears similarities to the birthday problem.)

71. A wealthy man passed away. He owned n equally valuable (and thus essentially indistinguishable) properties. His will stated that his properties were to be divided among his three daughters. If x_i represents the number of properties inherited by the ith daughter, with i = 1, 2, and 3, how many different ways could these properties be divided. Mathematically, this question reduces to: how many different integer-valued vectors (x_1,x_2,x_3) are there for which $x_i \geq 0$ for $i = 1,2,3$ and $x_1 + x_2 + x_3 = n$. (Hint 1: Draw a tree, with a stage dedicated to each of the daughters. Hint 2: A combinatorial argument can be made by imagining $n+2$ billiard balls is a row, choosing two of them, and noting that there are n left unchosen. How does this identify the number of properties each daughter got?)

72. Prove the following claim: the number of combinations of n objects taken r at a time, when

items are drawn with replacement (making repetitions possible), is given by

$$\binom{n+r-1}{r}. \tag{1.73}$$

(Hint: Try a combinatorial argument. Suppose you have $n+r-1$ indistinguishable marbles arranged in a row. Now, choose $(n-1)$ of them and mark them. Note that there are r unmarked marbles remaining. Show that your choice is equivalent to identifying r objects among n objects, with replacement.)

73. A certain engineering class had 8 male students and 12 female students. Four males and one female got A's as final grades. This apparent imbalance was called to the attention of the dean. The dean decided that if the distribution of A's in a class was found to be a "rare event" (that is, if it had less than a 5% chance of happening when grades were assigned randomly), the matter would be referred to the University Grade Change Committee. Assuming a random assignment of grades, with five students getting A's, calculate the probability that no more than one woman would get an A in this class. Do the grades actually assigned meet the dean's "rareness" criterion?

74. "Poker dice" is a game played by rolling a set of five balanced dice. (It's a popular game played in public houses (i.e., pubs). You can't miss it, because the noise associated with slamming down the leather cup which holds the dice is both loud and annoying.) But the game is fun to play. The possible outcomes are shown in the list below. Verify the stated probabilities.

    (a) P(five different numbers) = .0926
    (b) P(one pair) = .4630
    (c) P(two pairs) = .2315
    (d) P(three of a kind) = .1543
    (e) P(full house) = .0386
    (f) P(four of a kind) = .0193
    (g) P(five of a kind) = .0008.

    (Hint: It is helpful to think of the five rolls of your dice as a five-stage experiment. Imagine the tree that corresponds to it; it has 6^5 paths. Keep in mind that the paths of a tree are ordered (e.g., the path 65352 is different from the path 55623). In calculating the number of paths that yield a certain outcome, keep the possible orderings in mind.)

75. The Davis Philatelist Society has 15 members. (Obviously, some closet philatelists haven't seen fit to join.) The Society's By-Laws call for randomly selecting a 3-member group each week to do the necessary administrative and clerical work. The sampling each week involves all 15 members. (a) How many different groups can be formed to do the Society's work next week? (b) What is the probability that the groups formed in the next *two* weeks have exactly one member in common? Treat the sampling each week as independent and involving all members. (c) Give an expression, using permutation or combination symbols, for the probability that the groups formed in the next *three* weeks have no overlap, that is, involve nine different Society members?

76. Suppose that a bowl contains 5 white marbles and 10 red ones. Suppose that marbles are drawn at random, without replacement until all remaining marbles are the same color. What is the probability that all the red marbles are drawn before all the white marbles have been drawn?

77. In the game of bridge, each of the four players is dealt a 13-card hand at random. (a) Compute the probability that at least one of the four hands is missing at least one suit. (Hint: Use Theorem 1.3.7) (b) Compute the probability that each of the four players holds an ace in their hand.

78. Suppose that n pairs of people belong to a certain organization. These could be, for example, n couples attending a "Marriage Encounter" weekend, or it could be the 100 members of the U.S. Senate, each of whom is paired with another senator according to the state he/she represents. Suppose a subgroup of k individuals is to be chosen to oversee the performance of some task, where $k \leq n$. If these k individuals were chosen at random, what is that probability that the group contains both members of at least one pair? (Hint: Consider the complementary event.)

79. If n distinguishable balls are distributed at random into n urns so that all n^n arrangements are equally likely (i.e., ball i is equally likely to go into any urn), show that the probability p that there is exactly one empty box is

$$p = \frac{\binom{n}{2} n!}{n^n}. \tag{1.74}$$

80. Let k, m, n, and N be non-negative integers satisfying the constraints

$$\max(n,m) \leq N \text{ and } \max(m+n-N,0) \leq k \leq \min(n,m).$$

Consider the following combinatorial identity

$$\frac{\binom{n}{k}\binom{N-n}{m-k}}{\binom{N}{m}} = \frac{\binom{m}{k}\binom{N-m}{n-k}}{\binom{N}{n}}. \tag{1.75}$$

(a) Prove (1.75) algebraically. (b) Give a combinatorial proof of (1.75).

81. The game of Yahtzee is played with five standard dice. A turn consists of three rolls. The first roll uses all 5 dice. For the second roll, you can leave as many dice as you wish on the board and then roll the rest, trying for a particular poker-like objective. Same thing on the third roll. Suppose that you are on your last turn, and that you need to roll a "Yahtzee" (that is, "five of a kind") in order to win the game you are now playing. Assume that you rolled three "fours" on your first roll and that you left them on the board. You have two rolls left. What is the probability that you will roll a Yahtzee and win the game? (Hint: Use the probability tree to represent the several things that can happen (relative to rolling 4s) in the last two rolls of this turn. For example, if you roll 1 "four" on the second roll, you will leave it on the board and roll the one remaining die again. If you roll 2 "fours" on your second roll, the third roll is unnecessary.)

2

Discrete Probability Models

2.1 Random Variables

While the concept of a sample space provides a comprehensive description of the possible outcomes of a random experiment, it turns out that, in a host of applications, the sample space provides more information than we need or want. A simple example of this overly informative potential of the sample space is the 36-point space describing the possible outcomes when one rolls a pair of dice. It's certainly true that once a probability has been assigned to each simple event, we are in a position to compute the probability of any and all compound events that may be relevant to a particular discussion. Now suppose that the discussion *du jour* happens to be about the game of craps. The classic wager in the game of craps is based solely on the sum of the two digits facing upwards when two dice are rolled. When making this wager, we don't really care or need to know how a particular sum arose. Only the value of this sum matters. The sum of the digits facing up in two rolled dice is an example of a "random variable" in which we might be interested. This sum can only take on one of eleven possible values, so concentrating on the sum reduces the number of events of interest by more than 2/3. We proceed with our treatment of random variables with an appropriate definition and a collection of examples of the concept. For now, we will restrict attention to the discrete case, that is, to the case in which the sample space is either finite or countably infinite.

Definition 2.1.1. (Random Variables) A *random variable* is a function whose domain is the sample space of a random experiment and whose range is a subset of the real line.

This definition simply states that a random variable is a "mapping" that associates a real number with each simple event in a given sample space. Since different simple events can map onto the same real number, each value of the random variable actually corresponds to a compound event, namely, the set of all simple events which map on to this same number. If you were into linguistics, you would probably notice that the phrase "random variable" isn't an especially good name for the object just defined. You might insist that a random variable X is neither random nor is it a variable. It's actually just a function which maps the sample space S onto the real line R. In a mathematics book, you would find X described by the favored notation for "mappings" in that discipline, i.e., $X : S \to R$. Be that as it may, the language introduced above is traditional in Probability and Statistics, and we will stick with it. Before you know it, the term "random variable" will be rolling off your tongue and you will know precisely what it means simply because you will encounter and use the term with great frequency.

The values that a random variable X may take on will depend on exactly how it was defined. We will of course be interested in the values that X can take, but we will also be interested in the probabilities associated with those values. As mentioned above, each possible value of X corresponds to a compound event. It is thus natural for a particular value of X to inherit the probability carried by the compound event that it goes with. Suppose that A is the compound event consisting of every simple event "a" for which $X = k$, that is, suppose that $X = k$ if and only if $a \in A$. In mathematical function notation, we would write $X(a) = k$ if and only if $a \in A$. This correspondence allows us to identify the probability that must logically be associated with any given value of X. Thus, the probability model for the random experiment with which we started leads to a probability model for that random variable of interest. We often refer to this latter model as the probability distribution of X. In "small problems," this distribution is often displayed as a table listing the possible values of X and their corresponding probabilities. The following examples display the probability distributions of several random variables which arise in simple everyday problems.

Example 2.1.1. Suppose you toss a pair of *fair* coins and you observe a head or a tail on each of the coins. We may write the sample space of this random experiment as $S = \{HH, HT, TH, TT\}$. Let X be the random variable which keeps track of the number of heads obtained in the two tosses. Then clearly, $X \in \{0, 1, 2\}$. Since the coin is fair, each of the four simple events in S is equally likely, and the probability distribution of X may be displayed as

$X = x$	0	1	2
$p(x)$	1/4	1/2	1/4

Table 2.1.1. The probability distribution of X, the number of heads in two tosses of a fair coin.

■

The function p(x) in Table 2.1.1 is usually referred to as the *probability mass function* (or pmf) of X. We will use that phrase when describing the probabilities associated with any discrete random variable. The word "mass" refers to the weight (or probability) assigned to given values of X.

Example 2.1.2. Suppose that two *balanced* (6-sided) dice are rolled and we are interested in the digits facing upwards. If X is equal to the sum of the digits facing up, then it is clear that $X \in \{2, 3, 4, 5, 6, 7, 8, 9, 10, 11, 12\}$. The probability distribution of X is given by

$X = x$	2	3	4	5	6	7	8	9	10	11	12
$p(x)$	1/36	2/36	3/36	4/36	5/36	6/36	5/36	4/36	3/36	2/36	1/36

Table 2.1.2. The pmf of X, the sum of the digits facing up when two balanced dice are rolled.

■

Example 2.1.3. Let us suppose that five cards are drawn at random, without replacement, from a 52-card deck. Let X be the number of spades obtained. Clearly, X can only take the values in the set $\{0, 1, 2, 3, 4, 5\}$. The probability of each of the six possible values of X can be computed and then displayed in a table as in the two examples above. However, having studied some standard counting formulas in Chapter 1, we may turn to these formulas to write a general expression of the probability of any observed value of X. Recall that, when cards are sampled without replacement, all possible five-card poker hands are equally likely. In addition, note that there are

a) $\binom{52}{5}$ ways of picking 5 cards without replacement from the 52 cards,

b) $\binom{13}{x}$ ways of picking x spades without replacement from the 13 spades,

c) $\binom{39}{5-x}$ ways of picking $5-x$ cards without replacement from the 39 non-spades.

It thus follows from the basic rule of counting that the probability that $X = x$ is given by

$$P(X = x) = \frac{\binom{13}{x}\binom{39}{5-x}}{\binom{52}{5}} \text{ for } x = 0, 1, 2, 3, 4, 5.$$

This is the first occurrence of a very useful discrete probability model called the "hypergeometric" distribution. We will treat this distribution in some detail later in this chapter. ■

A probability histogram (PH) is a useful visual tool that may be employed in tandem with the table or formula that defines a probability mass function. The probability histograms associated with the pmfs displayed in the Examples 2.1.1–2.1.3 are shown in the figures below. The x-axis shows the possible values of a random variable X, and the graph shows a set of rectangles whose heights are the probabilities of the respective values of X.

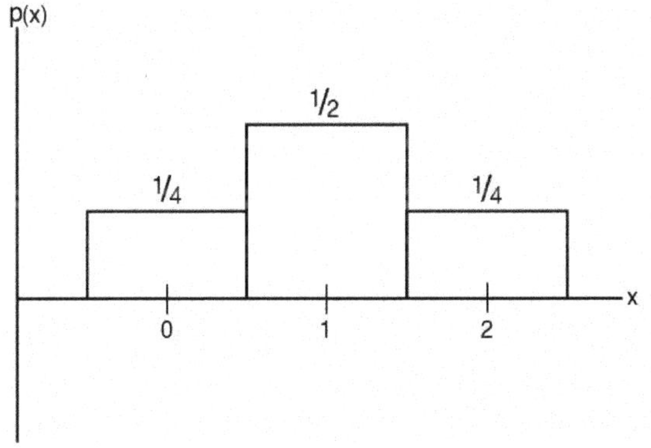

Figure 2.1.1. The PH for X = number of heads in 2 tosses of a fair coin.

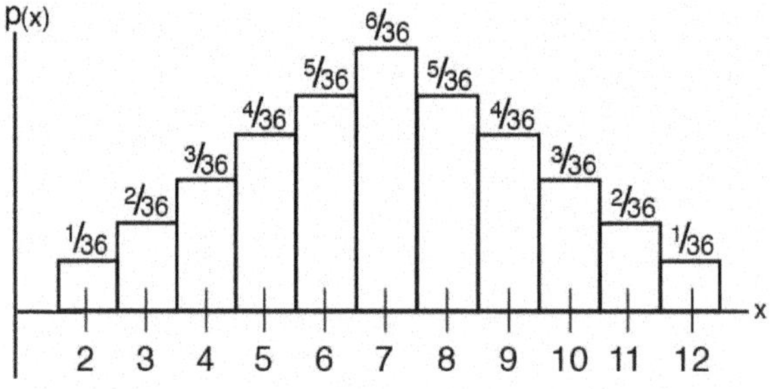

Figure 2.1.2. The PH for X = sum of digits facing up if 2 balanced dice are rolled.

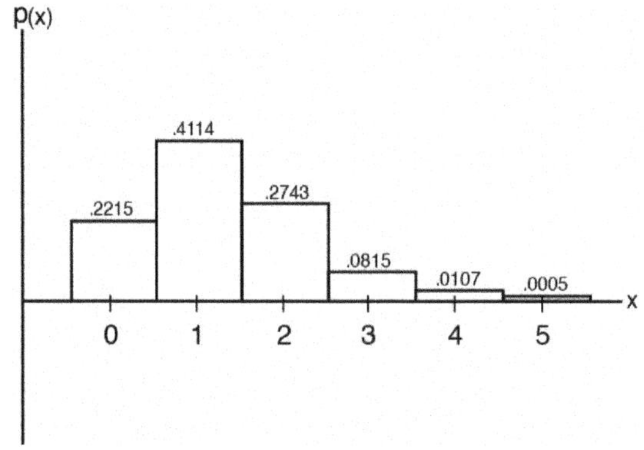

Figure 2.1.3. The PH for X = the number of spades in a random 5-card poker hand.

Exercises 2.1.

1. Suppose two balanced dice are rolled. Let X be the largest digit facing up. (If the digits facing up are both equal to the value c, then $X = c$.) Obtain the probability mass function of X.

2. Suppose a fair coin is tossed until the outcome "heads" (H) is obtained. Let X be the number of "tails" (T) obtained before the first H occurs, and let Y be the number of trials needed to get the first H. Find the probability mass functions $p_X(x)$ of X and $p_Y(y)$ of Y. What is the algebraic relationship between X and Y.

3. The Board of Directors of the MacArthur Foundation consists of 6 males and 4 females. Three members are selected at random (obviously without replacement) to serve on the Foundation's annual spring picnic committee. Let X be the number of females selected to serve on the committee. Obtain the p. m. f. of the random variable X.

4. Suppose that a fair coin is tossed four times. Let X be the number of heads obtained. Find the pmf of X. (Hint: Start with a probability tree.)

2.2 Mathematical Expectation

One numerical summary of the distribution of a random variable X that is quite meaningful and very widely used is the *mathematical expectation* of X. Often, this numerical summary is referred to as the *expected value* of X and is denoted by $E(X)$ or simply by EX. As the name suggests, this number represents the value we would "expect" X to be if the random experiment from which X is derived were to be carried out. We recognize, of course, that the actual value of the X that we observe might be larger or smaller than EX, but EX is meant to be interpreted as what we would expect X to be, "on the average," in a typical trial of the experiment. The expected value of any discrete random variable X may be computed by the following formula.

Definition 2.2.1. If X is a discrete random variable with probability mass function $p(x)$, then the expected value of X is given by

$$EX = \sum_{\text{all } x} xp(x). \tag{2.1}$$

Technically, the summation in (2.1) is taken over all x for which $p(x) > 0$, though the summation is also correct as written since the summand is zero for any x with $p(x) = 0$. The expected value of a random variable X is often called its mean and denoted by μ_X or simply by μ. Let us compute the mean or expected value of the variable X in Example 2.1.1. Recall that X is the number of heads obtained in two tosses of a fair coin. We thus have

$$EX = 0(1/4) + 1(1/2) + 2(1/4) = 1.$$

Intuitively, this answer seems appropriate since if we did do this experiment many times, we'd expect that get, on average, about 1 head per trial (that is, one head from the two coin tosses). This initial computation already suggests that we might be able to obtain EX "by inspection," that is, without any arithmetic, in certain kinds of problems. It is worth noting that the value taken on by EX need not be a value that the variable X can actually take on. For example, if a coin with $P(\text{heads}) = .4$ was tossed twice, it is easy to confirm that the expected value of X, the number of heads obtained, is $EX = .8$.

There are two common interpretations of EX that can occasionally be of help in computing EX and can also lend insight in certain types of applications (notably, situations involving games of chance). The first of these is the "center of gravity" interpretation. EX is supposed to be a sort of average value and, as such, should fall somewhere near the center of the values that the random variable X can take on. But as the formula in Equation (2.1) shows, it's not a pure average of the possible values of X. Rather, it is a weighted average of those values, with the weights based on how likely the various possible values of X are. Now, picture the probability histogram in Figure 2.1.1 as an object made out of wood, and ask where the balancing point would be if you were to try to lift the entire object with one finger. It turns out that if $EX < \infty$, then it will always be that balancing point. For the histogram in Figure 2.1.1, you could lift it with one finger if that finger was placed at the value 1 on the x-axis. If $EX < \infty$, the center of gravity of the probability histogram of a random variable X will be at the value EX. This interpretation (and property) of EX tells you something useful about computing EX. If the probability histogram is perfectly symmetric about a central point and if $EX < \infty$, then that central point is EX. The idea of the mean value of a random variable being infinite may seem difficult to imagine, but we'll see soon that this can happen. If it does, the center of gravity interpretation of EX goes out the window. Fortunately, this circumstance occurs fairly rarely.

If you look at the probability histogram in Figure 2.1.2, you can see that the mean EX of X is equal to 7 without actually computing it. You could verify this by applying the formula in (2.1) to the pmf given in Table 2.1.2. However, the symmetry of this pmf, together with the fact that $2 \leq X \leq 12$ is enough to assure you that EX is finite and that it is equal to the central value 7. You'll note, though, that a symmetry argument does not apply to the computation of EX for the random variable X in Example 2.1.3. There, you'll simply have to grind out the fact that $EX = 1.25$. (In Section 2.3, we'll derive a general formula for obtaining EX in probability models of this type.)

I mentioned above that there is a second interpretation of EX which has special utility. Consider a random experiment which yields a random variable X of interest, and suppose that this experiment is performed repeatedly (say, n times). Let $X_1, X_2, \dots, X_n$ be the observed values of the random variable X in the n repetitions of the experiment. Denote the average of these n values of X as

$$\bar{X}_n = \frac{1}{n}\sum_{i=1}^{n} X_i. \tag{2.2}$$

As n grows to ∞, the average or "sample mean" $\bar{X}_n$ will converge to the value EX. The sense in which $\bar{X}_n$ converges to EX can be made explicit and mathematically rigorous (with terms like "convergence in probability" or "convergence almost surely") and indeed this is done in Chapter 5, but for now, we will treat this convergence intuitively, simply writing $\bar{X}_n \to EX$ as $n \to \infty$, and interpreting this convergence as signifying, simply, that $\bar{X}_n$ gets closer and closer to EX as n grows. Interestingly, this holds true even when EX is infinite, in which case, the values of $\bar{X}_n$will grow without bound as $n \to \infty$.

This second interpretation of the expected value of X has some interesting implications when one considers "games of chance." Many games we play have a random element. For example, most board games rely on the spinning of a wheel or the roll of a die. The outcomes in the game of poker depend on random draws from a deck of cards. Indeed, virtually all the games one finds in Las Vegas (or all other) casinos depend on some random mechanism. Let us focus, for a moment, on games of chance in which a wager is offered. Let us refer to the two actors involved in the wager as the player and the house. A game involving a wager will typically involve some investment on the player's part (in dollars, let us say) and depending on the outcome of a chance mechanism, the player will either retain his/her investment and receive a payment from the house or will lose the investment to the house. In the end, a given play of this game will result in a random variable W that we will refer to as the player's winnings. The variable W may be *positive or negative*, with negative winnings representing losses to the player. Given this framework, we will evaluate the game by means of computing and interpreting EW. The expectation EW represents the long-run average winnings for the player after many repetitions of the game. (Technically, it is the limiting average winnings as the number of repetitions n tends to ∞.)

Definition 2.2.2. Consider a game between two adversaries—the player and the house—and suppose that the random variable W represents the player's winnings on a given play of the game. Then, the game is said to be a *fair game* if $EW = 0$. If $EW > 0$, the game is said to be *favorable to the player* and if $EW < 0$, the game is said to be *unfavorable to the player*.

Example 2.2.1. Years ago, I was sitting at home one evening watching the 11 o'clock news. My shoes had been kicked off a couple of hours earlier, and I was enjoying a glass of wine as the evening softly wound its way toward its inevitable end. Suddenly, the phone rang. It was a young man—a former student at Davis—who had a probability question for me. I asked him if he was a former student of mine in introductory statistics. He was impressively honest, admitting that actually, he had taken the course from another instructor and had called her first, but she wasn't answering her phone. Here was his dilemma. He was in a casino in South Lake Tahoe. He had been playing at the craps table for a while, and he was winning! He felt that he had discovered a bet that was favorable to the player, but before he invested too heavily in this wager, he thought he should verify his findings by talking to someone like me. As it happens, one of the wagers available at the craps table is *The Field Bet.* Now there are different versions of this bet out there, but most casinos use the version he described to me that night. The "field" is a particular collection of outcomes for the sum of the digits facing up when two dice are rolled. Specifically, the field F is equal to the set of outcomes $\{5,6,7,8\}$. In the field bet, you would put down a bet (say \$1.00) and would lose your bet if the outcome X was in the field. Otherwise, you would get to keep your bet and would also receive a payoff. Assuming a \$1.00 bet, your possible winnings, and the events associated with each, are shown below. As you can see, the house would sweeten the pot if you rolled a 2 or a 12.

Win W	-1	1	2	3
if	$X \in F$	$X \in \{3,4,9,10,11\}$	$X = 2$	$X = 12$

Table 2.2.1. The possible values of W in the Field Bet when \$1.00 is wagered.

The resulting probability distribution of W, assuming the dice used are balanced, is

$W = w$	-1	1	2	3
$p(w)$	20/36	14/36	1/36	1/36

Table 2.2.2. The probability distribution of W in the Field Bet.

After obtaining this distribution for my young friend, I calculated EW for him as

$$EW = \frac{20}{36}(-1) + \frac{14}{36}(1) + \frac{1}{36}(2) + \frac{1}{36}(3) = -\frac{1}{36}.$$

This finding was shocking and disappointing to the young man, especially because when he did this computation in his head earlier in the evening, he had gotten a different answer. But he had made an arithmetic mistake. In Table 2.2.2, he had mistakenly calculated $p(-1)$ as 19/36 and $p(1)$ as 15/ 36. As a result, what he thought of as a favorable bet was actually unfavorable to him. But there was an obvious silver lining. First, as luck would have it, he was ahead at the time. Secondly, he now knew that if he continued to make this bet repeatedly, he would lose, on average, about 3 cents per play and would eventually lose all his money. I felt good about saving him from this grim fate. He thanked me for my time, and no doubt resumed his determined search for a bet that would actually favor him. ■

It is perhaps worth mentioning that there are bets at all standard casinos that actually favor the player over the house. One example is the game of blackjack. As it happens, how well you do depends on your playing strategy. It is possible for a player using "the optimal strategy" to win, on the average, that is, to turn blackjack into a favorable game for the player. Good strategies in blackjack generally involve "card counting," that is, some form of accounting for the cards that have been seen as dealing proceeds from a shuffled deck of cards (or from a "boot" from which cards from several shuffled decks are dealt). It also pays to vary the size of your bets according to whether or not the current status of the deck favors you. You might wonder why casinos would offer a wager that is stacked against them. The answer is simple. Rather few players use anything close to the optimal strategy. Most players lose at blackjack, and casinos make tons of money from this game. So they don't mind losing a bit to the occasional "card counter," though most casinos reserve (and exercise) the right to ask you to leave the premises if they think you are "card counting" or otherwise take a dislike to you.

Since it's not my purpose here to turn my readers into gamblers, I'll fight off the temptation to tell you more. Instead, I'll give another example—a famous and rather intriguing example of a game in which a player's expected winnings are infinite, yet people who want to play the game are hard to find.

Example 2.2.2. (The St. Petersburg Paradox). Consider a game in which a fair coin is tossed repeatedly until it comes up heads. Let X be the number of trials that it takes to get the first heads. It is easy to confirm that the pmf of X is given by

$$P(X = x) = (1/2)^x \text{ for } x = 1, 2, 3, \ldots \tag{2.3}$$

Suppose that the house agrees to pay you $\$2^x$ if $X = x$ is the outcome in your play at the game. The probability distribution of your payoff Y in this game is:

$Y = y$	2	4	8	16	32	64	128	256	...
$p(y)$	1/2	1/4	1/8	1/16	1/32	1/64	1/128	1/256	...

Table 2.2.3. The distribution of the payoff Y in the St. Petersburg game.

The expected payoff from this game is thus

$$EY = \frac{1}{2}(2) + \frac{1}{4}(4) + \frac{1}{8}(8) + \frac{1}{16}(16) + \cdots = 1 + 1 + 1 + 1 + \cdots = \infty.$$

Given that the expected payoff is infinity, you should be willing to wager virtually all your money to play this game. Suppose that the house is willing to let you play for the piddling sum of \$1000.00, and that you happen to have access to that amount of cash. The house gets to keep your \$1000, whatever your outcome Y is. Would you take the bet? Remember that $\infty - 1000$ is still ∞, so your expected winnings would be infinite. Here's the problem, and the paradox. If you pay \$1000 to play this game, there is a very high probability that you'll lose money. Indeed the probability that the payoff will be larger than what you paid to play is $1/2^{10}$, or put another way, the chances that you lose money on your wager is $1023/1024 > .999$. Hey, what's up with that? ■

The idea of mathematical expectation is quite a bit more general than the ideas we have discussed thus far. The expected value of a random variable X tells us where the distribution of X is centered, but there is a good deal more about that distribution that we might be curious about. One thing we might ask about is the "spread" of the distribution—is it distributed quite tightly about EX or is the distribution more spread out. The term *dispersion* is often used in discussions about how spread out a probability distribution is. Also, one often says that one distribution is less diffuse than another if the first is more concentrated around its center than the second. In studying the dispersion of a random variable or of its distribution, we will consider the expectation of variables other than X itself; for example, we will be interested in the size of the expected value of X^2. At first glance, it would appear that calculating EX^2 requires that we obtain the probability distribution of the variable $Y = X^2$. If we have $p(y)$ in hand, then we can compute

$$EX^2 = EY = \sum_{\text{all y}} yp(y).$$

But this expectation can actually be computed more simply. It can be proven that one can obtain this expectation without actually identifying the distribution of X^2, that is, we may compute EX^2 as

$$EX^2 = \sum_{\text{all x}} x^2 p(x).$$

The theorem that justifies this simpler computation (for discrete random variables) is stated below without proof.

Theorem 2.2.1. Let X be a discrete random variable, and let $p(x)$ be its pmf. Let $y = g(x)$. If $Eg(X) < \infty$, then this expectation may be calculated as

$$Eg(X) = \sum_{\text{all x}} g(x)p(x).$$

We now define a number of other expected values of interest. The first involves the expectation of a useful class of random variables associated with a given random variable X — the class of linear functions of X. The following simple result will often prove useful.

Theorem 2.2.2. Let X be a discrete random variable with finite mean EX. Let a and b be real numbers, and let $Y = aX + b$. Then $EY = aEX + b$.

Proof. Using Theorem 2.2.1 to evaluate EY, we have

$$\begin{aligned} EY &= \sum_{\text{all x}} (ax+b)p(x) \\ &= \sum_{\text{all x}} axp(x) + \sum_{\text{all x}} bp(x) \\ &= a\sum_{\text{all x}} xp(x) + b\sum_{\text{all x}} p(x) \\ &= aEX + b. \end{aligned}$$

■

Definition 2.2.3. Let X be a discrete random variable with mean μ. The variance of X, denoted interchangeably by $V(X)$, σ_X^2 or when there can be no confusion, σ^2, is defined as the expected value

$$\sigma_X^2 = E(X-\mu)^2.$$

The variance of X (or of the distribution of X) is a measure of how concentrated X is about its mean. Since the variance measures the expected value of the squared distance of a variable X from its mean, large distances from the mean are magnified and will tend to result in a large value for the variance. Let us illustrate the computation of the variance using the simple random variable X treated in Example 2.2.1, that is, for $X =$ the number of heads in two tosses of a fair coin. For this particular variable X,

$$\sigma_X^2 = \sigma^2 = E(X-\mu)^2 = (0-1)^2\frac{1}{4} + (1-1)^2\frac{1}{2} + (2-1)^2\frac{1}{4} = \frac{1}{2}. \tag{2.4}$$

Besides the variance, which informs us about the spread of a probability distribution, there are a number of other expected values that provide meaningful information about the character or shape of a distribution. One class of such measures are called *moments* and are defined as follows:

Definition 2.2.4. For positive integers $k = 1,2,3\ldots$ the kth moment of a random variable X is the expected value

$$EX^k = \sum_{\text{all x}} x^k p(x).$$

As we discuss in more detail later, the moments of a random variable have a close connection to the shape of its distribution. The variance of a variable X can be obtained from its first two moments, as shown in the following result.

Theorem 2.2.3. For any discrete variable X for which $\sigma_X^2 < \infty$, the variance of X may be computed as

$$\sigma_X^2 = EX^2 - (EX)^2.$$

Proof. Write $EX = \mu$. Then

$$\begin{aligned}\sigma_X^2 &= \sum_{\text{all x}} (x-\mu)^2 p(x) \\ &= \sum_{\text{all x}} (x^2 - 2\mu x + \mu^2) p(x) \\ &= \sum_{\text{all x}} x^2 p(x) - 2\mu \sum_{\text{all x}} x p(x) + \mu^2 \sum_{\text{all x}} p(x) \\ &= EX^2 - 2\mu(\mu) + \mu^2 \\ &= EX^2 - \mu^2.\end{aligned}$$

■

The alternative formula given in Theorem 2.2.2 is useful when computing $V(X)$ using a hand-held calculator, as it requires fewer arithmetic operations than the original formula. One further note regarding the spread of a distribution is that the *standard deviation* σ_X, which is the square root of the variance, is the most frequently used measure of spread or dispersion. Its main virtue is that σ_X is a measure that is in the same units of measurement as the X values themselves. For example, if the Xs are measured in feet, then the standard deviation is measured in feet. Because variables are squared in the computation of the variance, this measure is not in the same units of measurement as the data themselves. For example, when we wish, later on, to discuss intervals of real numbers that are thought to contain the value of an unknown mean of a random variable with a certain specified probability, we will find it natural and convenient to use intervals of the form $X \pm k\sigma_X$; for example, if X is measured in feet, then such an interval says that the mean of X is thought to be within $k\sigma_X$ feet of the value X feet.

There are other expected values that are used to investigate specific aspects of a probability distribution. For example, one is sometimes interested in measuring the amount of asymmetry in a distribution. The term "skewness" is used to describe asymmetry. A discrete distribution is said to be *skewed to the right* if the right-hand tail of its probability histogram is longer and heavier than its left-hand tail, with the opposite circumstance described as skewness to the left. The third moment of a distribution is closely related to its skewness. The population skewness is defined $E(X-\mu)^3/\sigma^3$. The intuition behind this measure is that when the distribution of X is skewed to the right, say, the large the set of values to the right of the mean receive considerable weight because they are cubed, increasing their size, and receive more weight because the set of large values to the right of the mean tend to have higher probability of occurrence than the values at the same distance to the left of the mean. In a similar vein, the fourth moment of a random variable X provides some indication of the amount of "peakedness" of a distribution (the sharpness of the peak near its mean). A commonly used measure of peakedness is the so-called population "kurtosis" defined as $\kappa = E(X-\mu)^4/\sigma^4$.

Exercises 2.2.

1. Two balanced dice are rolled. Let X be the value facing up on the green die and Y be the value facing up on the red die. Let $Z = \max\{X, Y\}$, that is, the largest of these two values. Compute EZ and Var(Z).

2. The Dixon Stamp Collecting Club is throwing a fund raiser, and is offering eight door prizes—2 rare stamps worth \$20 each and 6 common stamps worth \$1 each. The number on your ticket has qualified you to pick two of the prizes at random. (You lucky dog!)

   a) Calculate the probability that the prizes you win are worth \$2, \$21, or \$40.

   b) Let X be the total value of your two prizes. Calculate $E(X)$.

3. NBA player Hunk Halloran is a great rebounder but a pretty average free-throw shooter. Suppose that X represents the number of free throws that Hunk will make out of 5 attempts. The probability distribution of X is given by:

X	0	1	2	3	4	5
$P(X)$	1/21	2/21	3/21	4/21	5/21	6/21

   Calculate the mean and variance of the random variable X.

4. You are dealt two cards without replacement from a standard 52-card deck. The dealer will pay you \$20 if the cards match, \$5 if the second card is larger than the first (with ace taken as "high"), and \$1 if the second card is smaller than the first. What should you be asked to pay to play this game if the game is intended to be a "fair game"?

5. Suppose a thumbtack is to be tossed in the air, and it has been determined that it lands with the point facing down (D) with probability .391. You are offered a wager in which you win \$3.00 if D occurs and you lose \$2.00 if D^c occurs. Determine, on the basis of your expected winnings, whether this wager is fair, favorable to you, or unfavorable to you.

2.3 The Hypergeometric Model

We can certainly construct discrete probability models any way we like. We only need to stipulate a finite or countably infinite set in which a particular random variable X will take on values (usually called the *support set* of the distribution of X) and a probability mass function $p(x)$ which gives every potential value x of X a specific probability of occurrence. Of course, adherence to the basic rules of probability is required. It is necessary that $\sum_{\text{all } x} p(x) = 1$. But among the countless possible distributions we could create, there are just a few that are sufficiently useful to merit a detailed discussion. We will treat the first of these in this section. The utility of this model, and the ones to be discussed in subsequent sections, derives from the fact that each closely mimics a real-life process and tends to yield an excellent approximation to the actual probabilities of outcomes of random experiments of a particular type. Indeed, the hypergeometric model considered here is the quintessential stochastic description of the process of sampling randomly from a finite population.

We've encountered the hypergeometric distribution already. In Section 2.2, we considered the probability distribution of the number X of spades drawn when five cards are drawn at random from a 52-card deck. We argued that the probability mass function of X is given by

$$P(X = x) = \frac{\binom{13}{x}\binom{39}{5-x}}{\binom{52}{5}} \quad \text{for } x = 0, 1, 2, 3, 4, 5.$$

We will now consider this same problem in a completely general context.

Let us suppose that a sample of size n is to be drawn at random, without replacement, from a population of size N. Suppose, further, that every member of the population can be classified into one of two types—type R individuals, and the rest. Let us denote the number of type R individuals in the population by r. There are thus $N - r$ individuals that are not of type R. A *hypergeometric random variable* is defined as a variable X having the following probability mass function:

$$P(X = x) = \frac{\binom{r}{x}\binom{N-r}{n-x}}{\binom{N}{n}} \quad \text{for } \max(0, n-N+r) \leq x \leq \min(r, n). \tag{2.5}$$

The logic of the formula above is simple. The denominator in (2.5) is simply the number of different samples of size n one can draw from a population of size N. Under the assumption of random sampling, each of these possible samples have the same probability. (Indeed, this latter property is the very definition of "simple random sampling" from a finite population.) The numerator in (2.5) is obtained by the basic rule of counting, as the process of drawing x items of type R and $(n - x)$ items that are not of type R are the two combinations symbols shown, and the number of ways in which one can do both is simply their product.

The odd-looking range of the possible x values in (2.5) may not be so transparent, but these two bounds make perfect sense! Since the number x of items of type R obtained in a sample of size n drawn from this population cannot exceed n, the size of the entire sample, nor can it exceed r, the total number of type R individuals that are available for sampling, it is clear that $x \leq \min(r,n)$. The lower bound on x is obtained by noting, similarly, that $(n-x)$, the number of individuals drawn that are not of type R, cannot exceed $(N-r)$ nor can it be less than zero. Usually, the range of x will actually turn out to be $0 \leq x \leq n$ simply because, in most applications, the population size N is much larger that both r and n and the sample size n is much smaller than r.

Perhaps the most frequent application of the hypergeometric model is to the quite pervasive practice of public opinion polling. The use of random sampling in the assessment of public opinion on political matters is familiar to all of us, perhaps even annoyingly so to some, as we tend to be besieged by the results and summaries of such polls on a regular basis. Polling on consumer preferences is behind many of the marketing decisions made by manufacturers and merchandisers here and abroad. Even what we see on television is largely driven by poll results. In all these cases, the information obtained from a sample is used to make inferences about the population from which the sample was drawn. Because the validity and value of the information obtained from a sample relies strongly on the extent to which the sample is representative of the population of interest, pollsters make a point of drawing the sample randomly. In a typical poll, and sample of size n is drawn at random from a population of size N, meaning that the sample is drawn randomly, without replacement, in a way that ensures that all possible samples of size n have the same chance of being chosen. If a particular yes/no question is of interest (e.g., "Do you favor candidate A?" or "Did you watch program B last night?", then r might be defined as the number of individuals in the population who would answer yes to this question. If X is the number of yes answers in the sample, then X is a hypergeometric variable with parameters N, r, and n. We will denote the hypergeometric model with these parameters by $HG(N,r,n)$.

For each probability model treated in this text, we will seek to determine certain properties of the model. Minimally, we would be interested in its mean and variance, since information about a distribution's center and its spread is a good starting point for fitting models to experimental data. We'll use the notation $X \sim D$ for the sentence "X is a random variable with the distribution D." For the random variable considered in this section, we will write $X \sim HG(N,r,n)$. As will be our custom, we will now seek to determine the expected value (or mean) of this variable. Obtaining an exact formula for EX is a mathematical exercise. Before embarking on that exercise, let us consider the matter intuitively. Ask yourself: given a sample of size n, a population size N and the fact that exactly r individuals in the population are of type R, how many type R individuals would I expect to see in the sample. Before reading the next sentence, close your eyes and think about it! Hopefully, what you thought went something like this: The fraction of the population that is of type R is r/N. This is the probability of drawing a type R individual in the ith draw (for any fixed i, without reference to what might be drawn in any other trial). If I draw n times without replacement from this population, I would expect to draw $n(r/N)$ individuals of type R. Intuitively, I would thus guess that $EX = nr/N$. This guess seems right, but is it? We will prove that it is right in the first theorem of this section. What we know at the moment is that our intuition points us toward this answer. We also know that it is correct in at least one instance. If $n = N$, that is, if I sampled the entire population, I would have, with probability one, exactly r individuals of type R in my sample. So then, $EX = N(r/N) = r$.

Before turning to a little mathematics, let's take a look at another situation in which the Hypergeometric model is helpful.

Example 2.3.1. How many fish are there in Lake Tahoe? An interesting question, but one that is seemingly hard to answer. The sampling of biological populations has been of interest for a great many years. The mark-recapture method of estimating the size of a population (usually called the Lincoln-Petersen method) works like this. Suppose N is the unknown size of the population of fish in the lake. An initial sample (taken with an eye toward randomness) results in r captured fish. They are marked and then returned to the lake. Time is allowed for the marked fish to disperse. Then, a random sample of size n is taken and the number X of marked fish in the second sample is observed. Note that $X \sim HG(N,r,n)$. Suppose $X = x$. Now assume for a moment that our guess at the expected value of X is right (which it is), that is, $EX = nr/N$. Since the randomly observed x is likely to be reasonably close to EX, we might ponder the equation $x = nr/N$, which estimates EX by x, and we might then solve the equation to obtain the estimate $\hat{N} = nr/x$. If two samples of 30 fish were drawn in this experiment, and there were 3 marked fish in the second sample, we would estimate there to be 300 fish in the lake. ■

We now proceed to a result establishing the expected value of a hypergeometric variable,

Theorem 2.3.1. If $X \sim HG(N,r,n)$, then $EX = nr/N$.

Proof. For simplicity, and because it is the case of principle interest, we will assume that the range of the variable X is the set $\{0, 1, \ldots, n\}$, that is, if $X = x$, then $0 \leq x \leq n$. The argument below is easily amended if a proof for the general case in (2.5) is desired.

$$
\begin{aligned}
EX &= \sum_{x=0}^{n} xp(x) = \sum_{x=1}^{n} xp(x) \text{ (since the summand is zero when x = 0).} \\
&= \sum_{x=1}^{n} x \frac{\binom{r}{x}\binom{N-r}{n-x}}{\binom{N}{n}} \\
&= \sum_{x=1}^{n} x \frac{\frac{r!}{x!(r-x)!}\binom{N-r}{n-x}}{\frac{N}{n}\binom{N-1}{n-1}} \\
&= \sum_{x=1}^{n} \frac{nr}{N} \frac{\binom{r-1}{x-1}\binom{N-r}{n-x}}{\binom{N-1}{n-1}} \quad \text{(by cancelling out } x \text{ and regrouping remaining terms)} \\
&= \frac{nr}{N} \sum_{y=0}^{n-1} \frac{\binom{r-1}{y}\binom{N-r}{n-1-y}}{\binom{N-1}{n-1}} \quad \text{(where the index } y \text{ represents } x-1\text{)} \\
&= \frac{nr}{N} \sum_{y=0}^{n-1} P(Y=y) \quad \text{(where } Y \sim HG(N-1, r-1, n-1)\text{)} \\
&= \frac{nr}{N} \qquad (2.6)
\end{aligned}
$$

since the last summation is simply the sum of all n of the $HG(N-1, r-1, n-1)$ probabilities, and is thus equal to 1. ■

Remark 2.3.1. The method of proof of the theorem above involves a mathematical "trick" that we will find very useful in many problems that will be discussed in the text or that will be relegated to section exercises or chapter problems. It thus seems worthwhile to call special attention to it. You will notice that as soon as we wrote down an expression for EX in the proof of Theorem 2.3.1, we were faced with a sum that we weren't sure how to evaluate. We know that $\sum p(x) = 1$, but we're not sure, without doing some work, what value $\sum xp(x)$ will take on. A direct proof of the theorem's claim would involve a serious attempt to evaluate $\sum xp(x)$. But you will note that the proof above doesn't do that. Instead, we pursued the possibility of rewriting this sum as a constant (nr/N) times a new summand whose form we immediately recognize. In the proof above, the final summation is recognizable as a sum of probabilities which add to 1, so that the constant we had factored out of the sum was the final answer in the calculation. Since we will make use of this approach often, I will tend to refer to it as "the trick." In the sequel, you will see, with some regularity, remarks like "let's use the trick" or "Hint: try the trick."

Obtaining the variance of the hypergeometric distribution is neither an easy intuitive exercise nor a particularly pleasant mathematical one, at least when one attempts a direct derivation. However, there is a "trick" that makes the calculation somewhat manageable. Because of the structure of hypergeometric probabilities, developments similar to those above make it possible to get an exact expression for the so-called "factorial moment" $A = E[X(X-1)]$. But since the latter expectation is equal to $EX^2 - EX$, we may use Theorem 2.2.2 to obtain the variance of X as $V(X) = A + EX - (EX)^2$. See Exercise 5 below for further details. The target result is recorded as:

Theorem 2.3.2. If $X \sim HG(N,r,n)$, then $V(X) = nr(N-r)(N-n)/N^2(N-1)$.

Exercises 2.3.

1. Twelve applicants (4 women and 8 men) for a beginning position with the Sacramento Police Department were judged to be fully qualified. The police chief has decided to hire four of them using a random lottery. Let X be the number of women hired. Calculate the probability $P(X = k)$ for $k = 0,1,2,3$ and 4 and calculate the expected number of women that will be hired.

2. The show *Dancing with the Stars* has been a huge hit on TV over the last several seasons. I read on a Hollywood-based blog that the show's three judges, each of whom is now a "star" in his/her own right, have decided to "move on," so there will be three new judges next year. The short list in the search for new judges consists of ten international champions in ballroom dancing. The show's producer has chosen three of the ten at random for a final interview. Totally independently, the show's director has also chosen three of them at random for a final interview. (They both thought it was their responsibility.) Let X = the number of people that these two "final threesomes" have in common. Clearly, $X \in \{0,1,2,3\}$. Obtain the probability distribution of X and calculate the expected value and variance of X.

3. The following lottery is offered in several states each week. A computer has been programmed to select six random numbers, without replacement, from among the integers $1,2,\ldots,49$, and is used to pick the six winning numbers each week. A player can buy a lottery ticket for \$1.00. If the player's ticket has fewer than three winning numbers, the player wins nothing. If the player's numbers include exactly three matches with the winning numbers, he/she wins \$5.00. The player wins quite a bit more for four or five

matches, but players who match all six winning numbers will win (or share with other players with 6 matches) the jackpot, which is always at least \$1 million and is occasionally much more. Let X be the number of matches on a particular player's lottery ticket. Calculate $P(X \leq 2), P(X = 3)$ and $P(X = 6)$.

4. Prove Theorem 2.3.1 without assuming that $0 \leq x \leq n$, that is, under the more restrictive assumption that $0 \leq \max(0, N-n+r) \leq x \leq \min(r,n) \leq n$.

5. Suppose $X \sim HG(N,r,n)$ and that $P(X=x) > 0$ for $0 \leq x \leq n$. Show that the "factorial moment" $E[X(X-1)]$ is equal to $n(n-1)r(r-1)/N(N-1)$. Then use the fact that the variance of X can be computed as $V(X) = E[X(X-1)] + EX - (EX)^2$ to obtain the expression $V(X) = nr(N-r)(N-n)/N^2(N-1)$.

2.4 A Brief Tutorial on Mathematical Induction (Optional)

Not everyone is totally enthralled with mathematical ideas, and some readers of this text may be less than excited by the thought of having to prove a mathematical claim. Well, the goals of this text do not include the plan of making its readers into mathematicians (though I must admit the thought is not abhorrent to me). One of my goals, however, is the much more modest one of taking some of the fear out of mathematics for those that have a bit of that fear in them. This section has the potential to help us toward that goal. Mathematical induction is an extremely useful tool in many fields of mathematics, but it is also a topic that is elementary enough to teach to virtually any college student (and in fact, it is often taught in a high school mathematical analysis class). Moreover, it allows us to take a close look at how mathematicians think and to see how mathematicians often prove things for themselves. The proofs we will discuss are completely rigorous (another phrase for "mathematically air tight"), and yet the simple logic that is involved can be understood and appreciated without great effort.

Let us suppose that we would like to determine if some particular property is true for all the positive integers, that is for 1, 2, 3, 4, etc. — an infinite collection of numbers. It is helpful to denote the property under consideration by $P(n)$. In other words, we will interpret "$P(n)$" as the version of the property for the positive integer n. We would therefore be interested in proving that $P(n)$ is true for any positive integer n. A concrete example of a proposition we might wish to prove is the following:

$$P(n): \text{ For any positive integer } n, \sum_{i=1}^{n} i = \frac{n(n+1)}{2}. \tag{2.7}$$

Now many people would be willing to believe this proposition without seeing a formal proof. After all,

$$1 = \frac{1(2)}{2}, \text{so it's true for } n = 1, \text{ and}$$
$$1+2 = \frac{2(3)}{2} = 3, \text{ so it's true for } n = 2, \text{ and}$$
$$1+2+3 = \frac{3(4)}{2} = 6, \text{ so it's true for } n = 3.$$

In fact, it's also true for $n = 4,5,6$, and 7. This may increase our confidence in the truth of $P(n)$ for all n, but how do we really know that there isn't an exception somewhere down the line? How do we know that $P(1,000)$ and $P(1,000,000)$ are true? This is a good example of where the logical power of mathematical induction really shines. In order to prove that $P(n)$ is true for all positive integers n, we need only verify two things:

(a) $P(1)$ is true and

(b) If $P(n)$ is true for a particular integer n, then $P(n+1)$ must be true.

Suppose these two steps have been accomplished. It then follows that $P(n)$ holds for all positive integers n. The logic of the argument stems from first establishing a foothold—that $P(1)$ is true—and then establishing that the truth of the property for one integer necessarily implies the truth of the property for the next integer. With step (a), we have that $P(1)$ is true. Once we've established step (b), it follows that the property is true for the successor of 1, namely 2. Of course we could have checked that directly, but the true richness of induction is that it can be applied for all future successors, one at a time. So because $P(2)$ is true, $P(3)$ is also true, which then implies the truth of $P(4)$ and so on, *ad infinitum*. Now the real trick in getting induction to work for you is establishing step (b). The assumption that $P(n)$ is true is called the "induction hypothesis." Given this assumption, one must show that it logically implies that $P(n+1)$ is true, There's nothing like a real example to demonstrate the process.

Theorem 2.4.1. For any positive integer n, $\sum_{i=1}^{n} i = n(n+1)/2$.

Proof. Let $P(n)$ be the proposition that $\sum_{i=1}^{n} i = n(n+1)/2$ holds for the positive integer n. First consider $P(1)$. Because both sides of the equation give the answer 1, we may conclude that $P(1)$ is true. Now suppose that $P(n)$ is true. Consider $P(n+1)$. Since the sum of the first $(n+1)$ positive integers may be written as the sum of the first n integers plus the number $(n+1)$, we have

$$\begin{aligned}
\sum_{i=1}^{n+1} i &= \sum_{i=1}^{n} i + (n+1) \\
&= \frac{n(n+1)}{2} + (n+1) \text{ (by the induction hypothesis)} \\
&= \left(\frac{n}{2}+1\right)(n+1) \\
&= \frac{(n+1)(n+2)}{2}.
\end{aligned}$$

This sequence of equalities shows that the $P(n+1)$ is true, from which it follows that the formula stated in the theorem must hold for any positive integer. ■

Now we know for sure that the property stated in 2.4.1 actually does indeed hold for $n = 1,000$ and $n = 1,000,000$!

Even though the problem above *looks like* a toy problem, it's actually a very useful result. We will, for example, make use of it later when discussing the large-sample properties of an important and widely used statistical procedure called the Wilcoxon Rank Sum Test. Here's another interesting example of a "proof by induction," this time involving a fundamental algebraic result that is at the core of the discrete probability model we will turn to next. The result is called the "binomial theorem." It derives its name from the fact that a sum of two elements, for example $x+y$, is called a "binomial" in algebra, the "bi" standing for "two."

Theorem 2.4.2. Let x and y be arbitrary real numbers, and let n be a positive integer. Then

$$(x+y)^n = \sum_{k=0}^{n} \binom{n}{k} x^k y^{n-k}. \tag{2.8}$$

Proof. Let us refer to the proposition in (2.8) above as $P(n)$. It is clear that $P(1)$ is true, as when $n = 1$, the right-hand side of $P(1)$ is the sum of precisely two terms, x and y, and thus matches the left-hand side of $P(1)$. Now, assume that $P(n)$ is true for some positive integer n. We must demonstrate that $P(n+1)$ is also true. To that end, let's rewrite $P(n+1)$ in a convenient way. Clearly, we may write $P(n+1)$ as

$$\begin{aligned}
(x+y)^{n+1} &= (x+y)(x+y)^n \\
&= (x+y)\sum_{k=0}^{n} \binom{n}{k} x^k y^{n-k} \text{ (by the induction hypothesis)} \\
&= \sum_{k=0}^{n} \binom{n}{k} x^{k+1} y^{n-k} + \sum_{k=0}^{n} \binom{n}{k} x^k y^{n+1-k} \\
&= \sum_{k=1}^{n+1} \binom{n}{k-1} x^k y^{n+1-k} + \sum_{k=0}^{n} \binom{n}{k} x^k y^{n+1-k} \\
&= 1 \cdot y^{n+1} + \sum_{k=1}^{n} \left(\binom{n}{k-1} + \binom{n}{k} \right) x^k y^{n+1-k} + 1 \cdot x^{n+1} \\
&= \sum_{k=0}^{n+1} \binom{n+1}{k} x^k y^{n+1-k},
\end{aligned}$$

the last equality being justified by the identity from Exercise 1.7.5 and the fact that $\binom{n+1}{0} = 1 = \binom{n+1}{n+1}$. ■

There is a version of mathematical induction that is equivalent to the version above but is occasionally more useful in a particular application. The alternative is generally called "complete induction." The goal remains the same: to prove that some particular property $P(n)$ is true for all positive integers n. Complete induction also relies on two steps.

(a′) $P(1)$ is true and

(b′) If $P(k)$ is true for all positive integers $k \leq n$, then $P(n+1)$ must be true.

The fact that complete induction accomplishes precisely that same goal as standard mathematical induction is quite transparent. Suppose steps (a′) and (b′) have been established. Then we have that $P(1)$ is true. Now, since $P(k)$ is true for all positive integers $k \leq 1$, $P(2)$ must be true. But this step may be repeated indefinitely, from which we conclude that $P(3)$ is true, $P(4)$ is true, and so on, *ad infinitum*.

The following result is attributed to the late great mathematician and humorist Edward Beckenbach.

Theorem 2.4.3. Every positive integer has an interesting property.

Proof. Let $P(n)$ represent the statement that the positive integer n has an interesting property. First, consider $P(1)$. Well, this statement hardly needs discussion. Not only is the number 1 the smallest positive integer, it is the "identity" element in multiplication and the only positive integer that is its own reciprocal. Of course $P(1)$ is true! Let us now suppose that $P(k)$ is true for all positive integers $k \leq n$. Consider $P(n+1)$. If $(n+1)$ did not have an interesting property, then it would be the smallest positive integer without an interesting property. Now, that's interesting!! ■

Exercises 2.4.

1. Prove by induction: For any positive integer n, $\sum_{i=1}^{n}(2i-1) = n^2$, that is, prove that the sum of the first n odd integers is equal to n^2.

2. Prove by induction: $\sum_{i=1}^{n} i^2 = n(n+1)(2n+1)/6$.

3. Prove the identity: $\sum_{k=0}^{n}\binom{n}{k} = 2^n$. (Hint: Apply Theorem 2.4.2.)

4. Prove the identity: $\sum_{k=0}^{n}(-1)^k\binom{n}{k} = 0$. (Hint: Apply Theorem 2.4.2.)

2.5 The Binomial Model

The binomial model is probably the most widely used discrete probability model. As with the hypergeometric model, it applies to populations which can be completely classified into two mutually exclusive groups. Experiments with just two types of possible outcomes are called *dichotomous*. Typical applications involve possible experimental outcomes like heads or tails, yes or no, male or female, success or failure, and so forth. A binomial experiment has three defining characteristics:

(a) There is a basic experiment with exactly two possible outcomes denoted, for convenience, as S or F for "success" or "failure."

(b) There are n independent repetitions of the basic experiment.

(c) The probability of success is constant, that is, $P(S)$ is the same for every trial.

The quintessential example of a binomial experiment is the process of tossing a coin n times in succession. The dichotomy of possible outcomes is present, we are usually willing to believe that the outcome of any toss is independent of the outcome of any other toss, and finally, the probability of heads is reasonably assumed to be the same in each coin toss. When something that looks like a binomial experiment isn't one, it is generally the independence assumption that has been violated. For example, if we draw 10 people, without replacement, from a population of 55 females and 45 males, we find that, even though the basic experiment (drawing and M or an F) is dichotomous, and the (unconditional) probability of drawing a female on any given trial is .55, the draws are not independent. If the first draw is a female, the probability of drawing a female on the second draw has diminished, that is $P(F_2|F_1) = 54/99 = .545$. The number of females drawn at random in 10 draws from this population has the hypergeometric distribution $HG(100, 55, 10)$.

The random variable of interest in a binomial experiment is the number of successes obtained in n independent trials of the basic experiment. We will denote this binomial variable by X. We will use the standard shorthand notation for the probabilities of success and failure: $p = P(S)$ and $q = P(F) = 1 - p$. The notation $X \sim B(n, p)$ will be used for the phrase "X has a binomial distribution based on n independent trials with probability p of success." The probability distribution of X is called the *binomial distribution*. The individual trials of the basic experiment on which a binomial experiment depends are generally referred to as *Bernoulli trials* in honor of the mathematician Jacob Bernoulli who first did extensive theoretical studies of their behavior in finite samples and asymptotically (that is, as $n \to \infty$). A random variable with distribution $B(1, p)$ is called a Bernoulli variable.

We now set ourselves to the task of obtaining a closed-form formula for the probability that a binomial variable X takes on a certain value, that is, a formula for $P(X = x)$, where x is an integer in the set $S = \{0, 1, 2, \ldots, n\}$. If $0 < p < 1$, each element of the sample space S is a possible value of the random variable X, and each value x of X has a positive probability of occurrence. To start, let's assume that x is an integer in the set $\{0, 1, \ldots, n\}$ that is both larger than 0 and smaller than n. Consider the following sequence of successes and failures that has exactly x successes:

$$\underbrace{SSS\ldots SSS}_{x}\underbrace{FFFF\ldots FFFF}_{n-x}, \tag{2.9}$$

where there are precisely x successes followed by precisely $(n - x)$ failures. Since each S in the sequence (2.9) contributes a p to the probability, and each F a contributes a q, the multiplication rule for independent events yields the product $p^x q^{n-x}$. Now consider a similar sequence of x Ss and $(n - x)$ Fs, but assume that they are now well mixed rather than clustered together as shown in (2.9). A little reflection will convince you that the probability of this new sequence is the same as the other. This follows from the way that the multiplication rule for independent events works; the calculation simply inserts a p in the product each time you encounter an S and inserts a q in the product each time you encounter an F. Since there are x Ss and $(n - x)$ Fs in the sequence, the product of these individual probabilities will again be equal to $p^x q^{n-x}$. What we can surmise from this is that every sequence with x Ss and $(n - x)$ Fs has the same probability. To complete our calculation of $P(X = x)$, we simply have to count how many ways there are to get exactly x Ss and $(n - x)$ Fs.

From our previous discussions about coin tossing, you will recognize the fact that an n-stage experiment consisting of n independent repetitions of a dichotomous experiment is an experiment which can be represented by a probability tree. The tree is generally too large to draw, but we can still imagine the tree involved. Because the number of branches in the tree doubles with each new trial, by the time the n trials are complete, there are 2^n separate paths in the tree. What we have done so far is find the probability of each path that has exactly x Ss and $(n-x)$ Fs. What we will do now is count the number of such paths. In the Table below, we picture an n-stage path of Ss and Fs as a collection of n spaces to be filled by Ss and Fs:

—	—	—	—	—	...	—	—	—
1	2	3	4	5		$n-2$	$n-1$	n

Table 2.5.1. Spaces $1,\ldots,n$ in which to place x Ss and $(n-x)$ Fs.

Suppose that we are interested in sequences that have exactly x Ss and $(n-x)$ Fs. How many such sequences are there? This is the same question posed above, but we can now readily see that the problem is a common counting problem for which we have a ready answer. It's really the same question as "How many ways can I choose a set of x things (the spaces to place the Ss in) from a collection of n things (all n spaces). The answer is clearly $\binom{n}{x}$. Once the Ss are put in place, there are no further choices to make since the Fs just go in the spaces that remain. Putting the two pieces above together, we arrive at the famous and very useful binomial formula, which we record as the following:

Theorem 2.5.1. Let $X \sim B(n,p)$. Then

$$P(X=x) = \binom{n}{x} p^x q^{n-x} \text{ for } x = 0,1,2,\ldots,n. \tag{2.10}$$

Don't you just love it when you get to a theorem and you don't have to prove it because you already have! But I must confess to one small sleight of hand in the argument above. Remember that I stated at one point that x was neither 0 or n. Well, actually, this was just for convenience, since I wanted to discuss a sequence of Ss and Fs, and I needed there to be at least one of each in order to do that. So all we need to do to make our argument air tight is to cover the two remaining cases. Now $P(X=0)$ is equal to the product of n consecutive $P(F)$s, that is, will be equal to q^n, while for similar reasons, $P(X=n) = p^n$. But the binomial formula in (2.10) gives precisely these answers when x is set equal to either 0 or n. Thus, the formula holds for the entire range of x values stated in (2.10).

We should feel confident that the binomial probabilities in (2.10) sum to 1. After all, we did a careful accounting, and the formula was developed by keeping track of every possible path in the probability tree with 2^n paths that represents the possible outcomes of a binomial experiment. But to confirm these calculations, let's check that fact that the binomial distribution with p.m.f. given by $p(x) = P(X=x)$ in (2.10) corresponds to a legitimate probability distribution. How do we know that $\sum_{x=0}^{n}\binom{n}{x}p^x q^{n-x} = 1$? Well, besides the fact that we trust the argument that led to the binomial distribution, we may invoke the binomial theorem (Thm. 2.4.2) to write

$$\sum_{x=0}^{n}\binom{n}{x}p^x q^{n-x} = (p+q)^n = 1^n = 1.$$

Example 2.5.1. A fair coin is tossed ten times. Let us find the probability of getting at least eight heads in the ten tosses. If X is the number of heads obtained, then $X \sim B(10, 1/2)$.

$$P(X \geq 8) = \binom{10}{8}\left(\frac{1}{2}\right)^8\left(\frac{1}{2}\right)^2 + \binom{10}{9}\left(\frac{1}{2}\right)^9\left(\frac{1}{2}\right)^1 + \binom{10}{10}\left(\frac{1}{2}\right)^{10}$$
$$= \frac{45+10+1}{2^{10}} = \frac{56}{1024} = .0547.$$

■

In Table A.1 in the Appendix, cumulative probabilities $P(X \leq x)$ are given for a binomial variable $X \sim B(n,p)$ for $x = 0,1,2,\dots,n$ for selected values of n and p. Specifically, Table A.1 covers most integers n between 2 and 25 and covers values of $p \in \{.05,.10,.15,\dots,.45,.50\}$). For the (n,p) pairs covered in Table A.1, one may obtain other probabilities of interest. For example, one may compute the probabilities $P(X > x) = 1 - P(X \leq x)$, $P(X = x) = P(X \leq x) - P(X \leq x-1)$ and $P(A \leq X \leq B) = P(X \leq B) - P(X \leq A-1)$. If $p > .5$, one may compute probabilities using Table A.1 by using the relationship $P(X \leq k|n,p) = P(X \geq n-k|n,1-p)$, a formula that simply recognizes that the probability of "k successes" in a binomial experiment with probability of success p will be equal to the probability of "$(n-k)$ failures" in a binomial experiment with probability of failure $(1-p)$. It should also be mentioned that individual and cumulative binomial probabilities, for arbitrary n and p, may be obtained directly from many of today's electronic devices.

We will now derive formulae for the mean and variance of a binomial variable. The mean of the $B(n,p)$ distribution is exactly what you would guess it would be. In n independent repetitions of an experiment in which the probability of success is p, you would "expect" to see, on average, np successes. This agrees with the natural intuition that if you toss a fair coin 10 times, you'd expect to see 5 heads, though of course you might well see more or less than 5 due to random variation. We now prove that our intuition is on the money. Besides establishing the formula of interest, the proof provides us with a second opportunity to use "the trick."

Theorem 2.5.2. If $X \sim B(n,p)$, then $EX = np$.

Proof.

$$EX = \sum_{x=0}^{n} x\binom{n}{x}p^x q^{n-x}$$
$$= \sum_{x=1}^{n} x\frac{n!}{x!(n-x)!}p^x q^{n-x} \text{ since the summand above is zero when } x = 0$$
$$= \sum_{x=1}^{n} \frac{n!}{(x-1)!(n-x)!}p^x q^{n-x} \text{ which, by using "the trick," we rewrite as}$$
$$= np\sum_{x=1}^{n} \frac{(n-1)!}{(x-1)!(n-x)!}p^{x-1} q^{n-x}, \text{ which, by letting } y = x-1, \text{ may be written as}$$

$$= np \sum_{y=0}^{n-1} \frac{(n-1)!}{y!(n-1-y)!} p^y q^{n-1-y}$$

$$= np \sum_{y=0}^{n-1} \binom{n-1}{y} p^y q^{n-1-y} \tag{2.11}$$

$$= np,$$

since the sum in (2.11) is the sum of all the probabilities of the $B(n-1,p)$ distribution and is thus equal to one. ■

Theorem 2.5.3. If $X \sim B(n,p)$, then $V(X) = npq$.

Proof. We'll first find $EX(X-1)$, then get EX^2 as $EX(X-1)+EX$ and then get $V(X)$ as $EX^2 - (EX)^2$.

$$EX(X-1) = \sum_{x=0}^{n} x(x-1)\binom{n}{x} p^x q^{n-x}$$

$$= \sum_{x=2}^{n} x(x-1)\frac{n!}{x!(n-x)!} p^x q^{n-x} \text{ since the summand above is zero for } x < 2$$

$$= \sum_{x=2}^{n} \frac{n!}{(x-2)!(n-x)!} p^x q^{n-x} \text{ which, by using "the trick," we rewrite as}$$

$$= n(n-1)p^2 \sum_{x=2}^{n} \frac{(n-2)!}{(x-2)!(n-x)!} p^{x-2} q^{n-x}, \text{ which, by letting } y = x-2,$$

$$= n(n-1)p^2 \sum_{y=0}^{n-2} \frac{(n-2)!}{y!(n-2-y)!} p^y q^{n-2-y}$$

$$= n(n-1)p^2 \sum_{y=0}^{n-2} \binom{n-2}{y} p^y q^{n-2-y} \tag{2.12}$$

$$= n(n-1)p^2,$$

since the sum in (2.12) adds all the probabilities of the $B(n-2,p)$ distribution. Thus, we have:

$$V(X) = E(X(X-1)) + EX - (EX)^2$$
$$= n(n-1)p^2 + np - n^2p^2 = np - np^2 = npq.$$

■

Upon reading about the hypergeometric and the binomial models in this chapter, it may well have occurred to you that the two models have some strong similarities. They both deal with dichotomous experiments. They both deal with n repetitions of a basic experiment. The obvious difference between them is the matter of the independence of successive trials. The easiest way to characterize the difference is that hypergeometric distribution is a model based on *repeated sampling, without replacement, from a finite population*. The binomial distribution, on the other hand, can be thought of as *repeated sampling from an infinite population*. The latter context would be appropriate in, say, repeated independent trials of a laboratory experiment where there is no change in the population from trial to trial. Another scenario where the binomial model is applicable is *repeated sampling, with replacement, from a finite population*. Here too, the population remains the same in every trial, thus yielding a constant probability of success.

One could imagine that, when a population is large, but not infinite, using either model for the probability distribution of the number of "successes" drawn might give quite similar answers. The hypergeometric model will give the exact answer, while the binomial model, with p set equal to r/N, should give quite a good approximation. This intuition is completely correct. Such a result is the subject of the "limit theorem" stated below. It is the first of a number of limit theorems we will encounter. All of them should be thought of as "approximation theorems," as what they really say is that as certain parameters get close to their limits, one model becomes a very good approximation for the other. Theorems such as these can be extremely useful when probability computation with the limiting model is straightforward while it is difficult (or completely intractable) with the other model.

Theorem 2.5.4. Let r and N be non-negative integers, with $0 \leq r \leq N$, and let $p \in (0,1)$. If $r \to \infty$ and $N \to \infty$ in such a way that $r/N \to p$, then for any fixed integers x and n, where $0 \leq x \leq n$,

$$\frac{\binom{r}{x}\binom{N-r}{n-x}}{\binom{N}{n}} \to \binom{n}{x} p^x q^{n-x}, \tag{2.13}$$

where $q = 1 - p$.

Proof. By canceling common elements of the numerators and denominators of the three combinations symbols on the left-hand side of (2.13), we may rewrite this fraction as

$$\frac{\frac{r(r-1)\cdots(r-x+1)}{x(x-1)\cdots 3\cdot 2\cdot 1} \cdot \frac{(N-r)(N-r-1)\cdots(N-r-n+x+1)}{(n-x)(n-x-1)\cdots 3\cdot 2\cdot 1}}{\frac{N(N-1)\cdots(N-n+1)}{n(n-1)\cdots 3\cdot 2\cdot 1}}. \tag{2.14}$$

Observing that the three remaining denominators in (2.14) are actually "whole factorials," the expression in (2.14) may be rewritten as

$$\binom{n}{x} \frac{r(r-1)\cdots(r-x+1)\cdot(N-r)(N-r-1)\cdots(N-r-n+x+1)}{N(N-1)\cdots(N-n+1)}. \tag{2.15}$$

Notice that the fraction in (2.15) contains the same number of terms (namely n) in its numerator as it does in its denominator. Upon dividing every term in this fraction by N, we obtain

$$\binom{n}{x} \frac{\frac{r}{N}\left(\frac{r}{N}-\frac{1}{N}\right)\cdots\left(\frac{r}{N}-\frac{x-1}{N}\right)\left(1-\frac{r}{N}\right)\cdots\left(1-\frac{r}{N}-\frac{n-x+1}{N}\right)}{1\left(1-\frac{1}{N}\right)\cdots\left(1-\frac{n-1}{N}\right)}. \tag{2.16}$$

As $r \to \infty, N \to \infty$, we have, by the theorem's hypothesis, that $r/N \to p$. We thus have that the ratio r/N, which occurs n times in the numerator of the fraction in (2.16), converges to the value p. Since n and x are fixed integers, every other ratio in (2.14) tends to zero as $N \to \infty$. It thus follows that, as $r \to \infty, N \to \infty$ and $r/N \to p$, the expression in (2.16) converges to the binomial probability $\binom{n}{x} p^x q^{n-x}$. ■

Remark 2.5.1. When he was five years old, my grandson Will picked up the popular phrase "What does that even mean?" somewhere, and he seemed to know precisely what the phrase's intended meaning was, as he used it with some frequency and always appropriately. The phrase comes to mind right now because the actual meaning of Theorem 2.5.4 may not be immediately clear. After all, when would we actually encounter a sequence of hypergeometric experiments in which N and r are growing, but at the same rate, so that their ratio actually tends to a constant p. In practice, the binomial model serves as a good approximation to the hypergeometric model when N is "close" to ∞ and r/N is "close" to the constant p. This just means that the binomial approximation of the hypergeometric works when N is large and we set $p = r/N$. Exercise 3 below will give you a bit of a feel for how and when the approximation works well.

Exercises 2.5.

1. The Academic Testing Association has published a study in which they assert that one of every six high school seniors will earn a college degree, and that one in every twenty will earn a graduate or professional degree. (a) If five high school seniors are selected at random, calculate the probability that exactly three of them will earn a college degree. (b) If 20 high school seniors are selected at random, what's the probability that at least two of them will earn a graduate or professional degree?

2. General Warren Peace, commander of the international force charged with preserving the peace in Bosnia, likes to toss military buttons in his quarters after hours. He claims nothing relaxes him more. Now military buttons are slightly rounded on one side and have a small loop on the other side. When flipped, these buttons have probability $p = .6$ of landing with the loop facing down. Suppose, on a given night, General Peace tosses a military button 25 times. If X is the number of times the button lands with the loop facing down, find the following probabilities: $P(X \leq 18)$, $P(X = 12)$, and $P(15 \leq X \leq 20)$.

3. A field biologist is using the "mark-recapture" method to count the number of fish in a pond. She tags an initial sample of r fish and releases them. In her second sample (this time of size n), she notes the number X of tagged fish. The variable X has will have the hypergeometric distribution $HG(N,n,r)$, where N is the number of fish in the pond. (a) If $N = 50$, $r = 10$, and $n = 10$, find $P(X \leq 2)$. (b) Compare the answer in (a) with the corresponding binomial approximation, that is, with the probability $P(Y \leq 2)$, where $Y \sim B(10,.2)$.

4. Let $X \sim B(n,p)$, and let k be a positive integer less than or equal to n. Show that

$$E\left(\frac{X(X-1)(X-2)\cdots(X-k+1)}{n(n-1)(n-2)\cdots(n-k+1)}\right) = p^k.$$

2.6 The Geometric and Negative Binomial Models

Let's examine the geometric model first. As the name suggests, the geometric model is related to the well-known geometric progression $1, A, A^2, A^3, \ldots, A^n, \ldots$. The sum of a finite geometric series, that is, the sum of the first k terms of this sequence for some given integer k, and the sum of an infinite geometric series are given in the following two results:

Lemma 2.6.1. For any number $A \neq 1, 1+A+A^2+A^3+\cdots+A^n = \frac{1-A^{n+1}}{1-A}$.

Proof. This result is easily proven by induction, but we will give a direct proof here. Note that

$$\begin{aligned}(1-A)(1+A+A^2+\cdots+A^n) &= (1+A+A^2+\cdots+A^n)\\ &\quad-(A+A^2+A^3+\cdots A^{n-1}\\ &= 1-A^{n+1}.\end{aligned} \tag{2.17}$$

Since $A \neq 1$, we may divide both sides of (2.17) by $(1-A)$ to obtain $1+A+A^2+\cdots+A^n = (1-A^{n+1})/(1-A)$. ■

The formula for summing an infinite geometric series follows easily from Lemma 2.6.1. We will only be interested in geometric series that have a stochastic interpretation, and we will thus restrict ourselves to positive values of A for which the infinite series converges. See Exercise 2.6.1 for more general results.

Theorem 2.6.1. For any $A \in (0,1)$, $\sum_{i=0}^{\infty} A^i = \frac{1}{1-A}$.

Proof. From Lemma 2.6.1, we have

$$\sum_{i=0}^{n} A^i = \frac{1-A^{n+1}}{1-A}. \tag{2.18}$$

It follows that

$$\lim_{n\to\infty} \sum_{i=0}^{n} A^i = \lim_{n\to\infty} \frac{1-A^{n+1}}{1-A} = \frac{1-0}{1-A} = \frac{1}{1-A}. \tag{2.19}$$

■

We are now in a position to introduce the geometric model on the positive integers. The model has a single parameter $p \in (0,1)$ and will be denoted by $G(p)$.

Definition 2.6.1. The random variable X has a *geometric distribution* on the set of positive integers $\{1,2,3,\dots\}$ if for some $p \in (0,1)$,

$$P(X=x) = p \cdot q^{x-1} \text{ for } x = 1,2,3,\dots, \tag{2.20}$$

where $q = 1-p$.

Clearly, $G(p)$ is a bona-fide probability model; by Theorem 2.6.1, the sum of the probabilities in (2.20) is $p(1/(1-q)) = 1$. The geometric model arises naturally in the same setting that gives rise to the binomial model. Imagine a sequence of Bernoulli trails with probability p of success. If instead of the number of successes obtained in n trials, we define the variable X to be the number of trials that are required in order to obtain the first success, then we have defined a variable that has no fixed upper bound. The variable X could, in theory, be larger than any predetermined number K, even though the probability of exceeding K might be extremely small. The probability mass function of X may be easily derived from the probability tree shown in Figure 2.6.1.

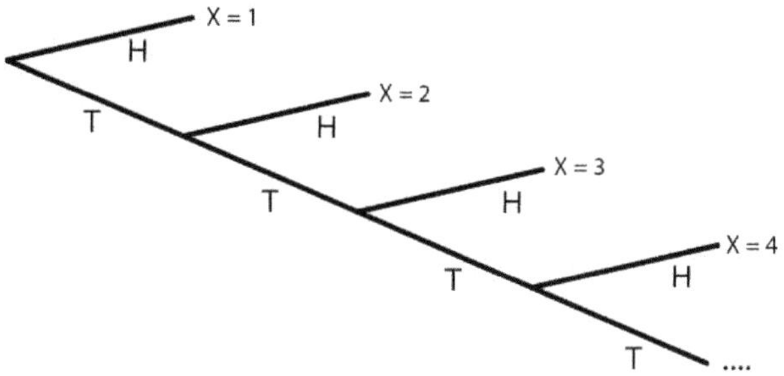

Figure 2.6.1. A probability tree corresponding to $X \sim G(p)$.

It is apparent from the tree above that the event $\{X = x\}$ will occur if and only if the outcomes of the first x Bernoulli trials consist of $(x-1)$ consecutive failures followed by a success. Under the assumptions of binomial sampling, such a sequence will occur with probability pq^{x-1}.

We will now turn to the problem of identifying the mean and variance of a geometric variable. While there are a variety of ways of obtaining the formulas for EX and $V(X)$ for $X \sim G(p)$, we will obtain these parameters of the model using elementary ideas that require no heavy mathematical artillery. The approach taken below has the added advantage of forcing us to play around with geometric series, and this practice helps cement the underlying ideas.

Theorem 2.6.2. If $X \sim G(p)$, then $EX = 1/p$.

Proof. By definition, $EX = \sum_{x=1}^{\infty} x \cdot p \cdot q^{x-1}$. Now, this is a sum that we don't recognize. It's certainly not the sum of a geometric series, though it was just before we placed that "x" in front of the probability element pq^{x-1}. Our strategy in evaluating this sum is going to entail breaking the sum down into a bunch of parts that we *can* evaluate. To this end, let us picture an element like "pq^k" as a brick. We are going to build a staircase with such elements at each step. Glancing at the sum $\sum_{x=1}^{\infty} x \cdot p \cdot q^{x-1}$, we immediately notice that the brick pq^{x-1} is counted x times. Consider the staircase displayed in the figure below.

				...
			pq^3	...
		pq^2	pq^3	...
	pq	pq^2	pq^3	...
p	pq	pq^2	pq^3	...

We may take the sum on all the values in this infinite staircase by repeatedly "peeling off" the top layer of the current staircase, adding its terms, and proceeding in a similar fashion with the top layer of each remaining staircase. From this, we see that our initial expression for EX may be rewritten as the sum of a collection of series that we'll be able to evaluate by repeated use of "the trick":

$$
\begin{aligned}
EX &= \sum_{x=0}^{\infty} pq^x + \sum_{x=1}^{\infty} pq^x + \sum_{x=2}^{\infty} pq^x + \ldots \\
&= \sum_{x=1}^{\infty} pq^{x-1} + q\sum_{x=1}^{\infty} pq^{x-1} + q^2\sum_{x=1}^{\infty} pq^{x-1} + \ldots \\
&= 1 + q + q^2 + \ldots \\
&= \frac{1}{1-q} = \frac{1}{p}.
\end{aligned}
$$

■

The variance of a geometric variable may be established with a similar argument. We won't pursue this here, as I'd prefer to show you a whole other way to get geometric moments, thereby adding to your problem-solving toolbox.

Theorem 2.6.3. Let $X \sim G(p)$. For any positive integer n,

$$E(X-1)^n = q \cdot EX^n \tag{2.21}$$

Proof.

$$
\begin{aligned}
E(X-1)^n &= \sum_{x=1}^{\infty} (x-1)^n pq^{x-1} \\
&= \sum_{y=0}^{\infty} y^n pq^y \text{ (where } y = x-1\text{)} \\
&= q\sum_{y=1}^{\infty} y^n pq^{y-1} \text{ (since the first term of the preceding sum is 0)} \\
&= q\sum_{x=1}^{\infty} x^n pq^{x-1} \text{ (by relabeling } y \text{ as } x\text{)} \\
&= q \cdot EX^n \text{ (recalling that } X \sim G(p)\text{)}.
\end{aligned}
$$

■

Corollary 2.6.1. If $X \sim G(p)$, then $V(X) = q/p^2$.

Proof. We know from Theorem 2.6.2 that $EX = 1/p$. (This result also follows, independently, from Theorem 2.6.3 with $n = 1$.) For $n = 2$, the latter theorem yields

$$E(X-1)^2 = q \cdot EX^2,$$

which may be rewritten as

$$EX^2 - 2EX + 1 = qEX^2,$$

or equivalently,

$$(1-q)EX^2 = 2EX - 1.$$

It then follows that

$$EX^2 = \frac{1+q}{p^2}.$$

From this, we obtain

$$V(X) = \frac{1+q}{p^2} - \frac{1}{p^2} = \frac{q}{p^2}.$$

■

The geometric distribution is a special case and simplest version of the negative binomial distribution. The latter model will be denoted by $NB(r, p)$, and is defined below.

Definition 2.6.2. Consider a sequence of Bernoulli trials with probability p of success. Let r be a positive integer and let X be the number of trials required to obtain the rth success. Then X has a negative binomial distribution with parameters r and p.

When $r = 1$, the negative binomial model is simply the geometric model. For larger r, one needs to do a tad bit of combinatorial work to obtain the exact probability mass function of X. This pmf is obtained in the following result.

Theorem 2.6.4. Let $X \sim NB(r, p)$. Then

$$P(X = x) = \binom{x-1}{r-1} p^r q^{x-r} \text{ for } x = r, r+1, \ldots. \tag{2.22}$$

Proof. As with the argument that established the binomial formula, we will obtain the pmf of X in two steps. Let us first note that the number of trials x needed to get the rth success can be no smaller than r (i.e., if $X = x$, then $x \geq r$). Now, suppose we have a sequence of Bernoulli trails in which there are r successes and $(x-r)$ failures. No matter the order in which the Ss and Fs appear, the probability of such a sequence, by the independence of the trials and the constant probabilities of S and F in each trial, must be equal to $p^r q^{x-r}$. If we were to imagine the negative binomial experiment as represented by a probability tree, then we could assert that every path in this tree with exactly r Ss and $(x-r)$ Fs will have the same probability of occurrence, namely, $p^r q^{x-r}$. It remains to determine how many such paths there are that correspond to the event $X = x$. Now the constraints on these paths are as follows: they must consist of x trials, r of which are Ss and $(x-r)$ of which are Fs; in addition, the outcome of the final (x^{th}) trial must be an S. We thus must determine how many ways there are to place the first $(r-1)$ Ss among the $(x-1)$ spaces that will contain the outcomes of the first $(x-1)$ trials. Clearly, there are $\binom{x-1}{r-1}$ ways this can be done. This number, times the common probability of each of the sequences with r Ss and $(x-r)$ Fs, results in the stated probability that $X = x$. ■

Example 2.6.1. Suppose that a fair coin is tossed until the fourth head occurs. Let's find the probability that no more than eight tosses are required. Letting X be the number of tosses needed to get the fourth head, we have that $X \sim NB(4, 1/2)$. Thus,

$$\begin{aligned}P(X \le 8) &= \sum_{x=4}^{8} P(X = x)\\ &= \binom{3}{3}\left(\frac{1}{2}\right)^4 + \binom{4}{3}\left(\frac{1}{2}\right)^5 + \binom{5}{3}\left(\frac{1}{2}\right)^6 + \binom{6}{3}\left(\frac{1}{2}\right)^7 + \binom{7}{3}\left(\frac{1}{2}\right)^8\\ &= \frac{163}{256} = .637.\end{aligned}$$

■

We state without proof the following properties of the $NB(r, p)$ model. We will return to prove this result in Chapter 4 when certain more powerful tools have been introduced.

Theorem 2.6.5. If $X \sim NB(r, p)$, then $EX = r/p$ and $V(X) = rp/q^2$.

An interesting connection between the geometric model and the negative binomial model is the fact that if $X \sim NB(r, p)$, then X has the same distribution as the sum $X_1 + X_2 + \cdots + X_r$, where $X_1, \ldots, X_r$ are random variables obtained from a sequence of independent and identical geometric experiments. Intuitively, this makes perfect sense; the number of trials needed to get the rth success must be the sum of the number of trials that were required to get each of the individual successes. This relationship is studied further in Section 4.6.

Exercises 2.6.

1. Show that the infinite sum in Theorem 2.6.1 (a) converges to the stated value for any $A \in (-1, 1)$, (b) does not converge if $A = -1$ (the successive finite sums vacillate between 1 and 0) or if $A < -1$ (different ways of grouping elements of the series give different answers), and (c) is equal to ∞ if $A = 1$.

2. Let $S_\infty = \sum_{i=0}^{\infty} A^i$. Confirm that the recursive relationship $1 + AS_\infty = S_\infty$ holds. Solve this equation for S_∞, providing an alternative proof of Theorem 2.6.1.

3. Evaluate the following infinite series:

   (a) $(1/3)^3 + (1/3)^5 + (1/3)^7 + \ldots$ and

   (b) $(1/2)(1/3)^2 + (1/2)^3(1/3)^3 + (1/2)^5(1/3)^4 + \ldots$.

4. After gambling all night in Lake Tahoe, you have \$2 left. You need \$3 to buy the "steak and eggs" breakfast you desperately want. You decide to risk your \$2. In a game where you either win or lose \$1 on each play, you bet repeatedly until you either reach \$3 or are completely broke. If successive trials are independent and the probability of winning on a single trial is .4, what's the probability that you'll be having steak and eggs for breakfast?

5. Suppose you make free throws in the game of basketball with probability $p = .8$. Your coach won't let any players go home from practice until they've made 25 free throws. What are the chances that you'll get to go home today having missed no more than five free throws?

2.7 The Poisson Model

The first thing we need to do is talk about pronunciation. "Poisson" is the French word for fish, and is also the name of a noted French scientist who happened to discover a particular discrete distribution with surprisingly broad applicability. The word "poisson" is pronounced (more or less) like "pwa–sõ", with the emphasis on the second syllable. Before discussing the Poisson distribution, let's take a moment to review another mathematical superstar, the number "*e*." This number plays an important role in many mathematical calculations, and we will encounter it frequently in future discussions. The precise definition of the number e is as the limit point of the sequence $(1+1/n)^n$ for $n = 1,2,\ldots$. More specifically, we define e as

$$e = \lim_{n\to\infty}\left(1+\frac{1}{n}\right)^n.$$

The number e is a "transcendental number" like π, meaning that it cannot be obtained as a root of a polynomial with rational coefficients. It is also irrational (i.e., cannot be expressed as a ratio of integers), but it is closely approximated by the decimal 2.718. Another limit of interest, for an arbitrary positive number λ, is

$$e^{\pm\lambda} = \lim_{n\to\infty}\left(1\pm\frac{\lambda}{n}\right)^n. \tag{2.23}$$

As is well known from the calculus, the function $f(x) = e^x$ is the unique real-valued function that is its own derivative, that is, for which $f'(x) = f(x)$. This fact is useful in obtaining the Maclaurin series expansion of $f(x) = e^x$. The Maclaurin series expansion of a function f is the Taylor series expansion of f about the point $x = 0$. (A peek at your old calculus book might be useful at this point.) Each of these expansions is an analytical representation of f as an infinite series, and each is valid for x values in a specific "radius of convergence" of the series. The Maclaurin series expansion of a function $f(x)$ is given by

$$f(x) = f(0)+xf'(0)+\frac{x^2}{2!}f''(0)+\frac{x^3}{3!}f'''(0)+\ldots.$$

The Maclaurin series expansion for the function $f(\lambda) = e^\lambda$ is given by

$$e^\lambda = 1+\lambda+\frac{\lambda^2}{2!}+\frac{\lambda^3}{3!}+\ldots, \tag{2.24}$$

a representation which holds for all real numbers λ. This expression implies that (upon multiplying both sides of (2.24) by $e^{-\lambda}$)

$$\sum_{x=0}^{\infty}\frac{\lambda^x e^{-\lambda}}{x!} = 1. \tag{2.25}$$

For $\lambda > 0$, we have thus identified a new probability model on the non-negative integers, as each summand in (2.25) is a positive value and the sum of all these values, as x ranges over all non-negative integers, is equal to 1. We may thus consider the value of the summand as the probability that $X = x$ for some random variable X taking values on the non-negative integers. This is a far cry from saying anything about the relevance or utility of this probability model; it only verifies that it meets the basic requirements for such models.

Definition 2.7.1. A random variable on the integers $\{0,1,2,\dots\}$ has a *Poisson distribution* with parameter $\lambda > 0$, if

$$P(X=x) = \frac{\lambda^x e^{-\lambda}}{x!} \text{ for } x = 0,1,2,\dots. \tag{2.26}$$

Our notation for this random variable and its distribution will be: $X \sim P(\lambda)$.

Example 2.7.1. Suppose the number of defectives among the light bulbs produced by General Eclectic, Inc. on any given day has a Poisson distribution with parameter $\lambda = 3$. What are that chances that no more than 2 light bulbs produced today will be defective? If X is the number of defectives, then $X \sim P(3)$, and

$$\begin{aligned} P(X \le 2) &= e^{-3} + 3e^{-3} + 3^2 e^{-3}/2! \\ &= e^{-3}(1+3+4.5) = (.049787)(8.5) = .423. \end{aligned}$$

■

In the Appendix, Table A.2 gives cumulative probabilities for selected Poisson distributions. Specifically, for selected values of λ from .1 to 16, $P(X \le x)$ is tabled for $X \sim P(\lambda)$. For example, if $X \sim P(4.4), P(X \le 6) = .8436$. One may obtain individual Poisson probabilities from the table by subtraction. Individual $P(\lambda)$ probabilities for values of λ not covered in Table A.2 may be easily obtained using a hand-held calculator.

Let us now turn our attention to identifying the mean and variance of a Poisson variable.

Theorem 2.7.1. If $X \sim P(\lambda)$, then $EX = \lambda$.

Proof. The trick proves useful again. By the definition of EX for $X \sim P(\lambda)$, we have

$$\begin{aligned} EX &= \sum_{x=0}^{\infty} x \cdot \frac{\lambda^x e^{-\lambda}}{x!} \\ &= \sum_{x=1}^{\infty} x \cdot \frac{\lambda^x e^{-\lambda}}{x!} \text{ (since the leading term of the series above is zero)} \\ &= \sum_{x=1}^{\infty} \frac{\lambda^x e^{-\lambda}}{(x-1)!} \\ &= \lambda \sum_{x=1}^{\infty} \frac{\lambda^{x-1} e^{-\lambda}}{(x-1)!} \\ &= \lambda \sum_{y=0}^{\infty} \frac{\lambda^y e^{-\lambda}}{y!} \text{ (where } y = x-1) \\ &= \lambda \text{ (since the last sum above adds all the probabilities of the } P(\lambda) \text{ model).} \end{aligned}$$

■

Theorem 2.7.2. If $X \sim P(\lambda)$, then $V(X) = \lambda$.

Proof. The proof of theorem 2.7.2 is left as Exercise 2.7.1. ■

Empirically, the Poisson distribution has been found to provide a very good approximation to the observed relative frequencies in a wide variety of counting problems. Consider the following (real) applications of the Poisson model. Poisson himself first postulated that the number of guilty verdicts reached by French juries of his day (roughly the first third of the nineteenth century) in a given period of time (say a month or year), has an approximate Poisson distribution. The model has also been used to describe or predict the number of insects on a random tobacco leaf, the number of calls entering a switch board from 9 to 10 in the morning, the number of tornadoes touching down in the state of Kansas in July, the number of plane crashes in a given year, and the number of deaths by horse kick in the Prussian army (circa 1800) in a given month. The rich variety of the phenomena that are well-modeled by a Poisson distribution at first seems quite surprising. But a second look may allow you to see some similarities among the scenarios just mentioned. It turns out it is often that case that two random variables which appear to come from quite different processes will have an essential feature in common. The Poisson distribution arises in situations in which we wish to count the number of occurrences of an event which has a rather small probability of happening in any given trial of an experiment, but we are interested in the number of times the event occurs in a large number of trials. Mathematically, the limit theorem that establishes that the Poisson distribution might serve as an approximation for the distribution of this discrete variable is called "the law of rare events." This law is stated and proven below.

Theorem 2.7.3. Let $X \sim B(n, p)$. If $n \to \infty$ and $p \to 0$ in such a way that $np \to \lambda$, where $\lambda > 0$, then for any non-negative integer x,

$$P(X = x) \to \frac{\lambda^x e^{-\lambda}}{x!}. \tag{2.27}$$

Proof. For convenience, in the argument that follows, we will use the symbol "lim" for $\lim_{n\to\infty, p\to 0, np\to\lambda}$, thus avoiding the repeated use of an awkward notation. We will thus write

$$\lim_{n\to\infty, p\to 0, np\to\lambda} \binom{n}{x} p^x q^{n-x} = \lim \binom{n}{x} p^x q^{n-x}.$$

Evaluating this limit, we may consider $np \approx \lambda$ since in the limit, the two are equal. We shall therefore make the replacement $p = \lambda/n$ in the expression above:

$$\begin{aligned} \lim \binom{n}{x} p^x q^{n-x} &= \lim \frac{n(n-1)\cdots(n-x+1)}{x!} \left(\frac{\lambda}{n}\right)^x \left(1 - \frac{\lambda}{n}\right)^{n-x} \\ &= \lim \frac{\lambda^x}{x!} \cdot \frac{n(n-1)\cdots(n-x+1)}{n^x} \left(1 - \frac{\lambda}{n}\right)^n \left(1 - \frac{\lambda}{n}\right)^{-x}. \end{aligned} \tag{2.28}$$

Examining the four terms in the product in (2.28), we see that each approaches a limit as $n \to \infty, p \to 0$ and $np \to \lambda$. We note that x is a fixed value at which the probability of interest is being calculated. The first term does not depend on n and p and thus remains equal to $(\lambda^x/x!)$. The second term tends to 1 (since it consists of a product of x items in both the numerator and denominator, and all of the individual ratios are approaching 1). The third term is a familiar sequence encountered in the first paragraph of this section, and it converges to $e^{-\lambda}$. Since the fourth term has a fixed exponent and $1-\lambda/n$ approaches 1 and $n \to \infty$, we see that the fourth term approaches 1. We may therefore conclude that, as $n \to \infty$, $p \to 0$ and $np \to \lambda$,

$$P(X = x) \to \frac{\lambda^x}{x!} \cdot 1 \cdot e^{-\lambda} \cdot 1 = \frac{\lambda^x}{x!} e^{-\lambda}.$$

■

The natural interpretation of the Law of Rare Events is that, in a binomial experiment with large n and small p (that is, in which "success" is a rare event but one has many chances to observe it), the number of times X that you will see a success is approximately distributed according to the Poisson distribution $P(np)$.

Exercises 2.7.

1. Prove Theorem 2.7.2 (Hint: Show that $EX(X-1) = \lambda^2$.)
2. Suppose that the number of defective widgets purchased from Wendell's Widget Warehouse on Saturday mornings is a Poisson variable with mean $\lambda = 6$. What is the probability that next Saturday morning, (a) no more than 3 defective widgets will be sold, (b) at least 10 defective widgets will be sold, (c) the number of defective widgets sold is an even integer. (By the way, zero is an even integer.) (Hint: Use table A.2 for all three parts of this problem.)
3. Kevin's Karpet Kompany offers their premium carpet at a bargain price. It has been discovered, however, that Kevin's premium carpets have a flaw (loose shag) in every 150 square feet, on average. Assuming that the number of flaws you will find in the premium carpet you buy from Kevin has a Poisson distribution, what's the probability that the 225 square feet of premium carpet you've just bought from Kevin will have at most one flaw?
4. Wee Willie Winkie lives in a two-story house, and just below the roofline on the tallest part of the house, there's a small round hole, apparently installed to allow heat to escape from the attic. Willie spends a good part of his weekends throwing tennis balls at this hole, trying to toss one into the attic. His dad told him that the chances of getting a ball in the hole were 1 in 500. If Willie were to toss a tennis ball at the hole 1000 times, what are the chances he'll get at least one ball in the hole?

2.8 Moment-Generating Functions

In this section, we will discuss a particular expected value that turns out to be extremely useful. Because of its unexpected form, you'd be unlikely to think of it yourself. While expectations like EX and $E(X-\mu)^2$ make some intuitive sense, why would you want to compute something like Ee^{tX} for some fixed real number t? Well, the answer is that this particular expectation turns out to be a valuable tool in answering a host of different questions about random variables and their distributions. Here, we will study this expected value when X is a discrete random variable. We'll encounter it again in other contexts, but the results treated here will allow us to use it to our advantage for discrete random variables such as those we have examined thus far.

Definition 2.8.1. Let X be a discrete random variable with pmf p(x). The *moment-generating function* (mgf) of X, denoted by m_X(t), is given by

$$m_X(t) = Ee^{tX} = \sum_{\text{all } x} e^{tx} p(x). \tag{2.29}$$

Among the important properties of moment-generating functions (mgfs) is the fact that, if they exist (i.e., if the expectation is finite), there is only one distribution with that particular mgf and there is, in this sense, a one-to-one correspondence between random variables and their mgfs. An immediate consequence is the fact that one may be able to recognize what the distribution of X is (in certain complex situations in which that distribution hasn't been explicitly stated) by recognizing the form of the mgf. But the utility of mgfs goes well beyond this fact. Its very name suggests a connection between the mgf and the moments of X. We'll show in this section precisely how to obtain EX^k directly from the mgf of X for any integer k for which the expectation exists. We will also see how these functions are useful tools when establishing certain asymptotic approximations. As an example of this, we will see how this tool can be used to provide a simple proof of the Law of Rare Events.

A quick aside: mathematicians love simple proofs! (Actually, everyone does, though perhaps for different reasons.). Professor Arnold Ross, a first-rate algebraist and an outstanding teacher (a former instructor of mine during the brief time I spent at The Ohio State University) once stated in class that mathematicians are lazy! What he meant was that they were naturally inclined to look for shortcuts, simple ways of demonstrating things. When you hear mathematicians describe something as "elegant," this is really what they are talking about. Dr. Ross noticed that one of the students in the class had a huge smile on his face after hearing his remark. Ross looked directly at him, raised his forefinger in the air, and added "Not *just* lazy." The class howled at Ross's postscript.

But I digress. Let's take a closer look at the object displayed in (2.29). The first thing to be acknowledged and understood is that the mgf is truly a function of the real number t, and as such, can be graphed and studied as a function. The figure below shows the graphs of several mgfs plotted in the plane.

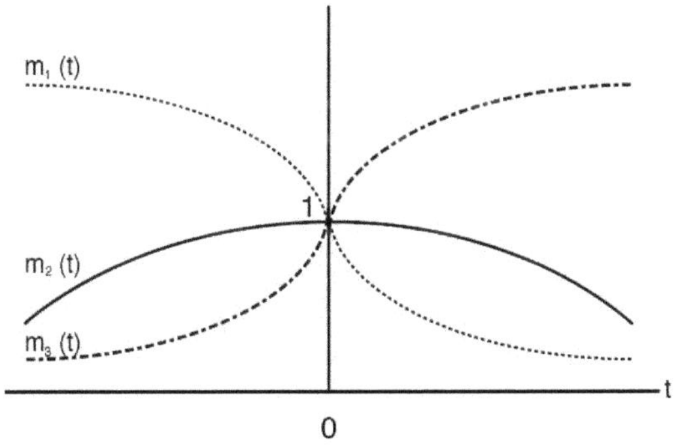

Figure 2.8.1. Several moment-generating functions.

You will note that the functions pictured above have some common characteristics. For example, each of these mgfs are strictly positive. This property is a consequence of the fact that a power of e, say e^y, is non-negative, being greater than 1 if $y > 0$ and taking a value between 0 and 1 if $y < 0$. A second common feature of all mgfs is the fact that, whatever X might be, $m_X(0) = 1$. This follows from the fact that

$$E(e^0) = E(1) = \sum_{\text{all } x} 1 \cdot p(x) = 1.$$

This fact also reveals that the moment-generating function of any discrete variable always "exists," and takes the value 1, at $t = 0$. There are random variables X for which $m_X(t)$ is defined and finite only at $t = 0$. As is common in the treatment of moment-generating functions, we will say that the mgf of a random variable *exists* if, and only if, it is defined and finite for some open interval of real numbers containing 0. The most notable property of an mgf $m_X(t)$, namely, the fact that it can be used to generate the moments $\{EX^k, k = 1, 2, 3, \dots\}$ depends on the precise characteristics of $m_X(t)$ for values of t near 0.

For many standard probability models, the moment-generating function is fairly easy to compute. We turn to several instructive examples.

Example 2.8.1. Suppose $X \sim B(1, p)$, that is, X is a Bernoulli variable with parameter $p \in (0,1)$, where $q = 1 - p$. Then

$$m_X(t) = Ee^{tX} = e^{t \cdot 0}p(0) + e^{t \cdot 1}p(1) = 1(q) + e^t(p) = q + pe^t.$$

■

Example 2.8.2. Suppose $X \sim B(n, p)$, that is, X is a binomial variable with parameters $n \geq 1$ and $p \in (0,1)$. Then

$$\begin{aligned} m_X(t) &= Ee^{tX} \\ &= \sum_{x=0}^{n} e^{tx} \binom{n}{x} p^x q^{n-x} \\ &= \sum_{x=0}^{n} \binom{n}{x} (e^t p)^x q^{n-x} \\ &= (q + pe^t)^n \text{ (by the binomial theorem).} \end{aligned}$$

■

The moment-generating function $m_X(t)$ in the following example is not defined for any arbitrary real number t, but rather, it is defined in a particular open interval of real numbers containing the value 0.

Example 2.8.3. Suppose $X \sim G(p)$, that is, X is a geometric variable with parameter $p \in (0,1)$. Then

$$m_X(t) = Ee^{tX}$$

$$= \sum_{x=1}^{\infty} e^{tx} p q^{x-1}$$

$$= pe^t \sum_{x=1}^{\infty} e^{t(x-1)} q^{x-1}$$

$$= pe^t \sum_{x=1}^{\infty} (qe^t)^{x-1}$$

$$= \frac{pe^t}{1-qe^t} \text{ (by Theorem 2.6.1, provided that } qe^t < 1).$$

If $qe^t \geq 1$, then the sum $\sum_{x=1}^{\infty} (qe^t)^{x-1}$ above is infinite. It is thus apparent that the mgf $m_X(t)$ of a geometric variable X exists only for $t < \ln(1/q)$. Since $q < 1$, we have that $\ln(1/q)$ is positive, so that $m_X(t)$ is defined and finite for t in the open interval of real numbers $(-\infty, \ln(1/q))$ which, of course, contains 0. ■

Example 2.8.4. Suppose $X \sim P(\lambda)$, that is, X is a Poisson variable with parameter $\lambda > 0$. Then

$$m_X(t) = Ee^{tX}$$

$$= \sum_{x=0}^{\infty} e^{tx} \cdot \frac{\lambda^x e^{-\lambda}}{x!}$$

$$= \sum_{x=0}^{\infty} \frac{(\lambda e^t)^x e^{-\lambda}}{x!}$$

$$= e^{-\lambda} e^{\lambda e^t} \sum_{x=0}^{\infty} \frac{(\lambda e^t)^x e^{-\lambda e^t}}{x!} \text{ (positioning ourselves to use our favorite trick)}$$

$$= e^{-\lambda + \lambda e^t} \text{ (since the terms in the sum above are all the } P(\lambda e^t) \text{ probabilities)}$$

$$= e^{\lambda(e^t - 1)}.$$

■

We state a simple but very useful property of mgf's, leaving its proof as an exercise.

Theorem 2.8.1. Let X be a discrete random variable with moment generating function $m_X(t)$, and suppose that $Y = aX + b$ where a and b are arbitrary constants. Then the mgf of Y is given by $m_Y(t) = e^{bt} m_X(at)$, for all t for which $m_X(at)$ exists.

Let's turn to some additional properties of moment-generating functions that will arise somewhat frequently in the sequel. One of the most useful is the *uniqueness property*. Now, it is true that, for some probability models, the moment-generating function does not exist. There is a more complex treatment available for such models involving what is called the characteristic function. Because the latter function is more mathematically complex (this being true quite literally, as it actually does involve complex numbers (like $a + bi$) rather than simply real numbers in their definition), it will not be discussed here. We will, for the most part, restrict attention to models of random variables X for which $m_X(t)$ does exist.

This does not involve much of a sacrifice for us since almost all of the commonly used univariate models treated in this book are in this latter class. For each such model, its mgf is uniquely associated with it, that is, no other random variable has this same mgf. The fact that the uniqueness property of mgfs is extremely useful may not be immediately evident. Its utility will become quite obvious as we encounter problems in which we can use it to our advantage.

To give you an idea of why we would ever make use of this property, one has to imagine encountering a random variable whose distribution you don't know but whose moment-generating function you are able to compute. Perhaps the variable was defined in a way that did not make its probability distribution apparent. A simple example of such a circumstance is the iterative modeling that one often encounters in situations involving two random variables. Here's an example that we will treat in more detail in Chapter 4. Suppose that $X \sim P(\lambda)$ and that given $X = x, Y \sim B(x, p)$. This is a modeling scenario that has been used to describe the number of traffic accidents X that an urban driver might have in a given three-year period and the number Y of those accidents that are his or her fault. Suppose we want to know the distribution of Y. It is not specified in the framework above, as the behavior of Y is only discussed under the condition that the value of X is known. A full discussion of this model requires that we treat the joint distribution of a pair (X,Y) of random variables and a collection of related ideas. Thus, we postpone a more detailed discussion for now. But the example should make evident the fact that it is possible for you to know quite a bit about a particular random variable Y without actually knowing its probability distribution. Suppose you find yourself in such a situation, and that, clever as you are, you can actually calculate the mgf of Y. Here's the interesting part. Suppose you find that the mgf of Y is

$$m_Y(t) = e^{\mu(e^t - 1)},$$

where the parameter μ is related in some way to λ and p. Then we can claim that Y has a Poisson distribution, simply because it has a Poisson-type mgf, even though you had not stipulated its distribution directly. The uniqueness property of mgfs can indeed help in identifying the probability of individual variables in complex stochastic models. It will prove especially useful in identifying the limiting distribution of a sequence of random variables, as is demonstrated in Theorem 2.8.4 below.

You're no doubt familiar with the phrase "naming rights," as it occurs all the time in connection to ball parks, stadiums, and other venues that get named after a sponsor who pays for the privilege. Sometimes, though, things are named after what they do or are, e.g., community park, city hall, or water tower. Well, moment-generating functions got their name from one of their distinctive properties. The property of being able to compute the moment EX^k, for any positive integer k, from the function $m_X(t)$ is special and quite useful, so the "mgf" name fits these functions well. To see precisely how the moment-generating feature of these functions works, let me remind you about the Taylor Series expansion of a continuously differentiable function f (that is, a function f for which the kth derivative of f exists for all positive integers k). If the real number "a" lies in the so-called radius of convergence of the Taylor series expansion of f, the function $f(x)$ may be represented as

$$f(x) = f(a) + f'(a)(x-a) + f''(a)(x-a)^2/2! + f'''(a)(X-a)^3/3! + \ldots$$

The requirement that all derivatives of the function f exist may be relaxed. If a fixed finite number of derivatives of f exist, then $f(x)$ may be written as a finite series known as a "Taylor Series with a Remainder." For many commonly used functions like, $\sum d_i x^i$, $ln(x)$, e^x and $\sin(x)$, the Taylor series expansion holds for all real numbers "a." As mentioned in Section 2.7, the Taylor series expansion about the number $a = 0$ is called the *Maclaurin* Series expansion of the function $f(x)$ and is written as

$$f(x) = f(0) + xf'(0) + \frac{x^2}{2!}f''(0) + \frac{x^3}{3!}f'''(0) + \ldots$$

Further, as noted earlier, the *Maclaurin* series expansion of the function $f(x) = e^x$ is

$$e^x = \sum_{k=0}^{\infty} \frac{x^k}{k!}.$$

We are now in a position to prove the moment-generating property of the mgf of a random variable X.

Theorem 2.8.2. Let X be a discrete random variable with moments $EX^k < \infty$ for $k = 1, 2, 3, \ldots$. Let $m_X(t)$ be the mgf of X, and assume that $m_X(t)$ exists and is finite in an open interval of real numbers containing 0. Then

$$\frac{\partial^k}{\partial t^k} m_X(t)|_{t=0} = EX^k. \tag{2.30}$$

Proof.

$$\begin{aligned}
m_X(t) &= Ee^{tX} \\
&= \sum_{\text{all x}} e^{tx} p(x) \\
&= \sum_{\text{all x}} \left(\sum_{k=0}^{\infty} \frac{(tx)^k}{k!} \right) p(x) \quad \text{(using the Maclaurin series for } e^{tx}\text{)} \\
&= \sum_{\text{all x}} \left(1 + tx + \frac{(tx)^2}{2!} + \ldots. \right) p(x) \\
&= \sum_{\text{all x}} p(x) + t \sum_{\text{all x}} xp(x) + \frac{t^2}{2!} \sum_{\text{all x}} x^2 p(x) + \ldots \\
&= 1 + t \cdot EX + \frac{t^2}{2!} \cdot EX^2 + \frac{t^3}{3!} \cdot EX^3 + \frac{t^4}{4!} \cdot EX^4 + \ldots
\end{aligned} \tag{2.31}$$

From (2.31), we obtain

$$\begin{aligned}
\frac{\partial}{\partial t} m_X(t)|_{t=0} \equiv m'(0) &= (0 + EX + t \cdot EX^2 + \frac{t^2}{2!} \cdot EX^3 + \frac{t^3}{3!} \cdot EX^4 + \ldots)|_{t=0} \\
&= EX.
\end{aligned}$$

Similarly,

$$\begin{aligned}
\frac{\partial^2}{\partial t^2} m_X(t)|_{t=0} \equiv m''(0) &= (EX^2 + t \cdot EX^3 + \frac{t^2}{2!} \cdot EX^4 + \ldots)|_{t=0} \\
&= EX^2.
\end{aligned}$$

It is clear that further derivatives of $m_X(t)$, when evaluated at $t = 0$, each yield the corresponding higher order moment of the variable X. ■

You perhaps noticed that, while the first and second moments of the binomial and Poisson models are reasonably easy to obtain directly from the corresponding series representations, this was not the case for the geometric model. We obtained these moments through a couple of "end runs," namely a "staircase" argument and a recursive relationship. The mgf of the geometric distribution provides a new vehicle for attacking this problem.

Example 2.8.5. Let $X \sim G(p)$. We have from Example 2.8.3 that the mgf of X is

$$m_X(t) = \frac{pe^t}{1-qe^t}, \quad \text{for } t < \ln\left(\frac{1}{q}\right).$$

An easy application of the calculus yields

$$m'(t) = \frac{pe^t}{(1-qe^t)^2} \tag{2.32}$$

and

$$m''(t) = \frac{pe^t(1-qe^t)^2 - 2(1-qe^t)(-qe^t)pe^t}{(1-qe^t)^4}. \tag{2.33}$$

From (2.32) and (2.33), we obtain

$$EX = m'(0) = \frac{p}{(1-q)^2} = \frac{1}{p} \text{ and } EX^2 = m''(0) = \frac{p^3+2p^2q}{(1-q)^4} = \frac{2-p}{p^2}.$$

Together, these latter equations yield

$$V(X) = \frac{2-p}{p^2} - \left(\frac{1}{p}\right)^2 = \frac{1-p}{p^2} = \frac{q}{p^2}.$$

■

We conclude our discussion of moment-generating functions with a discussion of the so-called *continuity property* of mgfs.

Theorem 2.8.3. Let $X_1, X_2, X_3, \ldots$ be a sequence of random variables with respective moment-generating functions $m_1(t), m_2(t), m_3(t), \ldots$. Suppose that the sequence $\{m_n(t)\}$ converges to a function $m(t)$, that is, suppose that as $n \to \infty$, $m_n(t) \to m(t)$ for all values of t in an open interval of real numbers containing 0. If $m(t)$ if the mgf of a random variable X, then as $n \to \infty$, $X_n \xrightarrow{D} X$, that is, the probability distribution of X_n converges to the probability distribution of X.

The convergence described in Theorem 2.8.2 is generally called "convergence in distribution." When convergence in distribution can be established in a particular setting, one may use the limiting distribution, that is, the distribution of X, as an approximation for the distribution of X_n for n sufficiently large. We have already encountered this type of convergence twice in this chapter. Theorem 2.5.4 gives conditions under which a sequence of hypergeometric variables will converge in distribution to a binomial variable. When these conditions are roughly satisfied, we can feel comfortable using binomial probabilities to approximate hypergeometric probabilities. The Law of Rare Events (Theorem 2.7.3) is another example of the convergence-in-distribution phenomenon. To show the power and utility of the continuity theorem for mgfs, I offer the following alternative proof of the "LORE."

Theorem 2.8.4. Let $X_n \sim B(n,p)$. If $n \to \infty$ and $p \to 0$ so that $np \to \lambda > 0$, then

$$X_n \xrightarrow{D} X, \text{ where } X \sim P(\lambda).$$

Proof. The notation $x_n \approx y_n$ means that $x_n/y_n \to 1$ as $n \to \infty$. If n is sufficiently large, we have, by assumption, that $np \approx \lambda$, or $p \approx \lambda/n$. Thus, for sufficiently large n,

$$(q+pe^t)^n \cong \left(1 - \frac{\lambda}{n} + \frac{\lambda}{n}e^t\right)^n.$$

It then follows from (2.23) that

$$\lim_{n\to\infty, p\to 0, np\to\lambda} (q+pe^t)^n = \lim_{n\to\infty}\left(1+\frac{\lambda(e^t-1)}{n}\right)^n = e^{\lambda(e^t-1)}.$$

Since $m_{X_n}(t) \to m_X(t)$, where $X \sim P(\lambda)$, it follows from the continuity theorem that $X_n \xrightarrow{D} X$. ■

Pretty cool proof, huh? If you liked that, you'll really be impressed when we prove the Central Limit Theorem in Chapter 5 using a similar (though a bit more involved) argument!

Exercises 2.8.

1. Find a closed-form expression for the moment-generating function of a discrete uniform random variable, that is, for the random variable X with pmf $p(x) = 1/n$ for $x = 1,2,3,\ldots,n$. (Hint: Use Lemma 2.6.1.)

2. Prove Theorem 2.8.1.

3. Let $m_X(t)$ be the moment-generating function of the random variable X, and let $R(t) = \ln m_X(t)$. Show that $R'(0) = EX$ and that $R''(0) = V(X)$.

4. Let $m_X(t) = (1-2t)^{-3}$ be the moment-generating function of the random variable X. Find the mean and variance of X.

5. Find the probability mass function of the random variable X with moment- generating function given by $m_X(t) = (.3)e^t + (.4)e^{3t} + (.3)e^{5t}$.

2.9 Chapter Problems

1. Let k be a positive integer. Find the constant c for which the function $p(x)$ below may serve as a discrete probability model on the integers $1,\ldots,k$.

$$p(x) = \frac{cx}{k^2} \text{ for } x = 1,2,\ldots,k.$$

2. Show that the function $p(x)$ below is a legitimate probability mass function:

$$p(x) = \frac{1}{x(x+1)} \text{ for } x = 1,2,3,\ldots.$$

3. Suppose that two evenly matched teams (say team A and team B), each with probability .5 of winning any given game, make it to the baseball World Series. The series ends as soon as one of the teams has won four games. Thus, it can end as early as the 4th game (a "sweep") or as late as the 7th game, with one team winning its fourth game after having lost three of the first six games. Calculate the probability that the series ends in 4 games, 5 games, 6 games, or 7 games. What is the expected value of X, the number of games it takes to complete the series?

4. Repeat problem 3 under the assumption that team A will win a given game with probability .6.

5. Suppose team A in problem 3 will play up to four games on its home field (game 1, 2, 6, and 7, the 5th, 6th, 7th games played only "if necessary"). The other games (game 3, 4, and, if necessary, game 5) will be played on Team B's home field. Assume that each team will win its home games with probability .6. What are that chances that team A wins the series?

6. Consider a bridge system in 5 components. The system is pictured in Figure 2.9.1 below.

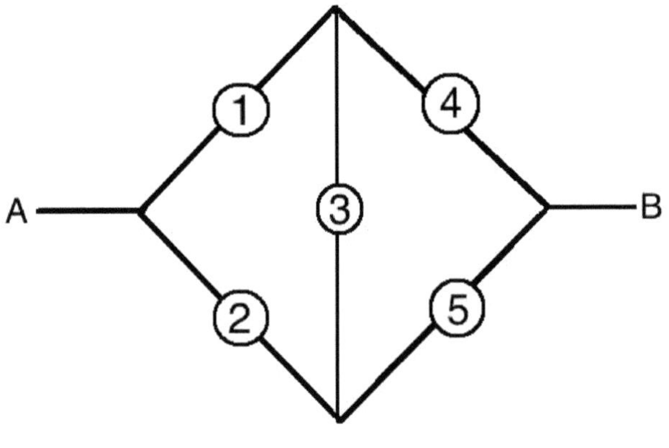

Figure 2.9.1. The bridge system.

The system works if there is a path consisting of working components which permits a transit from A to B. The minimal path sets of this system (that is, paths of working components that guarantee that the system works but which have no proper subsets with the same property) are $\{1,4\}$, $\{2,5\}$, $\{1,3,5\}$, and $\{2,3,4\}$. There are $5! = 120$ permutations of the numbers $1,2,3,4,5$. Let X be the discrete random variable defined as the index of the ordered component which causes the system to fail. For example, if components fail in the order $5,4,2,3,1$, then the system will fail upon the second component failure, so that $X = 2$. If the components of a system have independent lifetimes with a common distribution F, then the distribution of X is distribution free, that is, does not depend on F, but depends only on the distribution of the permutations of the component labels. The distribution of X in this setting is called the system's signature (see Samaniego, 2007). Derive the signature of the bridge system, that is, the distribution of X for the bridge system pictured above.

7. Let X have the discrete uniform distribution of the integers $1,2,\ldots,k$. Show that

$$EX = \frac{k+1}{2} \text{ and } V(X) = \frac{(k+1)(k-1)}{12}.$$

(Hint: Recall, from Section 2.4, that: $\sum_{i=1}^{k} i = \frac{k+1}{2}$ and $\sum_{i=1}^{k} i^2 = \frac{k(k+1)(2k+1)}{6}$.)

8. College playboy "Studley" Moore likes to throw lavish parties in his spacious Davis apartment. He has one quirk though. He insists that every guest, on their way in, play a game of "odd or even" with him. Each of the two sticks out one or two fingers, and Studley wins if the sum is *even*. Studley pays the guest \$5 when the sum is odd (i.e., there are *one* or *three* fingers showing), but he collects \$6 when the sum is *two* and \$4 when the sum is *four*. Suppose that Studley uses the strategy of sticking out one finger with probability 7/12 and two fingers with probability 5/12. Assume that each guest chooses one or two fingers with equal probability. Find the probability distribution for Studley's winnings W in a game of odd or even, and compute Studley's expected winnings in this game. Would always sticking out one finger be a better strategy for Studley (up until he gets found out)?

9. It is a tradition in almost every typing course in the country to have every student type, over and over, the sentence "The quick brown fox jumps over the lazy dog." Why, you ask? Well, it happens that this particular sentence contains all 26 letters of the alphabet. Suppose that you, just to pass the time, type a word chosen at random from the sentence alluded to above, Let X be its length. Obtain the probability distributions of X, and compute its mean and variance.

10. The State Lottery is offering a new game in which a digit from 0 to 9 is chosen at random, and your payoff is as shown below:

Outcome	0	$1-3$	$4-5$	$6-9$
Payoff	\$5	\$2	\$1	\$0

What should the State Lottery charge to play this game if it wishes to make a profit of 20 cents, on average, each time the game is played?

11. At a local fraternity's casino night, the following game is offered. You draw a single card from a 52-card deck. You win \$10 for drawing and ace, \$5 for drawing a face card (K,Q,J), and \$0 for drawing $2-10$. It costs \$2 to play this game. Is the game fair, or is it favorable or unfavorable to you?

12. Ten applicants (3 women and 7 men) for a beginning position with the Sacramento Police Department were judged to be fully qualified. The Police Chief has decided to hire three of them using a random lottery. Let X be the number of women hired. (a) Calculate the probability $P(X = k)$ for $k = 0, 1, 2, 3$. (b) Calculate the expected number of women that will be hired.

13. Suppose two balanced dice are rolled. Let U be the largest value facing up and let V be the smallest value facing up. Define $X = U - V$. Find the pmf of the variable X and compute its mean and variance.

14. Let X be a discrete random variable taking values in the set of non-negative integers $\{0,1,2,3,4,\ldots\}$. Assume $EX < \infty$. Show that the mean of X can be computed by the alternative formula $EX = \sum_{n=1}^{\infty} P(X \geq n)$.

15. Show that, for any discrete variable with a finite mean μ and variance σ^2,

$$E(X)(X-1) = \mu(\mu-1) + \sigma^2.$$

16. Let X be a discrete random variable with finite mean μ. Show that the average squared distance of X from the constant c, that is, the value $E(X-c)^2$, is minimized when $c = \mu$.

17. Let X be a discrete random variable and let m be a median of X (that is, m is a number with the property that $P(X \geq m) \geq 1/2$ and $P(X \leq m) \geq 1/2$. Show that the average distance between X and the constant c, that is, the value $E|X-c|$, is minimized when $c = m$.

18. Let $X_1, X_2, \ldots, X_k$ be k discrete random variables with pmfs $p_1(x)$, $p_2(x)$, $\ldots$, $p_k(x)$, respectively. Let $\{a_i, i = 1, \ldots, k\}$ be a set of nonnegative numbers such that $\sum_{i=1}^k a_i = 1$. Let X be the random variable that takes the value X_i with probability a_i for $i = 1, 2, \ldots, k$. Show that the pmf of X is $p(x) = \sum_{i=1}^k a_i p_i(x)$ and that the mean of X is $EX = \sum_{i=1}^k a_i EX_i$. The distribution of X is said to be a *mixture* of the distributions of $X_1, \ldots, X_k$, and the vector $\mathbf{a} = (a_1, \ldots, a_k)$ is called the *mixing distribution.*

19. A quality control engineer who works for Pep Boys. Inc. has established the following procedure for accepting a box of 12 spark plugs. Three spark plugs are tested, and if they all work, the box is accepted. What is the probability that a box containing two defectives will be accepted?

20. Field biologist Rayne Bo Trout uses the "mark-recapture" method to count the number of fish in a pond. Bo tags an initial sample of r fish and releases them. In his second sample (this time of size $n \leq r$), he notes the number X of tagged fish. It's clear that $X \sim HG(N, r, n)$, where N is the number of fish in the pond. If $X = x$ tagged fish are caught in the second sample, Rayne estimates the parameter N as $\hat{N} = (r+1)(n+1)/(x+1) - 1$. Calculate $E\left((r+1)(n+1)/(X+1) - 1\right)$. (Hint: Obtain $E\left((r+1)(n+1)/(X+1)\right)$ first. If r is large, this expectation is very close to $N+1$.)

21. If you toss a fair coin 10 times, what is the probability that you get more heads than tails?

22. If Rafa wins 60% of the tennis sets he plays against Roger, what is the probability that Roger will win more sets that Rafa in the next five sets they play?

23. Suppose that the random variables X and Y take values independently of each other. If $X \sim B(n, p)$ and $Y \sim B(n, q)$, where $q = 1 - p$, show that $P(X = k) = P(Y = n - k)$.

24. Suppose that $X \sim HG(N, r, n)$. Show that the mode of the distribution of X, that is, the most probable value of X, is the largest integer less than or equal to the real number $(n+1)(r+1)/(N+2)$.

25. A multiple choice test in Psychology 153, Impressions, Depressions and Couch Sessions, consists of 20 questions, each with four possible answers. (a) Suppose that student Claire Voyant forgot there was a test today, and since she hasn't read the relevant material, she simply picks an answer at random for each question. She will fail the test if she gets fewer than seven correct answers. (a) What is the probability that Claire fails the test? (b) Another student, I.B. Peeking, claims to have guessed a lot too, but he sat next to Claire and 10 of his answers are identical to hers. What is the probability that I.B. would match 10 or more of Claire's answers if he in fact guessed the answer to every question?

26. A 12-pack of D batteries has been lying around your garage for a couple of years. Suppose that 4 of the batteries are now dead and 8 of them still work. If you select 3 batteries at random, what is the probability that (a) exactly one of them is dead? or that (b) the first two batteries selected work but the third one is defective?

27. Perhaps you've heard of baseball's famous Giambi brothers? Well, they also like to play basketball. Jason makes a free throw with probability 1/2 and that Jeremy makes a free throw with probability 3/4. In a weekend tournament, Jason got to shoot three free throws and Jeremy got to shoot two. What's the probability that Jeremy *made* more free throws than Jason? (Hint: Let X and Y be independent binomial variables, with $X \sim B(3,1/2)$ and $Y \sim B(2,3/4)$. Compute $P(X < Y)$.)

28. A clinical trial is run with two groups of five patients. The first group is given a placebo, while the second group is given a newly developed pain medication. If the placebo succeeds in alleviating headache pain with probability .4 for each patient while the pain medication alleviates headache pain with probability .5, what is the probability that group 1 experiences more successes that group 2?

29. Let S be the sample space of a discrete random experiment, and let $A_1, \ldots, A_n$ be an arbitrary collection of compound events. Prove the Inclusion-Exclusion Rule stated in Theorem 1.3.7, that is, prove the identity in (1.19). (Hint: For $k = 1, \ldots, n$, show that if a simple event x is contained in exactly k of the events $\{A_i, i = 1, \ldots, n\}$, then its probability is counted exactly once on the RHS of (1.19).)

30. The game of roulette involves a spinning wheel with 38 numbers, each assumed equally likely to occur on a given spin. Of the 38 numbers, 18 are red, 18 are black and 2 are green. When you bet \$1.00 on red or on black, you will win \$1.00 if your color comes up (and, of course, you will also keep the \$1.00 you bet), and if your color does not come up, you simply lose your dollar. The chances that you win when either betting red or betting black is 18/38. Suppose you arrive at a casino with \$63.00 in your pocket. Consider the following betting strategy. You will always bet on red. Suppose that you approach the roulette table with the intention of playing five (potentially imaginary) sets of 6 bets in a row, starting with a bet of \$1.00 and repeatedly doubling your last bet if you don't win that bet. In reality, you'll only play the entire first set if you don't get a red in the first five plays and are forced to bet \$32.00 on the sixth play in the set. As soon as you get a red in a given set, the set is over and you have won \$1.00. If you win a set, you will start over with a new set using the same strategy. If you lose all six bets in any of the first five sets, you will have lost \$63.00 and the entire game will be over for you. Your strategy includes a provision for what to do if, at any point in time, you have just lost a bet and don't have enough money left to double your last bet. In that event, you lick your wounds, tip the roulette staff whatever cash you have left and hitch a ride home. But your plan is to play until you either lose or are \$5 ahead, exactly the price of a scrumptious hamburger with fries and a coke that you hope to buy before you head home. (a) Compute the probability that you will indeed leave the roulette table five dollars ahead. You'll see that your chances of leaving as a winner are pretty good! (Actually, about 90%). Who says that there's no such thing as a free lunch? (b) With the strategy above, you will leave the roulette table either \$5 ahead or \$63 behind. What are your expected winnings? (Hint: Calculate the probability that you lose a set. Now, calculate the probability that you win the first 5 sets. That's the probability that you walk away with a \$5.00 profit. The complementary probability goes with the outcome that you lose your entire stake of \$63.00.)

31. Let $X \sim B(n,p)$. Show that the mode of the distribution of X (the most probable value of X) is the greatest integer that is less than or equal to $(n+1)p$.

32. In the game of bridge, each player is dealt a 13-card hand at random. Suppose a bridge player plays 6 hands of bridge in a tournament. Find the probability that (a) he gets exactly two aces in exactly two of the hands and (b) he gets at least two aces in at least two of the hands.

33. Suppose that a sample of n students are selected at random (without replacement) from a group of r boys and $N-r$ girls. Let X be the number of boys in the sample. What choice of the sample size n maximizes the variance of X?

34. Suppose that $X \sim B(n,p)$, i.e., X has the binomial distribution based on n trials of a dichotomous experiment with $P(S) = p$. Show that

$$E2^X = (1+p)^n.$$

35. Characteristics inherited from one's parents are determined by carriers called genes. Suppose that genes can only take two forms, A and a. This leads to three possible genotypes: aa, aA, and AA. The genotypes aA and Aa are considered identical. The genotype of an offspring is determined by the selection of one gene at random from each parent. The selections for a given offspring are independent, and the genotypes for different offspring are also independent. Suppose both parents are of genotype aA. (a) What are the probabilities that an offspring will have genotypes aa, aA, or AA? (b) If two parents have six offspring, what is the probability that exactly three of the offspring will have genotype aA? (c) If two parents have six offspring, what is the probability that exactly two of the offspring will be of each genotype?

36. Let $X \sim B(n,p)$. Show that $V(X) \leq \frac{n}{4}$ for all p, and that $V(X) = \frac{n}{4}$ if and only if $p = 1/2$.

37. Moe, Larry, and Curly, our favorite stooges, take turns tossing a fair coin, continuing to toss the coin, in that order, until one of them gets "tails." The one who gets the first tails then gets to knock the other two guys' heads together (yuk, yuk, yuk). Find the probability that Curly gets to do the head knocking. What's the probability for Larry and for Moe? (Hint: Curly wins on trials like HHT, HHHHHT, etc.)

38. A fair coin is tossed until the fourth head is obtained. Find the probability that it takes at least seven tosses to obtain the fourth head.

39. Banker Willie Nillie has n keys on his key chain at work, and they all look pretty much the same. Each opens a bank deposit box, but only one will open the box he needs to open up today. Let X be the number of keys he must try before he actually finds the one key that will work. Find EX when (a) Willie randomly samples keys from the key chain without replacement and (b) Willie randomly samples from the key chain with replacement.

40. Suppose that $X \sim G(p)$. Show that, for any positive integer k, $P(X > k) = q^k$.

41. Show that the geometric distribution has the "memoryless property," that is, if $X \sim G(p)$, then $P(X > x+k|X > x) = P(X > k)$. (Note: The geometric distribution is the unique discrete distribution with this property.)

42. Anthony and Cleopatra are regulars at the Fox and Goose Bar and Restaurant in Sacramento, and they often play British-style dart games there. One evening, they are goaded by the other F&G regulars into a winner-takes-all match in which the first one to hit the (small) bull's eye in the center of the dart board wins \$100 from the other. Cleo is generally known to be better than Tony, and so Tony is allowed to go first. Tony's probability of hitting the bull's eye on a given toss is .1, while Cleo's probability is .2. Find the probability that Cleo hits the bull's eye before Tony does. (Hint: This will happen only on sequences that look like MH, MMMH, etc., where M = miss and H = hit.)

43. A biased coin is tossed until the fourth head is obtained. If $P(H) = .4$, what is the probability that the experiment ends on the tenth coin toss?

44. Let $X \sim G(p)$, and let $Y = \min(X, M)$, where M is an integer greater than 1. Find EY.

45. Let $X \sim G(p)$, and let $Y = \max(X, M)$, where M is an integer greater than 1. Find EY.

46. Accountant Ian Somnia's infamous napping at work has given rise to a challenge from colleague I. M. Tarde. At the company retreat, the two will take turns trying to stay awake during a sequence of 5-minute company training films. The probability that Somnia will fall asleep during a given film is 0.8, while the probability that Tarde does is 0.7. The first one to stay awake (as determined by a panel of alert judges) wins a \$1000 bonus. If Somnia and Tarde alternate watching films, with Somnia going first, what are the chances that Tarde wins the bonus? (Hint: A typical sequence of films for which Tarde wins the bonus is NNNY, where N = "not awake" and Y = "awake.")

47. Let $f(x) = \sum_{i=1}^{\infty} x^i$. Note that this function may also be written as $f(x) = \frac{x}{1-x}$. Obtain two expressions for $f'(x)$ and $f''(x)$, one from each representation of f. Now suppose that $X \sim G(p)$. Use these expressions to evaluate EX and $E[X(X-1)]$, and from these results, show that $V(X) = q/p^2$, where $q = 1 - p$.

48. In the 2010 gubernatorial race in California, politicos Meg and Jerry agreed to show up at one final face-off before the election. Instead of a debate, they participated in a Halloween pie-throwing contest. They stood 10 paces from each other and threw custard pies at each other until one scored a hit (H). Assume that all throws were independent and that the respective probabilities of a hit was .2 for Jerry and .1 for Meg. (a) If Meg threw first, what's the probability that she would score the first hit. (Note that this will happen only on sequences of the following type: H, MMH, MMMMH, etc., where M stands for "miss.") (b) If Jerry threw first, what's the probability that Meg would score the first hit?

49. Surreal Cereal Company is doing a promotion that they are sure will generate more business. Every family-size box of Surreal Cereal contains a valuable baseball card of a hall-of-fame baseball player in his rookie season. There are 10 different cards in all. Let X be the number of boxes of Surreal Cereal that you have to buy in order to collect all 10 cards. Find EX.

50. I love to play ping pong, don't you? It's ever so exciting! My friend Homer and I have this special game we play quite often, with the winner getting \$10.00 each time we play it. We simply play a sequence of points, and the first one to win two points in a row wins. I am a little better than Homer; I win 60% of the points played, irrespective of who's serving. Calculate the probability that I will win a given game. (Hint: Be sure to account for the two different ways I can win, for example, outcomes such as $WW, WLWW, \ldots$ and outcomes such as LWW. $LWLWW, \ldots$. It helps to treat them separately.)

51. Let $X \sim G(p)$. Write $EX = E(X - 1 + 1) = \sum_{y=0}^{\infty} ypq^y + 1 = qEX + 1$. From this, derive the fact that $EX = 1/p$.

52. You and your evil cousin Eddie each own a biased coin. Though neither of you know much about the other one's coin, it happens that your coin has a probability of heads of $P(H) = .6$, while Eddie's has $P(H) = .7$. Each New Year's Eve, you and Eddie play a game; the loser must serve as the designated driver as the two of you go off to take in the local party scene. Here's how the game works. You both toss your coins simultaneously. The game continues through a series of rounds. If you each have the same outcome (i.e., you both get H or you both get T), you toss again. The winner is the first of you to obtain a head while the other gets a tail. Find the probability that you'll be driving on New Year's Eve.

53. Suppose that X and Y are independent geometric variables with a common continuous parameter $p \in (0,1)$. Obtain an expression for $P(X > Y)$. (Hint: Write the probability $P(X > Y)$ as $\sum_{y=1}^{\infty} P(Y = y)P(X > y)$. Express this sum in terms of the parameters p and q, where $q = 1 - p$. The fact that $q^{2k-1} = q(q^2)^{k-1}$ may then be of use. Note: For checking your answer: if $p = 1/2$, $P(X > Y) = 1/3$.)

54. (The Coupon Collector Problem) Given a set of n coupons, how many coupons do you expect to draw, with replacement, in order to have drawn each coupon at least once. (Hint: Let X_i be the number of draws needed to draw the ith different coupon. Clearly, $X_1 = 1$ with probability 1. Each subsequent X_i is an independent geometric variable, and the desired answer may be obtained as $E\left(\sum_{i=1}^{n} X_i\right) = \sum_{i=1}^{n} E(X_i)$.)

55. Suppose that among all the widgets your company produces, the defective rate is .005. In the next batch of 1000 widgets produced, what is the probability that there are no more than three defectives? Find both the exact binomial probability and its Poisson approximation.

56. A radioactive substance emits α particles at the rate of $\lambda = .2$ per microsecond. What is the probability that there will be exactly 3 emissions during the next ten microseconds?

57. Suppose that a book has, on average, 1 misprint per page, and that X_i is the number of misprints on page i. Assuming that each X_i is an independent Poisson variable with mean $\lambda = 1$, find the probabilities

    (a) $P(X_1 = 0)$,

    (b) $P(X_1 + X_2 = 2)$

    (c) $P(\sum_{i=1}^{5} X_i = 1)$.

58. Suppose that $X \sim P(\lambda)$. Show that $E\left(\frac{1}{X+1}\right) = \frac{1}{\lambda}(1 - e^{-\lambda})$.

59. The probability that a given line of computer code has a "bug" is .02. If 100 lines of code are examined, what is the probability that these lines contain no bugs? What's the probability that they contain at least three bugs?

60. Let $X \sim P(\lambda)$. Show that the mode of the distribution of X (the most probable value of X) is the greatest integer that is less than or equal to λ.

61. Only once in every 5000 human births does a mother give birth to triplets. If 2000 mothers will give birth in Los Angeles this month, what is the probability the at least two sets of triplets will be born there?

62. Suppose X has a Poisson distribution with mean λ, that is, $X \sim P(\lambda)$. Prove Robbins' formula: For any function $f(x)$ for which $E(f(X)) < \infty$,

$$E[Xf(X-1)] = \lambda E[f(X)].$$

Note that if $f(x) = 1$ for all x, Robbins formula shows that $EX = \lambda$, and if $f(x) = x$ for all x, the Robbins formula shows that $E[X(X-1)] = \lambda^2$, which implies that $V(X) = \lambda$. There are, of course, tons of other possibilities for choosing f. Some are actually interesting!

63. My whole life, I've been trying to do a certain "magic trick" that's actually a total farce. When it finally works at some random gathering of people who don't know me, it will turn me into some sort of folk hero, and I will, thereafter, live on my laurels and never perform the "magic trick" again. Here's what I do. I take an ordinary deck of cards and have someone inspect it to confirm there's nothing special about it. I give it to someone to shuffle, and then ask him or her to pick a card and show it to one other person of his/her choice, but not to me. The card is then put back into the deck, and the deck is shuffled once more. I then hold the deck to my forehead and announce that the card that was selected was the eight of hearts. Since I have no real talent at this sort of thing, the chances that I am right are 1 in 52. If I'm wrong, I simply say "Oops, I guess I need to practice that trick a bit more." But one of these times, I'll get the card right. On that occasion, I'll say "Remind me to show you how I did that sometime" and then I'll triumphantly leave the gathering (after passing out my business card to everyone present). I have to admit that my lucky day hasn't come yet. But it will! I intend to try this "magic trick" 100 times over the next few years. If I do that, what's the probability that I achieve my goal?

64. Suppose that $X \sim P(\lambda)$. If $p(x)$ is the pmf of X, show that, for $x \geq 1$,

$$p(x) = \frac{\lambda}{x} \cdot p(x-1).$$

Use the recursive relationship above to compute $p(0), p(1), \ldots, p(9)$ for the Poisson distribution with parameter $\lambda = 3$. Compare your answers to those that may be obtained from the table of cumulative Poisson probabilities in Table A.2.

65. You will recall that we have shown that hypergeometric probabilities will converge to binomial probabilities under certain specific assumptions, and that binomial probabilities will converge to Poisson probabilities under certain specific assumptions. So it makes sense that, under the right assumptions, hypergeometric probabilities will converge to Poisson probabilities. Let $X \sim HG(N, r, n)$. Show that $P(X = x)$ converges to $\frac{\lambda^x e^{-\lambda}}{x!}$ if $N \to \infty$, $r \to \infty$, and $n \to \infty$ while $\frac{r}{n} \to 0$ and $\frac{rn}{N} \to \lambda > 0$. Explain why we should expect a Poisson limit under these conditions.

66. The truncated Poisson distribution is a useful model in situations in which an experimental outcome can only be observed if an event we are looking for happens at least once. (Otherwise, we can't be sure that the experiment was performed.) The pmf of the truncated Poisson distribution is given by

$$p(x) = \frac{\lambda^x e^{-\lambda}}{x!} / (1 - e^{-\lambda}) \text{ for } x = 1, 2, 3, \ldots$$

Suppose X has a truncated Poisson distribution with parameter $\lambda > 0$. Find the mean and variance of X.

67. Let X be a discrete random variable on the non-negative integers. The *probability-generating function* (pgf) of X is defined as $g_X(t) = Et^X = \sum_{x=0}^{\infty} t^x p(x)$.

    (a) Show that the function $g_x(t)$ generates individual probabilities for X through the relationship $\frac{\partial^x}{\partial t^x} g_X(t) = x! \cdot p_X(x)$ for $x = 0, 1, 2, \ldots$

    (b) Let $X \sim B(n, p)$. Show that the pgf of X is $g_X(t) = (q + pt)^n$.

    (c) Let $X \sim G(p)$. Show that the pgf of X is $g_X(t) = p/(1 - qt)$.

    (d) Let $X \sim P(\lambda)$. Show that the pgf of X is $g_X(t) = e^{\lambda(t-1)}$.

68. (See Problem 2.9.67) Suppose X is a discrete random variable whose pgf is $g_X(t) = \frac{1-t^{n+1}}{1-t}$. Find the pmf of X.

69. (See Problem 2.9.67) Let $g(t)$ represent the probability-generating function of a discrete random variable X with finite mean and variance. Show that $g'(t)|_{t=1} = EX$ and that $g''(t)|_{t=1} = EX(X-1)$.

70. (See Problem 2.9.67) Let X be the number facing up when a balanced six-sided die is rolled. Show that the probability-generating function of X is $g_X(t) = \frac{t^7 - t}{6(t-1)}$.

71. The logarithmic series distribution was introduced by R. A. Fisher (1942) in studies that sought to model the number of species represented by X individuals in sample taken from a biological population. Fisher applied it as a model in studies involving moths near a light source. The distribution is determined by a single parameter $\theta \in (0, 1)$ and has pmf given by

$$p(x) = \frac{\theta^x}{\alpha x} \text{ for } x = 1, 2, 3, \ldots$$

where $\alpha = \ln(1/(1-\theta))$. Suppose we denote the Logarithmic Series Distribution with parameter θ by $LS(\theta)$.

    (a) If $X \sim LS(\theta)$, show that the moment-generating function of X is given by $m_X(t) = \ln(1 - \theta e^t)/\ln(1 - \theta)$.

    (b) Show that the mean and variance of X are given by $EX = \alpha\theta/(1-\theta)$ and $V(x) = \alpha\theta(1 - \alpha\theta)/(1-\theta)^2$.

72. Identify (by name if possible, or by pmf) the discrete random variables with the following moment-generating functions if they exist, and obtain the mean and variance of each.

    (a) $m(t) = (.4 - .7e^t)^{10}$.

    (b) $m(t) = .2 + .8e^t$.

    (c) $m(t) = .5e^t/(1 - .5e^t)$, for $t < \ln 2$.

    (d) $m(t) = [.6e^t/(1 - .4e^t)]^3$, for $t < \ln(2.5)$.

    (e) $m(t) = e^{3(e^t - 1)}$.

    (f) $m(t) = \frac{1}{6}e^{2t} + \frac{2}{6}e^{4t} + \frac{3}{6}e^{6t}$.

    (g) $m(t) = \frac{1}{n}\sum_{j=1}^{n} e^{jt}$.

73. Suppose a fair coin is tossed repeatedly and independently until the pair HT occurs (i.e., a head followed by a tail). Let X be the number of trials required to obtain the ordered pair HT.

    (a) Show that the probability mass function of X is given by $p(x) = (x-1)/2^x$. (Hint: Draw a tree and note that at stage x, for $x \geq 2$, there are precisely $x-1$ paths in which an H was obtained at stage $x-1$.)

    (b) Show that the moment-generating function of X is $m(t) = e^{2t}/(e^t - 2)^2$.

74. Suppose that $X \sim NB(r, p)$. In this formulation of the Negative Binomial distribution, X represents the number of independent Bernoulli trials needed to obtain the r^{th}success. To obtain the limit theorem of interest here, it is useful to work with the pmf of the variable $Y = X - r$, which represents the number of failures which occur before the rth success in a sequence of independent Bernoulli trials.

    (a) Show that the pmf of the variable Y is given by

    $$p(y) = \binom{r+y-1}{r-1} p^r q^y \text{ for } y = 0, 1, 2, \ldots$$

    where $q = 1 - p$.

    (b) Show that $P(Y = y)$ tends to $\frac{\lambda^x e^{-\lambda}}{x!}$, the pmf of the $P(\lambda)$ distribution, as $r \to \infty, p \to 1$ and $rq \to \lambda > 0$.

75. IBM has devised a contest in which former world chess champions Bobby Fisher and Gary Kasporov will alternate playing chess against the latest version of BIG BLUE, IBM's state-of-the-art chess-playing computer program. The first one to beat BIG BLUE will win the \$1,000,000 prize. Suppose that Bobby and Gary have the same probability p of beating BIG BLUE in a given game. Bobby won the toss and will get to play first. Calculate Bobby's chances of winning the million dollars (as a function of p). Verify that this probability is an increasing function of p.

3

Continuous Probability Models

3.1 Continuous Random Variables

A random variable is continuous if it may take on any value in some interval of real numbers. While discrete random variables are typically associated with the process of counting (e.g., the number of times a particular event occurs in repeated trials of an experiment), continuous variables are generally the result of measurements rather than counts. Examples include the measurement of physical characteristics like height, weight, specific gravity, length of life, or intensity of sound. A variable is considered to be continuous if, in theory, it can take any value in an interval, even though its value may be "discretized" in practical applications. For example, we consider a person's height as a continuous variable even though we would often round that height off to the nearest inch.

The most common way to model a continuous random variable is by means of a curve called a *probability density function* (pdf) or simply a *density*. A probability density function $f(x)$ is a real valued function having two basic properties:

(i) $f(x) \geq 0$ for all x $\in$ ($-\infty$, ∞),

(ii) $\int_{-\infty}^{\infty} f(x)dx = 1$.

As a practical matter, we will wish to have $f(x)$ be an integrable function, as it is through the integration of f that we will be able to compute probabilities from the model. As defined above, a density has a natural interpretation as a probability model. If the random variable X has density $f(x)$, then for any real numbers $a < b$, the probability that $X \in (a,b)$ may be computed as

$$P(a \leq X \leq b) = \int_a^b f(x)dx. \qquad (3.1)$$

Conditions (i) and (ii) above guarantee that the integral in 3.1 is a real number between 0 and 1, inclusive, that is, it is a legitimate probability.

Example 3.1.1. Consider a spinner similar to ones you might encounter at a school carnival. Suppose that the perimeter of the circle within which the spinner revolves is indexed by the interval [0, 1] as in the following figure.

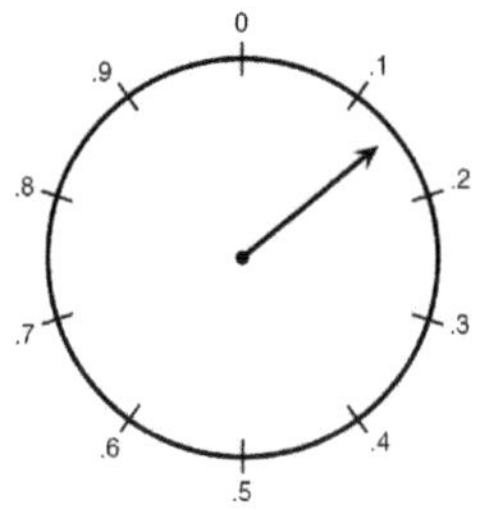

Figure 3.1.1. A spinner for random selection of a number in [0, 1].

If we believe that the spinner is well balanced and favors no particular subinterval between 0 and 1, then the "uniform density" on the interval [0, 1] would seem to be a reasonable model for the outcome X of a given spin. This density is specified as

$$f(x) = \begin{cases} 0 & \text{if } x \leq 0 \\ 1 & \text{if } 0 < x < 1 \\ 0 & \text{if } x \geq 0 \end{cases} \tag{3.2}$$

It is quite apparent that the probabilities computed from the density given above provide answers that match our natural intuition about this experiment. For example,

$$P(.3 \leq X \leq .5) = \int_{.3}^{.5} 1dx = .2.$$

Many of the densities we will encounter in the sequel involve different formulae for $f(x)$ on different subintervals of the real line. For convenience and simplicity, we will routinely employ "indicator function notation" in our specifications of density functions. Let A be a set of real numbers. We define the indicator function of the set A as

$$I_{\mathbf{A}}(x) = \begin{cases} 1 & \text{if } x \in A \\ 0 & \text{if } x \notin A \end{cases}$$

The uniform distribution on the interval $[0,1]$, which will be denoted by $U(0,1)$, has the density displayed in (3.2). This density may be written more compactly as

$$f(x) = I_{(0,1)}(x). \tag{3.3}$$

More generally, the uniform distribution on the interval (a,b), denoted by $U(a,b)$, has density

$$f(x) = \frac{1}{b-a} I_{(a,b)}(x). \tag{3.4}$$

Because of the shape of the density, some authors refer to this model as the rectangular distribution. ■

A quick aside about the interpretation of continuous probability models. Since probabilities computed from density functions are, basically, areas under curves, and there is no area associated with any individual real number, it follows that for a given real number A, $P(X = A) = 0$ for any continuous random variable X. Since the outcome of the associated experiment corresponds to a particular real number A, the fact that the model gave the observed value of X the probability zero of occurrence may at first seem disturbing. In truth, what the model is saying is that any individual value A, among the uncountably infinite potential values that X could take on, has no real chance of happening. Does this mean that if we measure the height of a random male, the probability that he is six feet tall is zero? Actually, yes it does! That's because being "six feet tall" really means that this person's height is 6.0000000000000000...., a decimal with an endless number of zeros, and we know that the chances of drawing such a person is nil. One characteristic of a continuous random variable X implied by this property is that $P(5 < X < 10) = P(5 \le X < 10) = P(5 < X \le 10) = P(5 \le X \le 10)$. All four probabilities are obtained as

$$\int_5^{10} f(x)dx,$$

where $f(x)$ is the density of X.

Example 3.1.2. Let X be a random variable with density

$$f(x) = \lambda e^{-\lambda x} I_{(0,\infty)}(x), \tag{3.5}$$

where $\lambda > 0$. This probability model is called the exponential distribution with failure rate λ and will be denoted by $Exp(\lambda)$. Later, we shall see where the parameter λ gets its name. For now, let us simply verify the $f(x)$ in (3.5) is a legitimate density. Note that

$$\int_{-\infty}^{\infty} f(x)dx = \int_0^{\infty} \lambda e^{-\lambda x} dx = -e^{-\lambda x}|_0^{\infty} = -e^{-\infty} - (-e^{-\lambda \cdot 0}) = 0 + 1 = 1.$$

■

Exercises 3.1.

1. For what value of k is the function $f(x) = kx^3 I_{[0,2]}(x)$ a probability density function?
2. Show that, whenever $0 < a < b$ and $f(x) = a_n x^n + a_{n-1} x^{n-1} + \cdots + a_1 x + a_0$ is a polynomial with positive coefficients $\{a_i\}$, the integral $\int_a^b f(x)dx$ is a positive constant (say k) so that the function $f(x) / k$ may serve as a probability density function on the interval (a,b).
3. Verify that the function $f(x) = k(x^2 + 3x + 1.25) I_{(.5,1.5)}(x)$ is a legitimate density function for some value of k. Identify that k.
4. Find the value k for which the function $f(x) = kxe^{-x^2} I_{(0,\infty)}(x)$ is a valid probability density function.
5. Find the value k for which the function $f(x) = \frac{k}{x^{\alpha+1}} I_{(\theta,\infty)}(x)$, where α and θ are positive parameters, is a valid probability density function on the interval (θ, ∞).

3.2 Mathematical Expectation for Continuous Random Variables

Definition 3.2.1. Let X be a continuous random variable with density $f(x)$. Then the *expected value* of X is given by

$$EX = \int_{-\infty}^{\infty} xf(x)dx. \tag{3.6}$$

As with discrete variables, we refer to EX here as the mean of X and sometimes denote it by μ_X or simply by μ. The number EX has the same interpretations for a continuous variable X that were ascribed to it in the discrete case. When $EX < \infty$, it represents the center of gravity of the probability density function, that is, the point at which a wooden model of the density function could be balanced on one finger. The condition that $EX < \infty$ is important since, as we shall see, there are continuous probability models for which either $EX = \infty$ or EX does not exist. Alternatively, EX may be viewed as the "long run average value" of X in many identical repetitions of the experiment associated with X. This latter property thus bestows the same interpretation as in the discrete case when applied to games of chance. If W is a continuous random variable representing the random winnings in a wager, then EW is equal to the expected winnings per play if this game was played repeatedly.

As with discrete random variables, there are a variety of other expected values that will be of interest in particular settings. Specifically, we will often make use of or refer to the moments $\{EX^k, k = 1,2,3,\dots\}$ of the random variable X, its variance and its moment-generating function. We will take these to be defined as follows:

$$\mu_k = EX^k = \int_{-\infty}^{\infty} x^k f(x)dx, \tag{3.7}$$

$$\sigma_X^2 \equiv V(X) = E(X-\mu)^2 = \int_{-\infty}^{\infty} (x-\mu)^2 f(x)dx \tag{3.8}$$

and

$$m_X(t) = Ee^{tX} = \int_{-\infty}^{\infty} e^{tx} f(x)dx. \tag{3.9}$$

The computational formula for the variance of a discrete random variable given in Theorem 2.2.2 holds as well for continuous variables. The same proof applies, the only difference being that if we were to write out any expectation here, it would be represented as an integral rather than as a sum. We record the continuous analog of Theorem 2.2.2 as:

Theorem 3.2.1. For any continuous random variable X for which $\sigma_X^2 < \infty$, the variance of X may be computed as $\sigma_X^2 = EX^2 - (EX)^2$.

Example 3.2.1. Let $X \sim U(0,\theta)$, the uniform distribution of the interval $(0,1)$. Then the mean of X may be calculated as

$$EX = \int_0^{\theta} x \cdot \frac{1}{\theta} dx = \frac{x^2}{2\theta}\Big|_0^{\theta} = \frac{\theta}{2}.$$

Since it is clear that EX exists for this model (and must lie somewhere between the lower bound 0 and the upper bound θ for X), the fact that $EX = \theta/2$ is obvious from the symmetry of the density about this value, which of course, implies that $\theta/2$ is the density's center of gravity. The second moment of X is

$$EX^2 = \int_0^\theta x^2 \cdot \frac{1}{\theta} dx = \frac{x^3}{3\theta}\Big|_0^\theta = \frac{\theta^2}{3},$$

so that the variance of X may be calculated as

$$\sigma_X^2 = \frac{\theta^2}{3} - \left(\frac{\theta}{2}\right)^2 = \frac{\theta^2}{12}.$$

■

Before moving on to a second example, let's briefly review the mechanics of the calculus technique known as *integration by parts*. The method is often used in evaluating an integral when the exact anti-derivative of the integrand is not available or known. There is no reason to use the technique otherwise, since if it is known that the function $g(x)$ has derivative $g'(x)$ (or alternatively, that $g'(x)$ has anti-derivative $g(x)$), then

$$\int_a^b g'(x)dx = g(x)|_a^b = g(b) - g(a).$$

This result is known as *The Fundamental Theorem of Calculus*, so you know it's got to be important. There are, however, many integrals that arise in practice to which this theorem is not usable, as the pairing (g, g') cannot be identified.

Let's suppose that we wish to evaluate an integral with respect to the variable of integration x, and that the anti-derivative of the integrand of interest is not known. Suppose, also, that the integral may be represented in the form $\int udv$, where u and v are two functions of the variable x, that is, $u = u(x)$ and $v = v(x)$. Then, as shown in most calculus texts, the following identity holds:

$$\int_a^b udv = uv|_a^b - \int_a^b vdu. \tag{3.10}$$

This manner of evaluating the integral $\int udv$ is known as integration by parts. To apply the method, one needs to identify the function u, which is going to be differentiated in the process to obtain du, and the "differential" dv, which will be integrated in the process to obtain v. By doing these differentiation and integration steps in parallel, one hopes to arrive at a simpler integral, one to which the Fundamental Theorem of Calculus applies. The precise way to separate the original integrand into the u and the dv parts is sometimes less than obvious. Often, however, the first or second thing you try will work and leads to a numerical value equal to the original integral. It should be noted that there are problems in which it takes a couple of applications of the process to get to a final answer. We will illustrate the use of integration by parts through the following example. (A quick aside first: I first learned integration by parts as a freshman at Loyola University (now named Loyola Marymount) from a crusty but spectacular teacher named Bert Wicker. I can still visualize Dr. Wicker telling us: "By the way, you'll find that if you use any other letters than u and v in applying this technique, you'll get it wrong. Only u and v work! Since he said it in his deadpan style, we believed him. I have to say, though, that I have found it to be sort of true. If you use other letters, you're just likely to confuse yourself!)

Example 3.2.2. Suppose X has an exponential distribution with failure rate parameter $\lambda = 1$, that is, suppose that $X \sim Exp(1)$. If we want to calculate the mean and variance of X, we will need to evaluate a couple of integrals:

$$EX = \int_0^\infty xe^{-x}dx \text{ and } EX^2 = \int_0^\infty x^2e^{-x}dx \tag{3.11}$$

Now, neither of these integrands have "perfect antiderivatives." We may or may not know this for sure, but what we do know is that we (you and I) can't immediately find functions whose derivatives are equal to the two integrands in (3.11). So naturally, we'll consider using integration by parts to evaluate these integrals. (This is not to say that this is the only tool we have for doing an end run in an integration problem, but it works often enough that a lot of people will at least give it a try first.)

Consider EX, that is, consider the first integral in (3.11). There are two obvious choices for the functions u and v: we could set $u = x$ and $dv = e^{-x}dx$ or we could set $u = e^{-x}$ and $dv = xdx$. For either choice, $xe^{-x}dx = udv$. However, one choice works beautifully, and the other doesn't (that is, leaves us with an integral that we still can't evaluate). The point of view we need to have in order to recognize the good choice is this: since we are going to differentiate u and integrate dv, which choice leads us to an integral $\int vdu$, which is simpler than $\int udv$, the integral we started with. Think about this carefully! Do it both ways if you like. You will see that the first choice is the one we want. Now, letting

$$u = x \text{ and } dv = e^{-x}dx,$$

we obtain, by differentiating the first and integrating the second,

$$du = dx \text{ and } v = -e^{-x}.$$

You can (and generally should) check this calculus by going backwards:

$$u = \int dx = x \quad \text{and} \quad \frac{dv}{dx} = \frac{d}{dx}(-e^{-x}) = e^{-x}, \text{ so that } dv = e^{-x}dx.$$

Now, applying (3.10), we have

$$\begin{aligned} EX &= \int_0^\infty xe^{-x}dx = \int_0^\infty udv = uv|_0^\infty - \int_0^\infty vdu \\ &= -xe^{-x}|_0^\infty + \int_0^\infty e^{-x}dx = 0 + (-e^{-x})|_0^\infty = 1. \end{aligned} \tag{3.12}$$

Finding the second moment of X requires two applications of the process of integration-by-parts. First, let $u = x^2$ and $dv = e^{-x}$. This leads to $du = 2xdx$ and $v = -e^{-x}$, which yields

$$\begin{aligned} EX^2 &= \int_0^\infty x^2e^{-x}dx = \int_0^\infty udv = uv|_0^\infty - \int_0^\infty vdu \\ &= -x^2e^{-x}|_0^\infty + \int_0^\infty 2xe^{-x}dx = 0 + 2\int_0^\infty xe^{-x}dx. \end{aligned}$$

What we have done thus far is to replace an integral we could not evaluate directly by another integral which we can't directly evaluate. Often, the evaluation of this latter integral requires a second integration by parts (and sometimes several more). Fortunately, in this instance, the integral that remains to be evaluated is precisely the integral we evaluated in (3.12). We may therefore identify EX^2 as 2 and compute

$$V(X) = EX^2 - (EX)^2 = 2 - 1^2 = 1.$$

An additional expected value that we will often wish to evaluate and utilize is the moment-generating function of a continuous variable. The definition of the mgf in this context is completely analogous to the discrete case, and is described in

Definition 3.2.2. Let X be a continuous random variable with pdf $f(x)$. The *moment-generating function* (mgf) of X, denoted by $m_X(t)$, is given by

$$m_X(t) = Ee^{tX} = \int_{-\infty}^{\infty} e^{tx} f(x)dx. \tag{3.13}$$

We will derive the moment-generating functions of many of the continuous variables used in this book. As an immediate example of the calculation involved, suppose that $X \sim Exp(\lambda)$, that is, X has the exponential density given in Equation (3.5). Then, using "the trick" in a continuous problem, we may compute

$$\begin{aligned} m_X(t) &= \int_0^{\infty} e^{tx}\lambda e^{-\lambda x}dx = \lambda \int_0^{\infty} e^{-(\lambda - t)x}dx \\ &= \frac{\lambda}{\lambda - t}\int_0^{\infty} (\lambda - t)e^{-(\lambda - t)x}dx = \frac{\lambda}{\lambda - t}, \quad \text{for} \quad t < \lambda. \end{aligned}$$

■

We record for future reference the continuous analog of Theorem 2.8.1 on the mgf of a linear function of a discrete random variable.

Theorem 3.2.2. Let X be a continuous random variable with moment- generating function $m_X(t)$, and suppose that $Y = aX + b$ where a and b are constants. Then the mgf of Y is given by $m_Y(t) = e^{bt}m_X(at)$ for all t for which $m_X(at)$ exists.

Exercises 3.2.

1. Let $X \sim U(0, \theta)$. Show that the moment-generating function of X is $m_x(t) = \left(e^{t\theta} - 1\right)/t\theta$ for $t \neq 0$, with $m_x(0) = 1$.
2. Suppose $X \sim U(a, b)$. Find the mean and variance of X.
3. Prove Theorem 3.2.2.
4. The integral $I = \int_0^1 x(1-x)^3dx$ may be evaluated directly by writing the integrand as a fourth-degree polynomial and integrating it term by term. Evaluate I, instead, using integration by parts with $u = x$ and $dv = (1-x)^3dx$. (Hint: $\frac{d}{dx}\left[-\frac{(1-x)^4}{4}\right] = (1-x)^3$.)
5. The Lopsided distribution $L(\alpha)$ has density $f(x) = kx^{\alpha}I_{(0,1)}(x)$, with $\alpha > 0$. It's been shown to be a good model for the proportion of *male* attendees at a random rooster fight. (a) Find the value of k for which $f(x)$ is a probability density function. (b) Suppose $X \sim L(\alpha)$; find EX. (c) Find the median of $L(\alpha)$, that is, the value m such that $\int_0^m f(x)dx = 1/2$.

6. Suppose X has the truncated exponential distribution on the interval (0, t). The density of X is given by
$$f(x) = \frac{\lambda e^{-\lambda x}}{1 - e^{-\lambda t}} I_{(0,t)}(x).$$
Obtain the mean of X.

3.3 Cumulative Distribution Functions

For any random variable X, the cumulative distribution function (cdf) of X is the function $F(x) = P(X \leq x)$. This function may be evaluated for any $x \in (-\infty, \infty)$ and is well defined for both discrete and continuous random variables. If X is discrete, then $F(x)$ is simply the sum of the probabilities of all values less than or equal to x that the variable X may take on. If X is continuous, then $F(x)$ is evaluated as the area under the density of X from $-\infty$ to x, that is,

$$F(x) = \int_{-\infty}^{x} f_X(t)dt. \tag{3.14}$$

Notice that the integrand in (3.14) uses the "dummy variable" t as the variable of integration, as one may not use the same symbol x in both the integrand and in the limits of integration. If we wished to use the variable x in the integrand, we would denote the argument of the distribution function F by a variable other than x. For example, we could write

$$F(x_0) = \int_{-\infty}^{x_0} f_X(x)dx. \tag{3.15}$$

Equations (3.14) and (3.15) are equivalent and interchangeable, as they both say that the cumulative distribution function F of the variable X, evaluated at a given real number, is simply the total probability that X is less than or equal to that number.

It should be clear from the discussion above that the cdf F of any random variable will have the following properties:

(1) For any $x \in (-\infty, \infty), 0 \leq F(x) \leq 1$. This is because, for any x, $F(x)$ represents the probability of a particular event, and thus it must take values in the interval [0, 1].

(2) *F(x) is a non-decreasing function of x.* This is because if $x < y$, the occurrence of the event $\{X \leq x\}$ implies the occurrence of the event $\{X \leq y\}$, so the probability of the former event cannot be larger than that of the latter event. (This is a consequence of Theorem 1.3.3, the monotonicity property of probabilities.)

(3) $F(x)$ has the following limits: $\lim_{x \to -\infty} F(x) = 0$ and $\lim_{x \to \infty} F(x) = 1$. This is so because, if it were not, the distribution of X would give probability less than 1 to the entire set of real numbers and thus, it would give positive probability to either ∞ or $-\infty$ or both. But random variables map sample spaces onto the real line so that the set of real numbers receives total probability 1.

One final property of note is best demonstrated graphically. A distribution function need not be a continuous function, but

(4) $F(x)$ is right-continuous, that is, $\lim_{x \to A^+} F(x) = F(A)$. This means that if a sequence of x values are approaching the number A from the right, then $F(x) \to F(A)$. It need not be true that $F(x) \to F(A)$ when a sequence of x values approach the number A from the left.

The two examples below display two cdfs, the first for a discrete random variable and the second for a continuous one. In the first case, $F(x)$ is right continuous but is not a continuous function. In the second case, $F(x)$ is a continuous function and is therefore both right-continuous and left-continuous.

Example 3.3.1. Suppose X is the discrete random variable with the pmf $p(x)$ shown below.

$X = x$	1	2	3
p(x)	1/3	1/3	1/3

Table 3.3.1. The pdf of a discrete random variable.

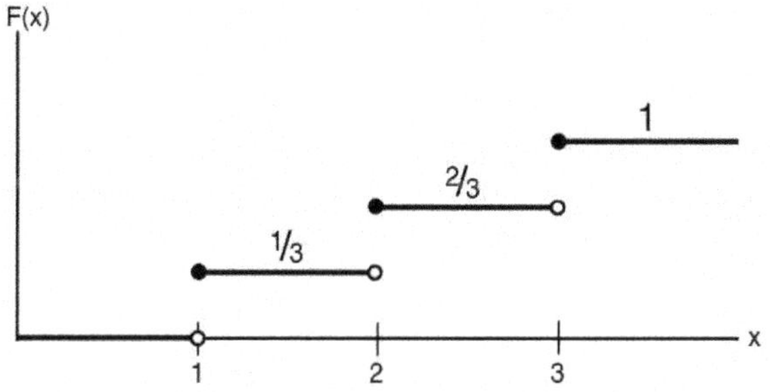

Figure 3.3.1. The cdf of a discrete variable.

■

Example 3.3.2. Suppose $X \sim Exp(1)$, the exponential distribution with density $f(x) = e^{-x} I_{(0,\infty)}(x)$. For $x > 0$, the cumulative distribution function $F(x)$ of X is given by $F(x) = \int_0^x e^{-t} dt = (-e^{-t})|_0^x = 1 - e^{-x}$. This cdf is pictured below.

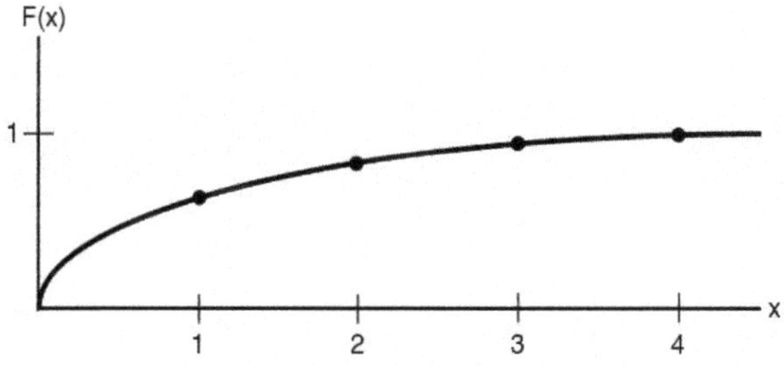

Figure 3.3.2. The cdf of an *Exp*(1) variable.

■

A stochastic model is often specified by stating its probability mass function or density, but it is completely specified as well if the cumulative distribution is given. For a discrete random variable, $F(x)$ has a jump at each point that receives positive probability, and the height of the jump at the point x_0 is precisely the value of $p(x_0)$. So the pmf can easily be recovered from the cdf, and, of course, vice versa. The same is true for the continuous case. Indeed, the Fundamental Theorem of Calculus tells us that if

$$F(x) = \int_{-\infty}^{x} f(t)dt,$$

then $f(x) = F'(x)$. Thus, the distribution function of a random variable X can be obtained from its density by integration, and the density of X may be obtained from the distribution function by differentiation. Stating either $F(x)$ or $f(x)$ completely specifies the probability model. Incidentally, so does stating the moment-generating function $m_X(t)$ of X. So far, we have three interchangeable ways of identifying the model that governs the behavior of a random variable X.

The probability $\bar{F}(x) = P(X > x) = 1 - F(x)$ is called the survival function of the random variable X. If X represents the lifetime of a system or a living being, then $\bar{F}(x)$ is simply the chances that this object survives beyond time x. In subfields of Statistics that are concerned with life length (for example, reliability theory, which studies engineered systems, or biostatistics, which typically studies human subjects), one is often interested in an index which measures a working or living object's propensity to fail at a given point in time. This measure is called the *failure rate* of the object, and is defined by

$$r(t) = \frac{f(t)}{\bar{F}(t)}. \tag{3.16}$$

The failure rate of a random variable X, or equivalently, the failure rate of its distribution, is often interpreted as the instantaneous rate of failure at time t. If one examines (3.16) carefully, one sees why this interpretation of $r(t)$ is reasonable. The fact that $r(t)$ is defined as a fraction suggests that it has something to do with conditional probability. It does indeed. The condition of interest is the event that the object has survived beyond time t. Let's consider how we would approximate the probability $P(t \le X \le t + \Delta t | X > t)$, where Δt represents some very small number. We are interested in the chances that the object will fail in the very near future, that is, in the time interval $(t, t + \Delta t)$, given that it has survived to time t. The actual value of this conditional probability is

$$P(t \le X \le t + \Delta t | X > t) = \frac{\int_t^{t+\Delta t} f(x)dx}{\bar{F}(t)} \cong \frac{f(t)\Delta t}{\bar{F}(t)} = r(t)\Delta t.$$

It follows that

$$\frac{P(X \in (t, t + \Delta t) | X > t)}{\Delta t} \to r(t) \text{ as } \Delta t \to 0,$$

a fact that justifies our thinking of the failure rate $r(t)$ as the instantaneous propensity to fail immediately following time t, given survival until time t.

The failure rate associated with a probability model is a feature of the model which has considerable interpretive value. Suppose we are modeling an engineered system. If the failure rate of the system is increasing, the system is getting weaker over time. Since systems tend to become less reliable as they age, an increasing failure rate might well be expected in many manufactured items. A decreasing failure rate is also possible, of course, and often happens in the initial phase of an object's lifetime (e.g., concrete grows stronger as it matures, wine often improves with age, human infants have an increasing likelihood of surviving to adulthood the longer they have survived beyond birth.) But aging invariably takes its toll, so that the failure rate of many systems or living things has a bathtub shape that decreases in early life, flattens out in midlife, and then increases from some point on. Interestingly, there is one probability model whose failure rate is a constant, irrespective of t. It can be shown that this property serves to characterize the distribution, i.e., $Exp(\lambda)$ is the unique continuous model with this property.

Example 3.3.3. Let $X \sim Exp(\lambda)$, the exponential distribution with density function

$$f(x) = \lambda e^{-\lambda x} I_{(0,\infty)}(x). \tag{3.17}$$

The cdf of X is given by $F(x) = 1 - e^{-\lambda x}$. Thus, the survival function of X is $\bar{F}(x) = e^{-\lambda x}$, and the failure rate of the exponential distribution is

$$r(t) = \frac{\lambda e^{-\lambda t}}{e^{-\lambda t}} = \lambda.$$

■

Exercises 3.3.

1. For the Lopsided distribution in Exercise 3.2.5, find the cumulative distribution function of the variable X.

2. Suppose that X is the proportion of his/her weekly allowance that a random eight-year-old will spend on candy. The density of X is known to have the form
$$f(x) = 20x^3(1-x) \text{ for } 0 < x < 1.$$
Obtain the distribution function $F(x)$ of X.

3. The Pareto distribution is a model for positive random variables. It has proven useful in modeling economic variables like income and in modeling the failure times of long-lived systems. The density function of the one-parameter Pareto model is given by
$$f(x) = [\theta/x^{\theta+1}] I_{(1,\infty)}(x),$$
where $\theta > 0$. Note that Pareto variables have an absolute lower bound, the bound being 1 in this particular version of the Pareto model. For an arbitrary value $x > 1$, find the distribution function $F(x)$ for the Pareto model. (Hint: for $s > 1$, $\int_a^b [1/x^s]dx = [-1/(s-1)x^{s-1}]|_a^b$.)

4. The "standard Cauchy model" has cumulative distribution function given by
$$F(x) = \frac{1}{2} + \frac{1}{\pi}\tan^{-1}(x) \text{ for } -\pi/2 < x < \pi/2.$$
(a) Confirm that $F(x)$ is a legitimate cdf. (b) Obtain the corresponding density $f(x)$. (c) Find the value x_0 such that $P(X > x_0) = .25$, where X has the distribution above.

5. Let $f(x)$ and $g(x)$ be probability density functions for two independent real-valued continuous random variables. Show that for any $c \in (0,1)$, $h(x) = cf(x) + (1-c)g(x)$ is a density function as well.

3.4 The Gamma Model

The so-called gamma integral, given by

$$\Gamma(\alpha) = \int_0^\infty x^{\alpha-1} e^{-x} dx, \tag{3.18}$$

where α is a positive constant, is a rather famous object in the history of mathematics. The celebrity status of $\Gamma(\alpha)$ among integrals of real-valued functions is due to the fact that it arises in a wide variety of mathematical applications. The importance of being able to evaluate the integral became increasingly clear over time. However, one of the distinguishing features of the integral is that there is no closed-form expression for it, that is, no formula that gives the value of the integral for arbitrary values of α. Today, there are tables which allow you to obtain an excellent approximation to the value of $\Gamma(\alpha)$ for any value of α of interest.

While the value of $\Gamma(\alpha)$ cannot be obtained analytically for most values of α, it can be expressed in closed form when α is a positive integer. For example, $\Gamma(1)$ is simply the integral of the $Exp(1)$ density and is thus equal to 1. For other positive integers α, the value of $\Gamma(\alpha)$ may be obtained from the following recursive relation satisfied by gamma integrals.

Theorem 3.4.1. For any $\alpha > 1, \Gamma(\alpha) = (\alpha-1)\Gamma(\alpha-1)$.

Proof. We may prove the claim by applying "integration by parts" to $\Gamma(\alpha)$. We have that

$$\Gamma(\alpha) = \int_0^\infty x^{\alpha-1} e^{-x} dx.$$

Now, let $u = x^{\alpha-1}$ and $dv = e^{-x}dx$. It follows that $du = (\alpha-1)x^{\alpha-2}dx$ and $v = -e^{-x}$.
Since

$$\int_a^b u dv = uv|_a^b - \int_a^b v du,$$

we may write

$$\Gamma(\alpha) = -x^{\alpha-1}e^{-x}|_0^\infty + \int_0^\infty (\alpha-1)x^{\alpha-2}e^{-x}dx = (0-0) + (\alpha-1)\Gamma(\alpha-1).$$

■

Since we know that $\Gamma(1) = 1$, we may calculate

$$\Gamma(2) = (2-1)\Gamma(1) = 1,$$
$$\Gamma(3) = (3-1)\Gamma(2) = 2\cdot 1 = 2,$$
$$\Gamma(4) = (4-1)\Gamma(3) = 3\cdot 2\cdot 1 = 6,$$

and so on. The general formula

$$\Gamma(n) = (n-1)! \quad \text{for all positive integers } n \tag{3.19}$$

follows easily from Theorem 3.4.1 using induction on n.

There are a few other values of α for which the value of $\Gamma(\alpha)$ is known exactly. One of these $\alpha = 1/2$. It can be shown that $\Gamma(1/2) = \sqrt{\pi}$ (See Problem 34 at the end of this chapter.) By the recursion in Theorem 3.4.1, it is thus clear that the value of $\Gamma(n/2)$ may be obtained exactly for any integer $n \geq 1$.

For any $\alpha > 0, \Gamma(\alpha)$ takes on a finite, positive value. Since the integrand of the gamma integral is positive for any positive value of the argument x, we may assert that the function

$$f(x) = \frac{1}{\Gamma(\alpha)} x^{\alpha-1} e^{-x} I_{(0,\infty)}(x)$$

is a legitimate probability density, that is, it is a non-negative function on the interval $(0, \infty)$ that integrates to 1 over that interval. This is a simple version of the gamma model. It is a probability model indexed by one parameter, the "shape" parameter $\alpha > 0$. That α determines the shape of the model is evident from the figure below.

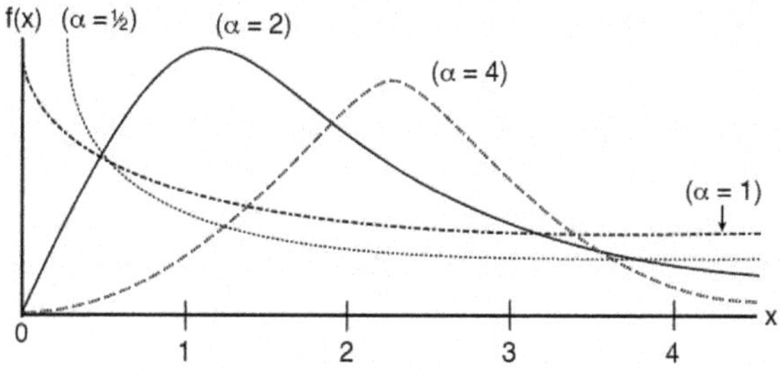

Figure 3.4.1. Shapes of the one-parameter gamma density.

The usual form of the gamma density includes a second parameter $\beta > 0$ called the "scale" parameter of the model. We will call the two-parameter model with density

$$f(x) = \frac{1}{\Gamma(\alpha)\beta^{\alpha}} x^{\alpha-1} e^{-\frac{x}{\beta}} I_{(0,\infty)}(x) \tag{3.20}$$

the *gamma model* and denote it by $\Gamma(\alpha, \beta)$. The fact that the integral of the function in (3.20) is equal to 1 may be easily verified using the "change of variable" theorem form calculus. As usual, the notation $X \sim \Gamma(\alpha, \beta)$ is shorthand for "X has the gamma distribution with shape parameter α and scale parameter β." It is worth mentioning that some authors use an *alternative parameterization* of the gamma distribution with $\lambda = 1/\beta$ as a basic parameter. We will occasionally represent the gamma model as $\Gamma(\alpha, 1/\lambda)$. The effect of the scale parameter is to stretch or contract the model without changing its basic shape. The effect of β is illustrated in the following figure.

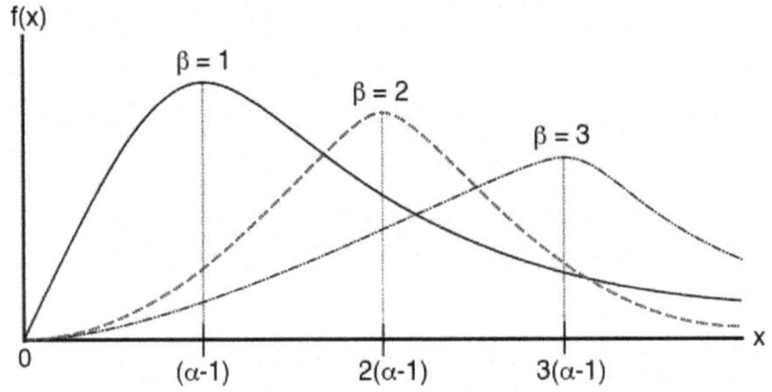

Figure 3.4.2. Three gamma densities with common shape α and varying scale β.

While there is no question that the gamma model $\Gamma(\alpha, \beta)$ is a legitimate probability distribution on the positive real line, the case for its utility as a model for positive random variables remains to be made. The gamma model has found application in many and diverse studies, including lifetime modeling in reliability studies in engineering and in survival analysis in public health, the modeling of selected physical variables of interest in hydrology and meteorology and the modeling the waiting time until the occurrence of k events when the number of events which occur in a time interval of length t is governed by a Poisson process. We will treat this latter application in more detail later in this section.

The gamma distribution also plays an important role in theoretical statistics; for example, we will encounter random variables which are necessarily positive (for example, the sum of squared measurements) and we will encounter particular situations in which the gamma model describes their theoretical behavior well. As we will see later, one specific instance is the fact that the sample variance of a random sample drawn from a "normal" or "Gaussian" distribution (treated in Section 3.5) can be shown to have a particular gamma distribution. Under certain specific conditions, the sums of squares that arise in the statistical areas of analysis of variance and regression analysis can be shown to have gamma distributions. So, for a variety of reasons, the gamma model turns out to be quite an important tool in statistical theory and practice.

Before examining the interesting connection between the gamma and Poisson models, we turn to three analytical results which will be useful in handling and interpreting the gamma distribution in practice. The following three theorems provide the answers to three questions we are by now accustomed to ask about a new model: What is its mean, its variance, and its moment- generating function?

Theorem 3.4.2. If $X \sim \Gamma(\alpha, \beta)$, then $EX = \alpha\beta$.

Proof. By definition,

$$EX = \int_0^\infty xf(x)dx = \int_0^\infty x \cdot \frac{1}{\Gamma(\alpha)\beta^\alpha} x^{\alpha-1} e^{-\frac{x}{\beta}} dx. \tag{3.21}$$

If the integrand in the final expression in (3.21) was a density, we would know the value of the integral, but this is clearly not the case. The most that can be said is that it is almost a gamma density. This fact suggests that some version of "the trick" may be helpful here. This will involve adjusting the integrand appropriately so that what we leave under the integral sign is in fact a perfect density. Rewriting (3.21), we obtain

$$EX = \int_0^\infty \frac{1}{\Gamma(\alpha)\beta^\alpha} x^\alpha e^{-\frac{x}{\beta}} dx$$

$$
\begin{aligned}
&= \frac{\Gamma(\alpha+1)\beta}{\Gamma(\alpha)} \int_0^\infty \frac{1}{\Gamma(\alpha+1)\beta^{\alpha+1}} x^\alpha e^{-\frac{x}{\beta}} dx \\
&= \frac{\Gamma(\alpha+1)\beta}{\Gamma(\alpha)} \text{ (as the integrand in the latter integral is the } \Gamma(\alpha+1,\beta) \text{ density)} \\
&= \frac{\alpha\Gamma(\alpha)\beta}{\Gamma(\alpha)} \text{ (using Theorem 3.4.1)} \\
&= \alpha\beta.
\end{aligned}
$$

■

Theorem 3.4.3. If $X \sim \Gamma(\alpha, \beta)$, then $V(X) = \alpha\beta^2$.

Proof. Using "the trick," we may evaluate the second moment of X as

$$
\begin{aligned}
EX^2 &= \int_0^\infty x^2 \cdot \frac{1}{\Gamma(\alpha)\beta^\alpha} x^{\alpha-1} e^{-\frac{x}{\beta}} dx \\
&= \int_0^\infty \frac{1}{\Gamma(\alpha)\beta^\alpha} x^{\alpha+1} e^{-\frac{x}{\beta}} dx \\
&= \frac{\Gamma(\alpha+2)\beta^2}{\Gamma(\alpha)} \int_0^\infty \frac{1}{\Gamma(\alpha+2)\beta^{\alpha+2}} x^{\alpha+2} e^{-\frac{x}{\beta}} dx \\
&= \frac{\Gamma(\alpha+2)\beta^2}{\Gamma(\alpha)} \text{ (since the integrand is the } \Gamma(\alpha+2,\beta) \text{ density)} \\
&= (\alpha+1)\alpha\beta^2. \text{ (using Theorem 3.4.1)}
\end{aligned}
$$

It then follows that

$$V(X) = EX^2 - (EX)^2 = (\alpha+1)\alpha\beta^2 - (\alpha\beta)^2 = \alpha^2\beta^2 + \alpha\beta^2 - \alpha^2\beta^2 = \alpha\beta^2.$$

■

Theorem 3.4.4. If $X \sim \Gamma(\alpha, \beta)$, then the mgf of X is $m_X(t) = (1-\beta t)^{-\alpha}$ for $t < 1/\beta$.

Proof. We may compute the mgf as

$$
\begin{aligned}
m_X(t) &= Ee^{tX} \\
&= \int_0^\infty e^{tx} \cdot \frac{1}{\Gamma(\alpha)\beta^\alpha} x^{\alpha-1} e^{-\frac{x}{\beta}} dx \\
&= \frac{1}{\Gamma(\alpha)\beta^\alpha} \int_0^\infty x^{\alpha-1} e^{-x\left(\frac{1}{\beta}-t\right)} dx \\
&= \frac{1}{\Gamma(\alpha)\beta^\alpha} \Gamma(\alpha) \left(\frac{\beta}{1-\beta t}\right)^\alpha \int_0^\infty \frac{1}{\Gamma(\alpha)\left(\frac{\beta}{1-\beta t}\right)^\alpha} x^{\alpha-1} e^{-x/\left(\frac{\beta}{1-\beta t}\right)} dx \qquad (3.22) \\
&= \frac{1}{\Gamma(\alpha)\beta^\alpha} \Gamma(\alpha) \left(\frac{\beta}{1-\beta t}\right)^\alpha
\end{aligned}
$$

(since the integrand in (3.22) is the $\Gamma(\alpha, \beta/(1-\beta t))$ density, when $t < 1/\beta$)

$$= \left(\frac{1}{1-\beta t}\right)^{\alpha}$$
$$= (1-\beta t)^{-\alpha} \quad \text{for } t < 1/\beta.$$

■

Let's now take a close look at the relationship between the gamma and Poisson distributions, as this particular connection explains why the gamma distribution might be a reasonable model for waiting times, that is, for the time it takes to observe an event of interest a certain number of times. We will first need to discuss the general notion of a stochastic process and the particular process of interest here—the Poisson process. We can think of a stochastic process as a collection of random variables $\{X(t), t \in T\}$. The variables $X(t)$ may be either discrete or continuous and the domain T may be a finite collection of numbers, a countably infinite collection (like the set of all integers) or an uncountably infinite collection (like the whole real line or the positive real line). For different values of t, say $t_1$and t_2, the random variables $X(t_1)$ and $X(t_2)$ are typically related to each other, and the probability model that describes their behavior must specify the distribution of $X(t)$ for every $t \in T$ as well as the joint stochastic behavior of any finite collection of them. The Poisson process is defined as a stochastic process $\{X(t), t \in T\}$ characterized by the following properties:

1. The domain of the process is $T = [0, \infty)$.
2. $X(t)$ is discrete, and represents the number of times an event of interest occurs in the time interval $[0,t]$. $X(0) = 0$ by definition.
3. The distribution of the variable $X(h+t) - X(h)$ is independent of h and depends only on t.
4. For a fixed constant $\lambda > 0$, and for any $t > 0, X(t) \sim P(\lambda t)$.
5. For any two non-overlapping intervals (s_1, s_2) and (t_1, t_2) of positive real numbers, the two differences $(X(s_2) - X(s_1))$ and $(X(t_2) - X(t_1))$ are independent random variables.

When a stochastic process has property 5 above, it is said to be a process with "independent increments."

The parameter λ of a Poisson process is referred to as the rate of the process per unit time. For example, at time $t = 1$, the expected number of occurrences of the event of interest is λ; the expected number of occurrences of this event by time $t = 2$ is 2λ. The Poisson process occurs in a wide variety of applications. No doubt many applications are a result of the applicability of the Law of Rare Events which gives specific conditions under which the Poisson distribution is a good approximation to the probability that a particular event will occur a certain number of times in a particular interval of time.

Note that since for any $s < t$, we may write $X(t) = X(s) + (X(t) - X(s))$, we see that $X(t)$ may be represented as the sum of the two independent variables U and V, where $U \sim P(\lambda s)$ and $V \sim P(\lambda(t-s))$.

Now, suppose that $\{X(t), t \in (0, \infty)\}$ is a Poisson process with rate parameter λ. Define

Y_k to be the waiting time until the kth occurrence of the event. Clearly, the event $\{Y_k \le t\}$ will occur if and only if $\{X(t) \ge k\}$. From this, it follows that

$$F_{Y_k}(y) = P(Y_k \le y) = 1 - P(X(y) < k) = 1 - \sum_{x=0}^{k-1} \frac{(\lambda y)^x e^{-\lambda y}}{x!}. \tag{3.23}$$

The density of Y_k may thus be obtained, for any $y > 0$, as

$$\begin{aligned}
f_{Y_k}(y) &= F'_{Y_k}(y) \\
&= \lambda e^{-\lambda y} - \sum_{x=1}^{k-1} \frac{x(\lambda y)^{x-1}\lambda}{x!} e^{-\lambda y} + \sum_{x=1}^{k-1} \frac{(\lambda y)^x \lambda}{x!} e^{-\lambda y} \\
&= \lambda e^{-\lambda y} - \sum_{x=1}^{k-1} \frac{(\lambda y)^{x-1}\lambda}{(x-1)!} e^{-\lambda y} + \sum_{x=1}^{k-1} \frac{(\lambda y)^x \lambda}{x!} e^{-\lambda y} \\
&= \lambda e^{-\lambda y} - \sum_{x=0}^{k-2} \frac{(\lambda y)^x \lambda}{x!} e^{-\lambda y} + \sum_{x=1}^{k-1} \frac{(\lambda y)^x \lambda}{x!} e^{-\lambda y} \text{ (by replacing } (x-1) \text{ by x)} \\
&= \lambda e^{-\lambda y} - e^{-\lambda y}\left(\lambda - \frac{(\lambda y)^{k-1}\lambda}{(k-1)!}\right)
\end{aligned}$$

(since the summands in the two sums above cancel each other out, with the exception of the leading term in the first sum and the last term of the second sum)

$$= \frac{\lambda^k}{\Gamma(k)} y^{k-1} e^{-\lambda y} \text{ for } y > 0. \tag{3.24}$$

The function in (3.24) is recognizable as the $\Gamma(k, 1/\lambda)$ density. Indeed, if we set $\beta = 1/\lambda$, the density of Y_k may be written in the usual form, namely,

$$f_{Y_k}(y) = \frac{1}{\Gamma(k)\beta^k} y^{k-1} e^{-y/\beta} I_{(0,\infty)}(y),$$

the density of the $\Gamma(k, \beta)$ distribution. We may thus conclude that in any setting in which the stochastic behavior of the number of occurrences of a given event over time is well approximated by a Poisson process, the distribution of the waiting time Y_k until the kth occurrence of this event is well approximated by a gamma distribution.

One final comment about the Poisson–gamma relationship described above. While the distribution function of a $\Gamma(\alpha, \beta)$ random variable cannot, in general, be evaluated in closed form for arbitrary values of α and β, the representation in (3.23) shows that this distribution function may be calculated exactly when the shape parameter α is a positive integer k; indeed, the integral of a $\Gamma(k, \beta)$ density over the interval $(0, t)$ may be computed in terms of k Poisson probabilities, that is, as the difference $1 - P(X < k)$, where $X \sim P(t/\beta)$.

The gamma distribution has been found to be very useful in empirical investigations of measurement data that are by nature positive-valued. Two specific versions of the gamma model are worth special mention. One is a special case of the $\Gamma(\alpha, \beta)$ model whose density is given in (3.20), and the other is a simple generalization of that model.

The model we have previously referred to as the exponential distribution, and denoted by $Exp(\lambda)$, is simply the $\Gamma(1, 1/\lambda)$ model. This special model has the intriguing memoryless property, that is, if $X \sim Exp(\lambda)$, then one may show that

$P(X > x+t|X > t) = P(X > x)$. (See Problems 31 and 32 at the end of the Chapter.) The mathematical appeal of this property is offset, to some extent, by the fact that it limits the practical utility of the exponential model. For example, when X represents the lifetime of a system, human or inanimate, stipulating an exponential model for X is the same as assuming that if the system is alive at a given time t, its prognosis for the future is precisely the same as when the system was new. Thus, the aging process is essentially nonexistent, as all working systems, whatever their age, have the same chance of surviving an additional x units of time. This is clearly a strong assumption and might be considered a reasonable assumption in only a rather narrow set of applications. It does seem to fit certain physical systems (like light bulbs and computer chips) that tend to fail from shocks to the system (like an electrical surge to a light bulb or a cup of hot chocolate spilled on a mother board). But in general, one would need a broader class of models to describe system lifetimes. The $\Gamma(\alpha,\beta)$ model contains the $Exp(\lambda)$ model as a special case, but also contains a subclasses of models that have a decreasing failure rate (DFR) and an increasing failure rate (IFR), respectively. Specifically, the $\Gamma(\alpha,\beta)$ distribution is DFR if $\alpha \leq 1$ and is IFR if $\alpha \geq 1$. For $\alpha = 1$, the failure rate is constant. Thus, $Exp(\lambda)$ is in both the IFR and the DFR classes, being a boundary member of each.

The phrase *support set* of a density function $f(x)$ is often used to describe the set of values of x for which $f(x)$ is positive. The support set of the $\Gamma(\alpha,\beta)$ model is the interval $(0,\infty)$. There are applications in which a different support set for a similar density is required. The three-parameter gamma distribution with shape parameter α, scale parameter β, and location parameter θ has density

$$f(x) = \frac{1}{\Gamma(\alpha)\beta^\alpha}(x-\theta)^{\alpha-1}e^{-\frac{(x-\theta)}{\beta}}I_{(\theta,\infty)}(x). \tag{3.25}$$

We will refer to this model as $\Gamma(\alpha,\beta,\theta)$. We note that its density looks exactly the same as the $\Gamma(\alpha,\beta)$ density, but it is shifted to have support set (θ,∞) instead of $(0,\infty)$. If $X \sim \Gamma(\alpha,\beta)$, then $Y = X+\theta \sim \Gamma(\alpha,\beta,\theta)$.

There are many occasions in which a particular type of gamma distribution is the appropriate model for a positive random variable arising in a statistical calculation. Often, the setting in which this happens involves a sum of squared variables which, of course, is positive by definition. We will refer to the $\Gamma(k/2,2)$ model as the chi-square distribution with k "degrees of freedom," and denote the distribution by χ^2_k. Clearly, it is a one-parameter model whose mean is k and whose variance is $2k$. The origins of the name "degrees of freedom" for the parameter of the distribution will become apparent in the sequel when various applications of the χ^2_k distribution arise. Because the chi-square distribution arises so frequently in statistical work, the critical values (or "cutoff points") of the χ^2_k distribution, that is, values x_α for which $P(X \geq x_\alpha) = \alpha$, where $X \sim \chi^2_k$, have been tabled for selected values of k and for commonly occurring probability levels α. See Table A.3 in the Appendix.

Exercises 3.4.

1. Let T be the lifetime of a randomly chosen 100-watt light bulb, measured in hours. Suppose that $T \sim Exp(\lambda)$, the exponential distribution with failure rate λ. Let N be the integer-valued variable defined as the smallest integer greater than or equal to T. (For example, if $T = 8.735$, then $N = 9$.) Thus, the light bulb fails during its Nth hour of service. For an arbitrary value of $n \in 1,2,3,\ldots$, evaluate $P(N = n)$ and identify the distribution of N by name.

2. Let $X \sim \Gamma(\alpha, \theta)$. Find $EX^{1/2}$.

3. Suppose that $X \sim \Gamma(5,2)$. Find the exact value of $P(X \leq 10)$. Also, find the value x_0 for which $P(X \leq x_0) = .95$.

4. Find the mode of the χ^2_k distribution, that is, the value x_0 which at which the χ^2_k density takes its maximal value.

3.5 The Normal Model

The most widely used continuous probability model is the so-called normal distribution. It is often called the "Gaussian distribution" in honor of the great German mathematician Karl Friedrich Gauss who introduced the "normal model" in a 1809 monograph as a plausible model for measurement error. Using this normal law as a generic model for errors in scientific experiments, Gauss formulated what is now known in the statistical literature as the "non-linear weighted least-squares" method. His work brought much attention to the model's utility in both theoretical and applied work. The normal distribution is entirely determined by two parameters, its mean μ and its variance σ^2. We will therefore denote the normal distribution by $N(\mu, \sigma^2)$. The probability density function of the $N(\mu, \sigma^2)$ distribution is given by

$$f(x) = \frac{1}{\sqrt{2\pi}\sigma} e^{-\frac{(x-\mu)^2}{2\sigma^2}} I_{(-\infty,\infty)}(x). \tag{3.26}$$

The graph of the normal density is a bell shaped curve whose location and spread depend on the parameters μ and σ^2. The $N(\mu, \sigma^2)$ density is symmetric about the mean μ, and the curve's concentration near μ is governed by its variance σ^2. The normal curve $N(0,1)$ is pictured in Figure 3.5.1 below. Figure 3.5.2 shows the effect of varying σ^2 on three normal curves with the same mean.

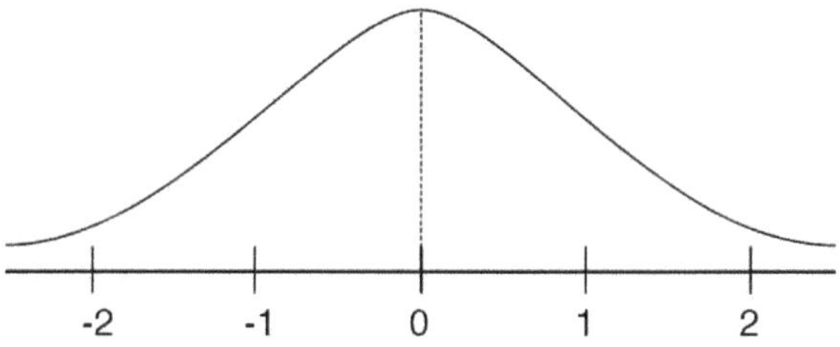

Figure 3.5.1. The density of the $N(0,1)$ distribution.

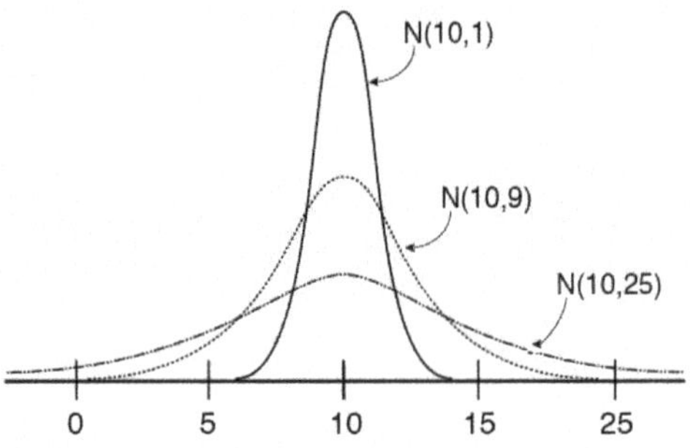

Figure 3.5.2. The densities of $N(10,1)$, $N(10,9)$ and $N(10,25)$.

The normal distribution occurs in a wide array of applications. Physical attributes like the heights and weights of members of certain subpopulations typically have distributions having bell-shaped densities. If we were to obtain a random sample of one hundred 20-year-old males attending a particular university, the histogram that we would construct for their heights or their weights (which graphically displays the relative frequency of occurrence of heights or weights in consecutive intervals of equal length) is very likely to look like an approximation to a bell-shaped curve. The measured characteristics of many natural populations have distributions that are approximately normal. As Gauss himself pointed out, the normal model often provides a good model for measurement or experimental errors. Another reason that the normal distribution arises in practice with some regularity is the fact that the distribution of numerical averages is often approximately normal, a consequence of the famous "Central Limit Theorem" which we will examine closely in Chapter 5.

The distribution function of the normal model does not have a closed form expression. No matter what values μ or σ^2 take, we cannot evaluate the integral

$$F(x_0) = \int_{-\infty}^{x_0} \frac{1}{\sqrt{2\pi}\sigma} e^{-\frac{(x-\mu)^2}{2\sigma^2}} dx$$

which represents the cdf of a $N(\mu, \sigma^2)$ random variable. But because the occurrence of the normal model is so pervasive in statistical work, it is essential that we be able to ascribe a value to this integral. This has been accomplished numerically for the particular normal curve $N(0,1)$, known as the "standard normal curve." We will denote the random variable with distribution $N(0,1)$ as Z, that is, $Z \sim N(0,1)$, and we will refer to Z as a *standard normal variable*. Table A.4 in the Appendix provides approximations, accurate to four decimal places, for the probability $P(0 \leq Z \leq z)$ for values of z among the decimals 0.00, 0.01, 0.02,..., 3.48, 3.49. Using Table A.4, one may compute $P(A \leq Z \leq B)$ for any real numbers A and B that are rounded off to two-decimal-place accuracy. For practical purposes, one may take the probabilities $P(Z \leq -3.49)$ and $P(Z \geq 3.49)$ as equal to zero.

Example 3.5.1. Let $Z \sim N(0,1)$. The following example of the use of Table A.2 includes cases which utilize the symmetry of the standard normal curve about the point $z = 0$ and that fact the $P(Z \leq 0) = 1/2 = P(Z \geq 0)$.

Certain probabilities may be read directly from Table A.3. For example,

$$P(0 \leq Z \leq 1.15) = .3749.$$

Obtaining other probabilities require using the symmetry of $N(0,1)$. For instance,

$$\begin{aligned} P(-0.25 \le Z \le 1.15) &= P(-0.25 \le Z \le 0) + P(0 \le Z \le 1.15) \\ &= P(0 \le Z \le 0.25) + P(0 \le Z \le 1.15) \\ &= .0987 + .3749 \\ &= .4936. \end{aligned}$$

The calculation $P(Z \ge 1.15) = .5000 - P(0 \le Z \le 1.15) = .5000 - .3749 = .1251$ utilizes the fact that the median of the $N(0,1)$ curve is 0. ∎

The fact that a table of probabilities for one particular normal curve will suffice in obtaining probabilities for any normal curve is by no means obvious. It is indeed fortuitous that knowing the probability that a standard normal variable Z takes values in any given interval (A,B) enables one to compute probabilities under any other normal curve. The intuition behind this fact is that one can transform a given normal curve into any other normal curve by changing the location and scale of the curve, that is, by shifting the curve to the left or right to give it a new center and by changing the scale of the new curve. Thus, it is possible to create a variable X having a $N(\mu,\sigma^2)$ distribution by taking a standard normal variable Z and setting $X = \sigma \cdot Z + \mu$. The mathematical justification of this relationship, and of the inverse relationship $(X-\mu)/\sigma \sim N(0,1)$, is given in the following theorem.

Theorem 3.5.1. Let $X \sim N(\mu,\sigma^2)$, and suppose the random variable Y is defined as $Y = aX + b$, where $a > 0$. Then $Y \sim N(a\mu + b, a^2\sigma^2)$.

Proof. We need to show that

$$P(Y \le y_0) = \int_{-\infty}^{y_0} \frac{1}{\sqrt{2\pi}a\sigma} e^{-\frac{(y-a\mu-b)^2}{2a^2\sigma^2}} dy, \tag{3.27}$$

that is, we need to show that cumulative probabilities for the random variable Y can be obtained by integrating the density function of the $N(a\mu + b, a^2\sigma^2)$ model. Now, since for $a > 0$, $P(Y \le y_0) = P(aX + b \le y_0) = P(X \le \frac{y_0-b}{a})$, we have that

$$P(Y \le y_0) = \int_{-\infty}^{\frac{y_0-b}{a}} \frac{1}{\sqrt{2\pi}\sigma} e^{-\frac{(x-\mu)^2}{2\sigma^2}} dx.$$

We now apply the "Change-of-Variable" theorem from calculus. Let $y = ax + b$. We wish to change $\int f(x)dx$ into $\int f(y)dy$. The Change-of-Variable theorem requires that we find the inverse transformation (i.e., find x in terms of y) and find the "Jacobian" $J = dx/dy$ which measures the rate of change of x as the variable y changes. Here, we have,

$$x = \frac{y-b}{a} \text{ and } J = \frac{dx}{dy} = \frac{1}{a}.$$

According to the Change-of-Variable theorem, the desired change from $\int f(x)dx$ to $\int f(y)dy$ is made by replacing, in the first integral, x by $(y-b)/a$ and dx by $|J|dy$ and by replacing $(-\infty, \frac{y_0-b}{a})$, the range of x in the limits of the first integral, by $(-\infty, y_0)$, the corresponding range of y. Making these replacements in the present context yields

$$P(Y \le y_0) = \int_{-\infty}^{\frac{y_0-b}{a}} \frac{1}{\sqrt{2\pi}\sigma} e^{-\frac{(x-\mu)^2}{2\sigma^2}} dx$$

$$= \int_{-\infty}^{y_0} \frac{1}{\sqrt{2\pi}\sigma} e^{-\frac{(\frac{y-b}{a}-\mu)^2}{2\sigma^2}} \frac{1}{a} dy$$

$$= \int_{-\infty}^{y_0} \frac{1}{\sqrt{2\pi}a\sigma} e^{-\frac{(y-a\mu-b)^2}{2a^2\sigma^2}} dy.$$

Thus, Equation (3.27) holds, and the proof is complete. ■

Remark 3.5.1. Theorem 3.5.1 holds for $a < 0$ as well. The proof is similar.

Corollary 3.5.1. If $X \sim N(\mu, \sigma^2)$, then $(X-\mu)/\sigma \sim N(0,1)$.

Proof. This result follows directly from Theorem 3.5.1 with $a = \frac{1}{\sigma}$ and $b = -\frac{\mu}{\sigma}$. ■

The process of transforming a $N(\mu, \sigma^2)$ variable X into $(X-\mu)/\sigma$ is called standardization, as it transforms X into a standard normal variable. Standardizing the variable X allows us to calculate probabilities for X, as the following example shows.

Example 3.5.2. Suppose $X \sim N(5,9)$. Then, from Table A.4, we may obtain

$$P(X \leq 8) = P(\frac{X-5}{3} \leq \frac{8-5}{3}) = P(Z \leq 1) = .8413$$

and

$$P(3.5 \leq X \leq 9.5) = P(-.5 \leq Z \leq 1.5) = .1915 + .4332 = .6247.$$

■

We will have occasion to use the standard normal table is a different way—something that I'll refer to as a *reverse calculation*. The calculation above computes a probability that corresponds to a numerical value or values of a normal random variable X. Suppose, instead, we need to find the numerical value of X corresponding to a given probability (for example, the numerical value of X which is larger than some given percentage of the normal population of interest). The following example shows how this reverse calculation is done.

Example 3.5.3. A dense jungle in Tanzania is famous for having a rather large number of resident gorillas. A particularly extroverted member of this clan goes by the name of Harry. Now it is known that the weight X of a random adult male gorilla in this region is normally distributed with mean $\mu = 650$ pounds and standard deviation $\sigma = 100$ pounds. It is also known that Harry the Ape weighs more than 67% of these adult male gorillas. Denote Harry's weight by H. To find H, we begin by writing

$$P(X \leq H) = .6700.$$

Standardizing X yields the equation

$$P\left(\frac{X-650}{100} \leq \frac{H-650}{100}\right) = .6700. \tag{3.28}$$

Now it may seem that the transformation above is pointless, since the value of the fraction $(H-650)/100$ is unknown, so little has been gained. But this view overlooks the fact that what we do know is the exact (standard normal) distribution of the fraction $(X-650)/100$. Indeed, we may rewrite (3.28) as

$$P(Z \leq \frac{H-650}{100}) = .6700,$$

where $Z \sim N(0,1)$. But from Table A.4, we may ascertain that

$$P(Z \leq 0.44) = .6700.$$

It follows that the unknown fraction $(H-650)/100$ must in fact be equal to 0.44. But if

$$\frac{H-650}{100} = 0.44,$$

then

$$H = 650 + 100(0.44) = 694.$$

■

As we get into problems of a statistical nature in Chapter 6 and beyond, we will have many opportunities to make use of numerical values that must be obtained by a reverse calculation of the type featured in Example 3.5.3. Since the normal distribution will occur in such problems with some regularity, we will take a moment here to carry out the typical reverse calculations involved with the standard normal distribution. We begin with the following.

Definition 3.5.1. Let $Z \sim N(0,1)$, and suppose that $0 < \alpha < 1$. Then Z_α is the number having the property that $P(Z > Z_\alpha) = \alpha$. We will refer to Z_α to as the *α-cutoff point* of the standard normal distribution.

The figure below provides a visual interpretation of the number Z_α.

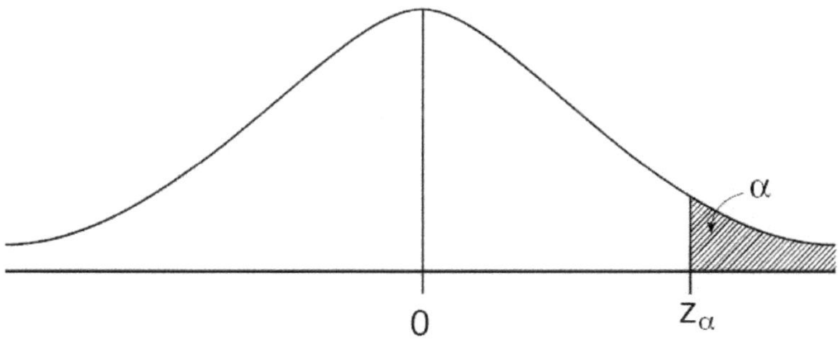

Figure 3.5.3. If $Z \sim N(0,1)$, then Z will exceed Z_α with probability α.

Calculating Z_α for particular values of α is a simple exercise which utilizes Table A.4

in a reverse fashion. We begin by specifying an area of interest, between 0 and 1, then use Table A.4 to find the number having that area to the right of it. For example, $Z_{.5}$ is the number having area .5 to its right under that standard normal curve. Since the $N(0,1)$ distribution is symmetric about 0, we know that $Z_{.5} = 0$. It is easy to verify from table A.4 that $Z_{.9236} = -1.43$, $Z_{.1587} = 1.0$, and $Z_{.0606} = 1.55$. There are certain values of α that will occur frequently in the sequel. The following table records these most commonly used α-cutoff points (rounded to two decimal point accuracy).

α	.1	.05	.025	.015	.01	.005
Z_α	1.28	1.64	1.96	2.17	2.33	2.58

Table 3.5.1. Values of Z_α for commonly used α values.

Now, for the mgf of a normal variable. The proof of the following result uses a familiar trick.

Theorem 3.5.2. If $Z \sim N(0,1)$, then the mgf of Z is $m_Z(t) = e^{t^2/2}$.

Proof. Starting with the definition of "moment-generating function," we have

$$\begin{aligned}
m_Z(t) &= \int_{-\infty}^{\infty} e^{tz} \frac{1}{\sqrt{2\pi}} e^{-z^2/2} dz \\
&= \int_{-\infty}^{\infty} \frac{1}{\sqrt{2\pi}} e^{-z^2/2+zt} dz \\
&= \int_{-\infty}^{\infty} \frac{1}{\sqrt{2\pi}} e^{-(z^2-2zt+t^2-t^2)/2} dz \\
&= \int_{-\infty}^{\infty} \frac{1}{\sqrt{2\pi}} e^{-(z-t)^2/2+t^2/2} dz \\
&= e^{t^2/2} \int_{-\infty}^{\infty} \frac{1}{\sqrt{2\pi}} e^{-(z-t)^2/2} dz \qquad (3.29) \\
&= e^{t^2/2},
\end{aligned}$$

where the last equality holds because the integrand in (3.29) is the density of the $N(t,1)$ distribution and thus integrates to 1. ■

Corollary 3.5.2. If $X \sim N(\mu, \sigma^2)$, then the mgf of X is $m_X(t) = e^{\mu t + (\sigma^2 t^2)/2}$.

Proof. The variable X has the same distribution as the variable $\sigma Z + \mu$. Using Theorem 3.2.2, along with Theorem 3.5.2, we may obtain the mgf of X as

$$m_X(t) = m_{\sigma Z+\mu}(t) = e^{\mu t} m_Z(\sigma t) = e^{\mu t} e^{\sigma^2 t^2/2} = e^{\mu t + \sigma^2 t^2/2}.$$

■

Remark 3.5.2. Thus far, we have made reference to the parameters μ and σ^2 of the $N(\mu, \sigma^2)$ distribution as its mean and variance, but we haven't actually derived these facts analytically. Given the symmetry of the $N(\mu,\sigma^2)$ density about μ, one could infer that, if $X \sim N(\mu,\sigma^2)$, then $EX = \mu$, provided that EX exists and is finite. Actually, the expected value of a normal variable is easy to evaluate directly. For example, if $Z \sim N(0,1)$, then

$$EZ = \int_{-\infty}^{\infty} z\frac{1}{\sqrt{2\pi}}e^{-\frac{z^2}{2}}dz = -\frac{1}{\sqrt{2\pi}}e^{-\frac{z^2}{2}}|_{-\infty}^{\infty} = 0-0 = 0.$$

Then, if $X \sim N(\mu,\sigma^2)$, we have that X has the same distribution as $\sigma Z+\mu$, so that $EX = E(\sigma Z+\mu) = \sigma EZ+\mu = 0+\mu = \mu$.

Proceeding similarly, EZ^2 can be shown to be equal to 1 by an easy integration by parts, from which we get that $V(Z) = 1-(0)^2 = 1$. Finally, for $X \sim N(\mu,\sigma^2)$, we have $V(X) = V(\sigma Z+\mu) = V(\sigma Z) = \sigma^2 V(Z) = \sigma^2$. Of course we may also now avail ourselves of the mgf of X, from which we get the confirmatory result

$$EX = m_X'(t)|_{t=0} = \frac{d}{dt}e^{\mu t+\sigma^2t^2/2}|_{t=0} = (\mu+\sigma^2 t)e^{\mu t+\sigma^2t^2/2}|_{t=0} = \mu.$$

Similarly, one may show that $EX^2 = m_X''(t)|_{t=0} = \sigma^2+\mu^2$, from which $V(X) = \sigma^2$ follows.

There is an important connection between the normal and gamma models. A special case of the relationship between these models is described in the following result. A proof of Theorem 3.5.3 is contained within Example 4.6.1.3 in the next chapter.

Theorem 3.5.3. If $Z \sim N(0,1)$, then Z^2 has a gamma distribution. More specifically, $Z^2 \sim \chi_1^2$, that is, Z^2 has the chi-square distribution with one degree of freedom.

The meaning of the term "degrees of freedom" begins to reveal itself in the result above. If one has a single normal variable Z, the distribution of Z^2 is a chi-square distribution with one degree of freedom. If one had two standard normal variables obtained from two independent experiments, it can be shown that the distribution of the sum $Z_1^2+Z_2^2$ has a chi-square distribution with "two degrees of freedom." Independence of the Zs plays an important role in this result. If the variables Z_1 and Z_2 are free to take values independently from each other, the sum of their squares has "two degrees of freedom." In Chapter 4, we will delve more deeply into such matters and will provide rigorous proofs of these claims.

One of the important applications of the normal distribution is as an approximation for other distributions for which the exact computation of probabilities is either cumbersome or analytically intractable. We will treat one such approximation here—the normal approximation of the binomial distribution.

Let us suppose that X is a binomial random variable. More specifically, let $X \sim B(n,p)$. We know that this distribution has mean np and variance npq, where $q = 1-p$. Suppose that p is neither particularly small nor particularly large. Indeed, for simplicity, suppose that $np \approx n/2$, that is, suppose $p \approx 1/2$. The probability histogram for the $B(n,p)$ distribution will then look approximately symmetric, much like the graph in the following figure. For convenience, we will assume here that n is an even integer.

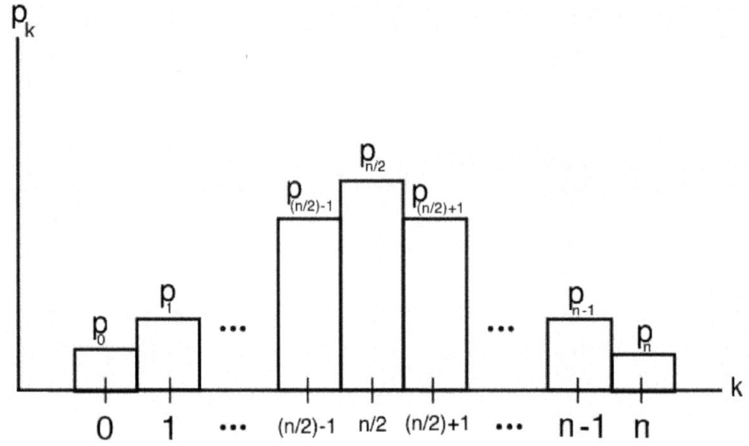

Figure 3.5.4. The probability histogram for $X \sim B(n, 1/2)$, with $p_k = P(X = k)$.

It should not be surprising that a normal curve might approximate the shape of this probability histogram well. Note that the histogram is symmetric about the point $x = n/2$, and that the total area under the histogram is exactly 1, just like the normal density. One might imagine that a normal curve centered at $n/2$, and with variance $n/4$ (the same variance as the $B(n, 1/2)$ model), would lay on top of the binomial histogram and fit pretty snugly. This intuition is entirely correct. In fact, the basic intuition applies quite a bit more broadly. Having perfect symmetry in the binomial histogram isn't really necessary for the approximation we will discuss here to work well.

Consider a binomial variable X obtained from a binomial experiment based on n trials with probability of success p, that is, suppose that $X \sim B(n, p)$. We might expect a normal curve to "fit" the probability histogram of a binomial distribution $B(n, p)$ well when the histogram is approximately bell shaped. This will happen if the mean of the distribution (that is, np) is not too close to either of the integers 0 or n. The standard rule of thumb for obtaining a good approximation of binomial probabilities is that $np > 5$ and $nq > 5$. The first condition ensures that p is not "too small" and the second that p is not "too large." When either of these conditions fail to hold, the binomial histogram will not resemble a normal curve and the approximation we discuss below can be extremely poor. An extreme example of this latter case is shown in Figure 3.5.5.

Under the conditions on p (and q) stated above, the normal curve with mean np and variance npq will look very much like a smooth, continuous version of the $B(n, p)$ probability histogram. The figure below shows normal curve $N(8, 4.8)$ curve superimposed on the probability histogram of the $B(20, .4)$ distribution. From this graph, it should be evident that these binomial probabilities can be reliably approximated by areas under the $N(8, 4.8)$ curve.

To make the guidelines for the normal approximation of the binomial distribution concrete, we'll need to discuss the so-called "continuity correction," an adjustment that is needed when we approximate probabilities for a discrete variable by areas under the density of a continuous variable. You will note that, in the probability histogram for a binomial variable X, the probability $P(X = k)$, for any k, is exactly equal to the area of the rectangle centered at the number k. This is because the height of this rectangle is $p_k = P(X = k)$

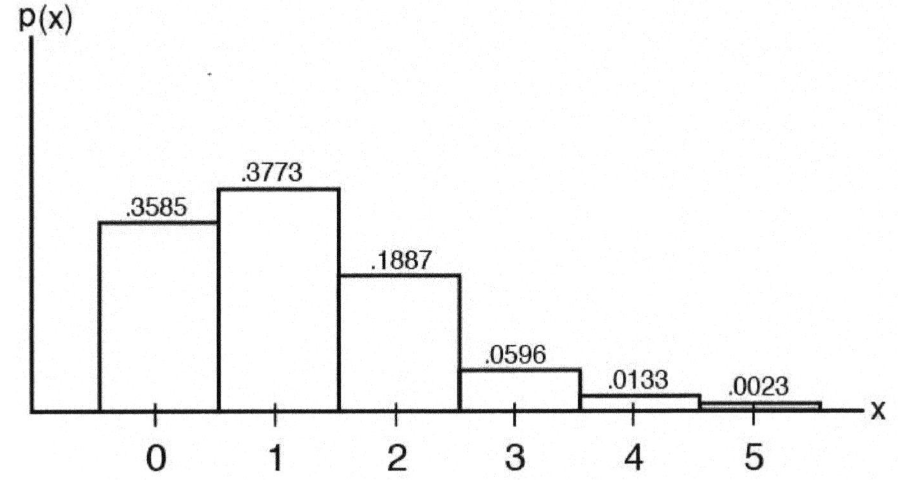

Figure 3.5.5. The probability histogram for $X \sim B(20, .05)$, with $p_k = P(X = k)$.

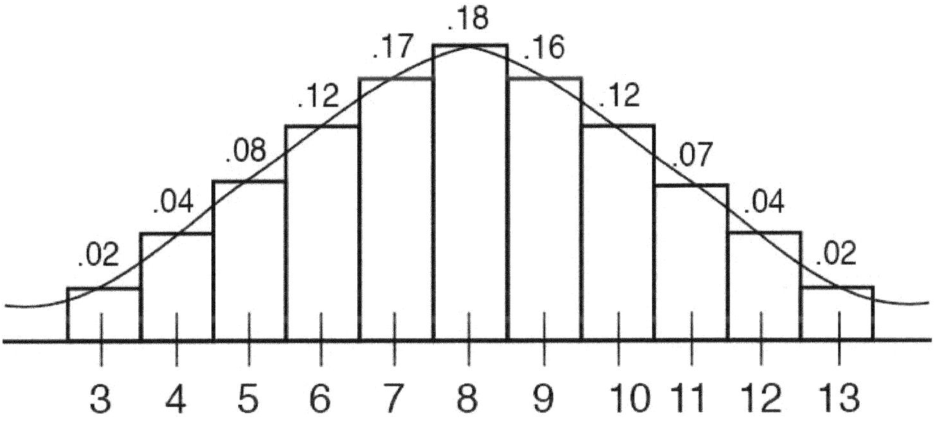

Figure 3.5.6. The probability histogram for $B(20, .4)$ and the $N(8, 4.8)$ curve.

and its width is 1. Since the rectangle associated with $X = k$ begins at $k - 1/2$ and ends at $k + 1/2$, the area under the associated normal curve that aptly approximates $P(X = x)$ is the area over the interval $(k - 1/2, k + 1/2)$. Let $X \sim B(n, p)$ and suppose $np > 5$ and $nq > 5$. If we set $Y \sim N(np, npq)$, then probabilities for X may be approximated as follows:

$$P(X = k) \approx P((k-1)/2 \leq Y \leq (k+1)/2),$$
$$P(X \leq k) \approx P(Y \leq (k+1)/2),$$
$$P(X \geq k) \approx P(Y \geq (k-1)/2)$$

and

$$P(A \leq Y \leq B) \approx P(A - 1/2 \leq Y \leq B + 1/2).$$

Example 3.5.4. Suppose that a fair coin is tossed 100 times. Let X be the number of heads obtained. Then $X \sim B(100, 1/2)$. Since $np = 50 = nq$ for this binomial distribution, the normal approximation of probabilities for X is well justified. Since the $B(100, 1/2)$ model has mean 50 and variance 25, the appropriate normal variable to use in this approximation is $Y \sim N(50, 25)$. Typical applications of the normal approximation here include the following:

$$\begin{aligned} P(X = 50) &\approx P(49.5 \leq Y \leq 50.5) \\ &= P\left(\frac{49.5 - 50}{5} \leq \frac{Y - 50}{5} \leq \frac{50.5 - 50}{5}\right) \\ &= P(-0.1 \leq Z \leq 0.1) = .0796. \end{aligned}$$

and

$$\begin{aligned} P(X \geq 55) &\approx P(Y \geq 54.5) \\ &= P\left(\frac{Y - 50}{5} \geq \frac{54.5 - 50}{5}\right) \\ &= P(Z \geq 0.9) = .1841. \end{aligned}$$

■

The accuracy of the normal approximation of the binomial distribution is excellent under the stated conditions on p. When the binomial histogram is symmetric (that is, when $p = 1/2$), the approximation even gives good answers if n is quite small. For example, if $X \sim B(16, 1/2)$, the exact value of $P(X = 8)$ is .1964 while the normal approximation is .1974. The normal approximation of the binomial complements the Poisson approximation of the binomial treated in Chapter 2 which applies when n is large and p is small, with $\lambda = np$ typically small.

Exercises 3.5.

1. Let $Z \sim N(0, 1)$. Find the following probabilities: $P(Z \leq 2.34)$, $P(Z \geq -1.26)$, $P(Z \geq 1.6)$, $P(Z \leq -0.35)$, $P(1.23 \leq Z \leq 2.23)$, and $P(-1.96 \leq Z \leq 1.96)$.

2. Let $X \sim N(10, 4)$. Find the following probabilities: $P(X \leq 12.32)$, $P(X \geq 8.36)$, $P(X \geq 13)$, $P(X \leq 9)$, $P(11.22 \leq X \leq 12.38)$, and $P(8.56 \leq X \leq 13.92)$.

3. Let $Z \sim N(0, 1)$. For $\alpha = 0.1, 0.05, 0.025, 0.015.0.01$, and 0.005, find the value Z_α with the property that $P(Z \geq Z_\alpha) = \alpha$. (Note: If the exact value of Z_α cannot be obtained from Table A.2, provide the value whose probability is closest to α.)

4. Harry the Ape has a friend Pee Wee (also an adult male gorilla) who weighs less than 98.5% of all adult male gorillas in Harry's jungle. How much does Pee Wee weigh? (See Example 3.5.3.)

5. Let $Z \sim N(0, 1)$. Write out EZ^2 as an integral and use integration by parts to show that $EZ^2 = 1$. (Hint: Choose $u = z$.)

6. Let $X \sim N(\mu, \sigma^2)$. Use the moment-generating function of X to show that the variance of X is σ^2.

7. Prove Theorem 3.5.3 using the change of variable technique from calculus. Specifically, write $P(Z^2 \leq x_0) = P(-\sqrt{x_0} \leq Z \leq \sqrt{x_0}) = 2\int_0^{\sqrt{x_0}} \frac{1}{\sqrt{2\pi}} e^{-\frac{z^2}{2}} dz$. Note that, in the latter integral, the variable z is positive, so that the transformation $x = z^2$ is one-to-one and may thus be inverted. Now, execute a change of variables, letting $x = z^2$ (so that $z = x^{1/2}$ and $J = (1/2)x^{-1/2}$).

3.6 Other Continuous Models

There are a host of other continuous distributions that have proven useful as models for univariate random variables X. In this section, I will provide you with some basic information on seven parametric models that have found widespread application. All of these models will be used in discussion, examples, and exercises in the sequel, and the present section will serve as a useful compilation of their basic properties.

3.6.1 The Beta Model

In Section 3.1, we introduced the uniform distribution on the interval $(0,1)$. This distribution is the idealized model for the selection of random decimals between 0 and 1, and it is seen as the model from which random decimals between 0 and 1 are generated, at least to the extent that a computer can be programmed to mimic such a process. A somewhat broader view of the $U(0,1)$ distribution is as a model for a randomly chosen proportion. After all, a random proportion will take values in the interval $(0,1)$, and if the proportion drawn was thought to be equally likely to fall in any subinterval of $(0,1)$ of a given length, the model would clearly be appropriate. But the uniform distribution is just one member of a large and flexible family of models for random proportions. The Beta family is indexed by a pair of positive parameters α and β. The model we will denote by $Be(\alpha,\beta)$ has density function

$$f(x) = \frac{\Gamma(\alpha+\beta)}{\Gamma(\alpha)\Gamma(\beta)} x^{\alpha-1}(1-x)^{\beta-1} I_{(0,1)}(x). \tag{3.30}$$

It is an easy exercise to show, using "the trick," that the following results hold.

Theorem 3.6.1. Let $X \sim Be(\alpha,\beta)$. Then

$$EX = \alpha/(\alpha+\beta) \quad \text{and} \quad EX^2 = (\alpha+1)\alpha/(\alpha+\beta+1)(\alpha+\beta).$$

It follows that $V(X) = \alpha\beta/(\alpha+\beta)^2(\alpha+\beta+1)$.

You might wonder where you would encounter the need for modeling a random probability p. One place in which it arises quite naturally is in the area of Bayesian inference. As we will see in Chapter 9, the Bayesian approach to the estimation of an unknown parameter is to model the parameter as a random variable having a "prior" probability distribution based on the experimenter's intuition or on his solicitation of an expert's opinion. Once experimental data is collected, the Bayesian statistician derives the "posterior" distribution of the random parameter, an updated version of the prior which takes account of the data and represents the statistician's revised opinion about the unknown parameter. When the unknown parameter is a probability p, as when the observed experimental data are obtained from a binomial experiment, the Beta model provides a flexible class of possible prior distributions for the unknown p. The varied shapes of different Beta densities are shown in the figure below. The range of available shapes for the densities in the Beta family is often seen as broad enough to allow the experimenter to represent his prior knowledge satisfactorily.

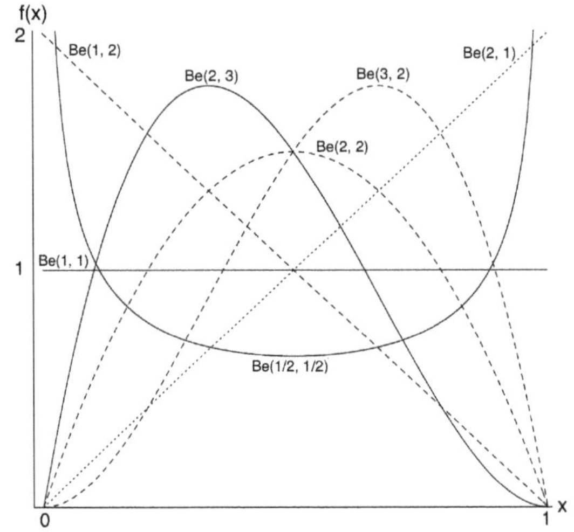

Figure 3.6.1. Graphs of the various Beta densities.

3.6.2 The Double Exponential Distribution

The double exponential ($DE(\lambda)$) density is obtained by cutting the height of the exponential density $Exp(\lambda)$ in half and reflecting it across the origin so that its support set is $(-\infty, \infty)$. When the exponential parameter λ is set equal to 1, the $DE(1)$ distribution has density

$$f(x) = (1/2)e^x \text{ if } x < 0 \text{ and } f(x) = (1/2)e^{-x} \text{ if } x \geq 0$$

The model is sometimes used as an alternative to the standard normal density when a heavier-tailed distribution is desired (for instance, if outliers beyond the typical range $(-3, 3)$ of a standard normal sample are anticipated.

Theorem 3.6.2. Let $X \sim DE(1)$. The moment-generating function of X is given by

$$m_X(t) = \frac{1}{1-t^2} \text{ for } |t| < 1.$$

Proof. First, we write $m_X(t)$ as

$$E(e^{tx}) = \int_{-\infty}^{0} e^{tx} f(x)dx + \int_{0}^{\infty} e^{tx} f(x)dx. \tag{3.31}$$

In the first integral in (3.31), change variables from x to $y = -x$; the inverse transformation is $x = -y$, the Jacobian is $J = -1$, and the range of y is $(0, \infty)$. We thus obtain

$$E(e^{tx}) = \int_{0}^{\infty} e^{-ty}(1/2)e^{-y}dy + \int_{0}^{\infty} e^{tx}(1/2)e^{-x}dx. \tag{3.32}$$

Using "the trick" on both integrals in (3.32) we obtain

$$E(e^{tX}) = (1/2)\left[\int_0^\infty e^{-y(1+t)}dy + \int_0^\infty e^{-x(1-t)}dx\right] \tag{3.33}$$
$$= \frac{1}{2(1+t)} + \frac{1}{2(1-t)}$$
$$= \frac{1}{1-t^2} \text{ for } |t| < 1.$$

It is clear that the mgf of X is undefined for $|t| \geq 1$ since, then, one of the two integrals in (3.33) is necessarily infinite. ∎

The more general version of the double exponential distribution that allows for varying location and scale is the model $DE(\mu, \lambda)$ with density function

$$f(x) = \frac{\lambda}{2}e^{-\lambda|x-\mu|}I_{(-\infty,\infty)}(x).$$

Using Theorem 3.2.2, it is easy to verify that $m_Y(t) = e^{\mu t}/(1+t^2/\lambda^2)$, for $|t| < \lambda$, is the mgf of the $DE(\mu, \lambda)$ distribution. The mean and variance of a variable $X \sim DE(\mu, \lambda)$ are $EX = \mu$ and $V(X) = 2/\lambda^2$, respectively.

3.6.3 The Lognormal Model

The lognormal distribution is a distribution on the positive real line that has proven useful as a model for lifetimes of selected animal species, for certain physical quantities such as the amount of airborne contaminants measured on a given day, and for economic variables that are generated through "multiplicative processes" (like compound interest). The model is derived via a transformation of a normal variable. If $X \sim N(\mu, \sigma^2)$, then $Y = e^X$ has a lognormal distribution. We denote the distribution of Y as $LN(\mu, \sigma^2)$. The density of the variable Y above can be shown to be

$$f(y) = (1/\sqrt{2\pi}\sigma)(1/y)e^{-(\ln y-\mu)^2/2\sigma^2}I_{(0,\infty)}(y). \tag{3.34}$$

The name of the $LN(\mu, \sigma^2)$ model comes from the simple fact that the natural logarithm of a lognormal variable Y has a normal distribution, that is, if $Y \sim LN(\mu, \sigma^2)$, then $\ln Y \sim N(\mu, \sigma^2)$. Keep in mind that, in our notation for the lognormal model, the distribution is indexed by the mean and variance of the associated normal distribution. It is easily shown (see Problem 54 at the end of this chapter) that the moments of the lognormal distribution are as claimed in the result below.

Theorem 3.6.3. Let $Y \sim LN(\mu, \sigma^2)$. Then $EY^k = \exp\{k\mu + k^2\sigma^2/2\}$.

One of the earliest references to the lognormal distribution appeared in a paper presented by the celebrated scientist Sir Francis Galton in 1879 to the Royal Society of London. At that time, various versions of the "Central Limit Theorem" were well known in scientific circles. This theorem states that under certain conditions, the distribution of the sample mean $\bar{X}$ of a random sample of size n from a fixed population would be approximately normally distributed if n was sufficiently large. Galton noted that this theorem implied that the geomantic mean $Y = (\prod_{i=1}^{n} X_i)^{1/n}$ of a random sample from a fixed population, again under suitable conditions, would have an approximately lognormal distribution when n was sufficiently large. This follows from the fact that $\ln Y$, being the arithmetic mean or average of the random variables $\{\ln X_i\}$, would typically be approximately normally distributed, and if $\ln Y$ is normal, then Y is lognormal. This observation suggests that we might expect to find the lognormal distribution arising in practice whenever the variables observed in an experiment can be thought of as having arisen from a multiplicative process, assuming the sample sizes are large.

3.6.4 The Pareto Distribution

The Pareto distribution is a model for positive random variables. It has proven useful in modeling economic variables like income. It has also been found useful as a lifetime model in applications in engineering and biostatistics, especially when a "heavy-tailed distribution" is called for. The density function of the one-parameter Pareto model is given by

$$f(x) = \alpha/x^{\alpha+1} \text{ for } x > 1, \tag{3.35}$$

where α is a positive parameter. Note that Pareto variables have an absolute lower bound, the bound being 1 in this particular version of the Pareto model.

A more useful version of the Pareto model is the two-parameter Pareto distribution with density

$$f(x) = \alpha\theta^{\alpha}/x^{\alpha+1} I_{(\theta,\infty)}(x). \tag{3.36}$$

The new parameter θ that has just been added is a positive scale parameter that permits one to stretch or shrink the one-parameter model to achieve greater flexibility. We will refer to the two-parameter Pareto models as $P(\alpha,\theta)$. The relationship between the two models above is that if $X \sim P(\alpha,\theta)$, then $X/\theta \sim P(\alpha,1)$, the one-parameter model. The model is named after Vilfredo Pareto, the Swiss-born economist of Italian heritage who first proposed the model for use in certain applications in economics and finance. In addition to providing a good fit to certain physical and financial variables, it has the additional appealing feature of being a very tractable model. The facts stated in the following theorem are easily verified.

Theorem 3.6.4. Let $X \sim P(\alpha,\theta)$. Then the cumulative distribution of X is given by

$$F(x) = 1 - \frac{\theta^{\alpha}}{x^{\alpha}}.$$

Further, if $\alpha > 1$, $EX = \frac{\alpha\theta}{\alpha-1}$ and if $\alpha > 2$, $V(X) = \frac{\alpha\theta^2}{(\alpha-1)^2(\alpha-2)}$.

Occasionally, it is convenient to use a different parameterization of the Pareto model. For example, suppose that $X \sim P(\alpha,1)$. Then the random variable $Y = X - 1$ takes values in the positive real line and has density

$$f(y) = \frac{\alpha}{(1+y)^{\alpha+1}} I_{(0,\infty)}(y). \tag{3.37}$$

The distribution of Y, in this parameterization, is referred to as the translated Pareto distribution.

3.6.5 The Weibull Distribution

In a paper published in 1939, the Swedish physicist Waloodi Weibull proposed a probability distribution, which today bears his name, as a model for the breaking strength of materials used in engineered systems. The distribution has since been used quite widely as a lifetime distribution in reliability and biostatistics. A theoretical result that suggests its potential applicability as a lifetime model is the fact that the Weibull distribution can be derived as a limiting distribution of a suitably standardized minimum of a sample of independent, identically distributed lifetimes. This suggests that the Weibull distribution might be an appropriate lifetime model for systems that tend to fail due to the failure of their weakest component.

The density function of the two-parameter Weibull distribution is

$$f(x) = \frac{\alpha x^{\alpha-1}}{\beta^{\alpha}} e^{-(x/\beta)^{\alpha}} I_{(0,\infty)}(x), \tag{3.38}$$

where $\alpha > 0$ and $\beta > 0$ are the model's "shape" and "scale" parameters, respectively. The Weibull distribution will be denoted by $W(\alpha, \beta)$. A closed form expression for the cdf of the $W(\alpha, \beta)$ model may be obtained by integration:

$$\begin{aligned} F(x) &= \int_0^x \frac{\alpha t^{\alpha-1}}{\beta^{\alpha}} e^{-(t/\beta)^{\alpha}} dt \\ &= -e^{-(t/\beta)^{\alpha}} \Big|_0^x \\ &= 1 - e^{-(x/\beta)^{\alpha}} \text{ for } x > 0. \end{aligned} \tag{3.39}$$

The family of Weibull distributions contains the exponential distribution as a special case, as is readily seen by setting $\alpha = 1$. For α near 1, the Weibull distribution can be viewed as a modest generalization of the exponential distribution which permits the failure rate to be non-constant. For arbitrary values of the shape and scale parameters, the failure rate of the Weibull distribution is easily obtained as

$$r(t) = \frac{f(t)}{1 - F(t)} = \frac{\frac{\alpha t^{\alpha-1}}{\beta^{\alpha}} e^{-(t/\beta)^{\alpha}}}{e^{-(t/\beta)^{\alpha}}} = \frac{\alpha t^{\alpha-1}}{\beta^{\alpha}}. \tag{3.40}$$

From (3.40), it is apparent that the failure rate $r(t)$ of the $W(\alpha, \beta)$ model is increasing in t if $\alpha > 1$, is a constant independent of t if $\alpha = 1$ and is decreasing in t if $\alpha < 1$. The modeling flexibility offered within the Weibull family of distributions is evident, as well, in the various shapes of the Weibull densities shown in Figure 3.6.2.

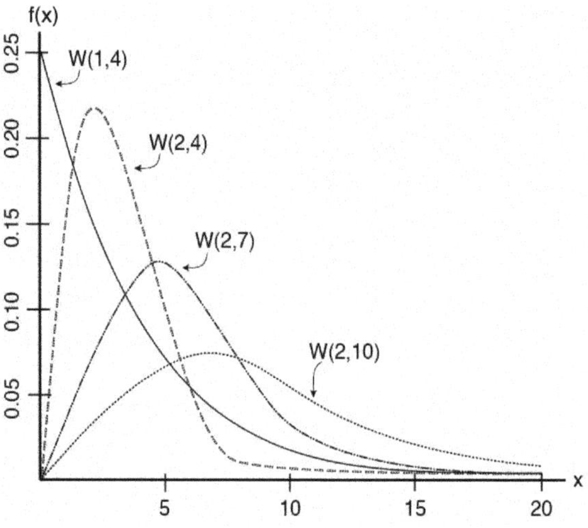

Figure 3.6.2. The $W(1,4)$, $W(2,4)$, $W(2,7)$, and $W(2,10)$ densities.

The following result provides a simple and useful connection between the general Weibull distribution and the exponential distribution.

Theorem 3.6.5. If $X \sim W(\alpha,\beta)$, then $X^\alpha \sim \Gamma(1,\beta^\alpha)$, that is, X^α has an exponential distribution with mean β^α.

Proof. We may derive the cdf of the random variable $Y = X^\alpha$ as follows.

$$P(Y \le y) = P(X^\alpha \le y) = P(X \le y^{1/\alpha}) = 1 - e^{-(y^{1/\alpha}/\beta)^\alpha} = 1 - e^{-y/\beta^\alpha}. \tag{3.41}$$

The last term in (3.41) is clearly the cdf of the $\Gamma(1,\beta^\alpha)$ distribution. ■

The mean and variance of a Weibull variable are stated below without proof.

Theorem 3.6.6. If $X \sim W(\alpha,\beta)$, then for any positive integer k, $EX^k = \beta^k\Gamma(1+\frac{k}{\alpha})$. In particular, $EX = \beta\Gamma(\frac{\alpha+1}{\alpha})$ and $V(X) = \beta^2[\Gamma(1+\frac{2}{\alpha}) - \Gamma^2(1+\frac{1}{\alpha})]$.

3.6.6 The Cauchy Distribution

The Cauchy model is ubiquitous in the statistical literature. It has a number of properties that make it the most widely cited and best-known model of its kind. To the naked eye, the graph of the Cauchy model with median m and scale parameter θ looks a good deal like a normal curve with mean value $\mu = m$ and some particular finite variance. In reality, the Cauchy density has heavier tails than any comparable normal density, that is, the Cauchy model assigns greater probability to values far from the center of the distribution.

We will denote this distribution by $C(m,\theta)$; its density function is given by

$$f(x) = \frac{1}{\pi\theta\left[1+\left(\frac{x-m}{\theta}\right)^2\right]} I_{(-\infty,\infty)}(x). \tag{3.42}$$

Interestingly, the integral

$$I = \int_{-\infty}^{\infty} |x| \frac{1}{\pi\theta\left[1+\left(\frac{x-m}{\theta}\right)^2\right]} dx$$

diverges, that is, I is equal to ∞. From this, it follows that the mean of the Cauchy distribution does not exist. Among other things, this implies that the famous Central Limit Theorem does not apply to random samples from a Cauchy distribution. In a nutshell, if $\bar{X}_n$ represents the mean of a random sample of size n from a Cauchy distribution, then $\bar{X}_n$ does not converge to the central value m as the sample size n tends to ∞. As we shall see in Chapter 5, the Central Limit Theorem says that, under certain mild conditions, the mean $\bar{X}_n$ of a random sample will not only converge to the population mean μ, but the probability distribution of $\bar{X}_n$ will be approximately normal when n is sufficiently large. In contrast to this, the mean $\bar{X}_n$ of a random sample from a Cauchy distribution will have a Cauchy distribution regardless of the size of n. The Cauchy model is the quintessential example of a probability distribution that does not satisfy the conditions under which the Central Limit Theorem holds. More on this later.

A special case of the Cauchy distribution arises in a certain geometric situation. The "standard Cauchy distribution" $C(0,1)$ with median 0 and scale parameter 1, has density function

$$f(x) = \frac{1}{\pi[1+x^2]} I_{(-\infty,\infty)}(x). \tag{3.43}$$

The integral of the function in (3.43) over any subinterval (a,b) of the real line may be obtained in closed form. Indeed, the cumulative distribution function F of the $C(0,1)$ model is given by

$$F(x) = \frac{1}{\pi}\left[\arctan x + \frac{\pi}{2}\right] \text{ for } x \in (-\infty,\infty),$$

where $\arctan x$ is the angle α, measured in radians, in the interval $(-\pi/2,\pi/2)$ for which $\tan\alpha = x$. This distribution is the correct model for the tangent X of the random angle $\alpha \in (-\pi,\pi)$ obtained when a well-balanced spinner is used to select a random uniformly distributed point on the circumference of a circle, as pictured in the figure below.

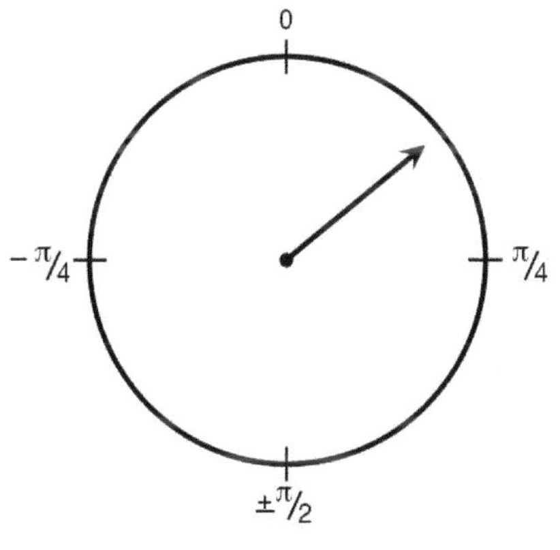

Figure 3.6.3. A spinner yielding a uniformly distributed random angle α.

A coincidental connection between the Cauchy and normal distributions is the following: If X_1, X_2 are independent random variables drawn from a $N(\mu,\sigma^2)$ population, and $Y = (X_1+X_2-2\mu)/|X_1-X_2|$, then Y has the standard Cauchy distribution. Later, we will identify the distribution of Y as a particular member of the family of "t distributions." More specifically, in Chapter 4, Y will be shown to have a t distribution with 1 degree of freedom, a special case of the fact that for random samples from a normal population, the fraction $Y = \sqrt{n}(\bar{X} - \mu)/s$ has a t-distribution, where $\bar{X}$ and s are, respectively, the mean and standard deviation of a random sample of size n from a $N(\mu, \sigma^2)$ population. A further connection: If X_1, X_2 are independent N(0,1), then X_1/X_2 has a Cauchy distribution.

3.6.7 The Logistic Model

A widely used model in applied statistical work is the logistic model. While many applications of the model occur in the context of regression analysis, the distribution is of interest in its own right. We will discuss the general probability model first and then briefly examine its important role in regression problems, a topic that we will take up in more detail in Chapter 11.

The logistic model is a probability model on the real line that is determined by two parameters, the mean $\mu \in (-\infty, \infty)$ of the distribution and the positive scale parameter β. The density of the logistic model is given by

$$f(x) = \frac{1}{\beta} \frac{e^{-(x-\mu)/\beta}}{[1 + e^{-(x-\mu)/\beta}]^2} I_{(-\infty,\infty)}(x). \tag{3.44}$$

Our notation for the logistic distribution with parameters μ and β will be $L(\mu, \beta)$. The logistic density function is integrable, and the integral of this density over any subinterval of the real line may be obtained in closed form. In particular, the cumulative distribution function $F(x)$ of $X \sim L(\mu, \beta)$ is given by

$$F(x) = \frac{1}{1 + e^{-(x-\mu)/\beta}} \text{ for } x \in (-\infty, \infty). \tag{3.45}$$

We state the following facts about the logistic model without proof.

Theorem 3.6.7. Let $X \sim L(\mu, \beta)$. Then the moment-generating function of X is given by $m_X(t) = e^{\mu t}\Gamma(1 - \beta t)\Gamma(1 + \beta t)$ for $|t| < \frac{1}{\beta}$. Further, the mean and variance of X are given by $EX = \mu$ and $V(X) = \frac{\pi^2\beta^2}{3}$.

Regression models are used in statistics as tools for understanding and explaining the possible relationship between a random variable Y and one or more "explanatory variables" that may be related to Y. A simple example would a model that might postulate that, subject to certain random fluctuations, a man's weight is linearly related to his height. We might well believe that this is true, on the average, since the taller a man is, the more weight he is likely to carry. We might think this even though we know that some short men are quite heavy and some tall men are very thin. Looking at the overall picture, it seems reasonable to postulate a model for the relationship between a man's weight Y and his height X as

$$Y = a + bX + \varepsilon, \tag{3.46}$$

where ε represents a random error which would account for random deviations above or below the "predicted weight" for a man whose height is X. The relationship in (3.46) stipulates that a straight line describes the average relationship between X and Y. The errors are often modeled as independent $N(0, \sigma^2)$ variables. There is much room for generalization here. For example, one could include other "independent variables" which might influence the value of Y. In general, multiple linear regression models provide a large class of flexible tools for exploring how a particular kind of measurement Y might be related to one or several X variables. But there are regression problems of considerable interest and importance which require a model quite different than that in (3.46). Suppose the "response variable" Y is a binary variable that represents the presence or absence of a particular trait (like lung cancer). In a regression setting, one might be interested in the possible relationship between Y and a continuous measurement X that records the number of cigarettes smoked per day. We know that a relationship like that in (3.46) is not likely to be very helpful. The model which is generally entertained in problems of this type recognizes that Y is actually a Bernoulli random variable, that is,

$$Y|X = x \sim B(1, p(x)),$$

so that what we are really trying to discover is how the value of the variable X influences the probability $P(Y = 1)$. The logistic curve is often used to model the relationship. For example, one might postulate that the relationship between X and Y has the following form:

$$p(x) = P(Y = 1|X = x) = \frac{1}{1 + e^{-(a+bx)}} \text{ for } x \in (-\infty, \infty). \tag{3.47}$$

Note that the logistic curve in (3.47) has the property that it takes only values between 0 and 1, tending to 0 as $x \to -\infty$ and tending to 1 as $x \to \infty$. It is thus a suitable model for describing the probability that $Y = 1$ as a function of the explanatory variable X when $P(Y = 1)$ is well modeled as an increasing function of X. When the data $\{(X_i, Y_i), i = 1, 2, \ldots, n\}$ are collected on a random sample of individuals, the parameter vector (a, b) may be estimated from data, leading to a particular logistic curve that provides insight into the relationship between smoking and lung cancer.

Remark 3.6.1. There are, of course, many other continuous models worthy of discussion. Some, like the t-distribution and the F-distribution arise in very specific statistical problems and will be discussed in detail when these problems are treated in later sections. Other models (like the inverse Gaussian distribution, extreme-value distributions, the Laplace distribution and the non-central chi square, t and F distributions) arise either as modeling alternatives or in particular statistical investigations, but are not utilized in this book and will thus not be specifically discussed. Readers who encounter the need to delve more deeply into the models treated in this chapter or to explore other continuous models not covered here should consult the comprehensive monograph by Johnson, Kotz, and Balakrishnan (1995) for detailed treatments.

Exercises 3.6.

1. Verify the claims in Theorem 3.6.1.

2. Let $X \sim DE(\mu, \lambda)$. Using the mgf of X, verify that $EX = \mu$ and $V(X) = \frac{2}{\lambda^2}$.

3. Suppose $Y \sim LN(\mu, \sigma^2)$. Find EY. (Hint: Write Y as e^X, where $X \sim N(\mu, \sigma^2)$, and use what you know about the mgf of the normal distribution.)

4. Verify the three claims in Theorem 3.6.4.

5. If a sample of independent variables $X_1, X_2, \ldots, X_n$ were drawn from a common distribution F, $X_{(1)} = \min\{X_1, X_2, \ldots, X_n\}$, that is, and $X_{(1)}$ stands for the smallest X, then the cdf $F_{1:n}(x)$ of $X_{(1)}$ would be given by $1 - P(X_1 > x, X_2 > x, \ldots, X_n > x) = 1 - (1 - F(x))^n$. Show that

the minimum of a sample of independent Pareto variables (that is, with $F = P(\alpha, 1)$) also has a Pareto distribution, but with a new parameter.

6. Prove Theorem 3.6.6 (Hint: Write X^k as $(X^\alpha)^{k/\alpha}$ and apply Theorem 3.6.5.)

3.7 Chapter Problems

1. The "More-Likely-Than-Not" distribution on the unit internal has density function $f(x) = kx^2(1-x)$ for $0 < x < 1$. It is the "go to" distribution that TV weatherpersons use when they need to state tomorrow's probability of rain on a given day during the months November–March. They draw a random decimal from this distribution and announce it as the required probability. (a) Find the constant k for which $f(x)$ is a valid density. (b) Compute the mean of the "MLTN" distribution. (c) Compute the variance of the distribution.

2. The probability X that a randomly chosen NFL football team will win its weekend game is a random variable taking values in the interval $[0,1]$ and having a density function of the form
$$f(x) = kx(1-x)I_{[0,1]}(x).$$
(a) For what value of k is $f(x)$ a legitimate density? (b) Since X has a density that is symmetric about $x = 1/2$, its mean is $EX = 1/2$ (which makes sense, since half the teams win and half of them lose). Find σ_X^2, the variance of X.

3. Suppose that the random variable X representing the probability that a random college student voted in the last student-body-president election has the density function $f(x) = kx^2(1-x)I_{(0,1)}(x)$. (a) Show that the unique value of k for which $f(x)$ is a legitimate density is $k = 12$. (b) Evaluate $P(.2 < X < .3)$. (c) Calculate the mean and variance of X, and also find $E(X/(1-X))$.

4. Find the value of k for which the function $f(x) = kx(2-x)I_{(0,2)}(x)$ is a valid probability density function. If X is a random variable having this density, find the mean and variance of X.

5. Let $X \sim U(0,1)$. X will partition the interval $(0,1)$ into two line segments. Find the probability that the shorter of these two segments is at least $1/3$ as long as the longer of the two segments.

6. Suppose the number b is chosen at random according to the uniform distribution of the interval $(-3,3)$. What's the probability that the quadratic equation $x^2 - bx + 1 = 0$ has two real roots?

7. Let X be a positive, continuous random variable with density function $f(x)$ and cumulative distribution function $F(x)$. Show that EX may be computed as
$$EX = \int_0^\infty \bar{F}(x)dx, \tag{3.48}$$
where $\bar{F}(x) = 1 - F(x)$. (Hint: Start by expressing $\bar{F}(x)$ as the integral $\int_x^\infty f(t)dt$.)

8. Let $X \sim Exp(\lambda)$. Compute EX using the identity in (3.48).

9. The identity in (3.48) applies to both continuous and discrete positive random variables. Obtain the mean of a geometric random variable ($X \sim G(p)$) by applying this identity.

10. Let X be a continuous random variable with support set equal to the whole real line. Show that
$$EX = \int_0^\infty P(X > x)dx - \int_0^\infty P(X < -x)dx. \tag{3.49}$$

11. Show that for an arbitrary non-negative random variable $X \sim F$,
$$EX^n = \int_0^\infty nx^{n-1}P(X > x)dx. \tag{3.50}$$
(Hint: Using (3.48), write $EX^n = \int_0^\infty P(X^n > x)dx = \int_0^\infty \bar{F}(x^{1/n})dx$, where $\bar{F}(x) = 1 - F(x)$. Then, change variables, letting $y = x^{1/n}$.)

12. Let $\bar{F}(x) = P(X > x)$. Show: for an arbitrary non-negative random variable X,
$$EX^2 = \int_0^\infty \int_0^\infty \bar{F}(x+y)dxdy. \tag{3.51}$$
(Hint: In the integral above, change variables to $u = x+y$ and $v = y$, and then apply the identity in (3.50).)

13. Determine whether or not there exist constants k for which the following functions are probability density functions. If there is, identify k.

    (a) $f(x) = (k/x)I_{(0,1)}(x)$
    (b) $f(x) = (k/(1-x)^{1/2})I_{(0,1)}(x)$
    (c) $f(x) = (k/(1+x)^{3/2})I_{(0,\infty)}(x)$.
    (d) $f(x) = (k/(1+x))I_{(0,\infty)}(x)$.

14. Let X and Y be continuous random variables obtained from two independent experiments. The density of the sum $Z = X+Y$ may be obtained from the following integral, known as the convolution of the densities of X and Y:
$$f_Z(z) = \int_{-\infty}^\infty f_X(x)f_Y(z-x)dx. \tag{3.52}$$
The intuition behind the integral is the fact that the weight associated with any value z of Z must clearly be an average of the weights on all x and y values which add up to z. Suppose that the X and Y of interest are independent exponential variables, both with the distribution $Exp(\lambda)$. Obtain the density of $Z = X+Y$. (Note: Here, the integrand in (3.52) is zero if x exceeds z.)

15. Suppose that X is a continuous random variable with density function given by $f(x) = (3/4)(1-x^2)I_{(-1,1)}(x)$. Obtain the cdf of X, and calculate its mean and variance.

16. Suppose X is a random variable with cumulative distribution function
$$F(x) = 1 - e^{-x^2}, \quad \text{for } x \geq 0.$$
(a) Evaluate $P(1 < X < 2)$. (b) Obtain the density function of X. (c) Identify the value m such that $P(X < m) = 1/2$.

17. Show that the function $F(x) = x^2/(x^2+1)$ is a legitimate cumulative distribution function for a random variable taking values in the positive real line.

18. A point is drawn at random from the interior of the unit circle C in the plane, that is, from the circle whose center is the point $(0,0)$ and whose radius is 1. Let X be the distance of the chosen point from the center of the circle. Assume that the probability of the chosen point being in any given region in C is proportional to the area of that region. Find the cdf of the variable X.

19. Suppose that the random variable X has density function given by

$$f(x) = xI_{(0,1]}(x) + (2-x)I_{(1,2)}(x).$$

Obtain the cumulative distribution function $F(x)$ of X. Use it to calculate the probabilities $P(.5 < X < 1.25)$ and $P(X > .8)$.

20. Suppose $X \sim U(0,1)$, and define $Y = [NX]^+$, where N is a fixed positive integer and $[x]^+$ is the smallest integer greater than or equal to x. Obtain the cdf of Y. This distribution is referred to, not too surprisingly, as the discrete uniform distribution.

21. Suppose $X \sim U(0,1)$. Find the cumulative distribution function of the random variable $Y = \sqrt{X}$. (Hint: Write $P(Y \le y)$ as $P(X \le y^2)$, for $0 < y < 1$.)

22. Suppose $X \sim U(-1,1)$. Find the cumulative distribution function of the random variable $Y = X^2$.

23. Suppose $X \sim U(-\pi/2, \pi/2)$. Obtain the cdf $F(y)$ and the pdf $f(y)$ of the random variable $Y = \tan X$.

24. Let X be a positive random variable with cdf F and density f. Distributions whose support set is the positive real line are good candidates, and are often utilized, as lifetime models in the field of Engineering Reliability. The function

$$R(x) = -\ln[\bar{F}(x)] \tag{3.53}$$

where $\bar{F}(x) = 1 - F(x)$, is called the *hazard function* corresponding to X or F. Its derivative $r(x)$ is called the *failure rate* or *hazard rate* of X. In Section 3.3, the failure rate of X was defined as $r(x) = f(x)/(1-F(x))$. Show that $\frac{d}{dx}R(x) = r(x)$.

25. Refer to Problem 24 above. The class of lifetime distributions having an increasing failure rate (that is, IFR class) is of special interest in engineering applications, as such models are appropriate when dealing with manufactured items that deteriorate as they age, an assumption that is very often considered tenable. Given F, f, R and r as above, show that r is increasing if and only if the hazard function R is convex.

26. Refer to Problems 24 and 25 above. Another nonparametric model of interest in Engineering Reliability is the "New is better than used" or NBU class. This class consists of models whose distributions obey the inequalities

$$\bar{F}(x+y) \ge \bar{F}(x)\bar{F}(y) \text{ for all } x > 0 \text{ and } y > 0, \tag{3.54}$$

where $\bar{F}(x) = 1 - F(x)$. A function g on $(0,\infty)$ is superadditive if it satisfies the inequalities

$$g(x+y) \leq g(x) + g(y) \text{ for all } x > 0 \text{ and } y > 0. \tag{3.55}$$

Show that F has the NBU property if and only if the hazard function $R(x)$ in (3.53) is superadditive.

27. Suppose U and V are $\Gamma(1,\mu)$ variables that take values independently of each other. Find the distribution of the variable $W = \min(U,V)$, and identify it by name. (Hint: Note that $W > w$ if and only if both $X > w$ and $Y > w$.) Can you identify the distribution of W if it were defined as the minimum of n independent $\Gamma(1,\mu)$ variables?

28. An energetic old Statistics Professor runs the Davis 10K race every year, come rain or come shine. He's never (yet) taken more than 2 hours to do it, but he's also never run it in less than 1 hour. A friend he runs with claims that the distribution with density

$$f(x) = 2(x-1) \qquad \text{for } 1 < x < 2,$$

is an excellent probability model for the time he'll take (in hours) on a random 10K run. (a) Show that $f(x)$ is a legitimate density, that is, that it integrates to 1. (b) Calculate EX, the professor's expected time on a random 10K run. (c) For $x_0 \in (1,2)$, obtain an expression for the cdf $F(x_0) = P(X \leq x_0)$ of the variable X with the density $f(x)$ above.

29. The lifetime of a 100-watt fluorescent light bulb is as exponential random variable with mean equal to 12000 hours. If a money-back guarantee is to be offered to users of such light bulbs lasting less than 1000 hours, what percentage of the light bulbs sold will be eligible for a full refund?

30. Let $X \sim \Gamma(1,\mu)$, the exponential distribution with mean μ. Show that, for any positive integer n, $EX^n = n!\mu^n$.

31. Prove the memoryless property of the exponential distribution, that is, prove that if $X \sim Exp(\lambda)$, then

$$P(X > x+t | X > x) = P(X > t). \tag{3.56}$$

A verbal interpretation of this property is that if an item having an exponentially distributed lifetime has survived x units of time, then its chances of surviving an additional t units of time is that same as when the item was new.

32. This exercise is aimed at showing that the exponential distribution is the *unique* continuous model with support on $(0,\infty)$ that enjoys the memoryless property defined in (3.56). Let's use the notation: $\bar{F}(x) = P(X > x)$. Then (3.56) may be written as

$$\bar{F}(x+t) = \bar{F}(x)\bar{F}(t). \tag{3.57}$$

Take (3.57) as given. The fact that $\bar{F}(x) = 1 - e^{-\lambda x}$ for some $\lambda > 0$ is to be proven.

(a) Show that for any $c > 0$ and positive integers m and n, (3.57) implies that

$$\bar{F}(nc) = [\bar{F}(c)]^n \text{ and } \bar{F}(c) = [\bar{F}(c/m)]^m. \tag{3.58}$$

(b) Show that $0 < \bar{F}(x) < 1$. (Hint: In other words, both of the identities $\bar{F}(x) = 1$ and $\bar{F}(x) = 0$ for any $x > 0$ contradict our assumptions.)

(c) Justify the claim that there exists $\lambda > 0$ such that $\bar{F}(1) = e^{-\lambda}$.

(d) Show that (3.58) implies that, for any positive integers m and n, $\bar{F}(n/m) = e^{-\lambda n/m}$, that is, that $\bar{F}(x) = e^{-\lambda x}$ for every rational non-negative number x.

(e) Use the right continuity of $\bar{F}$ (which follows from the right continuity of F) to conclude that $\bar{F}(x) = e^{-\lambda x}$ for all $x > 0$.

33. Let $X \sim \Gamma(\alpha, \beta)$. Derive an expression for $r(t) = f(t)/(1-F(t))$, the failure rate of X, and show that $r(t)$ is strictly increasing if $\alpha > 1$, is constant if $\alpha = 1$ and is strictly decreasing if $\alpha < 1$.

34. Show that $\Gamma(1/2) = \sqrt{\pi}$. (Hint: Write $\Gamma(1/2) = \int_0^\infty x^{-1/2} e^{-x} dx$ and then change variables to $y = \sqrt{2x}$. Apply the trick to conclude that $\Gamma(1/2)/\sqrt{\pi} = 1$.)

35. Suppose $X \sim Exp(\lambda)$. Show that a plot of $\ln(1 - F(x))^{-1}$ against x will be a straight line through the origin with slope λ. The empirical version of such a plot may be used to investigate whether or not experimental data appears to support the assumption that it was drawn from an exponential distribution.

36. Let X be a negative binomial variable; more specifically, let $X \sim NB(r, p)$. Recall that X represents the number of Bernoulli trials needed to obtain the rth success when the probability of success in a single trial is p. Now as $p \to 0$, the mean of X, that is, r/p, tends to infinity. But the "standardized" variable pX has mean r for all values of p, and thus should have a stable limit as $p \to 0$. Show that the distribution of pX converges to the $\Gamma(r, 1)$ distribution as $p \to 0$. (Hint: Use the continuity theorem, that is, show that $m_{pX}(t) \to m_Y(t)$ as $p \to 0$, where $Y \sim \Gamma(r, 1)$. This result provides a useful approximation for negative binomial probabilities when p is small (particularly in view of the connection between the gamma distribution and the Poisson process which provides a closed form expression for the approximation).

37. Prove that the density of the $\Gamma(\alpha, \beta)$ distribution is unimodal for arbitrary $\alpha > 0$ and $\beta > 0$.

38. Show that if $\alpha > 1$, the unique mode of the $\Gamma(\alpha, \beta)$ distribution is $(\alpha - 1)\beta$.

39. Let $X \sim \Gamma(1, \beta)$. Find the first, second, and third quartiles of X, that is, the values x_p for which $P(X < x_{.25}) = .25$, $P(X < x_{.50}) = .5$, and $P(X < x_{.75}) = .75$.

40. Let $X \sim \Gamma(\alpha, \beta)$, with $\alpha > 2$. Show that the random variable $Y = 1/X$ has mean $1/\beta(\alpha - 1)$ and variance $1/\beta^2(\alpha - 1)^2(\alpha - 2)$.

41. The diameter of a large pizza at Original Steve's restaurant in Davis, CA, is normally distributed with mean $\mu = 14$ inches and a standard deviation of $\sigma = 1$ inch. Gorilla Gonzales, star tackle on the UC Davis football squad, is delighted to have gotten an extraordinarily large pizza—one that's larger that 99.4% of all large pizzas they make. What's the diameter of this pizza?

42. In Section 3.5, we took it on faith that the integral of the $N(0, 1)$ "density function" $f(x) = \frac{1}{\sqrt{2\pi}} e^{-\frac{x^2}{2}} I_{(-\infty,\infty)}(x)$ is equal to 1. Prove that this is the case. (Hint: Let $I = \int_{-\infty}^\infty e^{-x^2/2} dx$, and write I^2 as $\int_{-\infty}^\infty \int_{-\infty}^\infty e^{-(x^2+y^2)/2} dxdy$. Then, execute a change of variables (to polar coordinates), replacing x by $r \sin\theta$, y by $r \cos\theta$ and $dxdy$ by $rdrd\theta$. Show that the resulting integral $I^2 = \int_0^\infty \int_0^{2\pi} re^{-r^2/2} d\theta dr$ is equal to 2π.)

43. My paper boy Ivan (the Terrible) Tosser manages to toss the daily newspaper onto my front porch only 25 % of the time. The rest of the time I have to go on an annoying scavenger hunt in my pajamas! What's the probability that I will find my newspaper on my porch at least 10 times in the month of November?

44. The annual rainfall X in Sacramento, California, is a normal random variable with a mean $\mu = 20.27$ inches and a standard deviation $\sigma = 4.75$ inches. Calculate the following probabilities (a) $P(16 \leq X \leq 21)$, (b) $P(X \geq 15)$, (c) $P(12 \leq X \leq 17)$, and (d) $P(X \leq 22)$. If the rainfall in Sacramento in a given year is below 10 inches, it is considered to be a drought year, and special water conservation measures are set in place. (e) What is the probability that next year will be a drought year?

45. If a normal random variable X with mean $\mu = 10$ will exceed the value 15 with probability .1, what is the variance of X (approximately)?

46. For obvious reasons, the probability distribution with density

$$f(x) = \frac{2}{\sqrt{2\pi}} e^{-\frac{x^2}{2}} I_{(0,\infty)}(x)$$

is called a folded-normal or half-normal distribution. Its density is obtained by folding the standard normal density over across the Y-axis so that the mass over the negative real line is 0 while the mass over the positive real line is 1. (a) Obtain the mean of the folded normal distribution. (b) Show, by a change of variable argument, that $Y = X^2$ has the $\Gamma(1/2, 2)$ distribution. Use this fact, together with part (a), to obtain the variance of the folded normal distribution.

47. Suppose that the heights of Irish males in their twenties is normally distributed with mean $\mu = 68$ inches and standard deviation $\sigma = 3$ inches. (a) What proportion of that population is over six feet tall? (b) What is this distribution of heights if the heights are measured in centimeters?

48. Give a direct proof, using the change of variable technique on the integral of the $N(\mu, \sigma^2)$ density, that if $X \sim N(\mu, \sigma^2)$, then $\frac{X-\mu}{\sigma} \sim N(0, 1)$.

49. Suppose that $X \sim \Gamma(\alpha, \beta)$. Show that if $\alpha \to \infty$ and $\beta \to 0$ at rates that assure that $\alpha\beta \to \mu$, where μ is a positive constant, then the distribution of X converges to the distribution of Y, a degenerate random variable that takes the value μ with probability 1. (Hint: Show that $m_X(t) \to e^{\mu t}$ for all real numbers t under that stated conditions on the model parameters.)

50. Derive the moment-generating function of the $N(\mu, \sigma^2)$ distribution directly by evaluating the integral

$$m_X(t) = \int_{-\infty}^{\infty} e^{tx} \frac{1}{\sqrt{2\pi}\sigma} e^{-\frac{(x-\mu)^2}{2\sigma^2}} dx.$$

51. The Weibull distribution is perhaps the most widely used parametric model for failure-time data in the field of engineering reliability. The density function of the Weibull model is given by

$$f(x) = \frac{\alpha x^{\alpha-1}}{\beta^\alpha} e^{-(x/\beta)^\alpha} I_{(0,\infty)}(x),$$

where $\alpha > 0$, $\beta > 0$. Denote this distribution as $W(\alpha, \beta)$. If $X \sim W(2, 3)$, calculate the probability $P(1 < X < 2)$.

52. A commonly used form of the one-parameter Pareto model has the density function

$$f(x) = \theta/[1+x]^{\theta+1} \text{ for } 0 < x < \infty,$$

where the parameter θ may take on any positive value. a) Let X be a variable having the Pareto distribution above. If $0 < \theta \leq 1$, then the expected value of X is ∞. Assuming that $\theta > 2$, evaluate EX and $V(X)$. (Hint: It's advisable to find first $E(1+X)^i$ for $i = 1$ and 2, and then obtain EX and $V(X)$ from these expressions.) b) Obtain an expression for $F(x)$, the cumulative distribution function of X, for a fixed but arbitrary value $x > 0$. (Hint: For any $a \neq -1$, $\int [1+x]^a = [1+x]^{a+1}/[a+1] + c$.)

53. The usual parameterization of the one-parameter Pareto density is as given in (3.35), namely,

$$f(x) = (\alpha/x^{\alpha+1})I_{(1,\infty)}(x).$$

(a) The kth moment EX^k of a Pareto variable is infinite if $k \geq \alpha$. Find an expression for EX^k when $k < \alpha$. (b) Suppose that X has the Pareto distribution above. Find the distribution function of the random variable $Y = \ln(X)$, and identify it by name. (Hint: Start by writing $F_Y(y) = P(Y \leq y) = P(\ln X \leq y)$.)

54. Suppose that $Y \sim LN(\mu, \sigma^2)$. Show that $EY^k = e^{k\mu + k^2\sigma^2/2}$.

55. Suppose $X \sim W(\alpha, \beta)$. Show that a plot of $y = \ln[\ln(1 - F(x))^{-1}]$ against $\ln x$ will be a straight line with slope α and y-intercept $-\alpha\ln(\beta)$. The empirical version of such a plot may be used to investigate whether or not experimental data appears to support the assumption that it was drawn from a Weibull distribution.

56. Let X be a logistic random variable, that is, let $X \sim L(\mu, \beta)$. Show that the mean and variance of X are given by

$$EX = \mu \text{ and } V(X) = \frac{\pi^2\beta^2}{3}.$$

57. Let X be a logistic random variable, that is, let $X \sim L(\mu, \beta)$. Show that the moment-generating function of X is given by $m_X(t) = e^{\mu t}\Gamma(1 - \beta t)\Gamma(1 + \beta t)$ for $|t| < \frac{1}{\beta}$.

58. Let $X \sim L(\mu, \beta)$. Show that the failure rate of X is given by $r(t) = F_X(t)/\beta$.

59. Show that if $\alpha > 1$ and $\beta > 1$, the density of the Beta (α, β) distribution is unimodal.

60. The density function of the general Cauchy distribution $C(m, \theta)$ was given in (3.39) as

$$f(x) = \frac{1}{\pi\theta\left[1 + \left(\frac{x-m}{\theta}\right)^2\right]} I_{(-\infty,\infty)}(x).$$

Let $X \sim C(m, \theta)$. Show that the parameter m of this distribution is the median of X, that is $P(X > m) = 1/2 = P(X < m)$, and that the values $m - \theta$ and $m + \theta$ are the lower and upper quartiles of X, that is, $P(X < m - \theta) = 1/4$ and $P(X > m + \theta) = 1/4$.

4

Multivariate Models

4.1 Bivariate Distributions

While a single random variable might be the sole focus in a particular experiment, there are many occasions in which two or more variables of interest occur together and are recorded simultaneously. Usually, the variables we record together are related in some natural way, and the univariate models (pmfs or pdfs) that we use to describe the variables separately are not sufficient to model the joint behavior of the variables. Take, for example, the recording of the heights and weights of a random collection of individuals. You might expect that, on the average, there is some sort of positive relationship between a person's height and weight (that is, on average, tall people are heavier than short people, notwithstanding the fact that there are plenty of heavy short people and skinny tall people around). The distribution of heights in the sampled population and the distribution of weights in that population don't carry all the relevant information about the random pair (height, weight). A *multivariate probability model* is a model describing the stochastic behavior of an n-dimensional vector $(X_1,X_2,\ldots,X_n)$ of random variables. In this section, we will treat *bivariate probability models* which describe the joint behavior of pairs (X,Y) of random variables. From such models, we will be able to calculate the values of probabilities like $P(X \leq x, Y \leq y)$, values which generally cannot be obtained from the individual models for X and Y alone. Consider the probability model shown below for the discrete random pair (X,Y).

$(X,Y)=(x,y)$	$(0,0)$	$(0,1)$	$(0,2)$	$(1,1)$	$(1,2)$	$(2,2)$
$p(x,y)$	$1/10$	$2/10$	$1/10$	$3/10$	$1/10$	$2/10$

Table 4.1.1. The probability mass function of a discrete random pair (X,Y).

It is apparent from the table above that the notation $p(x,y)$ for the joint pmf of the variables X and Y will be ambiguous in many situations; without indicating which variable is which, $p(a,b)$ might represent either $P(X=a,Y=b)$ or $P(Y=a,X=b)$. A more cumbersome, but unambiguous, notation for bivariate probabilities is the following. Let

$$p_{X,Y}(x,y) = P(X=x,Y=y).$$

In this and succeeding chapters, I will generally use subscripts in bivariate and univariate pmfs and pdfs, omitting them only when the context makes the meaning of the unsubscripted functions clear.

The individual probability distributions of each of the variables in the random pair (X,Y) are referred to as these variables *marginal distributions*. For a discrete random pair, the third axiom of probability (see Section 1.3) provides a vehicle for computing individual probabilities for either X or Y. For example, these calculations result in the individual probability mass functions of the variables X and Y. The marginal pmfs of the variables X and Y are defined as

$$P(X=x)=\sum_{\text{all } y} p_{X,Y}(x,y) \tag{4.1}$$

and

$$P(Y=y)=\sum_{\text{all } x} p_{X,Y}(x,y). \tag{4.2}$$

For (X,Y) with joint pmf in Table 4.1.1, we the marginal pmfs of the variables X and Y are given below:

$X=x$	0	1	2
$p_X(x)$	4/10	4/10	2/10

$Y=y$	0	1	2
$p_Y(y)$	1/10	5/10	4/10

Table 4.1.2. Marginal pmfs for random variables X, Y in Table 4.1.1.

From the calculations above, it is clear that the individual (univariate) distributions of the variables X and Y can be recovered from the bivariate distribution of the random pair (X,Y). The fact that this is not a two-way street should also be evident. If the variables X and Y happened to take on their values independently, we would have that $P(X=x,Y=y)=P(X=x)P(Y=y)$, so that the bivariate distribution for X and Y would have the pmf in Table 4.1.3 below.

$(X,Y)=(x,y)$	(0,0)	(0,1)	(0,2)	(1,0)	(1,1)	(1,2)	(2,0)	(2,1)	(2,2)
$p_{X,Y}(x,y)$	1/25	1/5	4/25	1/25	1/5	4/25	1/50	1/10	2/25

Table 4.1.3. The joint pmf of independent variables X and Y with marginal pmfs in Table 4.1.2.

The bivariate model in Table 4.1.3 will also yield the marginal models for X and Y shown in Table 4.1.2. Indeed, there are uncountably many bivariate distributions which will yield the marginal pmfs in Table 4.1.2, so the specific bivariate model governing the behavior of the pair (X,Y) carries much more information than do the marginal probability models for X and Y alone. For example, the model in Table 4.1.1 indicates that, while it is possible for X to take the value 1 and Y to take the value 0, the probability that they take these values simultaneously is zero.

A particularly useful piece of information that a bivariate model contains concerns the conditional behavior of one variable given that the value of the other variable is known. The probabilities $P(X=x|Y=y)$ and $P(Y=y|X=x)$ may both be derived from the joint pmf of X and Y. The conditional distributions of X given $Y=y$ and of Y given $X=x$ have pmfs given by

$$p_{X|Y}(x|y)=P(X=x|Y=y)=\frac{p_{X,Y}(x,y)}{p_Y(y)} \tag{4.3}$$

and

$$p_{Y|X}(y|x) = P(Y = y|X = x) = \frac{p_{X,Y}(x,y)}{p_X(x)}. \tag{4.4}$$

For the bivariate model in Table 4.1.1, we may calculate

$$P(X = 1|Y = 2) = \frac{p_{X,Y}(1,2)}{p_Y(2)} = \frac{1}{4}. \tag{4.5}$$

You can see that the subscripts on the pmfs in (4.5) are necessary for clarity, since, for example, the probability $p(2)$ is ambiguous without the subscript. The complete conditional pdf for X, given $Y = 2$, is shown in Table 4.1.4. below.

$X = x\|Y = 2$	0	1	2
$p_{X\|Y}(x\|2)$	1/4	1/4	1/2

Table 4.1.4. The conditional pmf of X, given $Y = 2$, for (X,Y) in Table **??**

The concept of "mathematical expectation" also arises in problems involving bivariate or multivariate distributions. It has the same interpretations as in the univariate cases treated in Chapters 2 and 3, namely, as the center of gravity of a pmf or as a long-run average value of the variable whose expectation is evaluated. The new wrinkle that arises here is that we will often need to take the expectation of functions of more than one variable. Let $g(x,y)$ be such a function. Our working definition of the expected value of $g(X,Y)$ of the discrete random pair (X,Y) will be

$$E(g(X,Y)) = \sum_{\text{all } (x,y)} g(x,y)p_{X,Y}(x,y). \tag{4.6}$$

An example of an expected value of interest in a bivariate problem is $E(XY)$, the "product moment" of X and Y. As we shall see, this particular expected value carries substantial information about the relationship between X and Y. For the bivariate distribution in Table 4.1.1, this expectation may be computed as

$$E(XY) = 0 \cdot \frac{1}{10} + 0 \cdot \frac{2}{10} + 0 \cdot \frac{1}{10} + 1 \cdot \frac{3}{10} + 2 \cdot \frac{1}{10} + 4 \cdot \frac{2}{10} = \frac{13}{10} = 1.3.$$

It can be shown that the computation of EX using the marginal pmf of X in Table 4.1.2 and the definition in (2.1.1) will give the same answer as the computation of EX using the joint distribution in Table 4.1.1 and the definition in (4.6).

Conditional expectations are treated in a similar fashion. For example, we define the conditional expectation of X, given $Y = y$, as

$$E(X|Y = y) = \sum_{\text{all } x} x p_{X|Y}(x|y). \tag{4.7}$$

More generally, the conditional expectation of a function $g(X)$ of X, given $Y = y$, is given by

$$E(g(X)|Y = y) = \sum_{\text{all } x} g(x)p_{X|Y}(x|y). \tag{4.8}$$

The conditional mean and variance of the variable X, given $Y = 2$, whose pmf is displayed in Table 4.1.4 may be computed as follows:

$$E(X|Y=2) = \sum_{\text{all } x} x p_{X|Y}(x|2) = 0 \cdot \frac{1}{4} + 1 \cdot \frac{1}{4} + 2 \cdot \frac{1}{2} = \frac{5}{4},$$

and

$$\begin{aligned}\sigma^2_{X|Y=2} &= E(X-(5/4))^2|Y=2) \\ &= \sum_{\text{all } x} (x-(5/4))^2 p_{X|Y}(x|2) \\ &= \frac{25}{16} \cdot \frac{1}{4} + \frac{1}{16} \cdot \frac{1}{4} + \frac{9}{16} \cdot \frac{1}{2} = \frac{11}{16}.\end{aligned}$$

Let us now consider these same notions—bivariate models, marginal and conditional distributions, and expected values—in the context of continuous random pairs (X,Y). In analogy with the univariate models treated in Chapter 3, where probabilities were obtained as areas under curves, we will now consider volumes under surfaces over the *xy*-plane and give them a probabilistic interpretation. We begin with the following.

Definition 4.1.1. A *bivariate (or joint) probability density function* of a random pair (X,Y) is an integrable function $f_{X,Y}(x,y)$ of two variables satisfying the conditions

(1) $f_{X,Y}(x,y) \geq 0$ for all $x \in (-\infty,\infty)$ and $y \in (-\infty,\infty)$, and

(2) $\int_{-\infty}^{\infty} \int_{-\infty}^{\infty} f_{X,Y}(x,y)dxdy = 1$.

A bivariate density can be pictured as a surface over (i.e., above or on) the *xy*-plane having a total volume of 1. Mathematically, the surface could be represented as the set $\{(x,y,z) \in (-\infty,\infty) \times (-\infty,\infty) \times [0,\infty)|z = f_{X,Y}(x,y)\}$. Since the surface takes on only nonnegative values and $f_{X,Y}(x,y)$ integrates to 1, the integral of this surface over any subregion of the plane is a nonnegative number between 0 and 1 and may thus be regarded as a probability. If the random pair (X,Y) is modeled as having the bivariate pdf $f_{X,Y}(x,y)$, then we may calculate probabilities for X and Y as integrals of $f_{X,Y}(x,y)$ over an appropriate subset of the Euclidean plane. More specifically, for any extended (i.e., possibly infinite) real numbers $a < b$, and $c < d$,

$$P(a \leq X \leq b, c \leq Y \leq d) = \int_c^d \int_a^b f_{X,Y}(x,y)dxdy. \tag{4.9}$$

Keep in mind that there are countless possibilities for such surfaces, and that the bivariate probability models used in practice tend to have manageable functional forms and certain intuitively interpretable shapes. But in theory, any surface satisfying the conditions of Definition 4.1.1 could qualify as a candidate model in a particular application.

Given a fixed joint density for the random pair (X,Y), we may obtain the marginal and conditional densities in the same fashion as we did in the discrete case, only now we will replace summations by integration. The marginal densities of X and Y are, respectively,

$$f_X(x) = \int_{-\infty}^{\infty} f_{X,Y}(x,y)dy \tag{4.10}$$

and

$$f_Y(y) = \int_{-\infty}^{\infty} f_{X,Y}(x,y)dx. \tag{4.11}$$

Remark 4.1.1. Two points are worth emphasizing when calculating marginal densities using (4.10) and (4.11). Note, first, that in each of these integrals, the integration is done with respect to one variable while the other variable is held fixed. In (4.10), for example, the computation requires your holding x fixed while you integrate with respect to y. Holding x fixed often influences the range of y values over which you integrate. This range should be set to precisely those values of y for which $f_{X,Y}(x,y) > 0$, with x viewed as a fixed constant. In general, while the computation of the marginal density may be appropriately represented by an integral over the entire real line $(-\infty,\infty)$, we will often replace these limits of integration by a smaller interval over which the integrand is positive.

Example 4.1.1. Consider a bivariate density function that resembles a shoe box. Let's take the pdf of the random pair (X,Y) to be

$$f_{X,Y}(x,y) = \frac{1}{2}I_{(0,2)}(x)I_{(0,1)}(y). \tag{4.12}$$

A slight simplification of indicator functions such as those in (4.12) will be used in selected bivariate problems. Here, we can rewrite the density above as

$$f_{X,Y}(x,y) = \frac{1}{2}I_{(0,2)\times(0,1)}(x,y). \tag{4.13}$$

The marginal densities of X and Y are, trivially, $f_X(x) = (1/2)I_{(0,2)}(x)$ and $f_Y(y) = I_{(0,1)}(y)$, that is, marginally, $X \sim U(0,2)$ and $Y \sim U(0,1)$. ■

In addition to the marginal densities associated with a bivariate density function, we will be interested, as we were in the discrete case, in the conditional distributions of each variable given that the value of the other variable is fixed and known. Given a continuous random pair (X,Y), the *conditional densities* of X given $Y = y$ and of Y given $X = x$, are defined as follows.

$$f_{X|Y}(x|y) = \frac{f_{X,Y}(x,y)}{f_Y(y)} \text{ for all values } x \text{ which are possible when } Y = y. \tag{4.14}$$

and

$$f_{Y|X}(y|x) = \frac{f_{X,Y}(x,y)}{f_X(x)} \text{ for all values } y \text{ which are possible when } X = x. \tag{4.15}$$

In Example 4.1.1, the conditional densities turn out to be quite special. In that example, it is easily verified that $f_{X|Y}(x|y) = f(x)$ and that $f_{Y|X}(y|x) = f_Y(y)$. While this would appear at this point to be a fact that diminishes the importance of conditional densities, it is, instead, conclusive evidence that the value of Y has no influence on the value that X will take on, and vice versa, a property called "independence" that we will study in some detail later in this chapter. In the following example, the dependence between X and Y will be immediately evident and is quite strong.

Example 4.1.2. Consider a random pair (X,Y) for which $0 < X < Y < 1$, so that the joint density is zero outside of the triangle below the diagonal line $x = y$ in the unit square. More specifically, suppose (X,Y) has the joint density

$$f_{X,Y}(x,y) = 2 \text{ for } 0 < x < y < 1. \tag{4.16}$$

We may visualize this density as being a surface of constant height 2 over the triangle $\{(x,y) : x \in (0,1), y \in (0,1), x < y\}$. It clearly has the required volume 1. Let us obtain the marginal and conditional densities for the variables X and Y. The marginals are easily obtained by integration:

$$\begin{aligned} f_X(x) &= \int_{-\infty}^{\infty} f_{X,Y}(x,y)dy \\ &= \int_x^1 2dy \quad (\text{since, for any fixed } x, f(x,y) > 0 \text{ only for } y \in (x,1)) \\ &= 2y|_x^1 = 2(1-x)I_{(0,1)}(x). \end{aligned} \tag{4.17}$$

Similarly, we have

$$f_Y(y) = \int_{-\infty}^{\infty} f_{X,Y}(x,y)dx = \int_0^y 2dx = 2x|_0^y = 2yI_{(0,1)}(y). \tag{4.18}$$

We may then easily obtain the two conditional densities. Given $Y = y$, we have

$$\begin{aligned} f_{X|Y}(x|y) &= \frac{f_{X,Y}(x,y)}{f_Y(y)} \text{ for } x \in (0,y) \\ &= \frac{2}{2y} \text{ for } x \in (0,y) \\ &= \frac{1}{y} I_{(0,y)}(x) \end{aligned}$$

and given $X = x$, we have

$$\begin{aligned} f_{Y|X}(y|x) &= \frac{f_{X,Y}(x,y)}{f_X(x)} \text{ for } y \in (x,1) \\ &= \frac{2}{2(1-x)} \text{ for } y \in (x,1) \\ &= \frac{1}{1-x} I_{(x,1)}(y). \end{aligned}$$

We conclude that $X|Y = y \sim U(0,y)$ and that $Y|X = x \sim U(x,1)$. ■

It should be noted that the definition of the conditional densities in (4.14) and (4.15) give rise to a useful representation of a bivariate probability model. From (4.15), for example, we see that we may write the joint density of a random pair (X,Y) as

$$f_{X,Y}(x,y) = f_X(x) f_{Y|X}(y|x). \tag{4.19}$$

The identity in (4.19) is a very useful modeling tool, allowing us to define a model iteratively by first specifying a univariate model for X and then, for any given value of X, specifying a univariate model describing the conditional behavior of Y, given $X = x$. You will encounter such iterative modeling in a variety of examples and applications in this book. In particular, it is a standard modeling device in the area of Bayesian inference, a topic which we will take up in Chapter 9. As a simple example of this modeling process, consider the following fully specified bivariate model: $X \sim N(\mu, \sigma^2)$ and $Y|X = x \sim N(x, \tau^2)$. The joint density of (X,Y) may be written as the product of the two normal densities. Later in this chapter, we'll identify this joint density as a special case of the so-called bivariate normal distribution.

Finally, we treat the analog of mathematical expectation for functions of a continuous random pair. If the pair (X,Y) has joint density $f_{X,Y}(x,y)$ and g is a function of two variables, then our working definition of $E(g(X,Y))$ is

$$E(g(X,Y)) = \int_{-\infty}^{\infty}\int_{-\infty}^{\infty} g(x,y) f_{X,Y}(x,y)\,dx\,dy. \tag{4.20}$$

As mentioned earlier in this section, the fact that the expectation $E(X)$ can be computed in two ways should cause no concern, since, by interchanging the order of integration, you can evaluate this expectation as

$$\begin{aligned} E(X) &= \int_{-\infty}^{\infty}\int_{-\infty}^{\infty} x f_{X,Y}(x,y)\,dx\,dy \\ &= \int_{-\infty}^{\infty}\left(x\int_{-\infty}^{\infty} f_{X,Y}(x,y)\,dy\right)dx \\ &= \int_{-\infty}^{\infty} x f_X(x)\,dx, \end{aligned}$$

which shows that the two definitions of $E(X)$ are equivalent. We will also be interested in conditional expectations for continuous random pairs. The relevant definitions are

$$E(X|Y=y) = \int_{-\infty}^{\infty} x f_{X|Y}(x|y)\,dx \tag{4.21}$$

and

$$E(Y|X=x) = \int_{-\infty}^{\infty} y f_{Y|X}(y|x)\,dy. \tag{4.22}$$

It is clear from these definitions that $E(X|Y = y)$ is simply the mean of the conditional distribution of X given $Y = y$ and $E(Y|X = x)$ is the mean of the conditional distribution of Y given $X = x$.

We conclude this section by returning to the computation of certain expected values for the random pair (X,Y) featured in Example 4.1.2. Specifically, we will calculate the product moment of the pair (X,Y) and the conditional expected value of X given $Y = y$. We may write

$$E(XY) = \int_{-\infty}^{\infty}\int_{-\infty}^{\infty} x \cdot y \cdot f_{X,Y}(x,y)\,dx\,dy$$

$$= \int_0^1 \int_0^y 2xydxdy$$

$$= \int_0^1 (yx^2)|_{x=0}^{y} dy$$

$$= \int_0^1 y^3 dy = \frac{y^4}{4}|_0^1 = \frac{1}{4}. \quad (4.23)$$

Further, we may calculate the conditional mean of Y, given $X = x$, as

$$E(Y|X = x) = \int_{-\infty}^{\infty} y f_{Y|X}(y|x) dy$$

$$= \int_x^1 y \frac{1}{1-x} dy \quad (4.24)$$

$$= \frac{1+x}{2} \text{ for any } x \in (0,1) \quad (4.25)$$

Since, for any $x \in (0,1)$, $Y|X = x \sim U(x,1)$, we could, of course, have predicted the value of the conditional expectation above from the "center of gravity" interpretation of the mean.

Remark 4.1.2. In the event that you are a bit rusty on setting limits for multiple integrals, I will offer here a brief tutorial. This isn't meant to substitute for a good hard look at your old calculus book, but it might get you by until you have time to do that. There are two situations that we will encounter quite often. I'll stick to the bivariate case, i.e., double integrals, since this is what you will be encountering with greatest frequency, but the routines described below are easily extended to n-fold integrals. Here are the scenarios you will need to deal with:

I. In setting limits for single integrals, as in obtaining a marginal density $f_X(x)$ from a joint density $f_{X,Y}(x,y)$, one variable is held fixed (x in computing $f_X(x)$) and you set the limits of the integral by asking "What values can y take on for this fixed value of x?" Answer this question by looking at the range of x and y in the bivariate density.

II. For setting limits of double integrals, for example, to obtain the product moment $E(XY) = \int_{-\infty}^{\infty}\int_{-\infty}^{\infty} xy f_{X,Y}(x,y) dxdy$, set the limits on the outside integral first (here, the limits are on y) by asking "What values are possible for y in this problem?" Let y range over all values that the variable Y can possibly take on. These values *must not* depend on x. For example, if $0 < x < y < 1$, then the "possible values" of y are all real numbers in the interval $(0,1)$. Once the outer limits are set, consider this variable fixed. One then sets the limits for the inner integral. For fixed y, ask "What can x be when $Y = y$?" This is similar to the scenario in item I above, and it is addressed in the same way.

Exercises 4.1.

1. Let (X,Y) be a discrete random pair with the following probability mass function:

(X,Y)	$(1,1)$	$(1,2)$	$(1,4)$	$(2,2)$	$(2,4)$	$(2,5)$	$(3,4)$	$(3,5)$
$p(x,y)$	1/10	1/10	2/10	1/10	1/10	2/10	1/10	1/10

Obtain the marginal pmfs for X and for Y, the conditional pmf of X, given that $Y=4$, and the expectations $E(XY)$, EX, EY, and $E(X|Y=4)$.

2. Let (X,Y) be a discrete random pair with pmf $p_{X,Y}(x,y)$ and marginal pmfs $p_X(x)$ and $p_Y(y)$. Show that EX, when obtained using $p_{X,Y}(x,y)$ and the Definition in (4.6), must yield the same answer as it does when obtained using $p_X(x)$ and the definition in (2.1). (Hint: When using Definition in (4.6), write $\sum_{\text{all }(x,y)}$ as $\sum_x \sum_y$ and evaluate the summation with respect to y first, holding x fixed.)

3. Set the limits for evaluating the integrals

$$E(XY) = \int_{-\infty}^{\infty}\int_{-\infty}^{\infty} xy f_{X,Y}(x,y)dxdy$$

and

$$\int_{-\infty}^{\infty}\int_{-\infty}^{\infty} xy f_{X,Y}(x,y)dydx$$

for densities $f_{X,Y}(x,y)$ that are nonzero only in the indicated sets of (x, y) pairs: a) $0 \le x \le y \le 1$; b) $0 \le y^2 \le x \le y \le 1$; c) $x \ge 0$, $y \ge 0$, $0 \le xy \le 1$; d) $0 \le x \le 1$, $0 \le y \le 1$, $0 \le x+y \le 1$; and e) $0 \le y \le 4$, $y \le x \le y+2$. In each case, plot the support set $\{(x,y)|f_{X,Y}(x,y) > 0\}$ of each density. Note that these plots can be quite helpful in setting limits of the two integrals above.

4. Let $f_{X,Y}(x,y) = 10xy^2$ for $0 < x < y < 1$ be the joint density function of the random pair (X,Y). (a) Obtain the marginal density $f(y)$ of Y. (b) Obtain the conditional density $f_{X|Y}(x|y)$ of X given $Y=y$. (c) Evaluate the conditional expectation of X, given $Y=y$.

5. Let X and Y be independent gamma variables, with $X \sim \Gamma(1,1)$ and $Y \sim \Gamma(1,2)$. Compute $P(X > Y)$. (Hint: Integrate the $f_{X,Y}(x,y)$ over $\{(x,y)|0 < y < x < \infty\}$.)

6. Consider the random pair taking values in the unit square according to the joint density

$$f(x,y) = kx^2y \quad \text{for } 0 < x < 1 \text{ and } 0 < y < 1.$$

(a) Determine the value of the constant k that makes $f(x,y)$ a legitimate bivariate density. (b) Using your answer to (a), evaluate the probability $P(X < Y)$. (Hint: Integrate $f(x,y)$ over the set $0 < x < y < 1$.)

4.2 More on Mathematical Expectation

I mentioned in the preceding section that bivariate probability models contain more information about the random pair (X,Y) than is captured in the marginal distributions of

X and Y alone. This happens because the variables X and Y are typically related in some way, so that the values that one variable takes on have some influence on the values that the other might take on. We will now explore the possible dependence between a pair of random variables and introduce a particular measure, the correlation between X and Y, that is often used to describe that dependence. We encountered the product moment $E(XY)$ of the random pair (X,Y) in Section 4.1. As we will shortly see, it is a key element of the dependence measure we will study here. The measure is well defined for arbitrary random variables, continuous or discrete, provided the expectations involved are finite.

Definition 4.2.1. The *covariance* between the random variables X and Y, denoted by $Cov(X,Y)$, is given by

$$Cov(X,Y) = E\left[(X-EX)(Y-EY)\right]. \tag{4.26}$$

The fact that $Cov(X,Y)$ measures a certain type of dependence, i.e., whether the variables X and Y vary directly or inversely, may be intuitively extracted from (4.26). If X tends to be above its mean when Y is above its mean, and X tends to be below its mean when Y is below its mean, then the products in (4.26) will tend to be positive (since the two components of the product have the same sign), while if X and Y vary inversely, the products in (4.26) will tend to be negative. The covariance between two variables may fail to recognize some other types of dependence (See Exercise 4.1.2). It is nonetheless a widely used tool in studying the dependence between variables. The following result provides a helpful computational formula for $Cov(X,Y)$.

Theorem 4.2.1. Let (X,Y) be a random pair with density or probability mass function $f_{X,Y}(X,Y)$. Denoting EX as μ_X and EY as μ_Y,

$$Cov(X,Y) = E(XY) - \mu_X\mu_Y. \tag{4.27}$$

Proof. We give a proof in the continuous case.

$$\begin{aligned}
Cov(X,Y) &= \int_{-\infty}^{\infty}\int_{-\infty}^{\infty}(x-\mu_X)(y-\mu_Y)f_{X,Y}(x,y)dxdy \\
&= \int_{-\infty}^{\infty}\int_{-\infty}^{\infty}(xy - x\mu_Y - y\mu_X + \mu_X\mu_Y)f_{X,Y}(x,y)dxdy \\
&= \int_{-\infty}^{\infty}\int_{-\infty}^{\infty}xyf_{X,Y}(x,y)dxdy - \int_{-\infty}^{\infty}\int_{-\infty}^{\infty}x\mu_Y f_{X,Y}(x,y)dxdy \\
&- \int_{-\infty}^{\infty}\int_{-\infty}^{\infty}y\mu_X f_{X,Y}(x,y)dxdy + \int_{-\infty}^{\infty}\int_{-\infty}^{\infty}\mu_X\mu_Y f_{X,Y}(x,y)dxdy \\
&= E(XY) - \mu_Y EX - \mu_X EY + \mu_X\mu_Y \\
&= E(XY) - 2\mu_X\mu_Y + \mu_X\mu_Y \\
&= E(XY) - \mu_X\mu_Y.
\end{aligned}$$

■

Remark 4.2.1. Later in this chapter, we will show that taking an expectation is a linear operation, that is, you may distribute an expectation through a linear function term by term. In other words, we will show that $E(\sum_{i=1}^{n} a_iX_i) = \sum_{i=1}^{n} a_iEX_i$. This property leads to a much shorter argument than the one above and handles the continuous and discrete cases simultaneously. The argument would go like this:

$$Cov(X,Y) = E(X-EX)(Y-EY) = E(XY - X\mu_Y - \mu_X Y + \mu_X \mu_Y) = E(XY) - \mu_X \mu_Y.$$

We have mentioned that $Cov(X,Y) > 0$ when X and Y vary together and that $Cov(X,Y) < 0$ when X and Y vary inversely. While the sign of $Cov(X,Y)$ carries important information about the relationship between two variables, the measure has a property that affects its interpretability. The covariance between two variables is not scale-invariant. This means that the size of $Cov(X,Y)$ depends on the scale in which X and Y are measured. Indeed, it is easy to show that for any constants a and b,

$$Cov(aX,bY) = abCov(X,Y).$$

Thus, if X and Y are measured in inches rather than in feet, the covariance between the variables will change by a factor of 144. This makes it difficult to judge whether a large positive or negative value of $Cov(X,Y)$ means that there is strong dependence between the variables. Fortunately, this deficiency can be remedied by a standardization process that creates a new measure that has the same sign as the covariance but is also scale-invariant, so that its size is meaningful in judging the strength of the dependence between variables.

Definition 4.2.2. The *correlation coefficient* $\rho_{X,Y}$ between the random variables X and Y is given by

$$\rho_{X,Y} = \frac{Cov(X,Y)}{\sigma_X \sigma_Y}, \tag{4.28}$$

where σ_X and σ_Y are the respective standard deviations of X and Y.

For computational purposes, the formula

$$\rho_{X,Y} = \frac{E(XY) - EXEY}{\sqrt{EX^2 - (EX)^2}\sqrt{EY^2 - (EY)^2}} \tag{4.29}$$

is generally used. The scale-invariance of the correlation coefficient may be easily confirmed using the latter formula. For arbitrary positive constants a and b,

$$\begin{aligned}
\rho_{aX,bY} &= \frac{EaXbY - EaX \cdot EbY}{\sqrt{E(aX - EaX)^2}\sqrt{E(bY - EbY)^2}} \\
&= \frac{ab(E(XY) - EX \cdot EY)}{\sqrt{a^2 E(X - EX)^2}\sqrt{b^2 E(Y - EY)^2}} \\
&= \pm \rho_{X,Y}.
\end{aligned} \tag{4.30}$$

Note that if $a > 0$ and $b < 0$ in (4.30), then $\rho_{aX,bY} = -\rho_{X,Y}$, so the size of the dependence $|\rho|$ is unchanged but the type of dependence is reversed.

To develop further properties of the correlation coefficient, we take a slight digression and treat the linearity property of expectation and establish, as well, a useful formula for the variance of a linear combination of random variables. Since the treatment below applies to random vectors of arbitrary dimension, we will assume we are dealing with multivariate probability distributions for the n-dimensional random vector $(X_1, X_2, \ldots, X_n)$.

Theorem 4.2.2. Suppose that $\mathbf{X} = (X_1, X_2, \ldots, X_n)$ has density or probability mass function $f_{X_1,\ldots,X_n}(x_1, \ldots, x_n)$ and that $EX_i < \infty$ for all i. For arbitrary real numbers $a_1, \ldots, a_n$,

$$E\left(\sum_{i=1}^{n} a_i X_i\right) = \sum_{i=1}^{n} a_i EX_i. \tag{4.31}$$

Proof. We give a proof in the continuous case. For a proof in the discrete case, simply replace the integration by summation.

$$\begin{aligned} E\left(\sum_{i=1}^{n} a_i X_i\right) &= \int_{-\infty}^{\infty} \cdots \int_{-\infty}^{\infty} \left(\sum_{i=1}^{n} a_i x_i\right) f_{X_1,\ldots,X_n}(x_1, \ldots, x_n)\, dx_1 \cdots dx_n \\ &= \sum_{i=1}^{n} a_i \int_{-\infty}^{\infty} \cdots \int_{-\infty}^{\infty} x_i f_{X_1,\ldots,X_n}(x_1, \ldots, x_n)\, dx_1 \cdots dx_n \\ &= \sum_{i=1}^{n} a_i EX_i. \end{aligned}$$

■

Corollary 4.2.1. Let $X_1, \ldots, X_n$ be a set of random variables with a common mean $\mu < \infty$. If $\bar{X} = (1/n)\sum_{i=1}^{n} X_i$, then $E\bar{X} = \mu$.

Theorem 4.2.3. Suppose that $\mathbf{X} = (X_1, X_2, \ldots, X_n)$ has density or probability mass function $f_{X_1,\ldots,X_n}(x_1, \ldots, x_n)$ and that $V(X_i) < \infty$ for all i. Then, for arbitrary real numbers $a_1, \ldots, a_n$,

$$V\left(\sum_{i=1}^{n} a_i X_i\right) = \sum_{i=1}^{n} a_i^2 V(X_i) + 2\sum_{i<j} a_i a_j Cov(X_i, X_j). \tag{4.32}$$

Proof. Again, we present the proof for continuous variables. For a proof in the discrete case, replace the integration by summation. Denoting EX_i by μ_i, we have

$$\begin{aligned} V\left(\sum_{i=1}^{n} a_i X_i\right) &= \int_{-\infty}^{\infty} \cdots \int_{-\infty}^{\infty} \left(\sum_{i=1}^{n} a_i x_i - \sum_{i=1}^{n} a_i \mu_i\right)^2 f_{X_1,\ldots,X_n}(x_1, \ldots, x_n)\, dx_1 \cdots dx_n \\ &= \int_{-\infty}^{\infty} \cdots \int_{-\infty}^{\infty} \left(\sum_{i=1}^{n} a_i (x_i - \mu_i)\right)^2 f_{X_1,\ldots,X_n}(x_1, \ldots, x_n)\, dx_1 \cdots dx_n \end{aligned}$$

$$\begin{aligned} &= \sum_{i=1}^{n} a_i^2 \int_{-\infty}^{\infty} \cdots \int_{-\infty}^{\infty} (x_i - \mu_i)^2 f_{X_1,\ldots,X_n}(x_1, \ldots, x_n)\, dx_1 \cdots dx_n \\ &+ \sum_{i \neq j} a_i a_j \int_{-\infty}^{\infty} \cdots \int_{-\infty}^{\infty} (x_i - \mu_i)(x_j - \mu_j) f_{X_1,\ldots,X_n}(x_1, \ldots, x_n)\, dx_1 \cdots dx_n \\ &= \sum_{i=1}^{n} a_i^2 \int_{-\infty}^{\infty} \cdots \int_{-\infty}^{\infty} (x_i - \mu_i)^2 f_{X_1,\ldots,X_n}(x_1, \ldots, x_n)\, dx_1 \cdots dx_n \\ &+ 2\sum_{i<j} a_i a_j \int_{-\infty}^{\infty} \cdots \int_{-\infty}^{\infty} (x_i - \mu_i)(x_j - \mu_j) f_{X_1,\ldots,X_n}(x_1, \ldots, x_n)\, dx_1 \cdots dx_n \\ &= \sum_{i=1}^{n} a_i^2 V(X_i) + 2\sum_{i<j} a_i a_j Cov(X_i, X_j). \end{aligned}$$

■

The bivariate cases of the two theorems above are worth stating separately.

Corollary 4.2.2. For the random pair (X,Y) for which the necessary (first and/or second) moments are finite, and for arbitrary real numbers a_1 and a_2,

$$E(a_1X_1 + a_2X_2) = a_1EX_1 + a_2EX_2 \tag{4.33}$$

and

$$V(a_1X_1 + a_2X_2) = a_1^2V(X_1) + a_2^2V(X_2) + 2a_1a_2Cov(X_1,X_2). \tag{4.34}$$

The results above are useful in a wide variety of applications, and you will find them employed in many of this book's examples and problems. Searching for the "best linear unbiased estimator" of an unknown parameter is one particular circumstance in which these results will prove useful. Here, we will make immediate use of Corollary 4.2.2 in proving a very important property of the correlation coefficient ρ.

Theorem 4.2.4. For any random pair (X,Y), each having finite second moments, the correlation $\rho_{X,Y}$ between X and Y satisfies the inequalities $-1 \leq \rho_{X,Y} \leq 1$.

Proof. The assumed finite second moments of X and Y guarantee that $\sigma_X^2 < \infty$ and $\sigma_Y^2 < \infty$. Let σ_X and σ_Y be the respective standard deviations of X and Y. Then, since

$$V\left(\frac{X}{\sigma_X}\right) = \frac{V(X)}{\sigma_X^2} = 1, V\left(\frac{Y}{\sigma_Y}\right) = \frac{V(Y)}{\sigma_Y^2} = 1 \text{ and } Cov\left(\frac{X}{\sigma_X}, \frac{Y}{\sigma_Y}\right) = \rho_{X,Y},$$

we have, by Corollary 4.2.2, that

$$V\left(\frac{X}{\sigma_X} + \frac{Y}{\sigma_Y}\right) = 1 + 1 + 2Cov\left(\frac{X}{\sigma_X}, \frac{Y}{\sigma_Y}\right) = 2 + 2\rho_{X,Y} \geq 0,$$

which implies that $\rho_{X,Y} \geq -1$. Similarly, by Corollary 4.2.2, we have that

$$V\left(\frac{X}{\sigma_X} - \frac{Y}{\sigma_Y}\right) = 1 + 1 - 2Cov\left(\frac{X}{\sigma_X}, \frac{Y}{\sigma_Y}\right) = 2 - 2\rho_{X,Y} \geq 0,$$

which implies that $\rho_{X,Y} \leq 1$, and completes the proof. ■

The fact that $|\rho_{X,Y}| \leq 1$ is very helpful in judging the strength of the dependence between X and Y. If X and Y are approximately linearly related, then $\rho_{X,Y}$ will be quite close to 1 or -1. It is easy to show that $|\rho_{X,Y}| = 1$ if and only if $Y = aX + b$, with $\rho_{X,Y}$ being equal to 1 if $a > 0$ and $\rho_{X,Y}$ being equal to -1 if $a < 0$.

Example 4.2.1. Consider the random pair (X,Y) of Example 4.1.2, that is, suppose that (X,Y) have the joint density

$$f(x,y) = 2 \text{ for } 0 < x < y < 1. \tag{4.35}$$

We wish to calculate the correlation $\rho_{X,Y}$ between X and Y. Recall that the marginal densities of X and Y are given by

$$f_X(x) = 2(1-x)I_{(0,1)}(x) \text{ and } f_Y(y) = 2yI_{(0,1)}(y).$$

The mean and variance of Y are thus easily computed:

$$EY = \int_0^1 y \cdot 2y dy = \frac{2y^3}{3}\Big|_0^1 = \frac{2}{3},$$

$$EY^2 = \int_0^1 y^2 \cdot 2y dy = \frac{2y^4}{4}\Big|_0^1 = \frac{1}{2} \text{ and}$$

$$V(Y) = \frac{1}{2} - \left(\frac{2}{3}\right)^2 = \frac{1}{18}.$$

Since the density of X is the mirror image of the density of Y about the line $x = 1/2$, it follows by symmetry that $EX = \frac{1}{3}$ and that $V(X) = \frac{1}{18}$. What remains in the calculation of $\rho_{X,Y}$ is the identification of the product moment $E(XY)$. This calculation was made in the last section (see Equation (4.23)). Using the known fact that $EXY = \frac{1}{4}$, we have

$$\rho_{X,Y} = \frac{E(XY) - EXEY}{\sigma_X \sigma_Y} = \frac{1/4 - (1/3)(2/3)}{\sqrt{1/18}\sqrt{1/18}} = \frac{1/36}{1/18} = 1/2.$$

A positive correlation between X and Y in this example should not be surprising, since the constraint $0 < X < Y < 1$ guarantees that Y will be large whenever X is large and that X will be small whenever Y is small. ■

There is a very useful result concerning the evaluation of an expected value that is appropriate for discussion only in the context of bivariate models. I've always called the result "Adam's Rule" because of the following story, told to me by my thesis advisor and dear friend, UCLA Professor Thomas S. Ferguson, many years ago. Tom was taking a course from the legendary Jerzy Neyman at the University of California, Berkeley, sometime in the 1950s. Tom was quite impressed with the broad utility of this particular formula for computing an expected value of a random variable X, and he asked Professor Neyman who deserved the credit for this neat result. Neyman responded, without a hint of a smile, that the result was due to Adam. Tom says he went to the library immediately after class to look up this statistician "Adam." Upon finding nothing, he realized that Neyman was talking about Adam of "Adam and Eve" fame. He was simply saying that the result was as old as the hills. That, of course, takes nothing away from its utility.

Theorem 4.2.5. (Adam's Rule) For any arbitrary random variables X and Y for which the expectations below are finite,

$$EX = E(E(X|Y)). \tag{4.36}$$

Remark 4.2.2. You might wonder why we should be interested in this result. On its face, it seems like it provides an alternative formula for computing EX when the standard definition of EX is simpler and easier to apply. But the problem with the usual definition of EX is that it requires that we know the distribution of X. In a bivariate problem (assuming continuous variables), the joint density of X and a second variable Y might be stipulated as $f_Y(y)f_{X|Y}(x|y)$, and if it is, then $f_X(x)$ is not known a priori and must be calculated first if it is to be used in obtaining EX using (2.1) or (3.6). So there are definitely problems in which the formula in (4.36) will be immediately applicable, while the direct computation of EX requires more work. A second question that arises from the statement of this theorem is: where does the Y come from? Can Y be any other random variable? The answer to these questions is that the right-hand side of (4.36) requires that the densities $f(y)$ and $f(x|y)$ be known, so that the Y in this theorem stands for a variable whose relationship with X is known, as would be the case when the joint density of X and Y is known.

Remark 4.2.3. In (4.36), we have used a common shorthand notation for a double expectation (that is, for an expectation with respect to the distribution of the two random variables X and Y). The expectation $E(E(X|Y))$ is meant to be interpreted as follows: The inner expectation is taken with respect to the conditional distribution of X given Y. It is therefore a function of Y. The outer expectation is taken with respect to the marginal distribution of Y, as it is simply the expected value of $E(X|Y)$, which is a function of Y alone. It is sometimes helpful to use subscripts on the expectation symbol E, which formally identifies the distribution with respect to which the expectation is being taken. We could, for example, express the expectation $E(E(X|Y))$ as

$$E(E(X|Y)) = E_Y\left[E_{X|Y=y}(X|Y=y).\right]$$

I will generally use the simpler notation for the expectation above, but I will occasionally add subscripts when needed for clarity or transparency. When expressing such expectations as a sum or integral, I will use subscripts to identify the relevant pmfs or pdfs.

Proof of Theorem 4.2.5. We give the proof for the continuous case. The proof in the discrete case is similar. We begin by noticing that $E(X|Y=y)$ is a function of y, and in order to evaluate its expectation, we will need to integrate it relative to the density of Y. We may therefore write

$$\begin{aligned}
E(E(X|Y)) &= \int_{-\infty}^{\infty} E(X|Y=y) f_Y(y)dy \\
&= \int_{-\infty}^{\infty} \left[\int_{-\infty}^{\infty} x f_{X|Y}(x|y)dx\right] f_Y(y)dy \\
&= \int_{-\infty}^{\infty}\int_{-\infty}^{\infty} x f_{X|Y}(x|y) f_Y(y)dxdy \\
&= \int_{-\infty}^{\infty}\int_{-\infty}^{\infty} x f_{X,Y}(x,y)dxdy \\
&= EX.
\end{aligned}$$

■

Example 4.2.2. Let's consider the problem of modeling traffic accidents over a fixed period of time (say a year) for a random driver. The following is a somewhat simplified version of a model that has actually been used in practice for this type of application. It seems reasonable that the "accident proneness" of a driver will vary in the population, as some drivers are very cautious and attentive and others are more aggressive and/or more easily distracted. If we draw a driver at random, his or her level of accident proneness will be a random variable, and we will denote it by Y. Let us model the number X of accidents that a driver with accident proneness $Y=y$ will actually have in the time period of interest as a Poisson variable. This seems pretty reasonable, since, for most people, accidents are rare events, but each driver tends to be faced with many opportunities to have one. Thus, we will set the conditional distribution of X as

$$X|Y=y \sim P(y). \tag{4.37}$$

The distribution of Y could be modeled in a variety of ways. Since we can think of Y as the average number of accidents per period as the driver type varies within the population, it should be modeled as a nonnegative variable. Let's consider a particularly simple choice of a model for Y; let

$$Y \sim Exp(\lambda). \tag{4.38}$$

The model we have proposed here is interesting for at least a couple of reasons. For one thing, it is the first example we've encountered that is actually revealed iteratively. The joint behavior of the pair (X,Y) is known through the specified distribution of Y and the conditional distribution of X, given Y. Secondly, the joint model is neither continuous nor discrete, but rather, it is a hybrid, with one continuous and one discrete element. This actually causes no conceptual or practical problems; to calculate a probability for the random pair (X,Y), we simply sum over the values of x of interest and integrate over the relevant values of y. The joint model, with its hybrid probability density-mass function, is fully specified by

$$f_{X,Y}(x,y) = \left(\lambda e^{-\lambda y}\right)\left(\frac{y^x e^{-y}}{x!}\right) \text{ for } y > 0 \text{ and for } x = 0,1,2,\ldots. \tag{4.39}$$

Suppose we were interested in the (marginal) distribution of X, the number of accidents per driver in the overall population. To find it, we would need to integrate the probability function in (4.39) with respect to the variable y. Proceeding with this calculation, we obtain the probability mass function of X as

$$\begin{aligned}
p_X(x) &= \int_0^\infty \frac{\lambda e^{-\lambda y} y^x e^{-y}}{x!} dy \\
&= \int_0^\infty \frac{\lambda}{\Gamma(x+1)} y^x e^{-(\lambda+1)y} dy \\
&= \frac{\lambda}{(\lambda+1)^{x+1}} \int_0^\infty \frac{(\lambda+1)^{x+1}}{\Gamma(x+1)} y^x e^{-(\lambda+1)y} dy \text{ (using your, by now, favorite trick)} \\
&= \frac{\lambda}{(\lambda+1)^{x+1}},
\end{aligned}$$

since the latter integrand is a perfect $\Gamma(x+1, 1/(\lambda+1))$ density). We may rewrite the pmf of X in a slightly more recognizable form:

$$p(x) = \left(\frac{1}{\lambda+1}\right)^x \frac{\lambda}{\lambda+1} \text{ for } x = 0,1,2,\ldots \tag{4.40}$$

This model is just another form of the geometric model $G(p)$ with $p = \lambda/(\lambda+1)$. While $G(p)$ is the distribution of the number of Bernoulli trials, say W, needed to obtain the first success (with $P(S) = p$), one could also count the number of failures X that occur before the first success. If $W \sim G(\lambda/(\lambda+1))$, then $X = W - 1$, and X will have the probability mass function displayed in (4.40). (For a generalization of the modeling above, see Problem 4.33.)

If we were interested in EX, the average number of accidents per person in the population during the period of interest, we could get it now by noting that, for the variable W above, $EW = \frac{\lambda+1}{\lambda}$, so that $EX = EW - 1 = \frac{1}{\lambda}$. If obtaining EX had been our main objective, though, we would have obtained it more quickly and easily from Adam's Rule, since we know that $Y \sim Exp(\lambda)$ and $X|Y = y \sim P(y)$. It follows that $EX = E[E(X|Y)] = E[Y] = \frac{1}{\lambda}$. ■

Adam's Rule is a handy formula that is useful in models that are specified iteratively. Models so specified are generally referred to as *hierarchical models*, and occur especially frequently in Bayesian statistical inference, as we'll see later. It is worth noting that Adam's rule can be applied, in two-stage hierarchical models, to the evaluation of the expectation of any variable X or any function of X of interest. We will encounter situations in the sequel in which it proves useful to apply Adam's rule to compute the moment-generating function of X as $EE(e^{tX}|Y)$ from an iteratively specified bivariate model for (X,Y). Note that such a computation stands to provide you with knowledge of the distribution of X as well as any other characteristics of X in which you might be interested.

Our final result in this section is a lovely theorem that is complementary to Adam's Rule, but deals with the computation of the variance of a random variable, rather than its mean, in the context of a two-stage hierarchical model. Not having a separate story for this particular identity, but recognizing it as being in the same family as Adam's Rule, I've chosen a rather natural name for it.

Theorem 4.2.6. (Eve's Rule): For any pair of random variables X and Y for which the expectations below are finite,

$$V(X) = E[V(X|Y)] + V[E(X|Y)]. \tag{4.41}$$

Proof. By definition, $V(X) = \mathbf{E}[X - EX]^2$. The bold "E" will denote the expectation with respect to the joint distribution of X and Y. In expressions with just one of the two variables (say $[X - EX]^2$, as in the expression above), we would integrate (or sum) first with respect to the distribution of $Y|X$, and since the integrand (or summand) doesn't depend on Y, we would then get the usual expectation, namely, $E[X - EX]^2$. More specifically,

$$\begin{aligned}\mathbf{E}[X - EX]^2 &= \int_{-\infty}^{\infty}\int_{-\infty}^{\infty}(x-\mu_x)^2 f(x,y)dydx \\ &= \int_{-\infty}^{\infty}(x-\mu_x)^2 f(x)dx = E[X - EX]^2.\end{aligned}$$

Now, let's expand the expression for $V(X)$ above by adding zero in the form of the difference "$E(X|Y) - E(X|Y)$." We may thus write

$$V(X) = \mathbf{E}[X - E(X|Y) + E(X|Y) - EX]^2. \tag{4.42}$$

Squaring the term on the right-hand side of (4.42), we get

$$V(X) = \mathbf{E}[X - E(X|Y)]^2 + \mathbf{E}[E(X|Y) - EX]^2 + 2\mathbf{E}[X - E(X|Y)][E(X|Y) - EX]. \tag{4.43}$$

The first term on the RHS of (4.43) may be written as

$$\mathbf{E}[X - E(X|Y)]^2 = E_Y\left[E_{X|Y}[X - E(X|Y)]^2\right] = E[V(X|Y)], \tag{4.44}$$

where the final expectation in (4.43) is with respect to the distribution of Y, since $V(X|Y)$

is a function of Y alone. Note that we have used subscripts on some of the expectations in (4.44) for clarity's sake. The second term on the RHS of (4.43) may be written as

$$\mathbf{E}[E(X|Y)-EX]^2 = E[E(X|Y)-E[E(X|Y)]]^2 \text{ (using Adam's rule on } EX) \\ = V[E(X|Y)]. \quad (4.45)$$

Thus, the theorem will hold if the third term in (4.43) is equal to zero. That this term is equal to zero follows from

$$\begin{aligned} &\mathbf{E}[X-E(X|Y)][E(X|Y)-EX] \\ &= E_Y E_{X|Y}[X-E(X|Y)][E(X|Y)-EX] && (4.46) \\ &= E_Y[[E(X|Y)-EX]E_{X|Y}[X-E(X|Y)]] && (4.47) \\ &= 0 && (4.48) \end{aligned}$$

where I have added subscripts to the expectations in these latter equations for clarity. Equation (4.47) is justified by the fact that $[E(X|Y)-EX]$ may be factored out of the inner expectation in (4.46) because this term depends only on Y, and the inner expectation in (4.46) conditions on Y, and thus $[E(X|Y)-EX]$ is a constant for any fixed value of Y. The equality in (4.48) follows from the fact that

$$E_{X|Y}[X-E(X|Y)] = E(X|Y)-E(X|Y) = 0.$$

This completes the proof. ■

Exercises 4.2.

1. Suppose a fair coin is tossed 3 times. Let X be the number of heads in the first 2 tosses and let Y be the number of heads in the last 2 tosses. Find (a) the joint probability mass function (pmf) of the pair (X,Y), (b) the marginal pmf of each, (c) the conditional pmf of X given $Y=1$ and also given $Y=2$ and (d) the correlation $\rho_{X,Y}$ between X and Y.

2. Let (X,Y) be a random pair which takes on the values $(0,0)$, $(0,1)$, $(1,-1)$, and $(1,2)$, each with probability $1/4$. Show that $cov(X,Y)=0$. Show that $P(Y=y|X=x) \neq P(Y=y)$ for all x and y.

3. Let X_1 and X_2 be the first two bids entered in a progressive auction involving an item that is worth θ dollars. Suppose that $X_1 \sim U(0,\theta)$ and that the conditional distribution of X_2, given X_1, is $X_2|X_1 = x_1 \sim U(x_1,\theta)$. Now both $2X_1$ and $4X_2/3$ are reasonable guesses at θ, but the guess $(-1/4)X_1+(3/2)X_2$ is an even better guess. Verify this claim by calculating and comparing the variances of these three variables. All three variables have expected value equal to θ. (Note: Corollary 4.22 comes in handy in solving this problem.)

4. Suppose that X is the number of hours that the University of Idaho's Coach Potato spends doing chores at home on a random weekend, and given $X=x$, let Y/x be the proportion of that time he spends whining while doing it. Assume that the joint density of the random pair (X,Y) is given by

$$f_{X,Y}(X,Y) = y^2 e^{-x/3}/3x^2, \text{ for } 0<y<x<\infty.$$

(a) Obtain the marginal density of X, including its range, and identify it by name. (b) Obtain the conditional density of Y, given $X = x$ (including the range of the variable Y). (Remark: Given $X = x$, Y/x has a beta distribution.) (c) Evaluate $E(Y|X)$, and use Adam's rule to obtain EY. (d) Use Eve's rule to evaluate $V(Y)$.

4.3 Independence

The notion of independence was treated in Chapter 1, and we could, if we wished to, use the concept defined there as the basis for defining the concept of independent random variables. Indeed, it is perfectly correct to declare the random variables X and Y to be independent if, for any sets A and B,

$$P(X \in A, Y \in B) = P(X \in A) \cdot P(Y \in B). \tag{4.49}$$

However, such a definition is not very useful operationally, since its use to verify independence would require our checking that (4.49) holds for all possible choices of A and B, seemingly quite a formidable task. The property highlighted in the alternative definition below is a quite a bit easier to check.

Definition 4.3.1. Two random variables X and Y are said to be *independent* if their ranges are independent, and, in addition, if

$$p_{X,Y}(x,y) = p_X(x) \cdot p_Y(y) \text{ for discrete } X \text{ and } Y \tag{4.50}$$

or

$$f_{X,Y}(x,y) = f_X(x) \cdot f_Y(y) \text{ for continuous } X \text{ and } Y. \tag{4.51}$$

It is clear that (4.50) and (4.51) are equivalent to the equality of conditional and unconditional distributions of either variable; for example, X and Y are independent if their ranges are independent and if $p(x|y) = p(x)$ or $f(x|y) = f(x)$. If you include the indicator functions for X and Y in the expression for the joint and marginal pmfs of densities of X and Y, then the identities in (4.50) and (4.51) will suffice for verifying independence, since the only way these identities can then hold is for the ranges to be independent. This is illustrated in the following example of a pair of variables X and Y that are shown to be independent.

Example 4.3.1. Suppose that the random pair (X, Y) had probability density function

$$f_{X,Y}(x,y) = 4xyI_{(0,1)}(x)I_{(0,1)}(y).$$

Applying Definition 4.3.1 to check the independence of X and Y requires that we obtain the marginal densities of the two variables. Here, we have, for $x \in (0,1)$,

$$f_X(x) = \int_0^1 4xydy = 2xy^2|_0^1 = 2x.$$

Similarly, we have, for $y \in (0,1)$,

$$f_Y(y) = \int_0^1 4xydx = 2x^2y|_0^1 = 2y.$$

In other words we have

$$f_X(x) = 2xI_{(0,1)}(x) \text{ and } f_Y(y) = 2yI_{(0,1)}(y).$$

Since we may write

$$f_{X,Y}(x,y) = 4xyI_{(0,1)}(x)I_{(0,1)}(y) = 2xI_{(0,1)}(x) \cdot 2yI_{(0,1)}(y) = f_X(x) \cdot f_Y(y),$$

we conclude that the random variables X and Y with the given joint density are independent variables. ■

In the example above, the calculation of the marginal densities of X and Y is a simple exercise, but you are no doubt willing to believe that in some problems, the integration (or, in the discrete case, summation) involved might be quite imposing. Wouldn't it be nice if you didn't actually have to do it? Wouldn't it be nice if you could just look at a joint density function and be able to tell "by inspection" that the variables involved are independent? The following theorem provides the means to do just that.

Theorem 4.3.1. (The Factorization Theorem) Suppose that X and Y are two random variables whose ranges are independent, that is, each variable takes its values in an interval that does not depend on the other variable. Then X and Y are independent variables if and only if there exist functions g and h such that the joint pmf or pdf of the random pair (X,Y) can be factored as

$$f_{X,Y}(x,y) = g(x) \cdot h(y). \tag{4.52}$$

Proof. For simplicity, we will consider only the continuous case. Since this result involves an "if and only if" statement, we must prove that independence implies factorization (the "only if" part) and that factorization implies independence (the "if" part). The "only if" part is immediate. If X and Y are independent, then their ranges are independent and, in addition,

$$f_{X,Y}(x,y) = f_X(x) \cdot f_Y(y).$$

Thus, identifying $g(x)$ as $f_X(x)$ and $h(y)$ as $f_Y(y)$, it is clear that (4.52) holds.

Now, assume that the ranges of X and Y are independent and that the factorization in (4.52) holds. Then the marginal densities may be computed as

$$f_X(x) = \int_{-\infty}^{\infty} g(x)h(y)dy = g(x)\int_{-\infty}^{\infty} h(y)dy = c \cdot g(x),$$

where c is a positive constant, and

$$f_Y(y) = \int_{-\infty}^{\infty} g(x)h(y)dx = h(y)\int_{-\infty}^{\infty} g(x)dx = d \cdot h(y),$$

where d is a positive constant. Moreover, since $f_{X,Y}(x,y)$ integrates to 1, we have

$$1 = \int_{-\infty}^{\infty}\int_{-\infty}^{\infty} g(x)h(y)dxdy = c \cdot d.$$

We thus see that

$$f_{X,Y}(x,y) = g(x) \cdot h(y) = c \cdot d \cdot g(x) \cdot h(y) = [c \cdot g(x)] \cdot [d \cdot h(y)] = f_X(x) \cdot f_Y(y),$$

showing that X and Y are independent. ■

The utility of Theorem 4.3.1 is immediately apparent. For the model in Example 4.3.1, we can see, by inspection, that X and Y are independent, as it is clear that the joint density may be factored as

$$f_{X,Y}(x,y) = [4xI_{(0,1)}(x)][yI_{(0,1)}(y)]. \tag{4.53}$$

It makes no difference that the two factors above are not actually the marginal densities of X and Y. If a factorization is possible, one factor will always be a constant multiple of $f_X(x)$ and the other factor will be a multiple of $f_Y(y)$. The fact that it is possible to factor $f_{X,Y}(x,y)$ into separate functions that, respectively, depend on x and y alone is sufficient to establish the independence of the variables involved. But a couple of practical matters are worth mentioning.

Remark 4.3.1. When the ranges of X and Y are dependent, that is "intermingled," you needn't even look at the possibility of factoring the joint density (or pmf) $f_{X,Y}(x,y)$. The possibility that X and Y are independent has already been excluded. For example, for the random pair (X,Y) with joint density

$$f_{X,Y}(x,y) = 2 \text{ for } 0 < x < y < 1,$$

the fact that Y must be larger than X already identifies the variables as dependent.

Remark 4.3.2. Theorem 4.3.1 gives us a useful tool for checking the independence of two variables X and Y. But you may have a nagging concern. What if a factorization exists but you simply don't recognize it? This is a real possibility, but it is not as large a worry as it may seem. You'll usually be able to factor $f_{X,Y}(x,y)$ when a factorization is possible. To do it, you may need to review some relevant algebra. Does the function e^{-x-y} factor? Yes it does: $e^{-x-y} = [e^{-x}][e^{-y}]$, by a well-known law of exponents. Does e^{-xy} factor? No it doesn't. We can write e^{-xy} as $[e^{-x}]^y$ by another law of exponents, but there's no way to tease out the y part from the x part. So if a joint density had such a term, multiplied by constants and perhaps other terms depending on x or y or both, a factorization would not be possible, and we could conclude that the X and Y involved were dependent. Does $(xy - 2y - 3x + 6)$ factor? It may not look like it at first, but $(xy - 2y - 3x + 6)$ may be written as $(x-2)(y-3)$, so that if the density for a random pair (X,Y) had the form

$$f_{X,Y}(x,y) = k \cdot (xy - 2y - 3x + 6) \cdot I_{(a,b)}(x) \cdot I_{(c,d)}(y)$$

for some constants k, a, b, c, and d, the variables X and Y would be independent.

The independence of two random variables has some important consequences. One implication of independence that we will find quite useful is the following.

Theorem 4.3.2. If X and Y are independent, then for any univariate functions r and s for which the expectations are finite,

$$E[r(X) \cdot s(Y)] = E[r(X)] \cdot E[s(Y)]. \tag{4.54}$$

Proof. We prove the claim for the continuous case. The discrete case is similar.

$$E[r(x) \cdot s(Y)] = \int_{-\infty}^{\infty}\int_{-\infty}^{\infty} r(x)s(y)f_{X,Y}(x,y)dxdy$$

$$
\begin{aligned}
&= \int_{-\infty}^{\infty}\int_{-\infty}^{\infty} r(x)s(y)f_X(x)f_Y(y)dxdy \\
&= \int_{-\infty}^{\infty} s(y)f_Y(y)\left[\int_{-\infty}^{\infty} r(x)f_X(x)dx\right]dy \\
&= E[r(X)]\int_{-\infty}^{\infty} s(y)f_Y(y)dy \\
&= E[r(X)]\cdot E[s(Y)].
\end{aligned}
$$

■

From Theorem 4.3.2, we obtain an important consequence of the property of independence. While the concepts of variables X and Y being independent and the variables being "uncorrelated" (that is, having $\rho_{X,Y} = 0$) are not equivalent, they are definitely related to each other, as evidenced by the following result.

Theorem 4.3.3. If X and Y are independent random variables, then $Cov(X,Y) = 0$. If, in addition, $\sigma_X > 0$ and $\sigma_Y > 0$, then X and Y are uncorrelated.

Proof. By Theorem 4.3.2, we have

$$Cov(X,Y) = E(X-EX)(Y-EY) = E(X-EX)\cdot E(Y-EY) = 0\cdot 0 = 0.$$

If $\sigma_X > 0$ and $\sigma_Y > 0$, it follows that $\rho_{X,Y} = \frac{Cov(X,Y)}{\sigma_X\sigma_Y} = 0$. ■

Remark 4.3.3. If, for example, X is degenerate (that is, $P(X = c) = 1$ for some constant c), then $\sigma_X = 0$, and $\rho_{X,Y}$ is undefined; it nonetheless follows that $Cov(X,Y) = 0$ for any random variable Y.

There are many statistical applications in which we will be interested in the behavior of sums of random variables, and often, in the context of interest, the assumption will be made that these variables are independent. Indeed, the "iid" assumption for a collection of n random variables will be made quite often. The acronym iid stands for "independent and identically distributed." Making the iid assumption is common for situations in which the data obtained are seen as coming from independent repetitions of the same experiment. We might make this assumption when tossing a coin n times or when performing the same laboratory experiment on n consecutive days. The data $X_1, X_2, \dots, X_n$ obtained in such a way can often justifiably be viewed as a random (iid) sample from a single underlying distribution F.

The following corollary of Theorem 4.3.3 provides a simplification of the variance formula in Theorem 4.2.3 under the additional assumption of independence.

Corollary 4.3.1. Assume that $X_1, \dots, X_n$ are independent random variables with finite variances. Then, for arbitrary real numbers $a_1, \dots, a_n$,

$$V\left(\sum_{i=1}^{m} a_i X_i\right) = \sum_{i=1}^{n} a_i^2 V(X_i).$$

Further, if $X_1, \dots, X_n \overset{\text{iid}}{\sim} F$, a distribution with variance $\sigma^2 < \infty$, and $\bar{X} = (1/n)\sum_{i=1}^{n} X_i$, then

$$V(\bar{X}) = \frac{\sigma^2}{n}.$$

We now turn our attention to moment-generating functions. As should be clear from our use of moment-generating functions in Chapters 2 and 3, these functions are valuable tools in the study of random variables and their exact or their limiting distributions. Our first result on mgfs considers the bivariate case and the sum of two random variables.

Theorem 4.3.4. If X and Y are independent, then for all values t for which the moment-generating functions exist,

$$m_{X+Y}(t) = m_X(t) \cdot m_Y(t). \tag{4.55}$$

Proof. By Theorem 4.3.2,

$$\begin{aligned} E(e^{t(X+Y)}) &= E(e^{tX} \cdot e^{tY}) \\ &= Ee^{tX} \cdot Ee^{tY} = m_X(t) \cdot m_Y(t). \end{aligned}$$

■

The following more general results are may be proven by similar arguments, and their proofs are left as exercises.

Theorem 4.3.5. Let $X_1, X_2, \ldots, X_n$ be a collection of n independent random variables. Then

$$m_{\sum_{i=1}^n X_i}(t) = \prod_{i=1}^n m_{X_i}(t). \tag{4.56}$$

The phrase "$X_1, X_2, \ldots, X_n$ are independent and identically distributed random variables with common distribution F" will typically be shortened in the sequel to "$X_1, X_2, \ldots, X_n \stackrel{\text{iid}}{\sim} F$." We use this shorthand notation for the first time in the following result.

Theorem 4.3.6. Suppose that $X_1, X_2, \ldots, X_n \stackrel{\text{iid}}{\sim} F$, a distribution with moment-generating function m. Then,

$$m_{\sum_{i=1}^n X_i}(t) = (m(t))^n. \tag{4.57}$$

Theorems 4.3.5 and 4.3.6 have a myriad of applications. Theorem 4.3.5 is often the best way to identify the distribution of a sum of independent variables. The main use of Theorem 4.3.6 is in identifying the exact distribution of a sum of iid variables having a known common distribution F. A sampling of the distributional results that may be obtained from applying the theorems above are given in the following example.

Example 4.3.2. (a) **Binomial.** Suppose that X and Y are independent variables, with $X \sim B(n,p)$ and $Y \sim B(m,p)$. Then the mgf of $X+Y$ is

$$m_{X+Y}(t) = (q+pe^t)^n(q+pe^t)^m = (q+pe^t)^{n+m},$$

where $q = 1-p$. Thus, $X+Y \sim B(n+m,p)$.

(b) **Binomial.** Suppose that $X_1, X_2, \ldots, X_n \stackrel{\text{iid}}{\sim} B(1,p)$. Then the mgf of $\sum_{i=1}^n X_i$ is

$$m_{\sum_{i=1}^n X_i}(t) = (q+pe^t)^n,$$

so that $\sum_{i=1}^n X_i \sim B(n,p)$.

(c) **Poisson.** Suppose that X and Y are independent variables, with $X \sim P(\lambda)$ and $Y \sim P(\mu)$. Then the mgf of $X+Y$ is

$$m_{X+Y}(t) = (e^{\lambda(e^t-1)})(e^{\mu(e^t-1)}) = e^{(\lambda+\mu)(e^t-1)},$$

so that $X+Y \sim P(\lambda+\mu)$.

(d) **Poisson.** Suppose that $X_1, X_2, \ldots, X_n \overset{\text{iid}}{\sim} P(\lambda)$. Then the mgf of $\sum_{i=1}^n X_i$ is

$$m_{\sum_{i=1}^n X_i}(t) = \left(e^{\lambda(e^t-1)}\right)^n = e^{n\lambda(e^t-1)},$$

so that $\sum_{i=1}^n X_i \sim P(n\lambda)$.

(e) **Exponential.** Suppose that $X_1, X_2, \ldots, X_n \overset{\text{iid}}{\sim} Exp(\lambda)$. Then the mgf of $\sum_{i=1}^n X_i$ is

$$m_{\sum_{i=1}^n X_i}(t) = \left(\left(1-\frac{t}{\lambda}\right)^{-1}\right)^n = \left(1-\frac{t}{\lambda}\right)^{-n},$$

so that $\sum_{i=1}^n X_i \sim \Gamma(n, 1/\lambda)$.

(f) **Gamma.** Suppose that X and Y are independent variables, with $X \sim \Gamma(\alpha_1, \beta)$ and $Y \sim \Gamma(\alpha_2, \beta)$. Then the mgf of $X+Y$ is

$$m_{X+Y}(t) = (1-\beta t)^{-\alpha_1}(1-\beta t)^{-\alpha_2} = (1-\beta t)^{-(\alpha_1+\alpha_2)},$$

so that $X+Y \sim \Gamma(\alpha_1+\alpha_2, \beta)$.

(g) **Normal.** Suppose that X and Y are independent variables, with $X \sim N(\mu_x, \sigma_x^2)$ and $Y \sim N(\mu_y, \sigma_y^2)$. Then the mgf of $X+Y$ is

$$m_{X+Y}(t) = e^{\mu_x t + \frac{\sigma_x^2 t^2}{2}} \cdot e^{\mu_y t + \frac{\sigma_y^2 t^2}{2}} = e^{(\mu_x+\mu_y)t + \frac{(\sigma_x^2+\sigma_y^2)t^2}{2}},$$

so that $X+Y \sim N(\mu_x+\mu_y, \sigma_x^2+\sigma_y^2)$.

(h) **Normal.** Suppose that $X_1, X_2, \ldots, X_n \overset{\text{iid}}{\sim} N(\mu, \sigma^2)$. Then the mgf of $\sum_{i=1}^n X_i$ is

$$m_{\sum_{i=1}^n X_i}(t) = (e^{\mu t + \frac{\sigma^2 t^2}{2}})^n = e^{n\mu t + \frac{n\sigma^2 t^2}{2}},$$

so that $\sum_{i=1}^n X_i \sim N(n\mu, n\sigma^2)$.

(i) **Normal.** Since $\bar{X} = \frac{1}{n}\sum_{i=1}^n X_i$ is a linear function of $\sum_{i=1}^n X_i$, we can identify the mgf of $\bar{X}$ as

$$m_{\bar{X}}(t) = m_{\sum_{i=1}^n X_i}(t/n) = e^{n\mu(t/n) + \frac{n\sigma^2(t/n)^2}{2}} = e^{\mu t + \frac{\sigma^2 t^2}{2n}},$$

so that $\bar{X} \sim N(\mu, \sigma^2/n)$.

■

Exercises 4.3.

1. Determine which of the random variables X and Y corresponding to density functions shown below are independent. Prove their independence when the property holds. If X and Y are not independent, show that $f(x) \neq f(x|y)$.

   (a) $f_{X,Y}(x,y) = 6e^{-2x-3y}I^2_{(0,\infty)}(x,y)$

   (b) $f_{X,Y}(x,y) = 24xyI_{(0,1)}(x)I_{(0,1-x)}(y)$

   (c) $f_{X,Y}(x,y) = \frac{6}{7}\left(x^2 + \frac{xy}{2}\right)I_{(0,1)}(x)I_{(0,2)}(y)$.

   (d) $f_{X,Y}(x,y) = e^{-2y}I_{(0,2)}(x)I_{(0,\infty)}(y)$.

2. Prove Theorem 4.3.5.

3. Prove Theorem 4.3.6.

4. Suppose that $X_1, X_2, \ldots, X_r \overset{\text{iid}}{\sim} G(p)$, the geometric distribution with parameter p. Then $Y = \sum_{i=1}^{r} X_i$ represents the number of trials needed to obtain the rth success. Thus, $Y \sim NB(r, p)$. Obtain the moment-generating function of Y.

4.4 The Multinomial Distribution

The multinomial model is a natural multivariate generalization of the binomial model. The binomial distribution models the number of occurrences X of a success in n independent repetitions of a basic experiment with two possible outcomes. In a binomial experiment, it suffices to keep track of X alone, the number of successes, since the number of failures will then necessarily be equal to $n - X$. In a multinomial experiment, the basic experiment has k outcomes, with $k > 2$, and the multinomial distribution keeps track of the number of occurrences $X_1, X_2, \ldots, X_k$ of each type of outcome in n independent repetitions of the basic experiment. As with the binomial case, the number X_k of occurrences of the kth outcome is completely determined when $X_1, \ldots, X_{k-1}$ are known. The three requirements that define a multinomial experiment are as follows:

(1) There is a basic experiment with k possible outcomes $O_1, O_2, \ldots, O_k$.

(2) The basic experiment is repeated independently n times.

(3) For each i, $P(O_i)$ is the same in every trial, and $\sum_{i=1}^{k} P(O_i) = 1$.

Let $p_i = P(O_i)$, and let X_i be the number of occurrences of the outcome O_i in the n trials of the basic experiment. Then the random vector $\mathbf{X}$ is said to have a multinomial distribution with parameters n and $\mathbf{p} = (p_1, p_2, \ldots, p_k)$. The fact that $\mathbf{X}$ has this distribution will be denoted as $\mathbf{X} \sim M_k(n, \mathbf{p})$. Let us derive the probability mass function of such an $\mathbf{X}$.

Consider a multinomial experiment consisting of n independent trials of a basic experiment with k possible outcomes. Using the notation above, let the vector $\mathbf{p} = (p_1, p_2, \ldots, p_k)$ represent the probabilities of the events $O_1, O_2, \ldots, O_k$ in an individual trial of the experiment. Now, if $x_1, x_2, \ldots, x_k$ are a set of k nonnegative integers for which $\sum_{i=1}^{k} x_i = n$, then we would like to evaluate the probability $P(X_1 = x_1, X_2 = x_2, \ldots, X_k = x_k)$. First we note that by the multiplication rule for independent events, any sequence of n outcomes in which outcome O_i occurs x_i times, for $i = 1, 2, \ldots, k$, will have the same probability, namely

$$\prod_{i=1}^{k} p_i^{x_i}.$$

To determine the probability of interest, we simply need to identify how many such sequences there can be. The answer is provided in Theorem 1.7.4 of Chapter 1. The number of ways in which n items can be divided into k groups of respective sizes $x_1, x_2, \ldots, x_{k-1}$, and x_k is given by the so-called multinomial coefficient

$$\binom{n}{x_1 \quad x_2 \quad \cdots \quad x_{k-1} \quad x_k} = \frac{n!}{\prod_{i=1}^{k} x_i!}.$$

It follows that

$$P(X_1 = x_1, X_2 = x_2, \ldots, X_k = x_k) = \frac{n!}{\prod_{i=1}^{k} x_i!} \prod_{i=1}^{k} p_i^{x_i} \tag{4.58}$$

for all vectors $\mathbf{x} = (x_1, x_2, \ldots, x_k)$ of integers $x_i \geq 0$ such that $\sum_{i=1}^{k} x_i = n$.

That the multinomial distribution is a natural generalization of the binomial distribution is confirmed by the fact that the marginal distribution of the component X_i, for any i, is $B(n, p_i)$. The validity of this claim is apparent from the fact that the occurrence or non-occurrence of the event O_i in any given trial is a dichotomous basic experiment, and X_i merely counts the number of occurrences of the outcome O_i in n independent trials in which $P(O_i) = p_i$. The three fundamental elements of a binomial experiment are thus satisfied, so that $X_i \sim B(n, p_i)$. Since sums of two or more X_is are also counts of the occurrence or non-occurrence of the corresponding O_is in n independent trials, it follows that $\sum_{i \in A} X_i \sim B(n, \sum_{i \in A} p_i)$. Finally, you might expect any two components of a multinomially distributed $\mathbf{X}$ would be negatively correlated since the sum of the X's is the fixed integer n, so that if one X is larger, other X's must be smaller. To confirm this intuition, we compute

$$\begin{aligned}
Cov(X_i, X_j) &= \left(\sum_{x_l \geq 0, l=1,\ldots,k; \sum x_l = n} x_i x_j \frac{n!}{\prod_{l=1}^{k} x_l!} \prod_{l=1}^{k} p_l^{x_l} \right) - ((np_i)(np_j)) \\
&= n(n-1) p_i p_j \left(\sum_{y_l \geq 0, l=1,\ldots,k; \sum y_l = n-2} \frac{(n-2)!}{\prod_{l=1}^{k} y_l!} \prod_{l=1}^{k} p_l^{y_l} \right) - ((np_i)(np_j)) \\
&\text{(where } y_s = x_s \text{ for } s \neq i \text{ or } j, y_i = x_i - 1 \text{ and } y_j = x_j - 1) \\
&= n(n-1) p_i p_j - (np_i)(np_j) \quad \text{(why?)} \\
&= -n p_i p_j.
\end{aligned}$$

Exercises 4.4.

1. Suppose that $\mathbf{X} \sim M_k(n, \mathbf{p})$, the multinomial distribution with parameters n and $\mathbf{p}$. Show that the vector $(X_1 + X_2, X_3, \ldots, X_k)$ has the multinomial distribution. Identify its parameters.
2. Suppose $\mathbf{X} \sim M_k(n, \mathbf{p})$. Calculate $V(X_i + X_j)$.
3. If a given traffic ticket during the work week is modeled as equally likely to be issued on any of the days Monday–Friday, what is the probability that, of the 15 tickets issued by Officer Krupke last week, 5 were issued on Monday, 5 on Wednesday, and 5 on Friday?

4.5 The Multivariate Normal Distribution

The multivariate normal distribution is the name given to a particular distribution of a random vector $\mathbf{X} = (X_1, X_2, \ldots, X_k)$, with $k \geq 2$, where each component X_i has a univariate normal distribution. There is more than one family of distributions that have this latter property, but there is one such model that has many other desirable properties and has thus gained exclusive ownership of the name. In this section, we will restrict our attention to the simplest case—the bivariate normal distribution—since our examination of this particular model will exhibit quite clearly the main characteristics of the general model.

The density of the bivariate normal distribution can be pictured as the surface you would get if you had a plastic object that looked like a normal curve and you stood it up on the floor and gave it a spin. You can imagine that the surface that was traced by the spinning of the upper boundary of the plastic curve would look like a bell in three dimensions. The bivariate normal density can look like that, but of course this is but a simple case. Bivariate normal curves needn't be symmetric, and horizontal cross sections of the density will generally look like ellipses rather than like circles. But if you think of a symmetrically shaped bell that has been distorted a bit by stretching its base along some line on the floor that goes through its central point (on the floor), then you will begin to see the possibilities in the shapes that these densities can have. A bivariate normal density appears on the book cover.

The bivariate normal (BVN) density is completely determined by five parameters. We will write

$$\begin{pmatrix} X \\ Y \end{pmatrix} \sim N_2(\mu, \Sigma) \tag{4.59}$$

where the mean vector μ and covariance matrix Σ are given by

$$\mu = \begin{pmatrix} \mu_x \\ \mu_y \end{pmatrix} \text{ and } \Sigma = \begin{bmatrix} \sigma_x^2 & \rho\sigma_x\sigma_y \\ \rho\sigma_x\sigma_y & \sigma_y^2 \end{bmatrix}. \tag{4.60}$$

We could write $N_2(\mu, \Sigma)$ as $BVN(\mu_x, \mu_y, \sigma_x^2, \sigma_y^2, \rho)$. The joint density of the $N_2(\mu, \Sigma)$ distribution is given by

$$f_{X,Y}(x,y) = \frac{1}{2\pi\sigma_x\sigma_y\sqrt{1-\rho^2}} e^{-\{(x-\mu_x)^2/2\sigma_x^2 + \rho[(x-\mu_x)/\sigma_x][(y-\mu_y)/\sigma_y] + (y-\mu_y)^2/2\sigma_y^2\}/(1-\rho^2)}, \tag{4.61}$$

for $-\infty < x < \infty$ and $-\infty < y < \infty$.

The fact that we refer to the distribution with the joint density given in (4.61) as the "bivariate normal distribution" suggests that the marginal distributions for both X and Y are univariate normal. Showing that this is the case is an easy but somewhat tedious exercise. If you integrate the joint density in (4.61) with respect to y or with respect to x, you will find that the marginal distributions of both X and Y are indeed normal; specifically,

$$X \sim N(\mu_x, \sigma_x^2) \text{ and } Y \sim N(\mu_y, \sigma_y^2). \tag{4.62}$$

These facts justify the symbols used to represent the four parameters μ_x, σ_x^2, μ_y, and σ_y^2 in the bivariate normal density. These parameters are, in fact, the means and variances of the component variables in the random pair (X,Y). You will notice that the fifth parameter in the bivariate normal density was given the name ρ, a name usually reserved for the correlation coefficient of the variables X and Y. You can verify by doing the requisite integration that

$$Cov(X,Y) = \rho\sigma_x\sigma_y, \tag{4.63}$$

and thus that the fifth parameter ρ of the bivariate normal model for (X,Y) is in fact the correlation coefficient $\rho_{X,Y}$.

The conditional distribution of one variable of the random pair (X,Y) given the value of the other variable is of special interest. Such distributions are not only univariate normal distributions, but they have a very interesting form. Given the facts stated above—the exact form of $f_{X,Y}(x,y)$, $f_X(x)$, and $f_Y(y)$—they are not difficult to derive. You will find that upon forming the appropriate ratios of these functions and simplifying, you'll obtain

$$X|Y = y \sim N(\mu_x + \rho\frac{\sigma_x}{\sigma_y}(y-\mu_y), \sigma_x^2(1-\rho^2)) \tag{4.64}$$

and

$$Y|X = x \sim N(\mu_y + \rho\frac{\sigma_y}{\sigma_x}(x-\mu_x), \sigma_y^2(1-\rho^2)). \tag{4.65}$$

If you have had some exposure to statistical regression analysis, the form of the conditional distribution in (4.65) will look hauntingly familiar. We will take a closer look at regression analysis in Chapter 11. For now, we will simply give a brief description that highlights its relationship to the conditional distribution above.

Regression analysis is a subfield of Statistics that seeks to examine and, if possible, quantify, the relationship between a random variable Y, often called the dependent variable, and one or more "explanatory" or independent variables. In the bivariate case, we are simply trying to identify the extent and form of the relationship between Y and a single explanatory variable X. We could examine many possible types of relationships in this case, but perhaps the most commonly investigated option is the possibility that the variables X and Y are related linearly. Now it would be too constraining, and generally unrealistic, to assume that Y is actually a linear function of X. Instead, the standard assumption made in "simple linear regression" is that X and Y are linearly related on the average. Specifically, the assumed model stipulates that

$$E(Y|X = x) = \beta_0 + \beta_1 x, \tag{4.66}$$

where β_0 and β_1 are unknown constants to be estimated from data, that is, from the observed experimental values $(x_1,y_1),\ldots,(x_n,y_n)$. The full model for the relationship may be written as follows: for any $i \in 1\ldots,n$, given $X_i = x_i$,

$$Y_i = \beta_0 + \beta_1 x_i + \varepsilon_i, \tag{4.67}$$

where $\varepsilon_i, i = 1,\ldots,n$ are uncorrelated random errors, often attributed to errors in measurement or errors of environmental origin. The most widely used assumption regarding the errors in (4.67) is quite a bit stronger: it is assumed that they are independent, normally distributed random variables with a common mean 0 and a common variance σ^2. Under this assumption, the simple linear regression model may be written as

$$Y|X = x \sim N(\beta_0 + \beta_1 x, \sigma^2). \tag{4.68}$$

Now, compare the display in (4.68) to the conditional distribution of $Y|X=x$ in (4.65) under the assumption the (X,Y) has a bivariate normal distribution. Note that the conditional distribution of Y, given $X=x$, in this latter case has a mean that is linear in x and has a variance that is a constant, independent of the value of X. In other words, if the bivariate normal model was an apt description of the stochastic behavior of the random pair (X,Y), then the standard simple linear regression model would describe the conditional behavior of Y, given X, perfectly. Now regression analysis is a useful method in situations well beyond the scope of the bivariate normal setting, but this latter setting is clearly one in which the fitting of straight lines to (x,y) data should prove fruitful for estimation and prediction purposes.

Example 4.5.1. Suppose that (X,Y) has a bivariate normal distribution with parameters $\mu_x=3$, $\sigma_x^2=4$, $\mu_y=5$, $\sigma_y^2=1$, and $\rho=1/2$. To compute the probabilities $P(4\leq X\leq 7)$ and $P(4\leq X\leq 7|Y=4)$, note first that from (4.62), that $X\sim N(3,4)$. It follows that $P(4\leq X\leq 7)=P(.5\leq Z\leq 2)=.2857$. From (4.65), we find that $X|Y=4\sim N(2,3)$. Thus, $P(4\leq X\leq 7|Y=4)\cong P(1.15\leq Z\leq 2.89)=.1232$. ■

The notion of variables X and Y being *independent* and the notion of X and Y being *uncorrelated* are similar and are clearly related. In Section 4.3, we saw that independent variables X and Y are necessarily uncorrelated. But the converse of this fact is not generally true. Uncorrelated random variables need not be independent. Exercise 4.5.4 represents an example of that fact for two continuous variables. In reality, the correlation between random variables X and Y is simply a measure of the strength of the *linear relationship* between them, it being positive if X and Y tend to be directly proportional and being negative if they tend to be inversely proportional. When $Cov(X,Y)=0$, neither of these descriptions apply.

This brings us to a treatment of a very special property of the bivariate normal distribution. The proof involves a lovely and fortuitous separation of parts that always reminds me of a famous biblical event.

Theorem 4.5.1. If (X,Y) has the bivariate normal distribution $N(\mu,\Sigma)$ and if $\rho=0$, then X and Y are independent.

Proof. (The Parting of the Red Sea.) Since X and Y clearly have independent ranges, we know that the independence of X and Y can be confirmed, under the stated condition, by showing that the joint density $f_{X,Y}(x,y)$ can be written as the product of the marginal densities of X and Y. The density of a bivariate normal random pair (X,Y) is given by

$$f_{X,Y}(x,y)=\frac{1}{2\pi\sigma_x\sigma_y\sqrt{1-\rho^2}}e^{-\{(x-\mu_x)^2/2\sigma_x^2+\rho[(x-\mu_x)/\sigma_x][(y-\mu_y)/\sigma_y]+(y-\mu_y)^2/2\sigma_y^2\}/(1-\rho^2)}.$$

When $\rho=0$, this density reduces to

$$\begin{aligned}f_{X,Y}(x,y)&=\frac{1}{2\pi\sigma_x\sigma_y}e^{-(x-\mu_x)^2/2\sigma_x^2+(y-\mu_y)^2/2\sigma_y^2}\\&=\frac{1}{\sqrt{2\pi}\sigma_x}e^{-(x-\mu_x)^2/2\sigma_x^2}\cdot\frac{1}{\sqrt{2\pi}\sigma_y}e^{-(y-\mu_y)^2/2\sigma_y^2}\\&=f_X(x)\cdot f_Y(y)\text{ for }(x,y)\in(-\infty,\infty)\times(-\infty,\infty).\end{aligned}$$

This factorization implies, by Theorem 4.3.2, that X and Y are independent. ■

Exercises 4.5.

1. Suppose the random pair (X,Y) has the bivariate normal density function displayed in (4.61). Verify that $X \sim N(\mu_x, \sigma_x^2)$.

2. Suppose the random pair (X,Y) has the bivariate normal density function displayed in (4.61). Confirm that $Cov(X,Y) = \rho\sigma_x\sigma_y$.

3. Suppose the random pair (X,Y) has the bivariate normal density function displayed in (4.61). Verify that the conditional distribution of Y, given $X = x$, is the normal distribution specified in (4.65).

4. Theorem 4.5.1 indicates that, at least when (X,Y) has a bivariate normal distribution, $Cov(X,Y) = 0$ implies that X and Y are independent. Examples abound of models in which that implication doesn't hold. Here's one involving two continuous variables. Suppose that the random pair (X,Y) has a uniform distribution on the circle $\{(x,y)|x^2+y^2 \le 1\}$. The density function of (X,Y) is given by

$$f_{X,Y}(x,y) = \frac{1}{\pi} \text{ for } -1 \le y \le 1 \text{ and } -\sqrt{1-y^2} \le x \le \sqrt{1-y^2}.$$

   Confirm that X and Y are not independent. Show that they are uncorrelated, that is, that $Cov(X,Y) = 0$.

4.6 Transformation Theory

The subject of Statistics is basically about the analysis and interpretation of data. A statistical study will often begin with the selection, and tacit assumption, of a general model for the experiment under study. This model will typically have one or more unknown parameters. When data (for example, a random sample of values X_i drawn from the population of interest) are collected, the statistical process of making inferences about the unknown parameters of the model begins. Typically, some function of the data (generally referred to as a "statistic") is identified and used as the main vehicle for the inferential process. For example, the sample mean $\bar{X}$ would typically be used in making inferences about the population mean μ and the sample variance s^2 would be used in making inferences about σ^2. This brings us to the main issue of this section, the process of finding the distribution of a function of one or more random variables from knowledge of the distributions of these variables. The area of study aimed at finding the distribution of a statistic based on the set of individual variables $X_1, X_2, \ldots, X_n$ is collectively referred to as "transformation theory." The three methods of handling transformations treated in the remainder of this section can be classified in terms of the basic tools on which they depend: i) the moment-generating function method, ii) the method of distribution functions, and iii) the "change of variable" technique. This section is divided into three subsections, one on each method.

4.6.1 The Method of Moment-Generating Functions

The fact that the moment-generating function can be a useful tool in finding the distribution of a statistic of interest is already apparent from the discussion and examples in Section 4.3. This "method" of finding the distribution of a particular statistic is clearly applicable when the statistic of interest is the sum of a collection of independent random variables, and it is especially useful when these variables are, in addition, identically distributed. Since random sampling from a large (or, conceptually, even infinite) population is typically modeled by the iid assumption (that is, $X_1, X_2, \dots, X_n \overset{\text{iid}}{\sim} F$, a distribution with mgf $m_X(t)$), and since $\sum_{i=1}^n X_i$ is a statistic that plays a key role in many statistical investigations, the moment-generating method will be relevant to many different modeling scenarios. The basic formula

$$m_{\sum_{i=1}^n X_i}(t) = (m_X(t))^n,$$

which holds under the iid assumption above, often provides the most efficient mechanism for discovering the distribution of $\sum_{i=1}^n X_i$. Example 4.3.2, which presents nine specific applications of the mgf method, aptly illustrates the broad reach of the approach.

While problems involving sums of independent random variables are no doubt the primary circumstance in which the mgf method is useful, the method can also be helpful in a variety of other settings. The examples below provide a glimpse into the broad applicability of this tool.

Example 4.6.1. Let $X \sim U(0,1)$, and let $Y = -\ln X$. We will identify the distribution of Y by computing its moment-generating function:

$$\begin{aligned} Ee^{tY} &= Ee^{-t\ln X} \\ &= EX^{-t} \\ &= \int_0^1 \frac{1}{x^t} \cdot 1 dx \end{aligned}$$

$$\begin{aligned} &= \frac{x^{-t+1}}{-t+1}\Big|_0^1 \quad \text{for } t \neq 1 \\ &= \frac{1}{1-t} \quad \text{for } t \neq 1. \end{aligned}$$

Thus, for $t < 1$, $m_Y(t) = (1-t)^{-1}$, a fact which implies that $Y \sim Exp(1)$. ■

Example 4.6.2. Let N be the number of bicycle accidents you have this year, and let X be the number of them that were your fault. Suppose that we model N as a Poisson variable and X, given N, as a binomial variable. Specifically, let

$$N \sim P(\lambda) \text{ and } X|N = n \sim B(n, p).$$

This model is well defined when $N = 0$, since then X is naturally taken to be equal to 0 with probability 1. To identify the marginal distribution of X, we will evaluate its moment-generating function via the application of Adam's rule. This calculation proceeds as follows:

$$
\begin{aligned}
Ee^{tX} &= E(E(e^{tX}|N)) \\
&= E[(q+pe^t)^N] \\
&= \sum_{n=0}^{\infty}(q+pe^t)^n\frac{\lambda^n e^{-\lambda}}{n!} \\
&= \sum_{n=0}^{\infty}\frac{[(q+pe^t)\lambda]^n e^{-\lambda}}{n!}
\end{aligned}
$$

$$
\begin{aligned}
&= e^{-\lambda}e^{\lambda(q+pe^t)}\sum_{n=0}^{\infty}\frac{[(q+pe^t)\lambda]^n e^{-\lambda(q+pe^t)}}{n!} \\
&= e^{-\lambda(q+p)}e^{\lambda(q+pe^t)} \\
&= e^{\lambda p(e^t-1)}.
\end{aligned}
$$

This calculation reveals that the marginal distribution of X is Poisson, or more specifically, that $X \sim P(\lambda p)$. ■

Remark 4.6.1. Note that this problem above derives the marginal distribution of the sum $X = \sum_{i=1}^{N} X_i$ of a random number of Bernoulli variables, a problem of interest in its own right. The answer obtained will of course depend on the way in which the variable N is modeled. When N is a Poisson variable, we see that X is also Poisson distributed.

Example 4.6.3. Suppose that $Z \sim N(0,1)$. Let us consider the random variable $Y = Z^2$. We will show that Y has a gamma distribution. More specifically, $Y \sim \chi^2_1$, the chi-square distribution with one degree of freedom. The mgf of the random variable $Y = Z^2$ can be calculated as

$$
\begin{aligned}
Ee^{tZ^2} &= \int_{-\infty}^{\infty} e^{tz^2}\frac{1}{\sqrt{2\pi}}e^{-z^2/2}dz \\
&= \int_{-\infty}^{\infty}\frac{1}{\sqrt{2\pi}}e^{-z^2(\frac{1}{2}-t)}dz \\
&= \int_{-\infty}^{\infty}\frac{1}{\sqrt{2\pi}}e^{-z^2/2(1-2t)^{-1}}dz
\end{aligned}
$$

$$
\begin{aligned}
&= (1-2t)^{-\frac{1}{2}}\int_{-\infty}^{\infty}\frac{1}{\sqrt{2\pi}(1-2t)^{-\frac{1}{2}}}e^{-\frac{z^2}{2(1-2t)^{-1}}}dz \qquad (4.69)\\
&= (1-2t)^{-\frac{1}{2}} \text{ for } t < 1/2,
\end{aligned}
$$

where the last equality holds (provided $t < 1/2$) because the integrand in (4.69) is the density of the $N(0,(1-2t)^{-1/2})$ distribution and thus integrates to 1. We can readily recognize the m.g.f. in (4.69) as that of the $\Gamma(\frac{1}{2},2)$ or χ^2_1 distribution. ■

Example 4.6.4. Let $Y_1, Y_2, \ldots \stackrel{\text{iid}}{\sim} B(1, p)$ be a sequence of independent Bernoulli trials, and let X_1 be the number of trials needed to obtain the first success. We know that $X_1 \sim G(p)$. Let X_2 be the number of additional trials needed to obtain the second success. If $X_1 = x$, then X_2 is a function of the Bernoulli trials $Y_{x+1}, Y_{x+2}, \ldots$ and these variables are independent of the Bernoulli trials $Y_1, \ldots, Y_x$. From this, it follows that X_2 is independent of X_1. The unconditional distribution of X_2 is clearly also $G(p)$, and $X_1 + X_2$, being the number of Bernoulli trials needed to obtain the second success, has the $NB(2, p)$ distribution. A similar argument establishes that for $r \geq 2$, $X = \sum_{i=1}^{r} X_i \sim NB(r, p)$. Since $X_1, \ldots, X_r \stackrel{\text{iid}}{\sim} G(p)$, it follows that the mgf of X is $m(t) = [pe^t/(1 - qe^t)]^r$. We may then identify $m'(0) = r/p$ and $m''(0) = (r^2 + r - rp)/p^2$, from which it follows that $EX = r/p$ and $V(X) = rq/p^2$. Of course, if all we wanted to know was the mean and variance of X, we could find them most easily using Theorem 4.2.2 and Corollary 4.3.1. ■

Exercises 4.6.1.

1. Suppose $X \sim \Gamma(\alpha, \beta)$, and let $Y = cX$, where c is a positive constant. What's the distribution of Y?

2. Let $X|Y \sim N(Y, 1)$, and suppose that $Y \sim N(0, 1)$. Show, by calculating the moment-generating function of X, that $X \sim N(0, 2)$.

3. The joint moment-generating function of the random pair (X_1, X_2) is defined as

$$m_{X_1, X_2}(t_1, t_2) = E\left(e^{t_1 X_1 + t_2 X_2}\right).$$

As in the univariate case, the joint mgf of a random pair (X_1, X_2), if it exists, may be shown to uniquely determine the joint distribution of (X_1, X_2). Suppose now that X and Y are independent $N(\mu, \sigma^2)$ variables. Show that the variables $X + Y$ and $X - Y$ are independent normal variables by showing that

$$m_{X+Y, X-Y}(t_1, t_2) = m_{X+Y}(t_1) \cdot m_{X-Y}(t_2).$$

From this, it can be seen that the joint distribution of $X + Y$ and $X - Y$ is the same as that of two independent normal variables. Identify the distributions of the variables $X + Y$ and $X - Y$. (Note: In the case of a sample of size 2 from $N(\mu, \sigma^2)$, this exercise shows that sample mean $(X + Y)/2$ and the sample variance s^2, which may be written as $(X - Y)^2/2$, are independent. The independence of the mean and variance of a random sample drawn from a normal population actually holds for samples of any size.)

4. Suppose that $X|p \sim B(n, p)$ and that $p \sim U(0, 1)$. (a) Find the moment-generating function of X and confirm that X has the discrete uniform distribution of the integers $0, 1, 2, \ldots, n$. (Hint: In evaluating the integral of $E(e^{tX}|p)$ with respect to p, expand this conditional expectation using the Binomial Theorem, and then apply the trick, term by term, using what you know about the beta distribution.) (b) Show that $p|X = x \sim Be(x + 1, n - x + 1)$. A historical aside: this problem is the very example used by the Reverend Thomas Bayes in putting forward his revolutionary proposal (published in 1763) for estimating an unknown proportion p based on n Bernoulli trials. The estimator he proposed was $\hat{p} = E(p|X = x) = \frac{x+1}{n+2}$. We'll return to the topic of Bayesian estimation in Chapter 9.

5. Suppose that $X \sim G(p)$, the geometric distribution on $\{1, 2, 3, \ldots\}$, and let $Y|X \sim NB(x, p)$,

the negative binomial distribution with $r = x$. Find the moment-generating function $m_Y(t)$ of Y. (Hint: First, recall that the mgf of the NB(r,p) distribution is $[pe^t/(1-qe^t)]^r$, where $q = 1-p$. Use Adam's rule to write $E[e^{tY}] = E[E(e^{tY}|X)]$. The inner expectation is just an NB mgf. The correct answer will not be a familiar mgf, but it qualifies as $m_Y(t)$ for $t < \ln[1/(q+pq)]$, where the latter number is necessarily positive since $(q+pq) < q+p = 1$.

4.6.2 The Method of Distribution Functions

In selected problems, the strategy of deriving an exact expression for the distribution function of a statistic T of interest works efficiently and well. We will refer to this approach as the "distribution function method." Since we often begin and end a problem involving a transformation with the density functions of the original and transformed variables, the distribution function method generally involves the additional step of obtaining the density of the transformed variable by differentiation. In a univariate problem in which the density $f_X(x)$ of X is known and the distribution function of a transformed variable $Y = g(X)$ is sought, the basic strategy is to write the distribution function of Y as

$$F_Y(y) = \int_{\{x:g(x)\le y\}} f_X(x)dx,$$

and then to obtain the density of Y as $f_Y(y) = F_Y'(y)$. The following example shows how the distribution function method works in a univariate problem.

Example 4.6.5. Suppose that $f_X(x) = 2xI_{(0,1)}(x)$, and let $Y = 8X^3$. For $y \in (0,8)$, the cdf of Y may be obtained as

$$\begin{aligned}
F_Y(y) &= P(Y \le y) \\
&= P(8X^3 \le y) \\
&= P(X \le \sqrt[3]{y}/2) \\
&= \int_0^{\sqrt[3]{y}/2} 2xdx \\
&= x^2\Big|_0^{\sqrt[3]{y}/2} \\
&= \frac{y^{2/3}}{4} \text{ for } y \in (0,8),
\end{aligned}$$

with $F_Y(y) = 0$ if $y \le 0$ and $F_Y(y) = 1$ if $y \ge 8$. From this, we obtain

$$f_Y(y) = F_Y'(y) = \frac{2}{3}y^{-1/3}\left(\frac{1}{4}\right)I_{(0,8)}(y) = \frac{1}{6y^{1/3}}I_{(0,8)}(y).$$

■

Among the most ubiquitous routines available on modern computers are those used to *generate random numbers*. These numbers are generally presented as a series of digits like 48351, say, with the understanding that each of the digits included in the string was generated in a way that gave every digit in the set $\{0,1,2,3,4,5,6,7,8,9\}$ the same likelihood of occupying the displayed digit's position. Such random digits are often provided by a computer without requiring any special software, but they are also often generated through a statistical package like Minitab, SAS, or R, or through a hybrid package like Excel. (By the way, some random number generators are known to be more reliable than others, and I recommend that you research the relevant literature before selecting one for use.) We will see momentarily that the standard mechanism for generating random variables having a given known distribution function F is derived from a straightforward application of the "method of distribution functions."

If you place a decimal point before the first digit in the strings of digits mentioned above, you can regard each outcome as a variable drawn from the distribution $U(0,1)$. The iid uniform variables you obtain from a random number generator may be used to create a random sample from any given distribution F. The process of *simulating* random variables from a given distribution F is widely used in both stochastic and statistical work in dealing with problems in which exact analytical solutions are cumbersome or even seemingly impossible. For example, if you wanted to know the likelihood that a certain Weibull variable X is larger than a particular logistic variable Y, where X and Y are independent, we would find that obtaining an exact answer isn't analytically feasible. However, the relative frequency of occurrence of the event $\{X > Y\}$ in 1000 simulated random (X,Y) pairs would provide a very good approximate answer. Simulations are also often used to assess the performance of a particular procedure, or to compare the performance of competing procedures, in addressing statistical problems of estimation, testing, or prediction. The transformation used to accomplish the transition from a variable having the continuous distribution F to a variable having the $U(0,1)$ distribution is called the *Probability Transform*. It is justified by the following:

Theorem 4.6.1. Suppose that $X \sim F$, where F is a continuous, strictly increasing probability distribution. Then $Y = F(X) \sim U(0,1)$. Conversely, if $Y \sim U(0,1)$, then $X = F^{-1}(Y) \sim F$.

Proof. Suppose, first, that $X \sim F$, and let $Y = F(X)$. Then

$$P(Y \leq y) = P(F(X) \leq y) = P(X \leq F^{-1}(y)) = F(F^{-1}(y)) = y,$$

where the existence of F^{-1} is guaranteed by our assumptions about F. It follows that $Y \sim U(0,1)$. Now, suppose that $Y \sim U(0,1)$. If $X = F^{-1}(Y)$, then

$$P(X \leq x) = P(F^{-1}(Y) \leq x) = P(Y \leq F(x)) = F(x),$$

which shows that $X \sim F$. ∎

The use of the Probability Transform F, in reverse, in generating unit exponential variables is illustrated below.

Example 4.6.6. Let $X \sim Exp(1)$, the unit exponential distribution with cdf

$$F(x) = 1 - e^{-x} \text{ for } x > 0.$$

To identify the function F^{-1}, we need to solve the equation

$$y = 1 - e^{-x}$$

for x in terms of y. This leads us to the expression

$$x = -\ln(1 - y).$$

The inverse function F^{-1} is thus given by

$$F^{-1}(y) = -\ln(1-y).$$

That F^{-1} is indeed the inverse of the function F is evident from the fact that

$$F^{-1}(F(x)) = -\ln(1-(1-e^{-x})) = -\ln(e^{-x}) = -(-x) = x,$$

that is, $F^{-1}F$ is the identity function. According to Theorem 4.6.1, if $Y \sim U(0,1)$, then $X = -\ln(1-Y) \sim Exp(1)$. We independently verify this conclusion below. For any $x > 0$, we have

$$\begin{aligned} F_X(x) &= P(X \le x) \\ &= P(-\ln(1-Y) \le x) \\ &= P(1-Y \ge e^{-x}) \\ &= P(Y \le 1-e^{-x}) \\ &= 1-e^{-x}. \end{aligned}$$

Thus, we see that $X \sim Exp(1)$, as claimed. ■

Let us now consider a bivariate example of the distribution function method.

Example 4.6.7. Let X and Y be independent random variables, each with an $Exp(1)$ distribution. The joint density of the pair (X,Y) is thus

$$f_{X,Y}(x,y) = e^{-x}I_{(0,\infty)}(x) \cdot e^{-y}I_{(0,\infty)}(y) = e^{-x-y}I_{(0,\infty)^2}(x,y).$$

Let $V = X+Y$. We already know the distribution of V, since the mgf of V is clearly $m(t) = [(1-t)^{-1}]^2 = (1-t)^{-2}$, the mgf of the $\Gamma(2,1)$ distribution. To obtain this result by the method of distribution functions, we write, for any $v > 0$,

$$\begin{aligned} F_V(v) &= P(X+Y \le v) \\ &= \iint_{\{(x,y):x+y\le v\}} e^{-x-y}dxdy \\ &= \int_0^v \int_0^{v-y} e^{-x-y}dxdy \\ &= \int_0^v e^{-y}\left[-e^{-x}\right]_0^{v-y} dy \\ &= \int_0^v e^{-y}(1-e^{-v+y})dy \\ &= \int_0^v (e^{-y}-e^{-v})dy \\ &= \left[-e^{-y}-ye^{-v}\right]_{y=0}^{v} \\ &= 1-e^{-v}-ve^{-v}. \end{aligned}$$

From this, we obtain, by differentiation,

$$f_V(v) = e^{-v} - e^{-v} + ve^{-v} = ve^{-v}I_{(0,\infty)}(v).$$

Thus, V has the $\Gamma(2,1)$ distribution, as determined earlier by other means. ■

Exercises 4.6.2.

1. Suppose X has a $Be(2,2)$ distribution with density $f_X(x) = 6x(1-x)I_{(0,1)}(x)$. Find the density of the random variable $Y = X/(1-X)$ using the method of distribution functions.

2. Let $X \sim W(\alpha, \beta)$, the two-parameter Weibull distribution with density function
$$f_X(x) = \frac{\alpha x^{\alpha-1}}{\beta^\alpha} e^{-(x/\beta)^\alpha} I_{(0,\infty)}(x),$$
where $\alpha > 0$ and $\beta > 0$ are the model's "shape" and "scale" parameters, respectively. The c.d.f. of the $W(\alpha, \beta)$ model is given by
$$F(x) = 1 - e^{-(x/\beta)^\alpha} \text{ for } x > 0.$$
Let $Y = X^\alpha$. Use the distribution function method to find the density of the variable Y, and identify the distribution of Y by name.

3. Suppose $X \sim W(\alpha, \beta)$ as in Exercise 4.6.2.2 above. Let $Y \sim U(0,1)$. Identify a function of Y that has the $W(\alpha, \beta)$ distribution.

4. Suppose $X \sim P(\alpha, \theta)$, two-parameter Pareto model with cdf $F(x) = 1 - (\theta/x)^\alpha$ for $x > \theta$. Let $Y \sim U(0,1)$. Identify a function of Y that has the $P(\alpha, \theta)$ distribution.

5. Let $X \sim Be(\theta, 1)$. Determine the unique function $Y = g(X)$ for which Y has the $U(0,1)$ distribution.

6. Let X and Y be independent "unit exponential" variables, that is, each have the $\Gamma(1,1)$ distribution. Let $U = X/Y$. Using the method of distribution functions, find the density of U.

4.6.3 The Change-of-Variable Technique

By far, the most broadly useful and widely used method of obtaining the density of a function $Y = g(X_1, \ldots, X_k)$ of one or more continuous random variables is based on the *Change-of-Variable* (CoV) technique from calculus. Some authors refer to it as the "substitution method." We'll introduce the method with a discussion of its application to functions of a single variable. Suppose that X is a random variable with known density $f_X(x)$ and that we wish to obtain the density function of the variable $Y = g(X)$, where g is a continuous function that is a *one-to-one* and *differentiable* mapping between its domain (the possible values of X) and its range (the set of all images $g(x)$ for x in the support set of X). The significance of these assumptions about the transformation g is that (1) they guarantee that g is invertible, that is, that there is a well-defined, unique function g^{-1} through which X may be expressed as $g^{-1}(Y)$ and (2) they specify that the function g^{-1} is differentiable, a smoothness condition that is required in the developments below. If the function g is defined on an interval (a,b) of real numbers, then g will be continuous and one-to-one in that interval if and only if g is strictly monotone (i.e., g is either strictly increasing or strictly decreasing for $x \in (a,b)$).

The CoV technique gets its name from the fact that it is inextricably associated with changing the variable of integration (let's call it x) in an integral such as $\int r(x)dx$ to a new variable of integration $y = g(x)$ in a new, but equivalent, integral $\int s(y)dy$. Given the density $f_X(x)$ of X and the 1:1, differentiable transformation g, with $Y = g(X)$, we can obtain the density $f_Y(y)$ of Y by executing the following four-step program:

(1) Solve for x in terms of y: $x = g^{-1}(y)$.

(2) Find the Jacobian $J = dx/dy$ of the inverse transformation:

$$J = \frac{dx}{dy} = \frac{d}{dy}g^{-1}(y).$$

(3) Identify the range B of the variable Y corresponding to the range (i.e., the support set) A of X, that is, identify the set $B = [y|y = g(x), x \in A]$.

(4) Write:

$$f_Y(y) = f_X(g^{-1}(y)) \cdot |J| \cdot I_B(y),$$

where $|J|$ is the absolute value of the Jacobian and B is the range of the variable Y.

The program above is precisely what you would implement if you were integrating a function $r(x)$ and were changing the variable x in the integrand $r(x)$ to the variable $y = g(x)$. Step 4 of this program changes this old integrand to $h(g^{-1}(y))$, changes the range of integration from the range A for x to the range B for y, and replaces the differential dx by $|J|dy$. The first of these changes simply makes the obvious substitution, replacing the variable x, wherever it is found in the integrand, by what x is when written as a function of y (i.e., replacing x by $g^{-1}(y)$). The change from the range of x to the range of y simply adjusts the limits of integration so that they reflect the values that the new variable y can take on. Finally, replacing dx by $|J|dy$ is required in order to adjust for the fact that the rate of change in y will be different than the rate of change in x. Even in the simplest case where $y = cx$, a unit change in x corresponds to a change of the amount c in y. The Jacobian of the inverse transformation is $J = 1/c$, and the fact that dx would be replaced by $|1/c|dy$ makes the differentials equivalent. In most applications, the Jacobian is not constant, but rather, it is a function of y. The absolute value of J is meant to account for the fact that, if we always evaluate integrals from the smaller limit to the larger limit, the integration will be correct when J is negative because we have simultaneously changed the sign of the differential while reversing the upper and lower limits of the integral.

I hope that you have found the brief "calculus review" above helpful. Of course, there's nothing more helpful than seeing the program *in action*. Let us consider a transformation problem that we solved in the previous subsection using a different method.

Example 4.6.8. Suppose that $f(x) = 2xI_{(0,1)}(x)$, and let $Y = 8X^3$. The density of Y may be obtained as follows:

(1) $y = 8x^3 \Rightarrow x = y^{1/3}/2$.

(2) $J = \frac{dx}{dy} = (1/3)y^{-2/3}/2 = 1/6y^{2/3}$.

(3) As x goes from 0 to 1, y goes from 0 to 8 (that is, $A = (0,1)$ and $B = (0,8)$).

(4) $f_Y(y) = 2 \cdot y^{1/3}/2 \cdot |1/6y^{2/3}| I_{(0,8)}(y)$,

which simplifies to

$$f_Y(y) = \frac{1}{6y^{1/3}} I_{(0,8)}(y).$$

■

Note that this is the same density derived in Example 4.6.5. You might sense that the CoV technique is slightly easier to implement than the method of distribution functions, although in this simple problem, both methods are quite straightforward. In more complex problems, the CoV method will almost always win, partly because the process is unambiguous and "automated," but also because no integration is required in obtaining the solution. Let's take another look at Example 4.6.2, this time applying the CoV technique.

Example 4.6.9. Suppose that $Z \sim N(0,1)$. Let us consider the random variable $Y = Z^2$. Now the transformation $g(z) = z^2$ is continuous but is not a one-to-one function. But since $|Z|^2 = Z^2$, we may take the original variable to be $X = |Z|$. The density of $|Z|$ is obtained by simply doubling the standard normal density on the positive real line, that is, X has the half normal distribution with

$$f(x) = \frac{\sqrt{2}}{\sqrt{\pi}} e^{-x^2/2} I_{(0,\infty)}(x).$$

(1) $y = x^2 \Rightarrow x = y^{1/2}$.

(2) $J = \frac{dx}{dy} = \frac{1}{2} y^{-1/2}$.

(3) as x goes from 0 to ∞, y goes from 0 to ∞.

(4) $f_Y(y) = \frac{\sqrt{2}}{\sqrt{\pi}} \cdot e^{-y/2} \cdot \left|\frac{1}{2} y^{-1/2}\right| I_{(0,\infty)}(y)$,

which we rewrite, after some cancelation and upon recalling that $\sqrt{\pi} = \Gamma(1/2)$, as

$$f_Y(y) = \frac{1}{\Gamma(1/2) \cdot 2^{1/2}} y^{(1/2)-1} e^{-\frac{y}{2}} I_{(0,\infty)}(y),$$

which is the $\Gamma(1/2,2)$or χ^2_1 density. ■

We now turn to the treatment of the bivariate case. Suppose we have a random pair (X_1, X_2) with known density function $f_{\mathbf{X}}(x_1, x_2)$, and we are interested in obtaining the density function of a random variable $Y = g(X_1, X_2)$. There is an inherent obstacle to the immediate application of the CoV method to obtaining the density of Y. In order to apply the method, it is necessary to have a one-to-one, invertible relationship between the original and the new variables. A mapping from the pair (X_1, X_2) to the single variable Y will not be invertible in any nontrivial problem. This difficulty is overcome by creating a second "Y variable" and mapping the original pair to the new pair. You might think that this doesn't accomplish our original goal of obtaining the density of the Y variable we had in mind, but this is a just a temporary illusion. What we will do, after identifying the joint density of the pair of Ys, is integrate the second Y over its range to obtain the marginal density of the first Y, precisely what we set out to find. The details of this process is the five-step program below, the fifth step being the integration needed to obtain the univariate density of interest.

Now, let $f_{X_1,X_2}(x_1, x_2)$ be the joint density of the random pair (X_1, X_2), and define

$$Y_1 = g_1(X_1, X_2)$$

and

$$Y_2 = g_2(X_1, X_2),$$

where $\mathbf{g} = (g_1, g_2)$ is a continuous function that is *one-to-one* and *differentiable*. The joint density of (Y_1, Y_2) may be obtained via the following five-step program:

(1) Invert the relationship between (Y_1,Y_2) and (X_1,X_2), that is, express X_1 and X_2 in terms of Y_1 and Y_2:

$$X_1 = g_1^{-1}(Y_1,Y_2)$$

and

$$X_2 = g_2^{-1}(Y_1,Y_2).$$

Note that g_1^{-1} and g_2^{-1} are not inverses of g_1 and g_2 in the usual sense, but are used here for notational convenience. In reality, they are the two functions that identify the X's in terms of the Y's.

(2) Find the Jacobian of the inverse transformation, that is, obtain the determinant of the matrix of partial derivatives below:

$$J = \det \begin{bmatrix} \frac{\partial x_1}{\partial y_1} & \frac{\partial x_1}{\partial y_2} \\ \frac{\partial x_2}{\partial y_1} & \frac{\partial x_2}{\partial y_2} \end{bmatrix}.$$

Note that the determinant of the matrix $\begin{bmatrix} a & b \\ c & d \end{bmatrix}$ is $ad - bc$.

(3) Identify the range B of the pair (Y_1,Y_2) which corresponds to the range A of (X_1,X_2), that is, identify the set

$$B = [(y_1,y_2)|(y_1,y_2) = (g_1(\mathbf{x}), g_2(\mathbf{x})), \mathbf{x} \in A].$$

(4) Write the joint density of (Y_1,Y_2) as

$$f_{\mathbf{Y}}(y_1,y_2) = f_{\mathbf{X}}(g_1^{-1}(y_1,y_2), g_2^{-1}(y_1,y_2)) \cdot |J| \cdot I_B(y_1,y_2).$$

(5) Obtain the density of Y_1: for any fixed value y_1 in the global range of Y_1, integrate the joint density $f_{\mathbf{Y}}(y_1,y_2)$ with respect to y_2 over all values that Y_2 may take on, when $Y_1 = y_1$.

While the development of the joint density of the two variables Y_1 and Y_2 is usually just a stepping stone on the way to obtaining the density of Y_1, the variable in which we are really interested, it is worth noting that their joint density itself will occasionally be of interest in its own right. There are problems in which both Y_1 and Y_2, and the relationship between them, are our main focus. For instance, we may be interested in whether Y_1 and Y_2 are independent, something that can be determined by inspection once we've obtained the joint density of the two variables. Further, there are occasions, in bivariate problems like the one treated above or in general multivariate problems, in which it is useful to identify the distribution of each of the transformed variables. We will encounter examples of both of these latter situations. However, since we will be principally occupied with the identification of the density of a single function of two or more of the original X variables, it is important to call attention to and, indeed, emphasize, an interesting and useful insight regarding the selection of the Y variables that will accompany the variable Y_1 of primary or sole interest. This insight is the fact that the precise way in which the auxiliary Ys are chosen actually makes no difference in the final answer, since to get the density of Y_1, all other Y variables will be integrated out of the problem. This implies that, as long as the choice of auxiliary variables provides a one-to-one, smooth (i.e., differentiable) relationship between the Xs and the Ys, they will suffice in carrying out the CoV technique. It will pay dividends to ask the question: what choice of auxiliary Y variables will simplify the CoV process? We will return to this question. Let us first examine a particular bivariate transformation problem with different choices of an accompanying variable Y_2.

Example 4.6.10. We will treat a now familiar problem using the CoV technique. Let $X_1, X_2 \overset{\text{iid}}{\sim} \Gamma(1,1)$ with joint density

$$f_{X_1,X_2}(x_1,x_2) = e^{-x_1-x_2} I_{(0,\infty)^2}(x_1,x_2),$$

and let

$$Y_1 = X_1 + X_2. \tag{4.70}$$

We have of course already established that $Y_1 \sim \Gamma(2,1)$. To obtain this result through our new method, we need to define a second variable Y_2 such that the mapping $(Y_1,Y_2) = \mathbf{g}(X_1,X_2)$ is one-to-one and differentiable. Let

$$Y_2 = \frac{X_1}{X_1 + X_2}. \tag{4.71}$$

This choice is motivated by the desire to simplify the inversion we need to perform to express the Xs in terms of the Ys. The fact that $Y_1 \cdot Y_2 = X_1$ suggests that the required inversion will be quite straightforward for this choice of Y_2. We now proceed with the five-step program.

(1) Invert the relationship between (Y_1,Y_2) and (X_1,X_2), that is, express X_1 and X_2 in terms of Y_1 and Y_2; given the formulae in (4.70) and (4.71), it's clear that

$$X_1 = Y_1 \cdot Y_2$$

and

$$X_2 = Y_1 - X_1 = Y_1 - Y_1 \cdot Y_2.$$

(2) Find the Jacobian of the inverse transformation:

$$J = \det \begin{bmatrix} \frac{\partial x_1}{\partial y_1} & \frac{\partial x_1}{\partial y_2} \\ \frac{\partial x_2}{\partial y_1} & \frac{\partial x_2}{\partial y_2} \end{bmatrix} = \det \begin{bmatrix} y_2 & y_1 \\ 1-y_2 & -y_1 \end{bmatrix} = -y_1 y_2 - y_1(1-y_2) = -y_1.$$

Since $P(Y_1 > 0) = 1$, we thus have $|J| = y_1$.

(3) Identify the range B of the pair (Y_1,Y_2) which corresponds to the range A of (X_1,X_2):

Since X_1 and X_2 can take arbitrary values in $(0,\infty)$, the range of $Y_1 = X_1 + X_2$ is also $(0,\infty)$. Suppose the value of Y_1 is fixed. Then $Y_2 = X_1/(X_1+X_2)$ clearly takes values between 0 and 1, and its range does not depend on the value of Y_1. We therefore conclude that the range of (Y_1,Y_2) is $(0,\infty) \times (0,1)$.

(4) Write the joint density of (Y_1,Y_2) as

$$f_{\mathbf{Y}}(y_1,y_2) = e^{-y_1} \cdot y_1 \cdot I_{(0,\infty)\times(0,1)}(y_1,y_2). \tag{4.72}$$

(5) Obtain the density of Y_1: Ordinarily, we would integrate the joint density in (4.72) with respect to Y_2 to obtain the desired univariate density. In this instance, we have our result by inspection, since it is clear from the factorization theorem that Y_1 and Y_2 are independent random variables. Indeed, writing their joint density as

$$f_{\mathbf{Y}}(y_1,y_2) = y_1 e^{-y_1} \cdot I_{(0,\infty)}(y_1) \times 1 \cdot I_{(0,1)}(y_2)$$

confirms both the independence of Y_1 and Y_2 and the fact that $Y_1 \sim \Gamma(2,1)$ while $Y_2 \sim U(0,1)$.

■

It is instructive to redo the solution above with a different auxiliary variable Y_2.

Example 4.6.11. With X_1 and X_2 as specified in Example 4.6.10, let us now define

$$Y_1 = X_1 + X_2. \tag{4.73}$$

and

$$Y_2 = X_2. \tag{4.74}$$

Choosing one of the original X variables as an auxiliary variable is the standard choice in most problems, as it immediately facilitates the inversion to express the Xs in terms of Ys, and it also guarantees, without giving the matter any extra thought, that the transformation from the Xs to the Ys is one-to-one. The five-step program now looks like this:

(1) Invert the relationship between (Y_1,Y_2) and (X_1,X_2), that is, express X_1 and X_2 in terms of Y_1 and Y_2; given the formulae in (4.73) and (4.74), it's clear that

$$X_1 = Y_1 - Y_2$$

and

$$X_2 = Y_2.$$

(2) Find the Jacobian of the inverse transformation:

$$J = \det \begin{bmatrix} \frac{\partial x_1}{\partial y_1} & \frac{\partial x_1}{\partial y_2} \\ \frac{\partial x_2}{\partial y_1} & \frac{\partial x_2}{\partial y_2} \end{bmatrix} = \det \begin{bmatrix} 1 & -1 \\ 0 & 1 \end{bmatrix} = 1.$$

We thus have $|J| = 1$. Not only is the inversion simpler in this scenario, the Jacobian is substantially simpler, as is often the case when the auxiliary Ys are set equal to original Xs.

(3) Identify the range B of the pair (Y_1,Y_2) implied by the range A of (X_1,X_2): Since X_1 and X_2 can take arbitrary values in $(0,\infty)$, the range of $Y_1 = X_1 + X_2$ is also $(0,\infty)$. Suppose the value of Y_1 is fixed. Then $Y_2 = X_2$ clearly can only take values that are smaller than Y_1. Thus, we conclude that the random pair (Y_1,Y_2) must obey the constraint $0 < Y_2 < Y_1 < \infty$. The range of (y_1,y_2) may be written as $(0,\infty) \times (0,y_1)$, or alternatively, as $(y_2,\infty) \times (0,\infty)$.

(4) Write the joint density of (Y_1,Y_2) as

$$f_{\mathbf{Y}}(y_1,y_2) = e^{-y_1} \quad \text{for } 0 < y_2 < y_1 < \infty.$$

(5) Obtain the density of Y_1:

$$f_{Y_1}(y_1) = \int_o^{y_1} e^{-y_1} dy_2 = e^{-y_1} [y_2]_0^{y_1} = y_1 e^{-y_1} I_{(0,\infty)}(y_1),$$

that is, $Y_1 \sim \Gamma(2,1)$. ■

It should be mentioned that the most challenging part of implementing the CoV technique in a bivariate problem is identifying the range of the transformed variables. Obviously, if that step is not executed correctly, there is no hope of obtaining the correct expression for the target univariate density, as the limits of integration in step 5 will be misspecified, leading to an error in evaluating the integral. The following example shows that the use of a visual tool can be helpful in getting the crucial step (3) right. In this and other examples in the sequel, we switch to a notation for the random variables involved in order to dispense with the use of subscripts.

Example 4.6.12. Let $X,Y \stackrel{\text{iid}}{\sim} U(0,1)$, and suppose we wish to obtain the density function of the variable

$$U = X + Y. \tag{4.75}$$

Let

$$V = X - Y. \tag{4.76}$$

The density of U may be obtained in the usual 5 steps.

(1) The inverse transformations are given by

$$X = \frac{U+V}{2}$$

and

$$Y = \frac{U-V}{2}.$$

(2) The Jabobian of the inverse transformation is

$$J = \det\begin{bmatrix} \frac{1}{2} & \frac{1}{2} \\ \frac{1}{2} & -\frac{1}{2} \end{bmatrix} = -\frac{1}{2}$$

so that $|J| = 1/2$.

(3) The random pair (X,Y) takes values in the unit square $(0,1)^2$. The corresponding range of (U,V) takes a little more work to identify. However, it's not difficult to confirm that the corresponding ranges of (X,Y) and (U,V) are shown in the figures pictured below.

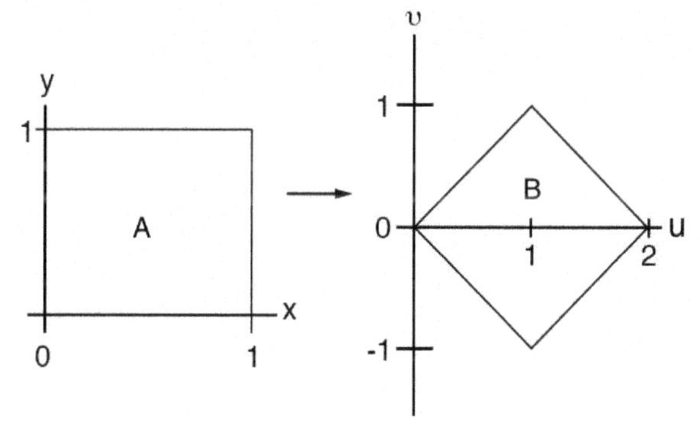

Figure 4.6.1. Ranges of (X,Y) and (U,V) in Example 4.6.12.

A method that is often used for obtaining the plot of the range of (U,V) in Figure 4.6.1 is to determine where the boundaries of the plot of the range of (X,Y) map onto in applying the transformations in (4.75) and (4.76). For example, the portion of the (X,Y) range A that lies on the y-axis, and thus contains values of the form $(0,y)$, maps onto the points $\{(y,-y), 0<y<1\}$ in the set B. This is the lower left-hand boundary of the plot showing the range B of (U,V). A similar analysis shows that the portion of the (X,Y) range that lies on the x-axis, and thus contains values of the form $(x,0)$, maps onto the points $\{(x,x), 0<x<1\}$. This is the upper left-hand boundary of the plot showing the range B of (U,V). The other two boundaries of the range A map onto the two remaining boundaries of the range B. Since the interior of the first plot maps onto the interior of the second plot, the range of the random pair (U,V) has been identified. Analytically, the range of the variables u and v may be expressed as:

$$\text{For } u \in (0,1), v \in (-u,u); \text{ for } u \in (1,2), v \in (u-2, 2-u). \tag{4.77}$$

The ranges in (4.77) can easily be confirmed to be in agreement with B in Figure 4.6.1.

(4) The joint density of U and V is given by

$$f_{U,V}(u,v) = 1 \cdot |-1/2| I_B(u,v) = (1/2) I_B(u,v).$$

(5) The density of U may then be obtained as follows. For, $0 \le u \le 1$,

$$f_U(u) = \int_{-u}^{u} 1/2 dv = \left[\frac{v}{2}\right]_{-u}^{u} = \frac{u}{2} - \left(-\frac{u}{2}\right) = u,$$

while for $1 < u \le 2$,

$$f_U(u) = \int_{u-2}^{2-u} 1/2 dv = \left[\frac{v}{2}\right]_{u-2}^{2-u} = \frac{2-u}{2} - \left(\frac{u-2}{2}\right) = 2-u.$$

This density is that of the (symmetric) triangular distribution on the interval $(0,2)$. It is pictured in the figure below.

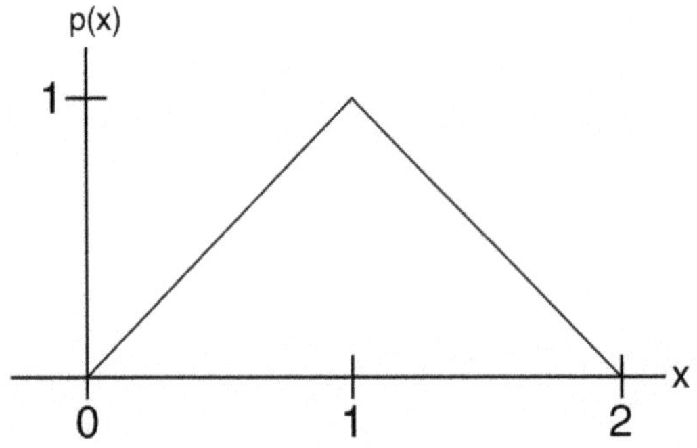

Figure 4.6.2. The density of the symmetric triangular distribution on $(0,2)$.

This concludes Example 4.6.12. ■

As my final example of the CoV method, I will derive the Student's t-distribution, a probability distribution that has a celebrated history in Probability and Statistics. William Gossett, a chemist turned statistician in the employ of the Guinness Brewing Company in Dublin, Ireland, in the early 1900s, found that he was constantly dealing with the analysis of small samples of product when examining the characteristics of interest in daily production of beer at the brewery. Let us recall (from Example 4.3.2 (part (i)) that the mean $\bar{X}_n$ of a random sample $X_1,\ldots,X_n$ from a normal population with mean μ and variance σ^2 has, itself, the normal distribution $N(\mu,\sigma^2/n)$. This fact was known in Gossett's day. Thus, the ratio $Z=(\bar{X}_n-\mu)/(\sigma/\sqrt{n})$ was known to have a standard normal distribution. The statistic was often used to test that the mean of the characteristic of the beer being measured had a certain required value μ_0. Since the population variance σ^2 was (as is typical) unknown, it was common at the time to use the statistic $t=(\bar{X}_n-\mu)\,/\,(s/\sqrt{n})$ as the test statistic, where s is the sample standard deviation, and to treat the statistic as having, approximately, a $N(0,1)$ distribution. Gossett was the first to notice that this approximation was quite poor when the sample size n was small. He set out to find the distribution of the statistic I have labeled as t above. He did find the distribution of t (and created useful tables for it) by a method that, today, would be described as "simulation." Guinness would not permit Gossett to publish his finding under his own name (for proprietary reasons); he published it under the admirably modest pen name "A Student." Ronald Fisher, at Gossett's request, studied the problem mathematically, and derived the density function using the approach taken in the example below. We present it as an abstract theorem in transformation theory. Its connection with Gossett's problem will be explained following the proof of the result.

Theorem 4.6.2. Let X and Y be independent random variables, with $X\sim N(0,1)$ and $Y\sim\chi^2_k$, the chi-squared distribution with k degrees of freedom (or the $\Gamma(\frac{k}{2},2)$ distribution). Let $T=X/\sqrt{Y/k}$. Then the statistic T has the "t-distribution with k degrees of freedom," denoted t_k, with density function

$$f_T(t)=\frac{\Gamma\left(\frac{k+1}{2}\right)}{\Gamma\left(\frac{k}{2}\right)\sqrt{\pi k}\left(1+\frac{t^2}{k}\right)^{\frac{k+1}{2}}}\quad\text{for } -\infty<t<\infty. \tag{4.78}$$

Proof. Let us first record the joint density of X and Y.

$$f_{X,Y}(x,y)=\frac{1}{\sqrt{2\pi}}e^{-x^2/2}\cdot\frac{1}{\Gamma(k/2)2^{k/2}}y^{k/2-1}e^{-y/2}\text{ for } x\in(-\infty,\infty)\text{ and } y\in(0,\infty).$$

Let $T=X/\sqrt{Y/k}$ and $V=Y$. Following the 5-step program above, we will first find the joint density of (T,V) and then obtain the density of T by integration. Proceeding as indicated,

(1) $X=T\sqrt{Y/k}=T\sqrt{V/k}$ and $Y=V$.

(2) The Jacobian of the inverse transformation is

$$J=\det\begin{bmatrix}\sqrt{v/k} & t/2\sqrt{vk}\\ 0 & 1\end{bmatrix}.$$

Since $V>0$ with probability 1, $|J|=\sqrt{v/k}$.

(3) Since $V=Y$, $V\in(0,\infty)$. For any fixed value $V=v$, it is clear that, as x takes values in $(-\infty,\infty)$, the range of $t=x/\sqrt{v/k}$ is also $(-\infty,\infty)$.

(4) The joint density of T and V is

$$f_{T,V}(t,v)=\frac{1}{\sqrt{2\pi}\Gamma(k/2)\,2^{k/2}}v^{k/2-1}e^{-t^2v/2k}e^{-v/2}\cdot\sqrt{v/k}\cdot I_{(-\infty,\infty)\times(0,\infty)}(t,v).$$

This joint density may be rewritten as

$$f_{T,V}(t,v) = \frac{1}{\sqrt{2\pi}\Gamma(k/2)\,2^{k/2}\sqrt{k}} v^{(k+1)/2-1} e^{-[v(1+t^2/k)]/2} \cdot I_{(-\infty,\infty)\times(0,\infty)}(t,v).$$

(5) To obtain the density of T, we will integrate the joint density of T and V with respect to v, using the trick to turn the integral into a constant (independent of v) times the integral of a certain gamma density. Toward that end, we write

$$f_T(t) = \frac{1}{\sqrt{2\pi}\Gamma(k/2)\,2^{k/2}\sqrt{k}} \cdot \Gamma\left(\frac{k+1}{2}\right)\left(\frac{2}{1+t^2/k}\right)^{\frac{k+1}{2}} \times$$
$$\int_0^{\infty} \frac{1}{\Gamma((k+1)/2)\,(2/(1+t^2/k))^{(k+1)/2}} v^{(k+1)/2-1} e^{-v/[2/(1+t^2/k)]} dv. \tag{4.79}$$

It is evident that the integrand in (4.79) is precisely the density function of the $\Gamma((k+1)/2, 2/(1+t^2/2))$ distribution and thus integrates to one. After some minor reorganization and cancellation, we obtain the desired result, that is,

$$f_T(t) = \frac{\Gamma\left(\frac{k+1}{2}\right)}{\Gamma\left(\frac{k}{2}\right)\sqrt{\pi k}\left(1+\frac{t^2}{k}\right)^{\frac{k+1}{2}}} \text{ for } -\infty < t < \infty. \quad \blacksquare$$

How does the result above provide an analytical solution to Gossett's problem? To see that it does, we begin by restating the conditions under which the result will be applied. Suppose that $X_1, X_2, \ldots, X_n \stackrel{\text{iid}}{\sim} N(\mu, \sigma^2)$. Then

$$X = \frac{(\bar{X}-\mu)}{\sigma/\sqrt{n}} \sim N(0,1).$$

This X will play the role of the numerator of the t-statistic displayed in the statement of Theorem 4.6.2. Secondly, it can be shown that the sample mean $\bar{X}$ and the sample standard deviation $s = [1/(n-1)]\sum_{i=1}^{n}(X_i - \bar{X})^2$ are independent. (This is proven for the special case $n = 2$ in Exercises 4.6.1.3 and 4.6.3.7, but the result holds for an arbitrary sample size n, and may be proven using the CoV method applied to the variables $\bar{X}$ and the $(n-1)$ differences $X_2 - X_1, X_3 - X_1, \ldots, X_n - X_1$. The independence of $\bar{X}$ and these differences can be shown by the factorization theorem, and since the sample variance is a function of the differences alone, it too is independent of $\bar{X}$.) Finally, since $\bar{X}$ and s^2 are independent, it follows from the algebraic identity

$$\sum_{i=1}^{n}\left(\frac{X_i-\mu}{\sigma}\right)^2 = \frac{(n-1)s^2}{\sigma^2} + \left(\frac{\bar{X}-\mu}{\sigma/\sqrt{n}}\right)^2, \tag{4.80}$$

using an mgf argument and the facts that the first term in (4.80) has a χ^2_n distribution and the third term has a χ^2_1 distribution, that

$$Y = \frac{(n-1)s^2}{\sigma^2} \sim \chi^2_{n-1}.$$

This Y is the chi-squared variable that appears in the denominator of the t statistic. Putting these elements together, we have that

$$\frac{\frac{(\bar{X}-\mu)}{\sigma/\sqrt{n}}}{\sqrt{\frac{(n-1)s^2}{\sigma^2}/(n-1)}} = \frac{\bar{X}-\mu}{s/\sqrt{n}} \sim t_{n-1}.$$

Exercises 4.6.3.

1. Suppose X has a $Be(2,2)$ distribution with density $f(x) = 6x(1-x)I_{(0,1)}(x)$. Find the density of the random variable $Y = X/(1-X)$ using the CoV technique.

2. The density function of the two-parameter Weibull distribution $W(\alpha,\beta)$ is

$$f_X(x) = \frac{\alpha x^{\alpha-1}}{\beta^{\alpha}} e^{-(x/\beta)^{\alpha}} I_{(0,\infty)}(x),$$

where $\alpha > 0$ and $\beta > 0$ are the model's "shape" and "scale" parameters, respectively. Use the CoV technique to find the density of the variable $Y = X^{\alpha}$.

3. The Cauchy distribution provides an example of a continuous probability model whose mean is not finite. The "standard" Cauchy variable X has a density function given by

$$f_X(x) = 1/\pi(1+x^2) \text{ for } -\infty < x < \infty.$$

Find the density of the random variable $Y = 1/X$. (Note: Y is also Cauchy. You may ignore the fact that Y is undefined when $X = 0$, since this is an event that occurs with probability 0.)

4. Let X and Y be independent "unit exponential" variables, that is, suppose that each has the $\Gamma(1,1)$ distribution. Let $U = X/Y$. Use the CoV technique to find the density of U. (Hint: The auxiliary transformed variable $V = Y$ will work well.)

5. Let X and Y be independent random variables with joint density $f_{X,Y}(x,y) = 8xy^3$ for $(x,y) \in (0,1)^2$. Find the density function of the variable $U = XY$. (Hint: Let $V = Y$. In obtaining $f_{U,V}(u,v)$ and $f_U(u)$, keep in mind that $U < V$, a consequence of the fact that $X < 1$.)

6. Let $X,Y \overset{\text{iid}}{\sim} N(0,1)$, and define $U = X+Y$ and $V = X-Y$. Find the joint density function of U and V. Show, by applying the factorization theorem, that U and V are independent. Conclude that both have the $N(0,2)$ distribution. (Remark: Note that this is essentially the same problem as Exercise 4.6.1.3, but the result is established with a different tool.)

7. Let (X,Y) have density $f_{X,Y}(x,y) = 6(x-y)$ for $0 < y < x < 1$. Find the density of $U = X-Y$. (Hint: Let $V = X$; then U and V satisfy the constraint $0 < U < V < 1$.)

4.7 Order Statistics

Suppose that F is a continuous univariate distribution function with density function f, and let $X_1, X_2, \ldots, X_n \overset{\text{iid}}{\sim} F$. In this section, we will be interested in the marginal and joint distributions of the "order statistics" associated with the sampled Xs. The order statistics corresponding to the Xs are defined as

$$X_{1:n} = \text{ the smallest X in the set } \{X_1, X_2, \ldots, X_n\}$$

$$X_{2:n} = \text{ the second smallest X in the set } \{X_1, X_2, \ldots, X_n\}$$

$$\ldots\ldots$$

$$X_{n:n} = \text{ the largest X in the set } \{X_1, X_2, \ldots, X_n\}.$$

For simplicity, we will use a shorthand notation for the order statistics from a sample of size n, one which assumes that the size of the sample under discussion is understood. We will denote the ith order statistic in a sample of size n by

$$X_{(i)} = X_{i:n} \qquad \text{for } i = 1, 2, \ldots, n.$$

The collection of order statistics consists of the same values as the original sample of Xs, with the additional constraint that they are ordered from smallest to largest:

$$X_{(1)} < X_{(2)} < \ldots \cdots < X_{(n)}.$$

We will display below the density function of the kth order statistic $X_{(k)}$ for arbitrary $k \in \{1, \ldots, n\}$. Throughout, we will assume, for simplicity, that the support set of the density $f(x)$ is the interval of real numbers (a, b). In most examples and problems you'll see, the intervals of interest will be $(-\infty, \infty), (0, \infty)$, or $(0, 1)$, but other intervals may also occur. Certain order statistics are especially easy to deal with. For example, we can derive the distribution and density functions of $X_{(n)}$, the maximum observation in the sample, by arguing as follows. For any $x \in (a, b)$,

$$\begin{aligned} P(X_{(n)} \leq x) &= P(X_1 \leq x, X_2 \leq x, \ldots, X_n \leq x) \\ &= \prod_{i=1}^{n} P(X_i \leq x) \text{ (by the independence of the } X\text{s)} \\ &= [F(x)]^n \text{ (since the } X\text{s have common distribution } F). \end{aligned}$$

From this, it follows that

$$f_{X_{(n)}}(x) = n(F(x))^{n-1} f(x) I_{(a,b)}(x). \tag{4.81}$$

The density of $X_{(1)}$, the smallest order statistic, can be obtained by a similar argument:

$$\begin{aligned} P(X_{(1)} \leq x) &= 1 - P(X_1 > x, X_2 > x, \ldots, X_n > x) \\ &= 1 - \prod_{i=1}^{n} P(X_i > x) \text{ (by the independence of the } X\text{s)} \\ &= 1 - (1 - F(x))^n \text{ (since the } X\text{s have common distribution } F). \end{aligned}$$

It thus follows that

$$f_{X_{(1)}}(x) = n(1 - F(x))^{n-1} f(x) I_{(a,b)}(x). \tag{4.82}$$

Obtaining the density of an arbitrary order statistic is conceptually similar but a bit more tedious algebraically. It is left as an exercise. However, since I am, generally speaking, a kind person, I will outline the steps to be taken before stating the result of interest and leaving the detailed derivation to you. To obtain the density of $X_{(k)}$ for $1 < k < n$, it helps to recognize the underlying binomial problem associated with order statistics: the event $\{X_{(k)} \leq x\}$ is equivalent to the event $\{Y \geq k\}$, where $Y \sim B(n, F(x))$. In other words, $X_{(k)}$

will be less than or equal to x precisely when k or more of the individual Xs are less than or equal to x. Thus, $P(X_{(k)} \leq x)$ is a sum of the $n-k+1$ binomial probabilities below:

$$P(X_{(k)} \leq x) = \sum_{i=k}^{n} \binom{n}{i} (F(x))^i (1-F(x))^{n-i}. \tag{4.83}$$

When a derivative of the sum in (4.83) is taken, using the product rule term by term, the entire expression can be recognized as a "telescoping sum" in which all terms except one cancel out. The remaining term is the desired expression, namely,

$$f_{X_{(k)}}(x) = \frac{n!}{(k-1)!(n-k)!}[F(x)]^{k-1}[1-F(x)]^{n-k} f(x) I_{(a,b)}(x). \tag{4.84}$$

The three densities above could have been derived as marginal densities obtained by integration from the joint density of $(X_1, X_2, \dots, X_n)$. Indeed, this is the preferred approach in delving further into the properties of order statistics. Since we make only moderate use of these more complex results, I will simply state certain useful facts here without proof, providing only a brief commentary on how they are derived. The most basic result characterizes the joint behavior of all n order statistics. The joint density of $(X_{(1)}, X_{(2)}, \dots, X_{(n)})$ is

$$f_{(X_{(1)},\dots,X_{(n)})}(x_{(1)}, \dots, x_{(n)}) = n! \prod_{i=1}^{n} f(x_{(i)}) \text{ for } a < x_{(1)} < \cdots < x_{(n)} < b. \tag{4.85}$$

The derivation of (4.85) is obtained through an interesting variation of the change-of-variable technique. When the range of a mapping can be partitioned into a collection of mutually exclusive sets B_i for which the transformation $\mathbf{g}$ of interest is one-to-one from the original variables to each B_i, then the density of the transformed variables may be obtained as a sum of the individual transformed densities. In the case of the transformation of the original iid data to the order statistics, the range space can be decomposed into $n!$ sets B_i, each representing one of the $n!$ ways in which the original Xs can be ordered. Since each of the components of the sum which constitutes the joint density of the order statistics consists of the product of the density f evaluated at the ordered x's (each with a different ordering), each component contributes the same product. The fact that there are $n!$ orderings leads to the multiple $n!$ in (4.85). A more detailed defense of the joint density in (4.85) may be found in Hogg, McKean, and Craig (2012). For our purposes, we will simply accept (4.85) as an established fact. To bolster your confidence that (4.85) is correct, let us derive, using this joint density, the density of the maximum order statistic $X_{(3)}$ in a sample of size 3. We may obtain the marginal density of $X_{(3)}$ as follows. (Compare it to the density in (4.81) when $n = 3$.)

$$\begin{aligned}
f_{X_{(3)}}(x_{(3)}) &= \int_a^{x_{(3)}} \int_a^{x_{(2)}} 3! f(x_{(1)}) f(x_{(2)}) f(x_{(3)}) dx_{(1)} dx_{(2)} \\
&= \int_a^{x_{(3)}} 3! \left[F(x_{(1)})\right]_a^{x_{(2)}} f(x_{(2)}) f(x_{(3)}) dx_{(2)} \\
&= \int_a^{x_{(3)}} 3! F(x_{(2)}) f(x_{(2)}) f(x_{(3)}) dx_{(2)} \\
&= 3! \left[\frac{F^2(x_{(2)})}{2}\right]_a^{x_{(3)}} f(x_{(3)})
\end{aligned}$$

$$= 3\left[F(x_{(3)})\right]^2 f(x_{(3)})I_{(a,b)}(x_{(3)}).$$

The joint density of a pair of order statistics $X_{(j)}$ and $X_{(k)}$, for $j < k$, may be obtained by integrating out all but those two order statistics from the overall joint density in (4.84). This process is facilitated by the fact that at each stage of the integration, the integrand may be viewed as constant multiples of terms of the form

(a) $[F(x_{(i)})]^k f(x_{(i)})$, which, when integrated with respect to $x_{(i)}$, has a perfect antiderivative $\frac{\left[F(x_{(i)})\right]^{k+1}}{k+1}$, or

(b) $[1-F(x_{(i)})]^k f(x_{(i)})$, which, when integrated with respect to x_i, has a perfect antiderivative $-\frac{\left[1-F(x_{(i)})\right]^{k+1}}{k+1}$, or

(c) $[F(x_{(j)})-F(x_{(i)})]^k f(x_{(i)})$, which, when integrated with respect to x_i, has a perfect antiderivative $-\frac{\left[F(x_{(j)})-F(x_{(i)})\right]^{k+1}}{k+1}$.

The result of such integration is the following: for $j < k$,

$$f_{(X_{(j)},X_{(k)})}(x_{(j)},x_{(k)}) = \frac{n!}{(j-1)!(k-j-1)!(n-k)!} \times [F(x_{(j)})]^{j-1}$$
$$[F(x_{(k)})-F(x_{(j)})]^{k-j-1} \times [1-F(x_{(k)})]^{n-k} f(x_{(j)})f(x_{(k)})I_{(a,b)}(x_{(j)})I_{(x_{(j)},b)}(x_{(k)}). \quad (4.86)$$

Remark 4.7.1. The smallest and largest order statistic occur together with some regularity in statistical applications, and in that case, the joint density in (4.86) takes the particularly simple form

$$f_{(X_{(1)},X_{(n)})}(x_{(1)},x_{(n)}) = n(n-1)[F(x_{(n)})-F(x_{(1)})]^{n-2}$$
$$\times f(x_{(1)})f(x_{(n)})I_{(a,b)}(x_{(1)})I_{(x_{(1)},b)}(x_{(n)}). \quad (4.87)$$

Remark 4.7.2. The univariate and joint densities above can be treated analytically only when both f and F for the model under study are known in closed form. Among the models we have studied, the uniform, exponential, Pareto, Weibull, Cauchy, and logistic models have this property.

Example 4.7.1. Let $X_1, X_2, \ldots, X_n \overset{\text{iid}}{\sim} U(0,1)$, and let $X_{(1)}, X_{(2)}, \ldots, X_{(n)}$ be the associated order statistics. Since $F(x) = x$ for $0 < x < 1$ and $f(x) = I_{(0,1)}(x)$, the density of $X_{(k)}$ reduces to

$$f_{X_{(k)}}(x) = \frac{n!}{(k-1)!(n-k)!}x^{k-1}(1-x)^{n-k}I_{(0,1)}(x);$$

thus, for this model, $X_{(k)} \sim Be(k, n-k+1)$, and $EX_{(k)} = k/(n+1)$ for $k = 1, \ldots, n$.

Also, the joint density of the smallest and largest order statistic in a random sample of n variables drawn from $U(0,1)$ may be written as

$$f_{(X_{(1)},X_{(n)})}(x_{(1)},x_{(n)}) = n(n-1)[x_{(n)}-x_{(1)}]^{n-2}I_{(0,1)}(x_{(1)})I_{(x_{(1)},1)}(x_{(n)}).$$

■

Exercises 4.7.1.

1. Prove the expression in (4.83) for the density of the kth order statistic $X_{k:n}$.

2. Let $X_1, X_2, \ldots, X_n \overset{\text{iid}}{\sim} P(\alpha, \theta)$, the two-parameter Pareto distribution. Find the density of the smallest order statistic and identify it by name.

3. Let $X_1, X_2, \ldots, X_n \overset{\text{iid}}{\sim} Exp(\lambda)$. Find the density of the smallest order statistic and identify it by name.

4. Let $X_1, X_2, \ldots, X_n \overset{\text{iid}}{\sim} U(\theta - 1, \theta + 1)$. Find the joint density of the largest and smallest order statistics. Derive an expression for $P(X_{1:n} < \theta < X_{n:n})$.

4.8 Chapter Problems

1. Let (X, Y) be a discrete random pair with probability mass function

$$p_{X,Y}(x,y) = (x+y)/27 \text{ for } x \in 0,1,2 \text{ and } y \in 1,2,3.$$

(a) Obtain the marginal pmfs of X and Y and the conditional pmfs of $X|Y = y$ and $Y|X = x$. (b) Evaluate the expectations $E(X)$, $E(XY)$, and $E(X|Y = 2)$.

2. Suppose that X is the proportion of his/her annual allowance that a random eight-year-old will spend on toys. The density of X is known to be

$$f_X(x) = 20x^3(1-x) \text{ for } 0 < x < 1.$$

Suppose that Y is the proportion of his/her annual allowance that is spent on candy. Assume that $Y|X = x \sim U(0, 1-x)$. (a) Obtain the marginal density of Y and the conditional density of X given $Y = y$. (b) Obtain an expression for $E(X|Y = y)$.

3. Geneticist D. N. Hay has been experimenting with two types of mice, light-haired mice cloned from a particularly hearty species and dark-haired mice that are direct descendants of a well-known mouse named Mickey. Assume that X and Y are, respectively, the proportion of each of these two types of mice that exit a specially designed maze within three minutes. Over many experiments with different batches of these mice, the variables X and Y have been found to be well-modeled by the joint density

$$f_{X,Y}(x,y) = kx^2(1-y) \text{ for } 0 < x < 1 \text{ and } 0 < y < 1.$$

(a) Find the value of k for which $f(x,y)$ is a valid bivariate density. (b) Find the expected value of the ratio Y/X. (c) Evaluate the probability that $X + Y \le 1$.

4. Suppose a fair coin is tossed 3 times. Let X = the number of heads in the first 2 tosses and let Y = the number of heads in the last 2 tosses. Find (a) the joint probability mass function (pmf) of the pair (X, Y), (b) the marginal pmf of each, (c) the conditional pmf of X given $Y = 1$ and also given $Y = 2$, and (d) the correlation $\rho_{X,Y}$ between X and Y.

5. Suppose that $Y \sim \Gamma(2,\theta)$ and $X|Y = y \sim U[0,y]$. The joint distribution of the random pair (X,Y) is thus given by
$$f_{X,Y}(x,y) = (1/\theta^2)e^{-y/\theta} \text{ for } 0 < x < y < \infty.$$
(a) Find the marginal density of X. (b) Evaluate $E(Y|X = x)$.

6. Let (X, Y) have joint density function
$$f_{X,Y}(x,y) = 10xy^2 \text{ for } 0 < x < y < 1.$$
(a) Evaluate the expected values of X, Y, XY, and of the ratio X/Y. (b) Calculate the correlation ρ between the variables X and Y.

7. Light of the World Industries, a religious organization based in the U.S. Bible Belt, is able to supplement the donations it receives by manufacturing high-quality light bulbs. The lifetimes of their light bulbs (measured in units of 1000 hours) are known to follow a $W(2,1)$ distribution, that is, the distribution with density
$$f(x) = 2xe^{-x^2}I_{(0,\infty)}(x).$$
Suppose that X and Y are the (independent) lifetimes of two random LWI light bulbs. Calculate the probability $P(X > 2Y)$. (Hints: You'll want to integrate the joint density of X and Y over the set $\{x,y\}|0 < 2y < x < \infty\}$. In evaluating this integral, you may find it helpful to take a good look at the general Weibull density in Equation (3.38) of the text. Be prepared to see a "β" that is less than 1.)

8. Suppose that $Y \sim \Gamma(\alpha, 1/\lambda)$ where $\alpha > 0$ and $\lambda > 0$. Further, let $X|Y = y \sim Exp(y)$ $= \Gamma(1, 1/y)$. The joint density of (X,Y) may thus be written as
$$f(x,y) = \frac{\lambda^\alpha}{\Gamma(\alpha)} y^\alpha e^{-y(\lambda+x)} I_{(0,\infty)}(x) I_{(0,\infty)}(y).$$
(a) Find the marginal density of X. (Hint: Keep the form of the gamma density in mind.) (b) Obtain the conditional density of Y, given $X = x$, and identify it by name. (c) Calculate the conditional expectation $E(Y|X = x)$ (Note: If you were able to observe $X = x$ but didn't know the value of Y, this expectation would be a reasonable guess at Y.)

9. The density of the random pair (X,Y) is positive in the triangle in the XY plane bounded by the lines $x = 0$, $y = 0$ and $x + y = 1$. The joint density of X and Y is given by
$$f_{X,Y}(x,y) = 24xy \text{ for } x > 0, y > 0 \text{ and } x + y < 1.$$
(a) Find the marginal density $f(y)$ of Y, including the range of y. (b) Obtain the conditional density of X, given $Y = y$, clearly specifying the range of X for a given value of y. (c) For an arbitrary $y \in (0,1)$, obtain an expression for $E\left(\frac{1}{X}|Y = y\right)$.

10. Suppose $\theta > 0$ and that the random pair (X,Y) has the joint density function
$$f_{X,Y}(x,y) = (1/\theta^2)e^{-y/\theta} \text{ for } 0 < x < y < \infty.$$
(a) Obtain $f_Y(y)$, the marginal density of Y (including its range), and identify it by name. (b) Obtain the conditional density $f_{X|Y}(x|y)$ (including its range), and identify it by name. (c) Show that $P(Y < 2X) = 1/2$. (Hint: Integrate the joint density above over the set $\{(x,y) : 0 < x < y < 2x < \infty\}$.)

11. Suppose that X is the lifetime, in hours, of the light bulb you use to illuminate your porch. When this light bulb fails, you replace it with the backup light bulb you keep handy. Suppose Y is the total number of hours that your porch stays illuminated when the two bulbs are used in succession. Assume that the joint density of the variables X and Y is given by
$$f_{X,Y}(x,y) = \lambda^2 e^{-\lambda y} \text{ for } 0 < x < y < \infty,$$
where $\lambda > 0$, with $f(x,y) = 0$ otherwise.

    (a) Obtain the marginal density of X, specifying clearly the range of the variable X. (b) Find the conditional density $f(y|x)$, including specification of the range of y in this density. (c) Evaluate $E(Y|X = x)$. (Hint: One approach is to use integration by parts. Another involves "recognizing" the conditional density in part (b).)

12. Let X and Y be independent random variables, each taking values in the interval $(0,1)$. The joint density of the random pair (X,Y) is
$$f_{X,Y}(x,y) = 4xy \text{ for } 0 < x < 1 \text{ and } 0 < y < 1.$$
Calculate the probability $P(X+Y \leq 1)$. (Hint: Integrate the joint density of X and Y over the set $\{(x,y) : 0 < x < 1, 0 < y < 1, x+y \leq 1\}$. You may find it helpful to plot this set before setting your limits of integration.)

13. Geologist Seymour Rocks has studied many of the crags in the Himalayas. He has proposed a quite novel probability model for the random pair (X,Y), where X and Y are standardized minimal and maximal azimuth measurements of the crag's midpoint (see *Azimuth Quarterly*, 2012, pp. 1021–1058)). By definition, X is always less than Y. The density of Rocks' distribution for (X,Y) is given by
$$f_{X,Y}(x,y) = \frac{9x^2}{y} \text{ for } 0 < x < y < 1.$$
(a) Show that $f(x,y)$ is a proper density, that is, that it integrates to 1 over the range $0 < x < y < 1$. (Hint: The integral $\int\int f(x,y)dxdy$ is easier to calculate than the integral $\int\int f(x,y)dydx$.) (b) Obtain the marginal density of X and the marginal density of Y, including the applicable range of each density. (Note: To get $f(x)$, recall that $\int \frac{1}{y}dy = \ln(y) + C$ and that $\ln(1) = 0$.) (c) Find the conditional expectation $E(X|Y = y)$.

14. Bulba Watson has made light bulbs at General Electric for years. Let X and Y be the lifetimes of the first and the second light bulb, respectively, that he makes on a given day. The joint density of X and Y is known to be
$$f(x,y) = \lambda x e^{-x(y+\lambda)} I_{(0,\infty)}(x) I_{(0,\infty)}(y),$$
where λ is a positive parameter. (a) Obtain the marginal density of X (including its range) and identify it by name. (Hint: Treat x as a fixed, known value and integrate $f(x,y)$ over the range of y.) (b) Obtain the conditional density of Y, given $X = x$, including its range, and identify it by name. (c) Calculate $E(X^2Y)$. (Hint: This can be done using integration by parts (several times), but it's easier to use your results from parts (a) and (b) and utilize Adam's rule by evaluating $E(X^2Y)$ as $E[E(X^2Y|X)] = E[X^2 \cdot E(Y|X)]$.)

15. Ambrose "Amby" Dextrous is famous for his amazing accuracy when using either hand to toss bags of peanuts or pretzels into the stands at Oakland A's home baseball games. Suppose X is the income (measured in thousands of dollars) that Amby will make selling peanuts during a random A's home game, and Y is the income during the same game from selling pretzels. The joint density of X and Y is

$$f_{X,Y}(x,y) = kx(1-x)y^2 \text{ for } 0 < x+y < 1.$$

(a) Determine the value of k that makes $f_{X,Y}(x,y)$ a valid density. (b) Find the marginal density of X, and identify it by name. (c) Calculate the expectations $E(Y)$, $E(XY)$, and $E(X|Y=y)$.

16. Two teachers at Way North High School in the upper Sacramento valley have made a wager regarding the average scores of their respective senior home room students on the December SAT test. Joe knows his class is somewhat weaker than Sue's, but the wager (which is based on the odds involved) has taken this difference into account. Let X and Y be the overall proportion of correct answers on the test for Joe's and Sue's classes, respectively. The joint distribution of X and Y is

$$f_{X,Y}(x,y) = 18x(1-x)y^2 \text{ for } x \in (0,1) \text{ and } y \in (0,1).$$

Calculate the probability that Joe wins the bet, that is, calculate $P(X > Y)$. (Hint: Integrate $f_{X,Y}(x,y)$ over the set $\{(x,y) : 0 < y < x < 1\}$.)

17. Medical researcher Luz Cannon is interested in methods of estimating an individual's propensity for "accidents" from data on their accident history. The number X of accidents a person has over the year for which she will study them will be modeled as a Poisson variable, i.e., $X \sim P(\lambda)$, where λ is their propensity for accidents. The propensity λ in the population is modeled as a $\Gamma(\alpha,\beta)$ variable. Thus, the joint probability function for (X,Λ) for a person chosen at random is given by

$$f_{X,\Lambda}(x,\lambda) = \frac{\lambda^x e^{-\lambda}}{x!}\frac{1}{\Gamma(\alpha)\beta^\alpha}\lambda^{\alpha-1}e^{-(\lambda/\beta)} \text{ for } x = 1,2,3,\dots \text{ and } \lambda \in (0,\infty).$$

(a) Obtain the marginal probability mass function of X, i.e., for fixed x, evaluate $\int_0^\infty f_{X,\Lambda}(x,\lambda)d\lambda$. (b) Obtain the conditional density $f_{\Lambda|X}(\lambda|x)$, and from it, obtain an expression for $\hat{\lambda} = E(\lambda|X=x)$. (This latter quantity is Luz's "Bayes estimate" of λ based on the accident datum $X = x$.)

18. Let $Y \sim \Gamma(\alpha, 1/\lambda)$, and suppose that given $Y = y$, $X_1,\dots,X_n \overset{\text{iid}}{\sim} Exp(y)$. Show that $Y|X_1 = x_1,\dots,X_n = x_n \sim \Gamma(\alpha+n, 1/(\lambda+\sum_{i=1}^n x_i))$.

19. Faculty who teach the course MAT 130: Math from A to Z, a course designed for unsuspecting non-majors, have long known that students always do better on the first midterm than on the second. The phenomenon is well-modeled by the density

$$f_{X,Y}(x,y) = 12y^2 \text{ for } 0 < y < x < 1,$$

where X and Y represent the proportions of the available points earned on the first and second tests, respectively. (a) Find the marginal density $f(x)$ of X and evaluate the expected score EX on the first midterm. (b) Specify the conditional density of Y given $X = x$. (c) Evaluate the conditional expectation $E(Y|X = .8)$.

20. Let (X,Y) be a random pair of continuous variables with joint density

$$f_{X,Y}(x,y) = x+y \text{ for } (x,y) \in (0,1)^2.$$

Evaluate $P(X > \sqrt{Y})$.

21. Let (X,Y) be a random pair of continuous variables with joint density

$$f_{X,Y}(x,y) = x+y \text{ for } (x,y) \in (0,1)^2.$$

Evaluate $P(X^2 < Y < X)$.

22. Let N be the number of bicycle accidents you have this year, and let X be the number of them that were your fault. Suppose that we model N as a Poisson variable and X, given N, as a binomial variable. Specifically, let

$$N \sim P(\lambda) \text{ and } X|N = n \sim B(n,p).$$

Take the distribution of X, given $N = 0$, as the degenerate distribution that gives probability 1 to the value $X = 0$. Identify the marginal distribution of X directly by evaluating its pmf as $p_X(x) = \sum_{n=x}^{\infty} p_N(n) p_{X|N}(x|n)$.

23. Suppose the random variables X and Y are uniformly distributed on the interior of the circle bounded by the curve $x^2 + y^2 = 1$. Find the marginal density of X and the conditional density of Y given $X = x$. (See Exercise 4.5.4.)

24. Let ρ represent the correlation coefficient between two variables X and Y with finite variances σ_Y^2 and σ_Y^2, respectively. Find the constant c for which the variables X and $Y - cX$ are uncorrelated.

25. Suppose X has a uniform distribution on the interval $(-c,c)$. Show that the variables X and X^2 are uncorrelated. Under what general conditions on X will this property hold?

26. Suppose $X_1, \ldots, X_m$ and $Y_1, \ldots, Y_n$ are random variables with the property that $Cov(X_i, Y_j) < \infty$ for all i and j. Show that, for any constants $a_1, \ldots, a_m$ and $b_1, \ldots, b_n$,

$$Cov\left(\sum_{i=1}^{m} a_i X_i, \sum_{j=1}^{n} b_j Y_j\right) = \sum_{i=1}^{m}\sum_{j=1}^{n} a_i b_j Cov(X_i, Y_j).$$

27. Let (X,Y) be a random pair of continuous variables with joint density

$$f_{X,Y}(x,y) = x+y \text{ for } (x,y) \in (0,1)^2.$$

Evaluate the variance of the variable $U = 3X - 2Y$.

28. Let X and Y be random variables with finite variances. Prove that if $|\rho_{X,Y}| = 1$, then there exist constants a, b, and c such that $aX + bY = c$ with probability 1.

29. If X and Y are random variables for which $\rho_{X,Y} \neq 0$, then if X can be observed before the value of Y is known, we might be able to predict the value of Y effectively from the observed value $X = x$. Show that the predictor $\hat{Y}(x)$ of Y that minimizes the squared error criterion function $E_{Y|X=x}(Y - \hat{Y}(x))^2$ is the predictor $\hat{Y}(x) = E(Y|X = x)$. (Hint: Use Adam's rule.)

30. Refer to Problem 4.8.29 above. Suppose that (X,Y) has joint density
$$f_{X,Y}(x,y) = x+y \text{ for } (x,y) \in (0,1)^2.$$
Find the best predictor of Y relative to the mean squared error criterion given that $X = 3/4$ has been observed.

31. Let the continuous random pair (X,Y) have, for some positive constant k, the joint density shown below
$$f_{X,Y}(x,y) = kxe^{(x+y)(x-y)} \text{ for } x \in (0,1) \text{ and } y \in (-\infty,\infty).$$
(a) Determine whether or not the variables X and Y are independent. (b) Find the constant k for which $f(x,y)$ integrates to one.

32. Consider an auction in which successively larger bids are tendered for an object whose value is known to be θ. Since the bidders know the object's value, no bidder will bid more than θ, but they are still interested in getting the object at the best price possible. Let's assume that the first bid, X_1, has a uniform distribution on $(0,\theta)$, that is $X_1 \sim U(0,\theta)$. For $i = 2,\dots,n$, $X_i|X_{i-1} \sim U(X_{i-1},\theta)$. Find the mean and variance of X_i for $i = 1,\dots,n$, and calculate $Cov(X_i,X_j)$ for $i \neq j$.

33. This problem extends the accident proneness model treated in Example 4.2.2. (a) Let $Z \sim \Gamma(r,\beta)$, where r is a positive integer, and let $Y|Z = z \sim P(z)$. Show that the marginal distribution of Y is a negative binomial distribution in the form associated with the number of failures occurring before the rth success. (b) Now, let $Z \sim \Gamma(r,\beta)$, where r is a positive integer, $Y|Z = z \sim P(z)$ and $X|Y = y \sim B(y,p)$. Show that the marginal distribution of Z is the same negative binomial distribution as found in part (a), but involving a different probability of success.

34. Let X and Y be uncorrelated variables with the same finite mean μ and variance σ^2. Show that the variables $X+Y$ and $X-Y$ are uncorrelated.

35. Suppose that the variables X, Y, and Z are iid, each having the $U(0,1)$ distribution. Calculate $P(X > YZ)$.

36. Let X and Y be independent Poisson variables, with $X \sim P(\lambda)$ and $Y \sim P(\mu)$. As shown in Example 4.3.2, $X+Y \sim P(\lambda+\mu)$. Identify the conditional distribution of X, given $X+Y = n$.

37. Assume that the random variables X and Y are independent. For convenience, take X and Y to be continuous. Show that the variables $U = g(X)$ and $V = h(y)$ are also independent variables, where you may take g and h to be one-to-one functions.

38. Let $X \sim U(0,1)$, and let $Y|X = x \sim N(x,x^2)$. Find the joint distribution of the variables $U = Y/X$ and $V = X$ and confirm that U and V are independent.

39. A parameter θ is a *location parameter* of the distribution of a random variable X if the density $f(x|\theta)$ can be written as a function of the difference $x-\theta$, that is, if $f(x|\theta) = g(x-\theta)$. A parameter σ is a *scale parameter* of the distribution of a random variable X if the density $f(x|\sigma)$ has the form $f(x|\sigma) = (1/\sigma)g(x/\sigma)$. The parameter pair (θ,σ) is a *location/scale parameter* for the distribution of a random variable X if the density $f(x|\theta,\sigma)$ has the form $f(x|\theta,\sigma) = (1/\sigma)g((x-\theta)/\sigma)$. Show that (a) if θ is a location parameter of the distribution

of X, then the random variable $Y = X - \theta$ has location parameter 0, (b) if σ is a location parameter of the distribution of X, then the random variable $Y = X/\sigma$ has scale parameter 1 and (c) if (θ, σ) is a location/scale parameter of the distribution of X, then the random variable $Y = (X - \theta)/\sigma$ has location parameter 0 and scale parameter 1.

40. Suppose that the conditional density of X, given Y, is

$$f_{X|Y}(x|y) = \frac{3x^2}{y^3} I_{(0,y)}(x),$$

and that Y is a continuous variable with an unspecified density $f(y)$ on $(0, \infty)$. Use the CoV method of transformations to show that the variables $U = X/Y$ and $V = Y$ are independent. (Interpretation: Y plays the role of the scale parameter of the density of X. Since the variable X/Y has scale parameter 1 and thus does not depend on the value of Y, the random variables X/Y and Y take values independently.)

41. Prove Minkowski's inequality for any real numbers a and b,

$$|a+b| \le |a| + |b|. \tag{4.88}$$

Now, use (4.88) to show that for any random variables with finite expected values,

$$E|X+Y| \le E|X| + E|Y|. \tag{4.89}$$

42. Ralph Lauren Inc. produces N polo shirts (a known number) on any given weekday. Historical records show that, on average, a known proportion p of these shirts are defective. Let X be the number of defective shirts in last Friday's batch. Assume that $X|N, p \sim B(N, p)$. Suppose that the Ralph Lauren store in Beverly Hills, CA, received a known number n of polo shirts from last Friday's batch. Let Y be the number of defective shirts received by this store. Assuming that on any particular day, the polo shirts delivered at a given store are selected at random from a population of N new shirts, the distribution of $Y|N, X, n$ is the hypergeometric distribution with probability mass function given by

$$P(Y = y|N, X, n) = \frac{\binom{x}{y}\binom{N-x}{n-y}}{\binom{N}{n}} \text{ for } 0 \le y \le \min(x, n).$$

(a) Show that the marginal distribution of Y, that is, the distribution of $Y|n, p$ is the $B(n, p)$ distribution. (Hint: Use the trick! Keep in mind that N, n, and p are fixed constants and that only X and Y are random. For fixed $Y = y$, the range of X is $y, y+1, \ldots, N-n+y$.)
(b) Identify $p_{X,Y}(x|y)$, the conditional pmf of X given $Y = y$. (Hint: The conditional pmf of $X - y|Y = y$ is more easily recognized than that of $X|Y = y$.)

43. Suppose X and Y have the joint density

$$f_{X,Y}(x, y) = \frac{1}{8}(x^2 - y^2)e^{-x} \text{ for } 0 < x < \infty \text{ and } -x < y < x.$$

Obtain the marginal densities of X and Y and the conditional densities of X, given Y, and of Y, given X.

44. Suppose a balanced six-sided die is rolled ten times. Find the probability that the numbers $1, 2, 3, 4, 5, 6$ faced up exactly $1, 2, 1, 3, 1, 2$ times, respectively.

45. Suppose a balanced six-sided die is rolled ten times. Find the probability that each of the six numbers faced up at least once.

46. Suppose $\mathbf{X} \sim M_k(n, \mathbf{p})$. Prove that, for $1 \leq i < j < k$, the triple (Y_1, Y_2, Y_3), where $Y_1 = \sum_{r=1}^{i} X_r, Y_2 = \sum_{r=i+1}^{j} X_r$ and $Y_3 = \sum_{r=j+1}^{k} X_r$, has a multinomial distribution. Identify its parameters.

47. Suppose $\mathbf{X} \sim M_k(n, \mathbf{p})$. Show that $E(X_i|X_j = x_j) = (n - x_j)p_i/(1 - p_j)$.

48. Let $\mathbf{X} \sim M_k(n, \mathbf{p})$. Since $\sum_{i=1}^{k} X_i = n$, the variable X_k may be replaced by $1 - \sum_{i=1}^{k-1} X_i$ in the pmf of $\mathbf{X}$. The moment-generating function of $X_1, \ldots, X_{k-1}$ is then well defined as $m(\mathbf{t}) = Ee^{\sum_{i=1}^{k-1} t_i X_i}$. Show that $m(\mathbf{t}) = \left(\sum_{i=1}^{k-1} p_i e^{t_i} + (1 - \sum_{i=1}^{k-1} p_i)\right)^n$.

49. Suppose X and Y have the following bivariate normal density:

$$f_{X,Y}(x,y) = \frac{1}{2\pi} e^{-\frac{x^2}{2} - \frac{y^2}{2}} \text{ for } (x,y) \in (-\infty, \infty)^2.$$

Define the random variables U and V as follows:

$$U = X^2 + Y^2 \text{ and } V = \frac{X}{\sqrt{X^2 + Y^2}}.$$

Show that U and V are independent. Can this result be interpreted geometrically?

50. The moment-generating function of the random vector $(X_1, \ldots, X_n)$ is defined as the expectation

$$m_{(X_1,\ldots,X_n)}(t_1, \ldots, t_n) = Ee^{\sum_{i=1}^{n} t_i X_i}.$$

Suppose that $(Z_1, Z_2) \sim BVN(0, 0, 1, 1, 0)$, where the five indicated parameters of the distribution are the values of $\mu_x, \mu_y, \sigma_x^2, \sigma_y^2$ and ρ, respectively. Show that the moment-generating function of this "standard BVN" distribution is

$$m_{(Z_1,Z_2)}(t_1, t_2) = e^{-\frac{t_1^2}{2} - \frac{t_2^2}{2}}.$$

51. See Problem 50. Obtain the mgf of the general BVN distribution.

52. Suppose that (X_1, X_2) has the $BVN(\mu_x, \mu_y, \sigma_x^2, \sigma_y^2, \rho)$ distribution. Show that, for arbitrary constants a, b, and c, with a or b or both non-zero, the random variable $aX_1 + bX_2 + c$ has a univariate normal distribution. Identify its mean and variance.

53. Let $Z_1, Z_2 \overset{\text{iid}}{\sim} N(0,1)$, and let $\mu_x, \mu_y, \sigma_x^2, \sigma_y^2$ and ρ be constants satisfying the conditions $-\infty < \mu_x < \infty$, $-\infty < \mu_y < \infty$, $\sigma_x > 0$, $\sigma_y > 0$, and $-1 < \rho < 1$. Define two random variables X and X_2 as follows:

$$X_1 = \sigma_1 Z_1 + \mu_1 \text{ and } X_2 = \sigma_2 \left[\rho Z_1 + (1 - \rho^2)^{1/2} Z_2\right] + \mu_2.$$

Show that (X_1, X_2) has the bivariate normal distribution with density displayed in (4.61).

54. In this and the following 2 problems, the multivariate normal distribution is developed without reference to the associated density function. This construction is equivalent to the standard development when the density exists, but applies as well to k-dimensional normal distributions for which the rank of the variance-covariance matrix Σ is less than k. **Definition**: A k-dimensional random vector $\mathbf{X}$ is said to have a k-dimensional (or multivariate)

normal distribution if for every k-dimensional vector $\mathbf{a} = (a_1, \dots, a_k)$ of real numbers, the random variable $Y_a = \sum_{i=1}^k a_i X_i$ has a (possibly degenerate) normal distribution on the real line. Suppose that $\mathbf{X}$ has a k-dimensional normal distribution. Prove that for any $n \times k$ matrix A of real numbers, the random vector $A\mathbf{X}$ has an n-dimensional normal distribution.

55. Using the definition in Problem 4.8.54, suppose $\mathbf{X}$ has a k-dimensional normal distribution. The moment-generating function of $\mathbf{X}$ is given by $m_X(\mathbf{t}) = Ee^{\sum_{i=1}^k t_i X_i}$. Note that this is the same as the mgf of the univariate random variable $\sum_{i=1}^k t_i X_i$.

    (a) Show that if $Z_1, \dots, Z_k \stackrel{\text{iid}}{\sim} N(0,1)$, then the moment-generating function of the vector $\mathbf{Z} = (Z_1, \dots, Z_k)$ is

    $$m_X(\mathbf{t}) = exp[\sum_{i=1}^k t_i^2/2].$$

    (b) Using the notation μ_i for EX_i and Σ for $[\sigma_{ij}]$, the $k \times k$ covariance matrix with $\sigma_{ii} = V(X_i)$ and $\sigma_{ij} = Cov(X_i, X_j)$, show that, if Σ is of full rank,

    $$m_X(\mathbf{t}) = \exp[t^T \mu - t^T \Sigma t].$$

56. Suppose $\mathbf{X}$ has a k-dimensional normal distribution with mean μ and with a covariance matrix Σ of full rank k. Show that the density of X is given by

    $$f(\mathbf{x}) = (2\pi)^{-1/2}(\det\Sigma)^{-1/2}\exp[-(\mathbf{x}-\mu)^T\Sigma^{-1/2}(\mathbf{x}-\mu)] \text{ for } \mathbf{x} \in (-\infty,\infty)^k.$$

57. Let X be a discrete random variable that may take on any of the values in the set $-3, -2, -1, 0, 1, 2, 3$, each with probability $1/7$. Obtain the probability mass function of the variable $Y = X^2 - |X|$.

58. Suppose that the probability distribution the random variable X has density

    $$f_X(x) = 3x^2 I_{(0,1)}(x).$$

    Find the density of the variable $Y = 1 - X^2$.

59. Suppose that the probability distribution the random variable X has density

    $$f_X(x) = (x/2) I_{(0,2)}(x).$$

    Find the density of the variable $Y = X(2 - X)$.

60. The logistic distribution is often used in modeling the regression function when a binary (or 0/1) response variable Y is modeled as a function of a continuous variable X. The standard logistic model, which has mean 0 and scale parameter 1, that is, the L(0, 1) model, has distribution function

    $$F(x) = \frac{1}{1+e^{-x}} \quad \text{for } -\infty < x < \infty.$$

    The general logistic distribution function is displayed in Equation (3.45). Logistic regression is based on the assumption that $P(Y = 1|X = x) = F(x)$.

    Suppose that you wish to simulate a L(0,1) variable. If the variable $U \sim U(0,1)$ is obtained from a random number generator, identify a random variable $X = g(U)$ whose distribution is L(0, 1). If the sampled value of U is .9000, what is the value of the corresponding logistic variable?

61. Explain how to use a random number generator (assumed to yield independent $U(0,1)$ random variables) to simulate a random sample of variables from a distribution with density function

$$f_X(x) = 6x^5 I_{(0,1)}(x).$$

Use any available computer program (e.g., Excel) to generate a sample of size 100 from this distribution. Draw a histogram for these data (using, for example, 5 intervals of length .2) and compare it to a graph of the density $f(x)$ above.

62. Explain how to use a random number generator (assumed to yield independent $U(0,1)$ random variables) to simulate a random sample of variables from a distribution with density function

$$f_X(x) = \frac{2}{(1+x)^3} I_{(0,\infty)}(x).$$

Use any available computer program (e.g., Excel) to generate a sample of size 30 from this distribution. Draw a histogram for these data (using, for example, a set of consecutive intervals of length 1, starting with $(0,1)$) and compare it to a graph of the density $f(x)$ above.

63. Recall Example 4.6.2. It was shown there, by a moment-generating function computation, that if $Z \sim N(0,1)$, then $X = Z^2$ has the χ^2_1 distribution, that is, X has the $\Gamma(1/2, 1/2)$ distribution. Here, we would like to obtain the same result using the distribution function technique. In applying this method, it is helpful to reformulate the problem into a version in which the relationship between the two variables is invertible. Note that the problem of interest is equivalent to finding the distribution of $X = Y^2$, where Y has the half-normal distribution with density

$$f(y) = \frac{2}{\sqrt{2\pi}} e^{-\frac{y^2}{2}} I_{(0,\infty)}(y).$$

Using this latter definition of X, find the density of X using the method of distribution functions.

64. Suppose that the random pair (X,Y) has the joint density given by

$$f_{X,Y}(x,y) = 30(y-x)^4 \text{ for } 0 < x < y < 1.$$

(a) Obtain the marginal density of Y, including a clear indication the range of Y. (b) Obtain the conditional density of X, given $Y = y$, including the range of X when $Y = y$. (c) Obtain the conditional density of $U = X/y$, given $Y = y$. (Hint: Transform X into U, using the conditional density $f(x|y)$ obtained in part (b), with y taken as fixed and known, as the appropriate density function for the variable X.) (d) Note from part (c) that $U|Y = y$ has a particular beta distribution. Use this to obtain $E(U|Y = y)$ and from that, obtain $E(X|Y = y)$.

65. The lognormal distribution is a distribution on the positive real line that has proven useful as a model for lifetimes of selected animal species, for certain physical quantities such as the amount of airborne contaminants measured on a given day, and for economic variables that are generated through "multiplicative processes," like compound interest. The model is derived via a transformation of a normal variable. If $X \sim N(\mu, \sigma^2)$, then $Y = e^X$ has a lognormal distribution. We denote the distribution of Y as $LN(\mu, \sigma^2)$. Show that the density of the variable Y is given by

$$f_Y(y) = (1/\sqrt{2\pi}\sigma)(1/y) e^{(1/2\sigma^2)(\ln y - \mu)^2} I_{(0,\infty)}(y).$$

66. Let U and V be independent chi-square variables, with $U \sim \chi_k^2$ and $V \sim \chi_n^2$. Define $X = \frac{U/k}{V/n}$ (a) Letting $Y = V$, obtain the joint density of (X,Y). (b) Show that X has the so-called "F-distribution with k and n degrees of freedom" with density given by

$$f_X(x) = \frac{\Gamma\left(\frac{k+n}{2}\right)(k/n)^{k/2}}{\Gamma(k/2)\Gamma(n/2)} \frac{x^{\frac{k}{2}-1}}{\left(1+\frac{kx}{n}\right)^{\frac{k+n}{2}}} I_{(0,\infty)}(x).$$

(Note: The F distribution with integer-valued parameters k and n called the distribution's "degrees of freedom" was named after Sir Ronald Fisher, one of the true geniuses of the 20th century, and the statistician who laid the foundations for the "analysis of variance" (ANOVA) where F tests are essential tools for statistical inference.)

67. The $F(k,n)$ distribution has density is given in Problem 4.8.66. Suppose that $X \sim F(k,n)$. Find the density of the variable $Y = \frac{kX}{n+kX}$. (It's a certain beta density.)

68. Let X and Y be independent unit exponential variables (that is, suppose they both have $\Gamma(1,1)$ distributions). Let $U = X/Y$ and $V = X+Y$. Are U and V independent random variables?

69. Consider the random pair with joint density given by

$$f_{X,Y}(x,y) = \frac{1}{x^2y^2} \text{ for } x > 1 \text{ and } y > 1.$$

Obtain the density of the random variable $U = X/Y$.

70. Let (X,Y) be a random pair of continuous variables with joint density

$$f_{X,Y}(x,y) = x+y \quad \text{for } (x,y) \in (0,1)^2.$$

Find the density function of the variable $U = XY$.

71. Let X and Y be independent random variables, and let $W = X+Y$. Suppose that $X \sim \Gamma(\alpha_1, \theta)$ and that $W \sim \Gamma(\alpha_2, \theta)$, where $\alpha_2 > \alpha_1$. Show that $Y \sim \Gamma(\alpha_2 - \alpha_1, \theta)$.

72. Consider a random pair (X,Y) having joint density given by

$$f(x,y) = 4xyI_{(0,1)}(x)I_{(0,1)}(y).$$

Let $U = XY$. Obtain the density function of U. (Hint: It's natural here to use the change of variable method (i.e., transformation method 3) and to define the companion variable in the transformation from (X,Y) to (U,V) to be $V = Y$. If you do, be careful to account for the fact that the ranges of U and V above are not independent. Also, don't panic if you get a minus sign in the density of U. Note that if $u \in (0,1)$, then $\ln(u) < 0$ and $-\ln(u) = \ln(1/u) > 0$.)

73. Let (X,Y) be a random pair of continuous variables with joint density

$$f_{X,Y}(x,y) = 4xy \text{ for } (x,y) \in (0,1)^2.$$

Find the joint density of the variables $U = X/Y$ and $V = XY$, and obtain the marginal densities of each. (Note that if $V = v \in (0,1)$, then $v \leq U \leq 1/v$ or, conversely, if $U = u \leq 1$, then $0 \leq V \leq u$ while if $U = u \geq 1$, then $0 \leq V \leq 1/u$. You can confirm this graphically by noting that the mapping $(X,Y) \to (U,V)$ is bounded by the lines $u = v$ and $v = 0$ and the upper branch of the hyperbola $uv = 1$.)

74. The two-parameter version of the Pareto model that has density function

$$f_X(x) = \alpha\theta^\alpha/x^{\alpha+1} \text{ for } x > \theta,$$

where α and θ are positive parameters. We will denote the distribution by $P(\alpha,\theta)$. (a) Let $X \sim P(\alpha,\theta)$. Show that the random variable $Y = X/\theta$ has the distribution $P(\alpha,1)$. (b) Let $Y \sim P(\alpha,1)$. Obtain the density of the variable $W = \ln Y$. Can you name it? (c) Suppose that $Y \sim P(\alpha,1)$. Show that the variable $U = Y - 1$ has the familiar one-parameter Pareto distribution with density

$$f(u) = \alpha/(1+u)^{\alpha+1} \text{ for } u > 0.$$

75. Sorry folks, we're going to take one more look at a celebrated model, and I don't mean Elle McPherson. I mean the one-parameter Pareto distribution having density

$$f_X(x) = \alpha/x^{\alpha+1} \text{ for } 1 < x < \infty,$$

with parameter $\alpha > 0$. And why shouldn't we take another look? After all, the model is pretty darned useful, having been applied to such diverse phenomena as the pricing of speculative investment offerings, the reliability of long-lived systems, and the clustering of errors in telephone networks. Suppose X and Y are independent Pareto variables having the density above, and let

$$U = XY \text{ and } V = Y.$$

(a) Find the joint distribution of U and V. (b) Find the marginal density of U. (A word of caution: since both X and Y are greater than 1, it follows that $U > V$.)

76. Suppose that the three random variables X, Y, and Z have joint density

$$f_{X,Y,Z}(x,y,z) = 8xyz \text{ for } (x,y,z) \in (0,1)^3.$$

Define the variables U, V, and W as $U = X$, $V = XY$, and $W = XYZ$. Find the joint density of U, V, and W.

77. Let (X,Y) be a random pair of continuous variables with joint density

$$f(x,y) = 8xy \text{ for } 0 \le x \le y \le 1.$$

Find the joint density of the variables $U = X/Y$ and $V = Y$. Are X and Y independent? Are U and V independent?

78. Let $X_1,\ldots,X_n \stackrel{\text{iid}}{\sim} U(0,1)$. Find the cdf of the variable $Y = \prod_{i=1}^n X_i$. (Hint: Recall that if $X \sim U(0,1)$, then $Y = -\ln X \sim Exp(1)$.)

79. Show that the density of the "t-statistic" in Theorem 4.6.2 can be obtained by the method of distribution functions by writing $F_T(t) = P(T \le t)$ as $P(X \le t\sqrt{Y/k})$, evaluating this probability by integrating the joint density of X and Y over the set $\{(x,y)|y \in (0,\infty) \text{ and } x \in (-\infty, t\sqrt{y/k})\}$ and then differentiating $F_T(t)$.

80. Suppose $X_1,\ldots,X_n \stackrel{\text{iid}}{\sim} \Gamma(1,1/\lambda)$, the exponential distribution with failure rate λ, and let $X_{(1)},\ldots,X_{(n)}$ be the corresponding order statistics. The "spacings" between order statistics, defined as $S_i = X_{(i)} - X_{(i-1)}$ for $i = 1,\ldots,n$, with $X_{(0)} \equiv 0$, have an interesting distribution theory. Show that (a) $S_i \sim \Gamma(1,(n-i+1)/\lambda)$ for $i = 1,\ldots,n$. (**Hint:** Start with the distribution of S_1. Then use the memoryless property to show that S_2 has the same distribution as the first spacing in a sample of size $n-1$ from $\Gamma(1,1/\lambda)$. Repeat the argument.) (b) Show that $S_1,\ldots,S_n$ are independent. (Hint: Obtain the joint density of $S_1,\ldots,S_n$ from the joint density of the order statistics (in the form of (4.84)) and show that it factors into the marginal distributions of $S_1,\ldots,S_n$. Note that the range of each S_i is $(0,\infty)$.) (c) Conclude that the normalized spacings $nS_1,(n-1)S_2,\ldots,(n-i+1)S_i,\ldots,S_n$ are iid with common distribution $\Gamma(1,1/\lambda)$. (d) Note that $X_{(i)} = \sum_{k=1}^{i} S_k$. From this relationship, obtain expressions for the mean and variance of $X_{(i)}$.

81. Let $X_1,\ldots,X_5 \stackrel{\text{iid}}{\sim} \Gamma(1,1)$, the unit exponential distribution, and let $X_{(1)},\ldots,X_{(5)}$ be the corresponding order statistics. Write the densities of $X_{(2)}$ and $X_{(5)}$ and the joint density of $(X_{(2)},X_{(5)})$. From these, compute the probabilities $P(X_{(2)} > 2)$, $P(X_{(5)} < 4)$, and $P(X_{(5)} - X_{(2)} > 3)$.

82. Suppose $X_1,\ldots,X_5 \stackrel{\text{iid}}{\sim} F$, a distribution with density $f(x) = 2xI_{(0,1)}(x)$ and let $X_{(1)},\ldots,X_{(5)}$ be the corresponding order statistics. Write the densities of $X_{(2)}$ and $X_{(5)}$ and the joint density of $(X_{(2)},X_{(5)})$. From these, compute the probabilities $P(X_{(2)} > .5)$, $P(X_{(5)} < .8)$, and $P(X_{(5)} - X_{(2)} > .3)$.

83. A stick of length 1 foot is broken into three pieces at random (that is, the three break points are iid, each having a $U(0,1)$ distribution). What is the probability that the pieces can be made into a triangle?

84. Suppose $X_1,\ldots,X_n \stackrel{\text{iid}}{\sim} F$, a continuous distribution on the real line with density function $f(x)$, and let $X_{(1)},\ldots,X_{(n)}$ be the corresponding order statistics. Show that the conditional joint density of the order statistics $X_{(2)},\ldots,X_{(n)}$, given $X_{(1)} = x_{(1)}$, is precisely equal to the distribution of the order statistics from a random sample of size $n-1$ from the distribution with density
$$f^*(x) = \frac{f(x)}{\bar{F}(x_{(1)})} \text{ for } x > x_{(1)},$$
where $\bar{F}(x) = 1 - F(x)$.

85. Suppose $X_1,\ldots,X_5 \stackrel{\text{iid}}{\sim} F$, a continuous distribution with density function $f(x)$ with support set on an interval (a,b) of the real line. Let $X_{(1)},\ldots,X_{(5)}$ be the corresponding order statistics. Show, by integrating the joint density in (4.85) with $n = 5$, that
$$f_{X_{(2)}}(x_{(2)}) = 20F(x_{(2)})[1 - F(x_{(2)})]^3 f(x_{(2)}) I_{(a,b)}(x_{(2)}),$$
in agreement with (4.83) for $k = 2$ and $n = 5$.

86. An engineered n-component system is said to be *coherent* if each of its components is relevant to its functioning and if the system is monotone in the sense that if a failed component is replaced by a working component, then the state of the system (i.e., its working or not) cannot deteriorate. Show that if the n components of a coherent system have iid

lifetimes with common continuous distribution F, then the survival function of the system lifetime T may be represented as

$$P(T > t) = \sum_{i=1}^{n} s_i P(X_{(i)} > t),$$

where $s_i = P(T = X_{(i)})$. The probability vector $\mathbf{s}$ is referred to as the signature of the system. Show that the expected lifetime of the system may be computed as

$$ET = \sum_{i=1}^{n} s_i EX_{(i)}.$$

For more detail on system signatures and their applications, see Samaniego (2007).

87. Given a random sample of size 5 is drawn from the $U(0,1)$ distribution, find the probability that the sample median $X_{(3)}$ is in the interval $(1/3, 2/3)$.

88. Let $X_1, \ldots, X_n \stackrel{\text{iid}}{\sim} W(\alpha, \beta)$, the Weibull distribution with shape and scale parameters α and β. Obtain the density function of $\min X_i$ and identify it by name.

89. (See example 1.7.11) As stated in the referenced example in which three students are each presumed to answer 12 particular multiple choice questions at random, it is possible for the students to get 1 common answer, 2 different answers among the three of them or 3 different answers. Verify that $P(1) = 1/25$, $P(2) = 12/25$ and $P(3) = 12/25$. Using these probabilities and the answers to Exercise 1.7.5, calculate the probability that the three students marked the same answer to at least 9 of the 12 questions under discussion.

5

Limit Theorems and Related Topics

5.1 Chebyshev's Inequality and Its Applications

Pafnuty Chebyshev, born in 1821, was a Russian mathematician of notable distinction. He specialized in the areas of probability, statistics, and number theory, and he is generally thought of as one of the founding fathers of Russian mathematics. Among his well-known students was the prolific probabilist Andrey Markov, celebrated for his contributions to stochastic processes and linear statistical models. Among Chebyshev's contributions to mathematics, he is perhaps best known for the inequality that bears his name. The French statistician Irénée-Jules Bienaymé (1796–1878), a friend and correspondent of Chebyshev, is sometimes credited with the inequality's discovery, and many modern books credit them both. Chebyshev's inequality is a simple but quite useful tool in probability, as it applies to any random variable whose variance is finite, and it leads to an immediate proof of a version of a famous result called the "weak law of large numbers" which gives conditions under which the mean of a random sample will converge to the mean of the sampled population as the sample size grows large. Another reason for the impact of Chebyshev's inequality on Statistics is that it showed quite dramatically the utility and relevance of the *variance* as a measure of dispersion. At first view, the quantity σ^2 seems like a reasonable but somewhat arbitrary measure of the "distance" of a random variable from its expected value. Chebyshev's inequality leaves no doubt that this measure says something quite meaningful about the dispersion of a distribution. The inequality applies to both theoretical and empirical distributions, that is, it can be applied when discussing the probability mass function or density of a random variable or a histogram based on discrete or continuous data.

Theorem 5.1.1. (Chebyshev's inequality) For any random variable X with finite mean μ and variance σ^2, and for any $\varepsilon > 0$,

$$P(|X-\mu| \geq \varepsilon) \leq \frac{\sigma^2}{\varepsilon^2}. \tag{5.1}$$

Proof. We give a proof in the continuous case. The proof of the discrete case is similar.

$$\begin{aligned}
\sigma^2 &= E(X-\mu)^2 \\
&= \int_{-\infty}^{\infty} (x-\mu)^2 f(x)dx \\
&\geq \int_{\{x:|x-\mu|\geq\varepsilon\}} (x-\mu)^2 f(x)dx \\
&\geq \int_{\{x:|x-\mu|\geq\varepsilon\}} \varepsilon^2 f(x)dx
\end{aligned}$$

$$= \varepsilon^2 P(|X - \mu| \geq \varepsilon).$$

Dividing the first and last term of the sequence above by ε^2 yields (5.1). ■

An alternative proof of Chebyshev's inequality may be obtained from the following seemingly simpler result, an inequality that has a variety of other applications that suggest it is worth including here (see Exercise 5.1.2).

Theorem 5.1.2. (Markov's inequality) For any random variable X and for any $a > 0$,

$$P(|X| \geq a) \leq \frac{E|X|}{a}. \tag{5.2}$$

Proof. Consider the indicator function of the event $\{|X| \geq a\}$, that is, the function $I_{[a,\infty)}(|X|)$. For any value of a, including $a > 0$ (the case of interest), we have that

$$aI_{[a,\infty)}(|X|) \leq |X|. \tag{5.3}$$

Note that $|X| < a \Rightarrow$ the RHS of (5.3) $= 0 \leq |X|$ while $|X| \geq a \Rightarrow$ RHS of (5.3) $= a \leq |X|$. Taking expectations in (5.3), we have

$$aP(|X| \geq a) \leq E|X|,$$

an inequality that may be rewritten as (5.2). ■

Chebyshev's inequality turns out to be a very useful tool in studying questions concerning the convergence of a sequence of random variables $\{X_n\}$ to a fixed constant. Such questions are of particular interest in applications in which an unknown parameter is estimated from data. We'd then be seeking to determine if the estimator selected for use has the "consistency" property, that is, if it converges to the true value of the parameter as the sample size increases. We consider such an application below, proving an early and famous version of a theorem concerning the limiting behavior of the mean $\bar{X}$ of a random sample. First, we define the relevant convergence concept.

Definition 5.1.1. Let $\{T_i, i = 1, 2, 3, \ldots\}$ be a sequence of random variables. Then T_n is said to converge to the constant c *in probability* (or to converge weakly to c), denoted $T_n \xrightarrow{p} c$, if for any $\varepsilon > 0$,

$$P(|T_n - c| \geq \varepsilon) \to 0 \text{ as } n \to \infty. \tag{5.4}$$

Theorem 5.1.3. (The Weak Law of Large Numbers) Let $X_1, X_2, \ldots, X_n, \ldots \overset{\text{iid}}{\sim} F$, where F is a distribution with finite mean μ and variance σ^2. Let $\bar{X}_n = (1/n)\sum_{i=1}^n X_i$ be the mean of a sample of size n. Then $\bar{X}_n \xrightarrow{p} \mu$ as $n \to \infty$.

Proof. Let $\varepsilon > 0$. By Chebyshev's inequality,

$$P(|\bar{X}_n - \mu| \geq \varepsilon) \leq \sigma^2_{\bar{X}}/\varepsilon^2 = \sigma^2/n\varepsilon^2 \to 0 \text{ as } n \to \infty.$$

■

I suppose that I should explain why the type of convergence referred to above is described as "weak." You guessed it! There are "stronger" ways that a sequence of variables can converge to a constant (that is, types of convergence that automatically imply convergence in probability). A commonly used alternative version of stochastic convergence is called "almost sure" convergence in probability theory and "convergence in measure" in mathematical fields like real analysis. Another form of convergence that is stronger than convergence in probability is "convergence in quadratic mean." These two modes of convergence are defined as follows.

Definition 5.1.2. Let $\{T_i, i = 1, 2, 3, \ldots\}$ be a sequence of random variables. Then T_n is said to converge to the constant c *almost surely* (or to converge strongly to c), denoted $T_n \stackrel{a.s.}{\rightarrow} c$, if for any $\varepsilon > 0$,

$$\lim_{N\to\infty} P(\sup_{n\geq N}|T_n - c| \leq \varepsilon) = 1, \tag{5.5}$$

where "sup" stands for "supremum," and is defined as the least upper bound of the values involved.

Definition 5.1.3. Let $\{T_i, i = 1, 2, 3, \ldots\}$ be a sequence of random variables. Then T_n converges to the constant c *in quadratic mean*, denoted $T_n \stackrel{q.m.}{\rightarrow} c$, if

$$E(T_n - c)^2 \to 0 \text{ as } n \to \infty. \tag{5.6}$$

Now, a thorough understanding of almost sure convergence involves some grounding in the mathematical field of measure theory, and this is well beyond the scope of this book. Still, it is worth knowing that there is a stronger (and thus more desirable) type of convergence than "$\stackrel{p}{\rightarrow}$". Some insight into why "$\stackrel{a.s.}{\rightarrow}$" is stronger can be gleaned from the fact that $|T_N - c| \geq \varepsilon$ implies that $\sup_{n\geq N}|T_n - c| \geq \varepsilon$, so that the latter event has a larger probability. Regarding the fact that $\stackrel{q.m.}{\rightarrow}$ is stronger than $\stackrel{p}{\rightarrow}$, this follows immediately from Markov's inequality (see Exercise 5.1.3).

A version of the "strong law of large numbers" due to A. N. Kolmogorov is stated below without proof. Using more powerful tools, Kolmogorov was able to obtain a stronger conclusion than that in Theorem 5.1.3 under weaker assumptions. Not only is a finite variance not required in the theorem below, the fact that the common mean μ of the Xs exists and is finite is a *necessary and sufficient condition* for the stated convergence to hold. Hats off to Kolmogorov!

Theorem 5.1.4. (The Strong Law of Large Numbers) Let $X_1, X_1, X_2, \ldots, X_n, \ldots \stackrel{\text{iid}}{\sim} F$. Then $\bar{X}_n \stackrel{a.s.}{\rightarrow} \mu$ as $n \to \infty$ if and only if $EX_i \equiv \mu < \infty$.

We close this section with a discussion of the practical interpretation and significance of Chebyshev's Theorem. First, it should be noted that the result is a very early contribution to "nonparametric statistics," that is, to the study of probability and statistics in contexts that make no assumptions about the form of the probability distribution or its pdf or pmf, be it continuous or discrete. The inequality holds for *all* distributions. Because of its breadth, the inequality will not provide, for many distributions, an accurate estimate of the probability that a variable will take a value within a certain distance of its mean. When we have a particular distribution in mind, we would typically be interested in computing probabilities such as that in (5.1) rather than in obtaining an upper bound for them.

The version of Chebyshev's inequality that is utilized most often states that for any $k > 0$,

$$P(|X-\mu|<k\sigma)\geq 1-\frac{1}{k^2}. \tag{5.7}$$

For example, regardless of the distribution of X, the chances that X will take a value within two standard deviations of the mean μ is at least .75, and there's at least an 89% chance that X will be within 3σ of the mean. The Six Sigma Society within the "Quality Control" community has as its fundamental tenet that any process that is "in control," i.e., is properly operating within its specifications, should display performance that reflects the high likelihood of staying within a band of width 6σ centered at the mean (with σ representing the standard deviation of the performance measure X being used). If one knows more about the distribution of the measures being used, one may want to use a tighter standard than 6σ. E.g., if $X\sim N(\mu,\sigma^2)$, then $P(|X-\mu|>2\sigma)<.05$ and if $X\sim U(a,b)$, $P(|X-\mu|>2\sigma)=0$. Exercise 5.1.5 shows that there are distributions for which the probability bound in Chebyshev's inequality can be sharp, that is, can hold with equality.

It is worth noting that empirical distributions, that is, step functions which place weight $1/n$ on each of the n observed data points, also obey Chebyshev's inequality. Thus, using $\bar{X}$ as the mean μ and $s^2=(1/n)\sum_{i=1}^{n}(X_i-\bar{X})^2$ as the variance σ^2, the inequality $P(|X-\bar{X}|\leq ks)\geq 1-(1/k)^2$ holds. The inequality also applies to relative-frequency histograms.

Exercises 5.1.

1. Prove Theorem 5.1.1 when X is a discrete random variable.
2. Show that Markov's inequality implies Chebyshev's inequality.
3. Use Markov's inequality to show that if $T_n\stackrel{q.m.}{\to}c$, then $T_n\stackrel{p}{\to}c$.
4. Compute $P(|X-\mu|\geq 2\sigma)$ for $X\sim P(\alpha,\theta)$, the two-parameter Pareto distribution, and compare this probability to the lower bound given by the version of Chebyshev's inequality given in (5.7).
5. Let X be a discrete variable with the pmf displayed below:

X	0	1/2	1
p(x)	.125	.75	.25

Show that $P(|X-\mu|<2\sigma)=3/4$, that is, that the Chebyshev inequality holds with equality in this particular instance.

5.2 Convergence of Distribution Functions

In Chapter 2, we encountered two different settings in which the probability mass functions of particular discrete distributions converge, as model parameters approach certain limits, to the probability mass function of a fixed limiting distribution. The first of these "limit theorems," involving the hypergeometric and binomial distributions, showed that

$HG(N,r,n)$ probabilities converge to $B(n,p)$ probabilities as $N \to \infty$, $r \to \infty$ and $r/N \to p$. The second limit theorem (the Law of Rare Events), involving the binomial and Poisson distributions, showed that $B(n,p)$ probabilities converge to $P(\lambda)$ probabilities as $n \to \infty$, $p \to 0$, and $np \to \lambda$. In both of these discrete settings, one could infer from the theorems we proved that the sequence of distribution functions $\{F_i(x), i = 1,2,3,\ldots.\}$ of the original variables would also converge to the distribution function of the limiting random variable. It is this latter type of convergence that is used to define *convergence in distribution*, this formulation applies equally to discrete and continuous variables and is a framework in which moment-generating functions can be easily applied. We will therefore define this mode of convergence in terms of the respective distribution functions.

Definition 5.2.1. The sequence of random variables $\{X_i, i = 1,2,3,\ldots\}$, with corresponding distribution functions $\{F_i, i = 1,2,3,\ldots\}$, is said to *converge in distribution* to the random variable X, with distribution function F, if, as $n \to \infty$, $F_n(x) \to F(x)$ at all continuity points of F, that is, for every x at which the function $F(x)$ is continuous. Convergence in distribution is denoted by $X_n \xrightarrow{D} X$.

If, in the definition above, the distribution function $F(x)$ is continuous at every x, then $X_n \xrightarrow{D} X$ if and only if $F_n(x) \to F(x)$ for all real numbers x. An example of convergence in distribution when F is a distribution with jump points is given below.

Example 5.2.1. Consider the sequence of random variables $\{X_n\}$ with respective distributions F_n, where $F_n(x) = I_{[1/n,\infty)}(x)$. Let $F(x) = I_{[0,\infty)}(x)$. Now $F_n(x) \to 0$ for $x \leq 0$ and $F_n(x) \to 1$ for $x > 0$, so that $F_n(x) \to F(x)$ for every x except for $x = 0$. Since $F_n(0) = 0$ for every n, we have that $\lim_{n\to\infty} F_n(0) = 0 \neq 1 = F(0)$. However, since 0 is not a continuity point of F, we still have that $X_n \xrightarrow{D} X$. ■

To give you a feeling for the connections between convergence in probability and convergence in distribution, I state below a useful collection of propositions which shed further light on the two concepts and their interplay. Proofs of the results in Theorem 5.2.1 may be found in Rao (1973, pp. 122–5).

Theorem 5.2.1. Let (X_n, Y_n) be a sequence of random pairs. Then

(i) If g is a continuous function and $X_n \xrightarrow{p} k$ for some constant k, then $g(X_n) \xrightarrow{p} g(k)$.

(ii) If g is a continuous function and $X_n \xrightarrow{D} X$, then $g(X_n) \xrightarrow{D} g(X)$.

(iii) If $X_n - Y_n \xrightarrow{p} 0$ and $X_n \xrightarrow{D} X$, then $Y_n \xrightarrow{D} X$.

(iv) (Slutsky's Theorem) If $X_n \xrightarrow{D} X$ and $Y_n \xrightarrow{p} k$, then: (a) $X_nY_n \xrightarrow{D} kX$, (b) $X_n + Y_n \xrightarrow{D} X + k$, and (c) if $k \neq 0$, then $X_n/Y_n \xrightarrow{D} X/k$.

Remark 5.2.1. Let me make a few comments about the various parts of the theorem above. Since the proofs are not included here, you might be more comfortable with using them if you have an intuitive appreciation for their validity. Parts (i) and (ii) of Theorem 5.2.1 have to do with the effect of applying a continuous transformation g to each member of a sequence of random variables. It is worth noting that these two results are exactly what we would expect. After all, the continuity of g simply means that if the variable x gets close to x_0, the values of $g(x)$ will get close to $g(x_0)$. In part (i) above, we know that for n sufficiently large, the probability that $|X_n - k|$ is small will be very close to 1. But the continuity of g then implies that $|g(X_n) - g(k)|$ is small with probability close to one. Since X_n approaches k as $n \to \infty$, it would be odd and surprising if $g(X_n)$ did not approach $g(k)$ as $n \to \infty$. The intuitive content of part (ii) is similar. To see (iii), note that because $X_n - Y_n \xrightarrow{P} 0$, X_n and Y_n are actually identical, in the limit, that is, they take on the same values with the same probabilities in the limit. It therefore should be predictable that the two variables have the same limiting distribution. Finally, take a look at Slutsky's theorem. Let's restrict our attention to part (a). Seeing some of the technical argument for part (a) will help you see why it's true. It suffices to take $k = 0$ in part (a) since the general case reduces to this special case by replacing Y_n by $Y_n - k$. For arbitrary $\varepsilon > 0$ and $m > 0$, write

$$P(|X_nY_n| > \varepsilon) = P(|X_nY_n| > \varepsilon, |Y_n| \le \varepsilon/m) + P(|X_nY_n| > \varepsilon, |Y_n| > \varepsilon/m),$$

a probability that is no greater than

$$P(|X_n| > m) + P(|Y_n| > \varepsilon/m).$$

Since the latter sum can be made arbitrarily small when m is sufficiently large and $n \to \infty$, it follows that $P(|X_nY_n| > \varepsilon)$ converges to 0 as $n \to \infty$.

Applications of Slutsky's theorem arise frequently, as it is often the case that the actual or asymptotic distribution of a particular statistic T_n are known but we are in fact interested in the asymptotic behavior of a slightly altered statistic T_n/Y_n, where Y_n tends to 1 in probability as $n \to \infty$. Slutsky's Theorem tells us that T_n and T_n/Y_n have the same limiting distribution. A case in point is the development of the t-statistic from a normal random sample. The following example complements the result in Theorem 4.6.2 in which the relationship between the t and the standard normal distributions is obtained for a fixed sample size n.

Example 5.2.2. As you know, the derivation of the t distribution was originally motivated by the fact that, given a $X_1, \dots, X_n \sim N(\mu, \sigma^2)$, the statistic

$$T_n = \frac{\bar{X} - \mu}{s/\sqrt{n}}$$

does not behave like a standard normal variable when n is small. But what can be said about the t statistic as n grows large? If you look over Table A.5 in the Appendix, you'll notice that α-cutoff points for the t_k distribution, for various frequently used values of α, are given for $k \le 30$, with the cutoff points of the standard normal distribution provided as approximations for larger values of k. This suggests that the t_k distribution approaches $N(0,1)$ as $k \to \infty$. This conclusion is justified by the following facts:

(1) If $X_1, X_2, \dots, X_n \overset{\text{iid}}{\sim} N(0, \sigma^2)$ and s_n^2 is the sample variance, then $(n-1)s_n^2/\sigma^2 \sim \chi^2_{n-1}$.

(2) From (1), it follows that $E(s_n^2) = \sigma^2$ and that $V(s_n^2) = 2\sigma^4/(n-1)$.

(3) By Chebyshev's inequality, we have $P(|s_n^2 - \sigma^2| \ge \varepsilon) \le 2\sigma^4/(n-1)\varepsilon^2 \to 0$ as $n \to \infty$, that is, $s_n^2 \xrightarrow{P} \sigma^2$, which implies that $s_n/\sigma \xrightarrow{P} 1$.

(4) Now, let $Y_n = \sqrt{n}(\bar{X} - \mu)/\sigma$ and $V_n = s_n/\sigma$. Since $Y_n \xrightarrow{D} Z \sim N(0,1)$ and $V_n \xrightarrow{p} 1$, it follows from part (c) of Slutsky's Theorem that $T_n = Y_n/V_n \xrightarrow{D} Z \sim N(0,1)$.

■

Remark 5.2.2. The result above uses a particularly simple form of Slutsky's Theorem since the variables Y_n in Theorem 5.2.1 are in fact iid with a standard normal distribution, so that the fact that $Y_n \xrightarrow{D} Z \sim N(0,1)$ is automatic, as $Y_n \sim N(0,1)$ for all n.

Perhaps the most obvious way to establish the convergence of a sequence of distributions $\{F_n\}$ to a limiting distribution F is to work directly with the form of the cdf $F_n(x)$ and show that it converges to $F(x)$ as n grows to infinity. Unfortunately, in many problems of interest, the exact distribution F_n of X_n, the relevant variable at the nth stage, is difficult to obtain in closed form. Indeed, this fact is usually the main motivation behind the study of convergence in distribution. If F_n is hard or impossible to identify, the fact that $X_n \xrightarrow{D} X$ justifies the use of F, the distribution of X, as an approximation to F_n when n is sufficiently large.

Given that the direct approach to demonstrating convergence in distribution is not generally available, it is fortuitous indeed that other feasible approaches exist. One approach that we have used to our advantage in the past, and will exploit further in the sequel, is the so-called continuity theorem. The result was mentioned in our discussion of moment-generating functions in Chapter 2. But because of its broad utility and its applicability beyond the discrete scenarios in which it was couched earlier, we restate it here, for emphasis, in full generality.

Theorem 5.2.2. (The Continuity Theorem). For $n = 1,2,3,\ldots$, let X_n be a random variable with moment-generating function $m_n(t)$ which exists and is finite in an open interval containing $t = 0$. If $X_n \xrightarrow{D} X$, then $m_n(t) \rightarrow m(t)$, where $m(t)$ is the mgf of X. Conversely, if $\{m_n(t), n = 1,2,3\ldots.\}$ and $m(t)$ are mgfs that exist in an open interval containing $t = 0$, and if $m_n(t) \rightarrow m(t)$, then $X_n \xrightarrow{D} X$, where X has mgf $m(t)$.

We conclude this section with an additional example of a sequence of discrete distributions converging to a limiting Poisson distribution. This result gives us one final opportunity to get warmed up for the use of the continuity theorem above in proving the Central Limit Theorem, without doubt the most glowing achievement among the ideas and results discussed in this chapter.

You will recall that we defined, in Section 2.6, a negative binomial variable $X \sim NB(r,p)$ to be the number of trials needed to obtain r successes in a sequence of independent Bernoulli trials having probability of success p. Here, we will consider a related variable $Y = X - r$, the number of failures obtained before the occurrence of the rth success. Now we have noted that the moment-generating function of the variable X is $m_X(t) = [pe^t/(1-qe^t)]^r$ for $t < \ln(1/q)$. Since, in general, the mgf of the linear function $aX + b$ of X is equal to $e^{bt}m_X(at)$, we may write the mgf of the variable Y as $m_Y(t) = e^{-rt}m_X(t) = [p/(1-qe^t)]^r$ for $t < \ln(1/q)$. We put this fact to immediate use in the following.

Theorem 5.2.3. Let $X_{r,p} \sim NB(r,p)$, and define $Y_{r,p} = X_{r,p} - r$. Suppose that $r \rightarrow \infty$ and

$q = 1 - p \to 0$ at rates such that $rq \to \lambda > 0$. Then $Y_{r,p} \xrightarrow{D} W \sim P(\lambda)$, that is, the distribution of $Y_{r,p}$ converges in distribution to the Poisson distribution with mean λ.

Proof. We will use the notation "$\overset{AE}{=}$" to mean that two terms connected by that symbol are "asymptotically equivalent"; if $a_n \overset{AE}{=} b_n$, then $a_n/b_n \to 1$ as $n \to \infty$.

Under the stated conditions, $q \overset{AE}{=} \lambda/r$ and $p \overset{AE}{=} 1 - \lambda/r$. Thus, for $t < \ln(1/q)$, we may write

$$\begin{aligned} \lim_{r\to\infty, q\to 0, rq\to\lambda} m_Y(t) &= \lim_{r\to\infty, q\to 0, rq\to\lambda} \left(\frac{p}{1-qe^t}\right)^r \\ &= \lim_{r\to\infty} \left(\frac{1-\frac{\lambda}{r}}{1-\frac{\lambda}{r}e^t}\right)^r \\ &= \lim_{r\to\infty} \frac{\left(1-\frac{\lambda}{r}\right)^r}{\left(1-\frac{\lambda e^t}{r}\right)^r} = \frac{e^{-\lambda}}{e^{-\lambda e^t}} = e^{\lambda(e^t-1)}. \end{aligned}$$

By the continuity theorem, we conclude that the shifted negative binomial variable $Y_{r,p}$ converges in distribution to a $P(\lambda)$ variable as $r \to \infty$, $q \to 0$ and $rq \to \lambda$. ■

Exercises 5.2.

1. Let $\{X_n\}$ be a sequence of random variables. (a) Suppose that $X_n \xrightarrow{p} X$. Prove that $X_n \xrightarrow{D} X$. (b) Prove that if $X_n \xrightarrow{p} c$, then $X_n \xrightarrow{D} \delta_c$, where δ_c a distribution that is degenerate at (that is, gives probability 1 to) the constant c.

2. Prove part (b) of Theorem 5.2.1 If $X_n \xrightarrow{D} X$ and $Y_n \xrightarrow{p} c$, then $X_n + Y_n \xrightarrow{D} X + c$.

3. Prove part (c) of Theorem 5.2.1 If $X_n \xrightarrow{D} X$ and $Y_n \xrightarrow{p} c$, where $c \neq 0$, then $X_n/Y_n \xrightarrow{D} X/c$.

4. Example 5.3.2 in the next section shows that, if $X_1, X_2, \ldots, X_n, \ldots \overset{\text{iid}}{\sim} P(\lambda)$, then
$$\frac{\sqrt{n}(\bar{X} - \lambda)}{\sqrt{\lambda}} \xrightarrow{D} Z \sim N(0,1),$$
where $\bar{X}$ is the sample mean. Assuming this result, show that
$$\frac{\sqrt{n}(\bar{X} - \lambda)}{\sqrt{\bar{X}}} \xrightarrow{D} Z \sim N(0,1),$$
so that in large samples, we may make the following approximate probability statement:
$$P(\bar{X} - \sqrt{\bar{X}/n}(1.96) < \lambda < \bar{X} + \sqrt{\bar{X}/n}(1.96)) \cong .95.$$

5.3 The Central Limit Theorem

If the world of lemmas, propositions, theorems, and corollaries was allowed to have its celebrities, the Central Limit Theorem would certainly be one of them. Its celebrity status could be justified by the "wonder" of the result alone, but it could also be justified on the basis of its broad applicability. The theorem basically gives conditions under which the average $\bar{X}$ of a random sample of size n will have an approximately normal distribution when n is large.

Recall that if $X_1, X_2, \ldots, X_n, \ldots \stackrel{\text{iid}}{\sim} N(\mu, \sigma^2)$, then the sample mean has a normal distribution, namely $N(\mu, \sigma^2/n)$. The latter distribution is often called the "sampling distribution of $\bar{X}$" due to the fact that its randomness is inherited from the process of random sampling. Were we to take many repeated samples of size n from the normal population $N(\mu, \sigma^2)$, we would find that variation among the different values of $\bar{X}$ one would get, as reflected in the histogram of $\bar{X}$ values obtained, would be quite consistent with the $N(\mu, \sigma^2/n)$ curve. The fact that the mean of a random sample from a normal population will have a normal distribution is not altogether surprising from an intuitive point of view. Further, the fact is quite easily proven (using mgfs), so whatever intuition was involved in our believing it is supported by a rigorous mathematical argument.

On the other hand, what the Central Limit Theorem states is really quite shocking: it turns out that this same outcome, i.e., $\bar{X} \sim N(\mu, \sigma^2/n)$, is approximately true in a wide array of situations, provided n is sufficiently large. The conditions under which this limit theorem holds are themselves surprisingly mild. All one needs is that the sample is taken from a population with a finite mean and variance. Since everything in world we live in is finite, it should not be surprising that the conditions under which the Central Limit Theorem holds arise in many applications of interest.

The Central Limit Theorem, or CLT, has a long history. There is general agreement that the first version of the CLT was put forth by the French-born mathematician Abraham de Moivre. In a paper published in 1733, he developed the normal approximation for a simple binomial distribution (with $p = 1/2$) associated with coin tossing. The reason that this approximation is an example of the CLT is that if the sum of n Bernoulli trials (i.e., the number of successes in n trials) is approximately normal, then so is the average number of successes, and vice versa, so theorems that find that one or the other is normal are really the same result. The famous French mathematician Pierre-Simon Laplace discovered de Moivre's long-ignored contribution and added refinements and generalizations in a celebrated treatise in 1812. Laplace is generally given credit for stating and treating the Central Limit Theorem in its modern, more general form.

Other early contributors to the development of the CLT include a veritable hall of fame of seventeenth- and eighteenth-century mathematicians including Cauchy, Bessel, and Poisson. Sir Francis Galton, writing about the CLT in his 1889 book *Natural Inheritance*, opined "The law would have been personified by the Greeks and deified, if they had known of it." The full significance of the CLT was recognized only in the 20th century. The sharpest formulation of the theorem can be found in a 1901 paper by the Russian mathematician Aleksandr Lyapunov. Because of the theorem's innate beauty, the many versions of the theorem that have surfaced over the years and the plethora of applications to which the CLT has been put, this celebrated theorem richly deserves to have been honored by William J. Adams in 1974 with its own biography entitled *The Life and Times of the Central Limit Theorem.*

Our proof of the CLT will require a fact from calculus that is slightly more general than the well-known result we cited in Section 2.7. Recall that the transcendental number "e" is defined in (2.23) as the limit of the sequence $\left(1+\frac{1}{n}\right)^n$ as $n \to \infty$. The following lemma, which includes the statement in (2.23) as a special case, will be needed in what follows.

Lemma 5.3.1. Let $\{a_i\}$ be a sequence of real numbers which converge to the real number a. Then

$$\lim_{n\to\infty}\left(1+\frac{a_n}{n}\right)^n = e^a. \tag{5.8}$$

Remark 5.3.1. In his book *A Course in Probability Theory* (1969, p. 157), Kai Lai Chung states essentially the same lemma without proof. (His version allows the a's to be complex numbers converging to complex number a). He refers to it as "an elementary lemma from calculus." In a footnote, he remarks: "Professor Doob kindly informed me that many students in his graduate class did not know how to prove this lemma properly." Curiously, Chung himself did not include a proof in his book. As it turns out, a published proof isn't that easily found. Here's a fairly straightforward proof of Lemma 5.3.1 (due to UCD Professor Sherman Stein) that relies only on elementary algebraic ideas.

Proof. Since $a_n \to a$, the sequence $\{a_n\}$ is bounded. Thus, there exists a positive real number b such that $|a_n| < b$ for all n, and there exists an integer N for which $|a_n/n| < 1$ for all $n > N$. We will now show that $\ln(1+a_n/n)^n \to e^a$. The Maclaurin series for $f(x) = \ln(1+x)$ is $f(x) = x - \frac{x^2}{2} + \frac{x^3}{3} - \frac{x^4}{4} + \ldots$; since the series is valid for $|x| < 1$, we may write, for $n > N$,

$$\begin{aligned}\ln\left(1+\frac{a_n}{n}\right)^n &= n\left(\frac{a_n}{n} - \frac{a_n^2}{2n^2} + \frac{a_n^3}{3n^3} - \frac{a_n^4}{4n^4} + \ldots\right)\\ &= a_n - \frac{a_n^2}{2n} + \frac{a_n^3}{3n^2} - \frac{a_n^4}{4n^3} + \ldots\\ &= a_n + A_n,\end{aligned}$$

where

$$A_n = -\frac{a_n^2}{2n} + \frac{a_n^3}{3n^2} - \frac{a_n^4}{4n^3} + \ldots$$

Since $|a_n| < b$ for all n, it follows that $|A_n| < B_n$, where

$$B_n = \frac{b^2}{n} + \frac{b^3}{n^2} + \frac{b^4}{n^3} + \ldots. = \frac{b^2}{n}\sum_{i=1}^{\infty}\left(\frac{b}{n}\right)^i = \frac{b^2}{n}\cdot\frac{1}{1-(b/n)}.$$

The fact that $B_n \to 0$ as $n \to \infty$ implies that $A_n \to 0$ as $n \to \infty$. We may thus conclude that

$$\ln\left(1+\frac{a_n}{n}\right)^n = a_n + A_n \to a + 0 = a \text{ as } n \to \infty.$$

Upon exponentiating, we have

$$\lim_{n\to\infty}\left(1+\frac{a_n}{n}\right)^n = e^a,$$

completing the proof. ■

Theorem 5.3.1. (The Central Limit Theorem) Let $X_1, X_2, \ldots, X_n, \ldots \overset{\text{iid}}{\sim} F$, a distribution with a finite mean μ and variance σ^2, and let $m_X(t)$ be the moment-generating function of F, assumed to exist in an interval (a,b) containing $t = 0$. Let $Y_n = (\bar{X}_n - \mu)/(\sigma/\sqrt{n})$, where $\bar{X}_n = (1/n)\sum_{i=1}^n X_i$. Then $Y_n \overset{D}{\to} Z \sim N(0,1)$.

Proof. For $t \in (a,b)$, we may write

$$m_{(X-\mu)}(t) = e^{-\mu t} m_X(t). \tag{5.9}$$

From this, we may easily obtain

$$m'_{(X-\mu)}(t) = e^{-\mu t} m'_X(t) - \mu e^{-\mu t} m_X(t), \tag{5.10}$$

and

$$\begin{aligned} m''_{(X-\mu)}(t) &= e^{-\mu t} m''_X(t) - \mu e^{-\mu t} m'_X(t) + \mu^2 e^{-\mu t} m_X(t) - \mu e^{-\mu t} m'_X(t) \\ &= e^{-\mu t} m''_X(t) - 2\mu e^{-\mu t} m'_X(t) + \mu^2 e^{-\mu t} m_X(t). \end{aligned} \tag{5.11}$$

Since X has mean μ and variance σ^2, we know that $m_X(0) = 1$, $m'_X(0) = \mu$ and $m''_X(0) = \sigma^2 + \mu^2$. It thus follows from (5.9)–(5.11) that $m_{X-\mu}(0) = 1, m'_{X-\mu}(0) = 0$ and $m''_{X-\mu}(0) = \sigma^2$.

The Maclaurin series expansion for $m_{(X-\mu)}(t)$, with a quadratic remainder term, may be written as

$$m_{(X-\mu)}(t) = m_{(X-\mu)}(0) + m'_{(X-\mu)}(0)t + \frac{m''_{(X-\mu)}(t^*)t^2}{2}, \tag{5.12}$$

where t* lies between 0 and t and $t \in (a,b)$. Making appropriate replacements in (5.12) and adding and subtracting the value $\frac{\sigma^2 t^2}{2}$ yields

$$m_{(X-\mu)}(t) = 1 + 0 + \frac{\sigma^2 t^2}{2} + \frac{[m''_{(X-\mu)}(t^*) - \sigma^2]t^2}{2}. \tag{5.13}$$

Now, let $Y_n = \sqrt{n}(\bar{X}_n - \mu)/\sigma = \sum_{i=1}^n (X_i - \mu)/\sqrt{n}\sigma$; we may then write, by the iid assumption,

$$\begin{aligned} m_{Y_n}(t) &= E(e^{tY_n}) \\ &= E\left(e^{t\sum_{i=1}^n (X_i-\mu)/\sqrt{n}\sigma}\right) \\ &= \prod_{i=1}^n E e^{t(X_i-\mu)/\sqrt{n}\sigma} \\ &= \left(m_{(X-\mu)}\left(t/\sqrt{n}\sigma\right)\right)^n \text{ by Thms 2.8.1, 3.2.2, and 4.3.6,} \end{aligned}$$

for $t \in (\sqrt{n}\sigma a, \sqrt{n}\sigma b)$. Then, using (5.13), we may write

$$m_{Y_n}(t) = \left(1 + \frac{t^2}{2n} + \frac{[m''_{(X-\mu)}(t^*/\sqrt{n}\sigma) - \sigma^2]t^2}{2n\sigma^2}\right)^n. \tag{5.14}$$

Using the fact that $m''_{(X-\mu)}(t)$ is continuous at $t = 0$ (a fact guaranteed by the assumption that $\sigma^2 < \infty$) and noting that $t^*/\sqrt{n}\sigma$ tends to 0 as n grows to ∞, it follows that $\lim_{n\to\infty}(m''_{(X-\mu)}(t/\sqrt{n}\sigma) - \sigma^2) = 0$. Thus, by Lemma 5.3.1, we have that $\lim_{n\to\infty} m_{(\bar{X}_n-\mu)/(\sigma/\sqrt{n})}(t) = e^{\frac{t^2}{2}}$. By the Continuity Theorem, this implies that

$$Y_n = \frac{\bar{X}_n - \mu}{\sigma/\sqrt{n}} \xrightarrow{D} Z \sim N(0,1).$$

■

There are many interesting and important applications of the CLT. The normal approximation of the binomial distribution $B(n,p)$ when n is sufficiently large is one that is particularly well known. Let us consider a second example of the CLT as a useful approximation theorem.

Example 5.3.1. Suppose that $X_1, X_2, \ldots, X_n, \ldots \stackrel{\text{iid}}{\sim} G(p)$, the geometric distribution based on independent Bernoulli trails with probability p of success. If such a geometric experiment is repeated many times, as it might be at a game booth at a school carnival, one might well be interested in the probability distribution of $\bar{X}_n$, the average number of trials needed to obtain a success. Now since we know that $X = \sum_{i=1}^{n} X_i$ has the negative binomial distribution $NB(n,p)$, probability computations for $\bar{X}_n$ can be made from the exact distribution of X. However, if n is moderate to large in size, the calculations can be quite challenging, even if they are computer aided. But the Central Limit Theorem provides an easy and accurate approximation of such probabilities. Suppose, for example, that $n = 200$ and $p = 1/2$, that is, we have 200 random draws from $G(1/2)$. Since the mean and variance of the parent distribution are $\mu = 2$ and $\sigma^2 = 2$, the CLT identifies the $N(2, .01)$ distribution as the approximate distribution of $\bar{X}_{200}$. We may thus compute

$$P(1.95 \le \bar{X}_{200} \le 2.05) \cong P(-0.5 \le Z \le 0.5) = 2(.1915) = .3830.$$

An exact calculation will show that this approximation is accurate to three decimal places. ■

Exercises 5.3.

1. A random sample of 64 high school seniors who have applied for admission to U. Betcha in northern Idaho had an average (math plus verbal) SAT score of 1020. From many years of experience, U. B. administrators feel that they know that the mean μ and standard deviation σ of their applicants' scores are, respectively, $\mu = 1000$ and $\sigma = 100$. Assuming that the U. B. administrators are correct about μ and σ, compute the approximate value of $P(\bar{X} \ge 1020)$.

2. Airlines usually oversell their flights so that they minimize the likelihood of their planes flying with empty seats. Suppose that a particular plane has 225 seats, but that the airline sells 250 tickets. If 10% of the people who buy an airline ticket fail to show up for the flight, what is the probability that the airline will be able to accommodate all ticket buyers who show up for the flight?

3. Professor Les Simpatico holds an office hour almost every day. The number of students X who drop in on a given day has the probability distribution:

$X = x$	0	1	2	3
$p(x)$	.5	.2	.1	.2

If he holds 140 office hours during the academic year, what's the probability that at least 120 students drop in on him?

4. Salesman I. M. Snooty at Nordtrom's department store in Seattle tends to be extremely rude. On a recent visit, I asked him to show me the cheapest suit in the store. He said to me "You're wearing it." It's no wonder he doesn't make many sales. In a given week, the number of suits X he sells has the probability distribution shown below.

$X = x$	0	1	2	3
$p(x)$	.5	.3	.1	.1

Approximate the probability that Mr. Snooty sells at most 21 suits in his next 30 days at work.

5. Sylvester (Sly) Fox has discovered an even wager at a casino in Reno, Nevada. On any given play, his winnings W on a \$1.00 bet are random with the probability distribution shown below:

$X = x$	-2	-1	1	2
$p(x)$	1/4	1/4	1/4	1/4

This morning, Sly played this game 90 times before breakfast. If $\bar{X}$ is his average winnings per play, find the approximate value of $P(\bar{X} \geq 1/3)$.

5.4 The Delta Method Theorem

For statistical work, the Central Limit Theorem is arguably the most important limit theorem in probability, as it has a myriad of applications and it applies to a process, namely "averaging," that arises in a great many quantitative problems. But there's another limit theorem that is also widely applicable and is extremely useful in many statistical settings. Its name in not nearly as impressive as that of the CLT; it's called the delta method theorem, getting its name from the technique used to prove the result. Basically, the theorem provides a straightforward way of identifying the asymptotic distribution of a function $g(T_n)$ of a variable T_n that is known to be asymptotically normal. As one might expect, there are requirements on the type of function g for which the theorem applies. We will assume that g is smooth enough for its first two derivatives to exist and be continuous. The δ-method theorem will come into play in statistical problems where we know how a particular statistic behaves but we're actually interested in a function of the statistic rather than the statistic itself. The following example illustrates the context in which the δ-method theorem turns out to be the go-to tool.

Example 5.4.1. Suppose that $X_1, X_2, \dots, X_n \overset{\text{iid}}{\sim} G(p)$, the geometric distribution with parameter p. A standard problem in statistical work involves generating a reliable guess of the value of an unknown parameter of a distribution from which a random sample has been drawn. The formal process of doing this falls within the domain of "estimation theory," a topic which we will take up in Chapters 6–8. If the model for the data is geometric, then the sample mean $\bar{X}$ would be a natural estimate of the population mean $1/p$. This suggests that $\hat{p} = 1/\bar{X}$ would make a sensible guess at the value of p. Here's the problem. We know exactly how $\bar{X}$ behaves, at least if the sample size is large. The Central Limit Theorem tells us that

$$\sqrt{n}(\bar{X} - \frac{1}{p}) \xrightarrow{D} Z \sim N(0, \frac{q}{p^2}). \tag{5.15}$$

In large samples, $\bar{X} \sim N(1/p, q/np^2)$, approximately, and we can carry the approximation further, since $\hat{p}$ should be a pretty precise estimate of p when n is large. We might say that there's about a 95% chance that $|\bar{X} - 1/p| < (1.96)\sqrt{\hat{q}}/\sqrt{n}\hat{p}$, where $\hat{q} = 1 - \hat{p}$.

But this is all about guessing the value of $1/p$. The δ-method is precisely the tool needed to make the transition to the problem of interest. The function $g(x) = 1/x$ is a smooth function that transforms our estimator of $1/p$ into an estimator of p. The δ-method will enable us to identify the large-sample distribution of $\hat{p} = 1/\bar{X}$ and to make statements about its precision as an estimator of p. ■

Theorem 5.4.1. (The δ-method theorem) Let $\{T_n\}$ be a sequence of random variables whose distributions depend on a real-valued parameter $\theta \in \Theta$, where Θ is an open interval of real numbers. Suppose that $T_n \xrightarrow{P} \theta$ and that

$$\sqrt{n}(T_n - \theta) \xrightarrow{D} X \sim N(0, \sigma^2(\theta)), \tag{5.16}$$

where $\sigma^2(\theta)$ represents the "asymptotic variance" of T_n and may depend on θ. Let g be a transformation with two continuous derivatives. Then

$$\sqrt{n}(g(T_n) - g(\theta)) \xrightarrow{D} Y \sim N(0, [g'(\theta)]^2\sigma^2(\theta)). \tag{5.17}$$

Proof. Using a first-order Taylor's Series expansion of g about the point θ, and using the Lagrange form of the remainder, we have

$$g(T_n) = g(\theta) + g\prime(\theta)(T_n - \theta) + g\prime\prime(\theta_n^*)(T_n - \theta)^2, \tag{5.18}$$

where θ_n^* lies between θ and T_n. We may thus write

$$\sqrt{n}(g(T_n) - g(\theta)) = g\prime(\theta)\sqrt{n}(T_n - \theta) + g\prime\prime(\theta_n^*)\sqrt{n}(T_n - \theta)^2. \tag{5.19}$$

Now, since $T_n \xrightarrow{P} \theta$, it follows that $\theta_n^* \xrightarrow{P} \theta$ and, by Theorem 5.2.1, part (i), we have that $g\prime\prime(\theta_n^*) \xrightarrow{P} g\prime\prime(\theta) < \infty$. This, in turn, implies that $g\prime\prime(\theta_n^*)\sqrt{n}(T_n - \theta)^2 \xrightarrow{P} 0$, and thus that

$$\sqrt{n}(g(T_n) - g(\theta)) \stackrel{AE}{=} g\prime(\theta)\sqrt{n}(T_n - \theta), \tag{5.20}$$

where the symbol "$X_n \stackrel{AE}{=} Y_m$" represents asymptotic equivalence, that is, X_n and Y_m have the same asymptotic distribution. We thus have from Theorem 5.2.1, part (iii),

$$\sqrt{n}(g(T_n) - g(\theta)) \xrightarrow{D} Y \sim N(0, [g'(\theta)]^2\sigma^2(\theta)), \tag{5.21}$$

as claimed. ■

Example 5.4.1. (continued) We are now in a position to identify the large-sample distribution of $\hat{p} = 1/\bar{X}$, where $\bar{X}$ is the mean of a random sample of size n from a geometric distribution $G(p)$. We have the asymptotic distribution of $\bar{X}$. The transformation of interest is $g(x) = 1/x$, and thus $g'(x) = -1/x^2$. By the δ-method theorem, we have that

$$\sqrt{n}\left(\frac{1}{\bar{X}} - p\right) \xrightarrow{D} Y \sim N\left(0, \left[-\frac{1}{1/p^2}\right]^2 \frac{q}{p^2}\right) = N\left(0, p^2 q\right). \tag{5.22}$$

■

To motivate one particularly important application of the δ-method, it is useful to give a brief preview of the notion of a *confidence interval* based on a random sample from a normal population. The topic will receive a more complete treatment in Chapter 8. If $X_1, X_2, \ldots, X_n, \ldots \stackrel{\text{iid}}{\sim} N(\mu, \sigma^2)$, then $\sqrt{n}(\bar{X} - \mu)/\sigma \sim N(0,1)$. Using the standard normal table, we may this write

$$P(-1.96 \leq \frac{\bar{X} - \mu}{\sigma/\sqrt{n}} \leq 1.96) = .95,$$

which is, of course, equivalent to the statement

$$P(|\bar{X} - \mu| \leq 1.96\sigma/\sqrt{n}) = .95.$$

If the population variance σ^2 were known, we could identify a random interval, namely the interval $(L,U) = (\bar{X} - 1.96\sigma/\sqrt{n}, \bar{X} + 1.96\sigma/\sqrt{n})$, as having a 95% chance of "trapping" the parameter μ. Once the data was in hand, the resulting numerical interval (l,u) is referred to as a 95% confidence interval for μ. The confidence level associated with this interval is linked to the probability that the random interval above will contain the parameter μ.

Now, suppose, again, that a normal random sample was available, but that the population variance was unknown. It is then clear that the random interval above is of no practical use, as its endpoints depend on an unknown parameter. This dilemma arises with some frequency when one wishes to construct an approximate confidence interval based on the asymptotic (normal) distribution of a particular statistic, an approximation that would be well justified if the sample size was sufficiently large. But if the asymptotic variance of that statistic depends on an unknown parameter, the random interval derived from its asymptotic distribution would not yield a usable confidence interval. One solution to this dilemma would be to estimate the unknown parameter in the asymptotic variance from the data. This, of course, adds a new level of approximation to an already-approximate process. An often used alternative is to use the δ-method to search for a transformed statistic whose asymptotic variance is free of unknown parameters. Let's examine how such a search would proceed.

Suppose we have a statistic T_n (based on a sample of size n) which serves as a good estimator for the parameter θ. Assume that

$$\sqrt{n}(T_n - \theta) \xrightarrow{D} X \sim N(0, \sigma^2(\theta)).$$

Now, recall that for any suitably smooth transformation g, the transformed statistic $g(T_n)$ is also asymptotically normal, with

$$\sqrt{n}(g(T_n) - g(\theta)) \xrightarrow{D} Y \sim N(0, [g'(\theta)]^2 \sigma^2(\theta)).$$

Our search thus focuses on finding a transformation g for which

$$[g'(\theta)]^2\sigma^2(\theta) = c^2, \tag{5.23}$$

where c is a positive constant independent of θ. Such a function g is called a *Variance Stabilizing Transformation* (VST). From (5.23), we are led to the abstract equation

$$g\prime(x) = \frac{c}{\sigma(\theta)}. \tag{5.24}$$

By the Fundamental Theorem of Calculus, this equation is equivalent to

$$g(\theta) = \int \frac{c}{\sigma(\theta)} d\theta. \tag{5.25}$$

The constant c in (5.25) is arbitrary and is generally chosen for convenience (for example, to make the asymptotic variance of $g(T_n)$ equal to 1 or to simplify $g(\theta)$). The following example illustrates the process of finding a VST and shows how it facilitates the development of a usable confidence interval.

Example 5.4.2. Suppose that we have a random sample from a Poisson distribution, that is, suppose that $X_1, X_2, \dots, X_n, \dots \overset{\text{iid}}{\sim} P(\lambda)$, where the mean λ is unknown. From the Central Limit Theorem, we have that the sample mean $\bar{X}$ is asymptotically normal, with

$$\sqrt{n}(\bar{X} - \lambda) \overset{D}{\rightarrow} U \sim N(0, \lambda).$$

Thus, if the sample size is large, $\bar{X} \sim N(\lambda, \lambda/n)$, approximately. This distribution does not lend itself directly to the development of a confidence interval for λ. To identify the variance stabilizing transformation in this problem, we need to solve the equation

$$g(\lambda) = \int \frac{c}{\sqrt{\lambda}} d\lambda. \tag{5.26}$$

A function g satisfies (5.26) if and only if it has the form $g(\lambda) = a\sqrt{\lambda} + b$, where a and b are arbitrary constants. Choosing $a = 1$ and $b = 0$, we identify $g(x) = x^{1/2}$ as a variance stabilizing transformation for $\bar{X}$. Then, since $g\prime(x) = 1/(2\sqrt{x})$, we have that

$$\sqrt{n}(\sqrt{\bar{X}} - \sqrt{\lambda}) \overset{D}{\rightarrow} V \sim N(0, \left(\frac{1}{2\sqrt{\lambda}}\right)^2 \lambda) = N(0, 1/4).$$

From (5.26), it follows that $(\sqrt{\bar{X}} - 1.96/2\sqrt{n},\ \sqrt{\bar{X}} + 1.96/2\sqrt{n})$ provides a 95% confidence interval for $\sqrt{\lambda}$. For large n, the lower bound of this interval will be a positive number close to $\sqrt{\bar{X}}$, so that squaring the lower and upper limits of this interval provides a 95% confidence interval for the original parameter λ. ■

Exercises 5.4.

1. Suppose that the statistic T_n based on a random sample of size n from the one-parameter model F_θ has asymptotic distribution shown below.

$$\sqrt{n}(T_n - \theta) \overset{D}{\rightarrow} Y \sim N(0, \theta^2).$$

Identify the asymptotic distribution of the random variable $U_n = \sqrt{T_n}$.

2. Suppose that the statistic T_n based on a random sample of size n from the one-parameter model F_θ has the asymptotic distribution shown below.

$$\sqrt{n}(T_n - \theta) \xrightarrow{D} Y \sim N(0, \theta^2).$$

Identify the asymptotic distribution of the random variable $V_n = \ln(T_n)$.

3. Suppose than the statistic T_n based on a random sample of size n from the one-parameter model F_θ has asymptotic distribution shown below.

$$\sqrt{n}(T_n - \theta) \xrightarrow{D} Y \sim N(0, \theta^2).$$

Identify the asymptotic distribution of $(T_n + 1)/T_n$.

4. Suppose that $X_1, \ldots, X_n, \ldots \overset{\text{iid}}{\sim} N(\mu, \sigma^2)$. Find the asymptotic distribution of $(\bar{X}_n)^3$, where $\bar{X}_n$ is the sample mean of the first n observations.

5. Confidence intervals for a population proportion p are generally based on the asymptotic distribution of the sample proportion $\hat{p}$ given below:

$$\sqrt{n}(\hat{p} - p) \xrightarrow{D} Y \sim N(0, p(1-p)) \text{ as } n \to \infty.$$

Show that the Variance Stabilizing Transformation for $\hat{p}$ is the transformation g given by

$$g(x) = \arcsin\sqrt{x}.$$

(You can confirm from an integral table that $g(x)$ is the antiderivative of the function $g'(x) = 1/\sqrt{x(1-x)}$.) Use the delta method to obtain the asymptotic distribution of $g(\hat{p})$. From this, obtain a 95% confidence interval for $g(p)$. From this interval, get a 95% confidence interval for p whose endpoints do not depend on the unknown parameter p.

5.5 Chapter Problems

1. Calculate $P(|X - \mu| < k\sigma) \geq 1 - 1/k^2$ for $X \sim U(0,1)$ and compare the result to the bound given by Chebyshev's inequality.

2. Let X be a discrete random variable taking integer values. Assume that the pmf of X is symmetric about 0 and that $P(|X| < k) = 1 - 1/k^2$. Show that the pmf of X is

$$p(x) = \frac{x + 1/2}{x^2(x+1)^2} \text{ for } x = 1, 2, 3, \ldots \text{ and}$$

$$p(x) = \frac{1/2 - x}{x^2(1-x)^2} \text{ for } x = -1, -2, -3, \ldots.$$

The symmetry of $p(x)$ implies that $EX = 0$. Calculate the variance σ^2 of X. Compare the exact value of $P(|X| < k\sigma)$ to the lower bound on that probability provided by Chebyshev's inequality.

3. Let $Z \sim N(0,1)$. Show that $P(|Z| > t) \le \sqrt{2/\pi}e^{-t^2/2}/t$ for $t > 0$. (Hint: Note that, when $t > 0$, $P(Z > t) \le \int_t^\infty (1/\sqrt{2\pi}(z/t)e^{-z^2/2}dz)$. Compare this upper bound to that of Chebyshev's inequality for the values $t = 2$ and $t = 3$.

4. A real-valued function $g(x)$ is *convex* if, for arbitrary values x and y and for any $\alpha \in [0,1]$, $g(\alpha x + (1-\alpha)y) \le \alpha g(x) + (1-\alpha)g(y)$. Geometrically, this means that the curve $y = g(x)$ will lie completely above any straight line $y = a + bx$ that is tangent to $g(x)$ at some point. Prove Jensen's inequality: If g is convex, then for any real-valued random variable, $E(g(X)) \ge g(EX)$. (Hint: Note that if $y = a + bx$ is the straight line that is tangent to $y = g(x)$ at the point $x = EX$, then the convexity of $g(x)$ guarantees that $g(x) \ge a + bx$ for all values of x.)

5. Calculate $P(|X - \mu| < k\sigma) \ge 1 - 1/k^2$ for $X \sim Exp(\lambda)$ and compare the result to the bound given by Chebyshev's inequality.

6. Prove the following alternative version of Chebyshev's inequality for any random variable X and any non-negative function g,

$$P(g(X) \ge c) \le \frac{Eg(X)}{c} \text{ for all } c > 0. \tag{5.27}$$

7. Let X be a random variable whose moment-generating function $m_X(t)$ exists and is finite in an interval $(-c,c)$ containing 0. Show that, for any real number d,

   (a) $P(X > d) \le e^{-dt}m_X(t)$ for $t \in (0,c)$ and

   (b) $P(X < d) \le e^{-dt}m_X(t)$ for $t \in (-c,0)$.

8. The "signal-to-noise ratio" r of a random variable with finite mean μ and variance σ^2 is defined as $r = |\mu|/\sigma$. The "relative deviation" of X from its signal is defined as $RD = |(X-\mu)/\mu|$. Show that

$$P(RD \le d) \ge 1 - \frac{1}{r^2d^2}. \tag{5.28}$$

9. See Problem 5.8. Compute the signal-to-noise ratio r for the random variables from the following distributions: (a) $P(\lambda)$, (b) $B(n,p)$, (c) $G(p)$, (d) $\Gamma(\alpha,\beta)$, (e) $W(\alpha,\beta)$, (f) $LN(\mu,\sigma^2)$, and (g) $P(\alpha,\theta)$, where $\alpha > 2$.

10. Let $\bar{X}$ and $\bar{Y}$ be the sample means from two independent samples of size n from a population with finite mean μ and variance σ^2. Use the Central Limit Theorem (treating it as exact) to approximate the smallest value of n for which $P(|\bar{X} - \bar{Y}| < \sigma/10) > .95$.

11. We have noted that $X_n \xrightarrow{qm} c$ implies that $X_n \xrightarrow{p} c$. Show that the converse does not hold. (Hint: Consider discrete variables X_n which take the values n and $-n$, each with probability $1/n$, and is equal to zero otherwise. Show that $X_n \xrightarrow{p} 0$, but that $X_n \xrightarrow{qm} \infty$.)

12. We have noted that $X_n \xrightarrow{as} X$ implies that $X_n \xrightarrow{p} X$. Show that the converse does not hold. (Hint: Fix the positive integer $k \ge 1$. Let $U \sim U(0,1)$. For each positive integer n such that $2^{k-1} \le n < 2^k$, let $j = n - 2^{k-1}$. Define

$$X_n = I_{(j/(2^{k-1}),(j+1)/(2^{k-1}))}(U) \text{ for } j = 0,1,\ldots,2^{k-1}-1.$$

Note that each X_n is equal to 1 on a subinterval of $[0,1]$; for n is such that $2^{k-1} \le n < 2^k$, this subinterval is of length $1/2^{k-1}$. For every k, $X_n = 1$ for exactly one n which satisfies $2^{k-1} \leq n < 2^k$ and $X_n = 0$ for all other n for which $2^{k-1} \leq n < 2^k$. Show that, for any $\varepsilon \in (0,1)$, $P(|X_n - 0| > \varepsilon) \to 0$ as $n \to \infty$, but that the event $\{X_n = 1\}$ occurs infinitely often, so that $P(\sup_{n \geq N} |X_n - 0| \leq \varepsilon) = 0$ for any $N > 0$.)

13. Let $X_1, \dots, X_n, \dots \stackrel{\text{iid}}{\sim} P(\alpha, \theta)$, the two-parameter Pareto distribution. Under what conditions on the parameters α and θ does Theorem 5.1.3, the Weak Law of Large Numbers, apply to the sequence of sample means $\{\bar{X}_n, n = 1, 2, 3 \dots\}$.

14. Let $X_1, X_2, X_3, \dots$ be a sequence of random variables. Show that $X_n \xrightarrow{p} X$ as $n \to \infty$ if and only if
$$E\left(\frac{|X_n|}{1 + |X_n|}\right) \to 0 \text{ as } n \to \infty.$$

15. Let $X_1, X_2, X_3, \dots$ be a sequence of independent Bernoulli random variables, with $X_i \sim B(1, p_i)$. (a) Identify a necessary and sufficient condition of the sequence $p_1, p_2, p_3, \dots$ for $X_n \xrightarrow{p} 0$ as $n \to \infty$. (b) Show that $\bar{X}_n \xrightarrow{p} p^*$, where $\bar{X}_n = \sum_{i=1}^n X_i$, if and only if $p_i \to p^*$.

16. Provide an example of a sequence $\{F_n, n = 1, 2, \dots\}$ of distribution functions which converge to the function $F^*(x) = 0$ for $-\infty < x < \infty$.

17. Construct an example of a sequence of random variables $X_1, X_2, X_3, \dots$ for which $X_n \xrightarrow{p} 0$ while $EX_n \to c \neq 0$.

18. Let $X_{(i)}$ be the ith order statistic from a sample of size n from the $U(0,1)$ distribution. Prove that (a) $X_{(1)} \xrightarrow{p} 0$ and (b) $U_n = nX_{(1)} \xrightarrow{D} V \sim \Gamma(1,1)$.

19. Suppose that $X_1, X_2, \dots, X_n, \dots \stackrel{\text{iid}}{\sim} F$, a distribution with finite mean μ and finite variance σ^2. Show that the sample mean converges to μ in quadratic mean, that is, $\bar{X}_n \xrightarrow{qm} \mu$.

20. Let $\{(X_n, Y_n), n = 1, 2, \dots\}$ be random pairs for which $(X_n - Y_n) \xrightarrow{qm} 0$. Show that if X is a variable for which $(X_n - X) \xrightarrow{qm} 0$, then $(Y_n - X) \xrightarrow{qm} 0$.

21. Let $X \sim P(\lambda)$ and let $Y = (X - \lambda)/\sqrt{\lambda}$. Use Theorem 5.2.2 (the Continuity Theorem) to prove that, as $\lambda \to \infty$, $Y \xrightarrow{D} Z \sim N(0,1)$, that is, show that $m_Y(t) \to m_Z(t)$ as $\lambda \to \infty$. (Hint: Make use of the Maclaurin series expansion of $e^{t\sqrt{\lambda}}$.)

22. Let $X_1, \dots, X_n, \dots \stackrel{\text{iid}}{\sim} F$, a distribution with mean μ. Suppose that, for some $\varepsilon > 0$, $E|X - \mu|^{1+\varepsilon} < \infty$. Establish an alternative form of the weak law of large numbers using Markov's inequality, that is, show that under these assumptions, $\bar{X}_n \xrightarrow{p} \mu$, where $\bar{X}_n = (1/n)\sum_{i=1}^n X_i$.

23. A random sample of n items are drawn from a distribution with mean $\mu < \infty$ and variance $\sigma^2 < \infty$. (a) Use Chebyshev's inequality to determine the smallest sample size n for which the following inequality holds:
$$P(|\bar{X}_n - \mu| \leq \sigma/4) \geq .99. \tag{5.29}$$
(b) Use the Central Limit Theorem to determine the smallest sample size for which (5.29) holds.

24. A certain Las Vegas casino offers one fair game. Players get to sit in comfortable overstuffed chairs overlooking the casino's food court and repeatedly put a \$1.00 chip in the slot on the chair's arm. On each turn, a player will win \$1.00 or lose \$1.00, each with probability 1/2. (By the way, the casino makes money from these players by wafting enticing aromas from the food court toward them as an encouragement to overeat.) Suppose that you show up with \$500 on a given day and you play this game 500 times. What's the probability that you'll be at least \$25.00 ahead at the end of the day?

25. True story. I found the following claim in the Appendix of a popular textbook on probability and mathematical statistics. Let $\mu < \infty$ and $\sigma^2 < \infty$. If $X \sim Be(\alpha,\beta)$ and if $\alpha \to \infty$ and $\beta \to \infty$ at such rates that $\alpha/(\alpha+\beta) \to \mu$ and $\alpha\beta/(\alpha+\beta)^2(\alpha+\beta+1) \to \sigma^2$, then $X \overset{D}{\to} Y \sim N(\mu,\sigma^2)$. Show that the claim is false, and that, actually, as $\alpha \to \infty$, $\beta \to \infty$ and $\alpha/(\alpha+\beta) \to \mu \in [0,1]$, the beta variable X above converges in probability to the constant μ. Still, it's true that the beta density does resemble a normal curve when α and β are large. Can you provide an explanation of why that should be so?

26. True story. I also found the following claim in the Appendix of a popular textbook on probability and mathematical statistics. Let $\mu < \infty$ and $\sigma^2 < \infty$. If $X \sim \Gamma(\alpha,\beta)$ and $\alpha \to \infty$ and $\beta \to 0$ at such rates that $\alpha\beta \to \mu$ and $\alpha\beta^2 \to \sigma^2$, then $X \overset{D}{\to} Y \sim N(\mu,\sigma^2)$. There is a germ of truth in this claim, that is, the gamma density does indeed look very much like a normal density when the shape parameter α is sufficiently large. However, the claim above can't be correct as stated because if $\alpha \to \infty$, $\beta \to 0$ and $\alpha\beta \to \mu$, then, necessarily, $\alpha\beta^2 \to 0$.

    (a) Show that, given $\alpha \to \infty$, $\beta \to 0$, and $\alpha\beta \to \mu$, the gamma variable X above converges in probability to the constant μ.

    Moving on ... it is worth noting that the normal approximation of the gamma distribution, when α is sufficiently large, can be justified theoretically by the Central Limit Theorem.

    (b) Let $X_1,\dots,X_n,\dots \overset{\text{iid}}{\sim} \Gamma(1,\beta)$. Identify the exact distribution of $S_n = \sum_{i=i}^{n} X_i$ and compare it to the approximate distribution of S_n based on the CLT applied to S_n/n, for $n = 30$ and $\beta = 2$. (Hint: The two relevant distributions for this comparison are the χ^2_{60} and the $N(60,120)$ distributions.)

27. Suppose you play roulette 1000 times and bet on red each time. As you know, you have the options of betting on red, black, or green, and the probability of red (or of black) on a given roll is $9/19$. What is the probability that you are even or ahead after 1000 trials? Assume that spins of the roulette wheel are independent, and assume, for concreteness, that you bet \$1.00 on red on every spin.

28. A fair coin is tossed until 100 tails occur. Find the probability that more than 220 coin tosses are required.

29. Let X be the number of servings that a randomly chosen meal-card holder will have for dinner at UC Davis's Segundo Dining Commons. The manager at Segundo has noticed that X has the distribution

x	0	1	2	3
$p(x)$	1/6	1/6	1/6	1/2

Approximate the probability that 96 random meal-card holders will consume at least 200 servings, that is, approximate $P(\bar{X} \geq 2.0833)$.

30. Stirling's formula provides a very useful approximation for $n!$ when n is large:

$$n! \cong \sqrt{2\pi}n^{n+.5}e^{-n}. \tag{5.30}$$

One way to derive this formula is to utilize the central limit theorem as follows. Given that $X_1, \ldots, X_n, \cdots \sim Exp(1)$, the CLT implies that

$$P\left(\frac{(\bar{X}_n - 1)}{1/\sqrt{n}} \leq z\right) \to P(Z \leq z), \tag{5.31}$$

where $Z \sim N(0,1)$. Differentiate both sides of (5.31) with respect to z to obtain

$$\frac{\sqrt{n}}{\Gamma(n)}(z\sqrt{n}+n)^{n-1}e^{-(z\sqrt{n}+n)} \cong \frac{1}{\sqrt{2\pi}}e^{-z^2/2}.$$

Setting $z = 0$, obtain (5.30).

31. Widgets weigh, on average, 12 ounces, with a standard deviation of 2 ounces. If 100 widgets are taken off the assembly line, what is the probability that their total weight is no greater that 1230 ounces?

32. Prove the Borel-Cantelli Lemma:

    (i) if $\{A_i, i = 1, 2, \ldots\}$ is a sequence of events such that $\sum_{i=1}^{\infty} P(A_i) < \infty$, then $P(A_i$ occurs for infinitely many $i) = 0$ and

    (ii) if $\{A_i, i = 1, 2, \ldots\}$ is a sequence of *mutually independent* events such that $\sum_{i=1}^{\infty} P(A_i) = \infty$, then $P(A_i$ occurs for infinitely many $i) = 1$.

    (Hint: Let $A = \bigcap_{r=1}^{\infty} \bigcup_{i=r}^{\infty} A_i$. To obtain (i), argue that since $A \subseteq \bigcup_{i=r}^{\infty} A_i$ for every r, countable subadditivity implies that $P(A) < \varepsilon$ for any arbitrary $\varepsilon > 0$. To obtain (ii), note that if the events $\{A_i\}$ are mutually independent, DeMorgan's Law, countable subadditivity and the multiplication rule may be used to show that $P(A^c) = 0$.)

33. A balanced six-sided die is rolled repeatedly until the sum of the digits rolled exceeds 360. Approximate the probability that at least 100 rolls are needed.

34. A seasoned college professor is quite sure that the distribution of scores on his final exam in the large freshman-level course he teaches every year has a mean of 70 and a variance of 64. (a) Obtain an upper bound for the probability that a random student's test score will exceed 80. (b) How small can the probability be that a random test score is between 70 and 90? (c) Use Chebyshev's inequality to determine how many test takers there must be to be 90% sure that the class average will fall between 72 and 88? (d) answer the question in part (c) under the assumption that the Central Limit Theorem provides an adequate approximation for the distribution of $\bar{X}_n$.

35. Suppose that 100 light bulbs are used successively, with a new bulb used to replace the current bulb as soon as the current light bulb fails. Assume that the lifetimes of these light bulbs are independent exponential variables with a mean of 10 hours. Approximate the probability that all the light bulbs have failed after 1050 hours.

36. An insurance company insures 20,000 automobiles. The distribution of claims each year from these policy holders has a mean value of \$550 with a standard deviation of \$1000. (This distribution is obviously heavily skewed to the right.) Approximate the probability that the total amount of the claims this year exceeds 11.2 million dollars.

37. Let $X_1, X_2, \dots, X_n \stackrel{\text{iid}}{\sim} F$, a distribution with finite mean μ, and suppose that the moment-generating function $m(t)$ of X exists for t in an open interval containing 0. Use Theorem 5.2.2 to show that $\bar{X} \xrightarrow{D} \delta_\mu$, the distribution degenerate at μ. (Hint: Write the mgf of $\bar{X}$ as displayed below, and apply Lemma 5.3.1.

$$m_{\bar{X}}(t) = (m_X(t/n))^n = \left(1 + (t/n) \cdot EX + (t^2/2!n^2) \cdot EX^2 + (t^3/3!n^3) \cdot EX^3 + \dots\right)^n.$$

38. Let $X_1, X_2, \dots, X_n \stackrel{\text{iid}}{\sim} \Gamma(1, \beta)$, the exponential distribution with mean β. The Central Limit Theorem implies that the sample mean $\bar{X}$ is asymptotically normal, or more specifically,

$$\sqrt{n}(\bar{X} - \beta) \xrightarrow{D} Y \sim N(0, \beta^2).$$

Find a variance stabilizing transformation g.

39. The California (CA) Department of Conservation assigns a "CA Redemption Value" (CRV) to recycled material which is due some reimbursement. The "commingled rate" R is defined as the following ratio for the population of recycled material of any given type: $R =$ weight of CRV material / total weight. The sampling distribution of R is not easy to obtain. Consider the following approach. Let $p =$ the proportion of CRV bottles in the population of all recycled bottles. If $\hat{p}$ is the proportion in a random sample of n bottles, then, by the CLT,

$$\sqrt{n}(\hat{p} - p) \xrightarrow{D} Y \sim N(0, p(1-p)) \text{ as } n \to \infty.$$

Now if A and B are known conversion factors (i.e., A = # of CRV bottles /lb and B = # of non-CRV bottles /lb) then we have $p = AR/[AR + B(1-R)]$. This relationship suggests (solving for R in terms of p) that we estimate R by

$$\hat{R} = B\hat{p}/[A(1-\hat{p}) + B\hat{p}].$$

Use the delta method theorem to obtain the asymptotic distribution of $\hat{R}$.

40. Suppose that $X_1, \dots, X_n \stackrel{\text{iid}}{\sim} N(0, \sigma^2)$. (a) Determine the asymptotic distribution of the reciprocal of the second sample moment, that is, of $Y_n = n/\sum_{i=1}^n X_i^2$. (b) Find a variance stabilizing transformation for the statistic $(1/n)\sum_{i=1}^n X_i^2$.

41. In simple linear regression, one typically estimates the correlation coefficient $\rho = Cov(X,Y)/\sigma_X\sigma_Y$ between X and Y by its sample analog r_n. It can be shown that

$$\sqrt{n}(r_n - \rho) \xrightarrow{D} Y \sim N(0, (1-\rho^2)^2).$$

R. A. Fisher recommended transforming r_n to the variable "Z_n" via the transformation $Z_n = g(r_n)$, where g is given by

$$g(x) = \frac{1}{2} ln\left(\frac{1+x}{1-x}\right).$$

The transformation was motivated by the fact that (a) its asymptotic variance does not depend on ρ and that (b) it converges more quickly that r does. This statistic is now called "Fisher's Z." Use the delta method to identify the asymptotic distribution of Z_n.

42. Suppose that $X_n \sim B(n,p)$, and let $\hat{p} = X/n$. The CLT implies that

$$\sqrt{n}(\hat{p}-p) \xrightarrow{D} Y \sim N(0, p(1-p)).$$

Use the δ-method theorem to identify the asymptotic distribution of the log odds $g(\hat{p}) = \ln(\hat{p}/(1-\hat{p}))$ as $n \to \infty$.

6

Statistical Estimation: Fixed Sample Size Theory

6.1 Basic Principles

In this chapter, we begin making the transition from the ideas and tools of probability theory to the theory and methods of statistical inference. It seems appropriate to begin with a discussion of how these two rather different fields actually fit together. But rather than thinking about how one glob, probability theory, can be joined (or glued) to another glob, statistics, I will concentrate here on how two important subfields of each of these areas—stochastic modeling and statistical estimation—are seamlessly connected. It is not difficult to make similar connections between other pairs of subfields of Probability and Statistics, though this will be left, for the most part, to your own initiative.

For simplicity, suppose that we have the data $X_1, X_2, \ldots, X_n$ in hand, where the Xs are assumed to be independent random variables with a common distribution F_θ indexed by the parameter θ whose value is unspecified. In our earlier discussions, we have used the notation $X_1, X_2, \ldots, X_n \stackrel{\text{iid}}{\sim} F_\theta$ to represent this scenario. In practice, we would often be willing to assume that the distribution F_θ has a density or probability mass function of some known form (a step which is the essence of "stochastic modeling"). For example, physical measurements might be assumed to be random draws from an $N(\mu, \sigma^2)$ distribution, counts might be assumed to be $B(n, p)$ or $P(\lambda)$ variables, and the failure times of engineered systems in a life-testing experiment might be assumed to be $\Gamma(\alpha, \beta)$ variables. Note that such assumptions do not specify the exact distribution that applies to the experiment of interest, but rather, specifies only its "type." Our initial modeling assumption stipulates only that the true model from which the data is drawn is a member of a class of models of a given form and is governed by one or more unspecified parameters. For now, we are setting aside the possibility of making "nonparametric" modeling assumptions like "the distribution F of the Xs has an increasing failure rate" or, something even less specific like "F is continuous," "F is symmetric," or "F can be anything." In such cases, the distributions cannot be indexed by a finite number of parameters. In this and the next two chapters, we will deal exclusively with statistical estimation for parametric models.

Suppose that a particular parametric model F_θ has been assumed to apply to the available data $X_1, X_2, \ldots, X_n$. It is virtually always the case in practice that the exact value of the parameter θ is not known. A naïve but nonetheless useful way to think of *statistical estimation* is to equate it with the process of guessing. The goal of statistical estimation is to make an *educated guess* about the value of the unknown parameter θ. What makes one's guess "educated" is the fact that the estimate of θ is informed by the data. A *point estimator* of θ is a fully specified function of the data which, when the data is revealed, yields a numerical guess at the value of θ. The estimator will typically be represented by the symbol $\hat{\theta}$, pronounced "theta hat," whose dependence on the data is reflected in the equation

$$\hat{\theta} = \hat{\theta}(X_1, X_2, \ldots, X_n).$$

Given the modeling assumption that has been made—namely, that the available data is drawn randomly from F_θ—and given the chosen estimator $\hat{\theta}$ of θ, one is now in a position to accomplish the primary goal of point estimation, that is, to refine the model by replacing its unknown parameters by estimates based on the data. While one starts one's analysis by postulating a general class of models $\{F_\theta, \theta \in \Theta\}$ which contain the true model for the observed data, we complete our analysis by refining our modeling and asserting that the estimated model $F_{\hat{\theta}}$ is our best guess at the true model.

Let us suppose, for example, that the experiment of interest involves n independent tosses of a bent coin, where n is a known integer. The parameter p, the probability of heads in a given coin toss, is treated as an unknown constant. It seems quite reasonable to assume that the experimental data we will observe is well described as a sequence of iid Bernoulli trials. Most people would base their estimate of p on the variable X, the number of successes in n iid Bernoulli trials; of course, $X \sim B(n, p)$. The sample proportion of successes, $\hat{p} = X/n$, is a natural estimator of p and is, in fact, the estimator of p that most people would use. If, in particular, $X \sim B(20, p)$ and we observe $X = 6$, then we would no doubt estimate p by $\hat{p} = .3$ and we would put forward the model $B(20, .3)$ as our best guess at the true underlying model.

The introductory remarks above provide a simple overview of the estimation process, but as with most simple overviews, it leaves a good many quite natural questions unanswered. Perhaps the most basic of these is the question of how one can make judgments concerning the quality of different estimators. Since we would naturally prefer to make good guesses rather than bad guesses, we will need to examine possible criteria by which different estimation strategies can be compared. In addressing this issue, it is natural to talk about two important characteristics of estimators that are helpful in making such judgments and comparisons—accuracy and precision.

Informally, you can think of an *accurate estimator* as one that is aimed correctly. If the parameter θ was equal to 5, then we would think of the estimator $\hat{\theta}$ of θ as being accurate if, when used repeatedly, the average value of $\hat{\theta}$ was fairly close to 5. Another description of an accurate estimation process is that it is "well calibrated"; it is aimed at the target (or at something very close by). Notice that the accuracy of $\hat{\theta}$, while an attractive, desirable property, is not, by itself, a sufficient basis to decide whether or not $\hat{\theta}$ is a good estimator. If $\theta = 0$ and your estimator always estimates it to be 100 or -100, each with probability .5, we would have to classify the estimator as "accurate," but we would also admit that it is unsatisfactory for most practical purposes. An "accurate" estimator that estimates θ to be 1 or -1, each with probability .5, would certainly be preferable. The reason we'd prefer the second estimator over the first is that it is still accurate, but it is also more precise.

An estimator may be thought of as a *precise estimator* if its variability is low in repeated trials. More simply, we will judge the precision of an estimator $\hat{\theta}$ by the size of its variance. A small variance indicates that the estimator gives approximately the same answer when used repeatedly. Precision seems like a more reliable criterion than accuracy, but it too has an Achilles' heel. If an estimator is very precise but is also always very far from the true value of the parameter, we would consider it to be defective.

From the discussion above, it should be clear that a truly good estimator should perform well in terms of both accuracy and precision. Estimators which fail in one or the other respect will generally be judged as inferior, and their average performance in repeated experiments can usually be improved by an alternative estimator. Besides the comparison with "guessing," an analogy between statistical estimation as "target practice" is also useful, particularly when picturing the concepts of accuracy and precision. The figures below display four possibilities of interest.

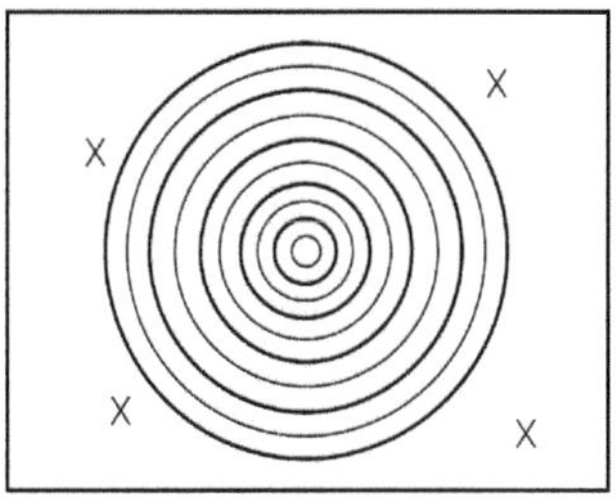

Figure 6.1.1. Accurate but not precise.

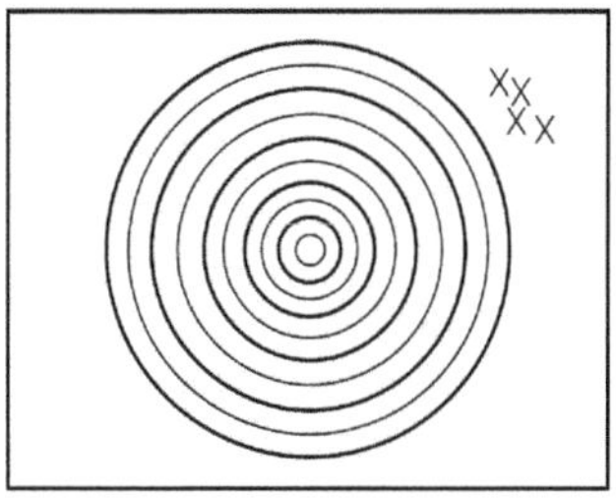

Figure 6.1.2 Precise but not accurate.

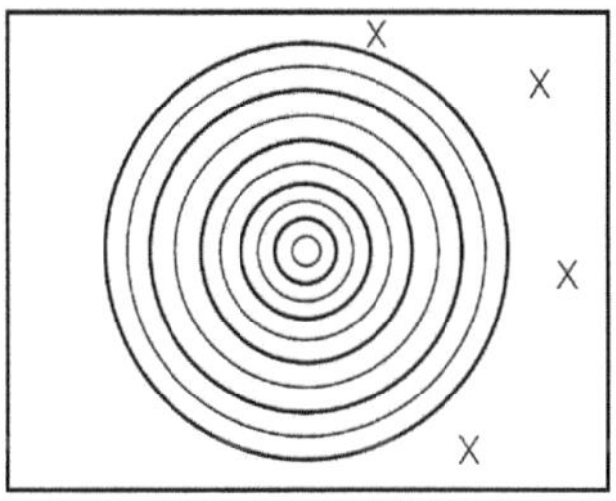

Figure 6.1.3. Neither accurate nor precise.

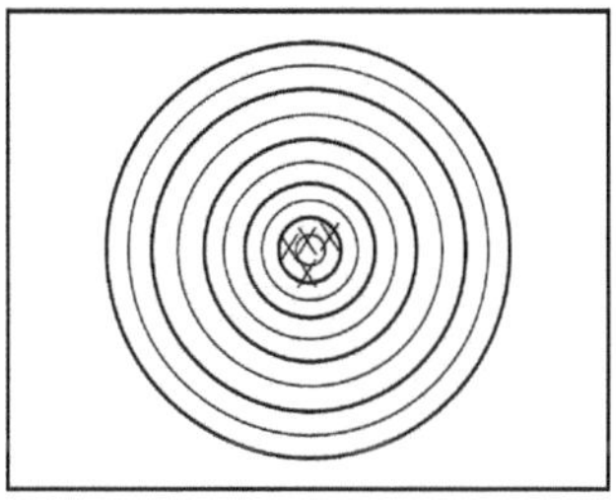

Figure 6.1.4. Accurate and precise.

We will now place the terms "accurate" and "precise" in the context of probability and statistics. Before proceeding, we record a remark worth emphasizing regarding the scope of the ideas we will develop in our treatment of estimation problems.

Remark 6.1.1. We will continue to assume that the estimation problems of interest are based on the random sample $X_1, X_2, \ldots, X_n \stackrel{\text{iid}}{\sim} F_\theta$. It should be noted that the properties of estimators to be examined here are of interest beyond this limited framework, that is, they also apply to estimators based on data which are modeled as dependent in some way and/or have different (marginal) distributions. Some of the examples, exercises, and problems in the sequel will call upon the reader to examine estimation problems in these more general scenarios. It should be noted, as well, that in much of what follows, we will take θ to be a scalar parameter, though we will generalize ideas and notation as needed when dealing with models with more than one parameter. ■

Let us assume that $X_1, X_2, \ldots, X_n \stackrel{\text{iid}}{\sim} F_\theta$, and that the estimator $\hat{\theta} = \hat{\theta}(X_1, \ldots, X_n)$ is to be used as a point estimator for θ. Let us first make note of the (perhaps transparent) fact that $\hat{\theta}$ is a random variable, its randomness being inherited from the randomness of the sample. If the random sampling process were repeated many times, the value of the estimator $\hat{\theta}$ would vary from sample to sample. The probability distribution of $\hat{\theta}$ could be obtained, in principle, using the transformation methods treated in Chapter 4. We turn now to an examination of a property of estimators that formally captures the notion of "accuracy."

Definition 6.1.1. The estimator $\hat{\theta}$ of θ is an *unbiased* estimator of θ if

$$E\hat{\theta} = \theta,$$

that is, if the mean of the distribution of $\hat{\theta}$ is θ.

As you know, $E\hat{\theta}$ is generally interpreted as the center (in the "center of gravity" sense) of the distribution of $\hat{\theta}$, so that if $\hat{\theta}$ is an unbiased estimator of θ, then we can think of the distribution of $\hat{\theta}$ as being well centered, or alternatively, we can think of $\hat{\theta}$ as being "aimed at the true value of θ." In many repetitions of the same experiment, we would expect the average value of $\hat{\theta}$ would be very close to θ.

One of the major themes of this chapter is the examination of ways in which the expected performance of two estimators can be compared. Eventually, we'll discuss a widely used criterion for comparing two arbitrary estimators based on a given random sample. We'll first consider the important special case in which the two estimators $\hat{\theta}_1$ and $\hat{\theta}_2$ of the parameter θ are both unbiased.

Using the language introduced earlier in this section, we would consider two unbiased estimators of a parameter θ to be equally "accurate" since both of their distributions are centered at the true value of θ. It thus remains to determine which estimator is more precise. We will judge the precision of an unbiased estimator $\hat{\theta}$ of θ by its variance $V(\hat{\theta}) = E(\hat{\theta} - \theta)^2$, a criterion function which measures the expected squared distance of the estimator from the true parameter value. We note, in passing, that alternative measures of precision, e.g., $E|\hat{\theta} - \theta|$, may be used; however, $V(\hat{\theta})$ is by far the most widely used measure in this context because of its interpretability as a measure of dispersion and because of the relative simplicity of computing it.

Now the variance of a random variable is, in general, a function of the parameters of its distribution, so the comparison alluded to above is a comparison between two functions. Since pairs of functions can cross each other one or more times, it is certainly possible that the variance of one unbiased estimator will be smaller than the variance of a second unbiased estimator for some values of the parameter and will be larger for other parameter values. In such cases, neither estimator can be declared to be superior to the other. On the other hand, given two unbiased estimators $\hat{\theta}_1$ and $\hat{\theta}_2$ of the parameter θ, we would consider $\hat{\theta}_1$ to be a better estimator of θ than $\hat{\theta}_2$ if $V(\hat{\theta}_1) \le V(\hat{\theta}_2)$ for all possible values of θ, with strict inequality ($<$) for at least one value of θ. We highlight this case in the following.

Definition 6.1.2. An unbiased estimator $\hat{\theta}_1$ of θ is *more efficient* than the unbiased estimator $\hat{\theta}_2$ of θ if

$$V(\hat{\theta}_1) \le V(\hat{\theta}_2) \text{ for all } \theta, \tag{6.1}$$

with

$$V(\hat{\theta}_1) < V(\hat{\theta}_2) \text{ for at least one value of } \theta. \tag{6.2}$$

The *relative efficiency* of $\hat{\theta}_1$ with respect to (wrt) $\hat{\theta}_2$ is simply the ratio of their variances, that is,

$$RE(\hat{\theta}_1 \text{ wrt } \hat{\theta}_2) = \frac{V(\hat{\theta}_2)}{V(\hat{\theta}_1)}. \tag{6.3}$$

If RE($\hat{\theta}_1$ wrt $\hat{\theta}_2$) ≥ 1 for all values of θ, then the estimator $\hat{\theta}_1$ is uniformly as good as or better than the estimator $\hat{\theta}_2$, regardless of the exact value of the parameter θ.

Let's now consider these ideas in a specific estimation problem.

Example 6.1.1. Let $X_1, X_2, \ldots, X_n \overset{\text{iid}}{\sim} U(0,\theta)$, with θ unknown. We will develop two particular unbiased estimators of θ and compare them side by side.

(a) Consider first an estimator of θ based on the sample mean $\bar{X}$. Recall from Corollary 4.2.1 that for a random sample from any probability model with a finite mean μ, the sample mean $\bar{X} = \frac{1}{n}\sum_{i=1}^{n} X_i$ has expected value μ, that is, $\bar{X}$ is an unbiased estimator of μ. Since the uniform model above has mean $\theta/2$ it follows that the estimator $\hat{\theta}_1 = 2\bar{X}$ is an unbiased estimator of θ. Using Corollary 4.3.1 and the fact that this uniform model has variance $\sigma^2 = \theta^2/12$ (see Example 3.2.1), we have that

$$V(\hat{\theta}_1) = V(2\bar{X}) = 4V(\bar{X}) = 4 \cdot \frac{\sigma^2}{n} = 4 \cdot \frac{\theta^2/12}{n} = \frac{\theta^2}{3n}. \tag{6.4}$$

(b) Next, let's consider estimating the parameter θ by a function of the order statistics $X_{(1)}, X_{(2)}, \ldots, X_{(n)}$ obtained from our sample from $U(0,\theta)$. One might guess that the largest order statistic $X_{(n)}$ gives the most information about θ. We will consider here only estimators of the form $\hat{\theta} = cX_{(n)}$. Later in this chapter, we will give theoretical support (on the basis of "sufficiency") for basing our inference about θ on the statistic $X_{(n)}$ alone. For now, we simply develop the estimator of interest and evaluate its variance. Note that, using the density of $X_{(n)}$ given in (4.81),

$$\begin{aligned} E(X_{(n)}) &= \int_0^\theta x_{(n)} \frac{n x_{(n)}^{n-1}}{\theta^n} dx_{(n)} \\ &= \left. \frac{n x_{(n)}^{n+1}}{(n+1)\theta^n} \right|_0^\theta \\ &= \frac{n\theta}{n+1}. \end{aligned}$$

It follows that

$$E\left(\frac{(n+1)X_{(n)}}{n}\right) = \theta,$$

that is, the estimator $\hat{\theta}_2 = \frac{(n+1)X_{(n)}}{n}$ is an unbiased estimator of θ.

The computations above may be simplified by employing some transformation theory from Section 4.6.3. First, we take note of the fact that if $Y \sim U(0,1)$, then $X = \theta Y \sim U(0,\theta)$. Recall from Example 4.7.1 that the order statistic $Y_{(k)}$ from a sample $Y_1, Y_2, \ldots, Y_n \overset{\text{iid}}{\sim} U(0,1)$ has a $Be(k, n-k+1)$ distribution. From this, we know that $E(Y_{(n)}) = n/(n+1)$. Since the variables $\theta Y_{(1)}, \theta Y_{(2)}, \ldots, \theta Y_{(n)}$ have the same distribution as the order statistics from a sample of size n from $U(0,\theta)$, we have that $E(X_{(n)}) = E(\theta Y_{(n)}) = n\theta/(n+1)$, from which we may infer that $\hat{\theta}_2$ above is an unbiased estimator of θ.

The transformation theory above is even more useful in obtaining the variance of $\hat{\theta}_2$. A direct derivation of this variance can of course be obtained by evaluating the second moment of $X_{(n)}$ as

$$E(X_{(n)}^2) = \int_0^\theta x_{(n)}^2 \frac{n x_{(n)}^{n-1}}{\theta^n} dx_{(n)}$$

and using it to obtain $V(X_{(n)})$ and then $V(\hat{\theta}_2)$. On the other hand, since we know that $X_{(n)}/\theta \sim Be(n,1)$, we have that

$$\begin{aligned} V(\hat{\theta}_2) &= V\left(\frac{(n+1)X_n}{n}\right) \\ &= \frac{(n+1)^2\theta^2}{n^2} V\left(\frac{X_{(n)}}{\theta}\right) \\ &= \frac{(n+1)^2\theta^2}{n^2} \cdot \frac{(n)(1)}{(n+1)^2(n+2)} \\ &= \frac{\theta^2}{n(n+2)}. \end{aligned} \tag{6.5}$$

What remains is to compare the variances of the two estimators. First, note that when $n = 1$, the two estimators are one and the same, i.e., the average of one observation is the same as the largest observation in a sample of size 1. So of course the variances in (6.4) and (6.5) are equal when $n = 1$. However, if $n \geq 2$, then

$$V(\hat{\theta}_2) = \frac{\theta^2}{(n+2)n} < \frac{\theta^2}{3n} = V(\hat{\theta}_1) \text{ for all } \theta > 0.$$

Thus, for $n \geq 2$, the estimator $\hat{\theta}_2$ is a more efficient estimator of θ than the estimator $\hat{\theta}_1$. For any sample size n, we have

$$RE(\hat{\theta}_2 \text{ wrt } \hat{\theta}_1) = \frac{n+2}{3}.$$

■

Exercises 6.1.

1. CIA agent Max Threat has been assigned to monitor e-mail correspondence between mild mannered terrorist "Bozo" and the angrier terrorist "Rambo." Mr. Threat has been counting the number of times the phrase "blow up" occurs per 1000 words that each of them writes. Max has determined that Rambo uses the phrase three times as often as Bozo. Threat has modeled their respective usage X and Y as independent Poisson variables, with $X \sim P(\lambda)$ for Bozo and $Y \sim P(3\lambda)$ for Rambo.

   (a) Show that $\hat{\lambda}_1 = (1/2)X + (1/6)Y$ is an unbiased estimator of λ.

   (b) Show that $\hat{\lambda}_2 = (1/10)X + (3/10)Y$ is also an unbiased estimator of λ.

   (c) Show that one of these estimators is a uniformly better estimator than the other.

2. Suppose you were given the choice of examining one of two sets of experimental data: (a) $X_1, \ldots, X_{10} \overset{\text{iid}}{\sim} N(\mu, 1)$ or (b) $Y_1, \ldots, Y_5 \overset{\text{iid}}{\sim} N(2\mu, 1)$. Both experiments would enable you to obtain an unbiased estimator of the parameter μ. Which data set would you prefer to examine. Justify your answer.

3. Suppose X_1, X_2 and $X_3 \overset{\text{iid}}{\sim} U(\theta - 1/2, \theta + 1/2)$.

(a) Show that both of the estimators $\hat{\theta}_1 = \bar{X}$ and $\hat{\theta}_2 = (X_{(1)} + X_{(3)})/2$, where $X_{(i)}$ is the ith order statistic in the sample, are unbiased estimators of θ. (Hint: Working with the random variables $Y_i = X_i - \theta + 1/2$, $i = 1,2,3$, which are iid $U(0,1)$ variables, should prove helpful. Note that $(Y_{(1)} + Y_{(3)})/2 = \hat{\theta}_2 - \theta + 1/2$.)

(b) Which one is the better estimator? (Hint: Again, work with the variables $\{Y_i\}$ mentioned in part (a). Verify that the random pair $(Y_{(1)}, Y_{(3)})$ has the joint density $f(y_1, y_3) = 6(y_3 - y_1)$ for $0 < y_1 < y_3 < 1$. Since, for arbitrary variables U and V, $Cov(U + a, V + b) = Cov(U,V)$ for all a and b, it follows that $Cov(Y_{(1)}, Y_{(3)}) = Cov(X_{(1)}, X_{(3)})$. All this facilitates finding $V(\hat{\theta}_2)$.)

4. Suppose X is the discrete variable with the distribution shown below. Consider two estimators of θ defined as (a) $T_1 = 0$ if $X = 0$ and $T_1 = 1$ if $X > 0$ and (b) $T_2 = 2X/3$. Show that both estimators are unbiased. Which is the better estimator?

$X = x$	0	1	2
$p(X)$	$1-\theta$	$\theta/2$	$\theta/2$

6.2 Further Insights into Unbiasedness

We noted in Chapter 4 that the mean $\bar{X}$ of a random sample $X_1, X_2, \ldots, X_n$ drawn from a distribution F with finite mean μ is an unbiased estimator of μ. The sample variance s^2, defined as

$$s^2 = \frac{1}{n-1} \sum_{i=1}^{n} (X_i - \bar{X})^2, \tag{6.6}$$

enjoys a similar property, that is, s^2 is an unbiased estimator of the population variance σ^2, provided that σ^2 is finite. This provides a partial, if not definitive, answer to the question: why do we divide the sum in (6.6) by $n - 1$ instead of by n, as we would do if we were simply taking an average of the n squared deviations of the Xs from the sample mean? The intuition behind dividing by something smaller than n lies in the fact that in trying to estimate the expected or average value of $(X - \mu)^2$, we find that $\sum_{i=1}^{n} (X_i - \bar{X})^2$ is necessarily smaller than $\sum_{i=1}^{n} (X_i - \mu)^2$, a claim that follows from the fact that $a = \bar{X}$ is the constant that minimizes $\sum_{i=1}^{n} (X_i - a)^2$. While $E\left[(1/n)\sum_{i=1}^{n} (X_i - \mu)^2\right] = \sigma^2$, this function is unusable as an estimator of σ^2 because it depends on the unknown parameter μ. Since $\sum_{i=1}^{n} (X_i - \bar{X})^2 < \sum_{i=1}^{n} (X_i - \mu)^2$ for any value $\mu \neq \bar{X}$, we see that $(1/n)\sum_{i=1}^{n} (X_i - \bar{X})^2$ tends to underestimate σ^2. That $(n - 1)$ is the "right" divisor of $\sum_{i=1}^{n} (X_i - \bar{X})^2$ is confirmed in the following result.

Theorem 6.2.1. Let $X_1, X_2, \ldots, X_n \overset{iid}{\sim} F$, a distribution with finite mean μ and variance σ^2. Then $s^2 = (1/(n-1))\sum_{i=1}^{n} (X_i - \bar{X})^2$ is an unbiased estimator of the parameter σ^2.

Proof. We may expand and reevaluate the expression $E(\sum_{i=1}^{n} (X_i - \bar{X})^2)$ as follows:

$$E\left[\sum_{i=1}^{n} (X_i - \bar{X})^2\right] = E\left[\sum_{i=1}^{n} (X_i - \mu + \mu - \bar{X})^2\right]$$

$$= E\left[\sum_{i=1}^{n}(X_i-\mu)^2+2(\mu-\bar{X})\sum_{i=1}^{n}(X_i-\mu)+n(\bar{X}-\mu)^2\right]$$

$$= E\left[\sum_{i=1}^{n}(X_i-\mu)^2-2n(\bar{X}-\mu)^2+n(\bar{X}-\mu)^2\right]$$

$$= E\left[\sum_{i=1}^{n}(X_i-\mu)^2-n(\bar{X}-\mu)^2\right]$$

$$= \sum_{i=1}^{n}E(X_i-\mu)^2-nE(\bar{X}-\mu)^2$$

$$= \sum_{i=1}^{n}\sigma^2-n\left(\frac{\sigma^2}{n}\right)=(n-1)\sigma^2.$$

Dividing the first and last terms in the sequence of identities above by $(n-1)$ establishes the claim. ■

The result above states that whenever one samples from a population in which $\sigma^2<\infty$, s^2 is an unbiased estimator for σ^2. In Exercise 6.6.3, we point out the fact that, in the commonly assumed scenario in which the Xs are randomly drawn from a normal population, the estimator s^2 can be improved upon, even though as we will see later, it is the unbiased estimator of σ^2 with the smallest possible variance.

We now move on to the interesting issue of "estimability."

Definition 6.2.1. Let $X_1,X_2,\dots,X_n \overset{\text{iid}}{\sim} F_\theta$. The parameter θ is said to be *estimable* if there exists an estimator $\hat{\theta}=\hat{\theta}(X_1,X_2,\dots,X_n)$ that is unbiased for θ.

From what you have seen so far, it may well appear to you that unbiased estimators were more or less identified in an *ad hoc* manner. It is natural to ask whether you would always be able to find one. Obviously, this is an even more basic question than that of finding out whether one unbiased estimator is better than another. The existence question has a somewhat unsettling answer. There do exist non-estimable parameters in certain modeling scenarios. Even more worrisome (since it plays upon an individual's mathematical insecurities) is the possibility of thinking that a parameter is not estimable just because one can't find an unbiased estimator when in fact one exists. Such worries can't be completely eliminated, but being familiar with some tried-and-true tools when searching for an unbiased estimator can help reduce the stress.

Here's the short-run and long-run agenda. In the short run, we will look into non-estimability more carefully, just to show that it is real. Then, we will discuss a couple of reliable methods for finding unbiased estimators. While these methods can't be guaranteed to work (after all, there might not be any unbiased estimators), you will at least have some tools to use in investigating the question, and you should be able to provide some evidence that you gave the matter the old college try. (Instructors like to see that!) You might not be able to find an unbiased estimator of a parameter of interest in a given problem. As we proceed, we'll consider other approaches to estimation whose performance tends to justify their use (sometimes, even over the best possible unbiased estimator). Since there are no unbiased estimators in some problems, it is sensible to take the positive approach espoused by the authors of a technical paper I once heard presented at a statistical conference. The title of their paper was: "We Still Can't Solve It, but Now We Don't Care!" In the long run, we'll show, in various ways, that unbiased estimators aren't really the gold standard in statistical estimation, and that there are many circumstances in which we would actually prefer a "good" biased estimator to the best unbiased one.

Let's begin with an example which shows that the matter of non-estimability is a not a figment of my imagination!

Example 6.2.1. Let X be a Bernoulli variable, i.e., let $X \sim B(1,p)$. The parameter p is of course estimable, since $EX = p$. But what about other functions of p? Is there an unbiased estimator of p^2? We'll show that p^2 is not estimable. Suppose there was a function g of X that satisfied the equation

$$E_p g(X) = p^2 \text{ for all possible } p \in (0,1). \tag{6.7}$$

For (6.7) to hold for all p, one must have

$$g(1)p + g(0)(1-p) = p^2 \text{ for all } p \in (0,1). \tag{6.8}$$

If Equation (6.8) holds for all p, then we would have a function f, namely

$$f(p) = p^2 + (g(0) - g(1))p - g(0),$$

that is quadratic in the variable p and is equal to zero for all p in the interval $(0,1)$. But there can be no such quadratic function, as quadratic functions (whose graphs are parabolas) can have at most two real roots, that is, can have at most two values of p for which $f(p) = 0$. It follows that a function g satisfying the property in (6.7) does not exist. ■

The example above makes it clear that we can have no assurance, in a given statistical problem, that the parameter of interest is estimable. Keep in mind that being non-estimable doesn't mean that we can't estimate the parameter; it simply means that an unbiased estimator of the parameter does not exist. In the example above, the parameter p is estimable but p^2 is not. We could, however, reason that if we have a trustworthy estimator $\hat{p}$ of p, then we should be willing to trust $\hat{p}^2$ as an estimator of p^2. (In the example above, this isn't very helpful since $\hat{p}^2 = \hat{p}$; there are other problems in which the strategy works better.) While a transformation g will not, in general, preserve the unbiasedness of a given estimator (see Exercise 6.2.1 for a special case in which it does), using $g(\hat{\theta})$ to estimate $g(\theta)$ when we trust $\hat{\theta}$ as an estimator of θ is often a reasonable thing to do.

Even though we cannot be sure of success, there are at least a couple of approaches which can prove fruitful in searching for an unbiased estimator. These are presented below. In a problem where neither of these methods lead us to an unbiased estimator, the possibilities that remain consist of (1) using an *ad hoc* approach that exploits the special features of the problem of interest and (2) the ever present possibility that the parameter of interest is not estimable. But before wrestling with such possibilities, let's take a look at the following approaches, both of which have proven useful in selected problems.

Method I: The Expectation Method. Typically, the expectation method involves three steps: (1) Evaluate the expectation of one or more random variables derived from the available data. If the data has been summarized by a single statistic T (as in the case of iid

Bernoulli variables that are summarized by their sum), the random variables whose expectation is calculated might be related variables like T, T^2, T^3,... (2) Try to express the parameter θ of interest as a linear function of these expected values, (3) Use the linearity property of expectation (Theorem 4.2.2) to write θ as the expected value of a linear function of these variables.

Consider the use of Method I in the following:

Example 6.2.2. Let $X \sim B(n,p)$, with $n \geq 2$. Let's find an unbiased estimator of the parameter $\theta = p^2$. In using Method I, we begin by evaluating expectations of powers of X:

$$EX = np \tag{6.9}$$

and

$$\begin{aligned} EX^2 &= V(X) + (EX)^2 \\ &= npq + (np)^2 \\ &= np + (n^2 - n)p^2. \end{aligned} \tag{6.10}$$

We could, of course, obtain additional moments of X, but we note that the parameter p^2 of interest has already arisen, so it makes sense to go to step (2) of the method. Using (6.9) and (6.10), we may write

$$EX^2 - EX = (n^2 - n)p^2,$$

from which we obtain, by the linearity property of expectation,

$$E\left(\frac{X^2 - X}{n^2 - n}\right) = p^2.$$

We thus have that

$$\hat{\theta} = \frac{X^2 - X}{n^2 - n} = \frac{X(X-1)}{n(n-1)}$$

is an unbiased estimator of $\theta = p^2$. ■

Method II. The Indicator Function Method. This method is applicable when the parameter θ of interest can be viewed as the probability of some event. Suppose that, in the random experiment under consideration, we can identify a statistic T, and an event A, such that $\theta = P(T \in A)$. Then the indicator function of the event A, that is,

$$\hat{\theta} = I_A(T), \tag{6.11}$$

is an unbiased estimator of θ. Recall that $I_A(t)$ can only take on the values 0 and 1, and $I_A(t) = 1$ if and only if $t \in A$. Therefore, we have that

$$E\hat{\theta} = 1 \cdot P(T \in A) + 0 \cdot P(T \notin A) = P(T \in A) = \theta.$$

Example 6.2.3. Let $X \sim B(n,p)$, and let $\theta = p^n$. In searching for an unbiased estimator of p^n using Method II, we would ask ourselves: Is p^n the probability of something? The answer is yes; in the binomial experiment of interest, we can recognize p^n as the probability of getting n successes in n trials. In other words, $p^n = P(X = n)$. Therefore, if we define $\hat{\theta} = I_{\{n\}}(X)$, i.e., if we set $\hat{\theta} = 1$ if $X = n$ and set $\hat{\theta} = 0$ if $X < n$, then $E\hat{\theta} = P(X = n) = p^n$. We thus have that $\hat{\theta}$ is an unbiased estimator of the parameter $\theta = p^n$. ■

There are estimation problems in which both methods are applicable. It's likely, in those cases, that one of them is easier to execute than the other. That would almost always be Method II. A case in point is Example 6.2.3. It is possible to obtain an unbiased estimator of $\theta = p^n$ using Method I. To do this, one would start with obtaining the first n moments of the variable X. This would result in n linear equations in the n "unknowns" $p, p^2, p^3, \ldots, p^n$. Solving these equations simultaneously would yield an expression for p^n as a linear combination of the moments $EX, EX^2, \ldots, EX^n$. Using Theorem 4.2.2, we would, in the end, find that the estimator

$$\tilde{\theta} = \frac{X(X-1)(X-2)\cdots(X-n+1)}{n(n-1)(n-2)\cdots(n-n+1)} \tag{6.12}$$

is an unbiased estimator of θ. Now, when we have two different unbiased estimators of a parameter θ, we would naturally turn to a comparison of their variances to determine if one is better than the other. Here, this next step is unnecessary. That's because our two estimators of p^n are actually one and the same. Notice that the estimator in (6.12) is equal to 0 whenever $X < n$, and is equal to 1 when $X = n$, that is, for all values of the variable X, $\tilde{\theta}(X) = \hat{\theta}(X)$.

The following derivation is a bit more than a "classroom example" as it deals with a real problem that arises in the area of engineering reliability.

Example 6.2.4. Reliability theory is a field of study on the interface of the mathematical sciences and engineering. The main focus of the field is to model the performance characteristics (for example, expected lifetime, reliability, average downtime) of engineered systems. Manufacturers will often perform "life testing experiments" on a newly developed system in order to learn about the properties of the population of such systems. Let's assume here that we are dealing with a simple system whose lifetime is well modeled by an exponential distribution with mean μ. Suppose that a random sample of system lifetimes is available for analysis. More concretely, suppose that $X_1, X_2, \ldots, X_n \stackrel{\text{iid}}{\sim} \Gamma(1, \mu)$. Since μ is the mean of the population, clearly $\bar{X}$ is an unbiased estimator of μ. Suppose, however, that we are not interested in μ itself, but rather, our aim is to find an unbiased estimator of the system's reliability $R(x) = P(X > x) = e^{-x/\mu}$ at some particular time x. While the estimator $\tilde{R}(x) = e^{-x/\bar{X}}$ might well come to mind, this estimator is not an unbiased estimator of $R(x)$. (See Problem 6.7.17 for details.) Later in this chapter, we will treat the general problem of finding the best unbiased estimator of $R(x)$ based on the available sample $X_1, X_2, \ldots, X_n$. Here, we will tackle the simpler problem of showing that $R(x)$ is an estimable parameter. Now, since $R(x)$ is clearly a probability, the estimator $\hat{R}_1(x) = I_{(x,\infty)}(X_1)$ is an unbiased estimator of $R(x)$. While this settles the question of the estimability of $R(x)$, we know that we can do better. We can immediately improve upon $\hat{R}_1(x)$ by an unbiased estimator obtained by averaging the similar unbiased estimators based on all the Xs, that is, by the estimator

$$\hat{R}(x) = \frac{1}{n}\sum_{i=1}^{n} \hat{R}_i(x), \tag{6.13}$$

where $\hat{R}_i(x) = I_{(x,\infty)}(X_i)$. The latter estimator reduces the variance of any of the individual estimators of $R(x)$ by a factor of $1/n$. Is this the best we can do? We will return to this problem in Section 6.4 where we show that there is an alternative unbiased estimator that is even better. There, with the help of more powerful tools, we will derive the unbiased estimator of $R(x)$ with the smallest possible variance. ■

Exercises 6.2.

1. Suppose that $T_1,\ldots,T_k$ are unbiased estimators of a parameter θ. Show that if $g(x_1,\ldots,x_k)=\sum_{i=1}^k c_i x_i$, where $\sum_{i=1}^k c_i = 1$, then $g(T_1,\ldots,T_k)$ is an unbiased estimator of θ.

2. Let $X \sim G(p)$, the geometric distribution with pmf $p(x) = pq^{x-1}$ for $x = 1,2,\ldots$ Find unbiased estimators of each of the following functions of p: p, q, $1/p$, q/p and q^5.

3. Let $X \sim P(\lambda)$, the Poisson distribution with mean λ. Find an unbiased estimator of λ^2.

4. Suppose that $X \sim P(\lambda)$. Obtain an unbiased estimator of $e^{2\lambda}$. (Hint: Think about utilizing the moment-generating function of X.)

5. The method of "randomized response" was devised to enable an experimenter to obtain sensitive information in a way that ensures each respondent's privacy. Sociologist I. B. High plans to ask a random sample of college freshmen whether they have used marijuana in the last 7 days. Let p be the proportion of the overall target population whose true answer to this question is "yes." To remove the possibility of self-incrimination, each participant will flip a coin and answer question A if they get heads and question B if they get tails. The outcome of their coin toss will be known only to them. Question A is "Does your mother's maiden name begin with the letters a–m?" It is reasonable to assume that the probability of a "yes" answer to question A is $1/2$. Question B is the marijuana question, which participants are asked to answer if their coin flip results in tails. Suppose a random sample of n freshmen participate in the experiment. Let X be the number of "yes" answers obtained in this experiment. What is the probability distribution of X? Find an unbiased estimator of p based on X.

6.3 Fisher Information, the Cramér-Rao Inequality, and Best Unbiased Estimators

Fisher Information. We begin this section with a discussion of a fundamental measure that will prove to be an important tool in the study of optimality in statistical estimation. Suppose we are dealing with data drawn from a one-parameter model F_θ with density (or pmf) f_θ for $\theta \in \Theta$. The measure of interest was introduced by Sir Ronald Fisher (1890–1962), one of the primary contributors to the foundations of mathematical statistics, and it is now called the Fisher information $I_X(\theta)$, a measure of the "information" contained in a single observation X from F_θ. One would wish such a measure to be large if X is very informative about θ as it would be, for example, if the density of X was centered at θ and had a very small variance, so that the observed value of X was likely to be very close to θ. The Fisher information of X is defined as

$$I_X(\theta) = E\left(\left[\frac{\partial}{\partial\theta}\ln f(X|\theta)\right]^2\right), \tag{6.14}$$

provided that the derivative in (6.14) is well defined and the expectation exists. Now the expectation in (6.14) is not a quantity that would occur to you, off hand, as a useful measure of the information content of one's data, and it is perhaps an indication of the genius of Ronald Fisher that he recognized it as such. Of course he came to this realization gradually, after encountering this same object arising in a variety of modeling contexts.

Later in this section, we will refer to a certain set of assumptions as "regularity conditions," and will speak of models which satisfy these conditions as "regular models." We will postpone a detailed discussion of regularity for now, save to say that they are a set of conditions on the density or probability mass function of the model of interest that guarantee a certain smoothness in the parameter θ that permits differentiation with respect to θ as well as certain convenient mathematical operations. As it happens, many of the probability models we have studied, both discrete and continuous, can be shown to satisfy the standard regularity conditions that are made in the development of the optimality theory for statistical estimation. Exceptions include models like $U(0,\theta)$ in which the support set $\{x|f(x|\theta) > 0\}$ of the model depends on the parameter θ. In the present context, it can be shown that, for a regular model, the Fisher information can be computed by an alternative formula, namely,

$$I_X(\theta) = -E\left(\frac{\partial^2}{\partial\theta^2}\ln f(X|\theta)\right). \tag{6.15}$$

As we will see, this alternative form of $I_X(\theta)$ will occasionally simplify the computation of this measure.

The measure $I_X(\theta)$ is best thought of in relative terms, that is, X is "more informative about θ" than Y is if $I_X(\theta) > I_Y(\theta)$. This view is supported, in part, by the following result which shows that Fisher information grows proportionately with the growth in the number of available observations. Since we'll usually be dealing with random samples of size n, this result will prove useful.

Theorem 6.3.1. If $X_1,\ldots,X_n \stackrel{\text{iid}}{\sim} F_\theta$, where F_θ is a regular one-parameter model, then $I_{X_1,\ldots,X_n}(\theta) = n\cdot I_X(\theta)$.

Proof. By the independence of the Xs, we have

$$\begin{aligned}
I_{X_1,\ldots,X_n}(\theta) &= -E\left(\frac{\partial^2}{\partial\theta^2}\ln f(X_1,\ldots,X_n|\theta)\right)\\
&= -E\left(\frac{\partial^2}{\partial\theta^2}\ln\prod_{i=1}^n f(X_i|\theta)\right)\\
&= \sum_{i=1}^n\left[-E\left(\frac{\partial^2}{\partial\theta^2}\ln f(X_i|\theta)\right)\right]\\
&= n\cdot I_X(\theta).
\end{aligned}$$

■

Although $I_X(\theta)$ in (6.14) looks somewhat imposing, it actually is rather easy to compute for many of the models with which we are familiar. Let's take a look at the normal model.

Example 6.3.1. Suppose X is a normal random variable with unknown mean μ and known variance $\sigma^2 = \sigma_0^2$, that is, $X \sim N(\mu,\sigma_0^2)$. We then have

$$f(x|\mu) = \frac{1}{\sqrt{2\pi}\sigma_0}e^{-\frac{(x-\mu)^2}{2\sigma_0^2}}, \quad -\infty < x < \infty,$$

so that

$$\ln f(x|\mu) = -\ln(\sqrt{2\pi}\sigma_0) - \frac{(x-\mu)^2}{2\sigma_0^2},$$

and

$$\frac{\partial}{\partial\mu}\ln f(x|\mu) = \frac{x-\mu}{\sigma_0^2}.$$

We may thus compute the Fisher information of X as

$$I_X(\mu) = E\left(\frac{X-\mu}{\sigma_0^2}\right)^2 = \frac{E(X-\mu)^2}{\left(\sigma_0^2\right)^2} = \frac{\sigma_0^2}{\left(\sigma_0^2\right)^2} = \frac{1}{\sigma_0^2}. \tag{6.16}$$

Noting that

$$\frac{\partial^2}{\partial\mu^2}\ln f(x|\mu) = -\frac{1}{\sigma_0^2},$$

we can also obtain the Fisher information of X using the alternative expression in (6.15):

$$I_X(\mu) = -E\left(-\frac{1}{\sigma_0^2}\right) = \frac{1}{\sigma_0^2}. \tag{6.17}$$

The fact that the Fisher information of a normal variable is the reciprocal of its variance supports the interpretation of $I_X(\mu)$ as measuring the amount of information in X about the parameter μ. The information content in X is inversely proportional to its variance. Extending this calculation to a normal random sample, if $X_1, X_2, \dots, X_n \stackrel{\text{iid}}{\sim} N(\mu, \sigma_0^2)$, we conclude that

$$I_{X_1,\dots,X_n}(\theta) = \frac{n}{\sigma_0^2}.$$

■

For discrete models, the Fisher information is calculated in the same fashion, with the density $f(x|\theta)$ representing the probability mass function of X.

Example 6.3.2. Suppose that $X \sim B(1,p)$. Then

$$f(x|p) = p^x(1-p)^{1-x} \qquad \text{for } x = 0, 1.$$

We thus have that

$$\ln f(x|p) = x\ln(p) + (1-x)\ln(1-p),$$

from which it follows that

$$\begin{aligned}\frac{\partial}{\partial p}\ln f(x|p) &= \frac{x}{p} - \frac{1-x}{1-p}\\ &= \frac{x(1-p)-p(1-x)}{p(1-p)}\\ &= \frac{x-p}{p(1-p)}.\end{aligned}$$

We may thus obtain the Fisher information of X as

$$I_X(p) = E\left(\frac{X-p}{p(1-p)}\right)^2 = \frac{E(X-p)^2}{(p(1-p))^2} = \frac{p(1-p)}{(p(1-p))^2} = \frac{1}{p(1-p)}.$$

Again, we see that the Fisher information of a variable X is the reciprocal of its variance. A similar computation shows that if $X \sim B(n,p)$, then $I_X(p) = n/p(1-p)$. This latter expression is also the Fisher information in a random sample of n Bernoulli variables, as they contain the same essential information about the parameter p as is contained in their sum X. ∎

The Cramér-Rao inequality. Given two unbiased estimators $\hat{\theta}_1$ and $\hat{\theta}_2$ of a parameter θ, we would typically seek to determine which is preferable by comparing their variances. If, for all values of θ, we confirm that $V(\hat{\theta}_1) \leq V(\hat{\theta}_2)$, then the estimator $\hat{\theta}_1$ is as good or better than $\hat{\theta}_2$, regardless of the true value of θ, and $\hat{\theta}_1$ could be recommended for use over $\hat{\theta}_2$. While a comparison such as this is satisfying, in that it helps us avoid using an inferior estimator, it offers no assurance that there isn't an alternative unbiased estimator with better performance than the best one discovered so far. Fortunately, it is possible, in certain circumstances, to confirm that a particular unbiased estimator is the best among all possible estimators in this class. When we are successful in doing so, we circumvent the need for comparing alternative unbiased estimators two at a time and we achieve that not-to-be undervalued feeling of being confident that there's no alternative unbiased estimator lurking around that improves on our estimator's performance. We will discuss two quite different avenues for finding best unbiased estimators, such estimators to be referred to, henceforth, as minimum variance unbiased estimators or MVUEs. The first approach, which is treated below, is to identify a lower bound on the variance of an unbiased estimator. Having that, we may establish that a particular unbiased estimator is better than any other if its variance is shown to achieve this lower bound for all possible values of the parameter of interest. We will treat an alternative method for identifying an MVUE in the next section. We turn now to the development of the famous Cramér-Rao lower bound which serves the purpose above.

A useful mathematical tool, both for the present goal of establishing the famous Cramér-Rao lower bound, and for many other problems in mathematical statistics, is the well-known inequality usually attributed to two prominent mathematicians who independently discovered versions of the result. Augustin-Louis Cauchy (1789–1857) was a leading French mathematician who was among the pioneers of the mathematical field of real analysis and Karl Hermann Amandus Schwarz (1843–1921) was a highly regarded German mathematician known primarily for his work in the mathematical field of complex analysis. The Cauchy-Schwarz inequality is stated separately below in the continuous and discrete cases. In the statement below, R represents the real line and R^n stands for n-dimensional Euclidean space. (R^2, for example, is the XY plane.)

Lemma 6.3.1. (The Cauchy-Schwarz inequality) Let f and g be real-valued functions of the vector-valued variable $\mathbf{y} \in R^n$ for which the integrals below exist and are finite. Then

$$\left(\int_{\mathbf{y}\in R^n} f(\mathbf{y})g(\mathbf{y})d\mathbf{y}\right)^2 \leq \int_{\mathbf{y}\in R^n} f^2(\mathbf{y})d\mathbf{y} \cdot \int_{\mathbf{y}\in R^n} g^2(\mathbf{y})d\mathbf{y}, \tag{6.18}$$

where

$$\int_{\mathbf{y}\in R^n} h(\mathbf{y})d\mathbf{y}$$

stands for the multiple integral $\int_{-\infty}^{\infty}\cdots\int_{-\infty}^{\infty} h(y_1,\ldots,y_n)dy_1\ldots dy_n$.

Proof. Note that for any real number $k \in R$, the function $h(k)$, defined as

$$h(k) = \int_{\mathbf{y}\in R^n} [kf(\mathbf{y}) + g(\mathbf{y})]^2 d\mathbf{y}, \tag{6.19}$$

is necessarily nonnegative. Expanding this integral as a quadratic in k, we have

$$h(k) = k^2 \int_{\mathbf{y}\in R^n} f^2(\mathbf{y})d\mathbf{y} + 2k\cdot\int_{\mathbf{y}\in R^n} f(\mathbf{y})g(\mathbf{y})dy + \int_{\mathbf{y}\in R^n} g^2(\mathbf{y})d\mathbf{y} \geq 0. \tag{6.20}$$

Since $h(k)$ is nonnegative for all values of k, it can have at most one real zero. Now, the potential zeros of $h(k)$ are given by the quadratic formula

$$k = \frac{-2\int f(\mathbf{y})g(\mathbf{y})d\mathbf{y} \pm \sqrt{\left(2\int f(\mathbf{y})g(\mathbf{y})d\mathbf{y}\right)^2 - 4\left(\int f^2(\mathbf{y})d\mathbf{y}\right)\left(\int g^2(\mathbf{y})d\mathbf{y}\right)}}{2\int f^2(\mathbf{y})d\mathbf{y}}. \tag{6.21}$$

The quadratic in (6.20) will have at most one real zero if and only if the discriminant D in (6.21), namely

$$D = \left(2\int_{\mathbf{y}\in R^n} f(\mathbf{y})g(\mathbf{y})d\mathbf{y}\right)^2 - 4\left(\int_{\mathbf{y}\in R^n} f^2(\mathbf{y})d\mathbf{y}\right)\left(\int_{\mathbf{y}\in R^n} g^2(\mathbf{y})d\mathbf{y}\right),$$

is non-positive. Rewriting the inequality $D \leq 0$ yields

$$\left(\int_{\mathbf{y}\in R^n} f(\mathbf{y})g(\mathbf{y})d\mathbf{y}\right)^2 \leq \int_{\mathbf{y}\in R^n} f^2(\mathbf{y})d\mathbf{y}\cdot\int_{\mathbf{y}\in R^n} g^2(\mathbf{y})d\mathbf{y},$$

completing the proof. ■

The discrete version of the Cauchy-Schwarz inequality is stated below. An efficient proof of this version of the inequality is suggested in Problem 6.7.28.

Lemma 6.3.2. Let $\{a_i\}$ and $\{b_i\}$ be arbitrary sequences of real numbers. Then

$$\left(\sum a_i b_i\right)^2 \leq \sum a_i^2 \times \sum b_i^2. \tag{6.22}$$

The inequality that we prove below will hold for problems in which the random sample on which an estimator is based satisfies the following "regularity conditions."

I. The set $A = \{x | f(x|\theta) > 0\}$ does not depend on θ.

II. For all $x \in A$, $\frac{\partial}{\partial\theta}\ln f(x|\theta)$ exists and is finite.

III. For any statistic $T(X)$ with $ET(X) < \infty$, $\frac{\partial}{\partial\theta^i}\int T(x)f(x|\theta)dx = \int T(x)\frac{\partial}{\partial\theta^i}f(x|\theta)dx$ for $i = 1,2$.

Models satisfying conditions I, II and III are said to be *regular*. Many of the models we've encountered in this book (for example, $B(n,p)$, $G(p)$, $P(\lambda)$, $NB(r,p)$, $N(\mu,\sigma_0^2)$, $N(\mu_0,\sigma^2)$ and $\Gamma(\alpha,\beta_0)$ and $\Gamma(\alpha_0,\beta)$) are regular one-parameter models. The $U(0,\theta)$ and $P(\alpha_0,\theta)$ models are examples of non-regular models.

We are now in a position to prove

Theorem 6.3.2. (The Cramér-Rao inequality) Let $X_1,\ldots,X_n \overset{\text{iid}}{\sim} F_\theta$, a regular model with density or probability mass function $f_\theta(x)$. If $T = T(X_1,\ldots,X_n)$ is an unbiased estimator of θ, then

$$V(T) \geq \frac{1}{nI_X(\theta)},$$

where $I_X(\theta)$ is the Fisher information in a single observation $X \sim F_\theta$.

Proof. We prove the theorem in the continuous case. We begin by making note of two identities implied by our assumptions:

$$\int_{\mathbf{x}\in R^n} f_\theta(\mathbf{x})d\mathbf{x} = 1 \tag{6.23}$$

and

$$\int_{\mathbf{x}\in R^n} T(\mathbf{x})f_\theta(\mathbf{x})d\mathbf{x} = \theta. \tag{6.24}$$

Differentiating both sides of each of these equations (employing regularity condition III twice) yields

$$\int_{\mathbf{x}\in R^n} \frac{\partial}{\partial\theta} f_\theta(\mathbf{x})d\mathbf{x} = 0 \tag{6.25}$$

and

$$\int_{\mathbf{x}\in R^n} T(\mathbf{x})\frac{\partial}{\partial\theta} f_\theta(\mathbf{x})d\mathbf{x} = 1. \tag{6.26}$$

Equation (6.25) may equivalently be written as

$$\int_{\mathbf{x}\in R^n} \theta\frac{\partial}{\partial\theta} f_\theta(\mathbf{x})d\mathbf{x} = 0. \tag{6.27}$$

Subtracting the LHS of (6.27) from the LHS of (6.26) yields

$$\int_{\mathbf{x}\in R^n} [T(\mathbf{x})-\theta]\frac{\partial}{\partial\theta} f_\theta(\mathbf{x})d\mathbf{x} = 1. \tag{6.28}$$

Preparing to apply the Cauchy-Schwarz inequality to (6.28), we rewrite the equation as

$$\int_{\mathbf{x}\in R^n} [T(\mathbf{x})-\theta]f_\theta(\mathbf{x})\frac{\partial}{\partial\theta}\ln f_\theta(\mathbf{x})d\mathbf{x} = 1,$$

an equation resulting from the fact that $(\partial/\partial\theta)\ln f_\theta(x) = (\partial/\partial\theta)f_\theta(x)/f_\theta(x)$. The latter equation is equivalent to

$$\int_{\mathbf{x}\in R^n} \left[(T(\mathbf{x})-\theta)\sqrt{f_\theta(\mathbf{x})}\right]\left[\sqrt{f_\theta(\mathbf{x})}\frac{\partial}{\partial\theta}\ln f_\theta(\mathbf{x})\right]d\mathbf{x} = 1.$$

From the Cauchy-Schwarz Inequality, we then have that

$$\begin{aligned} 1^2 &= \left(\int_{\mathbf{x}\in R^n} \left[(T(\mathbf{x})-\theta)\sqrt{f_\theta(\mathbf{x})}\right]\left[\sqrt{f_\theta(\mathbf{x})}\frac{\partial}{\partial\theta}\ln f_\theta(\mathbf{x})\right]d\mathbf{x}\right)^2 \\ &\leq \int_{\mathbf{x}\in R^n} (T(\mathbf{x})-\theta)^2 f_\theta(\mathbf{x})d\mathbf{x}\cdot\int_{\mathbf{x}\in R^n}\left(\frac{\partial}{\partial\theta}\ln f_\theta(\mathbf{x})\right)^2 f_\theta(\mathbf{x})d\mathbf{x} \end{aligned}$$

$$= V(T(\mathbf{X})) \cdot E\left(\frac{\partial}{\partial\theta}\ln f_\theta(\mathbf{X})\right)^2$$
$$= V(T(\mathbf{X})) \cdot n \cdot I_X(\theta),$$

the last step being justified by Theorem 6.3.1. It follows that

$$V(T(\mathbf{X})) \geq \frac{1}{nI_{\mathbf{X}}(\theta)}.$$

■

The following two examples show how the Cramér-Rao inequality may be used to verify that a given unbiased estimator of a particular parameter may be identified as the MVUE, while a third example shows that the Cramér-Rao Lower Bound (CRLB) may sometimes yield inconclusive results.

Example 6.3.1. (Continued) Suppose that $X_1, X_2, \ldots, X_n \overset{\text{iid}}{\sim} N(\mu, \sigma^2)$. The CRLB for the variance of an unbiased estimator of μ is equal to $1/I_{X_1,\ldots,X_n}(\mu) = \sigma^2/n$. Now, since the sample mean $\bar{X}$ is an unbiased estimator of μ, and $V(\bar{X}) = \sigma^2/n$, which is equal to the CRLB in this estimation problem, $\bar{X}$ has the smallest possible variance among unbiased estimators of μ, that is, $\bar{X}$ is the MVUE of μ. ■

Example 6.3.2. (Continued) Suppose that $X \sim B(n, p)$. Recall that X represents the number of successes in n independent Bernoulli trials with probability p of success. The Cramér-Rao Lower Bound for the variance of an unbiased estimator of p is $1/I_X(p) = p(1-p)/n$. Now, since the sample proportion $\hat{p} = X/n$ of successes is an unbiased estimator of p, and $V(\hat{p}) = p(1-p)/n$, it follows that $\hat{p}$ is the MVUE of μ. ■

Example 6.3.3. Let us consider the problem of estimating the variance of a normal population. Suppose that $X_1, X_2, \ldots, X_n \overset{\text{iid}}{\sim} N(\mu, \sigma^2)$. Let s^2 be the sample variance given by

$$s^2 = \frac{1}{n-1}\sum_{i=1}^{n}(X_i - \bar{X})^2.$$

As shown in Theorem 6.2.1, s^2 is, in general, an unbiased estimator of the population variance σ^2 (assuming only that $\sigma^2 < \infty$) and, of course, is an unbiased estimator of σ^2 in the particular case of a normal random sample considered here. As discussed in Chapter 4, the exact distribution of s^2 is known in the normal case. Specifically, we know that

$$Y = \frac{(n-1)s^2}{\sigma^2} \sim \chi^2_{n-1} = \Gamma\left(\frac{n-1}{2}, 2\right). \tag{6.29}$$

It is thus easy (though of course not necessary in light of Theorem 6.2.1) to confirm the unbiasedness of s^2:

$$E\left(\frac{(n-1)s^2}{\sigma^2}\right) = n-1, \tag{6.30}$$

from which $E(s^2) = \sigma^2$ immediately follows. The variance of the estimator s^2 is also easy to obtain from (6.29). For Y defined in (6.29), we may calculate $V(Y)$ as $[(n-1)/2] \cdot 2^2 = 2(n-1)$, from which it follows that

$$V(s^2) = \frac{2\sigma^4}{n-1}. \tag{6.31}$$

Next, we ask: Can Theorem 6.3.2 be used to demonstrate that s^2 is the MVUE of σ^2? Let's compute the CRLB. We have that

$$\ln f(x|\mu,\sigma^2) = -\frac{\ln(2\pi\sigma^2)}{2} - \frac{(x-\mu)^2}{2\sigma^2},$$

so that

$$\frac{\partial}{\partial\sigma^2}\ln f_{\mu,\sigma^2}(x) = -\frac{1}{2\sigma^2} + \frac{(x-\mu)^2}{2\sigma^4},$$

and

$$\frac{\partial^2}{\partial(\sigma^2)^2}\ln f_{\mu,\sigma^2}(x) = \frac{1}{2\sigma^4} - \frac{(x-\mu)^2}{\sigma^6}.$$

Now, recall that $(X-\mu)/\sigma \sim N(0,1)$ and thus that $(X-\mu)^2/\sigma^2 \sim \chi_1^2 = \Gamma(1/2,2)$. Using the fact that the first and second moments of this gamma distribution are 1 and 3, respectively, we have that

$$\begin{aligned}
I_X(\sigma^2) &= E\left(\frac{\partial}{\partial\sigma^2}\ln f_{\mu,\sigma^2}(X)\right)^2 \\
&= E\left(\frac{1}{4\sigma^4} - \frac{(X-\mu)^2}{2\sigma^6} + \frac{(X-\mu)^4}{4\sigma^8}\right) \\
&= \frac{1}{4\sigma^4} - \frac{\sigma^2}{2\sigma^6} + \frac{3\sigma^4}{4\sigma^8} \\
&= \frac{1}{2\sigma^4}.
\end{aligned}$$

The Cramér-Rao Lower Bound is thus given by $2\sigma^4/n$, a bound not achieved by the unbiased estimator s^2 of σ^2. In the next section, we will use an alternative approach to show that the sample variance s^2 of a normal random sample is the MVUE of σ^2, notwithstanding the fact that $V(s^2) >$ CRLB. ■

Exercises 6.3.

1. For each of the following models, obtain the Fisher information and the Cramér-Rao Lower Bound on the variance of an unbiased estimator of the parameter θ based on a random sample of size n from the model: (a) the geometric model $G(\theta)$ with $0 < \theta < 1$ and (b) the exponential distribution $\Gamma(1,\theta)$ with mean $\theta > 0$.

2. Suppose $X_1,\dots,X_n \overset{\text{iid}}{\sim} N(0,\theta)$, where the variance $\theta > 0$. (a) Find the CRLB for the variance of an unbiased estimator of θ. (b) Show that $\hat{\theta} = (1/n)\sum_{i=1}^n X_i^2$ is the MVUE of θ.

3. Show that under the regularity conditions I, II and III in Section 6.3, the Fisher information of the variable $X \sim F_\theta$, as defined in (6.14), may be computed by the alternative formula

$$I_X(\theta) = -E\left(\frac{\partial^2}{\partial\theta^2}\ln f(X|\theta)\right).$$

(Hint: Evaluate $\frac{\partial^2}{\partial\theta^2}\ln f_\theta(X)$ in two steps; using the quotient rule in the second step, show that $\frac{\partial^2}{\partial\theta^2}\ln f_\theta(X) = \frac{\frac{\partial^2}{\partial\theta^2}f_\theta(X)}{f_\theta(X)} - \left(\frac{\frac{\partial}{\partial\theta}f_\theta(X)}{f_\theta(X)}\right)^2$. Then, evaluate the expectation of both sides.)

4. Let X be the random variable associated with the randomized response experiment described in Exercise 6.2.5. Compute the Fisher information $I_X(p)$, and then compare the variance of your estimator to the CRLB. Can you claim that the unbiased estimator found in Exercise 6.2.5 is the MVUE of p?

5. Let $X_1, X_2, \dots, X_n \overset{\text{iid}}{\sim} F_\theta$, where F_θ is the "Paretoesque" distribution with density function

$$f_\theta(x) = \frac{1}{\theta} x^{-(\theta+1)/\theta} I_{(1,\infty)}(x).$$

   (a) Let $T = \prod_{i=1}^n X_i$. Show that if $X \sim F_\theta$, then $Y = \ln(X) \sim \Gamma(1, \theta)$. Using this fact, show that $\hat{\theta} = (\ln T)/n = (1/n)\sum_{i=1}^n \ln(X_i)$ is an unbiased estimator of θ.

   (b) By comparing $V(\hat{\theta})$ to the Cramér-Rao Lower Bound for the variance of unbiased estimators of θ, show that $\hat{\theta}$ is the best unbiased estimator, i.e., the MVUE, of θ.

6.4 Sufficiency, Completeness, and Related Ideas

Sufficient Statistics. Let us suppose that a random sample $X_1, \dots, X_n$ is drawn from F_θ, a distribution indexed by an unknown parameter θ. We are interested here in the idea that a particular real-valued statistic $T = h(X_1, \dots, X_n)$ might contain all the information about θ that the sample $X_1, \dots, X_n$ itself contains. If it does, then it would be appropriate to base any inferences we may wish to make about θ on the statistic T rather than on the entire sample, reducing the dimension of the problem from n to 1. We will call a statistic T a sufficient statistic for θ if it has such a property. We'll see that for many statistical models we have encountered, some sort of dimension reduction via sufficiency is indeed possible.

If one were to contemplate the issue of sufficiency from scratch, one might well decide to define the concept of sufficiency using a measure like Fisher information. For the statistic T alluded to above, if one found that $I_T(\theta) = nI_X(\theta)$, that is, the statistic T had precisely that same Fisher information as the sample $X_1, \dots, X_n$, one would feel justified in restricting attention to the statistic T in any statistical investigation concerning θ. While this is a reasonable way to approach the notion of sufficiency, it has certain limitations (for example, it is applicable only to models satisfying certain "regularity conditions"). We will use a different approach in defining the concept of sufficiency, one that is somewhat broader in scope, yet is also more elementary and more intuitive. The following is the most widely used definition of sufficiency and the one that we will carry forward to any further references to the concept in this book.

Definition 6.4.1. Suppose that $X_1, \dots, X_n \overset{\text{iid}}{\sim} F_\theta$. A statistic $T = h(X_1, \dots, X_n)$ is *sufficient* for θ if the conditional distribution of $X_1, \dots, X_n$, given $T = t$, does not depend on θ. If F_θ has density or probability mass function f_θ, then T is sufficient for θ if and only if $f_\theta(x_1, \dots, x_n | T = t)$ does not depend on θ.

If T is a sufficient statistic for θ, then when the value of T is known, the sample $X_1, \dots, X_n$ will provide no further information about θ. The conditional density or pmf of the data, given $T = t$, will formally depend on the parameter θ through the expression

$$f_\theta(x_1, \dots, x_n | t) = \frac{f_\theta(x_1, \dots, x_n, t)}{f_\theta(t)}, \tag{6.32}$$

but the sufficiency of T will be reflected in the fact that the apparent dependence of the ratio in (6.32) on θ will actually cancel out and the ratio will be free of θ altogether. An example of the use of the definition above to identify a particular statistic as sufficient for a parameter θ follows.

Remark 6.4.1. In (6.32) and elsewhere, I have used a shorthand notation "$f_\theta(x_1,\ldots,x_n|t)$" for the more cumbersome notation $f_{X|T=t,\theta}(x_1,\ldots,x_n|t)$. I will take this liberty whenever the context in which the shorthand notation is used makes the meaning clear.

Example 6.4.1. Let $X_1,\ldots,X_n \stackrel{\text{iid}}{\sim} B(1,p)$, that is, suppose that these Xs are Bernoulli random variables with probability p of "success." The Xs might represent, for example, the results of n independent tosses of a bent coin in which either a head ($X=1$) or a tail ($X=0$) was obtained. We might well suspect a priori that $T=\sum_{i=1}^n X_i$, the total number of heads obtained in the n trials is a sufficient statistic for p. To verify that it is, we will examine the conditional distribution of the Xs, given $T=t$. Since, if $T=t$, only certain types of vectors $\mathbf{x}=(x_1,\ldots,x_n)$ of 0s and 1s would be possible; we therefore may restrict the calculation to sets $\{x_i\}$ for which $\sum_{i=1}^n x_i=t$. For each i,

$$P(X_i=x_i)=p^{x_i}(1-p)^{1-x_i},$$

and by the independence of the Xs,

$$\begin{aligned}P(X_1=x_1,\ldots,X_n=x_n) &= \prod_{i=1}^n p^{x_i}(1-p)^{1-x_i}\\ &= p^{\sum_{i=1}^n x_i}(1-p)^{n-\sum_{i=1}^n x_i}\\ &= p^t(1-p)^{n-t}.\end{aligned}$$

Also, since $T\sim B(n,p)$, we have that

$$P(T=t)=\binom{n}{t}p^t(1-p)^{n-t}.$$

Thus, for vectors $\mathbf{x}$ of zeros and ones for which $\sum_{i=1}^n x_i=t$, it follows that

$$\begin{aligned}P(X_1=x_1,\ldots,X_n=x_n|T=t) &= \frac{P(X_1=x_1,\ldots,X_n=x_n)P(T=t|X_1=x_1,\ldots,X_n=x_n)}{P(T=t)}\\ &= \frac{p^t(1-p)^{n-t}\cdot 1}{\binom{n}{t}p^t(1-p)^{n-t}}\\ &= \frac{1}{\binom{n}{t}}.\end{aligned}$$

This calculation shows that, given $T=t$, every vector $\mathbf{x}$ of zeros and ones for which $\sum_{i=1}^n x_i=t$ has the same probability of occurrence, namely $1/\binom{n}{t}$. More importantly, it shows that these probabilities do not depend on p. We may thus conclude that the statistic $T=\sum_{i=1}^n X_i$ is a sufficient statistic for the parameter p. ■

It is perhaps worth noting that, in the context of Example 6.4.1,

$$I_T(\theta)=\frac{n}{p(1-p)}=I_{(X_1,\ldots,X_n)}(\theta),$$

as one would expect for a regular model such as $B(1,p)$. One can show, similarly, that if $X_1,\ldots,X_n \stackrel{\text{iid}}{\sim} N(\mu,1)$, then the statistic $T=\sum_{i=1}^n X_i$ is sufficient for μ, and if $X_1,\ldots,X_n \stackrel{\text{iid}}{\sim} U(0,\theta)$, then the maximum order statistic $X_{(n)}$ is sufficient for θ. It should be clear, however, from the binomial example above that finding a sufficient statistic for a given parameter using Definition 6.4.1 poses a couple of practical difficulties. First, it requires us to guess what statistic might be sufficient so that the definition can then be checked. Secondly, the calculation of the conditional probability of interest might turn out to be a nontrivial exercise. These considerations suggest that identifying a simpler and more practical way to find a sufficient statistic would be quite welcome. Well, two great pioneers in the development of the field of Statistics—Ronald Fisher and Jerzy Neyman—thought so too. These two brilliant statisticians were not on the best of terms, and in their later years, weren't even talking to each other. But each of them, independently, came up with the result below that now usually bears both of their names. The Fisher-Neyman factorization theorem substantially simplifies the search for a sufficient statistic. Indeed, it allows you to *identify* a potential sufficient statistic and *prove* its sufficiency simultaneously.

Theorem 6.4.1. (The Fisher-Neyman factorization theorem) Let $X_1,\ldots,X_n \stackrel{\text{iid}}{\sim} F_\theta$, a distribution with density or pmf f_θ, and let $T=h(X_1,\ldots,X_n)$. Then T is sufficient for the parameter θ if and only if there exist functions v and w such that

$$f_\theta(x_1,\ldots,x_n)=v(t,\theta)\cdot w(x_1,\ldots,x_n), \tag{6.33}$$

where $t=h(x_1,\ldots,x_n)$ and the function w does not depend on θ.

Proof. While we have often presented proofs of results that are true for both discrete and continuous models by treating just the continuous case, we will do the opposite here. The proof below is valid when the model F_θ is discrete. The proof for the continuous case involves some complexities that are beyond the scope of this book. The interested reader is referred to Lehmann (1986, pp. 20–21) for a discussion of the continuous case.

(The "only if" part.) Suppose that T is a sufficient statistic for θ. If $(x_1,\ldots,x_n)$ is such that $h(x_1,\ldots,x_n)=t$, then, since it is given that the conditional pmf of $\mathbf{X}|T=t$ does not depend on θ, we may write

$$f_\theta(x_1,\ldots,x_n|t)=w(x_1,\ldots,x_n), \tag{6.34}$$

an equation that may also be written as

$$\frac{f_\theta(x_1,\ldots,x_n)\cdot P(T=t|X_1=x_1,\ldots,X_n=x_n)}{f_\theta(t)}=w(x_1,\ldots,x_n),$$

Alternatively, since $P(T=t|X_1=x_1,\ldots,X_n=x_n)=1$, it may be written as

$$f_\theta(x_1,\ldots,x_n)=f_\theta(t)w(x_1,\ldots,x_n), \tag{6.35}$$

which shows that $f_\theta(x_1,\ldots,x_n)$ may be factored as in (6.33).

(The "if" part.) Suppose that there exist functions v and w such that

$$f_\theta(x_1,\ldots,x_n)=v(t,\theta)\cdot w(x_1,\ldots,x_n),$$

where $t = h(\mathbf{x})$. Then

$$f_\theta(t) = \sum_{(x_1,\ldots,x_n)\in A_t} v(t,\theta)w(x_1,\ldots,x_n) = v(t,\theta) \sum_{(x_1,\ldots,x_n)\in A_t} w(x_1,\ldots,x_n) = c\cdot v(t,\theta),$$

where $A_t = \{\mathbf{x}|h(\mathbf{x}) = t\}$ and c is a constant not depending on θ. It follows that

$$\begin{aligned} f_\theta(x_1,\ldots,x_n|t) &= \frac{f_\theta(x_1,\ldots,x_n)\cdot P(T=t|X_1=x_1,\ldots,X_n=x_n)}{f_\theta(t)} \\ &= \frac{v(t,\theta)w(x_1,\ldots x_n)\cdot 1}{cv(t,\theta)} \\ &= \frac{w(x_1,\ldots,x_n)}{c}, \end{aligned}$$

that is, it follows that $f_\theta(x_1,\ldots,x_n|t)$ does not depend on θ. ■

Example 6.4.1. (revisited) Let $X_1,\ldots,X_n \overset{\text{iid}}{\sim} B(1,p)$, and let $T = \sum_{i=1}^n X_i$. Then

$$P(X_1 = x_1,\ldots,X_n = x_n) = \prod_{i=1}^n p^{x_i}(1-p)^{1-x_i} = p^{\sum_{i=1}^n x_i}(1-p)^{n-\sum_{i=1}^n x_i} = p^t(1-p)^{n-t} = u(t,p)\cdot 1.$$

Here, the function $w(x_1,\ldots,x_n) \equiv 1$; T is thus sufficient for θ by Theorem 6.4.1. ■

Example 6.4.2. Let $X_1,\ldots,X_n$ be a random sample from a Poisson distribution with mean λ, and let $T = \sum_{i=1}^n X_i$. The joint pmf of the Xs may be written as

$$P(X_1 = x_1,\ldots,X_n = x_n) = \prod_{i=1}^n \frac{\lambda^{x_i}e^{-\lambda}}{x_i!} = \frac{\lambda^{\sum_{i=1}^n x_i}e^{-n\lambda}}{\prod_{i=1}^n x_i!} = \lambda^t e^{-n\lambda}\cdot\frac{1}{\prod_{i=1}^n x_i!} = v(t,\lambda)\cdot w(x_1,\ldots,x_n).$$

Again, by the factorization theorem, we conclude that T is sufficient for λ. ■

Example 6.4.3. Let $X_1,\ldots,X_n \overset{\text{iid}}{\sim} N(\mu,1)$ and let $T = \sum_{i=1}^n X_i$. The joint density of the Xs may be written as

$$\begin{aligned} f_\mu(\mathbf{x}) &= \prod_{i=1}^n \frac{1}{\sqrt{2\pi}}e^{-(x_i-\mu)^2/2} \\ &= \frac{1}{(2\pi)^{n/2}}e^{-\sum_{i=1}^n (x_i-\mu)^2/2} \\ &= \frac{1}{(2\pi)^{n/2}}e^{-\sum_{i=1}^n x_i^2+2\mu\sum_{i=1}^n x_i-n\mu^2/2} \\ &= \left[e^{2\mu\sum_{i=1}^n x_i-n\mu^2/2}\right]\left[\frac{1}{(2\pi)^{n/2}}e^{-\sum_{i=1}^n x_i^2/2}\right] \\ &= v(t,\mu)\cdot w(x_1,\ldots,x_n), \end{aligned}$$

where $t = \sum_{i=1}^n x_i$. We thus have that $T = \sum_{i=1}^n X_i$ is a sufficient statistic for μ. ■

The three examples above are all special cases of a general result that pertains to all probability models whose density or probability mass function takes on a certain form. A model is said to belong to a *one-parameter exponential family* if its density or pmf can be written in the form

$$f_\theta(x) = e^{A(\theta)B(x)+C(x)+D(\theta)} I_S(x), \tag{6.36}$$

where the support set S of the distribution does not depend on the parameter θ. Most of the models with which we are familiar are exponential families. We'll verify this for the Bernoulli, Poisson, and normal models treated in Examples 6.4.1–6.4.3:

(a) If $X \sim B(1,\theta)$, then $f_\theta(x) = \theta^x(1-\theta)^{1-x} = e^{x\ln\theta+(1-x)\ln(1-\theta)} = e^{x\ln\theta/(1-\theta)+\ln(1-\theta)}$ for $x \in \{0,1\}$, an exponential family with support $S = \{0,1\}$, $A(\theta) = \ln\theta/(1-\theta)$, $B(x) = x$, $C(x) = 0$ and $D(\theta) = \ln(1-\theta)$.

(b) If $X \sim P(\theta)$, then $f_\theta(x) = \theta^x e^{-\theta}/x! = e^{x\ln\theta-\ln x!-\theta}$ for $x = 0,1,2,3,\dots$, an exponential family with $S = \{0,1,2,\dots\}$, $A(\theta) = \ln\theta$, $B(x) = x$, $C(x) = -\ln x!$ and $D(\theta) = -\theta$.

(c) If $X \sim N(\theta,1)$, then $f_\theta(x) = \frac{1}{\sqrt{2\pi}}e^{-(x-\theta)^2/2} = e^{x\theta-x^2/2-\theta^2/2-\ln\sqrt{2\pi}}I_{(-\infty,\infty)}(x)$, an exponential family with $S = (-\infty,\infty)$, $A(\theta) = \theta$, $B(x) = x$, $C(x) = -x^2/2$ and $D(\theta) = -\theta^2/2 - \ln\sqrt{2\pi}$.

Now, suppose that $X_1,\dots,X_n \overset{\text{iid}}{\sim} F_\theta$, where $\{F_\theta\}$ is a one-parameter exponential family with density or pmf of the form in (6.36). Then the joint density or pmf of the sample of Xs is

$$\begin{aligned} f_\theta(\mathbf{x}) &= e^{A(\theta)\sum_{i=1}^n B(x_i)+\sum_{i=1}^n C(x_i)+nD(\theta)} = \left[e^{A(\theta)\sum_{i=1}^n B(x_i)+nD(\theta)}\right]\left[e^{\sum_{i=1}^n C(x_i)}\right] \\ &= v\Big(\sum_{i=1}^n B(x_i),\theta\Big)w(\mathbf{x}). \end{aligned} \tag{6.37}$$

The factorization in (6.37) shows that the statistic $T = \sum_{i=1}^n B(x_i)$ is a sufficient statistic for the parameter θ. If you take another look at the three examples above, you will be able to confirm quite easily that the sufficient statistic T in each of them can be identified as $\sum_{i=1}^n B(x_i)$ for each corresponding function $B(x)$.

Example 6.4.4. (A non-regular model) Let $X_1,\dots,X_n$ be a random sample from the uniform distribution $U(0,\theta)$. The joint pmf of the Xs may be written as

$$P(X_1 = x_1,\dots,X_n = x_n) = \frac{1}{\theta^n}\prod_{i=1}^n I_{(0,\theta)}(x_i). \tag{6.38}$$

Let $x_{(1)} < x_{(2)} < \cdots < x_{(n)}$ be the order statistics corresponding to the sampled Xs. The pmf in (6.38) may be rewritten as

$$\begin{aligned} P(\mathbf{X} = \mathbf{x}) &= \left[\frac{1}{\theta^n}I_{(0,\theta)}(x_{(n)})\right]\left[\prod_{i=1}^{n-1} I_{(0,x_{(n-i+1)})}(x_{(n-1)})\right] \\ &= v(x_{(n)},\theta)\cdot w(x_1,x_2,\dots,x_n) \end{aligned} \tag{6.39}$$

Thus, by Theorem 6.4.1, the maximum order statistic $X_{(n)}$ is sufficient for θ. ■

Example 6.4.5. (A two-parameter model) Let $X_1,\ldots,X_n \stackrel{\text{iid}}{\sim} N(\mu,\sigma^2)$ and let $T=\sum_{i=1}^n X_i$ and $S=\sum_{i=1}^n X_i^2$. It is easy to show that the joint density of the Xs may be written as

$$\begin{aligned} f_{\mu,\sigma^2}(\mathbf{x}) &= \frac{1}{(\sqrt{2\pi}\sigma)^n} e^{-\left(\sum_{i=1}^n x_i^2 - 2\mu\sum_{i=1}^n x_i + n\mu^2\right)/2\sigma^2} \\ &= v\left(t,s,\mu,\sigma^2\right)\cdot 1. \end{aligned}$$

By Theorem 6.4.1, we conclude that (T,S) are joint sufficient statistics for the parameter pair (μ,σ^2). Since the mapping $(T,S)\to(\bar{X},s^2)$ where $\bar{X}$ and s^2 are the sample mean and variance, is one-to-one, the pair $(\bar{X},s^2)$ is also sufficient for (μ,σ^2). ■

The Rao-Blackwell and Lehmann-Scheffé Theorems. It makes intuitive sense that, if $X_1,\ldots,X_n \stackrel{\text{iid}}{\sim} F_\theta$ and $T=h(X_1,\ldots,X_n)$ is a sufficient statistic for θ, then nothing is lost if we base our inferences about θ solely on T. After all, all the information about θ contained in the original sample is also contained in T. One advantage of restricting attention to T is, of course, that the dimension of our problem has been significantly reduced, often from n to 1. But its primary value is the fact that no information about θ has been lost. The following theorem shows that, in searching for the best unbiased estimator of a parameter θ, only functions of the sufficient statistic need be considered.

Theorem 6.4.2. (Rao-Blackwell) Suppose that $X_1,\ldots,X_n \stackrel{\text{iid}}{\sim} F_\theta$ and that the statistic $T=h(X_1,\ldots,X_n)$ is a sufficient statistic for θ. If S is an unbiased estimator of θ based on $X_1,\ldots,X_n$ and S is not a function of T, then the estimator $\hat{\theta}=E(S|T)$

a) is an unbiased estimator of θ, and

b) has as good or better precision than S, that is, $V(\hat{\theta})\le V(S)$.

Proof. It must first be noted that $\hat{\theta}$ is a legitimate estimator of θ. Since T is sufficient, the distribution of $X_1,\ldots,X_n$, given T, does not depend on the parameter θ, and thus the conditional distribution of S, given T, cannot depend on θ. Thus, $\hat{\theta}$ is a simply a function of T. The unbiasedness of $\hat{\theta}$ follows from an application of Adam's Rule (Theorem 4.2.5):

$$E(\hat{\theta})=E(E(S|T))=E(S)=\theta.$$

Further, from Eve's rule (Theorem 4.2.6), we have that

$$V(S)=V(E(S|T))+E(V(S|T))=V(\hat{\theta})+E(V(S|T))\ge V(\hat{\theta}).$$

■

To proceed further, we need to introduce a technical property that a statistic may have that, while being useful and having important consequences, is unfortunately quite devoid of intuitive content. The property is defined below.

Definition 6.4.2. Suppose that $X_1,\ldots,X_n \stackrel{\text{iid}}{\sim} F_\theta$. A statistic $T=h(X_1,\ldots,X_n)$ is a *complete sufficient statistic* for θ if it is sufficient for θ and if g is a real-valued function for which $E(g(T))=0$, then g must satisfy $g(t)=0$ for all t in the support set of the statistic T.

The following is an example of a proof of the completeness of a particular statistic.

Example 6.4.6. Let $X_1,\ldots,X_n \stackrel{\text{iid}}{\sim} B(1,p)$. The statistic $T = \sum_{i=1}^n X_i$ is sufficient for the parameter p. To see that it is complete, suppose that the function g has the property that $E(g(T)) = 0$. Since $T \sim B(n,p)$, we then have that

$$\sum_{t=0}^{n} g(t)\binom{n}{t}p^t(1-p)^{n-t} = 0 \text{ for all } p \in (0,1). \tag{6.40}$$

Let $y = p/(1-p)$. Since $p \in (0,1)$, $y \in (0,\infty)$. Upon dividing both sides of the equation in (6.40) by $(1-p)^n$, we way write this equation as

$$\sum_{t=0}^{n} g(t)\binom{n}{t}y^t = 0 \text{ for all } y \in (0,\infty). \tag{6.41}$$

Now the left-hand side of (6.41) is an nth degree polynomial in y and, as is well known, it can take the value zero for at most n different values of y. But (6.41) indicates that this polynomial is equal to zero at every positive value of y. This is only possible if all the coefficients of the polynomial are equal to zero. We thus conclude that

$$g(t) = 0 \text{ for } t = 0,1,\ldots,n,$$

that is, g is the zero function on the support set of the statistic T, proving that T is indeed a complete sufficient statistic. ■

Completeness checks are rather tedious, so the following theorem, stated without proof, is a useful fact to have in our hip pockets.

Theorem 6.4.3. Suppose that $X_1,\ldots,X_n \stackrel{\text{iid}}{\sim} F_\theta$, where F_θ is a member of a one-parameter exponential family of the form shown in (6.36). If T is a sufficient statistic for θ, then T is complete.

The relevance of the concept of completeness to unbiased estimation is evident in the following result.

Theorem 6.4.4. (Lehmann-Scheffé) Let $X_1,\ldots,X_n \stackrel{\text{iid}}{\sim} F_\theta$, and suppose that the statistic $T = h(X_1,\ldots,X_n)$ is a complete sufficient statistic for θ. If $\hat{\theta} = g(T)$ is an unbiased estimator of θ, then $\hat{\theta}$ is the unique minimum variance unbiased estimator (MVUE) of θ.

Proof. Because of the Rao-Blackwell theorem, we need only compare unbiased estimators of θ that are based on the statistic T. Suppose that, in addition to $\hat{\theta} = g(T)$, the function $k(T)$ is also an unbiased estimator of θ. Then

$$E(g(T) - k(T)) = \theta - \theta = 0.$$

The completeness of T implies that $g(t) = k(t)$ for all t in the support of T, that is, $\hat{\theta}$ and $k(T)$ are one and the same estimator. ■

We now look at a couple of applications of the theorems above.

Example 6.4.7. Let $X_1, \ldots, X_n \overset{\text{iid}}{\sim} P(\lambda)$. We know that $\bar{X}$ is the MVUE of the parameter λ, but let's suppose we are interested in finding the MVUE of $\theta = e^{-\lambda}$. Since $\theta = P(X = 0)$ when $X \sim P(\lambda)$, we have an immediate, albeit very simple, unbiased estimator of θ, namely

$$S = I_0(X_1),$$

the estimator that is equal to 1 if $X_1 = 0$ and is equal to 0 otherwise. By Theorem 6.4.3, $T = \sum_{i=1}^{n} X_i$ is a complete sufficient statistic for λ (and thus also for θ, which is a one-to-one function of λ). Thus, by the Rao-Blackwell theorem, $E(S|T)$ is an unbiased estimator of θ, and being a function of the complete sufficient statistic T, is the MVUE by the Lehmann-Scheffé theorem. To derive this estimator, we will first obtain the conditional distribution of X_1, given T. The marginal distributions of these two variables are, of course, $X_1 \sim P(\lambda)$ and $T \sim P(n\lambda)$. In the argument below, we will also utilize the fact that $\sum_{i=2}^{n} X_i \sim P((n-1)\lambda)$. For $0 \le k \le t < \infty$,

$$\begin{aligned}
P(X_1 = k|T = t) &= \frac{P(X_1 = k)P(T = t|X_1 = k)}{P(T = t)} \\
&= P(X_1 = k)P(\sum_{i=2}^{n} X_i = t - k)/P(T = t) \\
&= \frac{\lambda^k e^{-\lambda}}{k!} \cdot \frac{[(n-1)\lambda]^{t-k} e^{-(n-1)\lambda}}{(t-k)!} \Big/ \frac{(n\lambda)^t e^{-n\lambda}}{t!} \\
&= \frac{(n-1)^{t-k}}{k!(t-k)!} \Big/ \frac{(n)^t}{t!} \\
&= \binom{t}{k} \frac{(n-1)^{t-k}}{n^t} \\
&= \binom{t}{k} \left(\frac{1}{n}\right)^k \left(\frac{n-1}{n}\right)^{t-k}.
\end{aligned}$$

It is thus apparent that $X_1|T = t \sim B(t, 1/n)$. It follows that the MVUE of $\theta = e^{-\lambda}$ is

$$\begin{aligned}
E(S|T) &= 1 \cdot P(X_1 = 0|T) + 0 \cdot P(X_1 > 0|T) \\
&= \left(\frac{n-1}{n}\right)^T.
\end{aligned}$$

This estimator may look a bit odd, but one can see that it is in fact a reasonable estimator of $e^{-\lambda}$ by rewriting the estimator $E(S|T)$ as

$$\left[\left(1 - \frac{1}{n}\right)^n\right]^{\bar{X}},$$

and noting that, as n grows large, $\left(1 - \frac{1}{n}\right)^n$ converges to the constant e^{-1} while $\bar{X}$ converges the Poisson mean λ, so the estimator, being a continuous function of n and $\bar{X}$, converges to $e^{-\lambda}$. Of course the point of the example is the unbiasedness of the estimator. ■

Example 6.4.8. Let us further explore the problem of estimating the reliability function $R(x) = P(X > x)$ of a system based on a random sample of a system lifetimes assumed to have a common exponential distribution. More specifically, suppose that $X_1, X_2, \ldots, X_n \stackrel{\text{iid}}{\sim} \Gamma(1,\mu)$, the exponential distribution with mean μ. In Example 6.2.4, we noted that $\hat{R}_1(x) = I_{(x,\infty)}(X_1)$ is an unbiased estimator of the system's reliability function $R(x) = e^{-x/\mu}$ at the fixed time x. Since s^2 is a function of $(\sum X_i, \sum X_i^2)$ and $T = \sum_{i=1}^n X_i$ is a complete sufficient statistic for μ (and thus also for $R(x)$ at a fixed x), we may use the theoretical results of this section to identify the MVUE $E(\hat{R}_1(x)|T)$ of the system reliability $R(x)$. Our derivation of this estimator begins with the identification of the conditional distribution of X_1, given T. Note that $X_1 \sim \Gamma(1,\mu)$ and $T \sim \Gamma(n,\mu)$. Further, $T - X_1 = \sum_{i=2}^n X_i \sim \Gamma(n-1,\mu)$ and $T|X_1 = x_1$ has the three-parameter gamma distribution $\Gamma(n-1,\mu,x_1)$ with density given in (3.25). We may thus write, for $0 < x_1 < t < \infty$,

$$\begin{aligned} f_\mu(x_1|t) &= \frac{f_\mu(x_1)f_\mu(t|x_1)}{f_\mu(t)} \\ &= \frac{1}{\mu}e^{-x_1/\mu} \cdot \frac{1}{\Gamma(n-1)\mu^{n-1}}(t-x_1)^{n-2}e^{-(t-x_1)/\mu} \Big/ \frac{1}{\Gamma(n)\mu^n}t^{n-1}e^{-t/\mu} \\ &= (n-1)\frac{(t-x_1)^{n-2}}{t^{n-1}}. \end{aligned}$$

We can transform this density into a more convenient form using the CoV technique. Let $Y = X_1/t$, where $T = t$ and consider the distribution of Y, given $T = t$. We have that $X_1 = tY$ and $|J| = |\frac{\partial X_1}{\partial y}| = t$. The conditional density of $Y|T = t$ may be identified as

$$f(y|t) = (n-1)(1-y)^{n-2}I_{(0,1)}(y),$$

that is, $Y|T = t \sim Be(1, n-2)$. It follows that

$$\begin{aligned} E(\hat{R}_1(x)|T) &= P(X_1 > x|T = t) \\ &= P(Y > \frac{x}{t}|T = t) \\ &= \int_{x/t}^1 (n-1)(1-y)^{n-2}dy \\ &= -(1-y)^{n-1}\Big|_{x/t}^1 \\ &= \left(1 - \frac{x}{t}\right)^{n-1} \quad \text{for } 0 < x < t. \end{aligned}$$

For $x \geq t$, $E(\hat{R}_1(x)|T = t) = P(X_1 > x|T = t) = 0$. Thus, the MVUE of the system reliability $R(x) = e^{-x/\mu}$ at time x is

$$\hat{R}_T(x) = \begin{cases} (1 - x/T)^{n-1} & \text{if } x < T \\ 0 & \text{if } x \geq T, \end{cases}$$

where $T = \sum_{i=1}^n X_i$. ■

Example 6.4.9. In Example 6.3.3, we noted that the sample variance s^2 based on a random sample of size n drawn from an $N(\mu,\sigma^2)$ population is an unbiased estimator of σ^2 whose variance is larger than the Cramér-Rao Lower Bound. This fact leaves open the issue of whether or not s^2 is the MVUE of σ^2. The fact that it is can be deduced from the following considerations. First, we note that the result stated in Theorem 6.4.3 holds more generally. The k-parameter exponential family of distributions is defined in Problem 6.7.54. This problem asks you to verify that such a family has a k-dimensional sufficient statistic. The general version of Theorem 6.4.3 states that this k-dimensional sufficient statistic is *complete*. This implies, by the Lehmann-Scheffé Theorem, that an unbiased estimator of a parameter of the k-parameter exponential family based on the sufficient statistic is the unique MVUE of that parameter. Finally, we note that the $N(\mu,\sigma^2)$ distribution is a two-parameter exponential family and that the statistic $(\sum X_i, \sum X_i^2)$, and therefore also $(\bar{X}, s^2)$, is a complete sufficient statistic for (μ,σ^2). Since $E(s^2) = \sigma^2$, it follows that s^2 is indeed the minimum variance unbiased estimator of σ^2. ■

Exercises 6.4.

1. Let X and Y be independent Poisson variables, with $X \sim P(\lambda)$ and $Y \sim P(2\lambda)$. Of course X and $Y/2$ are each unbiased estimators of λ. Their respective variances are λ and $\lambda/2$, so the second is to be preferred over the first. But we ought to be able to do better!

   (a) Show that the 50–50 mixture of the two estimators above, namely $S = X/2 + Y/4$, is unbiased for λ and has a smaller variance that either of the estimators above. That's not bad, but the estimator surely isn't the best you can do because it's not a function of the sufficient statistic for λ.

   (b) Using the factorization theorem on $p(x,y)$, show that the statistic $T = X + Y$ is a sufficient statistic for λ.

   (c) Show that the estimator $T/3$ is unbiased for λ and is superior to the unbiased estimator S in part (a), that is, $V(T/3) < V(S)$ for all $\lambda > 0$.

2. Let $X_1,\ldots,X_n \overset{\text{iid}}{\sim} W(\alpha_0,\beta)$, the Weibull distribution with known shape parameter α_0. (a) Derive the Fisher information $I(\beta^{\alpha_0})$ of a single observation X from this model. (b) Write the density of the $W(\alpha_0,\beta)$ distribution as a one-parameter exponential family, and identify the complete sufficient statistic for β^{α_0}. (c) Use the Lehmann-Scheffé Theorem to identify the MVUE of β^{α_0}.

3. Let $X_1,\ldots,X_n \overset{\text{iid}}{\sim} Be(\theta,1)$. Write the $Be(\theta,1)$ density in the form of an exponential family. Find the complete sufficient statistic for θ, and obtain the minimum variance unbiased estimator of $1/\theta$. (Hint: Note that $Y = -\ln X \sim Exp(\theta)$.)

4. Let $X_1,\ldots,X_n \overset{\text{iid}}{\sim} U(0,\theta)$. Verify the fact that the maximum order statistic $X_{(n)}$ is a *complete* sufficient statistic for θ. Argue that $\hat{\theta} = [(n+1)/n]X_{(n)}$ is the MVUE of θ.

5. The Weibull distribution is the most widely used parametric distribution for modeling the failure time of an engineered system. It is also used in the fields of biostatistics and epidemiology as a lifetime model. One version of the Weibull model, denoted by $W(\theta)$, has density

$$f(x) = \frac{2x}{\theta}e^{-x^2/\theta}I_{(0,\infty)}(x)$$

   If $X \sim W(\theta)$, then $EX = \sqrt{\pi\theta}/2$ and $EX^2 = \theta$. Suppose that $X_1,\ldots,X_n \overset{\text{iid}}{\sim} W(\theta)$.

(a) Show that $T = \sum_{i=1}^{n} X_i^2$ is a sufficient statistic for the parameter θ.

(b) Show that if $X \sim W(\theta)$, then $Y = X^2 \sim \Gamma(1,\theta)$. Using this fact, show that $\hat{\theta} = T/n = \frac{1}{n}\sum_{i=1}^{n} X_i^2$ is an unbiased estimator of θ. Is $\hat{\theta}$ the best unbiased estimator (i.e., the MVUE) of θ?

6. Suppose that $X_1,\ldots,X_n \stackrel{\text{iid}}{\sim} U(\theta,\theta+1)$. (a) Show that the pair $(X_{(1)},X_{(n)})$, the smallest and largest order statistics from this sample, is a sufficient statistic for the parameter θ. (b) Show that $(X_{(1)},X_{(n)})$ is not a complete sufficient statistic for θ. (Hint: Show that $E(X_{(n)} - X_{(1)})$ is a constant that doesn't depend on θ.)

7. Let $X_1,\ldots,X_n \stackrel{\text{iid}}{\sim} \Gamma(\alpha,\beta)$. Find a two-dimensional sufficient statistic for the parameter pair (α,β).

6.5 Optimality within the Class of Linear Unbiased Estimators

In spite of the fact that we have presented, in the preceding sections, two techniques which can be helpful in identifying a particular estimator of a parameter θ as the best unbiased estimator of θ, there are many circumstances in which neither method succeeds and/or the identification of the MVUE proves impossible. Among the possibilities for such an outcome are situations in which the models or the data have complexities which render the search for an MVUE analytically intractable. In such situations, the following compromise is often considered.

Linear functions of the available data have a certain natural appeal because of their simplicity, the ease of analyzing them (because of Theorems 4.2.3 and 4.2.3, for example) and the fact that they are at least one way of taking advantage of all of the data in some reasonable fashion. Further, once the search for an MVUE has been abandoned, what to try next is often not clear, especially if some form of "finite-sample optimality" is sought. A particular instance in which an MVUE may well be out of reach is when the available data $X_1, X_2, \ldots, X_n$ is not a random sample from a common distribution F but rather constitutes the cumulative observations drawn from a variety of different sources, all depending on a common parameter θ. In this and other settings, it is not uncommon to turn to a class of estimators which share some desirable property in hopes of finding the best possible estimator within this class.

Suppose that each element X_i of the data set is drawn from a distribution F_i which depends on a single common parameter θ. One way of trying to capitalize on the information about θ contained in the observed Xs is to consider estimating θ by a linear function of the observations, that is, by a linear estimator of the form

$$\hat{\theta}_{\mathbf{a}} = \sum_{i=1}^{n} a_i X_i. \tag{6.42}$$

One might place some constraints on the coefficients $\{a_i\}$ in (6.42). A potential constraint that naturally comes to mind is to restrict attention to linear estimators that are unbiased estimators of θ. This restriction would simply take the form

$$\sum_{i=1}^{n} a_i E X_i = \theta. \tag{6.43}$$

One must recognize the fact that there may be no linear unbiased estimators, that is, there may be no constants $\{a_i\}$ for which the Equation (6.43) is satisfied. The fact that this is a possibility should not be too shocking in light of the fact that the parameter θ might not even be estimable, so that no unbiased estimator of θ can be found, linear or otherwise. Still, when the best unbiased estimator of θ is inaccessible, a search for the best linear unbiased estimator may well worth undertaking. The worst that can happen is that you can't find that estimator either. Taking a more optimistic view, this section is about the process of finding the *best linear unbiased estimator* of a parameter θ, an estimator that we will often refer to as the BLUE.

To begin the process of finding the BLUE in a particular problem, let's discuss how one might wish to compare two linear unbiased estimators. The desired criterion should be pretty obvious. Why not use the same criterion we've used for comparing arbitrary pairs of unbiased estimators—their respective variances? Since we are dealing with estimators that are aimed at the same target, namely the parameter θ, we would prefer to use the estimator with a smaller variance and, if possible, the smallest variance among all estimators under consideration. This leads to the following:

Definition 6.5.1. Let $X_1, X_2, \ldots, X_n$ be a collection of random variables whose joint distribution depends on a single scalar parameter θ. The estimator $\hat{\theta} = \hat{\theta}(X_1, X_2, \ldots, X_n)$ is the best linear unbiased estimator (BLUE) of the parameter θ if a) it is linear, that is, has the form in (6.42), b) it is unbiased, that is, satisfies (6.43) and c) it has the smallest variance among all linear unbiased estimators, that is,

$$V(\hat{\theta}) = \min_{\mathbf{a} \in A} V\left(\sum_{i=1}^{n} a_i X_i\right), \tag{6.44}$$

where A in (6.44) is the set of all vectors $(a_1, \ldots, a_n)$ satisfying (6.43).

The process of finding a BLUE is best understood by examining a concrete example. Since I don't work with concrete that much, I offer the following example, which I believe will do.

Example 6.5.1. Suppose that X and Y are independent normal variables, where $X \sim N(\mu, 1)$ and $Y \sim N(\mu, 4)$. One might think of these variables as the output of a common production process in two different factories, one of which has a better quality control program and thus greater precision in its product. Consider the class of linear estimators $\{\hat{\mu}_{a,b} = aX + bY\}$ of the common mean μ of these observations. Our first step in searching for the BLUE of μ is to identify the constraint placed on the constants a and b by the requirement that $\hat{\mu}_{a,b}$ be unbiased. Since $E(\hat{\mu}_{a,b}) = E(aX + bY) = (a+b)\mu$, we see that for $\hat{\mu}_{a,b}$ to be unbiased, the constraint $a + b = 1$ must be satisfied. We may thus set $b = 1 - a$ and consider the problem of minimizing the variance of an estimator of the form $\hat{\mu}_a = aX + (1-a)Y$. Since X and Y are independent, we have that

$$V(\hat{\mu}_a) = a^2 \cdot 1 + (1-a)^2 \cdot 4.$$

This variance is minimized by the unique solution to the equation

$$\frac{\partial}{\partial a} V(\hat{\mu}_a) = 2a - 8(1-a) = 10a - 8 = 0,$$

that is, by $a = 4/5$. This claim is substantiated by the "second derivative test," that is, by the fact that

$$\frac{\partial^2}{\partial a^2} V(\hat{\mu}_a) = 10 > 0.$$

We conclude that the estimator $\hat{\mu} = .8X + .2Y$ is the BLUE of the parameter μ. It is interesting and informative to note that, even though both X and Y are unbiased estimators of μ and X is notably better than Y, both of these "solo" estimators are inferior to the linear combination $\hat{\mu} = (4/5)X + (1/5)Y$. The more precise variable X does receive more weight than Y, which is less precise. ■

Remark 6.5.1. Constrained optimization problems are often solved by the method of *Lagrange multipliers*. As this is the first of several instances in which this alternative optimization tool will be helpful to us, I encourage you to look up the method in a calculus text. In the example above, the alternative solution would proceed as follows. The criterion function G one deals with will typically have the form: (the function to be optimized) $+\lambda\times$(the constraint set equal to zero). Here, after writing the constraint $a+b=1$ as $(a+b-1)=0$, we have

$$G(a,b,\lambda) = a^2 + 4b^2 + \lambda(a+b-1),$$

where a^2+4b^2 is simply $V(aX+bY)$ in Example 6.5.1. Differentiating G with respect to a, b and the Lagrange multiplier λ and setting these derivatives equal to zero yields three equations in three unknowns:

$$2a+\lambda = 0,$$

$$8b+\lambda = 0,$$

and

$$a+b-1 = 0.$$

The unique critical point of these equations is $(a,b,\lambda) = (4/5, 1/5, -8/5)$. Thus, the solution $(a,b) = (4/5,1/5)$ minimizes $V(\hat{\mu}_{a,b})$. ■

Exercises 6.5.

1. Find the BLUE of μ in Example 6.1.1 assuming that X and Y are dependent with $\rho_{X,Y} = 1/2$.

2. Suppose that $X_1,\dots,X_n \stackrel{\text{iid}}{\sim} F$, a distribution with mean $\mu < \infty$ and variance $\sigma^2 < \infty$. (a) What restrictions must be placed of the constants $a_1,\dots,a_n$ so that $T = \sum_{i=1}^n a_i X_i$ is an unbiased estimator of μ. (b) Show that the sample mean $\bar{X}$ is the BLUE of μ.

3. Suppose that the rate of infection by a particular virus is the same throughout the country, with $\lambda =$ expected number of infections per 100,000 people. Let $k_1,\dots,k_n$ be the population sizes, in multiples of 100,000, of the n largest cities in the country. For $i = 1,2,\dots,n$, assume that the number of infections X_i in a city i is a Poisson variable, that is, $X_i \sim P(k_i\lambda)$, with the number of infections in different cities assumed independent. Find the best linear unbiased estimator of λ based on the observations $X_1,\dots,X_n$.

4. A rare art object is to be sold at auction. Bidder 1 feels that its worth \$$\theta$ and bids the amount X, where $X \sim U[0,\theta]$. Bidder 2 feels that it's worth \$$2\theta$ and bids the amount Y, where $Y \sim U[0,2\theta]$. Find the best linear unbiased estimator (BLUE) of θ based on X and Y.

6.6 Beyond Unbiasedness

Although the property of unbiasedness clearly has intuitive appeal, it falls quite a bit short of being a moral imperative. It is, to be honest, a constraint that has been made for practical reasons, given that in any nontrivial statistical problem, there is no estimator that is uniformly better than all others. This latter point is worth a little elaboration. Why is it that there isn't a single estimator that beats all others? The simple answer is there are a host of estimators that, while being very poor overall, happen to be unbeatable for some particular value of the target parameter. To see this clearly, it helps to take a look at a specific estimation problem.

Suppose that you wish to estimate the mean μ of a normal population based on a random sample $X_1,\ldots,X_n$ drawn from that population. The time-honored estimator $\hat{\mu}_1 = \bar{X}$ would of course come to mind. We know that it is an unbiased estimator of μ, and in fact, it is the best possible unbiased estimator of μ. However, it cannot do better than the bonehead estimator $\hat{\mu}_2 = 5$ if the true value of μ happens to be 5. Even though $\hat{\mu}_2$ completely ignores the available data, and even though $\hat{\mu}_2$ exposes its user to a potentially huge estimation error if the true value of μ is quite distant from the number 5, there is no estimator that can improve on it, or even match it, when μ is actually equal to 5. One would never use an estimator such as this (or else one would refer to it as a guess rather than an estimate), but it is flat-out best, among all possible estimators, when μ happens to be equal to 5. If μ were equal to 10, then no estimator could do as well as the estimator $\hat{\mu}_3 = 10$. So the potential domination of the estimator $\hat{\mu}_2 = 5$ is obviously narrow and fleeting. Because there is no estimator that is universally best, looking for optimal estimators within some reasonable class of estimators would seem to be the next best thing. The MVUE is an example of such a compromise.

As we discussed in the first section of this chapter, there are two factors that we tend to examine carefully when assessing the performance of an estimator in a given problem—its *accuracy* and its *precision*. It is clear that estimators which are both inaccurate and imprecise are not generally worth consideration and would typically be dismissed from the competition straightaway. But the interplay between accuracy and precision needs to be treated more delicately. We will see that sacrificing a little accuracy for the sake of gaining greater precision may in fact be a rewarding strategy. Our discussion of the interplay alluded to above begins with the following definitions.

Definition 6.6.1. Let $X_1,\ldots,X_n \stackrel{\text{iid}}{\sim} F_\theta$, a distribution depending on a scalar parameter θ, and let $\hat{\theta} = \hat{\theta}(X_1,\ldots,X_n)$ be an estimator of θ. The difference

$$B(\hat{\theta}) = E\hat{\theta} - \theta \tag{6.45}$$

is called the *bias* of $\hat{\theta}$ as an estimator of θ.

It is clear from the definition above that, if $B(\hat{\theta}) > 0$, then $\hat{\theta}$ tends to overestimate θ on the average, while if $B(\hat{\theta}) < 0$, $\hat{\theta}$ tends to underestimate θ on the average. Obviously, when $\hat{\theta}$ is "unbiased," $B(\hat{\theta}) = 0$. The bias of an estimator can be thought of as a matter of "aim" or "calibration." If an estimator $\hat{\theta}$ of θ has a large bias when the parameter θ takes on a particular value, say θ_0, it clearly was not "aimed" at the target θ_0, that is, the center of its distribution was far from the true value θ_0 of θ. When considering an estimator for possible use, it is prudent to evaluate its bias. This assessment, together with an evaluation of its precision (measured, say, by its variance) will put us in a position to the judge the estimator's overall standing.

When judging an estimator's expected performance, we are naturally interested in its average distance from the target parameter, as measured by some reasonable metric. Certainly the absolute distance between $\hat{\theta}$ and θ, that is, the distance $d = |\hat{\theta} - \theta|$, is a natural metric for measuring estimation error. A more frequently used metric is squared distance, that is, $d^2 = (\hat{\theta} - \theta)^2$, which also keeps track of how far the estimator is from the target parameter, but is somewhat easier to handle in mathematical operations (like taking expected values). The squared error d^2 will be the metric that we focus on in our continuing investigations of an estimator's performance. Our primary tool for comparing estimators is defined as follows.

Definition 6.6.2. The *mean squared error* (MSE) of $\hat{\theta}$ as an estimator of θ is given by

$$MSE(\hat{\theta}) = E(\hat{\theta} - \theta)^2. \tag{6.46}$$

If $\hat{\theta}$ is an unbiased estimator of θ, then the mean squared error of $\hat{\theta}$ reduces to the estimator's variance, that is, if $E\hat{\theta} = \theta$, then $MSE(\hat{\theta}) = V(\hat{\theta})$. But the mean squared error is, in general, a function of both the bias and the variance of $\hat{\theta}$. The exact relationship is established in the following:

Theorem 6.6.1. The mean squared error of an estimator $\hat{\theta}$ of θ may be evaluated as

$$MSE(\hat{\theta}) = V(\hat{\theta}) + [B(\hat{\theta})]^2. \tag{6.47}$$

Proof.

$$\begin{aligned} MSE(\hat{\theta}) &= E[\hat{\theta} - \theta]^2 \\ &= E[\hat{\theta} - E\hat{\theta} + E\hat{\theta} - \theta]^2 \\ &= E[(\hat{\theta} - E\hat{\theta})^2] + 2E[(\hat{\theta} - E\hat{\theta})(E\hat{\theta} - \theta)] + E[(E\hat{\theta} - \theta)^2]. \end{aligned} \tag{6.48}$$

But, because $(E\hat{\theta} - \theta)$ is a constant that can be factored outside of the middle expectation in (6.48), we find that the expectation of the cross-product term above vanishes, that is,

$$E[(\hat{\theta} - E\hat{\theta})(E\hat{\theta} - \theta)] = (E\hat{\theta} - \theta) \cdot E(\hat{\theta} - E\hat{\theta}) = (E\hat{\theta} - \theta) \cdot 0 = 0.$$

Thus, we may evaluate the expectation in (6.48) as the sum of the first and last terms, leading us to the desired conclusion:

$$MSE(\hat{\theta}) = E(\hat{\theta} - E\hat{\theta})^2 + E(E\hat{\theta} - \theta)^2 = V(\hat{\theta}) + [B(\hat{\theta})]^2.$$

■

We will now take the view that the mean squared error of an estimator of an unknown parameter θ is the chosen criterion by which the performance of an estimator will be assessed and by which any two estimators will be compared. This choice takes as given that squared error is the preferred metric for accounting for estimation errors, and it also implies that the model that is assumed to generate the data is taken as correct (since the expectation to be evaluated explicitly relies on that model). An immediate issue that arises in the comparison of estimators by using their MSEs is the fact that, as an expected value, the

MSE of an estimator is not a constant, but rather, it is a function of the unknown parameter θ. It is thus quite possible, and perhaps even likely, that neither of the two estimators one wishes to compare will dominate the other. The first might have a smaller MSE for some values of θ, with the second having a smaller MSE for other values of θ. In such cases, neither estimator can be eliminated from consideration, at least not based on the MSE criterion. However, when one estimator does dominate (that is, is as good or better than the other) for all possible values of the parameter θ, then the latter estimator can indeed be set aside, for there is no circumstance in which it would be preferred. The following definition, borrowed from the statistical subfield of decision theory, makes these ideas more concrete.

Definition 6.6.3. Let $X_1, \dots, X_n \overset{\text{iid}}{\sim} F_\theta$, a distribution depending on a scalar parameter θ in the parameter space $\Theta \subseteq R$, and let $\hat{\theta}_1$ and $\hat{\theta}_2$ be two estimators of θ based on these data. If

$$MSE(\hat{\theta}_1) \le MSE(\hat{\theta}_2) \qquad \text{for all } \theta \in \Theta, \tag{6.49}$$

with a strict inequality in (6.49) for some θ, then $\hat{\theta}_2$ is *inadmissible* as an estimator of θ (relative to the squared error criterion).

An estimator is *admissible* if there is no estimator that dominates it in the sense of Definition 6.6.3. Admissibility is an "optimality" property of estimators, but only in a very weak sense. While we would not wish to use an estimator that was not admissible, we must acknowledge the fact that there are admissible estimators whose overall performance is completely unacceptable. Given a random sample of size n from a population with finite mean μ and variance σ^2, the estimator $\hat{\mu} = 5$ of μ is admissible, but its MSE is unbounded and tends to ∞ as the distance between μ and 5 grows large. On the other hand, the estimator $\bar{X}$ has the constant MSE equal to σ^2/n, independent of μ, and would no doubt be preferred to $\hat{\mu}$, even though it can't match the performance of $\hat{\mu}$ if μ is very close to 5.

While the admissibility of an estimator cannot be viewed as a badge of honor, its inadmissibility is quite rightly considered to be a fatal flaw. Interestingly, certain estimators that surface as best within some restricted class turn out, in the end, to be inadmissible. We conclude this section with a pair of examples of this phenomenon. Both examples demonstrate that a best unbiased estimator might well be improved upon by a biased estimator with a smaller variance.

Example 6.6.1. Consider the problem of estimating the mean of an exponential distribution. Suppose that $X_1, X_2, \dots, X_n \overset{\text{iid}}{\sim} \Gamma(1,\theta)$. Recall that $\sum_{i=1}^n X_i \sim \Gamma(n,\theta)$. It follows that the sample mean $\bar{X}$ is an unbiased estimator of θ, and that $V(\bar{X}) = \theta^2/n$. Further, since

$$\begin{aligned} I_X(\theta) &= E\left(\frac{\partial}{\partial\theta}\ln f_\theta(X)\right)^2 \\ &= E\left(\frac{\partial}{\partial\theta}[-\ln\theta - \frac{X}{\theta}\right)^2 \\ &= E\left(\frac{X-\theta}{\theta^2}\right)^2 = \frac{1}{\theta^2}, \end{aligned}$$

we have that the CRLB for unbiased estimators in this problem is θ^2/n. This implies that $\bar{X}$ is the MVUE of θ. However, $\bar{X}$ is inadmissible! Let us consider estimators of the form

$$\hat{\theta}_c = c\sum_{i=1}^n X_i.$$

We first compute the MSE for such estimators. We have

$$E\left(c\sum_{i=1}^{n} X_i\right) = cE\left(\sum_{i=1}^{n} X_i\right) = cn\theta,$$

from which we identify the bias of this estimator as $B(\hat{\theta}_c) = (cn\theta - \theta)$. Moreover,

$$V\left(c\sum_{i=1}^{n} X_i\right) = c^2 V\left(\sum_{i=1}^{n} X_i\right) = c^2 n\theta^2.$$

By Theorem 6.6.1, we have that

$$MSE\left(\hat{\theta}_c\right) = V\left(\hat{\theta}_c\right) + \left[B\left(\hat{\theta}_c\right)\right]^2 = c^2 n\theta^2 + (cn\theta - \theta)^2 = \left[c^2 n + (cn-1)^2\right]\theta^2. \tag{6.50}$$

To minimize the MSE in (6.50), we solve the equation $\partial/\partial c[MSE(\hat{\theta}_c)] = 0$, which leads to the unique minimizer

$$c = \frac{1}{n+1}.$$

Upon noting that $(\partial^2/\partial c^2)[MSE(\hat{\theta}_c)] > 0$, we conclude that the MVUE $\bar{X}$ is inadmissible as an estimator of θ, and that the estimator

$$\tilde{\theta} = \frac{1}{n+1}\sum_{i=1}^{n} X_i$$

is a uniformly better estimator than $\bar{X}$, that is, it has an MSE that is smaller than that of $\bar{X}$ for all $\theta > 0$. ■

Example 6.6.2. Let $X_1, \ldots, X_n \stackrel{\text{iid}}{\sim} U(0,\theta)$. The estimator $\hat{\theta} = \frac{n+1}{n} X_{(n)}$ is the MVUE of θ (see Exercise 6.4.4). Since $X_{(n)}/\theta \sim Be(n,1)$, we have that the

$$V(\hat{\theta}) = \left(\frac{n+1}{n}\right)^2 \cdot \frac{n\theta^2}{(n+1)^2(n+2)} = \frac{\theta^2}{n(n+2)}.$$

As in the previous example, we will consider the class of estimators which are constant multiples of $X_{(n)}$. Let $\hat{\theta}_c = cX_{(n)}$. Then

$$E\left(cX_{(n)}\right) = \frac{cn}{n+1}\theta \qquad \text{and } V\left(cX_{(n)}\right) = \frac{c^2 n\theta^2}{(n+1)^2(n+2)},$$

from which it follows that

$$MSE\left(\hat{\theta}_c\right) = V\left(\hat{\theta}_c\right) + B\left(\hat{\theta}_c\right) = \left[\frac{c^2 n}{(n+1)^2(n+2)} + \left(\frac{cn}{n+1} - 1\right)^2\right]\theta^2. \tag{6.51}$$

The equation $\partial/\partial c[MSE(\hat{\theta}_c)] = 0$ simplifies to

$$\frac{cn}{n+2} + cn^2 - n(n+1) = 0, \tag{6.52}$$

and the unique solution of (6.52) is

$$c = \frac{n+2}{n+1}.$$

Since $\partial^2/\partial c^2 MSE(\hat{\theta}_c) > 0$, we conclude that

$$MSE\left(\frac{n+2}{n+1}X_{(n)}\right) < MSE\left(\frac{n+1}{n}X_{(n)}\right),$$

that is, the MVUE $\hat{\theta} = [(n+1)/n]X_{(n)}$ is an inadmissible estimator of θ and the estimator $\tilde{\theta} = [(n+2)/(n+1)]X_{(n)}$ is a uniformly better estimator than $\hat{\theta}$, that is, it has an MSE that is smaller than that of $\hat{\theta}$ for all $\theta > 0$. ■

Exercises 6.6.

1. Epidemiologist Seymour Bacteria set up an experiment to estimate the incidence θ of a certain disease based on the sample $X_1, X_2, \ldots, X_n \stackrel{\text{iid}}{\sim} N(\theta, \theta^2)$. Seymour is inclined to use the sample mean $\bar{X}$ to estimate θ, but he has decided to turn to you for help. Yeah, you! Now as you know, insisting that an estimator be unbiased may not be the best thing to do. Let $\hat{\theta}_c = c\bar{X}$. Find the value of c that minimizes the mean squared error of $\hat{\theta}_c$. Seymour would so appreciate your improving upon his choice of $c = 1$!

2. Let $X \sim G(p)$, the geometric distribution with pmf $p(x) = p(1-p)^{x-1}$ for $x = 1, 2, \ldots$ Let $T = 1$ if $X = 1$, $T = -1$ if $X = 2$, and $T = 0$ if $X \geq 3$. Show that T is an unbiased estimator of p^2. Is T an admissible estimator of p^2 relative to squared error loss?

3. Suppose that $X_1, X_2, \ldots, X_n \stackrel{\text{iid}}{\sim} N(\mu, \sigma^2)$. The most widely used estimator of the population variance is the time-honored estimator $s^2 = [1/(n-1)]\sum_{i=1}^n (X_i - \bar{X})^2$, the sample variance. Show that s^2 is inadmissible as an estimator of σ^2. (Hint: Find the constant c for which $\hat{\sigma}_c^2 = c\sum_{i=1}^n (X_i - \bar{X})^2$ has the smallest possible mean squared error. You'll find that the value of c is not equal to $1/(n-1)$.)

4. During the U.S. lunar expedition in 1969, a single observation X was obtained on the depth of a gun shot into the lunar surface at point blank range. Experience on Earth suggests that $X \sim U(\theta, 2\theta)$. (a) Suppose that θ is to be estimated by a multiple of X. For what constant c is the estimator $T = cX$ an unbiased estimator of θ? (b) Is the estimator in part (a) the best estimator of the form $T = cX$, where c is a positive constant? Support your answer.

5. Suppose that $X_1, X_2, \ldots, X_n \sim N(\mu, 1)$. Using the method of expectations, find an unbiased estimator of μ^2. Is this estimator admissible?

6.7 Chapter Problems

1. The two-parameter Pareto model $P(k, \theta)$, where k is a positive integer and $\theta > 0$, has the following density, distribution function, and first and second moment:

   $f(x) = k\theta^k/x^{k+1}I_{(\theta,\infty)}(x), F(x) = 1 - \theta^k/x^k$ for $x > \theta$, $EX = k\theta/(k-1)$ for $k > 1$, $EX^2 = k\theta^2/(k-2)$ if $k > 2$.

   Now, let $X_1, \ldots, X_n \stackrel{\text{iid}}{\sim} P(3, \theta)$.

   (a) Find the value of c for which $c\sum_{i=1}^n X_i$ is an unbiased estimator of θ.

(b) Let $X_{(1)} = \min_{1 \le i \le n}\{X_i\}$, the smallest order statistic from this sample. Show that $X_{(1)}$ also has a Pareto distribution (see Section 4.7). Use this fact to find a constant d for which $\hat{\theta} = dX_{(1)}$ is an unbiased estimator of θ.

(c) Which of these estimators is better?

2. Let $X \sim B(n,p)$, and let $\hat{p} = X/n$. Find the constant c for which $c\hat{p}(1-\hat{p})$ is an unbiased estimator of $V(\hat{p}) = p(1-p)/n$.

3. Let $X_1,\dots,X_n \stackrel{\text{iid}}{\sim} \Gamma(1,\theta)$, and let $\lambda = 1/\theta$. Show that $\hat{\lambda} = (n-1)/\sum_{i=1}^n X_i$ is an unbiased estimator of λ.

4. Let $X_1,\dots,X_n,\dots$ be a sequence of Bernoulli trials with probability p of success, and suppose these trials continue until k successes have been obtained. Let N be the number of trials needed to obtain the kth success. Show that the estimator

$$\hat{p} = \frac{k-1}{N-1}$$

is an unbiased estimator of p.

5. Let $X,Y \stackrel{\text{iid}}{\sim} G(p)$. Use the expectation method, in combination with the indicator function method, to find an unbiased estimator of p^2.

6. Let $X \sim P(\lambda)$. Use X to devise an unbiased estimate of the parameter $\theta = e^{-2\lambda}$. (Hint: Start with the equation $E\hat{\theta}(X) = e^{-2\lambda}$. Multiply both sides of the equation by e^{λ}, and replace the right-hand side of the equation by the MacLaurin series expansion of $e^{-\lambda}$. Then, equate the coefficients of the two infinite series.) Note that your estimator is the unique unbiased estimate of θ based on the observation X. If so inclined, comment on what a terrible estimator it is!

7. Suppose that a sample of size n_1 is available from a population with mean μ_1 and variance σ^2 and a sample of size n_2 is available from a population with mean μ_2 and the same variance σ^2. For $i = 1$ and 2, let s_i^2 be the sample variance from the ith population. Show that the statistic

$$\hat{\sigma}^2 = \frac{(n_1-1)s_1^2 + (n_2-1)s_2^2}{n_1+n_2-2} \tag{6.53}$$

is an unbiased estimator of the common variance σ^2.

8. The Rayleigh distribution is a special case of the Weibull model that is frequently used in engineering applications. The density of the one parameter Rayleigh model is

$$f(x) = \frac{2}{\theta}xe^{-x^2/\theta}I_{(0,\infty)}(x). \tag{6.54}$$

Given a random sample $X_1,\dots,X_n$ from a population governed by this distribution, find an unbiased estimator of θ and calculate its variance. (Hint: Start by identifying EX^2 for a variable X with the density in (6.54).

9. Let $X_1,\dots,X_n \stackrel{\text{iid}}{\sim} F_\theta$, a distribution with density function given by

$$f_\theta(x) = \frac{(1+\theta x)}{2}I_{(-1,1)}(x),$$

where $\theta \in [0,1]$. Find the constant c for which $T = c\sum_{i=1}^n X_i$ is an unbiased estimator of θ.

10. Suppose that $X_1, \ldots, X \stackrel{\text{iid}}{\sim} F$, a distribution with finite mean μ and variance σ^2, and let $\bar{X}$ and s^2 be the sample mean and variance. (a) Show that the statistic $\bar{X}^2$ is not an unbiased estimator of μ^2. (b) Find a constant c for which the statistic $\bar{X}^2 - cs^2$ is an unbiased estimator of μ^2.

11. A parallel system with m components will work if and only if there is at least one working component. Suppose that components work independently of each other, and the probability that a given component is working at time t_0, the mission time for the system, is the constant $p \in (0,1)$. (a) Calculate the probability π that a parallel system in these m components will be working at time t_0. (b) Given that, in a random sample of n parallel systems, k of them were working at time t_0, find an unbiased estimator of π. (c) From the data obtained in part (b), how would you go about estimating p? Is your estimator of p unbiased?

12. Let $X_1, \ldots, X_n \stackrel{\text{iid}}{\sim} N(\mu, \sigma^2)$. We know from Theorem 2.6.1 that the sample variance s^2 is an unbiased estimator of σ^2 (with or without the assumed normality). However, the sample standard deviation s is not an unbiased estimator of σ. Find the constant c for which $E(cs) = \sigma$. (Hint: Use the fact that $(n-1)s^2/\sigma^2$ has a chi-square distribution.)

13. Let $X \sim \Gamma(\alpha, 1)$. Find functions of X that are unbiased estimates of α, $\alpha(\alpha+1)$ and α^2.

14. Suppose that a random sample of size n is taken from a finite population of size N in such a way that each of the possible $\binom{N}{n}$ samples of size n has an equal chance of being selected. Denote the measured population values by $X_1, X_2, \ldots, X_N$ and the n sampled values by $X_{i_1}, X_{i_2}, \ldots, X_{i_n}$. The sample and population means are denoted, respectively, by

$$\bar{X} = \frac{1}{n}\sum_{j=1}^{n} X_{i_j} \text{ and } \mu = \frac{1}{N}\sum_{i=1}^{N} X_i. \tag{6.55}$$

Show that $\bar{X}$ is an unbiased estimator of μ.

15. Suppose that $X \sim B(n,p)$. Show that, for any positive integer k, the parameter $\theta = 1/p^k$ is not estimable.

16. Under the sampling assumptions of problem 6.7.14, show that the sample variance $s^2 = \frac{1}{n-1}\sum_{j=1}^{n}\left(X_{i_j} - \bar{X}\right)^2$ is an unbiased estimate of $S^2 = 1/(N-1)\sum_{i=1}^{N}(X_i - \mu)^2$. (Note: The variance of X in a population of size N is usually defined as $(N-1)/NS^2$. However, Cochran (1977), the "bible" on finite population sampling, uses S^2 in all formulae referring to the variability in the population.)

17. Let $X_1, X_2, \ldots, X_n \stackrel{\text{iid}}{\sim} \Gamma(1, \mu)$ be the observed lifetimes on n independent systems and that one wishes to estimate the systems reliability $R(x) = P(X > x) = e^{-x/\mu}$ at some particular time x (sometimes referred to as the mission time). Since the sample mean $\bar{X}$ is an unbiased estimator of μ, the estimator $\tilde{R}(x) = e^{-x/\bar{X}}$ of $R(x)$ might come to mind. Show that this estimator is not an unbiased estimator of $R(x)$. (Hint: See Problem 5.5.4 (re: Jensen's inequality). Note that if g is a convex function and X is an arbitrary random variable, then $E(g(X)) \geq g(EX)$, and the inequality is strict when g above is strictly convex (that is, when $g'(x) > 0$ for all x in the domain of g). The present problem deals with the function $g(\mu) = e^{-x/\mu}$, which is a concave function of μ for $\mu > 0$; thus, $-g(\mu)$ is convex for $\mu > 0$. Jensen's inequality therefore implies that $E(g(\bar{X})) < g(\mu)$. Conclude that $Ee^{-x/\bar{X}} < e^{-x/\mu}$, showing that the estimator $\tilde{R}$ has a negative bias.)

18. Professor E. Z. Gradink is giving a True-False exam as Midterm I in his section of the course "Statistics Appreciation." Dr. Gradink's tests are indeed easy to grade, but there's a slight twist to the grades students receive on their tests. On an n-question test, a given student will know X answers for sure, where X has the binomial distribution $B(n,p)$. Dr. G. reasons that students will choose "True" or "False" at random (each with probability $1/2$) when they don't know an answer. Thus, the number of correct answers on a given student's exam can be modeled as $Y = X + (1/2)(n - X)$. Since Y is the observed outcome on a given test, Dr. G. will derive an unbiased estimate $\hat{p}$ of p based on the observed Y, and then assign the grade $100\hat{p}$ to the student. What is Dr. G's estimator of p?

19. Let $X_1, X_2, X_3 \stackrel{\text{iid}}{\sim} U(0,\theta)$, and let $X_{(i)}$ denote the ith order statistic. Show that the estimators $4X_{(1)}, 2X_{(2)}$, and $4X_{(3)}/3$ are all unbiased for θ. Calculate their variances and rank their performance.

20. Suppose that $X_i \sim U(0, i\theta)$ for $i = 1, 2, \ldots, n$, and that the Xs are independent. (a) Show that $\hat{\theta}_1 = (1/n)\sum_{i=1}^{n} X_i/2i$ is an unbiased estimator of θ. (b) Show that $\hat{\theta}_2 = [n + 1/n] \max_{1 \le i \le n}\{X_i/i\}$ is also an unbiased estimator of θ. (c) Which of these estimators has the smaller variance?

21. Let $X_1, \ldots, X_n \stackrel{\text{iid}}{\sim} U(\theta, 2\theta)$. Define two estimators of θ as

$$\hat{\theta}_1 = \frac{n+1}{2n+1} X_{(n)} \text{ and } \hat{\theta}_2 = \frac{n+1}{5n+4}(X_{(1)} + 2X_{(n)}).$$

(a) Show that $\hat{\theta}_1$ and $\hat{\theta}_2$ are both unbiased estimators of θ. (b) Show that one of these estimators has a variance that is uniformly smaller than the other.

22. Let $X \sim Be(\theta, 1)$. Derive the Fisher information $I_X(\theta)$ and the CRLB.

23. Let $X \sim NB(r, p)$. Derive the Fisher information $I_X(p)$ and the CRLB.

24. Let $X \sim G(p)$. Derive the Fisher information $I_X(p)$ and the CRLB.

25. Let $X \sim W(2, \theta)$, the Weibull distribution with shape parameter 2 and scale parameter θ. Derive the Fisher information $I_X(\theta)$ and the CRLB.

26. Suppose that $X \sim NB(r, p)$, the negative binomial distribution with parameters $r \in Z^+$ and $p \in (0,1)$. (a) Find an unbiased estimator of the parameter $\theta = 1/p$, and determine its variance. (b) Calculate the Cramér-Rao Lower Bound for the variance of unbiased estimates of $1/p$. Can you claim that the estimator found in part (a) is the MVUE of $1/p$?

27. Show that equality holds in the Cauchy-Schwarz inequality in (6.18) if $f(\mathbf{y}) = cg(\mathbf{y})$, where c is a constant independent of $\mathbf{y}$. Is this condition on the functions f and g also *necessary* for equality in (6.18) to hold?

28. The discrete version of the Cauchy-Schwarz inequality in Lemma 6.3.2 follows immediately from Lagrange's identity:

$$\sum a_i^2 \times \sum b_i^2 - \left(\sum a_i b_i\right)^2 = \sum_{i<j} (a_i b_j - a_j b_i)^2. \tag{6.56}$$

Prove the identity in (6.56). Also, use it to show that equality is attained in the (discrete) Cauchy-Schwarz inequality if and only if there exists a constant c such that $a_i = cb_i$ for all i.

29. Let $X_1,\ldots,X_n \stackrel{\text{iid}}{\sim} \Gamma(\alpha_0,\beta)$, where α_0 is a known positive number. (a) Show that the estimator

$$\hat{\beta} = \frac{1}{n\alpha_0}\sum_{i=1}^{n} X_i$$

is an unbiased estimator of β. (b) Calculate the Cramér-Rao Lower Bound for the variance of unbiased estimates of β, and show that $V(\hat{\beta})$ achieves the CRLB and is thus the MVUE of β.

30. Let $X_1,X_2,\ldots,X_n \stackrel{\text{iid}}{\sim} F_\theta$, where the family $\{F_\theta, \theta \in \Theta\}$ satisfies the Cramer-Rao regularity conditions. Suppose that $\hat{g}(\mathbf{X})$ is an estimator of $g(\theta)$ with bias given by $b(\theta) = (E\hat{g}(\mathbf{X}) - g(\theta))^2$. Assuming that g and b are differentiable functions, show that the variance of the estimator $\hat{g}(\mathbf{X})$ satisfies the inequality

$$V(\hat{g}(\mathbf{X})) \geq \frac{(g'(\theta)+b'(\theta))^2}{nI(\theta)}. \tag{6.57}$$

31. Let $X_1,X_2,\ldots,X_n \stackrel{\text{iid}}{\sim} P(\lambda)$, the Poisson distribution with mean λ. (a) Calculate the CRLB for this model. (b) Show that $V(\bar{X})$ is equal to the CRLB, that is, that $\bar{X}$ is the MVUE of λ.

32. Let $X_1,X_2,\ldots,X_n \stackrel{\text{iid}}{\sim} N(\mu,1)$. Show that the variance of the unbiased estimator $\bar{X}^2 - 1/n$ of μ^2 does not achieve the Cramér-Rao Lower Bound for the variance of unbiased estimators of μ^2. (Hint: Start by writing the normal density in terms of the new parameter $\theta = \mu^2$.)

33. Suppose T is the minimum variance unbiased estimator of a parameter θ, that is, among all unbiased estimators, $V(T)$ is minimal for all values of θ. Show that T is unique. (Hint: Suppose S is another unbiased estimator. Evaluate the variance of $U = (T+S)/2$, and show, using the Cauchy-Schwarz inequality, that $V(U) \leq V(T)$ for all θ. Given our assumption, this will imply that $V(U) = V(T)$ for all θ. Argue, again invoking Cauchy-Schwarz, that $U = T$ with probability 1.)

34. Let X be a Cauchy random variable with density

$$f_\theta(x) = \frac{1}{\pi(1+(x-\theta)^2)} \text{ for } -\infty < x < \infty.$$

If a random sample of size n is taken from this distribution, show that the CRLB for the variance of an unbiased estimate of θ is $2/n$.

35. Take Yourself Lightly, Inc., a company that makes florescent light bulbs, makes 100-watt light bulbs of three different types. The life length of the ith type of bulb is

$$X_i \sim \Gamma(1, i\theta) \text{ for } i = 1,2,3.$$

Based on iid samples $\{X_{i1},\ldots,X_{in_i}\}$ on the ith type of light bulb, for $i = 1,2,3$, find the best linear unbiased estimator of θ.

36. Suppose that $X_1,\ldots,X_m \stackrel{\text{iid}}{\sim} F_1$, a distribution with finite mean μ and variance σ^2, and suppose that $Y_1,\ldots,Y_n \stackrel{\text{iid}}{\sim} F_2$, a distribution with mean μ and variance $5\sigma^2$. Assume that the Xs are independent of the Ys. (a) Show that, for any constant c, the statistic $\hat{\mu}_c = c\bar{X} + (1-c)\bar{Y}$ is an unbiased estimator of μ. (b) Find the value of c for which $\hat{\mu}_c$ has the smallest possible variance.

37. Let $X_1,\ldots,X_n \overset{\text{iid}}{\sim} N(\theta,\theta^2)$. Note that the two statistics $\bar{X}^2 - s^2/n$ and s^2 are both unbiased estimators of θ^2. Find the best linear unbiased estimator of θ^2 based on these two estimators.

38. Let $X_1,\ldots,X_n$ be independent random variables, with $X_i \sim N(c_i\mu,\sigma^2)$. Derive the BLUE for estimating μ on the basis of these Xs.

39. Let $X_1,\ldots,X_n \overset{\text{iid}}{\sim} N(\mu,\sigma_i^2)$, where the variances $\{\sigma_i^2, i=1,\ldots,n\}$ are assumed known. Find the BLUE of μ.

40. Suppose that X and Y are independent binomial variables, with $X \sim B(n,p)$ and $Y \sim B(m,p)$. Derive the BLUE of p. Is your estimator actually the MVUE of p? Justify your answer.

41. A rare art object is to be sold at auction. The bidders are experienced art dealers and they all know what the object is worth, namely \$$\theta$. In the end, n (increasing) bids are submitted (as is the case in a "progressive auction"), and the object is bought by the last bidder. Suppose that the first bid is $X_1 \sim U[0,\theta]$ and that, for $i=1,\ldots,n-1, X_i|X_{i-1}=x_{i-1} \sim U(x_{i-1},\theta)$. Find the best linear unbiased estimate (BLUE) of θ, and obtain an expression for its variance.

42. Let X and Y be independent, with $X \sim \Gamma(2,\theta/2)$ and $Y \sim \Gamma(1,\theta)$. Find the best linear unbiased estimator (BLUE) of θ based on X and Y.

43. Let $T_1,\ldots,T_k$ be unbiased estimators of a parameter θ, and assume that $V(T_i)=\sigma_i^2$ and that $Cov(T_i,T_j)=0$ for $i\neq j$. Show that the estimator

$$T^* = \frac{\sum_{i=1}^k a_i T_i}{\sum_{i=1}^k a_i},$$

where $a_i = 1/\sigma_i^2$, is the best linear unbiased estimator of θ based on the Ts.

44. Suppose that $X \sim N(0,\sigma^2)$. Show that $|X|$ is a sufficient statistic for σ^2.

45. Suppose that $X_i \sim \Gamma(1,i\theta)$ for $i=1,2,\ldots,n$, and that the Xs are independent. Show that $\sum_{i=1}^n X_i/i$ is a sufficient statistic for θ.

46. Let $X_1,\ldots,X_n \overset{\text{iid}}{\sim} W(\alpha,\beta)$. Find a two-dimensional sufficient statistic for the parameter pair (α,β).

47. Let $X_1,\ldots,X_n \overset{\text{iid}}{\sim} Be(\alpha,\beta)$. Find a two-dimensional sufficient statistic for the parameter pair (α,β).

48. Let $X_1,\ldots,X_n \overset{\text{iid}}{\sim} F_\theta$, a one-parameter model with density function $f_\theta(x)$. Show that the order statistics $(X_{(1)},\ldots,X_{(n)})$ are sufficient statistics for the parameter θ.

49. Let $X_1,\ldots,X_n \overset{\text{iid}}{\sim} N(\theta,\theta^2)$, where θ is an unknown positive parameter. (a) Show that the pair $(\sum_{i=1}^n X_i, \sum_{i=1}^n X_i^2)$ is a sufficient statistic for the parameter θ. (b) Show that the pair $(\sum_{i=1}^n X_i, \sum_{i=1}^n X_i^2)$ is not a *complete* sufficient statistic for θ.

50. Let $X_1,\ldots,X_n \stackrel{\text{iid}}{\sim} F_\theta$, a one-parameter model with density function of the form

$$f_\theta(x) = g(\theta)h(x)I_{(0,\theta)}(x),$$

where θ is a positive parameter. Show that the maximum order statistic $X_{(n)}$ is a sufficient statistic for θ.

51. Suppose $(X_1,Y_1),\ldots,(X_n,Y_n) \stackrel{\text{iid}}{\sim} N_2(\mu,\Sigma)$, the bivariate normal distribution with means μ_x and μ_y, variances σ_x^2 and σ_y^2, and correlation coefficient ρ. Show that the 5-dimentional vector $(\Sigma X, \Sigma Y, \Sigma X^2, \Sigma Y^2, \Sigma XY)$ is jointly sufficient for the parameters μ_x, μ_y, σ_x^2, σ_y^2, and ρ.

52. Let $f(x,y|\theta_1,\theta_2,\theta_3,\theta_4)$ be the density of the uniform distribution on the rectangle in the plane with corners at the points $(\theta_1,\theta_2),(\theta_1,\theta_4),(\theta_3,\theta_2)$, and (θ_3,θ_4), where $\theta_1 < \theta_3$ and $\theta_2 < \theta_4$. Find a four-dimensional sufficient statistic for the parameter vector $(\theta_1,\theta_2,\theta_3,\theta_4)$.

53. Let $X_1,\ldots,X_n \stackrel{\text{iid}}{\sim} F_\theta$, a one-parameter distribution with density or pmf f_θ for $\theta \in \Theta$ and let $T = h(X_1,\ldots,X_n)$ be a sufficient statistic for θ. Suppose that the MLE $\hat{\theta}$ of θ uniquely maximizes the likelihood function $L(\theta|x_1,\ldots,x_n) = \prod_{i=1}^n f_\theta(x_i)$. Show that, if $g(x)$ is an arbitrary real-valued function with domain $X \equiv \Theta$, then $g(\hat{\theta})$ is the unique maximum likelihood estimator of $g(\theta)$.

54. This problem deals with the generalization of the one-parameter exponential family whose density function is displayed in (6.36) to two or more parameters and to random vectors $\mathbf{x} \in R^m$. Consider a family of distributions whose density of probability mass function $f(\mathbf{x}|\theta)$ depends on a k-dimensional parameter $\theta = (\theta_1,\ldots,\theta_k)$ taking values in the parameter space Θ. The family $\{f(\mathbf{x}|\theta)\}$ is called a *k-parameter exponential family* if $f(\mathbf{x}|\theta)$ can be written in the form

$$f(\mathbf{x}|\theta) = exp\{\sum_{i=1}^{k} A_i(\theta_1,\ldots,\theta_k)B_i(\mathbf{x}) + C(\mathbf{x}) + D(\theta_1,\ldots,\theta_k)\}I_S(\mathbf{x}), \tag{6.58}$$

where the support set S of $\mathbf{x}$ does not depend on θ. Suppose that $X_1 \ldots X_n$ is a random sample of size n from a distribution which belongs to a k-parameter exponential family. Show that the k statistics $\{B_1,\ldots,B_k\}$ defined by

$$B_i = \sum_{j=1}^{n} B_i(X_j) \text{ for } i = 1,\ldots,k, \tag{6.59}$$

are joint sufficient statistics for the parameter θ. (Note: It can be shown that the vector $(B_1,\ldots,B_k)$ is a *complete* sufficient statistic for θ.)

55. Let $X_1,\ldots,X_n \stackrel{\text{iid}}{\sim} N(\mu_0,\sigma^2)$, where μ_0 is a known number. Show that $T = \sum_{i=1}^n (X_i - \mu_0)^2$ is a sufficient statistic for σ^2.

56. (See Problem 6.7.54.) Show that the family of normal distributions $\{N(\mu,\sigma^2), \mu \in (-\infty,\infty), \sigma^2 > 0\}$ is a 2-parameter exponential family.

57. (See Problem 6.7.54.) Show that the family of multinomial distributions whose probability mass functions are given in (4.58) is a $(k-1)$-parameter exponential family.

58. (See Problem 6.7.54.) family of bivariate normal distributions whose density functions are given in (4.61) is a 5-parameter exponential family.

59. Let $X_1,\dots,X_n \overset{\text{iid}}{\sim} P(\lambda)$, the Poisson distribution with mean λ. Let $Y=\sum_{i=1}^n X_i$, the sufficient statistic for λ, and recall that $Y \sim P(n\lambda)$. (a) Prove that the parameter $\theta = 1/\lambda$ is not estimable, that is, an unbiased estimator based on Y does not exist. (Hint: Write out the expectation $E[\hat{\theta}(Y)]$ as an infinite series, set it equal to $1/\lambda$, and argue that this equation has no solution. (b) Show that the estimator defined as $\tilde{\theta} = n/(Y+1)$ is asymptotically unbiased for θ, that is, show that the bias of this estimator tends to zero as $n \to \infty$.

60. Let $X_1,\dots,X_n$ be a random sample from the Rayleigh distribution with density function

$$f(x) = \frac{2}{\theta} x e^{-x^2/\theta} I_{(0,\infty)}(x). \tag{6.60}$$

Show that the statistic $T=\sum_{i=1}^n X_i^2$ is a sufficient statistic for θ. Identify the MVUE of θ.

61. Let $X_1,\dots,X_n \overset{\text{iid}}{\sim} U(a,b)$. Identify a joint sufficient statistic (S,T) for the parameter pair (a,b), and obtain functions of S and T that are unbiased estimators of the mean $(a+b)/2$ and the range $(b-a)$.

62. Let $X_1,\dots,X_n \overset{\text{iid}}{\sim} F_\theta$, a distribution with density function

$$f(x) = \frac{1}{\theta} x^{(1-\theta)/\theta} I_{(0,1)}(x).$$

Show that the estimator $\hat{\theta} = -(1/n)\sum_{i=1}^n \ln X_i$ is an unbiased estimator of θ. Is it the MVUE of θ?

63. Suppose that $X_1,\dots,X_n \overset{\text{iid}}{\sim} P(\lambda)$. Find the MVUE of $\lambda e^{-\lambda}$.

64. The Rao-Blackwell theorem is usually stated as in Theorem 6.4.2, and shows that an unbiased estimator S of a parameter θ that is not a function of the sufficient statistic T for θ can be improved by the estimator $\hat{\theta} = E(S|T)$ that is also unbiased for θ and whose variance is no larger than $V(S)$ (and actually smaller when the loss function is strictly convex, as is the case with squared error loss). Another version of the Rao-Blackwell Theorem doesn't mention unbiasedness. Show that, for an arbitrary estimator S that is not a function of the sufficient statistic T, the estimator $\hat{\theta}$ has a smaller mean squared error than S.

65. Let $X_1,\dots,X_n \overset{\text{iid}}{\sim} F_\theta$, a distribution with pdf or pmf $f(x|\theta)$. Suppose that $\hat{\theta}$ is an unbiased estimator of θ and that $T=T(X_1,\dots,X_n)$ is a sufficient statistic for θ. Let $\tilde{\theta} = E(\hat{\theta}|T)$. Show that

$$E|\tilde{\theta}-\theta| \le E|\hat{\theta}-\theta| \text{ for all } \theta \in \Theta.$$

66. Let $X \sim G(p)$, the geometric distribution with parameter p. Let $\hat{p} = I_{\{1\}}(X)$, an estimator that is non-zero only when $X=1$. (a) Show that the geometric distribution is a one-parameter exponential family and conclude that X is a complete sufficient statistic for p. (b) Show that $\hat{p}$ is an unbiased estimator of p. (c) Argue that $\hat{p}$ is the MVUE of p.

67. Bioassay experiments are aimed at assessing the potency of a drug which is given to groups of animals at several different dose levels. Suppose k dose levels are tested. Denoting the k dose levels used as $x_1,\dots,x_k$, the probability of a positive response to the drug at dose level x_i is modeled as

$$p(x_i) = \frac{1}{1+e^{-(\alpha+\beta x_i)}}.$$

For $i = 1, \ldots, k$, let Y_i be the number of positive responses obtained from the ith group. Then $Y_i \sim B(n_i, p(x_i))$. (a) Obtain an expression for the joint distribution of $(Y_1, \ldots, Y_k)$. (b) Show that the statistic $\left(\sum_{i=1}^k Y_i, \sum_{i=1}^k x_i Y_i\right)$ is a joint sufficient statistic for (α, β). (c) Derive the minimum logit chi-square estimates of α and β, defined by Berkson (1955) as the values $\hat{\alpha}$ and $\hat{\beta}$ that minimize the function

$$\sum_{i=1}^{k} n_i \bar{Y}_i (1 - \bar{Y}_i)(\alpha + \beta x_i - \ln(\bar{Y}_i / (1 - \bar{Y}_i))^2,$$

where $\bar{Y}_i = 1/n_i \sum_{j=1}^{n_i} Y_{ij}$. (d) Show that the estimators $\hat{\alpha}$ and $\hat{\beta}$ in part (c) are not functions of the sufficient statistics in part (b). Note that this implies that these estimators may be improved by "Rao-Blackwellization" (as described in Problem 6.7.64.)

68. Suppose that $X_1, X_2, \ldots, X_n \stackrel{\text{iid}}{\sim} G(p)$. As in problem 6.7.66, let $\hat{p} = I_{\{1\}}(X_1)$. (a) Show that $S = \sum_{1=1}^n X_i$ is a sufficient statistic for p. (b) Since $\{G(p), p \in (0,1)\}$ is an exponential family, S is also complete. Conclude that $T = E(\hat{p}|S)$ is the MVUE of p. Derive a closed-form expression for T. (Hint: Identify the conditional distribution of X_1, given $S = s$.)

69. Suppose that $X_1, X_2, \ldots, X_n \stackrel{\text{iid}}{\sim} N(\mu, 1)$. Show that the estimator $\hat{\theta} = \bar{X} - 1/n$ is the MVUE of $\theta = \mu^2$. Is $\hat{\theta}$ admissible relative to the squared error criterion?

70. Suppose that $X_1, X_2, \ldots, X_n \stackrel{\text{iid}}{\sim} N(\mu, \sigma^2)$, where both parameters are unknown. Find the MVUE of the ratio μ/σ.

71. Suppose that $\{(X_i, Y_i), i = 1, \ldots, n\}$ is a random sample from a bivariate normal distribution with parameters μ_x, μ_y, σ_x^2, σ_y^2, and ρ. Find the MVUEs of $\mu_x \cdot \mu_y$ and of $\rho \cdot \sigma_x \cdot \sigma_y$.

72. Let $X_1, \ldots, X_n \stackrel{\text{iid}}{\sim} B(k, p)$, and let $\theta = kp(1-p)^{k-1}$. Find the minimum variance unbiased estimator of θ. (Hint: Since $\theta = P(X_1 = 1)$, $\hat{\theta} = I_{\{1\}}(X_1)$ is an unbiased estimator of θ. Let $X = \sum_{i=1}^n X_i \sim B(n, p)$, and show that

$$E(\hat{\theta}|X = x) = k \frac{\binom{k(n-1)}{x-1}}{\binom{kn}{x}}.$$

73. Suppose that X has uniform distribution $U(0, \theta)$. (a) Show that the estimator $\hat{\theta} = 2X$ is the best unbiased estimator of θ based on X. (b) Consider the class of estimators of the form $\hat{\theta}_c = c \cdot X$, where c is a positive constant. Calculate the mean squared error (MSE) of the estimator $\hat{\theta}_c$ of θ. (c) Show that $\hat{\theta}$ above is inadmissible as an estimator of θ (relative to squared error) by showing that there is an estimator in the class in part (b) that is better than $\hat{\theta}$ for all $\theta > 0$.

74. Let $X_1, \ldots, X_n \stackrel{\text{iid}}{\sim} \Gamma(\alpha, 1)$. Show that, while the sample mean $\bar{X}$ is an unbiased estimator of α, it is inadmissible, that is, find an alternative estimator of α whose MSE is uniformly smaller than that of $\bar{X}$.

75. Consider the setup in Problem 6.7.1, that is, let $X_1, \ldots, X_n \stackrel{\text{iid}}{\sim} P(1, \theta)$ and let $X_{(i)}$ be the ith order statistic of the sample. Suppose that $\hat{\theta}$ is the unbiased estimator based on $X_{(1)}$ that solves Problem 6.7.1. Is $\hat{\theta}$ the best estimator of θ of the form $cX_{(1)}$?

76. Let $X_1,\ldots,X_n \overset{\text{iid}}{\sim} U(0,\theta)$. Consider the estimators of $\theta : \hat{\theta}_1 = \bar{X}$ and $\hat{\theta}_2 = X_{(1)} + X_{(n)}$. Both are unbiased estimators of θ. Which is the better estimator, that is, has the smaller variance?

77. Let $X_1, X_2, X_3 \overset{\text{iid}}{\sim} \Gamma(1,\theta)$. Show that $\hat{\theta} = 5X_{(2)}/6$ is an unbiased estimator of θ, as is, of course, $\bar{X}$. Which of these estimators has the smaller variance?

78. Suppose that $X \sim B(n,p)$. For arbitrary positive constants α and β, consider the class of estimators of p of the form

$$\hat{p}_{\alpha,\beta} = \frac{X+\alpha}{n+\alpha+\beta}. \tag{6.61}$$

Calculate the mean, variance, and mean squared error of $\hat{p}_{\alpha,\beta}$.

79. Let $X_1,\ldots,X_n \overset{\text{iid}}{\sim} P(\lambda)$. As you know, the parameter λ is both the mean and the variance of the Poisson distribution. The sample mean and the sample variance are both unbiased estimators of λ. Which one has the smaller variance?

80. Let $\hat{p}_{\alpha,\beta}$ be the estimator of the binomial proportion p given in (6.61). Find the values of α and β for which the mean squared error of this estimator is a constant independent of p.

81. Suppose that $X_1,\ldots,X_n \overset{\text{iid}}{\sim} P(\lambda)$, the Poisson distribution with mean λ. For arbitrary positive constants a and b, consider the class of estimators of λ of the form

$$\hat{\lambda}_{a,b} = \frac{\sum_{i=1}^{n} X_i + a}{n+b}.$$

Calculate the mean, variance, and mean squared error of $\hat{\lambda}_{a,b}$.

7

Statistical Estimation: Asymptotic Theory

7.1 Basic Principles

The term "asymptotic" is the collegiate replacement for things that might earlier have been referred to as occurring "eventually" or "in the limit." This chapter is dedicated to the study of the asymptotic behavior of statistical estimators. The approaches to statistical estimation considered so far concentrate on the comparative performance of estimators based on a random sample of observations of some fixed size. It's not unreasonable to ask why one would wish to expand upon this framework. After all, doesn't the sample that we are working with in any given problem have some fixed size, perhaps dictated by the time or budget available to collect the data? Of course it does. So then why would one wish to explore something like what we are about to dive into—the performance of different estimators as the sample size is allowed to grow to infinity. There are several very good reasons to do this.

First, we must recognize that even when we succeed in deriving, under some specific restriction, an optimal estimator of a given parameter, there is no guarantee (as we saw in Section 6.6) that this estimator is good overall. Lurking in the background is the potential embarrassment of using or recommending an estimator that turns out to be inadmissible. Secondly, the fixed-sample-size methods that we have examined are somewhat limited in scope. We will expand upon this collection later in the book, adding the important method of least squares to the mix in our treatment of regression analysis in Chapter 11 and treating the topic of nonparametric estimation in Chapter 12. But even with this expanded collection of fixed-sample-size methods, there is only a modest array of stochastic models and scenarios for which an "exact analysis" (that is, the derivation of an estimator that is optimal in some restricted sense and the identification of its probability distribution) is analytically tractable. These potential difficulties motivate the search for alternative approaches that (a) relax the restrictions under which fixed-sample-size estimators are derived, (b) search for optimality under similar but more flexible criteria and (c) open the door to approximations, valid for sufficiently large samples, which bypass the need for exact solutions when analytic approaches are infeasible.

An asymptotic analysis of a statistical estimation problem involves the assessment of the limiting behavior of estimators as the sample size $n \to \infty$. The basic concepts to be employed in such an analysis have been presented in some detail in Chapter 5. Suppose $\hat{\theta}_n$ is an estimator of a parameter θ based on a random sample of size n from the distribution F_θ. Our first concern about $\hat{\theta}_n$ is whether or not it grows close to θ as $n \to \infty$. The convergence of $\hat{\theta}_n$ to θ can be described in a variety of ways. In Section 5.1, we defined three different ways in which a sequence of random variables can be said to converge to a constant. Of these, we will be especially interested, in this book, in "convergence in probability," a concept which is the subject of Definition 5.1.1. We will be interested in determining whether or not an estimator of interest converges in probability to the target parameter, that is, whether or not $\hat{\theta}_n \xrightarrow{p} \theta$. We will refer to such convergence as "consistency," that is, $\hat{\theta}_n$ is a *consistent estimator* of θ if $\hat{\theta}_n \xrightarrow{p} \theta$. Consistency constitutes a minimal prerequisite for an estimator that might be thought of as "asymptotically" competitive. We will not be interested here in studying estimators which don't satisfy this minimal property, although we will occasionally point out a potential estimator's inconsistency. Instead, our focus will be on finding consistent estimators that have other good properties and are, perhaps, best under other criteria which take an estimator's precision into account.

Let us first make clear that consistency itself can come in different flavors, even when speaking only of convergence in probability. For example, one consistent estimator of a parameter may converge more quickly than a second and would thus be preferred over the second. The convergence rate of an estimator is important! In Theorem 5.1.3, it was shown that, under certain fairly innocuous conditions, the mean $\bar{X}$ of a random sample is a consistent estimator of the population mean μ. In many situations, we would feel quite comfortable using $\bar{X}$ to estimate a population mean. Our first example shows that, on asymptotic grounds, there are at least a few occasions when that would not be a good idea.

Example 7.1.1. Let $X_1, X_2, \ldots, X_n \sim^{iid} U(0,\theta)$, with θ unknown. We have shown in Example 6.1.1 that both $\hat{\theta}_n^{(1)} = 2\bar{X}$ and $\hat{\theta}_n^{(2)} = (n+1)X_{(n)}/n$ are unbiased estimators of θ. Their respective variances are $V(\hat{\theta}_n^{(1)}) = \theta^2/3n$ and $V(\hat{\theta}_n^{(2)}) = \theta^2/n(n+2)$. It follows immediately from Chebyshev's inequality that both $\hat{\theta}_n^{(1)}$ and $\hat{\theta}_n^{(2)}$ are consistent estimators of θ. However, $\sqrt{n}(\hat{\theta}_n^{(1)} - \theta)$ has variance $\theta^2/3$, while $\sqrt{n}(\hat{\theta}_n^{(2)} - \theta)$ has variance $\theta^2/(n+2)$. So another application of Chebyshev's inequality shows that $\sqrt{n}(\hat{\theta}_n^{(2)} - \theta) \xrightarrow{p} 0$, while $\sqrt{n}(\hat{\theta}_n^{(1)} - \theta)$ does not converge to zero, in probability or in any other sense. We thus see that $\hat{\theta}_n^{(2)}$ converges to θ at a faster rate than $\hat{\theta}_n^{(1)}$. ■

The result in Example 7.1.1 above is not particularly surprising. After all, the estimator $\hat{\theta}_n^{(2)}$ is based on the sufficient statistic $X_{(n)}$ while its competitor is not. We have seen that $\hat{\theta}_n^{(2)}$ is better than $\hat{\theta}_n^{(1)}$ in fixed-sample-size comparisons, and we now see that it also has better asymptotic behavior. The fact that these two estimators converge to the target parameter at different rates leads naturally to a brief general discussion of rates of convergence and provides me with the opportunity of introducing some standard and quite useful notation for these rates. Let us first consider the meaning of a phenomenon like we see in the estimator $\hat{\theta}_n^{(1)}$ in Example 7.1.1. What exactly is happening when a sequence of random variables $\{T_n\}$ are such that T_n converges to θ in probability but the related variable $\sqrt{n}(T_n - \theta)$ has a constant variance independent of n? This phenomenon is telling us that, as n grows large, the difference $(T_n - \theta)$ is growing smaller at a rate that is precisely the reciprocal of the rate at which $\sqrt{n}$ is growing larger, so that their product neither tends to 0 nor to ∞. We might then say that $(T_n - \theta)$ is converging to zero at the rate $1/\sqrt{n}$. The shorthand notation that is typically used for describing the situation is

$$(T_n - \theta) = O_p(1/\sqrt{n}). \tag{7.1}$$

The term on the RHS of (7.1) is pronounced "Big O sub p of $1/\sqrt{n}$," and is meant to tell us precisely what multiplier of the variable $(T_n - \theta)$ stabilizes it so that it converges to a well-defined random variable rather than to 0 or to ∞. The "p" in the notation stands for "probability" and is used when the quantity on the left-hand side of (7.1) is a random variable. The symbol without the p is used for comparing two sequences of real numbers. We will make use of both "big O" and "little o" notation. Both are defined in the following:

Definition 7.1.1. Let $\{Y_n, n = 1,2,\ldots\}$ be a sequence of random variables and $\{a_n, n = 1,2,\ldots\}$ be a sequence of real numbers. Then

$$Y_n = O_p(a_n) \tag{7.2}$$

if and only if the sequence $\{Y_n/a_n, n = 1,2,\ldots\}$ is bounded in probability, that is, for any $\varepsilon > 0$, there exists a number $k = k(\varepsilon) < \infty$ such that $P\left(\left|\frac{Y_n}{a_n}\right| > k\right) < \varepsilon$ for all n.
Further,

$$Y_n = o_p(a_n) \tag{7.3}$$

if and only if $Y_n/a_n \xrightarrow{P} 0$.

When an estimator $\hat{\theta}$ is known to converge to the parameter θ at the rate $1/\sqrt{n}$, the estimator is generally said to be "root-n consistent." Of course this really means that $(\hat{\theta} - \theta) = O_p(1/\sqrt{n})$. In Example 7.1.1, it can be shown that each of these properties hold:

$$(\hat{\theta}_1 - \theta) = o_p(1), (\hat{\theta}_1 - \theta) = O_p\left(\frac{1}{\sqrt{n}}\right), (\hat{\theta}_2 - \theta) = o_p\left(\frac{1}{\sqrt{n}}\right) \text{ and } (\hat{\theta}_2 - \theta) = O_p\left(\frac{1}{n}\right).$$

Suppose we have two competing estimators, $\hat{\theta}_n^{(1)}$ and $\hat{\theta}_n^{(2)}$, of a parameter θ that converge to θ in probability at the same rate. For concreteness, let's assume that they both converge at the rate $1/\sqrt{n}$. What other asymptotic characteristics of these estimators would be deemed desirable? Well, of course, it would be convenient if we were able to identify the exact *asymptotic distribution* (AD) of each estimator, that is, the limiting distribution of $\sqrt{n}(\hat{\theta}_n^{(1)} - \theta)$ and $\sqrt{n}(\hat{\theta}_n^{(2)} - \theta)$. In dealing with these ideas, we will make use of the definitions and discussion in Section 5.2. We might prefer one estimator to the other if we had its AD in hand and couldn't obtain the AD of the other. Convenience aside, however, we would like to be able to select the estimator whose asymptotic distribution is more heavily concentrated about the target parameter than is the other's. Now in many of the statistical problems we will examine (in particular, problems involving models which we will refer to as "regular"), it will turn out that the methods we use to derive estimators with good asymptotic properties have these two features in common: the estimators they lead to will generally be root-n consistent, and the estimators, when properly scaled, will, asymptotically, have normal distributions. We will typically represent this situation as follows:

$$\sqrt{n}\left(\hat{\theta}_n - \theta\right) \xrightarrow{D} Y \sim N(0, AV). \tag{7.4}$$

We will refer to the variance of the *asymptotic normal distribution* of $\hat{\theta}$ as the *asymptotic variance* of the estimator (though, technically, it's the limiting variance of $\sqrt{n}\hat{\theta}_n$) and we will use "AV" as the shorthand notation for it. When random samples are drawn from regular models, root-n consistency is typically the best rate of convergence possible, and the estimator which satisfies (7.4) and has the smallest AV possible is referred to as the *best asymptotically normal* (or BAN) estimator. In the sections that follow, we turn our attention to the two most widely used methods of deriving consistent, asymptotically normal estimators.

Exercises 7.1.

1. Let $X_1, X_2, \dots, X_n \stackrel{\text{iid}}{\sim} U(\theta-1, \theta+1)$, and let $X_{(i)}$ be the ith order statistic, that is, the ith smallest X. Show that both $\hat{\theta}_n^{(1)} = \bar{X}$ and $\hat{\theta}_n^{(2)} = (X_{(1)} + X_{(n)})/2$ are consistent estimators of θ. Identify the rate of convergence of each.

2. Let $X_1, X_2, \dots, X_n \stackrel{\text{iid}}{\sim} G(p)$, the geometric distribution with parameter $p \in (0,1)$. Show that $\hat{p}_n = 1/\bar{X}_n$ is a consistent estimator of p.

3. The purpose of this exercise is to show that o_p is a stronger property than O_p. Specifically, show that if $Y_n = o_p(1)$, then $Y_n = O_p(1)$. (Hint: Argue as follows: if $Y_n = o_p(1)$, then $P(|Y_n| > 1) \to 0$. Thus, for any $\varepsilon > 0$, there exists an integer N such that $P(|Y_n| > 1) < \varepsilon$ for all $n > N$. Now, if k_0 is sufficiently large to ensure that $P(|Y_i| > k_0) < \varepsilon$ for $i = 1, \dots, N$, then for $k = \max(1, k_0)$, $P(|Y_n| > k) < \varepsilon$ for all n.)

7.2 The Method of Moments

The best-known and most highly regarded researcher in Statistics in the closing decades of the nineteenth century was Karl Pearson, a British mathematician who is credited with many of the early findings on probability modeling and various forms of statistical inference. Among his best-known accomplishments was his work with the correlation coefficient ρ which now bears his name, the chi-square goodness-of-fit test, his co-founding of the statistics journal *Biometrika* in 1901 (and service as its editor until his death in 1936), and his founding, in 1911, of the world's first university Department of Statistics at University College London. But more to the point here, he was an advocate of a quite broad and useful system of continuous univariate probability distributions which found many applications and continue to be used today (the standard gamma distribution being among them), and he introduced a complementary methodology aimed at estimating the unknown parameters of the varied collection of Pearson distributions from experimental data. Drawing from terms and ideas in common use in his day in the physical sciences, he called his estimation approach "the method of moments."

When the random sample one draws from a population is large, it is not unreasonable to expect that the characteristics of the population will be fairly well approximated by the characteristics of the sample. This type of intuition gives us some confidence that the sample mean $\bar{X}$ will be a reliable estimator of the population mean μ when the sample size n is large and, similarly, that the sample proportion $\hat{p}$ will be a reliable estimator of the population proportion p when n is large. Taking advantage of the expected similarity between a population and the sample drawn from it often lies behind the parameter estimation strategy employed. The method of moments is an approach to parameter estimation that is motivated by intuition such as this, and as we will show, it is a method that produces, under mild assumptions, consistent, asymptotically normal estimators of the parameters of interest.

Suppose that X is a random draw from a population of interest. Recall, from Sections 2.3 and 3.2, the definition of the "moments" of a random variable X. The kth moment of the population, or equivalently, the kth moment of X, is denoted by μ_k and is defined as

$$\mu_k = EX^k, k = 1, 2, \dots\dots \tag{7.5}$$

We have used the first two population moments repeatedly, as μ_1 is simply the population mean and the population variance σ^2 is usually computed as $\mu_2 - (\mu_1)^2$.

Given a random sample from a population of interest, that is, given the sample $X_1, X_2, \dots, X_n \stackrel{\text{iid}}{\sim} F$, let m_k denote the kth sample moment, a statistic that is defined as

$$m_k = \frac{1}{n}\sum_{i=1}^{n} X_i^k. \tag{7.6}$$

While μ_k is the expected (or average) value of X^k in the population, m_k is the average value of X^k in the sample. When the sample size is large, these averages are likely to be close to each other. Indeed, assuming that $EX^{2k} < \infty$ (a condition which guarantees that the variance of X^k is finite), Theorem 5.1.3 (the WLLN) yields that $m_k \xrightarrow{p} \mu_k$.

Let us now suppose that the model governing the available random sample depends on a finite dimensional vector of parameters $(\theta_1, \theta_2, \dots, \theta_r)$ for some $r \geq 1$. The population moments will typically depend on the parameters $\theta_1, \dots, \theta_r$ (though some may be equal to constants independent of the θs). We will refer to the vector of estimators $(\hat{\theta}_1, \hat{\theta}_2, \dots, \hat{\theta}_r)$ as "method of moments estimators" or MMEs if they are obtained by solving one or more equations of the form

$$m_k = \mu_k. \tag{7.7}$$

The equation in (7.7) is referred to as the kth moment equation. The LHS of (7.7) will be a number computed from the available data, while the RHS of (7.7) is a function of model parameters. When there are r unknown parameters in the model, obtaining the MMEs will involve finding the simultaneous solutions to r equations in the r unknowns. In theory, one is free to choose any collection of r moment equations, and different choices generally give rise to different MMEs. In the following example, we'll derive two MMEs, each of which is a valid consistent, asymptotically normal estimator of the target parameter.

Example 7.2.1. Suppose that $X_1, X_2, \dots, X_n \stackrel{\text{iid}}{\sim} U(0, \theta)$. Let us estimate the parameter θ by the method of moments. Using the first moment equation, that is, using

$$m_1 = \mu_1,$$

we have

$$\bar{X} = \frac{\theta}{2}. \tag{7.8}$$

Solving (7.8) for θ, we identify

$$\hat{\theta} = 2\bar{X}$$

as an MME of θ. Of course, we could also have used the second moment equation to obtain an MME of θ. The equation $m_2 = \mu_2$ may be written as

$$\frac{1}{n}\sum_{i=1}^{n} X_i^2 = \frac{\theta^2}{3},$$

from which we may identify an alternative MME of θ as

$$\hat{\hat{\theta}} = \sqrt{\frac{3}{n}\sum_{i=1}^{n} X_i^2}.$$

■

For the particular model in Example 7.2.1, all population moments are finite, and it can be shown that the two estimators obtained are both consistent at the same rate and are both asymptotically normal as, in fact, are all the MMEs that we could obtain from other moment equations. In practice, the MME that one would usually use is the one which is obtained from the moment equations of the lowest possible order. For a one-parameter problem, one would use the MME obtained from the first moment equation if in fact the equation yields an estimator. An example of a one-parameter model in which the second moment equation would have to be used is the $N(0,\sigma^2)$ model, as the first moment of the model is a constant independent of the parameter σ^2. Reasons for preferring low-order moment equations in the derivation of MMEs include the fact that the low-order MME's achieve asymptotic normality under weaker assumptions, they tend to be less variable and they are less dependent on the fine features of the assumed model and are thus somewhat more resistant to mild departures from that model (that is, to a modest amount of model misspecification). All this notwithstanding, a higher-order MME is sometimes preferable to a lower-order MME. (See Problem 7.7.15.)

Let us now consider the following two-parameter estimation problem.

Example 7.2.2. Let $X_1, X_2, \dots, X_n \stackrel{\text{iid}}{\sim} \Gamma(\alpha, \beta)$. To derive MMEs of the parameters α and β, we will solve the first two moment equations. The first equation results in

$$\alpha \cdot \beta = m_1, \tag{7.9}$$

and the second equation results in

$$\alpha \cdot \beta^2 + \alpha^2 \cdot \beta^2 = m_2, \tag{7.10}$$

where $m_1 = \bar{X}$ and $m_2 = 1/n\sum_{i=1}^n X_i^2$. From (7.9), we may write

$$\alpha = \frac{m_1}{\beta}. \tag{7.11}$$

Substituting (7.11) in (7.10) leads to an equation in β alone, namely,

$$\beta m_1 + m_1^2 = m_2,$$

from which we may identify the MME of β as

$$\hat{\beta} = \frac{m_2 - m_1^2}{m_1} = \frac{\sum_{i=1}^n (X_i - \bar{X})^2}{n\bar{X}}.$$

From (7.11), we then obtain the MME of α as

$$\hat{\alpha} = \frac{m_1}{\hat{\beta}} = \frac{n\bar{X}^2}{\sum_{i=1}^n (X_i - \bar{X})^2}.$$

■

Method of moments estimators tend to be easy to derive and are known to have, under appropriate assumptions, good asymptotic properties which include root-n consistency (where n is the sample size) and asymptotic normality. The fact that the method can provide more than one estimator based on the same data might, at first sight, cause some concern, but, as mentioned above, it is standard practice to utilize the MME based on the moment equations of lowest possible order yielding an MME. In any case, if two MMEs were seen as viable estimators in a given problem, one would choose the one which had the smaller asymptotic variance. We will now turn to a discussion of the asymptotic behavior of MMEs.

Suppose that $X_1, X_2, \dots, X_n \stackrel{\text{iid}}{\sim} F_\theta$, where the model parameter $\theta = (\theta_1, \dots, \theta_k)$ is a k-dimensional vector for some $k \geq 1$. The moments $\{\mu_i, i = 1, 2, \dots\}$ of the model are functions of θ. Method of moments estimators are obtained by solving a system of k moment equations (setting sample moments equal to population moments). We will assume, for simplicity, these equations are the following:

$$m_1 = \mu_1(\theta_1, \dots, \theta_k)$$
$$m_2 = \mu_2(\theta_1, \dots, \theta_k)$$
$$\dots\dots\dots$$
$$m_k = \mu_k(\theta_1, \dots, \theta_k).$$

Let us assume that the relationship between the vectors $\mathbf{m}$ and μ is one-to-one. Then an MME can be obtained by inverting the relationship above to obtain the θs as functions of the sample moments and then relabeling each θ_i as $\hat{\theta}_i$. This process of inversion begins by recognizing that the population moments $\mu_1, \dots, \mu_k$ are functions of the population parameters $\theta_1, \theta_2, \dots, \theta_k$ as represented below:

$$\mu_1 = \mu_1(\theta_1, \dots, \theta_k)$$
$$\mu_2 = \mu_2(\theta_1, \dots, \theta_k)$$
$$\dots\dots\dots$$
$$\mu_k = \mu_k(\theta_1, \dots, \theta_k)$$

By inverting this relationship, we obtain the relationship of interest, with the θs now expressed as functions of the population moments. Let us express these inverse relationships as

$$\theta_i = h_i(\mu_1, \mu_2, \dots, \mu_k) \text{ for } i = 1, 2, \dots, k \tag{7.12}$$

and

$$\hat{\theta}_i = h_i(m_1, m_2, \dots, m_k) \text{ for } i = 1, 2, \dots, k. \tag{7.13}$$

Under the assumption that the functions $\{h_i, i = 1, \dots, k\}$ are continuous (that is, for each i, $h_i(\mathbf{x_n}) \to h_i(\mathbf{x})$ whenever $\mathbf{x_n} \to \mathbf{x}$), we may obtain the consistency of the MMEs in (7.13) from the consistency of the sample moments. To obtain the asymptotic distribution of an MME, an additional assumption is required, namely, that the population moment μ_{2k} exists and is finite. This assumption guarantees that both the mean and the variance of each of the first k sample moments are finite. Since the first k sample moments are simply averages of iid variables that have a common finite mean and variance, Theorem 5.1.3 (the WLLN) applies, yielding the conclusion that $m_i \xrightarrow{p} \mu_i$ for $i = 1, \dots, k$ as $n \to \infty$. Then, using the (perfectly valid) multivariate version of Theorem 5.2.1, we obtain that

$$\hat{\theta}_i = h_i(m_1, m_2, \dots, m_k) \xrightarrow{p} h_i(\mu_1, \mu_2, \dots, \mu_k) = \theta_i \text{ for } i = 1, \dots, k. \tag{7.14}$$

To summarize, one can prove the consistency of MMEs under the following three assumptions: (a) the functional relationship between k population moments and the k model parameters is differentiable and one-to-one and thus invertible, (b) for $i = 1,\dots,k$, the function h_i expressing θ_i as a function of the k population moments is continuous, and (c) the moment μ_{2k} exists and is finite. The third assumption can be weakened to the condition that $\mu_k < \infty$, since Theorem 5.1.4 implies that the convergence in (7.14) holds under this weaker assumption. But we will need the stronger assumption in obtaining the asymptotic distribution of MMEs, and so we will consider assumptions (a), (b), and (c) above to be required for the full asymptotic theory of MMEs to hold. These three assumptions are reasonably easy to verify for many models to which the method of moments is applied. When the method fails to provide helpful results, it is usually because the model under study is somewhat "heavy tailed" and its higher-order moments are not finite. For example, the method of moments is useless in obtaining estimates of the parameters of the Cauchy distribution.

The asymptotic normality of an MME follows from the Central Limit Theorem (Theorem 5.4.1), together with the δ-method theorem. In problems with several parameters, one would need to invoke the multivariate versions of these theorems. The multivariate CLT gives conditions under which the vector of averages of the multivariate data $\{\mathbf{X_i} = (X_{1i},\dots,X_{ki}), i = 1,\dots,n\}$, that is, the vector $(\bar{X}_1,\bar{X}_2,\dots,\bar{X}_k)$ converges, when properly standardized, to a random vector $(Y_1,Y_2,\dots,Y_k)$ having a k-variate normal distribution. The multivariate δ-method theorem gives conditions under which k functions $\{g_1,g_2,\dots,g_k\}$ of a k-dimensional asymptotically normal vector like $(\bar{X}_1,\bar{X}_2,\dots,\bar{X}_k)$ will themselves have a multivariate normal distribution asymptotically. Precise statements of both of these theorems may be found in Lehmann and Casella (1998, pp. 343–4). For simplicity, we will restrict our treatment of the asymptotic distribution theory of MMEs to the estimation of a single (scalar) parameter θ. The first result is a direct application of the CLT and identifies, under specific assumptions, the marginal asymptotic distribution of the kth sample moment.

Theorem 7.2.1. Let $X_1,X_2,\dots,X_n \overset{\text{iid}}{\sim} F$, and let m_k represent the kth sample moment, where k is a positive integer. If the population moments μ_k and μ_{2k} exist and are finite, then

$$\sqrt{n}\,(m_k - \mu_k) \xrightarrow{D} Y \sim N\left(0, \mu_{2k} - \mu_k^2\right). \tag{7.15}$$

The result above requires no separate proof, as the statement is simply the CLT (Theorem 5.3.1) with $X_1,X_2,\dots,X_n$ replaced by $X_1^k,X_2^k,\dots,X_n^k$. Theorem 7.2.1 falls short of saying that the vector of sample moments $(m_1,\dots,m_k)$ have a multivariate normal distribution asymptotically, but it does say that the marginal asymptotic distribution of each individual sample moment is normal, provided the appropriate population moments are finite. To deal with an MME in a given one-parameter problem, we apply the δ-method theorem as follows.

Theorem 7.2.2. Let $X_1,X_2,\dots,X_n \overset{\text{iid}}{\sim} F_\theta$, and let m_k represent the kth sample moment, where θ is a scalar parameter and k is a positive integer. Suppose that

$$\sqrt{n}\,(m_k - \mu_k) \xrightarrow{D} Y \sim N\left(0, \mu_{2k} - \mu_k^2\right).$$

If $\hat{\theta} = h(m_k)$ is an MME of θ, and the function h is a transformation with two continuous derivatives, then

$$\sqrt{n}(h(m_k) - h(\mu_k)) \xrightarrow{D} W \sim N(0, [h'(\mu_k)]^2[\mu_{2k} - \mu_k^2]). \tag{7.16}$$

As with Theorem 7.2.1, the result above requires no proof, as it is a direct application of Theorem 5.4.1 with T_n and θ replaced by the kth moments of the sample and the population. We will now illustrate these two theorems with several examples.

Example 7.2.3. Let $X_1, X_2, \ldots, X_n \overset{\text{iid}}{\sim} Exp(\lambda)$, the exponential distribution with failure rate λ. Using the first moment equation $m_1 = \mu_1$ to obtain an MME, we have

$$\bar{X} = \frac{1}{\lambda},$$

an equation that leads to the MME

$$\hat{\lambda} = \frac{1}{\bar{X}}.$$

To find the asymptotic distribution of $\hat{\lambda}$ using MME theory, we first obtain, from the CLT, the AD of $\bar{X}$ for this model. We have that

$$\sqrt{n}\left(m_1 - \mu\right) \xrightarrow{D} Y \sim N\left(0, \mu_2 - \mu^2\right),$$

that is,

$$\sqrt{n}\left(\bar{X}_n - \frac{1}{\lambda}\right) \xrightarrow{D} Y \sim N\left(0, \frac{1}{\lambda^2}\right).$$

Let $h(x) = 1/x$. We will use the δ-method theorem to obtain the AD of the MME $\hat{\lambda} = h(\bar{X}_n)$. Note that

$$h'(x) = -\frac{1}{x^2},$$

and that we will apply h' to $x = 1/\lambda$. Thus, from Theorem 7.2.2, we have

$$\sqrt{n}\left(\frac{1}{\bar{X}_n} - \lambda\right) \xrightarrow{D} W \sim N\left(0, \left[-\frac{1}{(1/\lambda)^2}\right]^2 \frac{1}{\lambda^2}\right)$$

or

$$\sqrt{n}\left(\hat{\lambda}_n - \lambda\right) \xrightarrow{D} W \sim N\left(0, \lambda^2\right).$$

■

Example 7.2.4. Let $X_1, X_2, \ldots, X_n \overset{\text{iid}}{\sim} N(0, \theta)$, the normal distribution with mean 0 and variance $\theta > 0$. Note that, in this model, θ may also be expressed as EX^2. The first moment equation yields $\bar{X} = 0$, so it is of no use in estimating θ. But the second moment equation yields

$$\frac{1}{n}\sum_{i=1}^{n} X_i^2 = \theta,$$

an equation which immediately identifies the second sample moment as an MME of θ. In this case, the CLT suffices in identifying the asymptotic distribution of the MME. In order to identify the AD exactly, we will need to know the second and fourth moments of the $N(0,\theta)$ model. If $X \sim N(0,\theta)$, we know that the ratio X^2/θ has the χ_1^2 distribution. (See Theorem 3.5.3 and Example 4.6.3.) It follows that $\mu_2 = EX^2 = \theta$ and $\mu_4 = EX^4 = 3\theta^2$. Letting $\hat{\theta} = \left(\sum_{i=1}^n X_i^2\right)/n$, we may invoke Theorem 7.2.1 to get

$$\sqrt{n}\left(\hat{\theta}-\theta\right) \xrightarrow{D} W \sim N\left(0, 2\theta^2\right).$$

■

Example 7.2.5. Let $X_1, X_2, \ldots, X_n \stackrel{\text{iid}}{\sim} Be(\theta, 1)$, the Beta distribution with mean $\theta/(\theta+1)$ and variance $\theta/(\theta+1)^2(\theta+2)$. The first moment equation is given by

$$\bar{X} = \frac{\theta}{\theta+1},$$

from which it follows that

$$\hat{\theta} = \frac{\bar{X}}{1-\bar{X}}$$

is a method of moments estimator of θ. By the CLT, we have that

$$\sqrt{n}\left(\bar{X} - \frac{\theta}{\theta+1}\right) \xrightarrow{D} V \sim N\left(0, \frac{\theta}{(\theta+1)^2(\theta+2)}\right).$$

Let $h(x) = x/(1-x)$ for $0 < x < 1$, and note that $h'(x) = 1/(1-x)^2$. It thus follows from the δ-method theorem that

$$\sqrt{n}\left(\frac{\bar{X}}{1-\bar{X}} - \theta\right) \xrightarrow{D} W \sim N\left(0, \frac{\theta(\theta+1)^2}{(\theta+2)}\right).$$

■

While the method of moments is often easier to execute than other methods, it has a drawback which renders the method imperfect. Although its asymptotic behavior may be optimal for some models, there is no guarantee of optimality. Indeed, it is easy to find examples in which an MME is asymptotically inefficient, that is, has a larger asymptotic variance than some alternative estimator. Because of this, MMEs are often viewed as quick and easy initial estimators from which other estimators can be created. We will see in Section 7.5 that MMEs may be used as the seed in numerical algorithms aimed at iterating toward an asymptotically optimal estimator.

Exercises 7.2.

1. Suppose that $X_1, X_2, \ldots, X_n \stackrel{\text{iid}}{\sim} U(\alpha, \beta)$. Find estimators of the parameters α and β by the method of moments.

2. Suppose that $X_1, X_2, \ldots, X_n \stackrel{\text{iid}}{\sim} N(\mu, \sigma^2)$. Find estimators of the parameters μ and σ^2 by the method of moments.

3. Suppose that $X_1, X_2, \ldots, X_n \stackrel{\text{iid}}{\sim} L(\mu, \beta)$, the logistic distribution with parameters μ and β. Find estimators of the parameters μ and β by the method of moments. (Note: See Theorem 3.6.7.)

4. Suppose that $X_1, X_2, \ldots, X_n \overset{\text{iid}}{\sim} G(p)$, the geometric distribution with parameter p. Find an estimator $\hat{p}$ of p by the method of moments. Identify the asymptotic distribution of $\hat{p}$.

5. Suppose that $X_1, X_2, \ldots, X_n \overset{\text{iid}}{\sim} P(\alpha, 1)$, the one-parameter Pareto distribution with density given in (3.35). Assuming that $\alpha > 1$, identify an estimator $\hat{\alpha}$ of α by the method of moments. Assuming that $\alpha > 2$, identify the asymptotic distribution of $\hat{\alpha}$.

7.3 Maximum Likelihood Estimation

The method of maximum likelihood estimation has a somewhat more complex history than that of the method of moments. The basic idea of maximum likelihood appeared in the writings of the German mathematician Carl Friedrich Gauss (1777–1855), the French mathematician Pierre-Simon, Marquis de Laplace (1749–1827), Thorvald Nicolai Thiele (1838–1910), a Danish astronomer, actuary, and mathematician and Francis Ysidro Edgeworth (1845–1926), the Irish philosopher and political economist who made significant contributions to statistical theory, even though, as a mathematician, he was largely self-taught. Thus, the approach to estimation has traceable roots as far back as the 1700s and the main idea of the approach was reasonably well known by the early 1900s. But it is the British statistician and geneticist Sir Ronald A. Fisher (1890–1962) who named the method, made it the centerpiece of his approach to statistical estimation, and led to its widespread adoption as the preferred method of statistical estimation on the basis of its asymptotic behavior. In particular, he was able to demonstrate that the method of maximum likelihood was at least as good as, and often superior to, the method of moments. Karl Pearson did not accept this news graciously (although one could also make the case that Fisher did not deliver the news graciously), and the two were distinctly unfriendly to each other from that point on. In the end, Fisher won the battle of estimation approaches, and also established a compelling legacy in Statistics, as he is also given credit for many other creative and useful contributions to statistical theory and practice, not the least of which was the notion of sufficient statistics, Fisher information, the broadly used methodology known as the analysis of variance, and Fisher's various hypothesis testing methods in nonparametric settings. In his book on the history of mathematical statistics, Anders Hald (1998) referred to Fisher as "a genius who almost single-handedly created the foundations for modern statistical science." Interestingly, he is held in equally high esteem in the field of population genetics.

The basic idea behind maximum likelihood estimation is extraordinarily simple. (A former teacher of mine, Arnold Ross, whom I have mentioned before, liked to use this phrase often in his teaching. Over time, I grew a bit leery when I heard the phrase, as I came to think of "extraordinarily simple" as really meaning "very simple after giving the matter a couple of hours of thought." But in this instance, I'm using the phrase in its usual sense.) The term "maximum likelihood" is a perfect description for the meaning and intent of the approach. It simply answers the question: "What value of the parameter would make my experimental outcome most likely to occur?" It seems quite reasonable to expect that your guess at an unknown parameter should be as compatible as possible with the data on which your guess is based. Here's a way to quantify that intuition. For concreteness, I'll focus on a simple model with which you are familiar.

Example 7.3.1. Suppose your data consists of the single binomial observation X, that is, $X \sim B(n, p)$ is the result (and sufficient statistic) from a binomial experiment. How should the parameter p be estimated? The maximum likelihood estimator (MLE) of p is the value of the parameter that would make what you actually observed the *most likely*. If you got $x = 6$ heads in $n = 10$ tosses of a bent coin, you would know that $p = .1$ is not a good guess at p. If p was really .1, the chances of seeing what you've seen, namely $X = 6$, is extremely small. (It's actually equal to .00014.) What you have observed would have a larger likelihood of happening if p was larger, but not too much larger. For example, $P(X = 6|p = .9) = .01116$. What is the MLE here? It is obtained my maximizing $P(X = 6|p)$ with respect to p. Let's call this probability the *likelihood function*. Since the data is already known, it is a function only of the unknown parameter p. Since $P(X = 6|p)$ is a product of several terms, and some terms contain exponents, we'll do what is often done in such maximization problems, that is, we will maximize $\ln P(X = 6|p)$ rather than $P(X = 6|p)$ itself. Since the function $\ln x$ is a strictly increasing function of x for positive x, the same value of p that maximizes $\ln P(X = 6|p)$ will also maximize $P(X = 6|p)$. Now, since $\ln P(X = 6|p) = \ln \binom{10}{6} + 6\ln p + 4\ln(1-p)$, we wish to solve the equation

$$\frac{\partial}{\partial p} \ln P(X = 6|p) = \frac{6}{p} - \frac{4}{1-p} = 0. \tag{7.17}$$

Equation (7.17) reduces to

$$6(1-p) - 4p = 0$$

which results in the unique solution $\hat{p} = .6$, the maximum likelihood estimator of p. The second derivative test shows that the log likelihood ($\ln P$) is indeed maximized at this value of p. Not that we need to know this, but $P(X = 6|p = .6) = .25082$. This is not a huge probability, so it's not like you would definitely expect to see $X = 6$ if the probability of heads p for this bent coin was really .6. Still, if p was anything other than the value .6, the chance of seeing $X = 6$ would be even less. So the observation $X = 6$ encourages us to believe that the data came from a binomial model with $p = .6$. ■

The derivation of the MLE from data drawn from a continuous distribution is, operationally, pretty much the same as the above. However, the interpretation of the MLE must be altered a bit, since the data that one observes when the assumed model is continuous has probability zero of occurrence under the model. There is still a sense in which the parameter value that we will call the MLE in this case maximizes the likelihood of the occurrence of what we have seen. In the following example, we derive the MLE in a continuous setting.

Example 7.3.2. Suppose that $\bar{X}$ is the mean of a random sample of size 10 from the normal distribution $N(\mu, 10)$, a model in which we have, for simplicity, assumed that the population variance σ^2 is known to be equal to 10. It is thus known that $\bar{X} \sim N(\mu, 1)$. Suppose that $\bar{X} = 5$ is observed. How should μ be estimated? Now, no matter what the true value of the parameter μ is, the probability of observing $\bar{X} = 5$ is zero. So we cannot think in terms of maximizing that probability. If, on the other hand, we think about a small neighborhood of the number 5, say the interval $(4.99, 5.01)$, we can then ask: Which value of μ would give this interval the largest probability of occurrence? The value $\mu = 5$ is clearly the answer. Although $P(\bar{X} \in (4.99, 5.01)|\mu = 5) = .008$ is small, this value of μ gives the interval $(4.99, 5.01)$ the largest probability of occurrence, a consequence of the fact that the mean of the normal density is also its unique *mode* (its largest value) and the density is symmetric about its mean. In the sense above, we may consider the estimate $\hat{\mu} = 5$ to be the "maximum likelihood estimate" of μ. In practice, we dispense with the process of thinking about small neighborhoods containing the observed data, since the same result can be obtained by simply maximizing the density (or joint density) of the data with respect to the parameter. In the present example, the density of $Y = \bar{X}$ is given by

$$f(y|\mu) = \frac{1}{\sqrt{2\pi}} e^{-\frac{(y-\mu)^2}{2}},$$

so that solving the equation $\frac{\partial}{\partial \mu} \ln f(y|\mu) = 0$ leads to the equation

$$y - \mu = 0.$$

This, together with the fact $\frac{\partial^2}{\partial \mu^2} \ln f(y|\mu) < 0$, identifies $\hat{\mu} = \bar{X}$ as the MLE of μ. ■

Let us now develop a general treatment motivated by the two examples above. Suppose that a random sample is available from a one-parameter model (for simplicity of exposition), i.e., suppose that $X_1, X_2, \ldots, X_n \stackrel{\text{iid}}{\sim} F_\theta$. The likelihood $L(\theta)$ of the observed data $X_1 = x_1, X_2 = x_2, \ldots, X_n = x_n$ is a function of θ proportional to the joint density or probability mass function evaluated at the observed x's. We will in fact define the likelihood as the joint density or pmf, that is,

$$L(\theta) = f(x_1, x_2, \ldots, x_n|\theta) = \prod_{i=1}^{n} f(x_i|\theta), \tag{7.18}$$

by virtue of the iid assumption. In the discrete case, $L(\theta)$ is precisely the probability that we would observe $\mathbf{X} = \mathbf{x}$, the observations actually made. In the continuous case, $L(\theta)$ represents a weight that the parameter θ gives to the observed data, and a large value of $L(\theta)$ implies that individual densities in (7.18) took relatively large values at the observed data points.

Definition 7.3.1. Suppose that $X_1, X_2, \ldots, X_n$ is a random sample drawn from the model F_θ, where θ is scalar parameter in the parameter space Θ, and suppose that $X_1, X_2, \ldots, X_n$ take on the values $x_1, x_2, \ldots, x_n$. Then the estimator $\hat{\theta}$ is said to be a *maximum likelihood estimator* (MLE) of θ if

$$L(\hat{\theta}) = \max_{\theta \in \Theta} L(\theta), \tag{7.19}$$

where L is the likelihood function defined in (7.18).

Operationally, the MLE is typically obtained by (a) finding the solution $\hat{\theta}$ of the equation $\frac{\partial}{\partial \theta} \ln L(\theta) = 0$ and (b) verifying that $\frac{\partial^2}{\partial \theta^2} \ln L(\theta)|_{\theta=\hat{\theta}} < 0$. Of course, these two steps need not identify a unique MLE, as the "likelihood equation" in step (a) may have multiple solutions, and if so, several of them may satisfy the inequality in step (b), identifying all of these as associated with *local maxima* of $L(\theta)$. If that should occur, a unique MLE will usually be found by comparing the values of L at the various candidate values of θ. There are even problems, albeit rare ones, in which the equation in step (a) has no solutions, so that there is no maximizing θ. Still, in most of the MLE hunting problems you will face, the two-step process mentioned above will work. You'll find that when dealing with certain complex models, one has to do the MLE hunting numerically, but the approach is still based on the calculus. We'll treat this topic formally in the Section 7.5. Finding MLEs in "non-regular" problems is an exception requiring a different approach. We will treat both regular and non-regular models, by example, later in this section.

The MLE of a vector-valued parameter is generally found in a way that is analogous to steps (a) and (b) above. If θ is a k-dimensional vector, then the multivariate problem of maximizing the likelihood $L(\theta)$ requires two steps, though now, the steps are more difficult and/or time-consuming to execute. The steps above generalize to

(a) finding the solution $(\hat{\theta}_1, \hat{\theta}_2, \ldots, \hat{\theta}_k)$ of the simultaneous equations

$$\frac{\partial}{\partial \theta_i} \ln L(\theta) = 0 \text{ for } i = 1, 2, \ldots, k$$

and

(b) verifying that the "Hessian" matrix

$$H = \left[\frac{\partial^2}{\partial \theta_i \partial \theta_j} \ln L(\theta)|_{\theta=\hat{\theta}} \right]_{i,j=1}^{k}$$

is "negative definite" (meaning that for any k-dimensional column vector $\mathbf{a}$ of real numbers, the product $\mathbf{a}^T H \mathbf{a}$ is negative).

In the examples and problems on MLEs that you will encounter in this book, the dimension of the parameter vector of interest will never exceed 2. It will therefore be useful to provide a simple method for checking that a 2×2 matrix is negative definite. A symmetric matrix

$$\begin{bmatrix} u & v \\ v & w \end{bmatrix}$$

is negative definite if and only if $u < 0$, $w < 0$ and $uw - v^2 > 0$.

The connection between the MLE of a parameter θ and a statistic T that is sufficient for θ deserves some comment. It can be shown (see Moore, 1971) that if a random sample $X_1, X_2, \ldots, X_n$ is drawn from a model θ indexed by the parameter θ, and if $T = h(X_1, X_2, \ldots, X_n)$ is a sufficient statistic for θ, then, when the MLE of θ is *unique*, it must be a function of T. Many textbooks in mathematical statistics "prove" the seemingly stronger result (without requiring the uniqueness of the MLE) by arguing as follows. Using the Fisher-Neyman factorization theorem, the likelihood function $L(\theta)$ may be written as

$$L(\theta) = v(t, \theta) \cdot w(x_1, x_2, \ldots, x_n).$$

Thus, the maximization of $L(\theta)$ can only depend on the data through the value of the sufficient statistic. This argument has great intuitive appeal, but is technically incorrect. The problem is that when the set Θ^* of values of θ that maximize $L(\theta)$ consists of more than one value, an MLE may be chosen arbitrarily from Θ^*, and some of these MLEs may not be functions of T alone. If the MLE is not unique, the claim above must be weakened to "there exists an estimator among the collection of MLEs which depends on the data only through T." See Problem 7.7.55 for an example showing that the uniqueness of the MLE is necessary to ensure that the MLE is a function of T. The discussion above applies for any probability model, regular or non-regular.

Looking back at Examples 7.3.1 and 7.3.2, you will see that the underlying data were, in the first instance, iid Bernoulli variables and in the second instance, iid normal variables. But the derivation of an MLE in each of these examples dealt only with the pmf or density of the corresponding sufficient statistic. One will always get "an MLE" whether one maximizes the joint density (or pmf) of the sample or the density (or pmf) of the sufficient statistic. If the distribution of the sufficient statistic is known, using it in deriving the MLE of the parameter of interest is usually the simpler alternative.

We now turn to some further examples of the derivation of maximum likelihood estimators.

Example 7.3.3. In Example 7.2.3, we derived the method of moments estimator of the parameter of an exponential distribution. We'll now revisit this problem using a likelihood approach. So again, suppose that $X_1, X_2, \dots, X_n \stackrel{\text{iid}}{\sim} Exp(\lambda)$, the exponential distribution with failure rate λ. To obtain the MLE, we begin with the likelihood function of the data, which is given by

$$L(\lambda) = \prod_{i=1}^{n} \lambda e^{-\lambda x_i} = \lambda^n e^{-\lambda \sum_{i=1}^{n} x_i}.$$

Solving the equation $\frac{\partial}{\partial\lambda} \ln L(\lambda) = 0$ leads to the equation

$$\frac{n}{\lambda} - \sum_{i=1}^{n} x_i = 0,$$

which together with the fact that $\frac{\partial^2}{\partial\lambda^2} \ln L(\lambda) = -\frac{n}{\lambda^2}$, which is negative for all $\lambda > 0$, identifies the estimator

$$\hat{\lambda} = \frac{1}{\bar{X}}$$

as the unique MLE of the parameter λ. We note that since $T = \sum_{i=1}^{n} X_i$ is a sufficient statistic for λ, and $T \sim \Gamma(n, 1/\lambda)$, one would get the same result, perhaps a few nanoseconds faster, using the density of T as the likelihood. ■

Example 7.3.4. Let $X_1, X_2, \dots, X_n \stackrel{\text{iid}}{\sim} P(\lambda)$, the Poisson distribution with mean λ. Since $T = \sum_{i=1}^{n} X_i$ is a sufficient statistic for λ, and $T \sim P(n\lambda)$ (by a moment-generating function argument), we may find the MLE by maximizing

$$L(\lambda) = \frac{(n\lambda)^t e^{-n\lambda}}{t!}.$$

Solving the equation $\frac{\partial}{\partial\lambda} \ln L(\lambda) = 0$ is equivalent to solving the equation $t/\lambda - n = 0$. Since $\frac{\partial^2}{\partial\lambda^2} \ln L(\lambda) = -t/\lambda^2 < 0$ for all $t > 0$, it follows that $\hat{\lambda} = T/n = \bar{X}$ is the MLE of λ. ■

Example 7.3.5. How might one estimate the size of a biological population? This is a real application that wildlife biologists face with some frequency. This problem differs from those considered in our previous examples because the parameter of interest is an integer N, the unknown population size. One must immediately concede that this problem is not going to be solved by differentiating some likelihood with respect to N. In addressing the problem posed, quantitative biologists themselves invented a clever way of attacking the problem. The first challenge was to determine some way of collecting relevant data. Today, the sampling method used is called the *mark-recapture* method. To be concrete, let's suppose that N is the number of trout in a pond. The information to be collected involves a preliminary sample obtained by fishing in the pond, presumably in a random fashion that gives fish in all sectors of the lake a chance to be caught. Suppose that r fish are caught in this initial sample. These fish are then marked and released into the lake. Time is allowed for the marked fish to distribute themselves (presumably at random) throughout the population. Then a second fishing experiment is launched, and n fish are caught. The crucial, highly informative fact we will rely on heavily in estimating N is the number X of marked fish in the second sample. If you were to model the random variable X, what would you do? I encourage you to give this some thought before reading on.

I hope you would choose the hypergeometric model. Indeed, assuming the randomness mentioned above, the distribution of X, given N, r, and n, is $HG(N,r,n)$, that is, for integers x in the indicated range,

$$P(X=x)=\frac{\binom{r}{x}\binom{N-r}{n-x}}{\binom{N}{n}} \text{ for } \max\{0,r+n-N\}\leq x\leq \min\{r,n\}. \tag{7.20}$$

Now, let's treat the value x in (7.20) as the observed value of X in the two-stage fishing expedition described above. Then (7.20) is both the probability of observing that particular x and the likelihood function $L(N)$ associated with the observed data. In trying to maximize $L(N)$, the following hunch seems worth pursuing. Many likelihood functions have unimodal shapes, that is, they start out small, rise to its highest value (the likelihood's mode), and then decreases thereafter. If that were true of the hypergeometric likelihood (and fortunately for us, it is), we could find the maximum value of $L(N)$ by considering ratios of the form $L(N+1)/L(N)$, starting with the smallest possible value of N. Indeed, $L(N)$ is increasing when $L(N+1)/L(N)>1$ and is decreasing when $L(N+1)/L(N)<1$. Now this ratio is given by

$$L(N+1)/L(N)=\frac{\binom{r}{x}\binom{N+1-r}{n-x}}{\binom{N+1}{n}}\cdot\frac{\binom{N}{n}}{\binom{r}{x}\binom{N-r}{n-x}}.$$

After substantial cancellation, we find that the inequality $L(N+1)/L(N)\geq 1$ is equivalent to the inequality

$$\frac{Y-r}{Y-r-n+x}\leq\frac{Y}{Y-n}, \tag{7.21}$$

where $Y=N+1$. By cross-multiplying in (7.21), an operation which is valid for all candidate values of N, this latter inequality may be rewritten as

$$Y=N+1\leq\frac{rn}{x}. \tag{7.22}$$

The MLE of the parameter N may be easily deduced from (7.22). If rn/x takes on an integer value, then $L(N+1)/L(N)=1$ when $N=rn/x-1$, and the likelihood takes the same maximum value at the dual MLEs $\hat{N}=rn/x-1$ and $\hat{N}=rn/x$. If rn/x is not an integer, then $L(N+1)/L(N)>1$ if $N<rn/x-1$ and $L(N+1)/L(N)<1$ if $N>rn/x-1$. It therefore follows that $L(N)$ is maximized at the unique MLE $\hat{N}=[rn/x]$, where $[u]$ is the greatest integer less than or equal to u. ■

Remark 7.3.1. It is interesting to compare the (first-order) MME of the parameter N in Example 7.3.5 with the MLE just derived. The method of moments estimator is much easier to get, and is almost the same as the MLE. Its only deficiency, really, is that it's not necessarily an integer, while the target parameter is. Since our data consists of the observed value of the random variable X, the first moment equation results in

$$X = \frac{rn}{N}.$$

Solving for the unknown parameter N yields the MME

$$\hat{N} = \frac{rn}{x}.$$

If asymptotics made any practical sense in this problem, we would consider the two estimators to be equivalent.

Example 7.3.6. Let us now consider a two-parameter model—the normal model with unknown mean μ and variance σ^2. Suppose that $X_1, X_2, \dots, X_n \stackrel{\text{iid}}{\sim} N(\mu, \sigma^2)$. You may recall that in Example 6.4.5, we showed that the sample mean $\bar{X}$ and sample variance s^2 are joint sufficient statistics for the parameter pair (μ, σ^2). Further, in Section 4.6, we alluded to the fact that $\bar{X}$ and s^2, computed from a normal random sample, are independent random variables, with $\bar{X} \sim N(\mu, \sigma^2/n)$ and $s^2 \sim \Gamma\left(\frac{n-1}{2}, \frac{2\sigma^2}{n-1}\right)$. We may thus find the MLE of the parameter pair (μ, σ^2) by maximizing the joint density of $\bar{X}$ and s^2 which is given by

$$L(\mu, \sigma^2 | x, y) = \left(\frac{1}{\sqrt{2\pi\sigma^2/n}}\right) e^{-\frac{(x-\mu)^2}{2\sigma^2/n}} \times \frac{1}{\Gamma((n-1)/2)\,(2\sigma^2/(n-1))^{\frac{n-1}{2}}} (y)^{\frac{(n-3)}{2}} e^{-((n-1)y/2\sigma^2)},$$

where the variables x and y represent values of $\bar{X}$ and s^2, respectively. To facilitate your taking inventory of what is happening here, we will write out $\ln L(\mu, \sigma^2 | x, y)$ term by term:

$$\begin{aligned} \ln L = &-\frac{1}{2}\ln\left(\frac{2\pi}{n}\right) - \frac{1}{2}\ln\sigma^2 - \frac{n(x-\mu)^2}{2\sigma^2} - \ln\Gamma((n-1)/2) \\ &- \frac{n-1}{2}\left[\ln\frac{2}{n-1} + \ln\sigma^2\right] + \frac{n-3}{2}\ln y - \frac{(n-1)y}{2\sigma^2}. \end{aligned}$$

From this, we can derive the likelihood equations $\frac{\partial}{\partial\mu}\ln L = 0$ and $\frac{\partial}{\partial\sigma^2}\ln L = 0$ which will, hopefully, lead us to the MLE of (μ, σ^2). To that end, we have

$$\frac{\partial}{\partial\mu}\ln L(\mu, \sigma^2 | x, y) = \frac{2(x-\mu)}{2\sigma^2} = 0, \tag{7.23}$$

an equation that tells us that any critical point $(\hat{\mu}, \hat{\sigma}^2)$ of the likelihood equations must have $\hat{\mu} = \bar{X}$. We also have

$$\frac{\partial}{\partial\sigma^2}\ln L(\mu, \sigma^2 | x, y) = -\frac{1}{2\sigma^2} + \frac{n(x-\mu)^2}{2(\sigma^2)^2} - \frac{n-1}{2\sigma^2} + \frac{(n-1)y}{2(\sigma^2)^2} = 0. \tag{7.24}$$

Since, on account of (7.23), we may replace μ in (7.24) by x (which, of course, is really $\bar{X}$), Equation (7.24) reduces to

$$-\frac{n}{2\sigma^2} + \frac{(n-1)y}{2(\sigma^2)^2} = 0,$$

from which we identify a unique solution

$$\hat{\sigma}^2 = \frac{(n-1)y}{n} = \frac{(n-1)s^2}{n} = \frac{1}{n}\sum_{i=1}^{n}(X_i - \bar{X})^2.$$

We thus have that the pair $(\hat{\mu}, \hat{\sigma}^2) = \left(\bar{X}, \frac{1}{n}\sum_{i=1}^{n}(X_i - \bar{X})^2\right)$ is the unique critical point of the likelihood equations $\frac{\partial}{\partial\mu}\ln L = 0$ and $\frac{\partial}{\partial\sigma^2}\ln L = 0$. That $(\hat{\mu}, \hat{\sigma}^2)$ is actually the MLE of (μ, σ^2) follows from the fact that the Hessian Matrix of second partial derivatives is negative definite. Verifying this latter claim is left as Problem 7.7.53. It is of course true that the derivation of the MLE in this example could be accomplished by starting with the likelihood function for the full sample. ■

Example 7.3.7. Suppose that $X_1, X_2, \dots, X_n \stackrel{\text{iid}}{\sim} U(0, \theta)$, a model whose support set depends on the unknown parameter θ. For non-regular models such as this, the MLE is not found by a calculus argument, but rather by examining the effect that varying the parameter value has on the likelihood function. For the uniform model above, the likelihood function is given by

$$L(\theta, \mathbf{x}) = \prod_{i=1}^{n}\frac{1}{\theta}I_{(0,\theta)}(x_i) = \frac{1}{\theta^n}\prod_{i=1}^{n}I_{(0,\theta)}(x_i). \tag{7.25}$$

The MLE of θ is the value of θ that makes this likelihood as large as possible. Now the term $1/\theta^n$ in (7.25) is a decreasing function of θ, so that the height of the likelihood function increases as the value of θ decreases toward 0. However, since the indicator functions on (7.25) specify that $0 < x_i < \theta$ for $i = 1, \dots, n$, any value that θ can possibly take on must be as large or larger than every observed data point. From this, we deduce that the largest possible value of $L(\theta, x)$ occurs when θ takes the value of the largest observation. Thus, the MLE of θ is $\hat{\theta} = X_{(n)}$, the largest order statistic. ■

I mentioned above that neither the existence nor the uniqueness of an MLE can be guaranteed in general, though both its existence and uniqueness can be verified in most applications of interest. There is an additional property of maximum likelihood estimators that has not as yet been mentioned—the so-called *invariance property* of MLEs. Suppose that $X_1, X_2, \dots, X_n \stackrel{\text{iid}}{\sim} F_\theta$ and that $\hat{\theta}$ maximizes the likelihood function $L(\theta, \mathbf{x})$. Suppose that g is a one-to-one function of θ. Then the likelihood function, when parameterized in terms of the parameter $\eta = g(\theta)$, is maximized by the value $\hat{\eta} = g(\hat{\theta})$. In other words, based on the same sample of observations, if $\hat{\theta}$ is the MLE of θ, then $g(\hat{\theta})$ is the MLE of $g(\theta)$. The invariance of MLEs is not surprising since, if the likelihood is reparameterized in terms of $\eta = g(\theta)$, then it may be written simply as

$$L(\theta, \mathbf{x}) \equiv L(g^{-1}g(\theta), \mathbf{x}) \equiv L(g^{-1}(\eta), \mathbf{x}).$$

Now, since $g^{-1}(\hat{\eta}) = \hat{\theta}$, it is clear that $L(g^{-1}(\eta), \mathbf{x})$ is maximized by $\eta = \hat{\eta}$. Technically, the assumption that g is one-to-one isn't necessary, though the reference above to the inverse of the function g would not be valid. It would still be true that $g(\hat{\theta})$ is the MLE of $g(\theta)$ when $\hat{\theta}$ is the MLE of θ. The one-to-one assumption about g is generally made because it ensures that the reparameterization of the model results in the same model. An example of a function which violates the assumption is the function $g(\theta) = \theta^2$. If the original sample was drawn from the $N(\theta, 1)$ model, then the reparameterization would result in the model $N(\sqrt{\theta^2}, 1)$, where the mean is taken as the positive square root of θ^2. This is a different model than the original one in which the mean could be any value on the real line. However, since $\bar{X}$ is the MLE of θ, the invariance property still holds, that is, it is still true that $\bar{X}^2$ is the MLE of θ^2. You may recall that MVUEs are examples of estimators that don't, in general, enjoy a similar invariance property.

Let us now turn to a discussion of the asymptotic behavior of MLEs. Under specific regularity conditions, the method of Maximum Likelihood Estimation produces consistent estimators of model parameters. However, since there are alternative estimation approaches that also produce consistent estimators, and, in many problems, it is not difficult to construct a rather large collection of different consistent estimators, we are generally interested in estimators which have additional properties, including asymptotic normality with a relatively small (perhaps smallest possible) asymptotic variance. The following theorem asserts that, under certain conditions on the model for the data, MLEs have such additional properties.

Theorem 7.3.1. Let $X_1, X_2, \ldots, X_n \stackrel{\text{iid}}{\sim} F_\theta$, where F_θ is a regular one-parameter model (that is, obeys regularity conditions I, II and III stated in Section 6.3), and let $I(\theta)$ be the Fisher information for this model. If $\hat{\theta}_n$ is the unique MLE of θ, then as $n \to \infty$, $\hat{\theta}_n \stackrel{p}{\longrightarrow} \theta$ and

$$\sqrt{n}\,(\hat{\theta}_n - \theta) \stackrel{D}{\longrightarrow} Y \sim N(0, \frac{1}{I(\theta)}). \tag{7.26}$$

The proof of Theorem 7.3.1 is beyond the scope of this book. It is, however, a landmark development in the theory of estimation, and we will invest some time here understanding its meaning and applying the result in a variety of modeling scenarios. Let me first mention that this theorem is often stated without the requirement that the MLE is unique. A more general version of this result states simply that there exists a solution of the likelihood equation that is root-n consistent and whose asymptotic distribution is given in (7.26). In practice, uniqueness is not often an issue of concern, and the theorem, as stated above, will typically apply. This will be true, in particular, for the examples and problems you'll encounter in this book. Further, it is important to acknowledge that the theorem above extends to MLEs of vector-valued parameters. In that case, under suitable regularity conditions, a properly scaled vector of MLEs will be asymptotically normal with a covariance matrix equal to the inverse of the information matrix (for definitions and a formal statement of this extension, see Lehmann and Casella (1998, Section 6.5)).

Theorem 7.3.1 should be interpreted as identifying the large sample distribution of the MLE $\hat{\theta}_n$ when the sample size n is sufficiently large. There is, however, no theorem which provides guidance on how large n has to be for this approximation to be reliable. Experience suggests that, for many models and estimators, the approximation is good for sample sizes as small as 30, but it is known that for some models and estimators (for example, for $\hat{p}$ as an estimator of a binomial proportion p that is quite close to 0 or 1), sample sizes exceeding 100 or 1000 may be required. Let us assume that the approximation is reasonably accurate for a given sample size in a problem of interest. In that event, the theorem says that the large-sample distribution of the MLE $\hat{\theta}_n$ is given by

$$\hat{\theta}_n \sim N\left(\theta, \frac{1}{nI(\theta)}\right), \text{ approximately.} \tag{7.27}$$

If the distribution in (7.27) was exact, it would tell us that the MLE $\hat{\theta}_n$ is an unbiased estimator of θ, and further, that its variance is equal to the Cramér-Rao Lower Bound for the variance of unbiased estimators. Thus, when the sample size is large, the MLE behaves like the MVUE of the parameter θ. Since there are many problems in which the analytical derivation of the MVUE of a parameter of interest is difficult or impossible, it is notable that the MLE accomplishes the performance achieved by the MVUE to an extent that grows to perfection as the sample size grows to ∞.

For models that satisfy the standard regularity conditions alluded to in Theorem 7.3.1, the fastest possible rate of convergence is $O_p(1/\sqrt{n})$. Further, among estimators which converge at this rate and which are asymptotically normal, the MLE attains the smallest possible asymptotic variance. An estimator with these properties is referred to as the best asymptotically normal (BAN), and an estimator that achieves the smallest possible asymptotic variance would be referred to as asymptotically efficient. It would thus be appropriate to view the MLE as the gold standard, when n is larger, for estimators of the parameters of such models. If we are able to derive the MLE in closed form when estimating the parameter of a regular model, it is reasonable to use that estimator, or recommend it for use, when the sample size is considered to be "sufficiently large." The calculation of the estimator's asymptotic distribution is needed for further inferential work and would of course be executed in parallel. Other estimators need not be considered. When, for a regular model, the derivation of the MLE proves to be analytically intractable, it is common to use numerical methods to evaluate the estimator for the sample one has in hand. A widely used iterative method for finding the MLE numerically is treated in Section 7.5. To complete our discussion of MLE theory, we turn to several examples of the derivation of the asymptotic distribution of an MLE.

Example 7.3.8. Suppose that $X_1, X_2, \ldots, X_n \stackrel{\text{iid}}{\sim} Exp(\lambda)$. In Example 7.3.3, we showed that the maximum likelihood estimator of the failure rate λ is

$$\hat{\lambda} = \frac{1}{\bar{X}}.$$

This estimator is also the MME of λ, and its asymptotic distribution was obtained in Example 7.2.3 employing MME theory. Here, we'll obtain the AD of the estimator using MLE theory, as stated in Theorem 7.3.1. The derivation of the Fisher information for this model proceeds as follows. Since $\ln f(X|\lambda) = \ln\lambda - \lambda X$, it follows that

$$I(\lambda) = E\left(\frac{\partial}{\partial\lambda}\ln f(X|\lambda)\right)^2 = E\left(\frac{1}{\lambda} - X\right)^2 = V(X) = \frac{1}{\lambda^2}.$$

Therefore, in agreement with MME theory, we have that

$$\sqrt{n}\left(\hat{\lambda} - \lambda\right) \xrightarrow{D} W \sim N\left(0, \lambda^2\right).$$

■

Example 7.3.9. In Example 4.2.2, we encountered a version of the geometric distribution that differs slightly from the model we refer to as $G(p)$. The latter model is the distribution of the random variable Y, defined as the number of independent Bernoulli trails, each with probability of success p, that are required to obtain the first success. Here, we will be interested in the variable X, the number of failures that occur before the first success is obtained. The pmf of the variable X is given by

$$p(x) = p(1-p)^x \text{ for } x = 0, 1, 2, \ldots$$

Let's refer to this version of the geometric model as $G^*(p)$. If $X \sim G^*(p)$, then $EX = (1-p)/p$ and $V(X) = (1-p)/p^2$. Now, suppose that $X_1, X_2, \ldots, X_n \overset{\text{iid}}{\sim} G^*(p)$. To obtain the maximum likelihood estimator of p, we must solve the likelihood equation $\frac{\partial}{\partial p} \ln L(p) = 0$ which is given by

$$\frac{n}{p} - \frac{\sum_{i=1}^n x_i}{1-p} = 0. \tag{7.28}$$

The unique solution of (7.28) is

$$\hat{p} = \frac{n}{n + \sum_{i=1}^n x_i}.$$

The fact that $\frac{\partial^2}{\partial p^2} \ln L(p) = -\frac{n}{p^2} - \frac{\sum_{i=1}^n x_i}{(1-p)^2} < 0$ for all $p \in (0,1)$ confirms that the estimator

$$\hat{p} = \frac{1}{1+\bar{X}}$$

maximizes the log-likelihood function $\ln L(p)$ and is the unique MLE of p. Let us now derive the asymptotic distribution of $\hat{p}$. The Fisher information $I(p)$ for this model is obtained as

$$\begin{aligned} I(p) &= E\left(\frac{\partial}{\partial p} \ln f(X|p)\right)^2 \\ &= E\left(-\frac{X}{1-p} + \frac{1}{p}\right)^2 \\ &= \frac{1}{(1-p)^2} E\left(X - \frac{1-p}{p}\right)^2 \\ &= \frac{1}{(1-p)^2} V(X) \\ &= \frac{1}{(1-p)p^2}. \end{aligned}$$

The asymptotic distribution of the MLE $\hat{p}$ is thus given by

$$\sqrt{n}(\hat{p} - p) \xrightarrow{D} V \sim N(0, (1-p)p^2). \tag{7.29}$$

■

Exercises 7.3.

1. Suppose that $X_1, X_2, \dots, X_n \stackrel{\text{iid}}{\sim} G(p)$, the geometric distribution with parameter p. Find the maximum likelihood estimator $\hat{p}$ of p, and identify the asymptotic distribution of $\hat{p}$.

2. Suppose that $X_1, X_2, \dots, X_n \stackrel{\text{iid}}{\sim} N(0, \sigma^2)$. Find the maximum likelihood estimator $\hat{\sigma}^2$ of σ^2 and identify the asymptotic distribution of $\hat{\sigma}^2$.

3. Let $X_1, X_2, \dots, X_n \stackrel{\text{iid}}{\sim} P(\alpha, \theta)$, the Pareto distribution with power parameter α and scale parameter θ whose density is given in Equation (3.36). Find the MLE of the parameter pair (α, θ). (Note: For any fixed value of θ, find the MLE $\hat{\alpha}$ of α as you would in the one-parameter problem with θ known. Since $\hat{\alpha}$ does not depend on θ and maximizes the likelihood for each θ, one may substitute $\hat{\alpha}$ for α in the likelihood, and solve for the MLE of θ in the model $P(\hat{\alpha}, \theta)$, treating it as a one-parameter model.)

4. Let $X_1, X_2, \dots, X_n \stackrel{\text{iid}}{\sim} Be(\theta, 1)$, the Beta distribution treated in Example 7.2.5. Derive the maximum likelihood estimator of the parameter θ, and identify its asymptotic distribution. Compare the asymptotic variance of the MLE with that of the MME derived in Example 7.2.5. Verify that the MLE is superior asymptotically to the MME for all $\theta > 0$.

7.4 A Featured Example: Maximum Likelihood Estimation of the Risk of Disease Based on Data from a Prospective Study of Disease

In his classic text, Fleiss (1981) defines a prospective study as follows: "The comparative prospective study (also termed the *cohort* study) is characterized by the identification of the two study samples on the basis of the presence or absence of the antecedent factor and by the estimation for both samples of the proportions developing the disease or condition under study." We might, for example, be interested in the possible association between a particular disease and an individual's previous exposure to a suspected cause. Prospective studies may be designed in quite a few different ways, and could, in a complex study, involve multiple controls, use of matching or stratification of subjects and the accounting of a variety of confounding factors. We will treat a simple but useful version of a prospective study that foregoes the accommodation of these latter design elements. Let us assume that the data available for the prospective study of interest consist of two independent samples of Bernoulli random variables. Such a study would be appropriate in situations in which the sampled populations are seen as reasonably homogeneous, perhaps already controlled for potential differences by restricting the age, gender, and/or race of the subjects studied. This method of sampling differs from a *cross-sectional study*, which is based on a multinomial sample of a given size drawn from the population of interest, and also differs from a *retrospective (or case-control) study* in which one draws random samples of individuals who have a disease and of individuals who do not, and then one records the number of subjects in each sample who have been exposed to a particular antecedent factor. For more detail on retrospective studies, see Breslow and Day (1980).

As indicated above, we will assume that the data compiled for the study of interest here may be displayed as counts associated with the presence or absence of a certain type of previous exposure (E or E^c) and the presence or absence of a disease (D or D^c) of interest.

Our notation for such data is displayed in Table 7.4.1, where $n_{i\bullet}$ and $n_{\bullet j}$ represent row and column totals, respectively.

	E	E^c	
D	n_{11}	n_{12}	$n_{1\bullet}$
D^c	n_{21}	n_{22}	$n_{2\bullet}$
	$n_{\bullet 1}$	$n_{\bullet 2}$	$n_{\bullet\bullet}$

Table 7.4.1. Data from a simple prospective study.

The probabilities associated with each cell in the table above are denoted as in the table below, with the marginal probabilities of exposure or disease denoted in the standard dot notation.

	E	E^c	
D	p_{11}	p_{12}	$p_{1\bullet}$
D^c	p_{21}	p_{22}	$p_{2\bullet}$
	$p_{\bullet 1}$	$p_{\bullet 2}$	1

Table 7.4.2. Cell probabilities in a prospective study.

The notation above is the typical setup for drawing a random sample of size $n_{\bullet\bullet}$ from a multinomial population divided into four cells. However, we will now adapt this notation so that it is applicable when drawing two independent binomial samples (one each from the E and the E^c populations.) Henceforth, we will assume that separate and independent random samples were drawn from individuals with and without exposure, that is, we assume that

$$n_{11} \sim B(n_{\bullet 1}, p_{11}/p_{\bullet 1}) \quad \text{and} \quad n_{12} \sim B(n_{\bullet 2}, p_{12}/p_{\bullet 2}). \tag{7.30}$$

In Fleiss's language, disease is viewed as the outcome factor and exposure is viewed as the antecedent factor. The theoretical odds for getting the disease are

$$\frac{P(D|E)}{P(D^c|E)} = \frac{p_{11}/p_{\cdot 1}}{p_{21}/p_{\cdot 1}} = \frac{p_{11}}{p_{21}} \text{ for exposed subjects,}$$

and

$$\frac{P(D|E^c)}{P(D^c|E^c)} = \frac{p_{12}/p_{\cdot 2}}{p_{22}/p_{\cdot 2}} = \frac{p_{12}}{p_{22}} \text{ for unexposed subjects.}$$

The *odds ratio*, denoted by ω, is given by

$$\omega = \frac{P(D|E)/P(D^c|E)}{P(D|E^c)/P(D^c|E^c)} = \frac{p_{11}/p_{21}}{p_{12}/p_{22}} = \frac{p_{11}p_{22}}{p_{12}p_{21}}. \tag{7.31}$$

The odds ratio is a standard measure of association (i.e., dependence) between exposure and disease. Another measure, the relative risk r of contracting the disease depending on the presence or absence of exposure, is given as

$$r = \frac{P(D|E)}{P(D|E^c)} = \frac{p_{11}/p_{\cdot 1}}{p_{12}/p_{\cdot 2}} = \frac{p_{11}p_{\cdot 2}}{p_{12}p_{\cdot 1}}. \tag{7.32}$$

If the occurrence of the event D is unlikely, whether or not the subject has been exposed (that is, whether or not the event E has occurred)—an assumption that is called the "rare disease hypothesis"—then both $p_{21}/p_{\bullet 1}$ and $p_{22}/p_{\bullet 2}$ are very close to 1 and

$$\omega = \frac{p_{11}p_{22}}{p_{12}p_{21}} \approx \frac{p_{11}p_{\cdot 2}}{p_{12}p_{\cdot 1}} = r, \tag{7.33}$$

justifying the approximation of r by ω. Cornfield (1951), who introduced both measures, argued that the possibility of approximating r under the rare disease hypothesis provided strong motivation for the careful study of the parameter ω. But the odds ratio has interpretive value in other settings as well, and it is quite widely used today in the analysis of data from prospective studies.

The natural estimator of the odds ratio ω is the sample odds ratio $\hat{\omega}$ given by

$$\hat{\omega} = \frac{n_{11}n_{22}}{n_{21}n_{12}}. \tag{7.34}$$

Under the assumptions above, $\hat{\omega}$ is the maximum likelihood estimator of ω. This follows from the fact that the proportion $n_{ij}/n_{\bullet j}$ is the MLE of $p_{ij}/p_{\bullet j}$ for $i = 1,2$ and $j = 1,2$, and

$$\hat{\omega} \quad = g(n_{11}/n_{\bullet 1}, n_{21}/n_{\bullet 1}, n_{12}/n_{\bullet 2}, n_{22}/n_{\bullet 2}),$$

where $g(u,v,w,z) = uz/vw$. The first time I read Fleiss's book, I was a bit perplexed by the claim, stated without proof, that the standard error of the estimator $\hat{\omega}$ was approximately equal to

$$s.e.(\hat{\omega}) \quad = \quad \hat{\omega}\sqrt{\frac{1}{n_{11}} + \frac{1}{n_{12}} + \frac{1}{n_{21}} + \frac{1}{n_{22}}}, \tag{7.35}$$

provided that sample sizes $n_{\bullet 1}$ and $n_{\bullet 2}$ are sufficiently large. You may be feeling a little bewildered by this claim yourself! I will now provide justification for the approximate standard error in (7.34), albeit under the simplifying assumption that $n_{\bullet 1} = n = n_{\bullet 2}$. This assumption is not entirely innocuous, as it stipulates that our prospective study is balanced, that is, that we have included an equal number of exposed individuals (Es) and unexposed individuals (E^cs) in the study. (The unbalanced case is the subject of Problem 7.7.59.)

Let us now demystify (7.35) with some straightforward "δ-methoding." I'll begin by simplifying our notation. Let $\pi_1 = P(D|E)$ and $\pi_2 = P(D|E^c)$, and let $\hat{\pi}_1 = n_{11}/n$ and $\hat{\pi}_2 = n_{12}/n$; we may then express ω and $\hat{\omega}$, respectively, as

$$\omega = \frac{\pi_1/(1-\pi_1)}{\pi_2/(1-\pi_2)} \quad \text{and} \quad \hat{\omega} = \frac{\hat{\pi}_1/(1-\hat{\pi}_1)}{\hat{\pi}_2/(1-\hat{\pi}_2)}. \tag{7.36}$$

Further, we note that, by the Central Limit Theorem, as n grows to ∞,

$$\sqrt{n}(\hat{\pi}_1 - \pi_1) \xrightarrow{D} U_1 \sim N(0, \pi_1(1-\pi_1))$$

and

$$\sqrt{n}(\hat{\pi}_2 - \pi_2) \xrightarrow{D} U_2 \sim N(0, \pi_2(1-\pi_2)).$$

Now, for $x \in (0, 1)$, let g be the function defined as

$$g(x) = \ln\left(\frac{x}{1-x}\right).$$

Then $g'(x) = 1/x(1-x)$, and the δ-method theorem implies that, for $j = 1,2$,

$$\sqrt{n}\left(\ln\frac{\hat{\pi}_j}{1-\hat{\pi}_j} - \ln\frac{\pi_j}{1-\pi_j}\right) \xrightarrow{D} V_j \sim N\left(0, \frac{1}{\pi_j(1-\pi_j)}\right) = N\left(0, \frac{1}{\pi_j} + \frac{1}{1-\pi_j}\right). \quad (7.37)$$

If you sense that we're getting close, you're absolutely right! Note that we can write

$$\ln(\hat{\omega}) = \ln\left(\frac{\hat{\pi}_1}{1-\hat{\pi}_1}\right) - \ln\left(\frac{\hat{\pi}_2}{1-\hat{\pi}_2}\right).$$

Using the independence of $\hat{\pi}_1$ and $\hat{\pi}_2$, we may identify the asymptotic distribution of $\ln(\hat{\omega})$ as

$$\sqrt{n}\,(\ln(\hat{\omega}) - \ln(\omega)) = \sqrt{n}\left(\ln\left(\frac{\hat{\pi}_1}{1-\hat{\pi}_1}\right) - \ln\left(\frac{\pi_1}{1-\pi_1}\right)\right) - \sqrt{n}\left(\ln\left(\frac{\hat{\pi}_2}{1-\hat{\pi}_2}\right) - \ln\left(\frac{\pi_2}{1-\pi_2}\right)\right)$$

$$\xrightarrow{D} V_1 + V_2 = W \sim N\left(0, \frac{1}{\pi_1} + \frac{1}{1-\pi_1} + \frac{1}{\pi_2} + \frac{1}{1-\pi_2}\right),$$

or, equivalently (in our original notation), as

$$\sqrt{n}\,(\ln(\hat{\omega}) - \ln(\omega)) \xrightarrow{D} W \sim N\left(0, \frac{1}{p_{11}/p_{\cdot 1}} + \frac{1}{p_{21}/p_{\cdot 1}} + \frac{1}{p_{12}/p_{\cdot 2}} + \frac{1}{p_{22}/p_{\cdot 2}}\right). \quad (7.38)$$

So far, we have identified the asymptotic variance of $\ln(\hat{\omega})$. However, what we set out to obtain was the asymptotic variance of $\hat{\omega}$ itself. To obtain the latter, we apply the δ-method yet again. (Hey, your old favorite trick is getting jealous!) Consider the transformation $h(x) = e^x$. We have that $h'(x) = e^x$ and thus, by the δ-method theorem, it follows that

$$\sqrt{n}(\hat{\omega} - \omega) \xrightarrow{D} Y \sim N\left(0, \left(e^{\ln\omega}\right)^2\left(\frac{1}{p_{11}/p_{\cdot 1}} + \frac{1}{p_{21}/p_{\cdot 1}} + \frac{1}{p_{12}/p_{\cdot 2}} + \frac{1}{p_{22}/p_{\cdot 2}}\right)\right),$$

that is,

$$\sqrt{n}(\hat{\omega} - \omega) \xrightarrow{D} Y \sim N\left(0, \omega^2\left(\frac{1}{p_{11}/p_{\cdot 1}} + \frac{1}{p_{21}/p_{\cdot 1}} + \frac{1}{p_{12}/p_{\cdot 2}} + \frac{1}{p_{22}/p_{\cdot 2}}\right)\right). \quad (7.39)$$

It is clear from (7.39) that the standard error of $\hat{\omega}$, for large n, is approximately equal to

$$s.e.(\hat{\omega}) = \omega\sqrt{\frac{1}{np_{11}/p_{\cdot 1}} + \frac{1}{np_{21}/p_{\cdot 1}} + \frac{1}{np_{12}/p_{\cdot 2}} + \frac{1}{np_{22}/p_{\cdot 2}}}. \quad (7.40)$$

Finally, replacing ω and $p_{ij}/p_{\bullet j}$ in (7.40) by their MLEs $\hat{\omega}$ and $n_{ij}/n_{\bullet j}$ for $i = 1,2$ and $j = 1,2$, we arrive at the estimated standard error of $\hat{\omega}$ as given in (7.35), that is,

$$\widehat{s.e.}(\hat{\omega}) = \hat{\omega}\sqrt{\frac{1}{n_{11}} + \frac{1}{n_{21}} + \frac{1}{n_{12}} + \frac{1}{n_{22}}}. \quad (7.41)$$

Remark 7.4.1. $[P(E|D)/P(E^c|D)]/[P(E|D^c)/P(E^c|D^c)]$ is the odds ratio in retrospective studies, an expression that appears to differ from that in a prospective study. However, one can show algebraically that it is equal to the odds ratio ω as defined in (7.31). Thus, the odds ratio ω may be estimated from the data obtained in a retrospective study. The estimator $\hat{\omega}$ in (7.34), based on data from a retrospective study, is asymptotically normal with a standard error that may again be estimated as in (7.41).

Exercises 7.4.

1. The odds of getting a disease for the exposed group is

$$O_E = \frac{P(D|E)}{P(D^c|E)}, \tag{7.42}$$

while the odds of getting a disease for the unexposed group is

$$O_{E^c} = \frac{P(D|E^c)}{P(D^c|E^c)}. \tag{7.43}$$

Show that the odds ratios in (7.42) and (7.43) are equal if and only if $P(D|E) = P(D|E^C)$.

2. Show that the odds ratio ω in a prospective study exceeds 1 if and only if

$$P(D|E) > P(D|E^C).$$

3. Consider a retrospective study involving smoking and lung cancer. Independent random samples were drawn, one consisting of 200 lung cancer patients (C) and the other consisting of 200 individuals who are free of lung cancer. Each group was separated into smokers (S) and non-smokers; hypothetical data appears below.

	S	S^c	
C	140	60	200
C^c	50	150	200
	120	180	400

(a) Calculate the sample odds ratio and interpret it as a measure of the risk of smoking. (b) Calculate the estimated standard error of $\hat{\omega}$ and use it to provide a 95% confidence interval for ω.

4. The attributable risk r_E is defined as the fraction of all occurrences of a condition (e.g., D in Table 7.4.1) due to exposure to a specified risk factor (e.g., E in Table 7.4.1). The parameter r_E may be estimated from a retrospective study under the "rare disease assumption," since then, the odds ratio ω is approximately equal to the relative risk r and the conditional probability $P(D|E^c)$ is approximately equal to $P(D)$. (a) Show that, under the assumption above, r_E may be calculated as

$$r_E = \frac{P(D|E) - P(D|E^C)}{1 - P(D|E^C)}.$$

From the data displayed below, obtain an estimate of the risk r_E attributable to exposure.

	E	E^c	
D	10	10	20
D^c	150	330	480
	160	340	500

7.5 The Newton-Raphson Algorithm

Calculus is a powerful tool in optimization problems. We have, both in univariate and multivariate problems, a precise prescription for maximizing a function L of interest. The prescription often provides a closed form solution, that is, an explicit representation of a value of the argument $\mathbf{x}$ at which the function $L(\mathbf{x})$ attains a global maximum or a global minimum, whichever corresponds to one's goals. However, there are many problems, some of which arise even when the function L does not appear to be particularly complicated. The process of deriving the maximum likelihood estimator involves obtaining the solution(s) of one or more likelihood equations, a step which may itself be analytically intractable. It is no surprise that there are many systems of equations one could write down for which an exact analytical solution either doesn't exist or is exceedingly difficult to obtain. Even when dealing with problems that are analytically tractable, the verification that a particular solution maximizes the likelihood, globally or even locally, can itself be a substantial challenge. It is therefore useful to have some familiarity with numerical algorithms that can give increasingly accurate approximations of values at which a function is maximized. Among the "hill-climbing" methods in common use, perhaps the most widely known and regularly applied method is the *Newton-Raphson Algorithm.*

In its original form, the Newton-Raphson method (also sometimes referred to as Newton's method) was intended to find the roots or zeros of a function f, that is, the values of x for which $f(x) = 0$. It is instructive to examine the algorithm in that context. Consider a function $f(x)$ with derivative $f'(x)$. Take x_0 as an initial guess at a zero of $f(x)$. We then construct a series of iterations resulting in the updated values $x_1, x_2, x_3, \ldots$, where

$$x_i = x_{i-1} - \frac{f(x_{i-1})}{f'(x_{i-1})} \text{ for } i = 1,2,3,\ldots \tag{7.44}$$

To see why this algorithm might work, let's look at the iteration's first step. Suppose that x_0 is reasonably close to x^*, a value satisfying $f(x^*) = 0$. If $f(x)$ is increasing in an interval $(x^* - c, x^* + c)$ which contains x_0, and $x_0 < x^*$, then $f(x_0) < 0$ and $f'(x_0) > 0$, so that $x_1 > x_0$, that is, the iteration moves x_0 in the direction of x^*. If $x_0 > x^*$, then, for similar reasons, $x_1 < x_0$. Further, since $f(x^*) = 0$, $f(x)$ is close to zero when x is close to x^*. This indicates that, if x_i is close to x^*, the algorithm adjusts x_i only slightly, and if it actually reaches x^*, no further adjustment is made. In practice, the process is terminated when the difference $|x_i - x_{i-1}|$ becomes suitably small. The algorithm is known to converge to x^* at a quadratic rate, provided that the initial value x_0 is sufficiently close to x^* and the function f is sufficiently smooth. Convergence at a quadratic rate means that

$$|x_{i+1} - x_i| \le |x_i - x_{i-1}|^2.$$

A proof of the quadratic convergence rate of the algorithm above can be found on Wikipedia under the subject "The Newton-Raphson Algorithm." The following numerical example is drawn from that site.

Example 7.5.1. Consider the problem of finding the square root of the number 612. This is equivalent to finding a solution to the equation $x^2 = 612$, and it may also be thought of as the problem of finding zeros of the function $f(x) = x^2 - 612$. We know that there are two roots of the equation $f(x) = 0$, one positive and one negative. Suppose we're interested in the positive root. Let's start with the initial guess $x_0 = 10$ (as Wikipedia did—maybe they're not as smart as we all thought!). But we'll do it to show you that the Newton-Raphson method is pretty amazing. (OK, maybe that's what Wikipedia had in mind too.) Noting that $f'(x) = 2x$, we apply (7.44) for five iterations. The results are

$$x_0 = 10,$$

$$x_1 = 10 - \frac{10^2 - 612}{2 \cdot 10} = 35.6$$

$$x_2 = (35.6) - \frac{(35.6)^2 - 612}{2 \cdot (35.6)} = 26.395505618,$$

$$x_3 = (26.39\ldots) - \frac{(26.39\ldots)^2 - 612}{2 \cdot (26.39\ldots)} = 24.7906354025,$$

$$x_4 = (24.79\ldots) - \frac{(24.79\ldots)^2 - 612}{2 \cdot (24.79\ldots)} = 24.73868829408,$$

$$x_5 = (24.73\ldots) - \frac{(24.73\ldots)^2 - 612}{2 \cdot (24.73\ldots)} = 24.73863375377.$$

The fifth iteration is accurate to the tenth decimal place in approximating $\sqrt{612}$. ■

If $f(x)$ is a smooth function, searching for values of x that maximize (or minimize) $f(x)$ is generally done by finding the zeros of the function $f'(x)$. If we start with an initial value $x = x_0$, then the Newton-Raphson iterations will take the form

$$x_i = x_{i-1} - \frac{f'(x_{i-1})}{f''(x_{i-1})} \text{ for } i = 1, 2, 3, \ldots \tag{7.45}$$

As with the iteration in (7.44), the adjustment in (7.45) moves the current value toward a particular critical point of the function $f(x)$ provided that the initial value x_0 is sufficiently close to the critical point and the function f is sufficiently smooth.

We will now turn our attention to application of the Newton-Raphson algorithm for obtaining numerical approximations of maximum likelihood estimators. Suppose that $L(\theta)$ is the likelihood function in a statistical estimation problem involving a one-parameter model F_θ. Let $l(\theta) = \ln L(\theta)$, and let θ_0 be the initial guess at the MLE $\hat{\theta}$ of θ. It is worth mentioning that it is common to utilize the method-of-moments estimator of θ as an initial value, as it is likely to be close to the MLE, and its own consistency has some beneficial effects of the NR iterates that follow. The first Newton-Raphson iterate θ_1 is given by

$$\theta_1 = \theta_0 - \frac{l'(\theta_0)}{l''(\theta_0)}.$$

Interestingly, the first iteration of the Newton-Raphson algorithm has some notable properties of its own. It turns out that, if the initial value θ_0 of θ is sufficiently close to the MLE, and the likelihood function is sufficiently smooth, the first iterate θ_1 will behave, essentially, like the MLE itself. More specifically, when F_θ is a regular model and the initial value $\hat{\theta}_0(n)$ of θ (which is now assumed to be a function of the data $x_1, x_2, \ldots, x_n$) is a root-n

consistent estimator of θ, that is, if $\hat{\theta}_0(n) = O_p(\frac{1}{\sqrt{n}})$, then $\hat{\theta}_1(n)$ is asymptotically normal and asymptotically efficient

$$\sqrt{n}\left(\hat{\theta}_1(n) - \theta\right) \xrightarrow{D} Y \sim N\left(0, \frac{1}{I(\theta)}\right).$$

If one is determined to identify the MLE of θ, then, under the usual conditions on the smoothness of $L(\theta)$ and on the closeness of the initial value θ_0 to the MLE, the sequence of Newton-Raphson iterates will converge to the MLE at a quadratic rate. We conclude this section with a brief example of an application of the algorithm.

Example 7.5.2. Let $X_1, X_2, \ldots, X_n \stackrel{\text{iid}}{\sim} Be(\theta, 1)$, the Beta distribution treated in Example 7.2.5. The log likelihood function for this model is given by

$$\ln L(\theta) = n \ln \theta + (\theta - 1) \prod_{i=1}^{n} x_i.$$

In Example 7.2.5, we derived the MME $\hat{\theta} = \bar{X}/(1 - \bar{X})$ of θ. For this model, the MLE can be obtained in closed form (unlike the situation one faces with the two-parameter beta model). We will, nonetheless, apply the Newton-Raphson algorithm for illustration purposes, using the MME above as an initial value. The first Newton-Raphson iterate is

$$\begin{aligned}
\hat{\theta}_1 &= \theta_0 - \frac{\frac{n}{\theta_0} + \ln \prod_{i=1}^{n} x_i}{-\frac{n}{\theta_0^2}} \\
&= 2\theta_0 + \frac{\theta_0^2}{n} \cdot \ln \prod_{i=1}^{n} x_i \\
&= 2\left(\frac{\bar{X}}{1-\bar{X}}\right) + \frac{1}{n} \cdot \left(\frac{\bar{X}}{1-\bar{X}}\right)^2 \cdot \ln \prod_{i=1}^{n} x_i.
\end{aligned}$$

This estimator is a consistent and asymptotically efficient estimator of θ.

Code to do this in the R programming language might look like the following.

```
theta0 <- mean(xi) / (1- mean(xi))
l.prime <- function(theta, n, xi) { n/theta + log(prod(xi)) }
l.2prime <- function(theta, n) { -n / theta^2 }
newt.raph.sqrt <- function(theta, times){
  if (times==0) return(theta)
  newt.raph.sqrt(theta - l.prime(theta)/l.2prime(theta), times-1)
     }
newt.raph.sqrt(theta0, times)
```

The program sets an initial value, then runs through each iteration of the algorithm, decreasing the number of remaining iterations by one each time. When no iterations remain, the function returns the final value. ■

Exercises 7.5.

1. Use the Newton-Raphson algorithm to find the value of x for which the cosine of x is equal to x^3, that is, find the zero of the function $f(x) = cos(x) - x^3$. Since $|cos(x)| \leq 1$ for all x, $cos(x)$ and x^3 have opposite signs for $-1 < x < 0$ and $|x^3| > 1$ for $|x| > 1$, one may surmise that the solution $x \in (0,1)$. Take the initial value of x to be $x_0 = .5$.

2. It is well known that, if the initial value x_0 is not sufficiently close to a zero of a function $f(x)$, the Newton-Raphson algorithm might not converge. Let $f(x) = x^3 - 2x + 2$. Show that the Newton-Raphson algorithm for finding a zero of the function $f(x)$ using the initial value $x_0 = 0$ leads to an infinite loop that oscillates indefinitely between $x = 0$ and $x = 1$.

3. Let $X_1, \ldots, X_n \stackrel{\text{iid}}{\sim} \Gamma(\alpha, \beta)$. The maximum likelihood estimator of the parameter pair (α, β) cannot be obtained analytically, as there is no closed for expression of the derivative of $\Gamma(\alpha)$. In applying the Newton-Raphson algorithm for finding the MLE, one may reduce the problem to a one-parameter problem by noting that, given α, the MLE of the scale parameter is $\hat{\beta} = \bar{X}/\alpha$. (a) Write out an expression for the log-likelihood function $\ln L$ in this problem with the parameter β replaced by $\bar{X}/\alpha$. (b) Obtain expressions for the partial derivatives $(\partial/\partial\alpha)\ln L$ and $(\partial^2/\partial\alpha^2)\ln L$. (c) Obtain an explicit expression for the first iterate from the Newton-Raphson algorithm for estimating the parameter α using the method of moments estimator $\bar{X}^2/\left[1/n\left(\sum_{i=1}^n X_i^2 - n\bar{X}^2\right)\right]$, as the initial estimate α_0 of α. (Note: When applying Newton-Raphson iterations to data, tabled values of $\Gamma'(\alpha)$ and $\Gamma''(\alpha)$ are used.) (d) One thousand observations were generated from the $\Gamma(2, 1/3)$ model, resulting in the statistics $n = 1000$, $\ln \Pi X_i = -646.0951$, $\bar{X} = 0.6809364$ and $s^2 = 0.2235679$. Carry out a series of NR iterations which demonstrate convergence to the MLE $\hat{\alpha} = 2.060933$.

7.6 A Featured Example: Maximum Likelihood Estimation from Incomplete Data via the EM Algorithm

The title of this section (starting with "Maximum ...") is the same as the title of the much-cited research article by Dempster, Laird, and Rubin (DLR) (1977). In that paper, a lovely iterative technique for finding the MLE, which up to that time had been used in a variety of special cases, was shown to be quite general, remarkably broad in scope, and often quite easy to apply, even in seemingly difficult problems in which analytically deriving the MLE was basically impossible. The EM algorithm is an approach to solving a sequence of problems that are "similar" to the unsolvable problem you started with. Under specific conditions, the sequence of solutions converges to the MLE in the original problem—a surprising and highly satisfying outcome. The "E" in the EM algorithm stands for "expectation" and the "M" stands for "maximization." The introduction I provide here is just the tip of the iceberg, as the theory and application of the method has grown by leaps and bounds since the publication of the DLR paper. I will explain the EM algorithm through a particular application of its use. In the featured example, simple as it is, the main properties of the algorithm reveal themselves quite clearly.

The data to be treated was featured in the celebrated text by C. R. Rao (1965). The data consists of the classification of 197 animals into four categories. The model for these data is the multinomial model $M_4(197, \mathbf{p})$, where the probability vector $\mathbf{p}$ has a particular form, being a function of a single parameter which Rao denoted by π. The cell probabilities were determined by a particular genetic model which stipulated that the vector $\mathbf{p}$ was given by

$$\mathbf{p} = (1/2+\pi/4, (1-\pi)/4, (1-\pi)/4, \pi/4) \quad \text{for some } \pi \in [0,1].$$

The vector of observed frequencies is $\mathbf{y} = (125, 18, 20, 34)$. Given these data, the likelihood function, which is simply the multinomial probability of $\mathbf{y}$, is given by

$$L(\pi|\mathbf{y}) = \frac{197!}{125! \cdot 18! \cdot 20! \cdot 34!} \left(\frac{1}{2}+\frac{\pi}{4}\right)^{125} \left(\frac{1-\pi}{4}\right)^{18} \left(\frac{1-\pi}{4}\right)^{20} \left(\frac{\pi}{4}\right)^{34}. \tag{7.46}$$

It is clear that maximizing $\ln L(\pi|\mathbf{y})$ is not an easy matter. The troublesome term in the likelihood is the factor $(1/2+\pi/4)$; it is this particular term in $\ln L(\pi|\mathbf{y})$ that makes the equation $\partial/\partial\pi(\ln L(\pi|\mathbf{y})) = 0$ impossible to solve analytically. But an iterative solution is possible if we view the vector $\mathbf{y}$ as "incomplete data." To do this, we need to imagine what a model for the "complete data" looks like. Postulating such a model is no slam dunk in general, since there is often more than one complex model from which the vector $\mathbf{y}$ could have been obtained. But it is often the case that a simple model for the "complete data" can be identified. In the present case, let's suppose that the complete data was obtained from a larger multinomial model. Specifically, we will assume that the complete data, denoted by $\mathbf{x}$, is the observed outcome of the multinomial variable $\mathbf{X}$ specified as

$$\mathbf{X} \sim M_5(197, \mathbf{p}^*), \tag{7.47}$$

where $\mathbf{p}^* = (1/2, \pi/4, (1-\pi)/4, (1-\pi)/4, \pi/4)$. You might wonder why we would choose this particular model. The answer is that the model we started with can be obtained by combining the first two cells of the model in (7.47). In other words, the troublesome term in (7.46) is dealt with in the $M_5(197, \mathbf{p}^*)$ model by separating the first cell of the "incomplete data" into separate cells with the respective probabilities $1/2$ and $\pi/4$. It would not be an overstatement to refer to $M_5(197, \mathbf{p}^*)$ as the "dream model" in the present problem. If only we could have observed $\mathbf{x}$ instead of $\mathbf{y}$, we would be able to solve for the MLE of π in a flash. Given the likelihood function

$$L(\pi|\mathbf{x}) = \frac{n!}{x_1! \cdot x_2! \cdot x_3! \cdot x_4 \cdot x_5!} \left(\frac{1}{2}\right)^{x_1} \left(\frac{\pi}{4}\right)^{x_2} \left(\frac{1-\pi}{4}\right)^{x_3} \left(\frac{1-\pi}{4}\right)^{x_4} \left(\frac{\pi}{4}\right)^{x_5}, \tag{7.48}$$

it is evident that L is proportional to the core of a binomial likelihood, that is,

$$L(\pi|\mathbf{x}) \propto \pi^{x_2+x_5}(1-\pi)^{x_3+x_4}.$$

It follows that the MLE $\hat{\pi}(\mathbf{x})$ of π based on $\mathbf{X} = \mathbf{x}$ is

$$\hat{\pi}(x) = \frac{x_2+x_5}{x_2+x_3+x_4+x_5}. \tag{7.49}$$

The contrast in the ease of obtaining the MLE of π in the two scenarios examined above is striking. What remains is to explain the tool that allows us connect these two scenarios.

As with virtually all iterative methods and algorithms, one needs to have a seed, that is, an initial value, in this case a guess at π, that allows us to get the iteration started. Suppose we agree on the initial value $\pi^{(0)}$. The EM algorithm consists of a series of repetitions of two steps, the *estimation* step and the *maximization* step. Our routine will be to "estimate," at stage i, the complete data $\mathbf{x}^{(\mathbf{i})}$ from the incomplete data $\mathbf{y}^{(\mathbf{i})}$, together with the current estimate (or guess) $\pi^{(i-1)}$ of the parameter π. Looking at the models $M_4(197,\mathbf{p})$ and $M_5(197,\mathbf{p}^*)$ above, the expectation step "estimates," by $E(\mathbf{X}|\mathbf{y},\pi^{(i-1)})$, the value of $\mathbf{x}$ that we would expect to observe if we were able to sample from the model for the complete data. In the smaller model, we see that 125 animals were classified into a cell with probability $1/2+\pi^{(i-1)}/4$, utilizing our current estimate of π. If we were to split these 125 animals into separate cells with respective probabilities $1/2$ and $\pi^{(i-1)}/4$, the expected values on x_1 and x_2 (using well-known facts about the multinomial distribution) are

$$x_1^{(i)} = \frac{y_1^i(1/2)}{(1/2)+(\pi^{(i-1)}/4)} \text{ and } x_2^{(i)} = \frac{y_1^i(\pi^{(i-1)}/4)}{(1/2)+(\pi^{(i-1)}/4)}. \tag{7.50}$$

In the example under discussion here, it is clear that because of the matching cell probabilities in the $M_4(197,\mathbf{p})$ and $M_5(197,\mathbf{p}^*)$ models, we may assign values to the remaining x's as

$$x_3 = y_2, x_4 = y_3 \text{ and } x_5 = y_4. \tag{7.51}$$

Having estimated the complete data $\mathbf{x}$ from the expectation step, we proceed to the maximization step using the complete data model. But as we saw in (7.48) and (7.49), the maximization step for the complete data model is a piece of cake.

For concreteness, let's take the initial guess at π to be $\pi^{(0)} = .5$. Then, by (7.50) and (7.51), the complete data at stage 1 of the EM algorithm is

$$\mathbf{x} = (100, 25, 18, 20, 34).$$

The maximization step here is accomplished by using the general formula for the MLE in (7.49) to obtain

$$\pi^{(1)} = \frac{25+34}{25+18+20+34} = .608247423. \tag{7.52}$$

The level of accuracy shown in (7.52) simply reflects the fact that one would generally maintain considerable accuracy as the iteration proceeds. After enough iterations, we would hope for the iterates to approximate the MLE to 8–10 decimal places.

It can be shown, using the Newton-Raphson algorithm (see Problem 7.7.60), that the maximum likelihood estimate of π based on the given observed data $\mathbf{y} = (125, 18, 20, 34)$ drawn from the original model $M_4(197,\mathbf{p})$ is $\hat{\pi} = .6268214980$. In the table below, the first eight iterations of the EM algorithm, with the initial value $\pi^{(0)} = .5$, are shown. The highly satisfactory convergence of successive values of $\{\pi^{(i)}\}$ is reflected in the negligibly small difference $\pi^{(8)} - \hat{\pi}$.

i	$\pi^{(i)}$	$\pi^{(i)} - \hat{\pi}$
0	.500000000	.126821498
1	.608247423	.018574075
2	.624321051	.002500447
3	.626488879	.000332619
4	.626777323	.000044176
5	.626815632	.000005866
6	.626820719	.000000779
7	.626821395	.000000104
8	.626821484	.000000014

Table 7.6.1. Iterates of the EM algorithm for estimating a parameter π in (7.46).

Using the R programming language, an implementation of this example might look like this:

```
y <- c(125, 18, 20, 34); pi0 <- 0.5
EM <- function(y, pi, times){
  if (times == 0) return(pi)
  x1 <- (y[1] * .5) / (.5 + pi/4)
  x2 <- (y[1] * pi / 4) / (.5 + pi/4)
  pi <- (x2 + y[4]) / (x2 + y[2] + y[3] + y[4])
  EM(y, pi, times-1) }
EM(y, pi0, 8)
```

The function computes new values for x_1 and x_2 using the current best guess of π, then computes a new best guess using the new values of x_1 and x_2.

The following theoretical facts are known about the EM algorithm. The algorithm may be beneficially applied to a wide class of multi-parameter models, that is, the basic parameter θ of the model for incomplete data may be k-dimensional. Under mild conditions on the model for the incomplete data, the likelihood function increases monotonically as a function of successive iterations, that is, for all integers $i \geq 0$, $L(\theta^{(i)}|\mathbf{y}) \leq L(\theta^{(i+1)}|\mathbf{y})$. There are a variety of circumstances in which the convergence of the EM iterates $\{\theta^{(i)}. i = 0, 1, 2, \dots\}$ can be rigorously proven. Unfortunately, the convergence theorem presented in the original paper contained an unfixable error. Boyles (1983) provided a counterexample to that theorem and also stated and proved a convergence theorem under stronger assumptions. A widely cited reference on the convergence of the EM algorithm is the paper by Wu (1983). It is worth mentioning that the EM algorithm has proven useful in estimation problems in which the "incomplete data" feature of the original model is not present or is unimportant. One application of this alternative sort is to the problem of finding the mode of the posterior distribution in Bayesian analysis. More importantly, there are a good many sampling frameworks in statistical applications that can be fruitfully viewed as examples of incomplete data and may benefit from being treated through the use of the EM algorithm. Included among these frameworks are problems involving missing data, censoring, truncation, mixture models and latent variable models, among others. Bringing the attention of the Statistics community to such a broad array of applications of the EM algorithm constitutes one of the most valuable contributions of DLR (1977).

It is worth mentioning that the EM algorithm does involve a few troublesome features. There are number of published examples which show that the EM algorithm will diverge in certain situations. This fact underscores the need, in applications of the approach, to check that conditions under which the EM algorithm converges are satisfied. In problems in which the likelihood function has multiple maxima, it is entirely possible that the EM algorithm will converge to a local, non-global maximum. In such settings, the careless selection of the initial value of the parameter of the missing data model may lead to misleading conclusions. Running the algorithm with a variety of initial values will often offer protection against this potential problem. It should be noted that the expectation step need not produce a set of values of the complete data that can actually be taken on by real data from the complete data model. Indeed, in iterations 2–8 in the example above, the expectation step produces values of $x_1^{(i)}$ and $x_2^{(i)}$ that are not integers. Finally, it has been noted that the convergence of the EM algorithm can be, in some applications, painfully slow. Literature exists on methods of accelerating the convergence of the algorithm. The rate of convergence is known to be affected by the "complete data model" one selects. The Newton-Raphson algorithm, which is known to converge at a quadratic rate in the neighborhood of a maximum, will converge substantially more quickly than the EM algorithm when the initial value is judiciously chosen.

Exercises 7.6.

1. Suppose twelve observations are taken at random from a bivariate normal distribution with zero means. The data, with missing values marked as "*", is shown below.

X	1	1	−1	−1	2	2	−2	−2	*	*	*	*
Y	1	−1	1	−1	*	*	*	*	2	2	−2	−2

(a) Show that the likelihood function for these data has a saddle point at $\rho = 0$, $\sigma_1^2 = \sigma_2^2 = 5/2$ and two maxima at $\rho = \pm 1/2$ and $\sigma_1^2 = \sigma_2^2 = 8/3$. (b) Show that if the initial value of ρ is 0, the EM algorithm will converge to the saddle point. It can be shown that the EM algorithm will otherwise converge to one of the maxima, depending on the proximity on the starting point to each maximum.

2. Consider the 100 observations from following multinomial model with four cells governed by a parameter $p \in (0, .2)$

Cell	1	2	3
probability	$p+.2$	$p+.4$	$.4-2p$
frequencies	20	50	10

Use the EM algorithm to approximate the MLE of p. Take $p_0 = .1$ as an initial value.

7.7 Chapter Problems

1. Let $\{T_n\}$ be a sequence of estimators of the parameter θ, and suppose that

$$\text{(a) } V(T_n) \to 0 \text{ as } n \to \infty \text{ and } (b) ET_n \to \theta \text{ as } n \to \infty.$$

Show that $T_n \xrightarrow{p} \theta$.

2. Let $X_1, \ldots, X_n, \ldots$ be a sequence if iid $B(1, p)$ variables. Consider the statistic defined as

$$T_n = \frac{\sum_{i=1}^n X_i + \sqrt{n}/2}{n + \sqrt{n}}.$$

Show that $T_n \xrightarrow{p} p$. Can you identify the asymptotic distribution of $\sqrt{n}(T_n - p)$?

3. Let T_n be a consistent estimator of the parameter θ, and suppose that $\{a_n\}$ and $\{b_n\}$ are constants such that $a_n \to 1$ and $b_n \to 0$. Show that $a_n T_n + b_n \xrightarrow{p} \theta$.

4. Two sequences of estimators $\{S_n\}$ and $\{T_n\}$ are said to be asymptotically equivalent if $\sqrt{n}(S_n - T_n) \xrightarrow{D} 0$. Suppose that $\sqrt{n}(S_n - \theta) \xrightarrow{D} Y \sim N(0, 1)$. Show that each of the following sequences are asymptotically equivalent to $\{S_n\}$:

   (a) $b_n S_n + (1 - b_n)d$, for any constant d and any sequence $\{b_n\}$ such that $b_n \to 1$,
   (b) $(S_n)^{n/(n+k)}$, for any fixed real number k, and
   (c) $nS_n/(n+k)$, for any fixed real number k.

5. Show that, if $\sqrt{n}(S_n - \theta) \xrightarrow{D} Y \sim N(0, 1)$, then $S_n \xrightarrow{p} 0$.

6. Let $X_1, \ldots, X_n \overset{\text{iid}}{\sim} U(\theta, 2\theta)$ for some $\theta > 0$. (a) Show that $\hat{\theta}_1 = X_{(n)}/2$ is the maximum likelihood estimator of n. (b) Verify that $\hat{\theta}_1$ is a biased estimator of θ but that the estimator $\hat{\theta}_2 = (n+1)X_{(n)}/(2n+1)$ is unbiased for θ. (c) Show that the estimator $\hat{\theta}_3 = (n+1)(2X_{(n)} + X_{(1)})/(5n+4)$ is an unbiased estimator of θ with a uniformly smaller variance that $\hat{\theta}_2$. (d) Show that all three estimators above are asymptotically equivalent in the sense defined in Problem 7.7.4.

7. Let $X_1, \ldots, X_n \overset{\text{iid}}{\sim} N(\mu, \sigma^2)$, and let $T_n = \bar{X}_n + c/\sqrt{n}$. Show that the estimator T_n is a consistent estimator of μ but that it is asymptotically biased (that is, $E(T_n)$ does not converge to μ).

8. For each positive integer i, let $p_i \in (0, 1)$, and suppose that there exist a positive real number p such that $(1/n)\sum_{i=1}^n p_i \to p$. Let $\{X_i\}$ be a sequence of independent Bernoulli variables, with $X_i \sim B(1, p_i)$. Letting $\hat{p}_n = (1/n)\sum_{i=1}^n X_i$, show that $\hat{p}_n \xrightarrow{p} p$. Comment on the claim that this may be a more realistic model for repeated coin tosses than the usual one.

9. Let $X_1, \ldots, X_n \overset{\text{iid}}{\sim} F_{\alpha,\beta}$, a distribution with density

$$f(x) = \frac{\alpha x^{\alpha-1}}{\beta^\alpha} I_{(0,\beta)}(x).$$

(a) Find the joint method of moments estimator of the parameter pair (α, β). (b) Find the maximum likelihood estimators of α and β.

10. Let $X_1,\ldots,X_n \stackrel{\text{iid}}{\sim} U(\alpha,\beta)$, the uniform distribution on the interval (α,β). Obtain a method of moments estimator (MME) the maximum likelihood estimator (MLE) of the parameters α and β.

11. Australian immigrant Mel Bourne isn't very good at darts, but he just loves to throw darts at little star-shaped targets at the California State Fair. Suppose that X is the number of times Mel misses the targets before he finally hits one (and wins a stuffed Koala bear that will make him less homesick). The pmf of X is

$$p(x) = p(1-p)^x, x = 0,1,2,\ldots.$$

Note that if $Y \sim G(p)$ (the usual geometric), then $X = Y - 1$. Thus, $EX = (1-p)/p$ and $V(X) = (1-p)/p^2$. (a) Suppose that we wish to estimate p based on the observed value of X. As it turns out, the MME (from the 1st moment equation) and the MLE of p are identical. Derive this estimator $\hat{p}$ by one or the other method. (b) Find the asymptotic distribution of $\hat{p}$ using MME theory (i.e., the CLT, etc.). (c) Redo part (b), this time deriving the asymptotic distribution of $\hat{p}$ using MLE theory. (If you get different answers, check your work!)

12. Suppose $X_1, X_2,\ldots,X_n \stackrel{\text{iid}}{\sim} P(\theta,1)$, where $P(\theta,1)$ is the one-parameter Pareto distribution with density

$$f(x) = \theta/x^{\theta+1} \text{ for } 1 < x < \infty.$$

Assume that $\theta > 2$, so that the model $P(\theta,1)$ has finite mean $\theta/(\theta-1)$ and variance $\theta/(\theta-1)(\theta-2)^2$. (a) Obtain the MME $\hat{\theta}_1$ from the first moment equation and the MLE $\hat{\theta}_2$. (b) Obtain the asymptotic distributions of these two estimators. (c) Show that the MLE is asymptotically superior to the MME.

13. Suppose that $X_1, X_2\ldots,X_n \stackrel{\text{iid}}{\sim} Be(\theta,2)$, the beta distribution with density

$$f(x) = (\theta+1)\theta x^{\theta-1}(1-x) \text{ for } 0 < x < 1.$$

(a) Derive the method of moments estimator (MME) $\hat{\theta}_1$ of θ. (b) Identify the asymptotic distribution of $\sqrt{n}(\hat{\theta}_1 - \theta)$. (c) Identify the maximum likelihood estimator (MLE) $\hat{\theta}_2$ of θ as the positive root of a certain quadratic equation. (d) Identify the asymptotic distribution of $\sqrt{n}\,(\hat{\theta}_2 - \theta)$. (e) Show that the MLE of θ is (asymptotically) uniformly superior to the MME, i.e., show that $AV(\hat{\theta}_1) > AV(\hat{\theta}_2)$ for all $\theta > 0$, where "AV" stands for asymptotic variance.

14. Suppose that $X_1,\ldots,X_n \stackrel{\text{iid}}{\sim} F_\theta$, the "double exponential" distribution with mean 0 and scale parameter θ and with density

$$f(x) = \frac{1}{2\theta} e^{-\frac{|x|}{\theta}} I_{(-\infty,\infty)}(x),$$

where θ is a positive parameter. Take as given: $EX = 0$, $EX^2 = \theta^2$, $EX^3 = 0$, and $EX^4 = 24\theta^4$. Also, note that $Y = |X| \sim \Gamma(1,\theta)$. (a) Solve the "$2^{nd}$ moment equation" to obtain an "MME" of θ. (b) Identify the asymptotic distribution of $\sqrt{n}(\hat{\theta}_1 - \theta)$. (c) Find the MLE $\hat{\theta}_2$ of θ, verifying that it really maximizes the likelihood. (d) Identify the asymptotic distribution of $\sqrt{n}(\hat{\theta}_2 - \theta)$. (e) Show that the MLE of θ is (asymptotically) uniformly superior to the MME.

15. Let $W(\theta)$ be the one-parameter Weibull model with density

$$f(x) = \frac{2x}{\theta} e^{-x^2/\theta} I_{(0,\infty)}(x).$$

It is known that $EX = \sqrt{\pi\theta}/2$ and that $EX^2 = \theta$. Suppose that $X_1, \ldots, X_n \stackrel{iid}{\sim} W(\theta)$. (a) Using the "first moment equation," obtain a "method of moments estimator" (i.e., an MME) $\hat{\theta}_1$ of θ. (b) Identify the asymptotic distribution of $\sqrt{n}(\hat{\theta}_1 - \theta)$. (c) Find the maximum likelihood estimator $\hat{\theta}_2$ of θ, verifying that it really maximizes the likelihood. (d) Identify the asymptotic distribution of $\sqrt{n}(\hat{\theta}_2 - \theta)$. (e) Show that $\hat{\theta}_2$ is asymptotically superior to $\hat{\theta}_1$. Note, however, that $\hat{\theta}_2$ is also the solution to the"second moment equation," so that the MLE of θ is also an MME. This provides an example in which a higher-order MME is superior to a lower-order MME.

16. Let $X_1, \ldots, X_n \stackrel{iid}{\sim} \Gamma(1, \mu)$ and let $Y_1, \ldots, Y_n \stackrel{iid}{\sim} \Gamma(1, \nu)$, where the Xs are independent of the Ys. Suppose that

$$U_i = \min\{X_i, Y_i\} \text{ and } V_i = \begin{cases} 1 & \text{if } U_i = X \\ 0 & \text{if } U_i = Y. \end{cases}$$

Find the MLEs of μ and ν based on a collection of iid pairs (U_i, V_i).

17. Let $X_1, \ldots, X_n \stackrel{iid}{\sim} \Gamma(1, \beta, \theta)$, the three-parameter gamma distribution with density given in (3.24), with $\alpha = 1$. (a) Find the method of moments estimators of β and θ. (b) Find the maximum likelihood estimators of β and θ.

18. Let $X_1, \ldots, X_n \stackrel{iid}{\sim} N(\theta, \theta)$. Show that the MLE of θ is given by

$$\hat{\theta} = \frac{1}{2}\left[\left(1 + \frac{4}{n}\sum_{i=1}^{n} X_i^2\right)^{1/2} - 1\right].$$

19. Let $X_1, \ldots, X_n \stackrel{iid}{\sim} Be(\theta + 1, 1)$. Derive the MLE of θ, and identify its asymptotic distribution.

20. Let $X_1, \ldots, X_n \stackrel{iid}{\sim} Be(\theta + 1, 1)$. Derive the MLE of $\ln(\theta)$, and identify its asymptotic distribution.

21. Let $X_1, \ldots, X_n \stackrel{iid}{\sim} Be(2, \theta)$. Derive the MLE of θ, and identify its asymptotic distribution.

22. Let $X_1, \ldots, X_n \stackrel{iid}{\sim} B(k, p)$, where both k and p are unknown. Obtain estimators of k and p by the method of moments.

23. Let $X_1, \ldots, X_n \stackrel{iid}{\sim} F_\theta$. Suppose you wish to construct a consistent estimator of $E\left(1/\sqrt{X}\right)$. (a) If the form of F_θ is unknown, how would you go about estimating this expectation. (b) If F_θ is the $\Gamma(1, \theta)$ distribution, how would you estimate $E\left(1/\sqrt{X}\right)$.

24. Prove Stein's Lemma: If $X \sim N(\mu, \sigma^2)$ and g is a differentiable function with $E(g'(X)) < \infty$, then $E[g(X)(X - \mu)] = \sigma^2 E(g'(X))$. (Hint: Use integration by parts with $u = g(x)$ and $dv = (x - \mu)e^{-(x-\mu)^2/2\sigma^2} dx$. The fact that $\int_{-\infty}^{\infty} v du = 0$ follows from the assumption that $E(g'(X)) < \infty$.)

25. In Example 6.4.8, assuming the availability of a random sample of iid system lifetimes distributed according to an exponential distribution with mean μ, we derived the MVUE $(1-x/\bar{X})^{n-1}$ of the system reliability $R(x)=e^{-x/\mu}$, that is, of the probability that an individual system will survive beyond time x. Now, since $\bar{X}$ is the MLE of μ, we have, by the invariance property of MLEs, that $\hat{R}(x)=e^{-x/\bar{X}}$ is the MLE of $R(x)$. As it happens, the estimator $\hat{R}(x)$ is also the MME of $R(x)$. (a) Obtain the asymptotic distribution of $\hat{R}(x)$ from MLE theory. (b) Obtain the asymptotic distribution of $\hat{R}(x)$ from MME theory.

26. Let $X \sim N(\mu,\sigma^2)$. Use Stein's Lemma (see Exercise 7.7.24) to calculate EX^3 from the expression $EX^2(X-\mu)$. Now we know that the estimator $m_2-m_1^2$, derived from the first two moment equations, is an method of moments estimator of σ^2. Use the version of EX^3 obtained from Stein's Lemma to show that the estimator $(m_3-m_1m_2)/2m_1$ is also a method of moments estimator of σ^2. Is one superior to the other?

27. (Neyman and Scott) For each fixed $i=1,\dots,n$ and each $j=1,\dots,k$, suppose that $X_{ij} \sim N(\mu_i,\sigma^2)$. Show that the estimator

$$\hat{\sigma}^2=\frac{\sum_{i,j}(X_{ij}-\bar{X}_i)^2}{kn}$$

is the MLE of the parameter σ^2, where $\bar{X}_i=\sum_{j=1}^{k}X_{ij}$, and that, for any fixed value of k, it is inconsistent for σ^2 as $n\to\infty$.

28. Suppose the failure times of light bulbs are modeled as $X_1,\dots,X_n \overset{\text{iid}}{\sim} \Gamma(1,\mu)$. Suppose that data is observed under "type I censoring," that is, all the light bulbs are observed until time T at which time the experiment is terminated. The observed sample may thus be represented as $\{Y_i\}$, where $Y_i=\min(X_i,T\}$. Suppose that all n light bulbs last beyond time T. The likelihood of this event is

$$L=\prod_{i=1}^{n}e^{-T/\mu}=e^{-nT/\mu}.$$

Show that the maximum likelihood of μ does not exist.

29. Let $X_1,\dots,X_n \overset{\text{iid}}{\sim} U(\theta,\theta+1)$. Show that the MLE of θ based on these data is not unique, and there is an interval of real numbers $[a,b]$ such that every $\theta\in[a,b]$ is an MLE of θ.

30. Let $X_1,\dots,X_n \overset{\text{iid}}{\sim} P(\lambda)$. Show that the MLE of λ doesn't exist if every observed X is equal to zero. Assuming that at least one X_i is non-zero, find the MLEs of λ, $1/\lambda$, $e^{-\lambda}$, and $\lambda e^{-\lambda}$. Identify the asymptotic distribution of each of these estimators.

31. Let $X_1,\dots,X_n \overset{\text{iid}}{\sim} N(0,\sigma^2)$. Find the MLE of σ^2. Note that it is also an MME of σ^2. Find the asymptotic distribution of this estimator (a) by MLE theory and (b) by the central limit theorem directly.

32. Suppose that $X_1,\dots,X_n \overset{\text{iid}}{\sim} DE(\mu,1)$, the "double exponential" distribution with mean μ and scale parameter 1 and with density

$$f(x)=\frac{1}{2}e^{-|x-\mu|}I_{(-\infty,\infty)}(x).$$

Find an estimator of μ by the method of moments, and identify its asymptotic distribution.

33. Suppose that $X_1,\ldots,X_n \overset{\text{iid}}{\sim} DE(\mu,1)$, the "double exponential" distribution with mean μ and scale parameter 1 and with density

$$f(x) = \frac{1}{2}e^{-|x-\mu|}I_{(-\infty,\infty)}(x).$$

Find the MLE of μ. (Hint: Write the joint density of the data in terms of the ordered data $x_{(1)},\ldots,x_{(n)}$. Treat the cases of odd n and even n separately.)

34. Let $X_1,\ldots,X_n \overset{\text{iid}}{\sim} PS(\theta)$, the power series distribution with probability mass function

$$p(x) = c_x\theta^x/g(\theta) \text{ for } x = 0,1,2\ldots, \tag{7.53}$$

where $g(\theta) = \sum_{x=0}^{\infty} c_x\theta^x$. (a) Show that the MLE of θ is a root of the equation

$$\bar{x} = \frac{\theta g'(\theta)}{g(\theta)}, \tag{7.54}$$

where $\bar{x} = (1/n)\sum_{i=1}^{n} x_i$, (b) Show that the RHS of (7.54) is the mean of the $PS(\theta)$ distribution (proving that the MLE of θ for this class of models is also the MME of θ).

35. Let $(X_1,Y_1),\ldots,(X_n,Y_n) \overset{\text{iid}}{\sim} BVN(\mathbf{0},\mathbf{1},\rho)$, the bivariate normal distribution with $\mu_x = 0 = \mu_y$, $\sigma_x^2 = 1 = \sigma_y^2$ and correlation coefficient ρ. Find the maximum likelihood estimator of ρ and identify its asymptotic distribution.

36. Let X be a random variable that has distribution F_1 with probability p and has distribution F_2 with probability $(1-p)$, where F_1 and F_2 have respective densities

$$f_1(x) = I_{(0,1)}(x) \text{ and } f_2(x) = \frac{1}{2\sqrt{x}}I_{(0,1)}(x).$$

Given the single observation X, find the MLE of the mixing parameter p.

37. Let $(X_1,Y_1),\ldots,(X_n,Y_n) \overset{\text{iid}}{\sim} BVN(\mu,\sigma^2,\rho)$, the bivariate normal distribution with means μ_x and μ_y, variances σ_x^2 and σ_y^2, and correlation coefficient ρ. Find the maximum likelihood estimators of these 5 parameters.

38. Suppose that $X_1,X_2,\ldots,X_n \overset{\text{iid}}{\sim} P(\lambda)$, the Poisson distribution with mean λ. Since λ is both the mean and variance of the Poisson distribution, there are two natural MMEs for λ, $\hat{\lambda}_1 = m_1$ from the first moment equation and $\hat{\lambda}_2 = (-1+\sqrt{1+4m_2})/2$ from the second moment equation (since we're only interested in the positive root of this quadratic equation). Which one has the smaller asymptotic variance. (Hint: You'll need to find the fourth moment of a $P(\lambda)$ variable X. You can get it either from the mgf of X, or you can google the Poisson distribution. Either approach is fine.) The calculation of AVs aside, is there a good reason to suspect that one of these estimators is better than the other?

39. Suppose that $(X_1,\ldots,X_k) \sim M_k(n,(p_1,\ldots,p_k))$, the multinomial distribution with parameters n and $\mathbf{p}$. Show that the vector $(X_1/n,\ldots,X_k/n)$ is the MLE of $(p_1,\ldots,p_k)$. Verify that this estimator truly maximizes the likelihood function.

40. Let $X_1,X_2,\ldots,X_n \overset{\text{iid}}{\sim} C(m,1)$, the Cauchy distribution with median m (see (3.39)). The MLE of m is the value that minimizes $\prod_{i=1}^{n}[1+(x_i-m)^2]$. As this product is a polynomial of degree $2n$, the MLE cannot, in general, be found analytically. For the data: $-2,5,10,12,25$, find the MLE of θ using the Newton-Raphson algorithm.

41. Let $X \sim HG(N,r,n)$. Suppose that N and n are known. Find the maximum likelihood estimator of the parameter r.

42. Just for practice, consider applying the Newton-Raphson algorithm in obtaining a numerical approximation to the MLE of the parameter of the Pareto distribution with density

$$f(x) = \frac{\theta}{(1+x)^{\theta+1}} I_{(0,\infty)}(x).$$

Obtain an expression for the first Newton-Raphson iterate using the method of moments estimator of θ as a seed. You may assume that it is known that $\theta > 1$.

43. Let $X_1, X_2, \dots, X_n \overset{\text{iid}}{\sim} L(\mu, \beta)$, the logistic distribution with density given in (3.41). Take the scale parameter β to be known and equal to 1. Show that the likelihood equation for estimating μ may be written as

$$\sum_{i=1}^{n} \frac{1}{1+e^{(x_i-\mu)}} = \frac{n}{2}. \tag{7.55}$$

Show that the left-hand side of (7.55) is an increasing function of μ taking the value 0 at $\mu = -\infty$ and the value n at $\mu = \infty$. Conclude that the likelihood has a unique global maximum and, thus, that the MLE is unique. Find a suitable approximation to the MLE for the data below:

$$10, 25, 40, 45, 60, 80, 90.$$

(Hint: Try interval-halving.)

44. The variable X has a convoluted Poisson distribution if $X = Y + Z$, where Y and Z are independent discrete random variables, with $Y \sim P(\lambda)$. Suppose that Z has a geometric $(G(p))$ distribution with p known. Denote the distribution of X as $P + G(p)$. Let $X_1, X_2, \dots, X_n \overset{\text{iid}}{\sim} P + G(p)$. Show that the maximum likelihood estimator of λ is unique, and corresponds to the unique root of a decreasing function of λ. (Hint: Show that $(\partial/\partial\lambda)\ln L(\lambda) = 0$ may be written as

$$\sum_{i=1}^{n} \frac{P_\lambda(X_i = x_i - 1)}{P_\lambda(X_i = x_i)} = n \tag{7.56}$$

and that each of the fractions in (7.56) are decreasing in λ.

45. Suppose that a random sample of size n is drawn from a mixture distribution with density

$$f(x) = p \cdot e^{-x} + (1-p) \cdot (1/2)e^{-x/2}.$$

The density describes an experiment in which, on any given draw, you will obtain a $\Gamma(1,1)$ variable with probability p and a $\Gamma(1,2)$ variable with probability $1-p$. Note that the likelihood function is a polynomial in p of order n. Suppose the data obtained in a small experiment is: .3,.8.1.2,2.2,5. Use the Newton-Raphson algorithm to approximate the MLE of p. Can the EM algorithm be applied to this problem?

46. Hardy-Weinberg equilibrium describes the situation in which genotypes AA, Aa, and aa occur in a population with respective probabilities $(1-p)^2$, $2p(1-p)$, and p^2 for some $p \in (0,1)$. Suppose that in a particular population of n individuals, X_1 individuals have genotype AA, X_2 have genotype Aa, and X_3 have genotype aa. Find the maximum likelihood estimator of p as a function of X_1, X_2, and X_3.

47. The zero-truncated Poisson distribution has pmf

$$p(x) = \frac{\lambda^x e^{-\lambda}}{x!(1-e^{-\lambda})} \text{ for } x = 1, 2, \ldots \tag{7.57}$$

Let's denote this distribution by $P^*(\lambda)$. Suppose that $X_1, X_2, \ldots, X_n \stackrel{\text{iid}}{\sim} P^*(\lambda)$. The MLE of λ cannot be obtained analytically. Use the Newton-Raphson algorithm to obtain an expression for the first NR iterate using $\bar{X}$ as an initial estimate of λ.

48. Suppose that $X_1, \ldots, X_n \stackrel{\text{iid}}{\sim} F$, a distribution depending on two parameters α and β. Assume that the mean, second moment, and variance of the distribution are all functions of both parameters. Show that the method of moments estimators derived by the seemingly different estimating equations (a) $m_1 = \mu_1$ and $m_2 = \mu_2$ and (b) $m_1 = \mu_1$ and $s^2 = \sigma^2$ are the same estimators.

49. The cosine of the angle at which electrons are emitted in a decaying system has been modeled as a random variable X with density function

$$f(x) = \frac{1+\theta x}{2} I_{(-1,1)}(x) \text{ for } \theta \in (-1, 1).$$

Suppose that $X_1, \ldots, X_n \stackrel{\text{iid}}{\sim} F$, where F has the density above. (a) Obtain the method of moments estimator $\hat{\theta}_1$ of θ. (b) The MLE of θ cannot, in general, be obtained in closed form. Derive an expression for the first Newton-Raphson iterate using the MME as an initial estimator of θ.

50. Censored data is common in "life testing" experiments in the field of engineering reliability. Suppose n items are put "on test." The goal is to estimate the failure-time distribution of the items from the observed failures of the items on test. From a practical standpoint, waiting for all n items to fail is inconvenient, as the expected waiting time for the occurrence of the last of n failures tends to be rather large. Type II censoring (often called order-statistic censoring) is the process of terminating the experiment upon the occurrence of the rth item failure for some $r < n$. Let's suppose, henceforth, that individual item lifetimes are modeled as exponential variables, that is, suppose that $X_1, X_2, \ldots, X_n \stackrel{\text{iid}}{\sim} \Gamma(1, \theta)$.

    (a) By evaluating the integral $\int_{x_{(r)}}^{\infty} \cdots \int_{x_{(n-2)}}^{\infty} \int_{x_{(n-1)}}^{\infty} n! \prod_{i=1}^{n} f(x_{(i)}) dx_{(n)} dx_{(n-1)} \ldots dx_{(r+1)}$, show that the joint density, and thus the likelihood function, for the available exponential data is given by

    $$f_{X_{(1)},\ldots,X_{(r)}}(x_{(1)}, \ldots, x_{(r)}) = \frac{n!}{(n-r)!} \frac{1}{\theta^r} e^{-(\sum_{i=1}^{r} x_{(i)} + (n-r)x_{(r)})/\theta} \text{ for } x_{(1)} < \cdots < x_{(r)}. \tag{7.58}$$

    (b) Show that the sum $\sum_{i=1}^{r} x_{(i)} + (n-r)x_{(r)}$ is a sufficient statistic for θ.

    (c) For $i = 1, \ldots, r$, let $Y_i = X_{(i)} - X_{(i-1)}$, where $X_{(0)} \equiv 1$. Note that the sum $\sum_{i=1}^{r} x_{(i)} + (n-r)x_{(r)}$ may be written $\sum_{i=1}^{r} (n-i+1)Y_i$. Using transformation theory (note that $|J| = 1$), obtain the joint density of $(Y_1, \ldots, Y_r)$. Examine this density and conclude that the variables $nY_1, \ldots, (n-r+1)Y_r$ are iid $\Gamma(1, \theta)$ variables.

    (d) Derive the MLE $\hat{\theta}_{r,n}$ of the parameter θ. Expressing the ordered Xs in terms of the Ys, confirm that $\hat{\theta}_{r,n} \sim \Gamma(r, \theta)$.

(e) Show that the MLE is an unbiased estimator of θ and that the variance of $\hat{\theta}_{r,n}$ is equal to the CRLB for unbiased estimators of θ based on $X_{(1)},\ldots,X_{(r)}$, and is thus the MVUE of θ. (Hint: Obtain the CRLB using the fact that $(n-i+1)Y_i$, $i=1,\ldots,r$, are iid $\Gamma(1,\theta)$ variables.)

Remark 7.7.1. One of the lovely outcomes of the developments above is that the MLE of θ based on the first r order statistics form a sample of size n has exactly the same distribution as the MLE of θ derived from a complete sample of size r. So what, you ask? Well, think about this. If you place n items on test and only have to wait until the first r fail, your experiment will end earlier, perhaps substantially earlier, than if you have to wait for all r failures in a sample of size r. For example, you'll have to wait, on average, 8 times longer to observe the fifth failure in a sample of size 5 than to observe the fifth failure in a sample of size 20. Now if the exponential assumption is satisfied, the items that didn't fail are as good as new, and thus you've gained time, at essentially no cost, by performing an experiment with a type II censoring sampling design.

51. Let $X_1,\ldots,X_n \overset{\text{iid}}{\sim} W(\alpha,\beta)$, the two-parameter Weibull distribution with shape parameter α and scale parameter β. The maximum likelihood estimator of the parameter (α,β) cannot be obtained analytically. (a) Show that the likelihood equations for estimating α and β may be written as

$$g(\alpha)=\frac{\sum_{i=1}^n x_i^\alpha \ln(x_i)}{\sum_{i=1}^n x_i^\alpha}-\frac{1}{\alpha}=\frac{1}{n}\sum_{i=1}^n \ln(x_i)$$

and

$$\beta=\frac{1}{n}\sum_{i=1}^n x_i^\alpha.$$

Show that the equations above have a unique solution $\left(\hat{\alpha},\hat{\beta}\right)\in(0,\infty)^2$. (Note: Some useful hints on this problem are given in Lehmann and Casella (1998) Chapter 6, Example 6.1.)

52. Let $X_1,\ldots,X_n \overset{\text{iid}}{\sim} W(\alpha,1)$, the one-parameter Weibull distribution with shape parameter α. The maximum likelihood estimator of the parameter α cannot be obtained analytically. (a) Write down an expression for the log-likelihood function $\ln L$ in this problem as a function of α. (b) Obtain expressions for the partial derivatives $(\partial/\partial\alpha)\ln L$ and $(\partial^2/\partial\alpha^2)\ln L$. (c) Obtain an explicit expression for the first iterate from the Newton-Raphson algorithm for estimating the parameter α using the method of moments estimator as the initial estimate α_0 of α.

53. (An MLE in a two-parameter problem.) *Preliminaries*: Recall that a function $f(x,y)$ has a local maximum at the point (x_0,y_0) if x_0 and y_0 are solutions of the equations.

$$\frac{\partial}{\partial x}f(x,y)=0 \text{ and } \frac{\partial}{\partial y}f(x,y)=0, \tag{7.59}$$

and if the Hessian matrix

$$H(x,y)=\begin{bmatrix}\frac{\partial^2}{\partial x^2}f(x,y) & \frac{\partial^2}{\partial x\partial y}f(x,y)\\ \frac{\partial^2}{\partial x\partial y}f(x,y) & \frac{\partial^2}{\partial y^2}f(x,y)\end{bmatrix} \tag{7.60}$$

is "negative definite" when evaluated at the point (x_0, y_0). The matrix

$$\begin{bmatrix} a & b \\ c & d \end{bmatrix}$$

is negative if $a < 0$ and $d < 0$ and the determinant $ad - bc > 0$.

The Problem: Suppose that $X_1, X_2, \ldots, X_n \stackrel{\text{iid}}{\sim} N(\mu, \sigma^2)$, with both μ and σ^2 unknown. Prove that the maximum likelihood estimators of μ and σ^2, found in Example 7.3.6, actually maximize the likelihood. (Note that the likelihood equation has a unique critical point; thus, if the Hessian matrix is negative definite at that point, it must correspond to a *global* maximum.)

54. Let $X_1, \ldots, X_m \stackrel{\text{iid}}{\sim} N(\mu, \sigma_x^2)$, and let $Y_1, \ldots, Y_n \stackrel{\text{iid}}{\sim} N(\mu, \sigma_y^2)$, σ_x^2 and σ_y^2 are assumed known. Assume that the X sample is independent of the Y sample. (a) Show that the statistic

$$\hat{\mu} = \frac{(m/\sigma_x^2)\bar{X} + (n/\sigma_y^2)\bar{Y}}{m/\sigma_x^2 + n/\sigma_y^2}$$

is the unique maximum likelihood estimator of μ. (b) Show that $\hat{\mu}$ is asymptotically efficient. (c) Use part (d) of Theorem 5.2.1 (Slutsky's theorem) to show that the estimator

$$\hat{\hat{\mu}} = \frac{(m/s_x^2)\bar{X} + (n/s_y^2)\bar{Y}}{m/s_x^2 + n/s_y^2}$$

is also an asymptotically efficient estimator of μ when σ_x^2 and σ_y^2 are unknown.

55. Let $X_1, \ldots, X_n \stackrel{\text{iid}}{\sim} U(\theta - 1/2, \theta + 1/2)$. The order statistics $X_{(1)}$ and $X_{(n)}$ are jointly sufficient for θ. Show that any value $\tilde{\theta}$ in the interval $(x_{(n)} - 1/2, x_{(1)} + 1/2)$ maximizes the likelihood function. Show that the estimator

$$\hat{\theta} = x_{(n)} - 1/2 + (x_{(n-1)} - x_{(2)})(x_{(1)} - x_{(n)} + 1)$$

is a maximum likelihood estimator of θ, even though it does not depend solely on the sufficient statistics $X_{(1)}$ and $X_{(n)}$.

56. Let $X_1, \ldots, X_n \stackrel{\text{iid}}{\sim} LN(\mu, \sigma^2)$. Obtain the maximum likelihood estimator of the mean of this distribution.

57. Let $X_1, \ldots, X_n \stackrel{\text{iid}}{\sim} N(\mu, \sigma^2)$, and let $s^2 = (1/(n-1)) \sum_{i=1}^n (X_i - \bar{X})^2$ be the sample variance and let $S^2 = [(n-1)/n]s^2$. Show that these two statistics have distinctly different asymptotic distributions, that is, s^2 and S^2 are not asymptotically equivalent.

58. Let $X_1, \ldots, X_n \stackrel{\text{iid}}{\sim} F$. Show that the empirical distribution F_n, which places probability $1/n$ at each of the observed values $x_1, \ldots, x_n$, is the nonparametric maximum likelihood estimator of F. (Hint: Define the likelihood of the observed Xs as $L = \prod_{i=1}^n p_i$, where p_i is the weight placed on the observation x_i. Show that no vector $\mathbf{p}$ with zero elements can maximize L, and that the vector $\mathbf{p} = (1/n, \ldots, 1/n)$ is the maximizer of L. (See Remark 6.5.1. The method of Lagrange multipliers works nicely in this problem.)

59. Obtain the asymptotic distribution of the sample odds ratio $\hat{\omega}$ under the sampling framework in (7.30), but with an unbalanced sample. Specifically, assume that $n_{1\bullet} = n$ and $n_{2\bullet} \cong cn$ for some constant $c \neq 1$, and identify the limiting distribution of $\sqrt{n}(\hat{\omega} - \omega)$ an $n \to \infty$.

60. Use the Newton-Raphson algorithm to show that the maximum likelihood estimate $\hat{\pi}$ of π based on the observed data $\mathbf{y} = (125, 18, 20, 34)$ drawn from the multinomial model $M_4(197, \mathbf{p})$ is approximately equal to .6268214980.

61. Suppose $X_1, \ldots, X_n, \ldots \stackrel{\text{iid}}{\sim} F$, where F_θ is a member of a one-parameter exponential family, and let $\{\theta_i, i = 0, 1, 2 \ldots.\}$ be a sequence of iterations of the EM algorithm. Show that the likelihood function $L(\theta_i)$ is increasing in i.

62. Let $X_1, \ldots, X_n \stackrel{\text{iid}}{\sim} Be(1, \theta)$, the beta distribution with density $f_\theta(x) = \theta(1-x)^{\theta-1} I_{(0,1)}(x)$, where $\theta > 0$. If $X \sim Be(1, \theta)$, then $E(X) = 1/(1+\theta)$ and $V(X) = \theta/(\theta+1)^2(\theta+2)$. (a) Use the first moment equation to obtain an MME $\hat{\theta}_1$ of the parameter θ. (b) Obtain the asymptotic distribution of $\sqrt{n}(\hat{\theta}_1 - \theta)$. (c) Derive the MLE $\hat{\theta}_2$ of the parameter θ. (d) Calculate the Fisher information $I_X(\theta)$ for the $Be(1, \theta)$ model and identify the asymptotic distribution of $\sqrt{n}(\hat{\theta}_2 - \theta)$. (e) Show that $\hat{\theta}_2$ is asymptotically superior to $\hat{\theta}_1$.

8

Interval Estimation

8.1 Exact Confidence Intervals

While point estimates are very useful in identifying features of a population under study, and they are often what we are first interested in, we will invariably require further information. An investigator would no doubt also want to know the level of precision achieved by a given estimator. Let T be an estimator of the parameter θ. The standard deviation of the probability distribution of T, that is, σ_T, is generally referred to as the *standard error* of T. A small standard error is desirable, of course, since this will imply that the estimator is likely to be close to its expected value (by Chebyshev's inequality) and, if it is an unbiased estimator of the target parameter θ, or nearly so, it will then also be close to θ. For example, given a random sample of size n from a population with finite mean μ and variance σ^2, we know that the sample mean $\bar{X}$ is an unbiased estimator of μ, and that its standard error is $\sigma_{\bar{X}} = \sigma/n$. This tells us that if the sample size n is reasonably large, $\bar{X}$ will be quite close to the target parameter μ. Stating both the value of a point estimator and its exact or approximate standard error provides useful information; it provides our "best guess" at the parameter's value as well as some insight into the estimator's precision. In this section, we will describe a way of combining this information into a single procedure.

If we have in hand the exact or approximate distribution of a particular estimator, we may then be able to make assertions about the likelihood that the estimator will be within a certain distance of the target parameter. An interval estimate of a parameter θ is an interval (L,U) whose endpoints depend on an estimator of θ, but not on any unknown parameters, and for which the probability that the interval contains the unknown parameter θ can be explicitly derived or reliably approximated. When experimental data is available, the endpoints of the interval (L,U) can be computed, resulting in a numerical interval (l,u) called a *confidence interval* for the parameter θ. If the probability that the interval (L,U) contains the parameter θ is p, then the numerical interval (l,u) is referred to as a $100p\%$ *confidence interval* for θ. It may seem tempting to use the phrase "probability interval" in describing the interval (l,u), but this phrase would be inappropriate due to the fact that a given numerical interval either contains the true value of the parameter or it doesn't, so that the inequality $l < \theta < u$ has no immediate probabilistic meaning. Because the interval (l,u) is obtained as a realization of the interval (L,U), where $P(L < \theta < U) = p$, it is quite reasonable to say that we have $100p\%$ confidence that the numerical interval (l,u) will contain the true value of parameter θ.

There are several possible approaches to the development of a confidence interval for θ. For convenience, we will take θ to be a scalar parameter, although generalizations to higher dimensions cause no conceptual difficulties, and some such generalizations will be discussed later in the chapter. The approach we consider in this section is the one most commonly used, and it employs what we will refer to as a "pivotal statistic" for θ. This term is defined as follows.

Definition 8.1.1. Let $X_1, X_2, \dots, X_n \stackrel{\text{iid}}{\sim} F_\theta$. Let $S = g(\mathbf{X}, \theta)$ be a function of the data and the parameter θ. If the distribution of S does not depend on the parameter θ, then S is called a *pivotal statistic.*

Some simple examples of pivotal statistics are given below.

Example 8.1.1. (i) Suppose that $X_1, X_2, \dots, X_n \stackrel{\text{iid}}{\sim} N(\mu, 1)$, and let $S_1 = \bar{X} - \mu$. Since $S_1 \sim N(0, 1/n), S_1$ is a pivotal statistic.

(ii) Let $X_1, X_2, \dots, X_n \stackrel{\text{iid}}{\sim} N(\mu, \sigma^2)$, and define $S_2 = \sqrt{n}(\bar{X} - \mu)/\sigma$. Since $S_2 \sim N(0, 1), S_2$ is a pivotal statistic.

(iii) Let $X_1, X_2, \dots, X_n \stackrel{\text{iid}}{\sim} U(0, \theta)$. Then, since $X_i/\theta \sim U(0, 1)$, each such ratio is a pivotal statistic, as is their average $S_3 = \bar{X}/\theta$. Further, since order statistics of a $U(0, 1)$ sample have beta distributions, we have that $S_4 = X_{(n)}/\theta \sim Be(n, 1)$, so that S_4 is also a pivotal statistic for θ. It is clear from the latter development that there can be a multiplicity of pivotal statistics in a given modeling scenario.

■

You may have noticed that the phrase "pivotal statistic" constitutes a misuse of the word "statistic," since in the standard usage of the word, a statistic is a function of the data alone and does not depend on any unknown quantities. In spite of this, we will use the phrase as is, since it is by now in widespread use and it's really too late to turn the tide of history. Let us instead proceed with a discussion of the construction of confidence intervals based on a given pivotal statistic.

The "Pivotal Method" for constructing a confidence interval (CI) for a parameter θ begins with the identification of a pivotal statistic. As noted above, there can be more than one of them, so it is worth noting that the usual practice, a practice that is quite easy to defend, is to use a pivotal statistic that is a function of a sufficient statistic T for θ, assuming that (a univariate) one exists. Thus, in Example 8.1.1, part (iii), we would choose to work with the order statistic $X_{(n)}$ rather than with the sample mean $\bar{X}$. Now, given a particular pivotal statistic $S = g(T, \theta)$, where we are assuming, for convenience, that T is sufficient for θ, we would find cutoff points A and B such that

$$P(A \le g(T, \theta) \le B) = 1 - \alpha, \tag{8.1}$$

where α is some small value like .1, .05, or .01. The smaller the value of α used in (8.1), the more confidence we will have that the resulting interval will actually trap the true value of θ. It's not surprising, of course, that the smaller α is, the wider the interval will be. One can make the probability statement in (8.1) because, as a pivotal statistic, $g(T, \theta)$ has a distribution that is free of unknown parameters. We may therefore, either by the use of tables or by direct computation, identify constants A and B for which (8.1) is satisfied. It is worth mentioning that, typically, many different such (A, B) pairs may be found. While the choice among such pairs is arbitrary, it is common practice to select A and B so that

$$P(g(T, \theta) \le A) = \alpha/2 \text{ and } P(g(T, \theta) \ge B) = \alpha/2. \tag{8.2}$$

Sometimes, however, an investigator may wish to select the interval at the probability level α that has the shortest length $B-A$.

From (8.1), one can usually perform a series of algebraic steps to isolate the parameter θ in the center of the inequality, obtaining a probability statement of the form

$$P(L(T) \leq \theta \leq U(T)) = 1-\alpha. \tag{8.3}$$

Writing $L = L(T)$ and $U = U(T)$ for short, we have identified in (8.3) an interval, depending on the statistic T alone, that has probability $(1-\alpha)$ of capturing the true value of the parameter θ. Now, suppose that, from the observed data $x_1, \dots, x_n$, we find that $T = t$. Let $l = L(t)$ and $u = U(t)$. Then the numerical interval (l,u) is called a $100(1-\alpha)\%$ confidence interval for θ.

How should the "confidence" that one has in a confidence interval be interpreted. While an explicit probability statement should not be made about the interval (l,u), one can make the claim that $l \leq \theta \leq u$ with some confidence, that confidence stemming from the process by which the interval was generated. A confidence interval has the following "relative frequency" interpretation. For simplicity, let's take $\alpha = .05$, so that the interval (l,u) in the problem at hand is a 95% confidence interval. If one were to repeat the same experiment independently a large number of times, each interval we generated would have a 95% chance of trapping θ. If we were to perform the experiment 100 times, we would expect that about 95 of the confidence intervals we generated would contain the true value of θ, and that we would fail to capture θ about 5 times.

We now turn to some examples of the Pivotal Method in action.

Example 8.1.2. Suppose that $X_1, X_2, \dots, X_n \stackrel{\text{iid}}{\sim} N(\mu, \sigma_0^2)$, where the variance σ_0^2 is assumed known. Since the sample mean $\bar{X}$ is a sufficient statistic for μ, we will use $S = \sqrt{n}(\bar{X}-\mu)/\sigma_0$, a standard normal variable, as a pivotal statistic. We may then write

$$P\left(-Z_{\alpha/2} \leq \frac{\bar{X}-\mu}{\sigma_0/\sqrt{n}} \leq Z_{\alpha/2}\right) = 1-\alpha, \tag{8.4}$$

where $Z_{\alpha/2}$ is the $\alpha/2$-cutoff point of the standard normal distribution, as defined in Section 3.5. We may rewrite (8.4) as

$$P\left(-(\sigma_0/\sqrt{n})Z_{\alpha/2} \leq \bar{X}-\mu \leq (\sigma_0/\sqrt{n})Z_{\alpha/2}\right) = 1-\alpha, \tag{8.5}$$

or, equivalently, as

$$P\left(\bar{X}-(\sigma_0/\sqrt{n})Z_{\alpha/2} \leq \mu \leq \bar{X}+(\sigma_0/\sqrt{n})Z_{\alpha/2}\right) = 1-\alpha. \tag{8.6}$$

From (8.6), it follows that, given $\bar{X} = \bar{x}$, $\left(\bar{x}-(\sigma_0/\sqrt{n})Z_{\alpha/2}, \bar{x}+(\sigma_0/\sqrt{n})Z_{\alpha/2}\right)$ is a $100(1-\alpha)\%$ confidence interval for the population mean μ. If twenty-five observations were drawn randomly from a normal population with mean μ and known variance 4, and the sample mean $\bar{X}$ was equal to 12, then $(12-(2/5)(1.96), 12+(2/5)(1.96)) = (11.216, 12.784)$ is a 95% confidence interval for μ. ■

Example 8.1.3. Let $X_1, X_2, \dots, X_n \stackrel{\text{iid}}{\sim} U(0,\theta)$. Consider the pivotal statistic based on the sufficient statistic $X_{(n)}$, that is, let $S = X_{(n)}/\theta$. Since $S \sim Be(n,1)$, we may write

$$P(be_{n,1,1-\gamma/2} \leq X_{(n)}/\theta \leq be_{n,1,\gamma/2}) = 1-\gamma, \tag{8.7}$$

where the cutoff point $be_{\alpha,\beta,\gamma}$ is the value such that $P(X > be_{\alpha,\beta,\gamma}) = \gamma$ when the variable X has the $Be(\alpha,\beta)$ distribution. Now the $Be(n,1)$ distribution has density function $f(x) = nx^{n-1}I_{(0,1)}(x)$ and distribution function $F(x) = x^n$ for $0 < x < 1$. Since $F(be_{n,1,1-\gamma/2}) = \gamma/2$ and $F(be_{n,1,\gamma/2}) = 1-\gamma/2$, one finds that $be_{n,1,1-\gamma/2} = (\gamma/2)^{1/n}$ and $be_{n,1,\gamma/2} = (1-\gamma/2)^{1/n}$. We may thus rewrite the probability statement in (8.7) as

$$P(X_{(n)}/(\gamma/2)^{1/n} \leq \theta \leq X_{(n)}/(1-\gamma/2)^{1/n}) = 1-\alpha. \tag{8.8}$$

Now, suppose that $X_{(n)} = x_{(n)}$ is observed. Then, from (8.8), $(x_{(n)}/(1-\gamma/2)^{1/n}, x_{(n)}/(\gamma/2)^{1/n})$ is a $100(1-\gamma)\%$ confidence interval for θ. ■

We note that, unlike the normal case in Example 8.1.2, where the confidence interval for μ has the form $\bar{X} \pm C$, the confidence interval in Example 8.1.3 is not symmetric about the sufficient statistic $X_{(n)}$. In the normal case, the distribution of the pivotal statistic is symmetric about 0, so that one cutoff point is the negative of the other. The distribution of the pivotal statistic for the uniform model is quite skewed to the left and thus the simple relationship between the two cutoff points disappears.

Example 8.1.4. Let us now consider a more realistic version of Example 8.1.2. Suppose that $X_1, X_2, \ldots, X_n \stackrel{\text{iid}}{\sim} N(\mu,\sigma^2)$, where both μ and σ^2 are unknown. In Section 4.6, it was found that the t-statistic, given by

$$t = \frac{\bar{X}-\mu}{s/\sqrt{n}} \tag{8.9}$$

has a t distribution with $(n-1)$ degrees of freedom. Since this statistic has a distribution that does not depend on any unknown parameters, it may be used as a pivotal statistic in constructing a confidence interval for μ. If $t_{k,\alpha}$ is the α-cutoff point of the t distribution with k degrees of freedom, that is, if $P(X > t_{k,\alpha}) = \alpha$ when $X \sim t_k$, then we may write

$$P\left(-t_{n-1,\alpha/2} \leq \frac{\bar{X}-\mu}{s/\sqrt{n}} \leq t_{n-1,\alpha/2}\right) = 1-\alpha. \tag{8.10}$$

We may rewrite (8.10) as

$$P\left(-(s/\sqrt{n})t_{n-1,\alpha/2} \leq \bar{X}-\mu \leq (s/\sqrt{n})t_{n-1,\alpha/2}\right) = 1-\alpha, \tag{8.11}$$

or, equivalently, as

$$P\left(\bar{X}-(s/\sqrt{n})t_{n-1,\alpha/2} \leq \mu \leq \bar{X}+(s/\sqrt{n})t_{n-1,\alpha/2}\right) = 1-\alpha. \tag{8.12}$$

If $\bar{X} = \bar{x}$ is observed, we infer from (8.12) that $(\bar{x}-(s/\sqrt{n})t_{n-1,\alpha/2}, \bar{x}+(s/\sqrt{n})t_{n-1,\alpha/2})$ is a $100(1-\alpha)\%$ confidence interval for the population mean μ. ■

There are problems in which the pivotal method for constructing confidence intervals is not helpful. Such problems can arise when one simply can't identify a suitable pivotal statistic, but they can happen as well when the exact distribution of a candidate pivotal statistic cannot be determined analytically. It would thus be useful to consider an alternative approach to finding a confidence interval. A full description of the method I'll discuss here requires a basic understanding of the methodology of "hypothesis testing," and we will expand upon the present discussion in Chapter 10. For now, I'll give you an intuitive description and justification of the method which should suffice both for understanding it and being able to use it in practice. Let's first take another look at the notion of a confidence interval. What we are looking for are a set of possible values of the parameter of interest that are "believable," given the data that we have in hand. To anchor the discussion, let's reexamine the confidence interval for a normal mean μ in Example 8.1.2. In Equation (8.5), we noted that, given a sample of size n,

$$P\left(-(\sigma_0/\sqrt{n})Z_{\alpha/2} \leq \bar{X} - \mu \leq (\sigma_0/\sqrt{n})Z_{\alpha/2}\right) = 1 - \alpha.$$

This equation can also be written as

$$P\left(|\bar{X} - \mu| \leq (\sigma_0/\sqrt{n})Z_{\alpha/2}\right) = 1 - \alpha. \tag{8.13}$$

The latter equation admits to the following interpretation: for $\bar{X}$ and μ to be compatible with each other at the $(1 - \alpha)$ level of probability, the distance between them must not exceed $(\sigma_0/\sqrt{n})Z_{\alpha/2}$. This can be restated as: at the $(1 - \alpha)$ probability level, only values of μ that are close enough to $\bar{X}$ (that is, within $(\sigma_0/\sqrt{n})Z_{\alpha/2}$ of $\bar{X}$) can be considered compatible with the data. Given $\bar{X} = \bar{x}$, the values of μ in the interval $\left(\bar{x} - (\sigma_0/\sqrt{n})Z_{\alpha/2}, \bar{x} + (\sigma_0/\sqrt{n})Z_{\alpha/2}\right)$ are compatible with $\bar{X}$ at the $100(1 - \alpha)\%$ confidence level.

The following example illustrates this alternative approach which can be used for constructing confidence intervals in problems in which the pivotal method is unavailable. In Chapter 10, we'll refer to this method as "inverting a hypothesis test" about the parameter of interest.

Example 8.1.5. Suppose a coin is tossed 8 times and $X = 4$ heads are observed. We want to find a 90% confidence interval for p. While the distribution $B(8, p)$ is simple enough, there is no obvious statistic $S = g(X, p)$ whose distribution is completely known, that is, has no unknown parameters. Let's take the alternative approach suggested above. What we are looking for are a set of possible values of the parameter p that are "believable," given the data that we have in hand. We can eliminate certain possibilities immediately. For example, clearly $p \neq 0$, since we couldn't possibly observe $X = 4$ if $p = 0$. But we can make the same argument for other very small values of p. For example, if p were equal to 0.1, then we can calculate (or obtain from a binomial table) that $P(X \leq 3) = .995$. This means that anything as large as your observed data had a negligible chance of happening if p were equal to 0.1. This identifies the value $p = 0.1$ as incompatible with your data. We can make the same argument for values of p near 1. But as p grows closer to the observed proportion $\hat{p} = 1/2$, one finds that the chances of obtaining the observed value increases. In particular, $P(X \geq 4|p)$ increases from 0 to 1 as p goes from 0 to 1. So it makes sense to look for the smallest value of p for which the event $\{X \geq 4\}$ is still plausible. Suppose that we look for the value of p, call it p_L, for which $P(X \geq 4|p_L) = .05$. For any value of $p < p(L)$, the chances of observing $X \geq 4$ are too small to believe that the value $X = 4$ would have occurred if that value of p was the true parameter value. Similarly, we can find a value p_U of p for which $P(X \leq 4|p_U) = .05$. No value larger than p_U is credible either. What remains are the plausible values of p, given that we have observed $X = 4$. The interval (p_L, p_U) will serve as a 90% confidence interval for the true proportion p. Generally, one would write a small computer program to find the values of p_L and p_U, or perhaps use a computer package like R to compute these values. Having done the necessary computation, we would find that our 90% confidence interval for p is $(0.193, 0.807)$. It is easy to confirm that $P(X \geq 4|p = p_L) = .05$ and that $P(X \leq 4|p = p_U) = .05$ as I have claimed. ■

Exercises 8.1.

1. Suppose that $X_1, X_2, \dots, X_n \overset{\text{iid}}{\sim} N(0, \sigma^2)$. Find a pivotal statistic based on a sufficient statistic for the parameter σ^2, and develop a $100(1-\alpha)\%$ confidence interval for σ^2. (Hint: Recall that, for each i, $X_i/\sigma \sim N(0,1)$ and $X_i^2/\sigma^2 \sim \chi_1^2$.)

2. Suppose that $X_1, X_2, \dots, X_n \overset{\text{iid}}{\sim} N(\mu, \sigma^2)$. Note that $(n-1)s^2/\sigma^2$ is a pivotal statistic having the χ^2_{n-1} distribution. Develop a $100(1-\alpha)\%$ confidence interval for σ^2.

3. Suppose that $X_1, X_2, \dots, X_n \overset{\text{iid}}{\sim} \Gamma(1, \beta)$. Verify that $S = 2\sum_{i=1}^n X_i/\beta$ is a pivotal statistic having the χ^2_{2n} distribution. Obtain a $100(1-\alpha)\%$ confidence interval for β.

4. Suppose that $X_1, X_2, \dots, X_{10} \overset{\text{iid}}{\sim} P(\lambda)$, the Poisson distribution with mean λ. The sum $S = \sum_{i=1}^{10} X_i$ is a sufficient statistic for λ. Suppose $S = 14$ is observed. Use the method illustrated in Example 8.1.5 to obtain a 95% confidence interval for λ.

8.2 Approximate Confidence Intervals

We will first treat confidence intervals which use asymptotic approximations for the distribution of an estimator of a parameter of interest. It is not difficult to find examples in which a random sample of size n is drawn from a model F_θ and the exact distribution of a sufficient statistic T is not known. In such situations, obtaining an exact confidence interval for θ based on the statistic T will not be possible. However, if the sample size is large, one might be able to obtain an approximate confidence interval for θ from the asymptotic distribution of T. If T is the sum or average of the sampled data, then the Central Limit Theorem may be invoked to identify the approximate distribution of T, assuming of course that the population model has a finite variance. If the central limit theorem is not immediately applicable in the problem of interest, it's possible that the asymptotic theory of MMEs or MLEs can be used. Suppose that $\hat{\theta}_n$ is a $\sqrt{n}$- consistent, asymptotically normal estimator of the parameter θ, that is, suppose that, for sufficiently large n, the approximate distribution of $\hat{\theta}_n$ is given by

$$\hat{\theta}_n \sim N\left(\theta, \frac{AV(\theta)}{n}\right). \tag{8.14}$$

We may then construct the pivotal statistic

$$S = \frac{\hat{\theta}_n - \theta}{\sqrt{AV(\theta)/n}} \sim N(0,1). \tag{8.15}$$

It follows that the numerical interval

$$\left(\hat{\theta}_n - \sqrt{AV(\theta)/n} \cdot Z_{\alpha/2}, \hat{\theta}_n + \sqrt{AV(\theta)/n} \cdot Z_{\alpha/2}\right) \tag{8.16}$$

is an approximate $100(1-\alpha)\%$ confidence interval for θ. If the interval in (8.16) does not depend on θ (nor on any other unknown parameter), the interval is ready for immediate use as an approximate CI for θ. It is often the case that this interval does in fact depend on θ. In that case, it is standard practice to estimate θ (or other unknown parameters in (8.16)), obtaining an "approximate" approximate confidence interval for θ. In referring to the latter $100(1-\alpha)\%$ CI, given by

$$\left(\hat{\theta}_n - \sqrt{AV(\hat{\theta}_n)/n} \cdot Z_{\alpha/2}, \hat{\theta}_n + \sqrt{AV(\hat{\theta}_n)/n} \cdot Z_{\alpha/2}\right), \tag{8.17}$$

only one use of the word "approximate" is usually deemed necessary. The justification for plugging in an estimate for θ in the CI formula in (8.16) is that, when n is large, which must be the case for the interval in (8.16) to provide a good approximate CI, the estimator $\hat{\theta}_n$ should be quite close to the true value of θ so that $\text{AV}(\hat{\theta}_n)$ should be close to $\text{AV}(\theta)$. This of course isn't necessarily true, but it will be if the asymptotic variance of $\hat{\theta}_n$ is a reasonably smooth function of θ, which is often the case.

In the examples below, we will derive approximate, large-sample confidence intervals for parameters that one is often interested in.

Example 8.2.1. Suppose a random sample $X_1,\dots,X_n$ is available from a population with unknown mean μ and variance σ^2, both assumed finite. We will assume that the sample size n is sufficiently large to justify the application of the Central Limit Theorem. Specifically, let's assume that

$$\sqrt{n}(\bar{X}-\mu) \xrightarrow{D} Y \sim N\left(0,\sigma^2\right). \tag{8.18}$$

For large n, we may write

$$\bar{X} \sim N\left(\mu, \frac{\sigma^2}{n}\right) \text{ approximately.} \tag{8.19}$$

It follows that $\left(\bar{X} - \frac{\sigma}{\sqrt{n}}Z_{\alpha/2}, \bar{X} + \frac{\sigma}{\sqrt{n}}Z_{\alpha/2}\right)$ is an approximate $100(1-\alpha)\%$ confidence interval for μ. If the variance σ^2 were known, this interval would suffice, and we would use it with the specific probability level α of interest. Since σ^2 is virtually always unknown, it would typically be estimated by the sample variance s^2 given by

$$s^2 = \frac{1}{n-1}\sum_{i=1}^{n}(X_i-\bar{X})^2,$$

leading to the approximate $100(1-\alpha)\%$ confidence interval for μ given by

$$\left(\bar{X} - \frac{s}{\sqrt{n}}Z_{\alpha/2}, \bar{X} + \frac{s}{\sqrt{n}}Z_{\alpha/2}\right). \tag{8.20}$$

■

Example 8.2.2. Suppose that we have a random sample of Bernoulli trials, i.e., suppose that $X_1,\dots,X_n \stackrel{\text{iid}}{\sim} B(1,p)$. Then $X = \sum_{i=1}^{n} X_i$ has the $B(n,p)$ distribution, and the sample proportion $\hat{p} = \frac{X}{n}$ is an average whose asymptotic behavior is governed by the CLT. Thus, for n sufficiently large, we have that

$$\sqrt{n}(\hat{p}-p) \xrightarrow{D} Y \sim N(0, p(1-p)). \tag{8.21}$$

We may thus write that

$$\hat{p} \sim N(p, \frac{p(1-p)}{n}) \text{ approximately.} \tag{8.22}$$

It follows that

$$\left(\hat{p} - \left(\sqrt{p(1-p)}/\sqrt{n}\right) Z_{\alpha/2}, \hat{p} + \left(\sqrt{p(1-p)}/\sqrt{n}\right) Z_{\alpha/2}\right)$$

is an approximate $100(1-\alpha)\%$ confidence interval for p. Since this interval depends on the unknown p, we would opt for the approximate $100(1-\alpha)\%$ CI

$$\left(\hat{p} - \sqrt{\hat{p}(1-\hat{p})/n} Z_{\alpha/2}, \hat{p} + \sqrt{\hat{p}(1-\hat{p})/n} Z_{\alpha/2}\right).$$

■

An alternative to the double approximation involving the estimation of parameters in the asymptotic variance AV in the approximate confidence interval in (8.16) makes use of "variance stabilizing transformations," discussed in Section 5.4. As explained there, the VST in a particular problem may be found by applying the δ-method to an asymptotically normal estimator T_n of the parameter of interest. The VST g in a particular problem has the desirable property that $g(T_n)$ has an asymptotic variance that does not depend on the unknown parameter. The process of finding a VST as described in the abstract in Equations (5.23)–(5.26). The use of the VST in obtaining an approximate large-sample confidence interval for a Poisson parameter is illustrated in Example 5.4.2. Two further examples of this process are given below.

Example 8.2.3. Let us revisit the estimation problem treated in Example 7.3.8. Let $X_1, X_2, \ldots, X_n \stackrel{\text{iid}}{\sim} Exp(\lambda)$, and assume that n is "large." Then the estimator $\hat{\lambda} = 1/\bar{X}$ is both the MME and the MLE of the parameter λ and has the asymptotic distribution shown below:

$$\sqrt{n}\left(\hat{\lambda} - \lambda\right) \xrightarrow{D} W \sim N\left(0, \lambda^2\right). \tag{8.23}$$

Rather than settling for the approximate large sample $100(1-\alpha)\%$ confidence interval $\left(\hat{\lambda} - (\hat{\lambda}/\sqrt{n})Z_{\alpha/2}, \hat{\lambda} + (\hat{\lambda}/\sqrt{n})Z_{\alpha/2}\right)$, which involves the estimation of λ in the asymptotic variance of $\hat{\lambda}$, we search, instead, for the VST g in this particular problem. Taking $c = 1$ in Equation (5.26), we have that

$$g(\lambda) = \int \frac{1}{\lambda} d\lambda = \ln(\lambda).$$

From this, we deduce that

$$\sqrt{n}\left(\ln(\hat{\lambda}) - \ln(\lambda)\right) \xrightarrow{D} W \sim N\left(0, \left(\frac{1}{\lambda}\right)^2 \lambda^2\right) = N(0,1). \tag{8.24}$$

This latter equation leads to an approximate $100(1-\alpha)\%$ confidence interval for $\ln(\lambda)$:

$$(\ln(\lambda)_L, ln(\lambda)_U) = \left(\ln(\hat{\lambda}) - \frac{Z_{\alpha/2}}{\sqrt{n}}, \ln(\hat{\lambda}) + \frac{Z_{\alpha/2}}{\sqrt{n}}\right). \tag{8.25}$$

Since the logarithm is a strictly increasing function of its argument, we can immediately obtain from (8.25) an approximate $100(1-\alpha)\%$ confidence interval for λ itself, namely, $\left(e^{\ln(\lambda)_L}, e^{\ln(\lambda)_U}\right)$. ■

Example 8.2.4. Let's return to the scenario treated in Example 8.2.2, that is, let's reconsider confidence intervals (CIs) for a population proportion p based on the asymptotic distribution (AD) of the sample proportion $\hat{p}$. We have that

$$\sqrt{n}(\hat{p}-p) \xrightarrow{D} Y \sim N(0, p(1-p)). \tag{8.26}$$

Since the asymptotic variance in (8.26) depends on p, one would need to estimate p in the approximate standard error of $\hat{p}$ to obtain the approximate $100(1-\alpha)\%$ confidence interval

$$\left(\hat{p} - \frac{\sqrt{\hat{p}(1-\hat{p})}}{\sqrt{n}} Z_{\alpha/2}, \hat{p} + \frac{\sqrt{\hat{p}(1-\hat{p})}}{\sqrt{n}} Z_{\alpha/2}\right).$$

As an alternative that doesn't require the estimation of the asymptotic variance of $\hat{p}$, let's transform $\hat{p}$, using the delta-method, to get an AD whose variance doesn't depend on p. In the present context, the variance stabilizing transformation g is given by

$$g(x) = \arcsin\sqrt{x} \quad (\text{or } \sin^{-1}\sqrt{x}). \tag{8.27}$$

You can confirm from an integral table that g is the anti-derivative of

$$g'(x) = 1/\sqrt{x(1-x)}.$$

Using the delta-method to obtain the asymptotic distribution of $g(\hat{p})$, we get

$$\sqrt{n}\left(\arcsin\sqrt{\hat{p}} - \arcsin\sqrt{p}\right) \xrightarrow{D} V \sim N\left(0, \left(\frac{1}{\sqrt{p(1-p)}}\right)^2 p(1-p)\right) = N(0,1). \tag{8.28}$$

From this, we may obtain a $100(1-\alpha)\%$ confidence interval for $\arcsin\sqrt{p}$, namely

$$\left(\arcsin\sqrt{\hat{p}} - \frac{Z_{\alpha/2}}{\sqrt{n}}, \arcsin\sqrt{\hat{p}} + \frac{Z_{\alpha/2}}{\sqrt{n}}\right). \tag{8.29}$$

Now, if $a < \arcsin y < b$, then $\sin a < y < \sin b$. It thus follows from (8.29) that

$$\left(\left[\sin\left(\arcsin\sqrt{\hat{p}} - \frac{Z_{\alpha/2}}{\sqrt{n}}\right)\right]^2, \left[\sin\left(\arcsin\sqrt{\hat{p}} + \frac{Z_{\alpha/2}}{\sqrt{n}}\right)\right]^2\right)$$

is an approximate $100(1-\alpha)\%$ confidence interval for p. ■

Thus far, our discussion of interval estimation has been restricted to one-parameter models. For some multi-parameter models, some univariate and some multivariate, it is possible to identify regions depending on observable random variables for which explicit probability statements can be made. When a random sample has in fact been observed, the corresponding region based on the observed data may serve as a confidence region for the parameter vector, and it will have a similar frequency interpretation as in the one-parameter case. One important case in which the development of exact confidence regions for parameter vectors of interest is analytically tractable is in the estimating the mean of a multivariate normal distribution. We will not pursue such developments here. Instead, we discuss below a conservative but widely applicable method for generating simultaneous confidence intervals for two or more parameters of interest. The method to which we now turn is generally called the Bonferroni method and is based on the following inequality.

Theorem 8.2.1. (The Bonferroni inequality) Consider a random experiment with sample space S, and let $A_1, A_2, \ldots, A_k$ be k arbitrary events, i.e., subsets of S. Then

$$P\left(\bigcap_{i=1}^{k} A_i\right) \geq 1 - \sum_{i=1}^{k} P(A_i^c). \tag{8.30}$$

Proof. The inequality in (8.30) follows immediately from two results that we established in Chapter 1. We begin this proof by recalling these two facts. First, we have by DeMorgan's Law that for the k sets of interest here, we may write

$$\left(\bigcup_{i=1}^{k} A_i\right)^c = \bigcap_{i=1}^{k} A_i^c. \tag{8.31}$$

Since we are interested in the intersection of the sets $A_1, \ldots, A_k$ in (8.30), it is useful to reverse the role of each A_i and A_i^c and write DeMorgan's Law as

$$\left(\bigcup_{i=1}^{k} A_i^c\right)^c = \bigcap_{i=1}^{k} A_i. \tag{8.32}$$

In Section 1.3, we established the law of countable subadditivity." From that result, we know that

$$P\left(\bigcup_{i=1}^{k} A_i\right) \leq \sum_{i=1}^{k} P(A_i). \tag{8.33}$$

Together, these two facts, plus Theorem 1.3.1, justify the following claims:

$$\begin{aligned} P\left(\bigcap_{i=1}^{k} A_i\right) &= P\left(\left(\bigcup_{i=1}^{k} A_i^c\right)^c\right) \\ &= 1 - P\left(\bigcup_{i=1}^{k} A_i^c\right) \\ &\geq 1 - \sum_{i=1}^{k} P(A_i^c) \end{aligned}$$

■

Suppose one has a pair of 95% confidence intervals, one for parameter θ_1 and one for parameter θ_2. For the random intervals I_1 and I_2 on which they are based, we have $P(I_1) = .95$ and $P(I_2) = .95$. We may infer from the Bonferroni inequality that $P(I_1 \cap I_2) \geq .90$. Thus, the rectangular region in the XY plane forms by crossing the I_1 interval on one axis with the I_2 interval on the other axis, provides a confidence region for the pair (θ_1, θ_2) whose confidence level is at least .90. This process holds regardless of the possible dependence of the statistics upon which these intervals are based, and it thus provides a versatile tool which is applicable even in situations in which the joint distribution of the two statistics is not known. The following example illustrates the method in a familiar setting.

Example 8.2.5. Suppose that $X_1, X_2, \ldots, X_n \sim^{iid} N(\mu, \sigma^2)$, where both μ and σ^2 are unknown. We know that the t-statistic given by

$$t = \frac{\bar{X} - \mu}{s/\sqrt{n}},$$

where, as usual, $\bar{X}$ and s are the sample mean and standard deviation, is a pivotal statistic for μ and gives rise to $(\bar{x} - (s/\sqrt{n})t_{n-1,.025}, \bar{x} + (s/\sqrt{n})t_{n-1,.025})$ as a 95% confidence interval for μ. Similarly, the ratio

$$R = \frac{(n-1)s^2}{\sigma^2}$$

is a pivotal statistic for the population variance and, as noted in Exercise 8.1.3, gives rise to a confidence interval for σ^2. A 95% confidence interval for σ^2 is given by $\left(\frac{(n-1)s^2}{\chi^2_{n-1,.025}}, \frac{(n-1)s^2}{\chi^2_{n-1,.097}}\right)$. Using the Bonferroni inequality, we may assert that the rectangle $(\bar{x} - (s/\sqrt{n})t_{n-1,.025}, \bar{x} + (s/\sqrt{n})t_{n-1,.025}) \times \left((n-1)s^2/\chi^2_{n-1,.025}, (n-1)s^2/\chi^2_{n-1,.097}\right)$ in the (μ, σ^2) plane is at least a 90% confidence region for the true value of (μ, σ^2). ■

It is worth noting that, in Example 8.2.5, the independence of $\bar{X}$ and s^2 computed from a normal random sample allows us to identify the joint coverage probability of the two intervals, when used simultaneously, as $(.95)^2 = .9025$. The Bonferroni approximation is not far off. Its accuracy is often diminished when the intervals involved are dependent, and the approximate confidence level (which is actually a lower bound) can be quite inaccurate when many confidence statements are made simultaneously. However, the difficulty in getting exact results in multi-parameter problems has given the method both some practical value and moderate amount of popularity.

Exercises 8.2.

1. Suppose that $X_1, \ldots, X_n \overset{\text{iid}}{\sim} G(p)$, the geometric distribution with mean $1/p$. Assume that the sample size n is sufficiently large to warrant to invocation of the Central Limit Theorem. Use the asymptotic distribution of $\hat{p} = 1/\bar{X}$ to obtain an approximate $100(1-\alpha)\%$ confidence interval for p.

   Suppose that $X_1, \ldots, X_n \overset{\text{iid}}{\sim} N(0, \sigma^2)$ (a) Obtain the asymptotic distribution of the second sample moment $m_2 = (1/n)\sum_{i=1}^n X_i^2$. (b) Identify a variance stabilizing transformation $g(m_2)$. (c) Obtain a parameter-free approximate $100(1-\alpha)\%$ confidence interval for σ^2.

2. Suppose that $(X_1, Y_1), \ldots, (X_n, Y_n)$ is a random sample from a bivariate normal distribution and that we wish to obtain a confidence interval for the population correlation coefficient ρ. The sample correlation coefficient r_n, given by

$$r_n = \frac{1}{n} \frac{\sum_{i=1}^n (X_i - \bar{X})(Y_i - \bar{Y})}{\sum_{i=1}^n (X_i - \bar{X})^2 \sum_{i=1}^n (Y_i - \bar{Y})^2}$$

is $\sqrt{n}$ consistent and asymptotically normal as shown below:

$$\sqrt{n}(r_n - \rho) \xrightarrow{D} Y \sim N(0, (1-\rho^2)^2).$$

Verify that the statistic Z_n known as "Fisher's Z" and defined as $Z_n = g(r_n)$, where g is the function defined as

$$g(x) = \frac{1}{2}\ln(\frac{1+x}{1-x}),$$

is a variance stabilizing transformation for r_n, and obtain, from the asymptotic distribution of Z_n, an approximate $100(1-\alpha)\%$ confidence interval for $g(\rho)$ and from that, an approximate $100(1-\alpha)\%$ confidence interval for ρ.

8.3 Sample Size Calculations

Work as a statistical consultant is usually extremely rewarding. Because Statistics is widely applicable across the sciences, a statistical consultant will have many opportunities to delve into quantitative problems in other fields of study. Sometimes the problems encountered have rather straightforward solutions and can be solved using standard models and well developed statistical theory. Often, however, the problem will have quirks that distinguish it from "textbook problems" and will thus require new ideas in modeling or in statistical inference. There are many positives from working on such problems. The consultant has the opportunity to delve deeply into the subject matter from which the problem is drawn, to work toward an appropriate statistical solution to the problem posed and, if successful, to be a participant in the discovery of new scientific insights and in the advancement of science in a particular problem area. Why am I waxing about the joys of statistical consulting? Well, in truth, it's mostly because I think it's something you should know. Since you are studying Statistics at a fairly advanced level, you yourself may be approached sometime to provide some statistical advice. But the real reason that I wandered in this direction is to set the scene for answering the most frequently asked question that a statistical consultant hears: How large a sample do I need to obtain useful and reliable information about the population I'm studying? In this section, we will treat that question and provide some guidance that you yourself might someday be using as your stock response to this question.

The field of Statistics would not have developed into the science that it is if it were not true that estimates derived from random samples tend to improve in accuracy and precision as the sample size increases. That being the case, one might then imagine that if some particular level of precision was required, one might be able to solve the inverse problem of identifying a minimal sample size with which the desired level of precision can be achieved. We will examine two statistical paradigms with some care, as these two problems occur with some frequency and the solution proposed to each of them may be used as a template for dealing with other scenarios.

Let us suppose, initially, that we wish to estimate the mean μ of a normal population with a known variance σ_0^2. If we have available a random sample of size n from the population, and if $\bar{X}$ is the sample mean, then $\bar{X} \sim N(\mu, \sigma_0^2/n)$. A confidence interval for μ, based on the sufficient statistic $\bar{X}$ might be reasonably expected to be as good an inferential statement as can be made about the unknown parameter μ. Given that $\bar{X} = \bar{x}$, the interval $(\bar{x}-(\sigma_0/\sqrt{n})(1.96), \bar{x}+(\sigma_0/\sqrt{n})(1.96))$ is a 95% confidence interval for μ. Now suppose that $n=16$, $\bar{X}=110$, and $\sigma_0=40$, so that our 95% confidence interval reduces to $(90.4, 129.6)$. It is quite possible that a confidence interval whose width is nearly 40 would be deemed quite useless in an experiment such as this. If these data happened to be the measured IQs of a sample of 16 seventh-grade students, one might consider knowing that the average IQ of the population is somewhere between 90 and 130 only tells you that this group is somewhere between brilliant and severely challenged. How could we have designed a more useful experiment?

It is convenient to think of a $100(1-\alpha)\%$ confidence interval for the mean of a normal population as a set a values between a lower and an upper confidence limit. Let's express these two limits as the numbers $\bar{x} \pm (\sigma_0/\sqrt{n})Z_{\alpha/2}$. Now when an experiment is being designed, the experimenter generally has an idea of the type of precision that would be required or desired in the application at hand. In the case of IQs, one might wish to have sufficient precision to be 95% sure that $\bar{X}$ is within 5 units of μ. Another way of representing that goal is to stipulate that one wishes to have a 95% confidence interval of the form $\bar{x} \pm 5$. Once this stipulation is made, one can solve for the value of n for which this goal is approximately achieved. In general, one may find the required sample size by setting the real and idealized $100(1-\alpha)\%$ confidence intervals for μ side by side. We would like to have

$$\bar{x} \pm (\sigma_0/\sqrt{n})Z_{\alpha/2} = \bar{x} \pm E, \tag{8.34}$$

where the desired bound E on the estimation error, predetermined by the experimenter, is generally referred to as the "margin of error." To determine the required sample size, we solve the equation

$$(\sigma_0/\sqrt{n})Z_{\alpha/2} = E$$

to obtain

$$n = \frac{\sigma_0^2 Z_{\alpha/2}^2}{E^2}. \tag{8.35}$$

In the IQ example above, we would find that

$$n = \frac{(1600)(1.96)^2}{25} = 245.862.$$

We would conclude that a random sample of at least $n = 246$ would be required to ensure a margin of error no greater than 5 units at a 95% level of confidence.

Let us peel away one unrealistic assumption in the developments above. Let's assume that the population variance σ^2 is unknown. It then would seem that the formula in (8.35) is unusable. Fortunately, this is not entirely true. What needs to be done when σ^2 is unknown is to estimate it. Since we are at the early stage of wondering what size sample to take, this may seem like advice that's difficult to implement. But in practice, we proceed iteratively by first taking a preliminary sample whose size n_0 is large enough to provide a reasonably good estimate of σ^2 yet is not so large that we would expect the actual sample size needed to be substantially smaller than n_0. In the example above, suppose that we felt that it was almost certain that it would take at least 100 observations to get the precision we want. (Judgments like that become increasingly more reliable as one's experience with sample size computations grows.) Suppose that drawing a random sample of size $n_0 = 100$ yields a sample variance of $s^2 = 1200$. Substituting this estimated variance in (8.35) yields

$$n = \frac{(1200)(1.96)^2}{25} = 184.397. \tag{8.36}$$

We could then declare that a sample of size 185 will guarantee the required precision at the 95% level of confidence. If we pursue this thought, we would proceed to draw an additional 85 subjects at random from the remaining population to fill out the required sample size. But of course this initial calculation involved an approximation which was based on a moderate-sized, but not huge, sample of 100 subjects. It is thus reasonable to iterate the process further in hopes of getting a more reliable approximation of the desired n. One can iterate as many times as one wishes, but it would be quite rare in practice to repeat this calculation more than once, and it is often the case that the single calculation is deemed to be sufficient. If we were to repeat it in this example, we would next estimate the population variance from the full 185 random subjects selected thus far. Suppose that the sample variance of this larger sample was found to be $s^2 = 1500$. Then the calculation of a new required sample size would lead us to

$$n = \frac{(1500)(1.96)^2}{25} = 230.496. \tag{8.37}$$

This second pass would suggest that a total sample size of 231 will provide a 95% confidence interval for μ with a margin of error $E = 5$. We would probably trust this approximation of n because the estimated variance in the second calculation was based of 185 observations, a number that would give us some assurance that our estimate of the unknown variance is reasonably reliable.

Let us now consider another common paradigm in which sample size calculation arises. Suppose that a random sample of n Bernoulli trials will be available, that is, suppose that we will be able to observe $X_1, \dots, X_n \overset{\text{iid}}{\sim} B(1, p)$. Picture these observations as associated with n tosses of a bent coin, where $p = P(\text{heads})$. In estimating p from this sample, one may, as above, be interested in a specific level of precision in one's confidence interval. If n isn't too small (that is, if the usual rule of thumb "$np > 5$ and $n(1-p) > 5$" is satisfied), we can treat the sample proportion

$$\hat{p} = \frac{1}{n}\sum_{i=1}^{n} X_i$$

as having an approximate normal distribution, that is,

$$\hat{p} \sim N\left(p, \frac{pq}{n}\right), \quad \text{approximately,} \tag{8.38}$$

where $q = 1 - p$. From (8.38), one may obtain the $100(1-\alpha)\%$ confidence limits for p given by

$$\hat{p} \pm \sqrt{\frac{pq}{n}} Z_{\alpha/2}. \tag{8.39}$$

Ignore, for a moment, that the "confidence interval" above depends on the unknown parameter p. These limits may nonetheless be equated with the ideal $100(1-\alpha)\%$ confidence limits with the predetermined margin of error E, that is, we may solve for the sample size n for which

$$\sqrt{\frac{pq}{n}} Z_{\alpha/2} = E,$$

an equation that leads to the solution

$$n = \frac{pqZ^2_{\alpha/2}}{E^2}. \tag{8.40}$$

Now since the formula for n in (8.40) involves the unknown p, one must somehow come up with an approximation of p to render the equation useable. In this instance, there are two reasonable approaches to replacing p in (8.40) by a real number. The first is a conservative method which takes advantage of the known form of the variance of $\hat{p}$. A graph of the function $g(p) = p(1-p)$ is shown below.

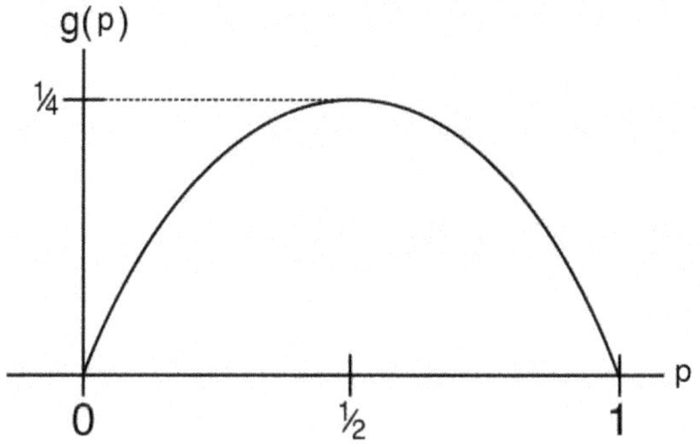

Figure 8.3.1. Graph of $g(p) = p(1-p)$ for $0 \leq p \leq 1$.

It is evident that, for $0 \leq p \leq 1$, the product $p(1-p)$ can never exceed $1/4$, its value when $p = 1/2$. Thus, substituting the value $1/2$ for p and for q in (8.40) results in a sample size n that is sure to be as large as required, regardless of the true value of the parameter p. Further, it is worthy of note that the function $g(p) = p(1-p)$ is quite a slowly varying function of p for p close to $1/2$. Thus, not much is lost if the true p is not exactly equal to $1/2$, but is actually in a neighborhood of $1/2$ like the interval $(.3, .7)$. For values of p in that interval, the product pq varies from 0.21 to 0.25.

When the true value of p is thought to be more distant from .5 than about 0.2, or if one wants to gain a little accuracy by estimating p rather than simply approximating it as .5, one can use an iterative process like the one used in obtaining the sample sizes shown in (8.36) and (8.37). Suppose that a preliminary sample of $n_0 = 100$ Bernoulli variables is drawn and that the preliminary estimate $\hat{p} = .33$ of p is computed from this sample. Substituting this estimated p into the formula in (8.40), with $\alpha = .05$ and $E = .06$, we have

$$n = \frac{(.33)(.67)(1.96)^2}{(.06)^2} = 235.938. \tag{8.41}$$

From this, one would conclude that a sample of size $n = 236$, is needed to obtain an approximate 95% confidence interval for p with margin of error $E = .06$. To complete the process, a random sample of 136 additional Bernoulli trials would be drawn and the confidence interval for p would be based on the resulting $\hat{p}$ based on the full sample. Of course this process could be repeated using the new estimate of p based on the sample of 236 observations, and doing so would either confirm that the sample of size $n = 236$ is adequate or that it should be increased to achieve the desired precision at the specified margin of error.

The treatment above for sample size calculations when estimating a population proportion is exact when the population may be thought of as infinite (as in a coin-tossing scenario), but requires some modifications when the sample is drawn from a finite population. As you know, the correct distribution for the number of observed "successes" in the latter case is the hypergeometric distribution. Now, two limit theorems with which you are familiar can usually be applied in the finite-population case: the binomial distribution as a limit of the hypergeometric and the normal distribution as a limit of the binomial distribution. Together, these two results will justify treating a sample proportion from a sample drawn from a finite population as a normal random variable, provided that both the population size N and the sample size n are sufficiently large. The difference in the corresponding formula for sample size computation in the finite-population case is that in this latter case, the sample proportion has greater precision than it has when the population is taken as infinite. The following fact is proven in most textbooks on sampling techniques (see, for example, Cochran (1977)).

Theorem 8.3.1. When a simple random sample of size n is drawn from a dichotomous population of size N, the sample proportion $\hat{p}$ of "successes" has expected value p, the true proportion of successes, and has a standard error given by

$$\sigma_{\hat{p}} = \sqrt{\frac{pq}{n}}\sqrt{\frac{N-n}{N-1}}. \tag{8.42}$$

The formula in (8.42) is interesting and merits some comment. The fraction $f = n/N$ is called the sampling fraction, and the fraction $(N-n)/(N-1)$ is called the "finite population correction." The latter value adjusts the standard error $\sqrt{pq/n}$ in the infinite-population case downward. As N $\to \infty$, the finite population correction vanishes, as it should. In practice, the finite population correction (fpc) is typically ignored when $f < .05$. When the fpc is taken into account, the formula for obtaining the required sample size for a $100(1-\alpha)\%$ confidence interval for p with margin of error E changes to the following:

$$n = \frac{\frac{pqZ^2_{\alpha/2}}{E^2}}{1 + \frac{1}{N}\left(\frac{pqZ^2_{\alpha/2}}{E^2} - 1\right)}. \tag{8.43}$$

The formula in (8.43) utilizes the fpc and will always result is a smaller value of n than is given by the formula in (8.40). In practice, p and q in (8.43) are replaced by $\hat{p}$ and $\hat{q}$ obtained from a preliminary sample.

Remark 8.3.1. Since I framed this section as guidance that a statistical consultant might give when asked about the sample size needed in a given experimental study, I should conclude with a comment about how this advice must be peppered with a dash of reality. It is often the case that a calculated sample size n obtained to address some pre-specified precision will appear to be shockingly large to the experimenter. For example, if a client is interested in conducting a survey in which population proportions may be estimated with a margin of error of .02, the formula in (8.40) will generate the answer $n = 2401$ when p is set equal to $1/2$ and a 95% confidence interval is required. To someone who had in mind a sample of about 200 subjects, this news can be disconcerting. The practical questions that need to be answered before calculations like this are made include (a) What precision is truly needed (not just what sounds nice)?, (b) Is there an interval of values that one can safely assume contain the true parameter value? (something that one might be comfortable stating on the basis of experience or knowledge with related studies), (c) How large is the target population? (d) What level of confidence would be minimally acceptable? and, the most practical question of all, (e) How large is your budget? Sometimes one has to work backwards from this last question and candidly tell the client the level of precision and confidence that his/her budget will allow. A former colleague of mine has a T-shirt with the statement "Up your sample size!" on the front. I don't think he meant to be rude. There may be more gentle ways of saying it, but the message is on target. In general, experimenters want to take as large a sample as they can afford, keeping in mind both the present study and future investigations that are planned or on the radar screen.

Exercises 8.3.

1. Marge N. O'Vera's polling firm has been retained to estimate the mean time to retirement, in years, of current state workers in California. (a) If Marge sampled 100 individuals and the mean and standard deviation of their answers were found to be 15 and 5 years, respectively, what would her 95% confidence interval for μ look like? (b) If a 95% confidence interval is required with a margin of error of $E = .5$ years, how large a sample would be needed to guarantee this level of precision? Use the sample standard deviation from part (a) as an estimate of σ.

2. A random sample of 100 registered voters in California was asked if they supported the governor's budget proposal, and 43 of them said yes. (a) If you assumed that the true population proportion in support of the governor's proposal is $p = 1/2$, and X represents the number of supporters in a random sample of size 100, obtain the normal approximation for $P(X \leq 43)$. (b) Based on the data collected above, obtain a 95% confidence interval for the unknown proportion p. Express your answer as an interval (L, U). (c) To obtain a 95% confidence interval with margin of error .05 (that is, an interval with confidence limits $\hat{p} \pm .05$), what total sample size n would be needed? First, use the sample proportion .43 from the preliminary sample above as an initial guess at the true p. Then, repeat the calculation under the assumption that p is completely unknown.

3. Ed "the head" Bedford is a whiz at school and doesn't have to study much. Consequently, he has a lot of extra time on his hands. Ed picked up a broken button on the sidewalk on his way home from school and noted that one side was smooth and one side was rough. Ed tossed the button 100 times, and the smooth side came up on top 58 times. (a) Find a 95% confidence interval for p, the probability that the smooth side will come up on a given toss. (b) If we were to assume, as Ed does, that nothing at all is known, *a priori*, about p, how many tosses would be needed to ensure that Ed's 95% confidence interval for p has a margin of error $E = .03$.

4. Derive the equation in (8.43) from the equation in (8.42).

8.4 Tolerance Intervals (Optional)

In this section, we will discuss a nonparametric approach to interval estimation that yields what is called a tolerance interval. Such an interval differs from a confidence interval for a population parameter in that the goal is to state with some assurance that a certain interval contains a fixed proportion of the population. A typical statement of this sort would say that "We are 95% confident that the interval contains 90% of the population." As this statement suggests, there will be two probability levels to keep track of here—the confidence we have in the procedure and the targeted proportion of the population we are trying to capture.

The method of tolerance intervals differs from the methods discussed above in that the approach is applicable very generally, that is, applies to any continuous, strictly increasing distribution F. The method is thus referred to as *distribution-free*, and it depends solely on the so-called probability transform discussed in Section 6.6. Recall that we showed there that if $X \sim F$, where F is a continuous strictly increasing distribution on some interval of the real line, then the random variable $Y = F(X)$ has a $U(0,1)$ distribution. We will now return to our discussion of order statistics, commenced in Section 4.7, and develop some additional results which allow us to exploit the fact that the distribution of $F(X)$ is known, even when the exact form of F is unknown.

Suppose that $X_1,\ldots,X_n \overset{\text{iid}}{\sim} F$, where F is assumed, henceforth, to be a continuous and strictly increasing cdf with density f. Since each variable $F(X_i)$ is independent and has the same distribution, we may then assert that $F(X_1),\ldots,F(X_n) \overset{\text{iid}}{\sim} U(0,1)$. In particular, the ordered values of $F(X_i)$ have the same distribution as the collection of order statistics drawn from the $U(0,1)$ distribution. In other words, we have, for example, that

$$F(X_{(i)}) \sim Be(i, n-i+1) \qquad \text{for } i = 1,2,\ldots,n. \tag{8.44}$$

For $i = 1,\ldots,n$, the differences $X_{(i)} - X_{(i-1)}$ between consecutive order statistics (where $X_{(0)}$ is defined as the greatest lower bound of the support set of the distribution F) are referred to as the "spacings" between order statistics. We will utilize the spacings between the $U(0,1)$ order statistics $F(X_1),\ldots,F(X_n)$ in constructing tolerance intervals with a suitably high likelihood of covering a desired percentage of the population. Let $S_i = F(X_{(i)}) - F(X_{(i-1)})$ for $i = 1,2,\ldots,n$. We will now obtain the joint probability distribution of $(S_1,\ldots,S_n)$. To do this, we will draw upon a result stated in Section 4.7, namely, the form of the joint density of the order statistics from the original sample from F, a distribution with support set (a,b). This density was given in (4.85), and is repeated here for convenience.

$$f_{(X_{(1)},X_{(2)},\ldots,X_{(n)})}(x_{(1)},x_{(2)},\ldots,x_{(n)}) = n!\prod_{i=1}^{n} f(x_{(i)}) \text{for } a < x_{(1)} < \cdots < x_{(n)} < b. \tag{8.45}$$

Now, let us use the notation $U_{(i)} = F(X_{(i)})$, $i = 1,2,\ldots,n$, in recognition of the fact that these ordered values are distributed like the order statistics from $U(0,1)$. Since the density of the $U(0,1)$ distribution is $f(x) = 1$ for $0 < x < 1$, we obtain from (8.45) that

$$f_{(U_{(1)},U_{(2)},\ldots,U_{(n)})}(u_{(1)},u_{(2)},\cdots < u_{(n)}) = n! \qquad \text{for } 0 < u_{(1)} < u_{(2)} < \ldots, u_{(n)} < 1. \tag{8.46}$$

We now apply the CoV technique to the one-to-one transformation that takes the n-dimensional random vector $\mathbf{U}$ to the n-dimensional vector of spacings $\mathbf{S}$. Recall that

$$S_1 = U_{(1)}, S_2 = U_{(2)} - U_{(1)}, \ldots, S_n = U_{(n)} - U_{(n-1)}.$$

Inverting this relationship yields

$$\begin{aligned} U_{(1)} &= S_1 \\ U_{(2)} &= S_1 + S_2 \\ &\cdots \\ U_{(n)} &= S_1 + S_2 + \cdots + S_n. \end{aligned}$$

The Jacobian of the inverse transformation has determinant 1, so that $|J| = 1$. Finally, the range of the variables $S_1,\ldots,S_n$ corresponding to the range of the original variables $0 < U_{1)} < U_{(2)} < \ldots, U_{(n)} < 1$ is the simplex $\{\mathbf{S} \in (0,1)^n | \sum_{i=1}^n S_i < 1\}$. It follows that the joint density of $S_1,\ldots,S_n$ is given by

$$f_{\mathbf{S}}(s_1,\ldots,s_n) = n! \qquad \text{for } \mathbf{s} \in (0,1)^n \text{ for which } \sum_{i=1}^n s_i < 1. \tag{8.47}$$

Given the joint density of the spacings in (8.47), some special scrutiny leads to an interesting and very useful conclusion. For example, we may deduce the fact that all sums of a set of r spacings have the same distribution. This claim is clear if we write

$$P(\sum_{i=1}^r S_{j_i} \le s) = \int_0^s \int_0^{s-s_{j_r}} \int_0^{s-s_{j_r}-s_{j_{r-1}}} \ldots \int_0^{s-\sum_{i=2}^r s_{j_i}} n! ds_{j_1} \ldots ds_{j_{r-2}} ds_{j_{r-1}} ds_{j_r}. \tag{8.48}$$

It is evident that the r variables of integration in (8.48) can be viewed as dummy variables, and that the value of the integral will be the same for any for any choice of indexes $j_1, j_2, \ldots, j_r$. A little reflection shows that this same conclusion may be obtained directly from the fact that the joint distribution is symmetric in the spacings $s_1,\ldots,s_n$ in the sense that the distribution remains the same for any permutation of the spacings. Since all sums of r spacings have the same distribution, it follows that, for $i < j$, the difference $U_{(j)} - U_{(i)} = \sum_{k=i+1}^j S_k$ has the same distribution as $U_{(j-i)} = \sum_{k=1}^{j-i} S_k$. Now, since $U_{(j-i)}$ is the $(j-i)^{th}$ order statistic from a sample of size n from the $U(0,1)$ distribution, we have that

$$U_{(j-i)} \sim Be(j-i, n-j+i+1).$$

Recalling that $U_{(k)} = F(X_{(k)})$, where $X_{(k)}$ is the kth order statistic from the original sample $X_1,\ldots,X_n$, we may write, for $i < j$, and for p between 0 and 1,

$$\begin{aligned} P\left(F(X_{(j)}) - F(X_{(i)}) \ge p\right) &= P\left(U_{(j)} - U_{(i)} \ge p\right) \\ &= P\left(U_{(j-i)} \ge p\right) \\ &= \int_p^1 \frac{\Gamma(n+1)}{\Gamma(j-i)\Gamma(n-i+j+1)} x^{j-i-1}(1-x)^{n-j+i} dx. \end{aligned} \tag{8.49}$$

Now the integral in (8.49) is a so-called incomplete beta integral. It can be shown by successive integration by parts (see Exercise 8.4.2) that this integral is equivalent to the partial binomial sum, that is,

$$P(F(X_{(j)} - F(X_{(i)}) \ge p) = \sum_{k=0}^{j-i-1} \binom{n}{k} p^k (1-p)^{n-k}. \tag{8.50}$$

For a predetermined value of p, one can determine the value of $(j-i)$ for which this probability in (8.50) is suitably large. For $0 < \pi < 1$, the statement $P(F(X_{(j)}) - F(X_{(i)}) \ge p) = \pi$ is equivalent to the statement that the interval $(X_{(i)}, X_{(j)})$ contains at least $100p\%$ of the population with probability π. The interval $(X_{(i)}, X_{(j)})$ is called a $100\pi\%$ tolerance interval for $100p\%$ of the population. The numerical interval that results from the observed value $X_{(i)} = x_{(i)}$ and $X_{(j)} = x_{(j)}$ has a frequency interpretation similar to that of confidence intervals. If the process which generated the numerical tolerance interval were repeated many times, the interval $(x_{(i)}, x_{(j)})$ would contain at least $100p\%$ of the population about $100\pi\%$ of the time. The following example illustrates the process described above.

Example 8.4.1. Suppose a random sample of 20 observations are drawn from a distribution that is assumed to be continuous and strictly increasing. Suppose that one wishes to obtain, approximately, a 90% tolerance interval for 70% of the population. From a binomial table with $n = 20$ and $p = 0.7$, one can determine that

$$\sum_{k=0}^{17} \binom{20}{k} (.7)^k (.3)^{20-k} = .893. \tag{8.51}$$

From this, we see that if $j - i = 18$, then the interval $(X_{(i)}, X_{(j)})$ provides a 89.3% tolerance interval for 70% of the population. Either of the intervals $(X_{(1)}, X_{(19)})$, or $(X_{(2)}, X_{(20)})$ may be chosen for this purpose. ■

Exercises 8.4.

1. Let $X_1, \ldots, X_{100}$ be a random sample from a continuous, strictly increasing distribution F. Verify that the interval $(X_{(1)}, X_{(100)})$ is a 95% tolerance interval for 97.05% of the population.

2. Use mathematical induction and repeated integration by parts to show that, for integers k with $0 \leq k \leq n$, the following identity holds:

$$\int_0^p \frac{\Gamma(n+1)}{\Gamma(k)\Gamma(n-k+1)} x^{k-1}(1-x)^{n-k} dx = \sum_{i=k}^{n} \binom{n}{i} p^i (1-p)^{n-i}. \tag{8.52}$$

It is helpful, in verifying (8.52), to begin by expressing the gamma functions as factorials. Note that the identity above implies the complementary identity

$$\int_p^1 \frac{\Gamma(n+1)}{\Gamma(k)\Gamma(n-k+1)} x^{k-1}(1-x)^{n-k} dx = \sum_{i=0}^{k-1} \binom{n}{i} p^i (1-p)^{n-i}$$

holds, as claimed in Section 8.4 regarding the integral in (8.49).

8.5 Chapter Problems

1. Let $X_1, X_2, \ldots, X_n \stackrel{\text{iid}}{\sim} P(1, \theta)$, the one parameter Pareto distribution with density given in (3.36) with the scale parameter θ. The first order statistic $X_{(1)}$ is a sufficient statistic for θ. (a) Show that $T = X_{(1)}/\theta$ is a pivotal statistic. (b) Obtain an expression for a $100(1-\alpha)\%$ confidence interval for θ. (c) If $X_{(1)}$ was observed to be 14, obtain a 95% confidence interval for θ.

2. Let $X_1, X_2, \ldots, X_n \stackrel{\text{iid}}{\sim} U(\theta - 1/2, \theta + 1/2)$. The order statistics $(X_{(1)}, X_{(n)})$ are sufficient statistics for θ. From the joint distribution of $(X_{(1)}, X_{(n)})$, derive the distribution of the statistic $T = (X_{(1)} + X_{(n)})/2$. Identify a pivotal statistic based on T and use it to obtain an expression for a 100(1 - α)% confidence interval for θ.

3. Let $X_1, X_2, \ldots, X_n \stackrel{\text{iid}}{\sim} N(\mu_x, \sigma_1^2)$ and $Y_1, \ldots, Y_m \stackrel{\text{iid}}{\sim} N(\mu_y, \sigma_2^2)$ be two independent samples from different normal populations. Assume that the two population variances are known. Identify the distribution of the difference $\bar{X} - \bar{Y}$ of sample means and use it to form a pivotal statistic whose distribution does not depend on the population means. Use that statistic to obtain a $100(1-\alpha)\%$ confidence interval for $\mu_x - \mu_y$.

4. Suppose that $X_1, X_2, \ldots, X_n \stackrel{\text{iid}}{\sim} F_\theta$, where F_θ is a distribution on the positive real line with density function
$$f_\theta(x) = e^{-(x-\theta)} I_{(\theta,\infty)}(x).$$
(a) Use the Pivotal Statistic Method to obtain an expression for a $100(1-\alpha)\%$ confidence interval for θ based on the sufficient statistic $X_{(1)}$. (b) Show that the shortest $100(1-\alpha)\%$ confidence interval of the form $(X_{(1)} - A, X_{(1)} - B)$ is the interval $(X_{(1)} - \chi^2_{2,\alpha}/2n, X_{(1)})$.

5. Let $X_1, X_2, \ldots, X_n \stackrel{\text{iid}}{\sim} W(\alpha, \beta)$, the two-parameter Weibull distribution. Suppose that the parameter α is known. Show that $T = (1/\beta)\sum_{i=1}^n X_i^\alpha$ is a pivotal statistic, and use it to obtain an expression for a $100(1-\alpha)\%$ confidence interval for β.

6. Suppose that $X_1, X_2, \ldots, X_n \stackrel{\text{iid}}{\sim} Be(\theta, 1)$. Use the Pivotal Statistic Method to obtain an expression for a $100(1-\alpha)\%$ confidence interval for θ. (Hint: Recall that if $X \sim Be(\theta, 1)$, then $-\ln X \sim Exp(\theta)$, the exponential distribution with failure rate θ.)

7. Suppose that $X_1, X_2, \ldots, X_n \stackrel{\text{iid}}{\sim} L(1, \beta)$, the one-parameter logistic distribution with scale parameter β. Use the Pivotal Statistic Method to derive a $100(1-\alpha)\%$ confidence interval for β.

8. Suppose that $X_1, X_2, \ldots, X_n \stackrel{\text{iid}}{\sim} L(\mu, 1)$, the one-parameter logistic distribution with mean μ. Use the Pivotal Statistic Method to derive a $100(1-\alpha)\%$ confidence interval for μ.

9. Suppose that $X_1, X_2, \ldots, X_n \stackrel{\text{iid}}{\sim} DE(\mu, 1)$, the one-parameter double exponential distribution with mean μ. Use the Pivotal Statistic Method to derive a $100(1-\alpha)\%$ confidence interval for μ.

10. Suppose that $X_1, X_2, \ldots, X_n \stackrel{\text{iid}}{\sim} DE(1, \lambda)$, the one-parameter double exponential distribution with scale parameter λ. Use the Pivotal Statistic Method to derive a $100(1-\alpha)\%$ confidence interval for λ.

11. Suppose that $X_1, X_2, \ldots, X_n \stackrel{\text{iid}}{\sim} LN(\mu, 1)$, the one-parameter lognormal distribution for which $E(\ln X_i) = \mu$. Use the Pivotal Statistic Method to derive a $100(1-\alpha)\%$ confidence interval for μ.

12. Suppose that $X_1, X_2, \ldots, X_n \stackrel{\text{iid}}{\sim} LN(0, \sigma^2)$, the one-parameter lognormal distribution for which $V(\ln X_i) = \sigma^2$. Use the Pivotal Statistic Method to derive a $100(1-\alpha)\%$ confidence interval for σ^2.

13. On the surface of the planet Mars, the vehicle Curiosity made carefully selected probes to measure the hardness of Martian soil. These probes were considered to be risky because of the non-negligible risk that the instrument to make the measurement would break. In a particular region under study, the resistance of the surface to the probe has been modeled as a Weibull variable. Because of risk considerations, only a single measurement X was taken, with $X \sim W(\theta, 1)$. (a) Define a statistic $T = g(X, \theta)$ whose distribution does not

depend on θ and use it as a pivotal statistic to obtain a $100(1-\alpha)\%$ confidence interval for θ. (b) If $X = 1087$ pounds was the measured resistance, obtain a 95% confidence interval for θ.

14. Suppose that $X_1, X_2, \ldots, X_n \overset{\text{iid}}{\sim} C(m,1)$, the one-parameter Cauchy distribution with median m. Use the Pivotal Statistic Method to derive a $100(1-\alpha)\%$ confidence interval for m. (Hint: Use the fact that the cumulative distribution F of the $C(\mu,1)$ model is given by
$$F(x) = \frac{1}{\pi}\left[\arctan(x-\mu) + \frac{\pi}{2}\right] \text{ for } x \in (-\infty, \infty).)$$

15. Suppose that $X_1, X_2, \ldots, X_n \overset{\text{iid}}{\sim} \Gamma(k,\beta)$, the gamma distribution with known integer-valued shape parameter k. Use the Pivotal Statistic Method to derive a $100(1-\alpha)\%$ confidence interval for β. (Hint: The chi-square distribution will come into play.)

16. Suppose that $X_1, X_2, \ldots, X_n \overset{\text{iid}}{\sim} U(0,\theta)$, the uniform distribution on the interval $(0,\theta)$. The maximum order statistic $X_{(n)}$ is sufficient for θ. (a) Use the Pivotal Statistic Method to derive $100(1-\alpha)\%$ confidence interval for θ of the form $(AX_{(n)}, BX_{(n)})$. (b) Determine the values of A and B which yield the $100(1-\alpha)\%$ confidence interval with the shortest length.

17. A researcher in the health sciences reported that the mean response time to a certain drug she administered to 9 patients was 25 seconds, with a standard deviation $s = 6$ seconds. She reported the confidence interval $(21.28, 28.72)$ for the mean μ of the corresponding population of patients, but neglected to mention the level of confidence. Find the confidence level associated with this interval. (Assume the data is normally distributed.)

18. Let $X_1, \ldots, X_n \overset{\text{iid}}{\sim} U(\theta - 1/2, \theta + 1/2)$. (a) Calculate the confidence level of the confidence interval $(X_{(i)}, X_{(j)})$ for θ, where $i < n/2 < j$. (b) How large a sample size must be taken so that the confidence level of the interval $(X_{(1)}, X_{(n)})$ is at least 95%?

19. Suppose that the sequence of random variables $\{T_n\}$ is consistent for a real-valued parameter θ and is asymptotically normal, with $\sqrt{n}(T_n - \theta) \xrightarrow{D} X \sim N(0,\theta)$. Use a variance stabilizing transformation to develop an approximate $100(1-\alpha)\%$ confidence interval for θ.

20. Let $X_1, \ldots, X_n \overset{\text{iid}}{\sim} Exp(\lambda)$, the exponential distribution with failure rate λ. We showed in Sections 7.2 and 7.3 that $\hat{\lambda} = 1/\bar{X}$ is both the MME and the MLE of λ, and that its asymptotic distribution is given by
$$\sqrt{n}\left(\hat{\lambda} - \lambda\right) \xrightarrow{D} W \sim N\left(0, \lambda^2\right). \tag{8.53}$$
Use the normal distribution in (8.53) to obtain, via a variance stabilizing transformation, an approximate $100(1-\alpha)\%$ confidence interval for λ.

21. Suppose that $X_1, X_2, \ldots, X_n, \ldots \overset{\text{iid}}{\sim} P(\lambda)$, the Poisson distribution with mean λ. From the Central Limit Theorem, we have that the sample mean $\bar{X}$ is asymptotically normal, with
$$\sqrt{n}(\bar{X} - \lambda) \xrightarrow{D} V \sim N(0,\lambda).$$
(a) Find the asymptotic distribution of $g(\bar{X})$, where $g(x) = \sqrt{x}$, and use it to obtain a $100(1-\alpha)\%$ confidence interval for $\sqrt{\lambda}$. (b) Transform this interval to obtain a parameter-free $100(1-\alpha)\%$ confidence interval for λ.

22. Suppose that $X_1, X_2, \ldots, X_n \overset{\text{iid}}{\sim} F_p$, where F_p is a discrete distribution with probability mass function

$$p(x) = p(1-p)^x \text{ for } x = 0, 1, 2, \ldots.$$

(See Example 7.3.9.) The MLE $\hat{p} = 1/(1+\bar{X})$ has the asymptotic distribution below:

$$\sqrt{n}(\hat{p} - p) \xrightarrow{D} Y \sim N(0, (1-p)p^2). \quad (8.54)$$

Use the normal distribution in (8.54) to obtain, via a variance stabilizing transformation, an approximate $100(1-\alpha)\%$ confidence interval for p.

23. Let $X_1, \ldots, X_n, \ldots \overset{\text{iid}}{\sim} B(1, p)$, and let $\hat{p}_n$ be the sample proportion based on the first n observations. The Central Limit Theorem implies that

$$\sqrt{n}(\hat{p}_n - p) \xrightarrow{D} Y \sim N(0, p(1-p)) \text{ as } n \to \infty.$$

Note that the asymptotic variance of $g(\hat{p}_n) = \arcsin\sqrt{\hat{p}_n}$ does not depend on p. Use the latter distribution to obtain a 95% confidence interval for $g(p)$. From this interval, get a 95% confidence interval for p whose endpoints do not depend on the unknown parameter p.

24. Suppose that $X_n \sim B(n, p)$, and let $\hat{p}_n = X/n$. The CLT implies that

$$\sqrt{n}(\hat{p}_n - p) \xrightarrow{D} Y \sim N(0, p(1-p)).$$

(a) Let $l_n = \ln[\hat{p}_n/(1-\hat{p}_n)]$, an estimator of the parameter $\ln[p/(1-p)]$ known as the log odds. Obtain the asymptotic distribution of the sample log odds $\hat{l}_n = g(\hat{p}_n) = \ln(\hat{p}_n/(1-\hat{p}_n))$ as $n \to \infty$. (b) Find a variance stabilizing transformation for $\hat{l}_n$, and use it to obtain a parameter-free approximate $100(1-\alpha)\%$ confidence interval for l.

25. Let $(X_1, Y_1), \ldots, (X_n, Y_n), \ldots$ be a random sample from a bivariate distribution with finite means, variances, and covariance. The sample correlation coefficient r_n is generally used to estimate the population correlation coefficient $\rho = \frac{Cov(X,Y)}{\sigma(X)\sigma(Y)}$. It can be shown that

$$\sqrt{n}(r_n - \rho) \xrightarrow{D} Y \sim N(0, (1-\rho^2)^2).$$

R. A. Fisher noted that the transformation $Z_n = g(r_n)$, where g is given by

$$g(x) = \frac{1}{2}\ln\left(\frac{1+x}{1-x}\right),$$

stabilized the variance of r_n. (a) Use the asymptotic distribution of Z_n to obtain an approximate $100(1-\alpha)\%$ confidence interval for $g(\rho)$. (b) Transform this interval into a parameter-free approximate $100(1-\alpha)\%$ confidence interval for ρ.

26. Sunkist, Inc. has managed to cross a cotton bush with an orange tree, producing a tree (or perhaps, more accurately, a "trush") whose fruit has the taste and consistency of orange flavored cotton candy. One hundred Sunkist customers were polled, and 65 indicated that they were positive toward the new product. (a) Obtain an approximate 98% confidence interval for the true proportion p of Sunkist customers that would be positive toward this product. (b) Consider the 100 customers above as a preliminary sample. How many additional customers should be polled if Sunkist wishes to have a 98% confidence interval with margin of error $E = .05$.

27. Rework part (b) of Problem 26 if it is known that the target population for this experiment is the collection of owners of cotton candy stands at state and county fairs in the US, a population of known size $N = 743$.

28. Suppose that you have been retained as a consultant in a sampling experiment involving measuring steel rods (in inches). A sample $X_1, X_2, \ldots, X_n \stackrel{\text{iid}}{\sim} N(\mu, \sigma^2)$ will be obtained, where the variance σ^2 is known to be 70. If a 95% confidence interval of total width 3 inches is required (that is, a confidence interval with limits $\bar{X} \pm 1.5$), what sample size will you recommend to your client?

29. The national mean score on the SAT test (including verbal, quantitative, and written parts) seems to vary from year to year, so that last year's mean score is not a reliable predictor of this year's mean score. Anyway, these means would be inapplicable to students at Carnegie High (CH) in Pittsburgh, a school which only admits kids whose parents have IQs above 160. As it turns out though, the variability in SAT scores of CH seniors has been shown to be pretty constant (at or around $\sigma = 150$) over time. Let's assume that the population standard deviation of SAT scores for this year's CH seniors is known to be $\sigma = 150$. A random sample of 25 seniors at Carnegie High had an average SAT score of 1956 this year. (a) Obtain a 95% confidence interval for the overall mean SAT score μ of Carnegie High School seniors. (b) How large a sample would be required to obtain a 95% confidence interval for μ with margin of error $E = 40$?

30. Sixty of the 200 undergrads at Sacramento City College who were interviewed last week have part-time jobs. (a) Obtain a 95% confidence interval for the proportion p of all SCC undergrads who have part-time jobs. (b) Using the conservative assumption that the true value of p is $1/2$, how many additional undergrads should be interviewed in order to obtain a 95% confidence interval with margin of error $E = .05$? (c) If the true value p was estimated from the "preliminary" sample of 200 mentioned above, how many additional undergrads should be interviewed in order to obtain a 95% confidence interval with margin of error $E = .05$?

31. Suppose a bent coin is to be tossed n times in order to estimate the probability of heads p. (a) If the coin is tossed $n = 20$ times and comes up heads 6 times, obtain an approximate 95% confidence interval (or 95% confidence limits) for p. (b) Since the initial experiment above does not have sufficient precision in estimating p, assume that the true p is not known. Find the sample size n which guarantees that the largest possible margin of error for a 95% confidence interval for p is $E = .04$, regardless of the true value of p.

32. The Amazing Hairballi, a cat made famous by its appearance on the David Letterman Show's segment "Stupid Animal Tricks," is now being studied by Animal Scientist U. R. Maaaavelous. In 400 trials, Hairballi hit (with a hairball) the safe area on a rotating human target 90% of the time. (a) Obtain 95% confidence limits for p, the probability that Hairballi will hit the safe area in a random trial. (b) Suppose that Dr. Maaaavelous would like to obtain a 95% confidence interval for p with margin of error $E = .02$. What sample size would be needed? (You may assume that the value .9 is as an excellent approximation for the true value of p.)

33. Gray Matter, a noted Sacramento brain surgeon, is trying to perfect a gene-implant technique that stands to improve people's IQs. While lab mice generally take 20 seconds to go through a standard maze, 30 mice in Matter's lab had a mean completion time of 16.8 seconds, with a standard deviation of 4.2 seconds. You may assume normality as needed.

(a) Obtain 95% confidence limits for μ, the post-implant population mean. (b) Assume the population standard deviation σ of the maze-completion time for Matter mice is well approximated by the number 4.2 above. Calculate the sample size needed to obtain a 95% confidence interval for μ with maximum allowable error $E = .6$.

34. Phillip A. Coin picked up a bent coin on the sidewalk and noted that one side was smooth and one side was rough. (a) Suppose Phil tossed the coin 100 times, and the smooth side came up on top 70 times. Find 95% confidence limits for p, the probability that the smooth side will come up on a given toss. (b) If we were to assume, as Phil does, that nothing at all is known about p, how many tosses would be needed to ensure that Phil's 95% confidence interval for p has maximum allowable error $E = .03$.

35. Suppose a random sample of size 6 is drawn from a continuous distribution F. Show that the interval $(X_{(1)}, X_{(6)})$ is a 34% tolerance interval for 80% of the population.

36. Let $X_1, \dots, X_n$ be a random sample from a continuous distribution F. How large must n be so that $(X_{(1)}, X_{(n)})$ is a 95% tolerance interval for 50% of the population.

37. Let $X_1, \dots, X_n$ be a random sample drawn from the $U(0,1)$ distribution, and let R be the range of the sample, that is, let $R = X_{(n)} - X_{(1)}$. (a) Show that, for any positive integer n, $P(R > p) = 1 - np^{n-1} + (n-1)p^n$. (b) Apply this expression to solve Problem 8.5.36.

38. Let $X_1, \dots, X_n$ be a random sample from a continuous distribution F. Suppose that you need a 95% upper tolerance limit for 90% of the population, that is, you'd like to ensure that $P(F(X_{(n)}) > .9) \geq .95$. What is the smallest sample size that guarantees such a result?

9

The Bayesian Approach to Estimation

9.1 The Bayesian Paradigm

In this section, I will describe the way in which a "Bayesian statistician" thinks about an estimation problem. You will soon see that the approach is substantially different from the classical approach described in Chapters 6–8. The two approaches start in the same place, but then diverge rather strikingly. As you know, the classical (or frequentist) approach to statistical inference begins with the stipulation of a probability model that is thought of as a reasonable description of the form of the likelihood of the possible outcomes in a random experiment. While we have concentrated on the iid framework for the available data, this is a choice that we have made partly for convenience in presenting the main ideas of probability and mathematical statistics and partly because it is a framework that often occurs in practice and is often adopted in both theoretical and applied statistical studies. But it is worth mentioning that the model one might select at the beginning of a statistical study can be as complex as one deems appropriate, allowing for dependencies among the data and for possible singularities (e.g., $P(X_i = X_j) > 0$, with $i \neq j$), even when X_i and X_j are continuous variables. We will nonetheless assume throughout this discussion that our inferences about an unknown parameter θ are based on the sample $X_1, X_2, \ldots, X_n \stackrel{\text{iid}}{\sim} F_\theta$. When possible, the frequentist will restrict attention to a low-dimensional statistic which is a sufficient statistic for the unknown parameter(s) of the model.

Having adopted a specific probability model for the data and reduced the data via sufficiency, the frequentist statistician would then seek to identify an estimator of the unknown parameter(s) having some desirable (frequency-based) performance properties. Regarding the modeling of the data and the possible data-reduction via sufficiency, the Bayesian and the frequentist approaches are identical. At this point, however, the Bayesian statistician takes off in a markedly different direction.

The Bayesian approach to statistics involves both a *philosophy* and a *technique*. One essential tenet of the philosophical underpinnings of Bayesian analysis is the notion that one must deal with "uncertain events" through the calculus of probability. (This conclusion follows from axiom systems which form the basis for Bayesian inference. See, for example, DeGroot (1970).) The practical implications of this tenet are quite substantial. In any estimation problem, the target to be estimated is an unknown quantity. In a one-parameter model F_θ for the data, the parameter θ is unknown. It follows that for any interval (a,b) of real numbers, the "event" $\{\theta \in (a,b)\}$ is an uncertain event, and thus must be treated probabilistically. The Bayesian statistician deals with this mandate by postulating a probability distribution for θ. At first glance, this may seem like a bold, and perhaps even dangerously inappropriate, move. After all, in almost any estimation problem we can think of, the parameter of interest is simply an unknown constant. One can reasonably ask: What can possibly be gained by treating the parameter as a random variable? The answer is generally surprising (to the novice) and also quite intriguing. What can often, though certainly not always, be gained is better inference about the unknown parameter (in the sense of tending to get closer to the true value of the target parameter than one would otherwise). In the next section, I will provide some intuition about why this assertion might be true, but there my primary focus will be on the technical mechanics of the Bayesian approach to estimation. In Section 9.3, I will present some empirical evidence that Bayesian estimation can be quite efficacious. I will present some theoretical justification of the claim above in Section 9.4.

Let's discuss the intuitive justification of Bayesian estimation first. A Bayesian statistician regards the probability distribution G placed on an unknown parameter θ as representing his personal, subjective opinion about the likelihood that θ will take on certain values in the parameter space Θ, the collection of all values that θ can possibly take. The distribution G is referred to as the *prior distribution* of θ. Because it is subjective, there is no inherent logical problem with the fact that different people would postulate different priors for θ. Indeed, because different people would generally have different experience, opinions, and knowledge regarding the parameter θ (or about the random experiment for which θ is an index), we expect this to be the case. Next, we should acknowledge that there are many situations in which a statistician (or the relevant expert with whom he consults) has real and useful prior knowledge about the unknown parameter. The types of prior probability modeling which *you* personally might feel comfortable engaging in might range from the mundane (the chances that you get heads when you toss a coin), to the practical (the chances that it will rain today), to the semi-serious (that chances that you don't break any equipment in your lab experiment this week) to the truly important (the chances that at least one of the 10 scholarships you've applied for comes through). Sometimes you have substantial confidence in your intuition, sometimes you realize that your intuition is a bit weak. Depending on what your state of knowledge is, you may decide to consult an "expert" before finalizing your prior distribution. Be that as it may, it often turns out that when you know something about the experiment you are about to witness or perform, it is beneficial to quantify your knowledge and incorporate it into the statistical inference you will carry out. Following the Bayesian paradigm, you would select a fully specified "prior" probability distribution for the unknown parameter, then use the data that becomes available to update your prior opinion and, finally, derive an estimate of the unknown parameter based on the "posterior" distribution that the data leads you to. We now turn our attention to describing more carefully the main elements of the Bayesian approach.

Consider an experiment involving an outcome X whose distribution depends on a scalar parameter θ. For simplicity of exposition, I will treat both X and θ as continuous variables (leading to certain integrals rather than sums), but the process is essentially the same when either X or θ or both are discrete variables. Suppose that your prior opinion about θ has been modeled by the density $g(\theta)$, while the experimental outcome (which, again for simplicity, we treat as a univariate variable X—perhaps a sufficient statistic for θ) is assumed to have the probability density $f(x|\theta)$. From these models, one can derive the *posterior density* of θ, that is, the conditional density $g(\theta|x)$, our shorthand notation for $g(\theta|X=x)$. Specifically, we can write the joint likelihood of the random pair (X,θ) as

$$f(x,\theta) = g(\theta)f(x|\theta). \tag{9.1}$$

From this joint likelihood, one may obtain the desired posterior probability model for θ

using the standard calculus of probability. Specifically,

$$g(\theta|x) = f(x,\theta)/f(x), \tag{9.2}$$

where the marginal density $f(x)$ of X is obtained as

$$f(x) = \int_{-\infty}^{\infty} f(x,\theta)d\theta. \tag{9.3}$$

Note that I have used a shorthand notation for densities in (9.1)–(9.3) for ease of exposition. Since $g(\theta)$ represents your knowledge or opinion about the value of θ prior to seeing the experimental data, it may be viewed as associated with stage 1 in a two-stage process. Drawing the experimental data is then viewed as stage 2 of the process. It is thus clear that the derivation of the conditional distribution of the stage 1 outcome θ, given that $X = x$ occurred at stage 2, is an application of Bayes' Theorem, thereby justifying the name that this particular approach to statistical inference carries.

It remains to determine what we should do with the posterior distribution of θ. This of course depends on what our statistical goals are. Since we are presently focusing on point estimation, what we want to do is make a defensible guess at the true value of θ based on our prior knowledge about θ together with the information we have obtained from the available experimental data. We will now discuss how a Bayesian does this. To understand the process, it is worth emphasizing that, in general, the Bayesian statistician will base whatever statistical inference is of interest on the posterior distribution of the unknown parameter(s). It is fair to say the posterior distribution of the model parameter(s) carries all the information a Bayesian statistician needs or wants in a given application, and the inference that follows is drawn completely from this one source. For estimation problems, one needs to introduce, before proceeding, some appropriate criterion for judging the quality of any given estimator. In the frequentist approach to estimation, an estimator's mean squared error is often used in assessing its worth and in comparing one estimator with another. To explain what a Bayesian does will require a brief digression into some basic ideas of the statistical subfield of decision theory.

The decision-theoretic approach to statistics was largely developed in the 1940s and 1950s. Abraham Wald (1950) was a major contributor to its foundations, and important contributions made a group of notable statisticians and economists a bit too numerous to mention. An excellent introduction (and quite thorough treatment) of the decision-theoretic approach to Statistics may be found in the well-known text by Thomas Ferguson (1967). In brief, the setup we are interested in goes like this. An estimation problem is viewed as a game between "nature," a player who will choose the parameter value θ from the parameter space Θ, and the statistician, a player who will choose an estimator (generally called a "decision function" in this setup) which is simply a mapping from the sample space X which contains the experimental outcome to the parameter space Θ. One can think of the statistician's estimator of θ simply as his/her best guess at θ, given the available data. The statistician's goal is, of course, to get as close to the true parameter value as possible. There is the need to specify a criterion for judging how well one does in problems such as this. The criterion function, which is generally referred to as the "loss function," is a mapping that associates a nonnegative real number with every possible pair (θ, a) representing the parameter value "θ" chosen by nature and the estimate "a" chosen by the statistician. In the usual mathematical notation, we may write $L : \Theta \times A \rightarrow [0, \infty)$, where Θ is the parameter space and A is the set of values ones' estimator might take on. In estimation problems, A is taken to be equal to Θ. The loss $L(\theta_0, a_0)$ represents the penalty that the statistician incurs if the true value of the parameter is θ_0 and the statistician's guess at the parameter is the value a_0. All commonly used measures share the property that one suffers a positive loss only when one's guess is incorrect, that is, when $a \neq \theta$. The most widely used loss function in estimation problems is "squared error loss," that is,

$$L(\theta, a) = (\theta - a)^2, \tag{9.4}$$

a measure of goodness that is also often used in frequentist studies. Two reasonable alternatives are "absolute error loss," that is,

$$L(\theta, a) = |\theta - a| \tag{9.5}$$

which is, of course, a natural way to measure the distance between a guess and a target value, and "linex loss," that is,

$$L(\theta, a) = \exp c(\theta - a) - c(\theta - a) - 1, \tag{9.6}$$

a measure that is intentionally asymmetric, assigning a greater loss for overestimating θ than for underestimating θ when c is set to be negative, and vice versa when c is set to be positive, with the requirements of the problem at hand determining the sign of c. In our developments in this chapter, we will focus, primarily, on squared error loss.

Once we have selected a loss function, we are in a position to specify our overall criterion for comparing estimators. When estimators are being viewed as "decision rules," one could use to common notation of decision theory and consider an estimator to be a mapping $d : X \to \Theta$. Thus, when one observes $X = x$, our estimate of θ would be denoted as $d(x)$ and the loss we would incur would be $(\theta - d(x))^2$. In what follows, we will use the more common notation of $\hat{\theta}$ or, when necessary for clarity, $\hat{\theta}(x)$ for the estimator of θ based on the observation $X = x$.

An essential decision-theoretic tool for comparing estimators is the risk function, defined as

$$R(\theta, \hat{\theta}) = E_{X|\theta} L\left(\theta, \hat{\theta}(X)\right). \tag{9.7}$$

When squared error loss is used, the risk function

$$R(\theta, \hat{\theta}) = E_{X|\theta}(\theta - \hat{\theta})^2 \tag{9.8}$$

is nothing other than the mean squared error (MSE) of the estimator $\hat{\theta}$. In general, $R(\theta, \hat{\theta})$ is a function of the parameter θ, and can be used, as we have seen, to compare one estimator with another. You will recall that if the risk function in (9.8) of $\hat{\theta}_1$ is no greater than that of $\hat{\theta}_2$, and it is smaller for at least one value of θ, then $\hat{\theta}_1$ is a better estimator than $\hat{\theta}_2$, and the latter estimator is "inadmissible" relative to the MSE criterion.

As we saw in Chapter 6, the comparison of curves like the MSEs of a pair of estimators might well be inconclusive. It's often the case that one estimator is better than a second estimator for some values of the unknown parameter θ but will be worse than the second estimator for other values of θ. And because the risk of the trivial estimator $\hat{\theta} = \theta_0$ for any arbitrary value θ_0 is zero when $\theta = \theta_0$, there can be no universally best estimator relative to any of the standard risk functions. This ambiguity can, however, be resolved if the worth of every estimator were to be determined by a single real number or score related to the risk function. Then, for example, we could declare the estimator with the smallest score to be the best estimator relative to this particular scoring system. One criterion that yields a *total ordering* (which guarantees that any two members of the class of interest can be ranked as $<$, $=$ or $>$) among estimators is the *minimax criterion.* According to this criterion, each estimator is assigned the score equal to the maximum value of its risk function (which may well be $+\infty$ for some estimators). The estimator $\hat{\theta}_{mm}$ is the minimax estimator of θ if

$$R(\theta,\hat{\theta}_{mm}) = \min_{\hat{\theta}}\max_{\theta}R(\theta,\hat{\theta}). \tag{9.9}$$

The minimax criterion is generally considered to be too conservative a measure of the worth of an estimator (since it only offers protection against the worst possible circumstance, identifying the estimator corresponding to the best "worst case"), and it is not often used in practice. But the minimax criterion does occasionally lead to a quite reasonable estimator; for example, the sample mean based on a random sample drawn from a normal population can be shown to be a minimax estimator of the population mean μ. Although it is not a criterion that we will emphasize, we will encounter the minimax criterion in some of the discussion and problems later in this chapter.

A widely used alternative is the approach referred to as Bayesian estimation. As we have mentioned, the Bayesian approach to Statistics involves a prior distribution G on the parameter θ. Given a prior distribution on θ, one can entertain the possibility of averaging the risk function with respect to G. The *Bayes risk* of the estimator $\hat{\theta}$ relative to the prior G is given by

$$r(G,\hat{\theta}) = E_{\theta}E_{X|\theta}L(\theta,\hat{\theta}), \tag{9.10}$$

where the outer expectation is taken with respect to the prior distribution G on θ. For squared error loss, the Bayes risk of the estimator $\hat{\theta}$ is given by

$$r(G,\hat{\theta}) = E_{\theta}\left[E_{X|\theta}(\theta-\hat{\theta})^2\right]. \tag{9.11}$$

The Bayes risk takes a non-negative value for each estimator we might consider, and these "scores" allow us to compare any two estimators and declare one better than the other according to the Bayes risk criterion. The Bayes estimator with respect to the prior distribution G (and squared error loss), to be denoted by $\hat{\theta}_G$, is the estimator that achieves the smallest possible value of $r(G,\hat{\theta})$ among all estimators for which the expectation in (9.11) exists. We may thus write

$$r(G,\hat{\theta}_G) = \min_{\hat{\theta}}E_{\theta}E_{X|\theta}(\theta-\hat{\theta})^2. \tag{9.12}$$

Since the class of all estimators of interest is a large, amorphous collection that is difficult to describe analytically, one might think that finding the Bayes estimator of θ in a particular problem can be quite imposing. But a slight reformulation of the problem will lead us to a relatively simple and direct way of identifying the estimator $\hat{\theta}_G$. Notice that the order taken in evaluating the double expectation in (9.10) or (9.11) is to take the (inner) expectation with respect to X first, holding θ fixed, and then to take the expectation with respect to θ. Suppose we reverse this order, writing the Bayes risk in (9.11), instead, as

$$r(G,\hat{\theta}) = E_X\left[E_{\theta|X}(\theta-\hat{\theta})^2\right]. \tag{9.13}$$

You might think that playing ring-around-the-rosy like this could make little difference to us in minimizing $r(G,\hat{\theta})$, but if you do, you'd be mistaken. It actually helps a lot. Looking at the interior expectation in (9.13), we see that, in finding the Bayes estimator, our job now is to minimize the so-called "posterior expected loss," given by

$$E[L(\theta,\hat{\theta}(x))|X=x] = E_{\theta|X=x}(\theta-\hat{\theta}(x))^2. \tag{9.14}$$

Problem 16 in Chapter 2 asks you to do just that—minimize an expected squared distance between a random variable and a constant. If you didn't work that problem, please take a moment and do it now! You'll find that the mean μ_X of the distribution of X is the constant c that minimizes $E(X-c)^2$. (By the way, the same proof works in the discrete and continuous cases.) Proving the result is an easy calculus problem, but it can also be shown algebraically by first writing $(X-c)^2$ as $[(X-\mu_X)+(\mu_X-c)]^2$ and squaring out its expectation.) In the present context, what we can infer from this is that the value $\hat{\theta}(x)$ that minimizes $E_{\theta|X=x}(\theta-\hat{\theta}(x))^2$ is simply the mean $E(\theta|X=x)$ of the conditional distribution of θ, given $X=x$, or in other words, is the mean of the posterior distribution of θ, given $X=x$. So far, we have that

$$\hat{\theta}_G(x) = E_G(\theta|X=x) \tag{9.15}$$

minimizes the interior expectation in (9.13). However, we get something more, something quite important, and it comes without any extra effort. Since for every value of x, the posterior expected loss $E_{\theta|X=x}(\theta-\hat{\theta}(x))^2$ is minimized by $\hat{\theta}_G(x)$, it follows that, when $E_{\theta|X=x}(\theta-\hat{\theta}(x))^2$ is averaged over x (by taking the outer expectation in (9.13)), this average of minimized conditional expectations yields the smallest possible Bayes risk. Thus, the Bayes estimator may be written, as a function of the experimental input X, as

$$\hat{\theta}_G(X) = E_{G(\theta|X)}(\theta|X). \tag{9.16}$$

When the experimental outcome $X=x$ is in hand, the numerical value of the Bayes estimator is as in (9.15).

I can't resist telling you about some advice offered by the celebrated statistician Herbert Robbins in a seminar presentation I heard him give at Berkeley years ago. Herb was doing some computation which involved a double integral and he paused to make the following remark. He said "When I was a student, I was told that switching the order of integration was not always mathematically valid, and so before ever doing it, one should check that the conditions which justify the switch are satisfied. This sounded perfectly reasonable. But as I moved on in my mathematical studies and career, I discovered that no one seemed to know what these conditions were. This bothered me for a while, but it doesn't anymore. Now, when I encounter a double integral, the first thing I do is switch the order of integration and check to see if I get anything interesting! If I do, I then scrounge around for a justification." Robbins was deservedly famous for his statistical accomplishments, but he was also renowned for his stimulating and amusing lectures. In this instance, though, Robbins's remark was more comedic than factual, as there's a very nice result called Fubini's theorem that is pretty specific about when switching the order of integration is valid. I should nevertheless add that Robbins' advice on this matter has some real value. You can indeed discover some interesting things by switching the order of integration.

It is evident from the discussion above that the Bayes estimator of a parameter θ with respect to the prior G minimizes both the posterior expected loss and the Bayes risk relative to G. One of the other interesting tenets of the Bayesian school of statistics is the dictum that one's statistical inference must depend solely on the observed experimental value (for us, the observed value of X) and not otherwise on the random quantity X. The formal statement of this rule is referred to as the *likelihood principle*. The motivation for the principle is the idea that our inference about θ should depend only on what we actually observed, not on what we *might have observed*, but didn't. If one adheres to the likelihood principle, then the idea of averaging over the sample space (as when one takes an expectation with respect to the distributions of X, or of $X|\theta$, is anathema (that is, "not OK"). Since the Bayes risk of an estimator involves such averaging, a strict Bayesian would not consider it as an appropriate measure for choosing an estimator. One interesting thing about the developments above is the fact that one obtains the same solution, that is, the best that one can do under the assumption that $\theta \sim G$, whether one minimizes the posterior expected loss, which depends only on the observed value of x, or minimizes the Bayes risk with respect to G, which is a measure which involves averaging over x.

Within the Bayesian community, a statistical procedure that violates the likelihood principle has been given the onerous label "incoherent," chosen in part, some claim, to draw more statistical practitioners into the Bayesian camp. After all, who wants to be thought of as incoherent! Many frequentist procedures are considered incoherent by orthodox Bayesians, including criteria (like variance or MSE) that average over the entire sample space, the construction of confidence intervals (which are justified by the likelihood of certain events in yet-to-be-performed future experiments). To read more about the likelihood principle, see Berger and Wolpert (1984).

I've mentioned that there are strong philosophical differences between the Bayesian and frequentist schools of thought. These differences are important, as they are often highly influential in determining a statisticians' choice between embracing the Bayesian approach or rejecting it. The frequentists have one major criticism about the Bayesian approach, and it's a difficult one for Bayesians to counter. It seems quite valid to ask: "What in the world gives a statistician the right to interject his subjective opinion into a statistical investigation?" The criticism is especially piercing if one is talking about a scientific investigation in which, for example, the measurement of some physical quantity is involved. Scientists are trained to try to eliminate all possible forms of bias from the design and analysis of an experiment. The integration of subjective judgments into the final analysis of one's data seems totally antithetical to (again, not in agreement with) the scientist's goals. The Bayesian has a reasonably good answer, but not one which protects him against occasionally making gross mistakes. The answer is that when one has useful prior information in a situation in which experimental data will be made available, one is ill-advised to ignore it, since using it carefully can quite often lead to better inference (for example, to estimators that tend to be closer than the best frequentist estimator to the true value of the parameter of interest). Whole tomes have been written on this and related topics, and space does not permit much further elaboration here. For further discussion, I refer the reader to Samaniego (2010), my recent book on comparative inference. In Section 9.3, I will treat Bayesian and frequentist estimation from a comparative perspective, thereby introducing you to some of the main ideas in the aforementioned monograph.

Exercises 9.1.

1. Suppose that $X \sim B(n,p)$. Let $\hat{p} = X/n$ be the sample proportion. Taking squared error as the loss criterion, derive the risk function $R(p,\hat{p})$ of the estimator $\hat{p}$.

2. Suppose that $X_1,\ldots,X_n \overset{\text{iid}}{\sim} N(\mu,\sigma^2)$, and let $\hat{\mu} = \bar{X}$, the sample mean. Show that the risk

function of $\hat{\mu}$ relative to squared error loss is a constant not depending on μ. (Its value does depend on σ, of course.) An estimator whose risk function R is a (known or unknown) constant is called an *equalizer rule*.

3. Let $L(\theta, a)$ be a non-negative loss function which takes the value 0 when $\theta = a$. Show that if an estimator $\hat{\theta}$ is an equalizer rule and is a Bayes rule with respect to the prior probability distribution G, then $\hat{\theta}$ is a minimax estimator of θ.

4. Suppose that $X \sim B(n, p)$. Let $\hat{p} = X/n$ be the sample proportion. Let the loss function be $L(p, a) = (p-a)^2/p(1-p)$. Verify that (a) the risk function of $\hat{p}$ is a constant independent of p and (b) $\hat{p}$ is the Bayes estimator with respect to the prior distribution $G = U(0,1)$. Conclude (from the result in Exercise 9.1.3 above) that $\hat{p}$ is a minimax estimator of p under the assumed weighted squared error loss function.

9.2 Deriving Bayes Estimators

In this section, my intent is to carry out some full-blown Bayesian analyses with a view toward shedding light on how the Bayesian approach to estimation is applied (with a shortcut or two tossed in at no extra charge). We'll also examine how the results of such analyses are meant to be interpreted. We begin with the familiar and frequently occurring problem of estimating a population proportion p. We'll make the simplifying assumption that the data on which our inference is based consist of n independent Bernoulli trials with probability p of success. Of course we recognize that when sampling without replacement from a finite population, the number of successes in one's sample has a hypergeometric rather than a binomial distribution. If the population can be thought of as infinite (for example, it consists of the collection of all potential repetitions of a particular laboratory experiment), then individual draws would indeed be Bernoulli trials, and their sum, the number of successes in n trials, would be a binomial variable. When the population size is large, we know from Chapter 2 that the binomial model is a very good approximation to the hypergeometric model. We'll assume we are dealing with a problem of one of the latter two types. The model also applies, of course, to sampling with replacement from a finite population where p is the relevant population proportion.

If $X_1, \dots, X_n \stackrel{\text{iid}}{\sim} B(1, p)$, and if $X = \sum_{i=1}^{n} X_i$, then $X \sim B(n, p)$. Now, in estimating p using the Bayesian approach, a particular family of prior distributions is often used. The family of beta distributions has a number of properties that recommend its use in this problem. First, the support set of each beta distribution is the interval $(0,1)$. Since the unknown parameter p is a probability and thus takes values in $(0,1)$, the beta family lends itself quite naturally as a distribution for a random probability. Secondly, the beta family has densities taking a variety of shapes—bell-shaped, U-shaped, flat, strictly increasing or decreasing—and this flexibility is helpful in prior modeling. But the nicest property of the beta–binomial pairing is something that is usually called the *conjugacy* property (a certain type of "model preservation"). Although this term is widely used for this and other pairs of models, it doesn't uniquely define the families of priors involved. We'll explore the matter further in the paragraph that follows. While acknowledging the lack of uniqueness of probability distributions with the conjugacy property, we will proceed, without any great loss, with the usual treatment of conjugacy in the examples, exercises, and problems in this chapter.

The property generally called *conjugacy* amounts to the simple phenomenon that the posterior distributions belong to the same family of distributions as the priors do. While most of the priors we'll use in this text satisfy this property of conjugacy under Bayesian updating, it is known that there is often a large collection of prior families which share the property in a given application. (Diaconis and Ylvisaker (1979) point out that if the family $\{g(\theta|\alpha), \alpha \in A\}$ has the "conjugacy" property when sampling from $F_{X|\theta}$, then the family $\{c \cdot h(\theta)g(\theta|\alpha), \alpha \in A\}$, where the constant $c = (\int_{-\infty}^{\infty} h(\theta)g(\theta|\alpha)d\theta)^{-1}$, is a conjugate family as well, provided $c < \infty$.) We will thus refer to the conjugate prior distributions we discuss and use in this text as "the standard conjugate priors," as these are the conjugate priors that are routinely used in applications of the Bayesian approach to estimation. In the present case, we are seeking a Bayes estimator of the unknown proportion p and we shall see that when a beta prior distribution is postulated for p, the posterior distribution is also a beta distribution. It should be recognized that this choice of prior is a *convenience* rather than a *necessity*. In many applications, one could use any prior one likes and obtain the exact posterior distribution or a suitable approximation of it numerically or through an iterative technique. As we shall see, conjugate priors do facilitate dealing with inferences based on the posterior distribution. For example, the posterior mean can then often be identified by inspection rather than by a potentially tedious integration or an approximation procedure. While the use of a beta prior is by no means necessary in estimating an unknown proportion, it remains true that, because the beta family is fairly rich in terms of possible shapes for capturing prior opinion, a particular beta prior can often be justified as an appropriate quantification of the prior information available to the experimenter.

Let us take, as a prior distribution for p, a $Be(\alpha, \beta)$ distribution, where the parameters α and β are specific numerical values chosen to reflect our prior opinion about p. This prior has density function

$$g(p) = \frac{\Gamma(\alpha+\beta)}{\Gamma(\alpha)\Gamma(\beta)} p^{\alpha-1}(1-p)^{\beta-1} I_{(0,1)}(p).$$

The joint likelihood of the pair (X, p) may be written as

$$f(x,p) = \binom{n}{x} p^x (1-p)^{n-x} \frac{\Gamma(\alpha+\beta)}{\Gamma(\alpha)\Gamma(\beta)} p^{\alpha-1}(1-p)^{\beta-1} \text{ for } x = 0,\ldots,n \text{ and } p \in (0,1),$$

an expression which we can simplify to

$$f(x,p) = \binom{n}{x} \frac{\Gamma(\alpha+\beta)}{\Gamma(\alpha)\Gamma(\beta)} p^{\alpha+x-1}(1-p)^{\beta+n-x-1} \text{ for } x = 0,\ldots,n \text{ and } p \in (0,1). \quad (9.17)$$

Evaluating the marginal pmf of the variable X seems like a nasty piece of business, but in reality, it falls out rather easily by a simple application of "the trick." For $x = 0, 1, \ldots, n$, we may write the pmf of X as

$$f(x) = \int_0^1 \binom{n}{x} \frac{\Gamma(\alpha+\beta)}{\Gamma(\alpha)\Gamma(\beta)} p^{\alpha+x-1}(1-p)^{\beta+n-x-1} dp. \quad (9.18)$$

From (9.18), we can immediately recognize the integrand as being "almost a beta density."

Applying the trick yields

$$\begin{aligned} f(x) &= \binom{n}{x}\frac{\Gamma(\alpha+\beta)}{\Gamma(\alpha)\Gamma(\beta)}\frac{\Gamma(\alpha+x)\Gamma(\beta+n-x)}{\Gamma(\alpha+\beta+n)} \\ &\quad\times \int_0^1 \frac{\Gamma(\alpha+\beta+n)}{\Gamma(\alpha+x)\Gamma(\beta+n-x)}p^{\alpha+x-1}(1-p)^{\beta+n-x-1}dp \\ &= \binom{n}{x}\frac{\Gamma(\alpha+\beta)}{\Gamma(\alpha)\Gamma(\beta)}\frac{\Gamma(\alpha+x)\Gamma(\beta+n-x)}{\Gamma(\alpha+\beta+n)} \quad \text{for } x = 0,1,\dots,n. \end{aligned} \tag{9.19}$$

The discrete distribution in (9.19) is, quite appropriately, known as the beta-binomial distribution, and it has arisen as a model of interest in a variety of applied studies. When $\alpha = \beta = 1$, (9.19) reduces to the discrete uniform distribution on the integers $0,1,\dots,n$. Our main focus here, of course, is on the posterior distribution of p, given $X = x$. From (9.17) and (9.19), it is evident that the posterior density of p is given by

$$g(p|x) = \frac{f(x,p)}{f(x)} = \frac{\Gamma(\alpha+\beta+n)}{\Gamma(\alpha+x)\Gamma(\beta+n-x)}p^{\alpha+x-1}(1-p)^{\beta+n-x-1}I_{(0,1)}(p), \tag{9.20}$$

i.e., $p|X = x \sim Be(\alpha+x, \beta+n-x)$. We thus see that the Beta family enjoys the conjugacy property when coupled with the binomial distribution. It is the standard conjugate family employed in this context. The Bayes estimator of p with respect to the prior $G = Be(\alpha,\beta)$ and squared error loss is the mean of the posterior distribution of p given by

$$\hat{p}_G = E(p|X = x) = \frac{\alpha+x}{\alpha+\beta+n}. \tag{9.21}$$

Before commenting on the interpretation of the Bayes estimator in (9.21), we point out a useful shortcut in its computation that is quite helpful in problems like the one above in which a conjugate prior has been employed. Looking back at the joint likelihood function displayed in (9.17), and reflecting on the process of deriving the posterior density $g(p|x)$ from it, we note that the desired posterior will necessarily be equal to a constant multiple of $f(x,p)$. This is because the posterior is obtained by dividing the joint likelihood $f(x,p)$ by $f(x)$, a function that is independent of p, while the conditional density $g(p|x)$ can be properly thought of as a constant times some function of p times an indicator function specifying the support set of p given $X = x$. It follows from (9.17) that, in the beta-binomial problem,

$$g(p|x) = \frac{c}{f(x)}p^{\alpha+x-1}(1-p)^{\beta+n-x-1}I_{(0,1)}(p) \propto p^{\alpha+x-1}(1-p)^{\beta+n-x-1}I_{(0,1)}(p), \tag{9.22}$$

where the symbol "$\propto$" is read "is proportional to." From (9.22), we can identify $g(p|x)$ without any further calculation (that is, we can identify $g(p|x)$ "by inspection") since we know that the function in (9.22) is the core of a beta density, and that the only multiplicative constant k that will make this function of p into a density is the constant

$$k = \frac{\Gamma(\alpha+\beta+n)}{\Gamma(\alpha+x)\Gamma(\beta+n-x)}.$$

It follows that $p|X = x \sim Be(\alpha+x, \beta+n-x)$.

You will notice immediately that the Bayes estimator of p in (9.21) is quite similar to the standard frequentist estimator $\hat{p} = x/n$. Indeed, if the prior parameters α and β are near zero, the difference between $\hat{p}_G$ and $\hat{p}$ might be considered negligible. However, for other choices of these parameters, the difference between the two estimators can be substantial. It is instructive to re-express the Bayes estimator $\hat{p}_G$ as

$$\hat{p}_G = \frac{\alpha+\beta}{\alpha+\beta+n} \cdot \frac{\alpha}{\alpha+\beta} + \frac{n}{\alpha+\beta+n} \cdot \frac{x}{n}. \tag{9.23}$$

From this latter equation, it is apparent that the Bayes estimator of p is a *convex combination* of the mean $p_G = \alpha/(\alpha+\beta)$ of the prior distribution of p and the standard frequentist estimator $\hat{p} = x/n$, which is, simultaneously, the MVUE, the MME, and the MLE of the parameter p. A natural and quite valid interpretation of this estimator emerges. The Bayes estimator of p is a mixture of my best "guess" p_G of p before any data is available and the estimator $\hat{p}$, the standard "guess" at p that any frequentist would make based on the data alone. Looking more closely at (9.23), we see that the weight placed on the components p_G and $\hat{p}$ of the Bayes estimator depend on the relative sizes of n and $\alpha+\beta$. If $\alpha+\beta = n$, the two components receive equal weight, while p_G receives greater weight than $\hat{p}$ when $\alpha+\beta > n$ and less weight than $\hat{p}$ when $\alpha+\beta < n$. Because of the role that the sum $\alpha+\beta$ plays in the Bayes estimator above, it is often referred to as the "prior sample size." One can think of the value a Bayesian chooses for $\alpha+\beta$ as the number of observations he believes his prior information is worth.

One further elaboration on the notion of prior sample size is worthy of mention. Since there are many statistical problems you might encounter in which you feel that the prior information available to you is somewhat limited, the topic of the "elicitation" of prior information from experts with whom you can consult naturally arises. Elicitation usually proceeds by asking the expert a series of questions. There is no unique way of proceeding in this area, but when inference relative to a conjugate prior distribution is contemplated, as in the beta-binomial pairing discussed above, proceeding as follows is quite natural. For the beta distribution, what needs to be determined are the values of the parameters α and β. As you can no doubt imagine, asking someone to specify values of the parameters α and β of the beta prior distribution is not a good approach, as the meaning and operational role that these parameters have in the prior and posterior distributions of p is not widely understood nor easily explained. Some Bayesian statisticians like to pose their questions in terms of the quantiles of the prior that an expert might be able to specify; for example, what are the .1 and the .9 quantiles of the prior distribution (that is, the values $p_{.10}$ and $p_{.90}$ such that $P(p < p_{.1}) = .1$ and $P(p < p_{.9}) = .9$)? Two easier questions to answer, for most people, are (1) What's your best guess at p? and (2) Given that you will be obtaining a sample of size n, how much weight would you place on the standard estimator $\hat{p}$, with the rest of the weight going to your prior guess? The answer to the first question is just the prior mean p_G, and equating that answer to the mean $\alpha/(\alpha+\beta)$ of the $Be(\alpha,\beta)$ distribution gives you one equation to be solved. The second answer, say η, can be set equal the weight $n/(\alpha+\beta+n)$ that the Bayes rule places on the estimator $\hat{p}$, providing a second equation that the desired α and β must satisfy. In summary, if p_G and η are the expert's answers to questions (1) and (2) above, then simultaneously solving the equations

$$p_G = \alpha/(\alpha+\beta) \text{ and } \eta = n/(\alpha+\beta+n)$$

for α and β identifies the beta prior consistent with these answers. If, for example, $n = 30$, $p_G = 1/2$, and $\eta = 3/4$, then the beta prior that has been specified is $Be(5,5)$.

As mentioned earlier, one of the persistent criticisms of Bayesian inference concerns the worrisome use of subjective information in inferential problems where "objectivity" is highly valued. The popular phrase "let the data speak for themselves" describes this position well. But there's another side to this issue. The fact that the use of prior knowledge or opinion can often improve one's inferences, and can occasionally save one from abject disaster, is well illustrated in the following simple example.

Example 9.2.1. Suppose a freshly minted coin is tossed ten times, and we wish to estimate the probability p that it comes up heads in any single toss. If we were to obtain 10 heads in 10 tosses, the standard, universally recommended frequentist estimator of p, the sample proportion $\hat{p}$, would take the value 1. But of course we know that p is no doubt very close to 1/2. If we were to do a Bayesian analysis, we would probably use a beta prior, perhaps the prior $G = Be(100, 100)$, which is quite heavily concentrated around its mean $1/2$. Given our vast experience in tossing coins over the years (together with our intuition), the prior sample size of 200 seems well justified! When we observe 10 heads in 10 tosses of the coin, we update our prior opinion, and we estimate p on the basis of the posterior distribution $Be(110, 100)$. With respect to squared error loss, our Bayes estimate of p is the mean of this posterior distribution, which is $\hat{p}_G = .5238$. While we thought, initially, that the coin was fair, we are in fact affected by the surprising result of the experiment. Now, we no longer believe that the coin is fair, as our posterior opinion properly moderates our initial thoughts, resulting in a small estimated bias in the coin. As simple as this example is, it reveals an essential truth. A Bayesian analysis *here* doesn't introduce questionable subjectivity, intentional mischief, or any other form of arbitrary alteration of the data. What it does is use some pretty solid prior knowledge to save us from the embarrassment of making a ridiculously poor inference. This example suggests that the use of prior information can be extremely useful. The challenge in more complex estimation problems is to determine whether or not useful prior information is in fact available. ■

You will find it helpful to strengthen your hold on the ideas and the process used above in estimating a population proportion by thinking through a second Bayesian analysis. Let's consider a second example, this time with the other ubiquitous statistical model in common use, the normal distribution. Suppose that, under the squared error loss criterion, we wish to derive a Bayes estimator of the mean μ of a normal population based on a random sample $X_1, \ldots, X_n \overset{\text{iid}}{\sim} N(\mu, \sigma_0^2)$, where the variance of the population is assumed to be the known number σ_0^2. This makes the model of interest a one-parameter model comparable to the binomial model we dealt with above. I should mention that the Bayesian approach is implemented in precisely the same way as above in estimating vector-valued parameters, but I will forego providing the details of such applications for now.

Our first order of business here is to specify a family of prior distributions which provides the statistician with an adequate vehicle for expressing his prior opinion about the parameter μ. Since the normal distribution is generally used as a model for physical measurements, one might think of μ as a single measurement (treated as unknown to the experimenter) which happens to coincide with the population average. It's a short leap from there to the thought that a normal distribution for μ might make a reasonable prior distribution. Let us suppose that we select $G = N(\nu, \tau^2)$ as the prior model for the unknown mean μ. In deriving the posterior model for μ, we will replace the available sample by $\bar{X}$, the sufficient statistic for μ. We can, of course, obtain the posterior distribution by deriving the posterior density $g(\mu|\bar{x})$ as we did in (9.1.2) for the beta-binomial problem. Instead, we will take advantage of the shortcut mentioned above and seek to identify $g(\mu|\bar{x})$ by inspection.

Note first that core of the likelihood function for the available sample may be written as

$$f(\mathbf{x}|\mu) \propto \exp\left[-\frac{1}{2\sigma_0^2}\sum_{i=1}^{n}(x_i-\mu)^2\right]. \tag{9.24}$$

where $exp[A] = e^A$. Now, we know that $\sum_{i=1}^{n}(x_i - \bar{x}) = 0$; this fact justifies the following rewriting of the sum in (9.24):

$$\sum_{i=1}^{n}(x_i-\mu)^2 = \sum_{i=1}^{n}(x_i-\bar{x}+\bar{x}-\mu)^2 = \sum_{i=1}^{n}(x_i-\bar{x})^2 + \sum_{i=1}^{n}(\bar{x}-\mu)^2, \tag{9.25}$$

since, in the cross product term, $\sum_{i=1}^{n}(x_i-\bar{x})(\bar{x}-\mu) = (\bar{x}-\mu)\sum_{i=1}^{n}(x_i-\bar{x}) = (\bar{x}-\mu)(0) = 0$. We may thus write

$$f(\mathbf{x}|\mu) \propto \exp\left[-\frac{1}{2\sigma_0^2}\left\{\sum_{i=1}^{n}(x_i-\bar{x})^2 + n(\mu-\bar{x})^2.\right\}\right]. \tag{9.26}$$

Now, since the prior distribution on μ has density satisfying

$$g(\mu) \propto exp\left[-\frac{1}{2\tau^2}(\mu-\nu)^2\right],$$

we have that the posterior density of μ, given the sufficient statistic $\bar{X} = \bar{x}$, is a function of μ that satisfies the proportionality relation

$$g(\mu|\bar{x}) \propto exp\left\{-\frac{1}{2}\left[\frac{n}{\sigma_0^2}(\mu-\bar{x})^2 + \frac{1}{\tau^2}(\mu-\nu)^2\right]\right\}. \tag{9.27}$$

The final step in identifying, by inspection, the posterior density as a familiar model is to rewrite the exponent in (9.27). It is left as Exercise 9.2.3 to confirm that, by combining the two terms in the exponent and completing the square in μ, the bracketed element of the exponent in (9.27), namely,

$$\frac{n}{\sigma_0^2}(\mu-\bar{x})^2 + \frac{1}{\tau^2}(\mu-\nu)^2 \tag{9.28}$$

may be rewritten as

$$\left(\frac{\sigma_0^2\tau^2}{(\sigma_0^2+n\tau^2)}\right)^{-1}\left(\mu - \frac{\sigma_0^2\nu+n\tau^2\bar{x}}{\sigma_0^2+n\tau^2}\right)^2 + \left(\frac{n}{(\sigma_0^2+n\tau^2)}\right)(\bar{x}-\nu)^2. \tag{9.29}$$

Since the second term in (9.29) does not depend on μ, we have, finally, that the posterior density of μ satisfies

$$g(\mu|\bar{x}) \propto exp\left[-\left(\frac{2\sigma_0^2\tau^2}{(\sigma_0^2+n\tau^2)}\right)^{-1}\left(\mu - \frac{\sigma_0^2\nu+n\tau^2\bar{x}}{\sigma_0^2+n\tau^2}\right)^2\right]. \tag{9.30}$$

From (9.30), we see that the posterior distribution of μ is a normal distribution; more precisely,

$$\mu|\bar{X} = \bar{x} \sim N\left(\frac{\sigma_0^2 \nu + n\tau^2 \bar{x}}{\sigma_0^2 + n\tau^2}, \frac{\sigma_0^2 \tau^2}{\sigma_0^2 + n\tau^2}\right). \tag{9.31}$$

This confirms that the normal prior distribution we have employed has the conjugacy property discussed above. It is the standard conjugate prior when sampling from a normal population with an unknown mean. The mean of the posterior distribution, and thus the Bayes estimator of μ under squared error loss, is clearly

$$\mu_G = \frac{\sigma_0^2 \nu + n\tau^2 \bar{x}}{\sigma_0^2 + n\tau^2}. \tag{9.32}$$

This estimator can be rewritten as the following convex combination of the prior mean and the standard frequentist estimator $\bar{x}$:

$$\mu_G = \frac{\sigma_0^2}{\sigma_0^2 + n\tau^2}\nu + \frac{n\tau^2}{\sigma_0^2 + n\tau^2}\bar{x}. \tag{9.33}$$

The amount of weight that one places on the two "guesses" at μ—the prior guess and the data-driven guess—is controlled by the prior variance τ^2. If τ^2 is small, the prior is quite concentrated about the prior mean ν, representing the statistician's strong confidence in his prior guess. This is reflected in the weight that he gives to his prior guess, a weight that approaches 1 as $\tau^2 \to 0$. If τ^2 is large, the prior is quite diffuse, reflecting rather little confidence in the prior guess; in that event, the Bayes estimator places more weight on the sample mean, that weight tending to 1 as $\tau^2 \to \infty$.

The convenience of conjugacy is quite apparent from the developments above. Several other conjugate pairs will be explored in the exercises and chapter problems that follow. The collection of pairs of models which enjoy the conjugacy property includes the following: the gamma model $\Gamma(\alpha, \beta)$ is the standard conjugate prior for the Poisson model $P(\lambda)$, the beta model $Be(\alpha, \beta)$ is the standard conjugate model for the negative binomial distribution $NB(r, p)$, the gamma model $\Gamma(\alpha, \beta)$ is the standard conjugate prior for the exponential distribution $\Gamma(1, \mu)$, the gamma model $\Gamma(\alpha, \beta)$ is the standard conjugate prior for the one-parameter normal model $N(\theta_0, \theta)$ parameterized in terms of a *known* mean θ_0 and a positive parameter $\theta = 1/\sigma^2$ referred to as the "precision" of the distribution, and the Pareto model $P(\alpha, \theta_0)$ is the standard conjugate prior for the uniform distribution $U(0, \theta)$.

It is not uncommon to encounter situations in which the standard conjugate family is inadequate (as when a bimodal prior is deemed desirable when estimating a particular population proportion) or no simple conjugate family is known to exist. The use of non-conjugate priors causes no conceptual difficulties, but they do require additional work. One will need to do the calculus required to obtain the marginal distribution needed for identifying the posterior distribution of the target parameter. The integration involved is sometimes difficult, and sometimes, it is analytically intractable. But a Bayesian analysis is generally still possible, as good approximations of posterior distributions can usually be obtained either through numerical integration or through iterative techniques such as Markov Chain Monte Carlo schemes. A good reference on modern methods of Bayesian computation is the text by Gamerman and Lopes (2006). Of course a Bayesian analysis using a non-conjugate prior doesn't always require special numerical or computational tools, as the following simple example shows.

Example 9.2.2. Suppose that $X|\theta \sim U[0,\theta]$ and that θ has the gamma prior distribution $G = \Gamma(2,1)$, that is, the prior distribution with density

$$g(\theta) = \theta e^{-\theta} I_{[0,\infty]}(\theta).$$

The joint distribution of X and θ is thus

$$f(x,\theta) = e^{-\theta}, \text{ for } 0 < x < \theta < \infty.$$

The marginal density of X is that of the exponential (or $\text{Exp}(1) \equiv \Gamma(1,1)$) distribution with mean 1; thus, the posterior density of θ is

$$f(\theta|x) = e^{-(\theta-x)} I_{[x,\infty]}(\theta),$$

the density of the "translated exponential distribution" or, alternatively, the density of $x+Y$, where $Y \sim Exp(1)$. Since the mean of Y is 1, it follows that, under squared error loss, the Bayes estimate of θ with respect to G is $\hat{\theta}(x) = x+1$. ■

When prior information appears to be rather limited, one can opt for using a prior distribution with a considerably large dispersion, the result of which would be a Bayes estimator which places very little weight on the prior mean and quite a large amount of weight on the data-based estimator. An alternative approach is to seek a prior distribution which is "non-informative," that is, a prior which places minimal weight on prior information and maximal weight on the data. Such priors are generally referred to as "objective priors" (or "generalized priors") and are usually found to be limiting cases of conjugate priors as the prior parameters approach certain limits. The resulting "distribution" of the parameter is typically an improper distribution, that is, it actually gives infinite weight to the parameter space. As an illustration, consider the estimation of the mean of a normal population treated above, where $X \sim N(\mu,\sigma_0^2)$ and $\mu \sim N(\nu,\tau^2)$. As the prior variance τ^2 is allowed to grow to ∞, the density of the prior distribution grows very flat and has the property that all intervals of the same length receive approximately the same weight under the prior. One can envision a general measure on the real line that has exactly these properties. Lebesgue measure can be described as a measure that gives infinite weight to the whole real line, with any interval of length L receiving weight L. Lebesgue measure is often used as an "objective" prior for a location parameter (like the mean μ of a normal distribution). The "generalized Bayes estimator" of a normal mean with respect this "prior" is the estimator $\hat{\mu}$ which minimizes the generalized posterior expected loss

$$\int_{-\infty}^{\infty} (\mu-\hat{\mu})^2 g(\mu|\bar{X}=\bar{x})d\mu, \tag{9.34}$$

is $\hat{\mu} = \bar{x}$. (In the normal case, the integral in (9.34) exists and is finite when $g(\mu) = 1$ by virtue of the fact that the posterior distribution of μ is a proper probability distribution even though the prior distribution of μ is not.) It is apparent that the use of the "objective" Lebesgue prior represents an analysis devoid of subjective input. The approach has the appeal of permitting the derivation of an estimator using the Bayesian approach which is not subject to the criticism of introducing potential bias through the use of a subjective prior. But the "objective Bayesian" approach does draw certain criticisms.

First, the idea of allowing a prior distribution placing infinite mass on the parameter space has no clear interpretation. The notion of prior mean is absent, and the possibility of placing infinite mass on a part of the parameter space that is considered impossible in the application of interest (like the complement of the interval (0 feet, 10 feet) when estimating the average height of a human subpopulation) is disturbing. Thus, the averaging in (9.34) is difficult to justify. The derivation of an estimator using an improper prior violates a basic tenet of Bayesian inference in that the uncertainty one has concerning the unknown parameter is not treated through the use of a legitimate probability distribution. Orthodox Bayesians would describe the so-called "objective Bayesian inference" as incoherent. But perhaps the most important criticism of the routine use of the objective Bayes approach is that it intentionally foregoes the opportunity to take advantage of the prior information that may be available. It is precisely when useful prior information is available to the statistician that Bayesian inference is most beneficial. Failure to use that information may result in inferior statistical performance.

In spite of such criticisms, objective Bayes methods are in use today, and they seem to be growing in popularity. For a statistician who is open to using either Bayesian or frequentist methods, depending on what a particular application seems to require, one might consider the discussion above as involving semantics rather than substance. It is true that "objective Bayesian methods" are not really Bayesian. They simply resemble Bayes procedures due to the mechanics of obtaining them. In that sense, they are simply misnamed, and should be classified as frequentist procedures obtained in a new way. The fact that one gets the perfectly respectable estimator $\hat{\mu} = \bar{X}$ in the problem of estimating a normal mean suggests that the objective Bayesian approach may give good answers in an array of applications, and that they may be especially useful in problems in which alternative frequentist approaches are either intractable or unacceptable.

The fact that the estimator $\hat{\mu} = \bar{X}$ of a normal mean is a *generalized Bayes estimator* relative to Lebesgue measure under squared error loss but cannot be a Bayes estimator with respect to any proper prior is a special case of the general property of Bayes estimators treated in the following result.

Theorem 9.2.1. Consider a statistical estimation problem based on the non-degenerate random variable $X \sim F_\theta$. Suppose that X is unbiased for θ, that is, $E(X|\theta) = \theta$. Then, under squared error loss, the estimator X is not a Bayes rule with respect to any proper prior distribution G.

Proof. Let $\theta \sim G$, and assume that X is the Bayes estimator with respect to the prior G. Then X can be realized as the mean of the posterior distribution of θ, that is, $X = E(\theta|X)$. Let us now compute the Bayes risk of the estimator X with respect to the prior G. We have

$$r(G,X) = E_{X,\theta}(\theta - X)^2 = E_\theta\theta^2 - 2E_{X,\theta}(\theta X) + E_X X^2. \tag{9.35}$$

But since X is unbiased for θ, it follows from Adam's Rule that

$$E_{X,\theta}(\theta X) = E_\theta E_{X|\theta}(\theta X|\theta) = E_\theta \theta E_{X|\theta}(X|\theta) = E_\theta\theta^2. \tag{9.36}$$

On the other hand, since X is the Bayes estimator of θ, Adam's Rule also yields

$$E_{X,\theta}(\theta X) = E_X E_{\theta|X}(\theta X|X) = E_X X E_{\theta|X}(\theta|X) = E_X X^2. \tag{9.37}$$

Because of (9.36) and (9.37), we may replace $2E_{X,\theta}(\theta X)$ in (9.35) by the sum $E_\theta\theta^2 + E_X X^2$, yielding

$$r(G,X) = E_\theta\theta^2 - E_\theta\theta^2 - E_X X^2 + E_X X^2 = 0. \tag{9.38}$$

However, in reality,

$$\begin{aligned} r(G,X) &= E_{X,\theta}(\theta - X)^2 \\ &= E_\theta E_{X|\theta}\left[(X-\theta)^2|\theta)\right] \\ &= E_\theta E_{X|\theta}(X - E(X|\theta))^2 \\ &= E_\theta V(X|\theta). \end{aligned} \tag{9.39}$$

Since X is non-degenerate, by assumption, the latter expectation is positive or infinite, contradicting (9.38). This shows that X cannot be a Bayes estimator. ■

Exercises 9.2.

1. Suppose that $X|\lambda \sim P(\lambda)$, the Poisson distribution with mean λ. Assuming squared error loss, derive that Bayes estimator of λ with respect to the prior distribution $\Gamma(\alpha,\beta)$, first by explicitly deriving the marginal probability mass function of X, obtaining an expression for the posterior density of λ and evaluating $E(\lambda|x)$ and secondly by identifying $g(\lambda|x)$ by inspection and noting that it is a familiar distribution with a known mean.

2. Suppose that $X|\theta \sim U(0,\theta)$, the uniform distribution on the interval $(0,\theta)$. Assuming squared error loss, derive that Bayes estimator of θ with respect to the prior distribution $P(\alpha,\theta_0)$, the two-parameter Pareto model specified in (3.36), first by explicitly deriving the marginal probability mass function of X, obtaining an expression for the posterior density of θ and evaluating $E(\theta|x)$ and secondly by identifying $g(\theta|x)$ by inspection and noting that it is a familiar distribution with a known mean.

3. Show that the expression in (9.28) is algebraically equivalent to the expression in (9.29).

4. For an arbitrary real-valued random variable X, show that $E|X-c|$ is minimized by $c = m$, where m is a median of the distribution of X, that is, m is any value such that $P(X \geq m) \geq 1/2$ and $P(X \leq m) \geq 1/2$.

5. Find the Bayes estimate of θ in Example 9.2.2 under the absolute error loss function $L(\theta,a) = |\theta - a|$. (See Exercise 9.2.4.)

9.3 Exploring the Relative Performance of Bayes and Frequentist Estimators

There are many professional statisticians who make a more or less permanent choice to stick with procedures of one type or of the other—either Bayesian or frequentist. Their decisions are sometimes based on the orientation of the academic program in which they studied, but they are often based on other factors. For example, it may be that the statistical methodology they utilize most often, or perhaps almost exclusively, is simply more extensively developed in one paradigm or in the other. For example, most statisticians whose work largely involves linear or generalized linear models tend to be inclined toward the frequentist approach to inference. On the other hand, it is not uncommon to choose between the Bayesian and frequentist approaches to inference based on philosophical or logical grounds. There are certainly those who would dismiss the Bayesian approach on the grounds that they cannot accept the Bayesian's use of subjective information as a legitimate form of statistical inference. On the other hand, there are those who believe strongly in the foundations of Bayesian inference and feel that ignoring prior information poses a greater risk than does using it.

This section is devoted to a very practical question which is not covered by any of the reasons mentioned above for adopting or rejecting either statistical approach. By now, you may have come up with this question yourselves: "How does the performance of Bayes estimators compare to that of frequentist estimators in real-life problems?" If, for example, you were to discover that Bayes estimators outperform frequentist estimators with impressive regularity (according to some criterion you considered reasonable), then you might be more inclined to use a Bayes estimator in practice. If, on the other hand, Bayes estimators only did well under quite narrow conditions (for example, only when your prior guess is highly concentrated on the true value of the parameter), then you would probably be less inclined to opt for a Bayesian estimator with any regularity. The content of this section is totally empirical. In this and the next section, we will try to give some useful guidance on how the question "When should I be a Bayesian?" can be addressed.

In the paragraphs that follow, I will describe to you a real experiment in which 99 college students were asked to construct a Bayes estimator in a particular statistical problem. The estimators they provided, and how well these estimators performed, compared to performance of the standard frequentist estimator used in the same problem, will be discussed in detail. Theoretical developments that help understand the findings from our experiment are treated in the section that follows.

I'll begin with a discussion of the real experiment I will refer to as the "word-length experiment." Ninety-nine students in an elementary statistics class at the University of California, Davis, were asked to participate in an experiment involving an observed binomial variable with an unknown probability p of "success." The population of interest was the collection of "first words" on the 758 pages of a particular edition of Somerset Maugham's 1915 novel *Of Human Bondage*. Ten pages were to be sampled randomly, *with replacement*, and the number X of long words (i.e., words with six or more letters) was to be recorded. After a brief introduction to the Bayesian approach to the estimation of a population proportion, each student was asked to provide a Bayes estimate of the unknown proportion p of long words. The elicitation of the students' beta priors was accomplished by obtaining each student's best guess p^* at p and the weight η he or she wished to place on the sample proportion $\hat{p} = X/10$, where X is the number of long words found in a sample of size 10 drawn with replacement from the population of "first words" on 758 pages of Maugham's novel. The remaining weight $(1-\eta)$ was placed on the prior guess p^*. The prior specifications $\{(p^*, \eta)\}$ obtained from the students are displayed in Figure 9.3.1. (The (p^*, η) values needed for this plot are given in Table 9.3.1.

As Figure 9.3.1 suggests, the 99 students who participated in the word-length experiment had rather diverse views about this author's word usage, and the confidence these students showed in their prior guesses at p was quite variable as well. The scatter plot above might be described as "a shotgun blast into the unit square." So it's natural to ask:

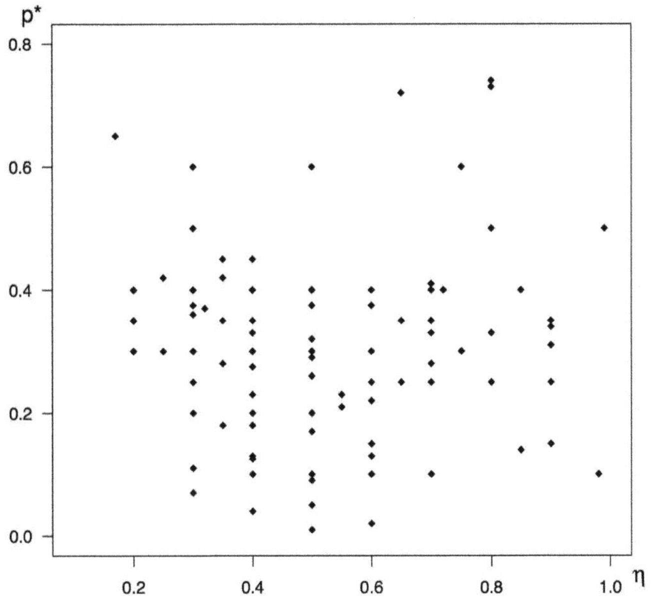

Figure 9.3.1. Scatter plot of (p^*, η) values in the word-length experiment.

How many of these nouveau Bayesians would tend to be closer to the true value of p than a statistician using the sample proportion $\hat{p}$ as an estimator for p? It may surprise you to learn that about 90% of these young Bayesians have an advantage over a frequentist who uses $\hat{p}$ to estimate p. In the paragraph below, I will provide a few additional details about this unexpected domination. In the next section, we'll revisit the word-length experiment after a theoretical development which sheds light on why this domination might have been expected.

Let me now reveal that the true proportion of long words in the target population was found to be $p = 228/758 = .3008$. That's right, I actually counted them! (That's why the target population wasn't defined as all the words in the book.) In Table 9.3.1, each students' prior specification (p^*, η) is displayed, along with the prior sample size ω implied by his/her choice of (p^*, η) and a score (to be explained later) which measures the relative performance of the Bayes and frequentist estimators of p. For now, you may interpret scores that are less than 1 as indicating that the Bayes estimator outperforms $\hat{p}$ (in terms of expected closeness to the true p), while scores that are larger than 1 indicate the opposite. Table 9.3.1 shows that the Bayes estimators are classified as superior in 88 of 99 cases. Since these 99 Bayesians seem to have rather shaky prior knowledge, Table 9.3.1 begs the question: Why do a large fraction of the Bayes estimates outperform the time-honored estimator $\hat{p}$? We'll examine this question theoretically in the next section. Before reading on, see if you can determine from Table 9.3.1 what makes a Bayes estimator good.

ID	p^*	η	ω	score	ID	p^*	η	ω	score
1	0.35	0.35	18.571	0.171	50	0.1	0.6	6.667	0.667
2	0.15	0.9	1.111	0.821	51	0.25	0.7	4.286	0.501
3	0.15	0.6	6.667	0.533	52	0.18	0.4	15	0.41
4	0.375	0.5	10	0.315	53	0.28	0.7	4.286	0.492
5	0.26	0.5	10	0.27	54	0.15	0.6	6.667	0.533
6	0.2	0.5	10	0.371	55	0.5	0.8	2.5	0.715
7	0.4	0.2	40	0.34	56	0.29	0.5	10	0.251
8	0.35	0.2	40	0.114	57	0.13	0.6	6.667	0.582
9	0.03	0.2	40	2.271	58	0.2	0.3	23.333	0.327
10	0.03	0.4	15	1.415	59	0.36	0.3	23.333	0.172
11	0.04	0.4	15	1.324	60	0.42	0.35	18.571	0.408
12	0.25	0.3	23.333	0.15	61	0.25	0.8	2.5	0.645
13	0.4	0.4	15	0.328	62	0.28	0.35	18.571	0.131
14	0.4	0.3	23.333	0.319	63	0.34	0.9	1.111	0.811
15	0.4	0.2	40	0.34	64	0.31	0.9	1.111	0.81
16	0.45	0.4	15	0.541	65	0.13	0.4	15	0.659
17	0.35	0.65	5.385	0.437	66	0.3	0.5	10	0.25
18	0.6	0.5	10	1.314	68	0.375	0.6	6.667	0.402
19	0.02	0.6	6.667	0.96	69	0.73	0.8	2.5	0.99
20	0.33	0.7	4.286	0.494	70	0.1	0.7	4.286	0.663
21	0.05	0.5	10	0.998	71	0.33	0.8	2.5	0.642
22	0.3	0.3	23.333	0.09	72	0.3	0.5	10	0.25
23	0.4	0.5	10	0.367	73	0.21	0.55	8.182	0.382
24	0.25	0.9	1.111	0.811	74	0.01	0.5	10	1.255
25	0.4	0.85	1.765	0.733	75	0.3	0.4	15	0.16
26	0.1	0.4	15	0.85	76	0.125	0.4	15	0.689
27	0.1	0.5	10	0.729	77	0.18	0.35	18.571	0.416
28	0.32	0.5	10	0.254	78	0.4	0.2	40	0.34
29	0.41	0.7	4.286	0.541	79	0.23	0.55	8.182	0.351
30	0.22	0.6	6.667	0.41	80	0.4	0.4	15	0.328
31	0.09	0.5	10	0.778	81	0.375	0.3	23.333	0.218
32	0.35	0.2	40	0.114	82	0.4	0.5	10	0.367
33	0.74	0.8	2.5	1.007	83	0.4	0.4	15	0.328
34	0.65	0.17	48.824	4.023	84	0.3	0.6	6.667	0.36
35	0.25	0.6	6.667	0.38	85	0.6	0.75	3.333	0.829
36	0.275	0.4	15	0.171	86	0.4	0.7	4.286	0.532
37	0.5	0.3	23.333	1.015	87	0.4	0.3	23.333	0.319
38	0.1	0.98	0.204	0.961	88	0.35	0.4	15	0.201
39	0.3	0.75	3.333	0.563	89	0.4	0.72	3.899	0.555
40	0.14	0.85	1.765	0.75	90	0.23	0.4	15	0.246
41	0.42	0.25	30	0.443	92	0.35	0.9	1.111	0.811
42	0.35	0.7	4.286	0.5	93	0.2	0.5	10	0.371
43	0.25	0.65	5.385	0.438	94	0.07	0.3	23.333	1.331
44	0.5	0.99	0.101	0.98	95	0.4	0.5	10	0.367
45	0.3	0.25	30	0.063	96	0.33	0.4	15	0.175
46	0.11	0.3	23.333	0.938	97	0.26	0.5	10	0.27
47	0.72	0.65	5.385	1.446	98	0.2	0.4	15	0.334
48	0.6	0.3	23.333	2.176	99	0.17	0.5	10	0.453
49	0.45	0.35	18.571	0.57					

Table 9.3.1. Performance of Bayes and frequentist estimators in the word-length experiment.

9.4 A Theoretical Framework for Comparing Bayes vs. Frequentist Estimators

Let us consider the possibility of comparing two estimators on the basis of their proximity to the true value of the parameter being estimated. Since it is clear that what a statistician really wants from an estimator is that it tends to be close to the target parameter's true value, it would be difficult to argue that this basis for comparison is not relevant. The difficulty with such criteria is that we generally don't know the truth (which is, of course, the very reason we attempt to estimate population parameters). But here's a revolutionary idea—let's model the truth as a random variable. Suppose that we observe $X \sim F_\theta$, and that the unknown θ has a probability distribution G_0. It is reasonable to have some concern about the fact that G_0 is not known or the possibility that the parameter θ is actually just an unknown constant rather than a random quantity. Regarding the second issue, there is no real problem since if θ is just the constant θ_0, the situation is covered (like a glove!) by the framework above, since in that case, the distribution G_0 is simply the degenerate distribution which places probability 1 on the value θ_0. The fact that G_0 is not known or knowable seems like a serious obstacle to progress, but surprisingly, it really isn't. Accounting for the "true prior" distribution G_0 turns out to be quite helpful in allowing us to identify the basic principles which lead to Bayes estimators that outperform their frequentist counterparts (and vice versa).

We now turn our attention to the technical underpinnings of a comparative analysis of Bayesian and frequentist estimation. We will assume that two statisticians, one from each camp, are faced with the same problem of estimating an unknown parameter θ based on a random sample $X_1, X_2, \ldots, X_n \stackrel{\text{iid}}{\sim} F_\theta$ and a fixed loss function $L(\theta, a)$. Suppose that the frequentist statistician is prepared to estimate the unknown parameter θ by the estimator $\hat{\theta}$ and that a Bayesian statistician is prepared to estimate θ by the estimator $\hat{\theta}_G$, the Bayes estimator relative to his chosen prior distribution G. We'll take the true value of the target parameter to be a random variable with a (possibly degenerate) distribution G_0 which we will refer to as the "true prior" distribution. In many estimation problems, the true θ is just an unknown constant. In other settings, as when θ is the proportion of defective items in a particular day's production lot (a value which varies from day to day), it may be appropriate to consider G_0 to be non-degenerate. In either case, we take G_0 to be a description of "what is," the actual (random or fixed) state of nature in the experiment we have at hand. Think of G_0 as God's prior, unknown to us and to the two statisticians who are trying to estimate the parameter θ. Accounting for the unknown state of nature in this way gives neither statistician any particular advantage, as the exact form of G_0 is unknown and unknowable in any real estimation problem.

No comparative analysis is possible without the specification of a criterion on which comparisons will be based. Let us consider the possibility of using the Bayes risk of an estimator, relative to the true prior G_0, as a criterion for judging the superiority of one estimator over another. For a fixed loss function L, the Bayes risk of an estimator $\hat{\theta}$ with respect to the true prior G_0 is given by

$$r(G_0, \hat{\theta}) = E_\theta E_{\mathbf{X}|\theta} L(\theta, \hat{\theta}(\mathbf{X})),$$

where the outer expectation is taken with respect to G_0. For the sake of clarity and simplicity, I will provide a defense of this criterion for squared error loss, that is, for $L(\theta,a) = (\theta - a)^2$. After all, the use of squared error loss in assessing the performance of estimators is ubiquitous in statistical work. It will be apparent that the arguments given can be easily generalized.

Three questions naturally arise at this point, and they need to be addressed satisfactorily before we, or anyone else, would feel comfortable proceeding with the proposed comparisons. First, does the criterion measure something interesting and important? Second, does the criterion favor one of the statisticians over the other? And finally, does the fact that the distribution G_0, whether it is degenerate or not, is an unknown quantity make the comparison impossible to resolve? I will answer each of these questions in turn, but will do so briefly, summarizing the more detailed discussion provided in Samaniego (2010).

So, what are we measuring here? The risk function $R(\theta,d)$ of the decision rule or estimator d is simply the mean squared error of d. This is hardly a new quantity in statistical work. The new wrinkle here is that we intend to focus on the risk function averaged over the distribution G_0. Suppose for a minute that G_0 is degenerate at the point θ_0. Then $r(G_0,d)$ is actually just the mean squared error of d evaluated at the true value of θ. The importance of this quantity in a given estimation problem is transparent. In this case, $r(G_0,d)$ just measures the average squared distance between the estimator and the true value of θ. An impartial third party would no doubt judge the estimator for which $r(G_0,d)$ is smaller to be the better estimator. The fact that the Bayesian statistician chooses to minimize the measure $r(G,d)$ relative to his personal prior G shows that the proposed criterion measures something that the Bayesian in interested in. The fact that the frequentist statistician seeks an estimator with a small mean squared error at arbitrary values of θ suggests that a small mean squared error at the true value of θ would be especially desirable to the frequentist. Setting aside the fact that the truth is unknown, a criterion that rewards the statistician that is able to get closer to the truth must be regarded as both relevant and appropriate. The relevance of this criterion when G_0 is non-degenerate may be supported similarly.

Does the criterion function favor either the Bayesian of the frequentist statistician? The Bayesian statistician is using an estimator that minimizes the Bayes risk with respect to his operational prior G. Since the true prior G_0 is unknown, there is no built-in advantage for the Bayesian. One would expect the Bayesian to do well only if G resembles G_0 in some sense, but this depends on the quality of his choice of prior, not on the criterion itself, since the criterion simply rewards the statistician who manages to get close to the unknown truth. The Bayesian's estimation process is not driven by the true prior G_0, but there can be no question that an impartial adjudicator would be interested in $r(G_0, \hat{\theta})$ rather than in $r(G,\hat{\theta})$, as it is the former measure, rather than the latter, which pertains to how well the Bayesian did in estimating the parameter θ. The frequentist statistician is letting the data lead him to the true value of θ. A criterion function which rewards a frequentist for getting close to the true θ is compatible with his goals, but since the truth is not known, the frequentist is neither favored nor disadvantaged by the use of the proposed criterion. The frequentist will do well only when his estimator effectively hones in on the truth, whatever that truth may be.

Finally, let's consider the fact that G_0 is unknown. The fact that this doesn't matter is perhaps the most surprising feature of the approach I've proposed. If our goal was to raise the hand of one of the two statisticians and proclaim a clear winner in a given estimation problem, we would have to acknowledge that this task is beyond us. What *is* possible, even though G_0 is unknown, is to identify the characteristics of Bayes estimators in an estimation problem that have strong performance characteristics relative to the performance of the best (or at least good) frequentist estimators. This claim is hard to digest in the abstract. Theorem 9.4.1, the main result of this section, together with our general discussion of this theorem and its relevance in interpreting the results of the word-length experiment, will serve to make the point more clearly. We shall see that, in many estimation problems of interest, it is possible to detect when a Bayesian might have the advantage over the frequentist. We will also see that the collection of "good Bayes estimators" is a broader class than it is generally thought to be, that is, there is a certain natural robustness in the specification of a prior distribution that permits Bayesians to have the advantage even when their priors are quite imperfect. The general principles of good Bayesian estimation will be identified, as will the type of estimation problem in which the Bayesian approach is not very promising.

I now will define the "threshold problem" in a general way and then describe the version of it whose solution will be presented here. For simplicity and clarity, I will assume that we are interested in estimating a scalar parameter θ. Also, assume that a loss function L has been specified. As proposed above, I will use the Bayes risk $r(G_0,\hat{\theta})$ of a point estimator $\hat{\theta}$ with respect to the true prior distribution G_0 as the ultimate measure of the estimator's performance. Let $\mathcal{G} = \{G\}$ represent the class of prior distributions that a Bayesian statistician will draw from in deriving a Bayes estimator $\hat{\theta}_G$ of θ. Let $\hat{\theta}$ be the estimator of θ selected by the frequentist statistician. Then, the "threshold problem" may be defined as follows: determine the boundary which divides the class $\mathcal{G}$ into the subclass of priors G for which

$$r(G_0,\hat{\theta}_G) < r(G_0,\hat{\theta}), \tag{9.40}$$

from the subclass of priors G for which

$$r(G_0,\hat{\theta}_G) > r(G_0,\hat{\theta}). \tag{9.41}$$

There are several reasons why the threshold problem, as formulated above, may seem impenetrable: (1) the class $\mathcal{G}$ of all possible priors is large and, in the context of studying the inequality in (9.40), it seems quite unmanageable analytically, (2) it appears that one must solve multiple threshold problems, one for each estimator the frequentist might consider for use, and (3) the true prior G_0 is completely unknown, so that assessing an estimator's worth by means of a criterion that is a function of G_0 seems infeasible. To make the threshold problem tractable, I will make some simplifying assumptions. Let's assume that our data consists of a *random sample* from a distribution indexed by θ, that is, that $X_1, X_2, \ldots, X_n \overset{\text{iid}}{\sim} F_\theta$, where the distribution F_θ belongs to a *one-parameter exponential family*. Further, take L to be *squared error loss*, and let $\mathcal{G}$ be the class of *standard conjugate priors* corresponding to the distribution F_θ. As will be seen, the fact that G_0 is unknown will not prove to be a major obstacle. In the problems of primary interest, it is often reasonable to assume that G_0 is degenerate at a single point θ_0, and this assumption makes drawing some conclusions from our analysis not only feasible but especially clear. These simplifying assumptions are not absolutely necessary to make the threshold problem well defined and manageable, but as we will see, they are sufficient.

Under the assumptions in the preceding paragraph, we will show that the determination of the threshold representing the boundary between the subclasses defined by (9.40) and (9.41) is a problem that is amenable to an analytical treatment. First, we will be able to restrict attention to just one frequentist estimator, the estimator $\hat{\theta}$ that I will refer to as the "best frequentist estimator." Our ability to restrict attention to $\hat{\theta}$ is a consequence of the fact that exponential families have complete sufficient statistics, and for the usual target parameters of interest, MVUEs generally exist. Not only are these the typical frequentist estimators of choice in such problems, but all of the reputable alternative estimators are one and the same; that is, the same estimator arises whether one approaches the problem by finding the best unbiased estimator, the method of moments estimator, the maximum likelihood estimator, the best linear unbiased estimator, or the least-squares estimator of θ. Thus, one can consider the apparent multiplicity of threshold problems defined by (9.40) and (9.41) to be reducible to a single basic problem. Further, the standard conjugate families to exponential families of sampling distributions are indexed by a fixed number of parameters. Thus, the characterization of conjugate priors for which (9.40) holds reduces to a search over a finite-dimensional space of prior parameters. Thirdly, under squared error loss, the identification of Bayes estimators and the evaluation of the Bayes risks for the Bayes or frequentist estimators are generally straightforward. Under the simplifying assumptions we have made, it is not reasonable to expect that the threshold problem will admit to an analytical solution.

The theorem below is proven under assumptions that are less restrictive than those satisfied by the primary intended applications. The settings which motivated the result below are (1) estimating a population proportion from iid Bernoulli trials and (2) estimating a normal mean based on a random sample from a normal population (with variance σ_0^2 assumed known). The common characteristic of these two estimation problems is that the available sample is drawn from a distribution F_θ which belongs to a one-parameter exponential family. We will be interested in estimating a scalar parameter θ under squared error loss. We posit the existence of a statistic $\hat{\theta}$ that is sufficient for θ and is an unbiased estimator of θ. In the applications of primary interest, $\hat{\theta}$ can be thought of as the minimum variance unbiased estimator (MVUE) of θ. We will refer to $\hat{\theta}$ as "the best frequentist estimator," since in the contexts of primary interest, the standard alternative frequentist estimators will also be equal to $\hat{\theta}$.

I will make a fairly standard assumption in distinguishing between Bayes and frequentist estimators. By the term "Bayes estimator," I mean estimators that are Bayes with respect to proper prior probability distributions, that is, that minimize the posterior expected loss relative to a prior distribution that assigns total mass 1 to the real line. This restriction is common in "subjective" Bayesian analysis, where the prior model is taken as a representation of the Bayesian's *a priori* assessment of his uncertainty about the value of the unknown parameter θ. The orthodox application of the Bayesian theory of estimation requires that the measure placed on the parameter space be a probability measure. We know (from Theorem 9.2.1) that an unbiased estimator cannot be a Bayes rule with respect to a proper prior distribution and squared error loss. Thus, under squared error loss, the estimator $\hat{p}$ of a population proportion of p and the estimator $\bar{X}$ of the mean μ of a normal distribution are estimators that are available to the frequentist but unavailable to the Bayesian. The specific prior distribution G used by the Bayesian will be referred to as his operational prior. Although not formally included as an assumption in the theorem below, typical applications of the theorem would involve conjugate prior distributions, as it is then that Bayes estimates of the parameter of interest necessarily have the form stipulated in the theorem.

As explained above, the Bayes risk of an estimator relative to the "true prior" G_0 will be taken as the criterion for judging the performance of either estimator. We will seek to compare $r(G_0, \hat{\theta}_G)$, the Bayes risk of the Bayesian's estimator $\hat{\theta}_G$, with respect to the true prior G_0, to the corresponding Bayes risk $r(G_0, \hat{\theta})$ of the frequentist's estimator $\hat{\theta}$. Having put forward the case for using the Bayes risk of an estimator, relative to the true prior distribution G_0, as an appropriate and relevant measure of its quality, I now present this section's main result.

Theorem 9.4.1. (Samaniego and Reneau, 1994) Assume that a random sample $X_1,\ldots,X_n$ is drawn from a distribution F_θ. Let $\hat{\theta}_G$ be the Bayes estimator of θ under squared error loss, relative to the operational prior G. If $\hat{\theta}_G$ has the form

$$\hat{\theta}_G = (1-\eta)E_G\theta + \eta\hat{\theta}, \tag{9.42}$$

where $\hat{\theta}$ is a sufficient and unbiased estimator of θ and $\eta \in [0,1)$, then for any fixed distribution G_0 for which the expectations exist,

$$\mathbf{r}(G_0,\hat{\theta}_G) \leq \mathbf{r}(G_0,\hat{\theta}) \tag{9.43}$$

if and only if

$$V_{G_0}(\theta) + (E_G\theta - E_{G_0}\theta)^2 \leq \frac{1+\eta}{1-\eta}\mathbf{r}(G_0,\hat{\theta}). \tag{9.44}$$

Proof. For a fixed but arbitrary θ, the mean squared error of the Bayes estimator $\hat{\theta}_G$ may be written as

$$E_{F_\theta}(\hat{\theta}_G - \theta)^2 = E_{F_\theta}[\eta(\hat{\theta}-\theta) + (1-\eta)(E_G\theta - \theta)]^2. \tag{9.45}$$

Squaring out the RHS of (9.45), we find that the cross-product term

$$2\eta(1-\eta)E_{F_\theta}(\hat{\theta}-\theta)(E_G\theta-\theta)$$

vanishes due to the fact that $\hat{\theta}$ is assumed to be unbiased. We may thus rewrite (9.45) as

$$E_{F_\theta}(\hat{\theta}_G - \theta)^2 = \eta^2 E_{F_\theta}(\hat{\theta}-\theta)^2 + (1-\eta)^2(E_G\theta-\theta)^2. \tag{9.46}$$

Taking the expectation of both sides of (9.46) with respect to the distribution G_0, we have

$$r(G_0,\hat{\theta}_G) = \eta^2 r(G_0,\hat{\theta}) + (1-\eta)^2 E_{G_0}(\theta - E_G\theta)^2. \tag{9.47}$$

Viewing $E_{G_0}(\theta - E_G\theta)^2$ as the mean squared error of the random variable θ as an estimator of $E_{G_0}\theta$, we may replace $E_{G_0}(\theta - E_G\theta)^2$ in (9.47) by

$$V_{G_0}(\theta) + (E_G\theta - E_{G_0}\theta)^2. \tag{9.48}$$

It then follows that the inequality in (9.43) is equivalent to

$$(1-\eta)^2(V_{G_0}(\theta) + (E_G\theta - E_{G_0}\theta)^2) \leq (1-\eta^2)\mathbf{r}(G_0,\hat{\theta}), \tag{9.49}$$

an inequality that may equivalently be written as

$$V_{G_0}(\theta) + (E_G\theta - E_{G_0}\theta)^2 \leq \frac{1+\eta}{1-\eta}\mathbf{r}(G_0,\hat{\theta}).$$

■

Theorem 9.4.1 is quite revealing about the characteristics of Bayes estimators which will tend to outperform the frequentist estimator $\hat{\theta}$. Note, first, that the left-hand side (LHS) of (9.44) can be made equal to zero, while the right-hand side (RHS) of (9.44) is necessarily positive. When the LHS of (9.44) is zero, the Bayesian will win the contest with certainty. Further, note that the variance on the LHS of (9.44) is the variance of the true prior, not the variance of the operational prior, a fact which suggests that in typical estimation problems (where $V_{G_0}(\theta)$ may be considered to be zero), the Bayes estimator $\hat{\theta}_G$ looks especially promising. Finally, since the weight η placed on $\hat{\theta}$ resides in the interval

$[0,1)$, the multiplier $(1+\eta)/(1-\eta)$ on the RHS of (9.44) only takes on values in the interval $[1,\infty)$. Not only can it never be smaller than 1, it can be made arbitrarily large by taking η sufficiently close to 1. Let us reflect on the implications of these three revelations. First, when the true prior G_0 is taken to be a degenerate distribution on θ_0, the "true value of θ," we see that the inequality in (9.44) goes well beyond our natural intuition on when Bayes estimators should be good. Intuitively, one would expect the Bayesian to outperform the frequentist in estimating an unknown constant θ whenever the mean of the operational prior is close to the true value of θ *and* the operational prior has a small variance. This intuition is articulated, for example, in following statement by Diaconis and Freedman (1986): "A statistician who has sharp prior knowledge of these parameters (sic) should use it, according to Bayes' Theorem ... On this point, there seems to be general agreement in the statistical community." But notice that the LHS of (9.44) makes no mention of the variance of the operational prior G. The variance of G enters the picture only through the value of η on the RHS of (9.44). The statement quoted above is thus seen to be unduly conservative. The inequality in (9.44) suggests, instead, that the Bayesian will outperform the frequentist whenever the mean of the operational prior is "sufficiently close" to the true value of θ (or more generally, is close to the mean of G_0) and the *true* prior has a small variance. That's both interesting and highly relevant! In most problems of practical interest, the true prior has variance zero. This brings into focus the role of the "mean correctness" of an operational prior, a property that is defined as follows:

Definition 9.4.1. In the context discussed above with G and G_0 being, respectively, the "operational" and "true" priors of the random parameter θ, the operational prior G is said to be *mean correct* if $E_G\theta = E_{G_0}\theta$.

The following corollary follows immediately from Theorem 9.4.1.

Corollary 9.4.1. Under the hypotheses of Theorem 9.4.1, a Bayes estimator with respect to a mean-correct prior G has a smaller Bayes risk than the best frequentist estimator $\hat{\theta}$ if and only if

$$V_{G_0}(\theta) \leq \frac{1+\eta}{1-\eta}\mathbf{r}(G_0,\hat{\theta}). \tag{9.50}$$

Further, if the true prior distribution G_0 is degenerate at a point, any Bayes estimator with respect to a mean-correct operational prior is superior to the best frequentist estimator.

It is thus clear that Theorem 9.4.1 goes well beyond the "general agreement" mentioned above about "sharp" (that is, accurate and precise) priors. For example, the uniform distribution on the interval $[0,1]$ is not generally regarded as a sharp prior distribution on a population proportion p, and a U-shaped beta prior like $Be(.1,.1)$ is anything but sharp, but if they happen to be mean correct (that is, the true p is equal to $1/2$), the Bayes estimators with respect to either of these priors will outperform the best frequentist estimator $\hat{p}$. Note also that (9.44) shows that mean correctness isn't necessary for Bayesian superiority. When the true prior is degenerate, there is clearly an interval (a,b) containing the true value of θ (or, more generally, $E_{G_0}\theta$) such that, when $E_G\theta \in (a,b)$, the Bayes estimator is necessarily superior to the frequentist estimator $\hat{\theta}$. Finally, even when $E_G\theta$ lies well outside that interval, the Bayes estimator will still be superior to the best frequentist estimator $\hat{\theta}$ provided that the weight η that the Bayes estimator places on $\hat{\theta}$ is not too small. This finding is quite counterintuitive and thus constitutes an interesting and useful insight. An example of this circumstance will help underscore the point. Suppose that you use a Bayes estimator of a binomial proportion p, and that your prior distribution has a mean near 0 when in fact the true value of p is close to 1. Your prior could hardly be any worse! It seems reasonable to expect that the price to be paid for this poor prior modeling would diminish if the weight you place on your prior guess is suitably small. But why should a convex combination of $\hat{\theta}$ and a terrible guess at θ ever be superior to the estimator $\hat{\theta}$ alone? Theorem 9.4.1 tells us precisely when this will happen: the Bayes estimator will outperform $\hat{\theta}$ whenever η, the weight the Bayes estimator places on $\hat{p}$, is sufficiently close to 1. The reason things work this way is explored in a little more depth in the next paragraph.

The inequality (9.44) identifies two particular mistakes that a Bayesian can make in estimating the parameter θ, and it shows what their consequences will be. The first mistake is a poor specification of the mean of his prior. The second is placing too much weight on his prior mean (that is, having too much confidence that his prior mean is a good guess at the true value of θ). Interestingly, neither of these mistakes will necessarily, by itself, cause the Bayesian to lose his advantage. If, for example, the heights of a particular human population are modeled as normally distributed, the Bayes estimator, relative to a normal prior distribution with mean 1000 ft will actually outperform the frequentist estimator $\bar{X}$ if the weight placed on the prior mean is sufficiently small. Such phenomena can be understood by examining the damping effect of the weight $(1-\eta)$ placed on the prior mean, as seen in the following expression (obtained from (9.47) when G_0 is degenerate at θ_0) for the mean squared error of the Bayes estimator $\hat{\theta}_G$ at the true value $\theta = \theta_0$:

$$MSE_{\theta_0}(\hat{\theta}_G(\eta)) = \eta^2 MSE_{\theta_0}(\hat{\theta}) + (1-\eta)^2(\theta_0 - E_G\theta)^2. \tag{9.51}$$

Note that the RHS of (9.51) is a parabola in η that is bounded below by 0, takes the value $MSE_{\theta_0}(\hat{\theta})$ (which is the same as $V_{\theta_0}(\hat{\theta})$ due to the unbiasedness of $\hat{\theta}$) when $\eta = 1$, and has a positive derivative at $\eta = 1$. It follows that there exists a value $\eta^* \in [0,1)$ such that if $\eta > \eta^*$, then $MSE_{\theta_0}(\hat{\theta}_G(\eta)) < MSE_{\theta_0}(\hat{\theta})$. Thus, the Bayes estimator will outperform $\hat{\theta}$ when η is sufficiently large.

While the artifact of assuming the existence of a true prior distribution G_0 has served a useful purpose, namely, helping to clarify how the various elements of a statistical model affect the comparative performance of Bayes and frequentist estimators, it is possible to develop the comparison of these estimators under the assumption that the parameter θ is simply an unknown constant θ_0. In that case, Theorem 9.4.1 may be rewritten as

Corollary 9.4.2. Under the hypotheses of Theorem 5.1, a Bayes estimator with respect to the prior G is closer, on average, to the true parameter value θ_0 than the best frequentist estimator $\hat{\theta}$ if and only if

$$(E_G\theta - \theta_0)^2 \leq \frac{1+\eta}{1-\eta} MSE_{\theta_0}(\hat{\theta}). \tag{9.52}$$

From either Theorem 9.4.1 or Corollary 9.4.2, the Bayesian's winning strategy clearly reveals itself. A Bayesian who hopes or expects his estimator to outperform the best frequentist estimator $\hat{\theta}$ should choose a prior distribution with the following two guidelines in mind. First, the prior mean, the Bayesian's best guess at θ prior to the examination of data, should be chosen with great care. It is clear that a judicious choice of the prior mean stands to give the Bayesian a strong advantage, and might in fact ensure the his estimator will outperform $\hat{\theta}$, on the average. This step is so important that the choice of the prior mean should be given considerable thought, perhaps involving a little research regarding similar past experiments and about general knowledge concerning the experiment. When the stakes are really high, consulting with an expert on the subject matter in question would no doubt be advisable. Secondly, it is clear from (9.44) or (9.51) that overstating one's confidence in a prior guess can lead to inferior performance, so conservative prior modeling is indicated; if one is to err in specifying the weight η one places on the frequentist estimator $\hat{\theta}$, it's better to err on the high side, thereby understating the confidence associated with one's prior guess. The recipe for successful Bayesian estimation is thus clear: careful specification of the prior mean $E_G\theta$, and the placement of only a modest amount of weight on the prior mean (protecting against overconfidence). Failing in just one of these guidelines is not necessary a fatal flaw, but failing at both generally will be. The prospects for success for a Bayesian who is both *misguided* and *stubborn* are rather poor.

A final consequence of Theorem 9.4.1, and no doubt the most important, is that in the context of exponential families, conjugate prior distributions, and squared error loss, it provides an explicit solution to the threshold problem. The characterization of Bayesian superiority is contained in the following:

Corollary 9.4.3. Under the hypotheses of Theorem 9.4.1, the Bayes estimator $\hat{\theta}_G$ and the frequentist estimator $\hat{\theta}$ have the same Bayes risk with respect to the true prior G_0 for any operational prior G corresponding to the prior parameters (Δ, η) satisfying the hyperbolic equation

$$\Delta\eta + \eta(r(G_0, \hat{\theta}) + V_{G_0}(\theta)) - \Delta + (\mathbf{r}(G_0, \hat{\theta}) - V_{G_0}(\theta)) = 0, \tag{9.53}$$

where $\Delta = (E_G\theta - E_{G_0}\theta)^2$ and $\eta \in [0, 1)$ is the weight parameter specified in (9.42). Further, if $\hat{\theta}_G = (1-\eta)\theta^* + \eta\hat{\theta}$ is the Bayes estimator of θ with respect to the prior G, where $\theta^* = E_G\theta$ and $\eta \in [0, 1)$, then

(i) for fixed θ^*, there exists a constant $\eta^* \in [0, 1)$ such that $\hat{\theta}_G$ is superior to the best frequentist estimator $\hat{\theta}$ if and only if $\eta > \eta^*$, and

(ii) for fixed η, there exists a constant $\Delta^* \in [0, \infty)$ such that the Bayes estimate $\hat{\theta}_G$ is superior to the estimator $\hat{\theta}$ if and only if $\Delta = (\theta^* - E_{G_0}\theta)^2 < \Delta^*$.

In Section 9.3, I introduced you to the word-length experiment. Recall that the experiment involved 99 students who were asked to provide Bayes estimates, relative to their individually elicited beta priors, of the proportion p of "long words" among the first words appearing on the 758 pages of a copy of Somerset Maugham's novel *Of Human Bondage*. The data that these students would be expecting to see, when implementing their estimation of p, was a random sample of 10 Bernoulli trials (that is, 10 draws, with replacement, from the population of 758 words). From Figure 9.3.1, it is evident that the 99 students had rather diverse opinions about the value of p and also differed substantially in the amount of confidence they had in their prior guesses. As Table 9.3.1 indicates, a seemingly surprising fraction of these students (namely, 8/9) constructed Bayes estimators of p that outperformed the sample proportion $\hat{p}$. We now examine this experiment in light of the theoretical results above. The true proportion of long words in the population was determined to be $p = 228/758 = .3008$. In Table 9.3.1, each student's prior specification (p^*, η) is shown, together with the prior sample size $\omega = (n/\eta) - n$ of the corresponding beta prior (with $n = 10$) and a score associated with the comparison of each Bayes estimator with the sample mean. The specifics of the score function were not identified in Table 9.3.1, but it was in fact that natural score function based on the inequality (9.40) which defines the class of Bayes estimators of p that outperform the sample proportion $\hat{p}$. The score for each Bayes estimator is simply the ratio of the Bayes risk of the Bayes estimator relative to the true prior distribution, to the Bayes risk of $\hat{p}$, relative to the true prior distribution, the true prior being the distribution giving probability 1 to the true value .3008 of p. Thus, a score less than 1 indicates Bayesian superiority. Bayes estimators are superior in 88 of 99 cases. A further indication of the extent of the domination of Bayes estimators over the sample proportion is the fact that, for the 38 students for whom $|\ln[r(G_0,\hat{\theta}_G)/r(G_0,\hat{\theta})]| > 1$, that is, for whom one Bayes risk was (roughly) three or more times as large as the other, the Bayes estimator outperformed the sample proportion 37 times.

While the strong domination of the Bayes over the frequentist estimators apparent from Table 9.3.1 seems, at first view, surprising, let us examine this domination carefully in light of the theoretical developments above. We do, of course, have the rare luxury here of knowing the true prior distribution G_0; it is simply the degenerate distribution at the point .3008. Now, from the inequality (9.44), it follows that the Bayes estimator based on the prior mean p^* (and irrespective of the prior weight parameter η) will be superior to $\hat{p}$, relative to our Bayes risk criterion, whenever

$$(p^* - .3008)^2 \leq \mathbf{r}(G_0, \hat{p}) = .02103. \tag{9.54}$$

This shows that a Bayesian has a fairly generous window of opportunity for outperforming the frequentist; the Bayes estimator will prevail if the Bayesian's prior mean is within .145 of the true value .3008 of p. With a little reflection, the fact that 66 of the 99 Bayes estimators dominated the best frequentist estimator on the basis of a "good" prior guess might well be anticipated. For the Bayes estimators for which $|p^* - .3008| > .145$, a Bayes estimator will be superior to the best frequentist estimator precisely when

$$(p^* - .3008)^2 < \frac{1+\eta}{1-\eta}(.02103). \tag{9.55}$$

For each value of p^* in this latter class, Bayesian superiority involves a direct relationship between the distance $|p^* - .3008|$ and the prior weight η that the Bayesian places on the sample proportion $\hat{p}$. Of the 33 Bayes estimators for which the prior means p^* that were far enough from the true mean .3008 so that the value of η actually plays a role in determining the direction of superiority, 22 of them chose a value of η that was large enough to satisfy the inequality (9.55). In total, 89% of the Bayes estimators outperformed the sample proportion $\hat{p}$.

The empirical evidence provided by the word-length experiment suggests, by itself, that Bayesian inference can be quite forgiving. The theoretical developments of this section help us understand why. The Bayesian will be penalized when he is *misguided*, as the Bayes risk $r(G_0, \hat{\theta}_G)$ is an increasing function of the distance between the prior mean and the true value of θ (or the value of $E_{G_0}\theta$). The Bayesian will also suffer for being *stubborn* about his prior opinion, a characteristic that, in our context, is reflected in the fact that the Bayes estimator places an unduly high weight on the prior mean. A Bayesian who has one, but not both, of these characteristics still has a fighting chance at outperforming the frequentist. In the word-length experiment, Student 85, who was clearly misguided but wasn't stubborn about it, outperformed the frequentist. Student 8, exhibited the opposite combination—a very good prior guess but, it seems, quite an inflated amount of confidence in that guess. Student 8 also outperformed the frequentist. On the other hand, a misguided and stubborn Bayesian such as Student 34 had virtually no chance of success in this experiment.

Figure 9.4.1, the threshold between "superior" and "inferior" Bayes estimators in the word-length experiment is displayed. The curve shown in Figure 9.4.1 is the collection of priors $G(p^*, \eta)$, where $p^* = E_G\theta$, for which the Bayes estimator wrt G has the same Bayes risk with respect to the true prior G_0 as the sample proportion $\hat{p}$, that is, for which

$$(p^* - .3008)^2 = \frac{1+\eta}{1-\eta}(.02103). \tag{9.56}$$

The graph shows quite vividly that the collection of Bayes estimators that are superior to the frequentist estimator $\hat{p}$ constitutes a non-negligible fraction of the unit square (being all points (p^*, η) above the threshold).

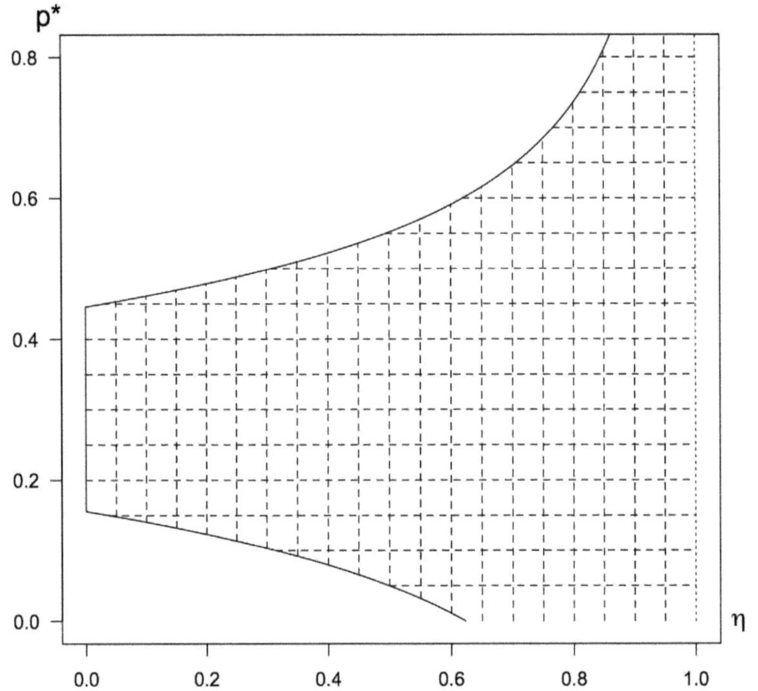

Figure 9.4.1. Graph of the threshold (9.56) and the region of Bayesian superiority in the word-length experiment.

The graph above, when expanded to include the entire unit square, shows that all Bayes estimators for which the prior mean $p^* \in (.1558, .4458)$ are necessarily superior to $\hat{p}$, regardless of the weight η the Bayesian places on $\hat{p}$, and that all Bayes estimates which place weight $\eta > .9601$ are necessarily superior to $\hat{p}$, regardless of the prior mean p^* selected by the Bayesian. The proportion of the unit square taken up by prior specifications (p^*, η) corresponding to Bayes estimates that are superior to $\hat{p}$ is .55. The proper interpretation of this proportion is that, if a Bayesian were to pick a prior specification (p^*, η) completely at random according to a uniform distribution on the unit square (something akin to the aforementioned shotgun blast), the Bayes estimator would outperform $\hat{p}$, the frequentist estimator, 55% of the time. For Bayesians who have "useful" prior information available about the parameter p, one would expect even better performance, on the average. The fact that 89% of the students participating in the word-length experiment specified Bayes estimators that outperformed $\hat{p}$ suggests that most of them did have such information, even though their prior distributions could hardly be referred to as "sharp."

It is only fair to mention the "other side" of the story. While the performance of Bayes estimators in the word-length experiment is eye-catching, you should not lose sight of the fact that, both in theory and in practice, the frequentist estimator $\hat{p}$ sometimes beats the Bayes estimator $\hat{p}_G$. In settings in which Theorem 9.4.1 applies, neither the Bayes estimator nor the frequentist estimator will be uniformly superior. The natural conclusion one comes to is that you need to be careful in selecting the approach you'll use on a given problem. Perhaps the real take-home lesson of this section is that statisticians should remain open to using one approach or the other, depending on which promises to provide better performance.

It may have occurred to you that the small sample size in the word-length experiment might play an important role in whether the Bayesian or the frequentist has better performance. After all, when the sample size is very small, you might argue that even weak prior information might in fact be useful and might lead to a better result than placing all your confidence on weak experimental evidence. It is possible to study this question and get some definitive conclusions about it. For example, in the word-length experiment, you could identify the beta prior distribution elicited from each of the 99 students involved, and then see if a Bayes estimate using that prior would outperform $\hat{p}$ when the sample size is allowed to grow. I have done exactly that for the information in Table 9.3.1. In the table below, I display the percentage of Bayes estimators that would be superior to the frequentist estimator $\hat{p}$ for the word-length experiment with different sample sizes. The true prior G_0 remains the same, as does the Bayes risk criterion, relative to G_0, for comparing the Bayes and frequentist estimators.

Sample size n	Number of SBEs	Percentage of SBEs
1	99	100%
5	93	94%
10	88	89%
50	83	84%
100	81	82%
500	80	81%
1,000	80	81%
5,000	79	80%
10,000	79	80%
∞	79	80%

Table 9.4.1. The percentage of Bayes estimators (SBEs) that are superior to the sample proportion in the word-length experiment, as a function of n.

It is clear from Table 9.4.1 that the size of the experiment has relatively little impact on the comparisons we have made. Indeed, the outcomes in this table represent yet one more surprise in our investigations. Imagine, if you were given a sample of $10,000$ words, drawn with replacement, from the 758 words in the population we've been dealing with, the estimator $\hat{p}$ would be a stunningly precise estimator of p. Indeed, the distribution of $\hat{p}$ would be beautifully approximated by a normal distribution with mean $p = .3008$ and variance $pq/10,000 = .00002132$. The standard error of the estimator $\hat{p}$ is .004586, which means that we would have about 95% confidence that $\hat{p}$ is within .009 of the true value .3008 of p. Now take a look at the prior guesses of the 99 rag-tag Bayesians who participated in the word-length experiment. With such an accurate estimator $\hat{p}$ of p, who would guess that most students (actually 80%) would be closer to the true value of p, on average, with the Bayes estimators based on their elicited priors. Is there something magical going on here? Actually, the identity in (9.51) is really what's going on. The weight parameter η in that equation converges to 1 as n grows, and it converges fast enough to allow most Bayesians to retain their edge.

It is pretty clear that the frequentist approach to estimation took quite a trouncing in the word-length experiment. This experiment, and the theoretical developments that followed it, seem to suggest that the Bayesian approach to estimation is worth considering. That's a fairly conservative conclusion, really, since a 90% rate of domination is pretty impressive. These final paragraphs have two purposes. First, I have a few caveats to mention. Second, it is true that, in the problem (and problem type) I have discussed here, Bayesians rule! But there is clearly room for the frequentist, even in this arena, and even more so in the problems I will now discuss. So I will close this section with what might be thought of as a "pep talk" for frequentists.

Let's do the caveats first. There are certain characteristics of the problems covered in this section that makes these scenarios special. I've been dealing with one-parameter problems, I've been utilizing a nice symmetric loss function, and I've restricted attention to a collection of models (exponential families with conjugate priors) in which analytical work is feasible and for which nice things tend to occur. What happens when we expand our horizons to more complex scenarios? The first thing I would say is that the framework of positing the existence of a true prior distribution and searching for the threshold separating good priors from bad priors carries over to every extension I've examined. What doesn't carry over, in general, is the facility in getting exact analytical results and characterizations of what the threshold will look like. In Samaniego (2010), a variety of alternative modeling scenarios are examined. I will mention two in particular—the problem of estimating a k-dimensional parameter vector for $k \geq 3$, treated in Chapter 7 of that book, and the problem of estimating a scalar or vector-valued parameter with asymmetric loss, treated in Chapter 8. The higher dimensional problem that is treated there is the estimation of the mean of a multivariate normal distribution, and the comparison pursued concerns the relative performance of a frequentist "shrinkage estimator" (the James-Stein estimator) and the Bayes estimator with respect to a multivariate normal (MVN) prior and generalized squared error loss. By the way, the Bayes estimator is also a "shrinker," but it shrinks the vector of sample means toward the mean of the prior distribution rather than to a pre-specified constant. What does the comparison look like? Well, as you might imagine, Bayesian estimation of a high dimensional parameter is not easy. This becomes very apparent as the dimension grows. In a nutshell, specifying a prior distribution of a vector-valued parameter is a very delicate matter, and a lot can go wrong. So as the dimension increases, the collection of "good priors" becomes a smaller and smaller subset of the class of the available MVN priors. Bottom line: when estimating the mean of a multivariate normal distribution, you would be well advised to use the James-Stein estimator (or one of its cousins) instead of a Bayes estimator unless you have prior information about the unknown parameter vector that you can reasonably consider as highly reliable.

For insights into the estimation with asymmetric loss, the problems of estimating a multivariate normal mean and also a linear function of regression parameters under generalized

Linex loss are examined in Samaniego (2010). There, it is found that Bayes estimators with respect to normal priors tend to be competitive when compared to the maximum likelihood estimator. Examples of findings in this work are: (1) for arbitrary values of the true and operational prior means, the Bayes estimator will outperform the MLE if the operational prior is sufficiently diffuse and (2) if the operational prior is mean-correct and the true prior is degenerate at a point, the Bayes estimator is universally superior to the MLE.

The discussion above supports the premise that neither Bayes nor frequentist estimators will be universally optimal in any nontrivial estimation problem. This leads me to suggest that the real question you should ask is not "Should I be a Bayesian" but rather "When should I be a Bayesian". In every estimation problem, there will be good and bad priors. Deciding which side of the threshold your prior is in is a serious challenge. While it is difficult to proceed without risk, your task when dealing with an estimation problem is to determine whether the prior information you have about model parameters is useful. You should recognize that it needn't be perfect to be useful. But when the prior information is of dubious value (seemingly not much better than taking a wild guess), a frequentist estimator like the MLE is probably your best bet. If I myself were a Bayesian, I wouldn't be writing such things. In my "Bayes book," as I call it, I admit that I am not a Bayesian, but I also admit to being a "Bayesian sympathizer." I find the Bayesian approach attractive, intellectually satisfying, and quite often, very effective. It's an approach and a methodology that I feel should be in every statistician's toolbox.

Exercises 9.4.

1. Suppose that $X_1,\dots,X_n \stackrel{\text{iid}}{\sim} P(\lambda)$, the Poisson distribution with mean λ, and let the prior distribution G be specified as $\lambda \sim \Gamma(\alpha,\beta)$. Show that, under squared error loss, the Bayes estimator of λ with respect to the prior G, may be written as a convex combination of the prior mean $\lambda^* = \alpha\beta$ and $\bar{X}$, the MVUE of λ, that is, show that the Bayes estimator satisfies the convexity condition in (9.42).

2. Suppose that $X_1,\dots,X_n \stackrel{\text{iid}}{\sim} Exp(\lambda)$, the Poisson distribution with mean λ, and let the prior distribution G be specified as $\lambda \sim \Gamma(\alpha,\beta)$. Show that, under squared error loss, the Bayes estimator of λ with respect to the prior G, may be written as a convex combination of the prior mean $\lambda^* = \alpha\beta$ and $\bar{X}$, the MVUE of λ, that is, show that the Bayes estimator satisfies the convexity condition in (9.42).

3. Prove Corollary 9.4.3.

4. Let $I(\theta)$ be the Fisher information in a single observation X from the distribution F_θ. Suppose the hypotheses of Theorem 5.1 hold and that, in addition,

   (i) for some fixed positive number ω and for any fixed n, $\eta = n/(n+\omega)$,

   (ii) the model F_θ satisfies the Cramer-Rao regularity conditions, and

   (iii) the estimator $\hat{\theta}$ is an efficient estimator of θ.

   Show that the Bayes estimator $\hat{\theta}_G$ is superior to the estimator $\hat{\theta}$ as $n \to \infty$ if and only if $V_{G_0}(\theta) + (E_G\theta - E_{G_0}\theta)^2 \le \frac{2}{\omega} E_{G_0} I^{-1}(\theta)$.

9.5 Bayesian Interval Estimation

As you know, it is often of interest to go beyond the process of point estimation by providing a numerical interval that is highly likely, in some sense, to contain the true value of the unknown parameter. The frequentist approach to interval estimation was treated in Chapter 8. As we have seen, the frequentist uses the probability model for the data to make a probability statement about a data-based random interval that will contain the unknown parameter θ with probability $1-\alpha$ for some small positive value α. Realizations of this random interval based on the observed data yield a numerical interval called a $100(1-\alpha)\%$ confidence interval for θ. In many replications of the process that generates such intervals, the relative frequency of the event that the interval obtained will contain the true value of the parameter is expected to be close to $1-\alpha$. A Bayesian statistician would object to this method of interval estimation as incoherent, both because it violates the likelihood principle and because it fails to assign probabilities to uncertain events like $\theta \in (a,b)$ for two given real numbers $a < b$. The method is also occasionally criticized on certain practical grounds. The following example illustrates the difficulties one might encounter in practice.

Example 9.5.1. Let $X_1, X_2, X_3 \stackrel{\text{iid}}{\sim} U(\theta-1, \theta+1)$, and let $X_{(1)} < X_{(2)} < X_{(3)}$ be corresponding order statistics. Since $P(X_{(3)} < \theta) = 1/8$ and $P(X_{(1)} > \theta) = 1/8$, it follows that

$$P(X_{(1)} < \theta < X_{(3)}) = 3/4.$$

Therefore, the interval $(X_{(1)}, X_{(3)})$ is a 75% confidence interval for the parameter θ. Now suppose you observe a sample for which $X_{(1)} = 4.0$ and $X_{(3)} = 5.2$. We then know that the 75% confidence interval $(4.0, 5.2)$ contains the unknown θ with certainty (as happens, of course, whenever $X_{(3)} - X_{(1)} > 1$). On the other hand, if your sample yielded $X_{(1)} = 4.0$ and $X_{(3)} = 4.1$, you would have very little confidence that the interval $(4.0, 4.1)$ contains the unknown θ. It is clear from these considerations that the 75% coverage probability for the random interval $(X_{(1)}, X_{(3)})$ translates, in practice, into actual coverage in 75% of replications of the experiment, but the 75% figure may be meaningless or misleading in any particular replication. ■

The Bayesian approach to interval estimation differs quite significantly from the frequentist approach. The Bayesian approach is simpler, and it manages to avoid potential conflicts with the observed data. It is, however, firmly based on a subjective assessment of the statistician's uncertainty about the unknown parameter θ, and thus does not rely exclusively on the observed data. This renders the approach vulnerable to the usual criticism of the Bayesian approach—the fact that subjective judgments about θ may result in biased or misleading conclusions. The Bayesian camp has a stock response: it may also often lead to more reliable and useful inferences.

The posterior distribution of the parameter θ comprises the basis for all Bayesian inference about θ. The Bayesian counterpart of a confidence interval for θ is called a *credible interval* for θ, and it is obtained from the posterior distribution by selecting an interval whose posterior probability is equal to the desired probability level. For example, any interval (θ_L, θ_U) for which

$$\int_{\theta_L}^{\theta_U} g(\theta|\mathbf{x})d\theta = 1-\alpha \tag{9.57}$$

is a $100(1-\alpha)\%$ credible interval for θ, where $g(\theta|\mathbf{x})$ is the posterior density of θ, given $\mathbf{X}=\mathbf{x}$. The credible interval used most often is the central one in which the limits θ_L and θ_U are chosen to satisfy

$$\int_{-\infty}^{\theta_L} g(\theta|\mathbf{x})d\theta = \alpha/2 = \int_{\theta_U}^{\infty} g(\theta|\mathbf{x})d\theta. \tag{9.58}$$

A credible interval represents a statistician's posterior judgment about intervals that contain θ with a given probability. Since credible intervals are based on the data only through the likelihood function $L(\theta,\mathbf{x})$, they conform with the requirements of the Likelihood Principle. While the integrals in (9.57) and (9.58) may not be amenable to analytical calculations, they can, in most situations of interest, be well approximated using numerical or iterative methods. In the following example, we obtain a 95% credible interval for a positive parameter θ in a problem in which the calculation is tractable.

Example 9.5.2. As in Example 9.5.1, suppose that $X|\theta \sim U[0,\theta]$ and that θ has the gamma prior distribution $G=\Gamma(2,1)$, that is, the prior distribution with density

$$g(\theta) = \theta e^{-\theta} I_{(0,\infty)}(\theta),$$

The joint distribution of X and θ is thus

$$f(x,\theta) = e^{-\theta}, \text{ for } 0 < x < \theta < \infty.$$

We know that the posterior density of θ is

$$f(\theta|x) = e^{-(\theta-x)} I_{(x,\infty)}(\theta),$$

the density of the "translated exponential distribution," or alternatively, the density of $x+Y$, where $Y \sim Exp(1)$. It follows that, given $X=x$, we may obtain a 95% credible interval for θ as the interval $(x+a, x+b)$, where a and b may be identified from the posterior distribution of θ as constants satisfying the equations

$$P(\theta < x+a|X=x) = .025 \text{ and } P(\theta > x+b|X=x) = .025. \tag{9.59}$$

Since $\theta - x|X=x \sim Exp(1)$, the constants a and b in (9.59) must satisfy

$$\int_a^{\infty} e^{-y}dy = .975 \text{ and } \int_b^{\infty} e^{-y}dy = .025. \tag{9.60}$$

From (9.60), we obtain

$$a = -\ln(.975) = .0253 \text{ and } b = -\ln(.025) = 3.6889.$$

We thus have that the interval $(x+.0253, x+3.6889)$ is a 95% credible interval for the parameter θ. ■

You're probably asking the natural question at this point—can anything be said about the comparative performance of Bayesian credible intervals and frequentist confidence intervals? If this question occurred to you, then you really did digest the main point if Section 9.4, and I am (or would be, if I only knew about it) so very pleased! Anyway, whether you're itching to know or not, I'm going to make a few comments about such a comparison. It's actually quite an interesting story, including a surprise or two. I'll only treat the tip of the iceberg, though, in deference to the fact that college students do need their sleep.

In what follows, I will restrict my attention to the problem of estimating a binomial proportion. I will assume that the available data is the binomial variable X, where $X \sim B(n,p)$.

I'm doing this, partly for ease of exposition, but the main reason is that we have the word-length experiment in hand, and taking another look at that from the point of view of interval estimation will be quite helpful. The story begins with the good old standard frequentist large-sample confidence interval for p, the interval with confidence limits

$$\hat{p} \pm \frac{\sqrt{\hat{p}(1-\hat{p})}}{\sqrt{n}} Z_{\alpha/2}, \tag{9.61}$$

where $\hat{p} = X/n$. We know that, if the binomial model is correct, the random interval from which the interval in (9.61) is calculated will contain p with approximate probability $1-\alpha$ when n is sufficiently large. Interestingly, a growing number of statisticians recommend *never* using this interval! Instead, they recommend using the "plus-four" confidence interval for p. The plus-four confidence interval is a hybrid interval estimator that combines some Bayesian thinking with some frequentist thinking. To compute the plus-four interval for p, one simply replaces $\hat{p}$ in (9.61) by the alternative estimator $\tilde{p}$ of p given by

$$\tilde{p} = \frac{X+2}{n+4}, \tag{9.62}$$

yielding the "preferred" interval with confidence limits

$$\tilde{p} \pm \frac{\sqrt{\tilde{p}(1-\tilde{p})}}{\sqrt{n}} Z_{\alpha/2}. \tag{9.63}$$

Let's first explain why the word "hybrid" aptly describes this interval. First, notice that $\tilde{p}$ in (9.62) is actually the Bayes estimator of p, under squared error loss, with respect to the $Be(2, 2)$ prior. However, the interval corresponding to (9.63) is by no means a credible interval for p. Indeed, it really has nothing to do with probabilities calculated from the $Be(x+2, n-x+2)$ distribution, the posterior distribution of p given the $Be(2,2)$ prior. Instead of referring to this posterior distribution, the interval in (9.63) is constructed by mimicking what the frequentist does, using the same form as the frequentist confidence interval, but with a substitute for $\hat{p}$. So the procedure might be thought of as being half-Bayesian and half-frequentist. The reasons that the hybrid confidence works better than the standard confidence interval are not very well understood, but there's a ton of documentation that it does. The usual reasons given for the problems experienced with the standard interval have to do with the discreteness of X. It turns out that the discreteness of X results in a rather unsteady probability of coverage (usually below the advertized coverage probability $1-\alpha$) as the sample size varies. Why the hybrid estimator fixes this problem is not entirely clear, though it is thought that it has to do with the fact that $\tilde{p}$ "shrinks" $\hat{p}$ toward 1/2. Shrinking an estimator toward a fixed constant was first shown to be an effective estimation strategy by Charles Stein, the statistician who proved, in the 1950s, that the mean of a sample from a k-dimensional normal distribution, with $k \geq 3$, was inadmissible, and that an estimator that shrinks the mean toward the origin $\mathbf{0}$ was actually better. The estimator $\tilde{p}$ has an additional advantage. Because $\tilde{p}$ is closer to 1/2 than $\hat{p}$, the confidence interval based on $\tilde{p}$ is actually slightly wider than the interval based on $\hat{p}$, so that is casts a wider net in trying to trap the unknown parameter p. Theoretical musings aside, the empirical evidence in support of $\tilde{p}$ is really quite overwhelming. One quick indication: researchers, using simulated experiments, have investigated the question: How large does n have to be to guarantee that the actual probability that a 95% confidence interval actually covers the true value of the parameter p is at least .94 for all samples of size n or larger? Suppose p happens to be .1. For that value of p, the required n for this guaranteed performance by the standard confidence limits in (9.61) is $n = 646$. Remarkably, the n required for this guaranteed performance by the plus-four confidence limits in (9.63) is $n = 11$. This is just the tip of the iceberg. The evidence in favor of the plus-four interval is even more impressive than that. It has been shown, again via simulation, that the plus-four interval even tends to give better results when compared to exact intervals based on a binomial computation rather than a normal approximation. Case closed!

Or is it? Hey, there's another player here!! Let's call the interval estimates in (9.61) and (9.63) F and H, respectively, for "frequentist" and "hybrid." Where do Bayesian credible intervals fit in? Well, let's now consider B, the Bayesian credible interval. The comparisons between F and H mentioned above are impressive, but they are not airtight. One can do lots of simulations and get a pretty good idea of what's going on, but together, they will still fall short of what we would like to have—definitive instructions about what to do. Well, the comparisons between B and F and B and H are even more challenging. It goes without saying that what I will tell you here isn't going to be definitive either. But I do think you'll find it interesting. (I'd say "fascinating," but I don't expect you to be at that stage of self-actualization yet; maybe next year!)

I'm going to share with you how F, H, and B did in the word-length experiment. You'll recall that I elicited the prior distributions of 99 rag-tag Bayesians trying to estimate the proportion p of long words in a Somerset Maugham novel. Each of these Bayesians would be able to put forward a central credible interval for p. In order to ensure that the interval estimators F and H were not disadvantaged (by an inappropriate normal approximation) in a comparison with a given B, I (and my colleague D. Bhattacharya) increased the sample size n at which the comparison would be made to $n = 50$.

In 1000 simulations in which the known $p = .3008$ was used to generate $X \sim B(50, p)$, we recorded the coverage probability P for each of the 99 intervals of type F, H, and B and also recorded the width W of the resulting interval. Now, neither of these two measures is appropriate, by itself, to serve as a criterion for comparing interval estimates. For example, we prefer a higher to a lower value of P—a higher probability of coverage is better—but if this was the only basis for comparison, why not just use the interval $[0, 1]$, which covers the true p with certainty. We can't do better than 100% coverage! On the other hand, a tighter interval is preferred over a broader one, but if that's all we cared about, the interval $(\hat{p} - .001, \hat{p} + .001)$ would look pretty good. A criterion that makes more sense than either of these is the ratio W/P. We should be happy with intervals that have a small value of that ratio. If one procedure has a larger coverage probability P than another but also has a larger W, which should be judged better? If W/P is smaller for the first procedure, then we may conclude that it is better than the other because, proportionately, the increased width of the first was smaller than that of the second (that is, $W_1 < (P_1/P_2)W_2$.) Given this as a preamble, here's how the comparisons came out. (Drum roll, please!)

How did the hybrid interval do against the frequentist interval? On the basis of the criterion W/P, we found that H beat F 93 out of 99 times. Interestingly, H beat F in 100% of the cases when judged in terms of coverage probability alone. These results are perhaps expected, given the discussion above. How did the Bayes interval do against the frequentist interval? On the basis of the criterion W/P, we found that B beat F 66 out of 99 times. This domination was not quite as easy to predict as the first. While the Bayesian point estimators tend to outperform their frequentist counterparts quite dramatically in the word-length experiment, the success of credible intervals, which draw much more heavily on the posterior distribution, is not a slam dunk. The fact that $2/3$ of the Bayesian intervals outperform the standard frequentist interval is a new and (for the Bayesian) quite encouraging finding. So the contest now reduces to the world series of interval estimation, the contest between H and B. If you were inclined to make a wager at this point, which way do you think that comparison would come out? Although B beat F, H beat F quite a bit more convincingly. This suggests that the smart money would be backing H in this duel. Well, surprise of surprises, on the basis of the criterion W/P, we found that B beat H 65 out of 99 times. While this result might not be classified as a monumental triumph for the Bayesian, it is nonetheless a victory one cannot ignore. After all, H is roundly accepted these days as the king of the hill! My own interpretation of these findings is that F is highly suspect, that H is clearly good, and that B is also good and ought to be considered more seriously, as it obviously has some promise. Bayesian credible intervals have been used rather sparingly in practice. Perhaps the study I've reported on here will convince a few more people that, on the basis of the performance-related criterion of the sort we have discussed, Bayesian interval estimates seem to merit some consideration.

Exercises 9.5.

1. Let $X_1, X_2 \stackrel{\text{iid}}{\sim} U(\theta, \theta+2)$. Argue that $P(X_{(1)} < \theta + 1 < X_{(2)}) = 1/2$. Conclude that the interval $(X_{(1)} - 1, X_{(2)} - 1)$ is a 50% confidence interval for θ. Discuss the interpretation of this confidence interval in the case in which the observed value of $X_{(2)} - X_{(1)}$ exceeds 1.
2. Suppose $X|p \sim B(1, p)$ and $p \sim Be(5, 1)$. Construct a central 95% credible interval for the parameter p.
3. Let $X_1, \dots, X_5|\mu \stackrel{\text{iid}}{\sim} N(\mu, 1)$ and let the prior distribution of μ be $N(0, 1)$. Suppose the observed sample mean is $\bar{x} = 1.2$. Construct a 95% credible interval for μ. Compare this interval to a 95% confidence interval for μ.

9.6 Chapter Problems

1. It is well documented that lie detector tests have proven useful is the context of criminal investigations. Take it as known that when a criminal suspect gives an answer to a relevant question, he/she answers truthfully (T) only 50% of the time. Suppose, further, that a lie detector test will classify a truthful answer as a lie (L) 5% of the time, and will classify an untruthful answer (T^c) as a lie 90% of the time. On a given question that the lie detector test classifies as a lie, what's the probability that the suspect is telling the truth?
2. Ultrasound tests after the first trimester of pregnancy are often used to predict the gender of a fetus. The tests are known to be perfectly accurate in identifying a female fetus as a girl, but they have only a 75% accuracy rate in predicting a male fetus as a boy. Given that 48% of all births are girls, what is the probability that a child who was predicted to be a girl will in fact be a boy?
3. (The concavity of minimum Bayes risks.) Let G_1 and G_2 be two proper prior distributions on the parameter space Θ. Show that for any number $\alpha \in (0, 1)$,
$$\mathbf{r}^*(\alpha G_1 + (1-\alpha)G_2) \geq \alpha \mathbf{r}^*(G_1) + (1-\alpha)\mathbf{r}^*(G_2), \tag{9.64}$$
where $\mathbf{r}^*(G) = \inf_d \mathbf{r}(G, d)$, that is, $\mathbf{r}*(G)$ is the Bayes risk of the Bayes rule d^* with respect to the prior G.

4. Let $X|p \sim B(n,p)$, and let the prior distribution of p be the $Be(\alpha,\beta)$ model. As shown in Section 9.2, the Bayes estimator of p relative to squared error loss is

$$\hat{p}_{\alpha,\beta} = \frac{X+\alpha}{n+\alpha+\beta}. \tag{9.65}$$

   (a) Show that the risk function of the estimator in (9.57) is a quadratic function of p.

   (b) Identify the unique values α_0 and β_0 for which $R(p,\hat{p}_{\alpha,\beta})$ is a constant independent of p. (Note that these values will depend on the sample size n.)

   (c) Is the Bayes rule with respect to the $Be(\alpha_0,\beta_0)$ prior a minimax estimator of p? Defend your answer.

5. Prove the following claim: If an estimator $\hat{\theta}$ of a parameter θ is an equalizer rule and if $\hat{\theta}$ is admissible, then $\hat{\theta}$ is a minimax estimator of θ.

6. Suppose that the Bayes estimator $\hat{\theta}$ of a parameter θ with respect to the proper prior G and loss function L is unique, that is, $\hat{\theta}$ uniquely minimizes the Bayes risk r with respect to G. Show that the estimator $\hat{\theta}$ is admissible.

7. Suppose that $X_1,X_2,\ldots,X_n|\theta \sim^{iid} U[0,\theta]$, the uniform distribution on $(0,\theta)$, and let $\theta \sim P(\alpha,\theta_0)$, the one-parameter Pareto distribution with density

$$g(\theta) = \alpha\theta^{\alpha-1}/\theta_0^{\alpha} \text{ for } \theta > \theta_0,$$

   where θ_0 is known.

   (a) Verify that the posterior density of θ, given $x_1,x_2,\ldots,x_n$, is the Pareto distribution $P(\alpha+n,\theta_0^*)$, where $\theta_0^* = \max(\theta_0,x_1,\ldots,x_n)$.

   (b) Find the Bayes estimator of θ relative to squared error loss.

8. Let $X_1,X_2,\ldots,X_n|p \overset{\text{iid}}{\sim} NB(r,p)$, the negative binomial distribution on the integers $x = r,r+1,\ldots,$ where r is a (known) positive integer and $p \in (0,1)$. Suppose that $p \sim Be(\alpha,\beta)$, the standard conjugate prior for the negative binomial model.

   (a) Derive the posterior distribution of $p|\mathbf{X}$.

   (b) Identify the Bayes estimator of p relative to squared error loss. (You may simplify your derivations by restricting attention to $T = \sum_{i=1}^{n} X_i$, the sufficient statistic for λ in this problem.)

9. Let $X_1,X_2,\ldots,X_n|\lambda \overset{\text{iid}}{\sim} Exp(\lambda)$, the exponential distribution with failure rate λ. Suppose that $\lambda \sim \Gamma(\alpha,\beta)$, the conjugate prior for the exponential model. (a) Derive the posterior distribution of $\lambda|\mathbf{X}$. (b) Identify the Bayes estimator of λ relative to squared error loss.

10. Let $X_1,X_2,\ldots,X_n|\theta \overset{\text{iid}}{\sim} N(\mu_0,1/\theta)$, the normal distribution with known mean μ_0 and precision parameter $\theta > 0$, where θ is the reciprocal of the variance σ^2. Suppose that $\theta \sim \Gamma(\alpha,\beta)$, the standard conjugate prior for the normal when parameterized in this way. (a) Derive the posterior distribution of $\theta|\mathbf{X}$. (b) Identify the Bayes estimator of θ relative to squared error loss.

11. Suppose that $X_1,X_2,\ldots,X_n|\theta_1,\theta_2 \overset{\text{iid}}{\sim} U(\theta_1,\theta_2)$, the uniform distribution on the interval (θ_1,θ_2). Assume that the prior distribution on (θ_1,θ_2) is the bivariate, bilateral Pareto with parameters $r_1 < r_2$ and $\alpha > 0$ and density

$$g(\theta_1,\theta_2|r_1,r_2,\alpha) = \frac{\alpha(\alpha+1)(r_2-r_1)^{\alpha}}{(\theta_2-\theta_1)^{\alpha+2}} \text{ for } \theta_1 < r_1 \text{ and } \theta_2 > r_2. \tag{9.66}$$

   Show that the posterior density of (θ_1,θ_2), given $X_1,X_2,\ldots,X_n$, is a bivariate, bilateral Pareto density, and identify its parameters.

12. The *Dirichlet distribution* is a multivariate generalization of the beta distribution whose marginal distributions are in fact beta distributions. The probability vector $\mathbf{p} = (p_1, \ldots, p_k)$ has the Dirichlet distribution with parameter $\alpha = (\alpha_1, \ldots, \alpha_k)$, where $\alpha_i > 0$ for $i = 1, \ldots, k$, if the probability density function of $\mathbf{p}|\alpha$ is given by

$$f(\mathbf{p}|\alpha) = \frac{\Gamma(\alpha_1 + \cdots + \alpha_k)}{\Gamma(\alpha_1)\cdots\Gamma(\alpha_k)} p_1^{\alpha_1 - 1} \cdots p_k^{\alpha_k - 1} \text{ for } \mathbf{p} \in [0,1]^k \text{ for which } \sum_{i=1}^{k} p_i = 1. \quad (9.67)$$

We'll use $\mathbf{p} \sim D(\alpha)$ as shorthand notation for this. Since $P(\sum_{i=1}^{k} p_i = 1) = 1$, the function $f(\bullet|\alpha)$ is positive only on the $(k-1)$ dimensional simplex $\mathbf{P}$ given by

$$\mathbf{P} = \{\mathbf{p}|p_i \geq 0 \quad \forall i \text{ and } \sum_{i=1}^{k} p_i = 1\}.$$

For example, for $k = 2$, $p_1 \sim Be(\alpha, \alpha_2)$ and $p_2 \equiv 1 - p_1$. Show that the Dirichlet distribution is a conjugate prior for the multinomial distribution, that is, show that if $\mathbf{X} \sim M_k(n, \mathbf{p})$ and $\mathbf{p} \sim D(\alpha)$, then $\mathbf{p}|\mathbf{X} = \mathbf{x}$ has a Dirichlet distribution with parameter $(\alpha_1 + x_1, \ldots, \alpha_k + x_k)$.

13. Shirt Off My Back, Inc. (popularly known as "Zombie") Inc. produces N polo shirts (a known number) on any given weekday. Historical records show that, on average, a known proportion p of these shirts are defective. Let X be the number of defective shirts in last Friday's batch. Assume that $X|N, p \sim B(N, p)$. Assume that the value of X is unknown. Suppose that the Zombie store in Key West, Florida, received a known number n of polo shirts from last Friday's batch. Let Y be the number of defective shirts received by the store. Assuming a random distribution scheme, Y has the hypergeometric distribution, that is, $Y|N,x,n \sim HG(N,X,n)$ with probability mass function (pmf) given by

$$P(Y = y|N,x,n) = \frac{\binom{x}{y}\binom{N-x}{n-y}}{\binom{N}{n}} \text{ for } 0 \leq y \leq x.$$

Your aim is to derive the Bayes estimator of the unknown (parameter) x with respect to squared error loss based on the observed value $Y = y$. (a) Show that $Y|n, p \sim B(n, p)$. (b) Identify $p(x|y)$, the conditional pmf of X given $Y = y$. (Hint: Note that if $Y = y$, the range of X is $y \leq x \leq N - n + y$. The probability mass function of $X - y|Y = y$ is more easily recognized than that of $X|Y = y$.) (c) Find $E(X|Y = y)$, the Bayes estimator of X based on the observation $Y = y$.

14. In general, Bayes estimators do not enjoy an "invariance" property that would guarantee that $h(\hat{\theta})$ is a Bayes estimator of $h(\theta)$ when $\hat{\theta}$ is the Bayes estimator of θ. Let $X|p \sim B(n, p)$. Demonstrate this lack of invariance by showing that $\hat{p}_B^2$ is not the Bayes estimator of p^2 with respect to the uniform prior $U[0,1]$ on p and squared error loss, where $\hat{p}_B = (X + 1)/(n + 2)$ is the Bayes estimator of p with respect to the uniform prior and squared error loss.

15. Let $X|\theta$ have density $f(x|\theta)$ and let θ have the prior distribution G with density $g(\theta)$. Consider the estimation of θ relative to the Linex loss function given by

$$L(\theta, a) = \exp\{c(\theta - a)\} - c(\theta - a) - 1, \quad (9.68)$$

where c is chosen to be negative when overestimating θ is considered an especially serious error and c is chosen to be positive when underestimating θ is seen as especially serious. Show that the Bayes estimator of θ with respect to G and the Linex loss function is given by

$$\hat{\theta}(x) = -\frac{1}{c} \ln\{E_{\theta|X=x} e^{-c\theta}\}.$$

(You may assume that the mgf of $\theta|X=x$ exists and is finite for t in an open interval containing 0.

16. Consider the models in Example 9.2.2, that is, suppose that $X|\theta \sim U[0,\theta]$ and that θ has the gamma prior distribution G with density
$$g(\theta) = \theta e^{-\theta} I_{[0,\infty]}(\theta).$$
Find the Bayes estimator of θ relative to the Linex loss function given by
$$L(\theta,a) = exp\{c(a-\theta) - c(a-\theta) - 1\}.$$

17. Let $X|\theta \sim F_\theta$ with density or probability mass function $f_\theta(x)$, and suppose that θ has the prior distribution G such that $G'(\theta) = g(\theta)$. Show that if G is an improper prior, then the marginal distribution of X is also improper. Verify the general result by examining the special case where $X \sim B(n,\theta)$ and $g(\theta) = [1/\theta(1-\theta)]I_{(0,1)}(\theta)$.

18. There have been a number of approaches proposed for modeling the state of "prior ignorance." Intuitively, one might think that a uniform distribution on a bounded parameter might serve as a "non-informative prior." Consider such an approach in the problem of estimating a binomial parameter p. Suppose our prior model for p assumes that $p \sim U(0,1)$. Derive the distribution of p^2. Note that if one is ignorant about p in the sense above, then one is not ignorant about p^2. As these facts are not logically compatible, this approach to modeling prior ignorance must be judged to be inadequate.

19. Jeffreys (1961) proposed an alternative formulation of prior ignorance which has two attractive features: it is relatively easy to compute and it is invariant under smooth transformations. Let $X_1, X_2, \ldots, X_n|\theta \overset{\text{iid}}{\sim} F_\theta$. The Jeffreys prior on a parameter θ is given by
$$g(\theta) \propto [I_X(\theta)]^{1/2}, \tag{9.69}$$
where $I(\theta)$ is the Fisher information of a single observation from F_θ. Suppose that $\lambda = h(\theta)$, where h is a one-to-one differentiable transformation. Show that
$$[I_X(\lambda)]^{1/2} = [I_X(h^{-1}(\lambda))]^{1/2} \cdot \left|\frac{d\theta}{d\lambda}\right|. \tag{9.70}$$

20. See Problem 9.6.19. Show that the Jeffreys prior of a location parameter is constant.

21. See Problem 9.6.19. Let $X_1, \ldots, X_n \overset{\text{iid}}{\sim} N(\mu,\sigma^2)$. Assuming that σ^2 is known, show that the Jeffreys prior on μ is $g_1(\mu) \propto 1$. Assuming μ is known, show that the Jeffreys prior on σ is $g_2(\sigma) \propto 1/\sigma$.

22. See Problem 9.6.19. Let $X \sim B(n,p)$. Show that the Jeffreys prior on p is the $Be(1/2,1/2)$ distribution.

23. Suppose that $L(\theta,a)$ is an arbitrary non-negative function which takes the value 0 when $\theta = a$. An estimator $\hat{\theta}$ is called an *extended Bayes rule* if for any $\varepsilon > 0$, there is a prior probability distribution G_ε on θ such that
$$r(G_\varepsilon,\hat{\theta}) \leq \min_{\hat{\theta}} r(G_\varepsilon,\hat{\theta}) + \varepsilon.$$
(This means that for arbitrary ε, no matter how small, one can find a prior distribution G_ε such that $\hat{\theta}$ has Bayes risk that is within ε of the smallest Bayes risk possible with respect to G_ε.) Prove "by contradiction": If the risk function R of an estimator $\hat{\theta}$ of θ is a constant independent of θ (that is, if $\hat{\theta}$ is an equalizer rule), and $\hat{\theta}$ is an extended Bayes rule, then $\hat{\theta}$ is a minimax estimator of θ. (Remark: In the setting of Exercise 9.1.2, it can be shown that the estimator $\hat{\mu} = \bar{X}$ is an extended Bayes estimator of μ and is thus a minimax estimator of μ.)

24. Suppose that $X_1, \dots, X_n \stackrel{\text{iid}}{\sim} F_\theta$, where F_θ has density or pmf f_θ, and suppose that θ had the prior distribution G with density g. The *predictive density* of a future observation X_{n+1} is given by

$$f(x_{n+1}|x_1, \dots, x_n) = \int f(x_{n+1}|\theta, x_1, \dots, x_n) g(\theta) d\theta. \tag{9.71}$$

Show that the predictive density may be evaluated via the posterior density $g(\theta|x_1, \dots, x_n)$, that is, show that

$$f(x_{n+1}|x_1, \dots, x_n) = \int f(x_{n+1}|\theta) g(\theta|x_1, \dots, x_n) d\theta. \tag{9.72}$$

25. See Problem 9.5.23. Let $X_1, \dots, X_n \stackrel{\text{iid}}{\sim} B(1, p)$ and suppose that $p \sim Be(\alpha, \beta)$. Derive the predictive distribution of the future observation X_{n+1} given $X_1, \dots, X_n$. If $\alpha = \beta = 1$ and the observed value of $X = \sum_{i=1}^n X_i$ is x, what is the predictive distribution of X_{n+1}?

26. Suppose that a medical school research team has obtained a grant to fund a clinical trial to compare a new treatment for esophageal cancer with the currently favored treatment. For a patient who receives the new treatment, let p be the probability of "success" (that is, remission of the cancer for a specified period of time). Suppose the first such patient has a successful outcome. If the prior distribution of p is taken to be the $U(0, 1)$ distribution, obtain the Bayes estimate of p under squared error loss and under absolute error loss.

27. Let $X|p \sim B(n, p)$, and let the prior distribution of p be the $Be(\alpha, \beta)$ model. Treating α and β as known constants, show that the Bayes estimator of p relative to squared error loss is (a) a consistent estimator of p and (b) asymptotically normal with the same asymptotic distribution as the sample proportion $\hat{p}$.

28. Let p be the proportion of defective items in a lot of widgets. Suppose that 25 widgets are sampled at random from this lot and exactly one defective widget is found. If the prior distribution of p is taken to be $Be(10, 90)$, find the Bayes estimate of p under squared error loss.

29. Suppose that $X_1, X_2, \dots, X_n \stackrel{\text{iid}}{\sim} N(\mu, \sigma_0^2)$, the normal distribution with known variance σ_0^2. Let G be an improper prior distribution which assigns equal mass to all intervals of the real line having the same length (or, equivalently, let $g(\mu) = 1 for -\infty < \mu < \infty$. Show that $\hat{\mu} = \bar{X}$ is the "generalized Bayes" estimator of μ relative to squared error loss, that is, show that $\hat{\mu}$ minimizes the generalized posterior expected loss

$$E(L(\mu, a_{\bar{x}})|\bar{X} = \bar{x}) = \int_{-\infty}^{\infty} (\mu - a_{\bar{x}})^2 f(\bar{x}|\mu) g(\mu) d\mu$$

with respect to $a_{\bar{x}}$ at each fixed value of the observed sample mean $\bar{x}$.

30. Consider estimating the parameter θ under the weighted squared error loss function $L(\theta, a) = w(\theta)(\theta - a)^2$. Show that if the Bayes estimator $\hat{\theta}_B$ of θ with respect to the prior G is unbiased and if $Ew(\theta) < \infty$, then the Bayes risk $r(G, \hat{\theta}_B)$ of $\hat{\theta}_B$ must be equal to 0.

31. Suppose that $X|\mu \sim N(\mu, 1)$, and that the prior distribution of μ is taken to be $N(0, \sigma^2)$. (a) Show, by integrating the joint density $f(x, \mu)$, that the marginal distribution of X is $N(0, 1 + \sigma^2)$. (b) Show that the posterior distribution of μ, given X, is $N(x\sigma^2/(1 + \sigma^2), \sigma^2/(1 + \sigma^2))$ and conclude that the Bayes estimator of μ relative to squared error loss is given by $\hat{\mu} = x\sigma^2/(1 + \sigma^2)$.

32. Let $X \sim B(n, p)$. Suppose two statisticians provide Bayes estimators of p, the first using the prior $Be(a, b)$ and the second using the prior $Be(c, d)$. If the resulting Bayes estimators are denoted by $\hat{p}_1$ and $\hat{p}_2$, respectively, show that $\hat{p}_1 - \hat{p}_2 \xrightarrow{p} 0$ as $n \to \infty$.

33. Recall that maximum likelihood estimators have the very nice invariance property if $\hat{\theta}$ is the MLE of θ and g is a one-to-one function, then $g(\hat{\theta})$ is the MLE of $g(\theta)$. Bayes estimators do not enjoy this property. To illustrate this, let $X \sim B(1,p)$ and suppose that $p \sim U(0,1)$. Let $\hat{p}$ be the Bayes estimator of p. Show that $\hat{p}^2$ is not the Bayes estimator of p^2. (Incidentally, the function $g(p) = p^2$ is a one-to-one function when p is restricted to the interval $[0,1]$.)

34. Suppose that $X \sim F_\theta$ and that $\theta \sim G$. For $c > 0$, define the loss function $L(\theta,a)$ as

$$L(\theta,a) = \begin{cases} c|\theta - a|, & \text{if } \theta < a \\ |\theta - a|, & \text{if } \theta \geq a. \end{cases} \tag{9.73}$$

Assume that θ is modeled as a continuous variable. Show that any $(1/(1+c))$ quantile of the posterior distribution of θ is a Bayes estimator of θ.

35. Prove or disprove: If $\hat{\theta}$ is the Bayes estimator of θ with respect to the proper prior G and squared error loss $(\theta - a)^2$, and c and d are fixed constants, then $\hat{\eta} = c\hat{\theta} + d$ is the Bayes estimator of $\eta = c\theta + d$ with respect to the prior G on θ and squared error loss $(\eta - a)^2$.

36. (Refer to Problem 9.6.12) Let $X \sim B(n, p_1 + p_2)$, where $p_1 > 0$, $p_2 > 0$ and $p_1 + p_2 < 1$. The parameter pair (p_1, p_2) is an example of a *nonidentifiable* parameter, as there are multiple values of the pair (in this present case, infinitely many values) that have the same sum and thus give rise to the same distribution for X. There are no classical statistical methods for estimating nonidentifiable parameters. However, the Bayesian paradigm provides a logical mechanism for the estimation of such parameters. For estimating the pair (p_1, p_2), suppose that $(p_1, p_2) \sim D(\alpha_1, \alpha_2, \alpha_3)$, the Dirichlet distribution with parameter α. (a) Show that the posterior distribution of (p_1, p_2), given $X = x$, is a mixture of Dirichlet distributions of the form

$$f(p_1, p_2 | X = x) = \sum_{i=0}^{x} c_i D(\alpha_1 + i, \alpha_2 + x - i, \alpha_3 + n - x),$$

and identify the constants $\{c_i\}$. (b) Show that the Bayes estimator relative to generalized squared error loss

$$L(\mathbf{p}, \mathbf{a}) = (p_1 - a_1)^2 + (p_2 - a_2)^2$$

has the form

$$(\hat{p}_1, \hat{p}_2) = \left(\sum_{i=0}^{x} c_i \frac{\alpha_1 + i}{\alpha_1 + \alpha_2 + \alpha_3 + n}, \sum_{i=0}^{x} c_i \frac{\alpha_2 + x - i}{\alpha_1 + \alpha_2 + \alpha_3 + n} \right).$$

37. This exercise serves as an introduction to the *empirical Bayes* approach to statistics. The EB framework assumes that there is a sequence of independent but similar experiments from which data is available, and it seeks to "borrow strength" from past experiments to develop an effective estimator of the parameter value in the current experiment. Suppose that $(X_1, \lambda_1), \ldots, (X_k, \lambda_k), (X_{k+1}, \lambda_{k+1})$ are a collection of independent random pairs, where $\lambda_1, \ldots, \lambda_k, \lambda_{k+1} \overset{\text{iid}}{\sim} G$, where G is taken as the true but unknown (prior) distribution of the λs. For $i = 1, \ldots, k+1$, assume that $X_i | \lambda_i \sim P(\lambda_i)$, the Poisson distribution with mean λ_i. (a) Show that the Bayes estimator of λ with respect to G and squared error loss and based on the observation $X \sim P(\lambda)$ may be written as

$$\hat{\lambda}_G(x) = \frac{(x+1) p_G(x+1)}{p_G(x)},$$

where $p_G(x) = \int p(x|\lambda) dG(\lambda)$ is the marginal pmf of X. (b) Now, let $p_k(x)$ be the empirical pmf based on the first k Xs. Suppose that $X_{k+1} = x_{k+1}$. Then the estimator $\hat{\lambda}_{EB}$ of λ is given by

$$\hat{\lambda}_{EB}(x_{k+1}) = \frac{(x_{k+1}+1)p_k(x_{k+1}+1)}{p_k(x_{k+1})},$$

is an "empirical Bayes" estimator of λ_{k+1}. Show that, for any integer $x \geq 0$, $\hat{\lambda}_{EB}(x) \to \hat{\lambda}_G(x)$ as $k \to \infty$. (Note: The really interesting thing about this result is that, for large k, you can approximate the Bayes estimator $\hat{\lambda}_G(x)$ even though G is unknown. On a philosophical level, you should note that the EB approach is not truly Bayesian, as the method neither requires nor uses any subjective input about the unknown parameters or about the unknown G.)

38. Repeat Problem 9.5.37 with the Poisson distribution $P(\lambda)$, for $\lambda > 0$, replaced by the geometric distribution $G^*(\lambda)$, with pmf $p(x) = (1-\lambda)\lambda^x$, $x = 0,1,2,\ldots$ for $0 < \lambda < 1$. In this case, note that the Bayes estimator of λ with respect to G and squared error loss and based on the observation $X \sim G(\lambda)$ may be written as

$$\hat{\lambda}_G(x) = \frac{p_G(x+1)}{p_G(x)},$$

where $p_G(x) = \int p(x|\lambda)dG(\lambda)$ is the marginal pmf of X. Find an empirical Bayes estimator of λ_{k+1} which uses past data (from experiments $1,\ldots,k$) to estimate $p_G(x)$.

39. Let θ be a real-valued parameter and let $X \sim F_\theta$. Suppose the loss function L is given by

$$L(\theta,a) = \begin{cases} 0 & \text{if } |\theta - a| < c \\ 1 & \text{if } |\theta - a| \geq c. \end{cases} \tag{9.74}$$

Show that the Bayes estimator of θ with respect to the prior G is equal to the midpoint of the modal interval of length $2c$, that is, the interval of length $2c$ with the highest probability, under the posterior distribution of θ, given $X = x$.

40. Suppose the posterior distribution of a parameter θ, given the available sample, is $\Gamma(2,1)$. Obtain a 95% central credible interval for θ, that is, an interval (L,U) for which the posterior probability below L and the posterior probability above U are both equal to .025.

41. Let $X|\theta \sim U(0,\theta)$ and let $\theta \sim P(\alpha,\theta_0)$, the one-parameter Pareto distribution with density $g(\theta) = \alpha\theta^{\alpha-1}/\theta_0^\alpha$ for $\theta > \theta_0$, where θ_0 is known. Identify the posterior distribution of θ, given $X = x$, and derive the central 95% credible interval for θ.

42. Let $X_1, X_2, \ldots, X_n | \theta \overset{\text{iid}}{\sim} F_\theta$, and let $\hat{\theta}$ be a sufficient and unbiased estimator of θ. Suppose that G^* is the prior distribution of θ and let $\theta^* = E_G\theta$. Then the Bayes estimator $\hat{\theta}_G$ wrt G under squared error loss is said to be *self-consistent* if

$$E(\theta|\hat{\theta} = \theta^*) = \theta^*.$$

Show that a Bayes estimator that satisfies condition (9.42) of Theorem 9.4.1 is self-consistent.

43. (See Problem 9.5.42). Under squared error loss, the standard Bayes estimates of a binomial proportion p or of a normal mean μ are self-consistent. This exercise shows that the concept extends beyond exponential families and their standard conjugate priors. Let $X \sim U(0,2\theta)$ and let $\theta \sim \Gamma(2,1)$. Show that the Bayes estimator of θ under squared error loss is self-consistent.

10

Hypothesis Testing

10.1 Basic Principles

In ordinary language, a hypothesis is basically just a guess or a hunch about an event or process. Statements like: "It will rain tomorrow" or "At least 10 percent of the automobiles driven on U.S roads today are hybrid (gas/electric) vehicles" would qualify as hypotheses we might entertain at a given moment in time. In statistical settings, we adopt a narrower view of the term "hypothesis," a view that relates directly to the stochastic model that might be used to describe a given random experiment. If we have assumed a particular parametric model F_θ, say, for a random experiment, then the types of hypotheses that might well arise typically have to do with the possible values of θ. Given a hypothesis of this sort, for example the hypothesis that θ is no larger than 5, and given a random sample drawn from F_θ, we could test this hypothesis on the basis of the observed data and reach some (perhaps tentative) conclusion about whether or not the hypothesis is true. In this chapter, we will explore the statistical process of testing hypotheses, identifying the basic characteristics of such problems, and examining the principles which one uses in constructing "good" tests and in determining when one way of testing a particular hypothesis is superior to some alternative approach.

In problems in which a particular parametric model has been adopted as a general description of an experiment of interest, we will use, for the sake of clarity, the following working definitions:

Definition 10.1.1. A *hypothesis* is a statement about a population parameter. A hypothesis is said to be *simple* if it specifies a fixed numerical value of the parameter. Hypotheses which specify that the parameter belongs to some collection of values is said to be *composite*. A *hypothesis test* is a procedure for choosing, on the basis of observed experimental data, which of two competing hypotheses is to be considered "correct." We will use the terms *accept* a hypothesis or *reject* a hypothesis in reference to the decision we make regarding a particular hypothesis as an appropriate description of the conclusion reached about the experiment we are concerned with.

When two hypotheses are being compared, it is common to refer to one as H_0 and the other as H_1. I know, this choice seems like the obvious one. By the way, the first of these hypotheses is pronounced "H naught" (sounds like "H knot"), "naught" being a word the British occasionally use in place of "zero." But besides the fact that these labels are the most obvious ones, there's another little known (historical) reason for their use. The foundations of the theory of hypothesis testing were laid out in England in the 1920s and 1930s. The two main architects of the theory in use today were Jerzy Neyman and Egon Pearson, two statisticians who worked together in England for about a decade. As the story goes, when they started working together on hypothesis testing problems, Neyman said to Pearson "Let's call this hypothesis H." Pearson simply said "Not!" and the name stuck. In applications, H_0 is often referred to as the "null hypothesis."

The thought process of statistical hypothesis testing is most easily explained through a detailed example.

Example 10.1.1. Let us suppose that an automobile manufacturer claims that the latest model of a particular car gets 40 miles per gallon (mpg) of gasoline when driven under normal highway conditions. The consumer advocacy group CAG, which publishes its experimental findings in a glossy magazine every other month, decides to test this particular claim. Suppose that CAG purchases 9 new cars of the model of interest and has each of them driven for 1000 miles under "normal highway conditions." For simplicity, we will assume that the miles per gallon for these 9 cars are normally distributed; indeed, if X_i is the mpg of the ith car, let's assume that

$$X_1, \dots, X_9 \overset{\text{iid}}{\sim} N(\mu, 9). \tag{10.1}$$

The assumption that the variance of the distribution in (10.1) is known to be 9 is made for simplicity and for convenience. (For example, we will be working with the sufficient statistic $\bar{X}$ which, under the assumption in (10.1), has the $N(\mu, 1)$ distribution.) We will soon leave simplifying assumptions such as this behind. The manufacturer's claim will play the role of the hypothesis of primary interest, that is, we will set $H_0 : \mu = 40$. Since claims about the performance of automobiles are often overstated, the hypothesis $H_1 : \mu < 40$ would be the alternative that we might consider most reasonable if H_0 is false. Let's agree that these specifications of H_0 and H_1 have been made.

Now, suppose that CAG finds that, for their sample of 9 cars, $\bar{X} = 37.5$. This result seems to discredit the manufacturer's claim, but some further analysis of its implications is needed. After all, when one drives this car under the stated conditions, one doesn't expect to get exactly 40 mpg. Naturally, there will be some variability in the results from car to car (and even in repeated tests with the same car), so we will need to determine whether outcome $\bar{X} = 37.5$ is indeed a rare event under the assumption that $\mu = 40$. What we need is a measure of "rareness" that will give us some guidance as to whether we should believe the manufacturer's claim that $\mu = 40$ or reject it in favor of the alternative that $\mu < 40$. Since the measure $P(\bar{X} = 37.5|\mu = 40) = 0$, it is clearly of no use. A reasonable alternative measure is the probability $P(\bar{X} \le 37.5|\mu = 40)$. This measure tells us the chances that we would see a result as rare as, or rarer than, what we actually observed based solely on the inherent randomness of the observations. Computing this measure yields

$$\begin{aligned} P(\bar{X} \le 37.5|\mu = 40) &= P\left(\frac{\bar{X} - 40}{1} \le \frac{37.5 - 40}{1}\right) \\ &= P(Z \le -2.5) = .0062. \end{aligned} \tag{10.2}$$

Upon seeing the probability .0062, we would probably reason as follows: There's less than a 1 in 100 chance that we would see a value of $\bar{X}$ this small or smaller if the mean mpg was really 40. So it seems quite reasonable to conclude that the manufacturer's claim was inflated and that, in reality, $\mu < 40$. If, on the other hand, the experimental outcome had been $\bar{X} = 38.5$, we might take a different view. Of course this latter outcome also discredits the manufacturer's claim, to an extent. But carrying out our "rareness" computation, we would find that

$$P(\bar{X} \leq 38.5|\mu = 40) = P\left(\frac{\bar{X}-40}{1} \leq \frac{38.5-40}{1}\right)$$
$$= P(Z \leq -1.5) = .0668. \tag{10.3}$$

The probability .0668 is still small, but now we cannot be as confident about reaching the same conclusion as above. After all, there is nearly a 7% chance of getting an outcome like that merely by chance. In our daily lives, we see events with probability .07 happen often enough that we are not that shocked to see them. For example, if you pulled out a quarter and tossed it four times, would you conclude that it was not a fair coin if you happened to toss four heads in a row? Of course not. But the probability of that happening with a fair coin is 1/16 = .0625! Reaching a conclusion in which we are really confident would require greater evidence than this. Regarding the car manufacturer, we might still have doubts about the claim that $\mu = 40$, but it's not likely that we could confidently declare it as absolutely false based on the outcome in (10.3). The probabilities calculated in (10.2) and (10.3) are generally referred to as p-*values*. Informally, one can think of p-values as the chances, under the null hypothesis, of obtaining an experimental outcome that is at least as rare as what was actually observed. Here, we see that the p-value of the observation $\bar{X} = 37.5$ is .0062 and the p-value of the observation $\bar{X} = 38.5$ is .0668. ■

The example above illustrates one of the standard approaches to hypothesis testing. The approach can be summarized as follows: Choose a threshold probability level α (like $\alpha = .05$, for example), then calculate the p-value of the "test statistic" (usually a sufficient statistic for the parameter of interest) and compare the two. If the p-value $< \alpha$, then H_0 is rejected. Otherwise, it is not rejected (meaning that the evidence in support of rejecting H_0 was not strong enough).

There is an alternative, but equivalent, approach to carrying out a hypothesis test, namely: Choose a threshold probability level α (like $\alpha = .05$, for example), then determine an appropriate threshold value (large or small, depending on the context) which the test statistic will exceed with exactly probability α. If $\alpha = .05$ in Example 10.1.1, we would be looking for the constant c for which $P(\bar{X} < c|\mu = 40)$. Note that this is the same measure of rareness which we utilized in Example 10.1.1. To find c, we compute

$$P(\bar{X} < c|\mu = 40) = P\left(\frac{\bar{X}-40}{1} < \frac{c-40}{1}\right)$$
$$= P(Z < c-40) = .05. \tag{10.4}$$

You will recognize what we are doing here as a "Harry the Ape" type of calculation (see Example 3.5.3)—we are given the probability level and we want to find the cutoff point corresponding to it. From Table 3.5.1, we know that $P(Z < -1.64) = .05$. This means that

$$c - 40 = -1.64,$$

that is,

$$c = 38.36.$$

This calculation identifies the values of the test statistic $\bar{X}$ for which H_0 should be rejected. If we reject H_0 if and only if $\bar{X} < 38.36$, there will be only a 5% chance of rejecting H_0 when H_0 is true. The following definitions explain the jargon that is usually used in summarizing results using the latter approach to testing.

Definition 10.1.2. In testing the hypotheses H_0 versus H_1 using the test statistic T, the set of values of T for which H_0 is rejected is called the *critical region* of the test. The probability α that H_0 is rejected, assuming that H_0 is true, is called the *significance level* of the test.

The two approaches for carrying out a test of H_0 vs. H_1 may thus be summarized as follows:

Approach 1: Choose a significance level α. Determine an appropriate test statistic T, and compute the p-value of that statistic. The context of the problem will usually make it obvious what values of T would be considered rare when H_0 is true. If the p-value of the observed value of T is smaller than α, reject H_0 at significance level α.

Approach 2: Choose a significance level α. Determine an appropriate test statistic T, and compute the critical region of a test with significance level α. The context of the problem will usually make it obvious what the critical region should look like (i.e., T too large, T too small, T either too large or too small). If the observed value of the test statistic lands in the critical region of the test, reject H_0 at significance level α.

If I can ask for a bit more of your patience, I will return to the well one more time. Let's reexamine Example 10.1.1 once again, but this time with a new wrinkle. Let's take, as before, the "null hypothesis" (as H_0 is generally called) to be $H_0: \mu = 40$, but let's suppose that CAG wishes to test H_0 against the alternative $H_1: \mu = 38$ (last year's average mpg). This gives us the opportunity to look at two specific curves, the two normal densities under the null and alternative hypotheses. These curves are pictured below.

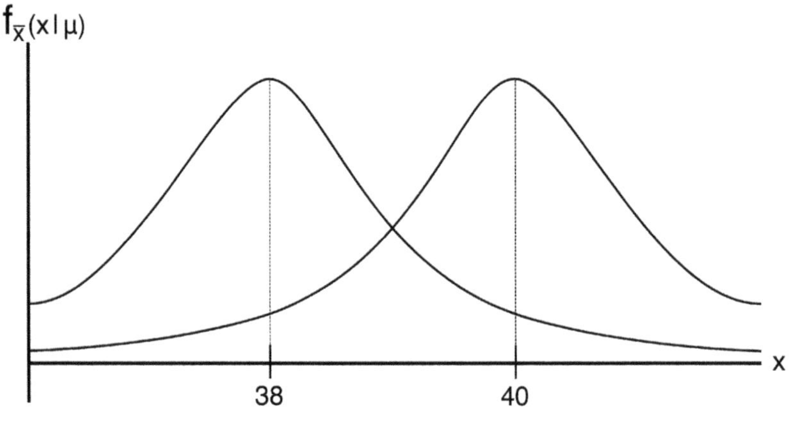

Figure 10.1.1. The distributions of $\bar{X} \sim N(\mu, 1)$ under $H_0: \mu = 40$ and $H_1: \mu = 38$.

Consider resolving the hypothesis test using a critical region of the form "$\bar{X} < c$". Since large values of $\bar{X}$ favor H_0 and small values of $\bar{X}$ favor H_1, this is certainly the natural type of critical region to use. Now, no matter what value of c you choose, you'll note that there are two possible errors which might be committed, the error resulting when $\bar{X} < c$ but $\mu = 40$, and the error resulting when $\bar{X} \geq c$ but $\mu = 38$. In general, the four possibilities we face in any given test of hypotheses are displayed in the following table:

The truth ↓ / The Decision →	Accept H_0	Reject H_0
H_0 is true	Correct	Type I error
H_1 is true	Type II error	Correct

Table 10.1.1. The four possible consequences of a given testing procedure.

It is apparent from Figure 10.1.1 that the probabilities of each of the errors in Table 10.1.1 will be positive, no matter what threshold c is used in defining the critical region. It should also be apparent that reducing the probability of making one type of error necessarily increases the probability of making the other type of error. Clearly, reducing the size of c in the critical region which rejects H_0 when $\bar{X} < c$ will decrease the probability of making a type I error while increasing the probability of making a type II error.

The standard notation for the probabilities on the errors in Table 10.1.1 is

$$\alpha = P(\text{type I error}) = P(\text{Rejecting } H_0 | H_0 \text{ is true}) \tag{10.5}$$

and

$$\beta = P(\text{type II error}) = P(\text{Accepting } H_0 | H_1 \text{ is true}). \tag{10.6}$$

When the critical region of a test has been fixed, we can always compute, in theory, the error probabilities α and β. Let us return to the automobile manufacturer problem to make some computations of this sort.

Example 10.1.2. Let's suppose that we wish to test $H_0 : \mu = 40$ against $H_1 : \mu = 38$ based on a test statistic $\bar{X}$ which has a $N(\mu, 1)$ distribution. A novice at hypothesis testing might feel that H_0 should be rejected if and only if $\bar{X} < 39$. After all, values of $\bar{X}$ that are less than 39 favor H_1 over H_0, just as values of $\bar{X}$ that are greater than 39 favor H_0 over H_1. The consequences of choosing such a critical region are that both α and β are quite large. Specifically,

$$\alpha = P(\bar{X} < 39 | \mu = 40) = P(Z < -1) = .1587$$

and

$$\beta = P(\bar{X} \geq 39 | \mu = 38) = P(Z \geq 1) = .1587.$$

The figure below displays these two error probabilities.

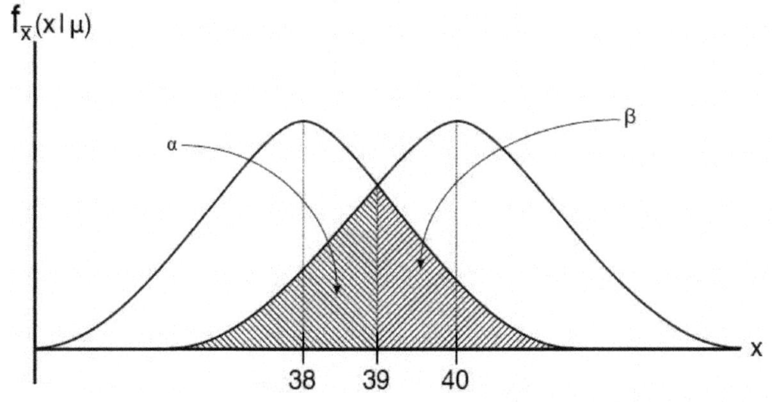

Figure 10.1.2. Probabilities α and β for the critical region "Reject H_0 if $\bar{X} < 39$" when testing $H_0 : \mu = 40$ vs. $H_1 : \mu = 38$.

Were we to reject H_0 precisely when $\bar{X} < 39$, we couldn't be very sure of making the correct decision, as there is about a 16% chance of rejecting H_0 when it is true. There's also about a 16% chance of accepting H_0 when it is false.

Consider, instead, using the critical region: Reject H_0 if $\bar{X} < 38.36$. With this new critical region, the probabilities of type I and type II errors change as follows:

$$\alpha = P(\bar{X} < 38.36|\mu = 40) = P(Z < -1.64) = .05$$

and

$$\beta = P(\bar{X} \geq 38.36|\mu = 38) = P(Z \geq .36) = .3594.$$

These two error probabilities are displayed in the figure below.

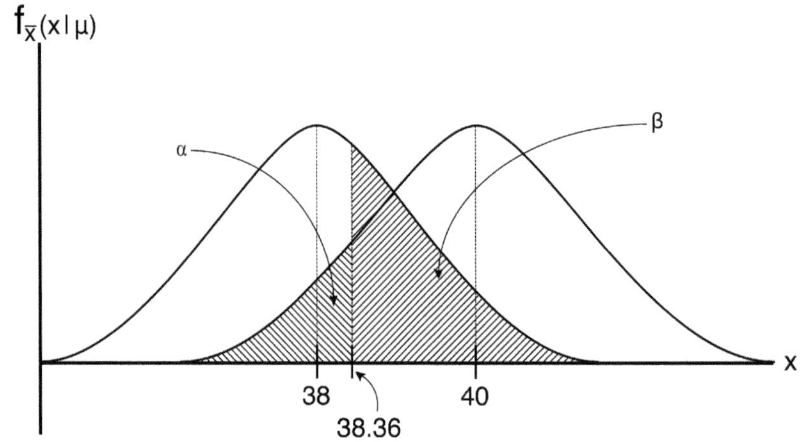

Figure 10.1.3. Probabilities α and β for the critical region "Reject H_0 if $\bar{X} < 38.36$."

■

Since one can't control both α and β simultaneously (given a sample of a fixed size), it is standard practice to seek to control the value of α, setting it at a pre-specified small value, and then simply accept whatever value of β comes with it. In this way, at least one of the error probabilities is small. In particular, if α is suitably small (say $\alpha = .05$ or .01), then we can have substantial confidence that we have made the correct decision if our data indicates that H_0 should be rejected. But since the corresponding value of β may be quite large, the strategy of controlling α but not β has some important implications. It definitely influences the way in which we set up a testing problem.

Let's assume that we plan to operate at the 5% significance level. This means that whatever statement we designate as H_0, we will set up the test so that, when H_0 is true, there's only a 5% chance that we would reject it. In most circumstances, we would be comfortable rejecting H_0 when our test statistic lands in our prescribed critical region, reasoning as follows: If H_0 were true, the event we've just witnessed would hardly ever happen. So we can interpret the event in one of two ways: (1) H_0 is true, and a very rare event has occurred or (2) H_0 is not true, which means that the event we have observed, which is no doubt more compatible with H_1 than with H_0, isn't necessarily a rare event at all, but an event that could well have happened under H_1. In the scientific community, rejecting a null hypothesis at the 5% significance level is a widely used and generally accepted standard for reaching the decision that the null hypothesis is false. There are settings in which a higher standard of proof is deemed necessary for a null hypothesis to be rejected. A 1% significance level is often used when stronger evidence of the inconsistency between the null hypothesis and the observed data is desired or required.

The reasoning above leads to some natural guidelines that are helpful when we are setting up a test for deciding between two hypotheses. Consider the following strategies, both motivated by the fact that we will be carefully controlling α, the probability of falsely rejecting H_0, by ensuring that the critical region of our test is a set of experimental outcomes that has a very small probability of occurrence if H_0 happens to be true.

(1) Let H_0 be the statement that you hope your data will disprove,

or

(2) Choose H_0 so that the error of falsely rejecting that hypothesis is the more serious of two possible types of errors.

The first of these guidelines simply reflects the fact that the rejection of a null hypothesis at some small significance level like .05 is generally accepted as "proof," that is, strong evidence, that the null hypothesis is false. Thus, if there is something that you would like to prove, let that be the alternative hypothesis, and set the null hypothesis to be what you hope to disprove. In this way, if your test statistic indicates that you should reject H_0, you can reject it with a fair amount of confidence, and as a consequence, you can assert that H_0 has been "proven false" (in a statistical sense) and H_1 may thus be considered true. Note that it is not possible to make assertions of this sort about decisions in the other direction. If our test statistic does not land in the critical region of the test, then we are not able to reject H_0. Even though we will use the phrase "accept H_0" when we fail to reject the null hypothesis, this is not equivalent to the assertion that H_0 is true. Indeed, because we have not been able to control the probability of type II error, there could be a non-negligible probability that H_0 is false but that the evidence against H_0 provided by our data was not striking enough to allow us to reject it. If, on the other hand, our data calls for the rejection of H_0, then the fact that this would happen very rarely if H_0 is true gives us confidence when we declare that H_0 is false. The following examples of the application of the guidelines above help to understand how they might come into play in deciding how to assign the labels H_0 and H_1 to two competing hypotheses.

Example 10.1.3. Suppose that you work for a pharmaceutical company that has developed a new medication for a common medical condition. Let p be the proportion of the population of people with this condition who will react favorably to the new medication. If the proportion of this population which reacts favorably to the best medication currently available is p_0, then the two hypotheses that your company will wish to test are

$$p < p_0 \text{ and } p > p_0. \tag{10.7}$$

Which of these two statements should play the role of H_0? The first guideline above provides an immediate answer. It is clear that you would like to prove that the new medication is better that the old one, that is, you would like to prove that $p > p_0$. This statement should thus be chosen to play the role of H_1. The statement we wish to disprove will then play the role of H_0. Thus, the desired setup is

$$H_0 : p \leq p_0 \text{ and } H_1 : p > p_0. \tag{10.8}$$

(Note that, as is customary, I have made the equality of p and p_0 part of the null hypothesis.) If H_0 is rejected when the significance level α is suitably small, you can reasonably claim that the new medication is superior to the old. The FDA would probably take notice! ■

Example 10.1.4. Suppose you work for a company that makes hand grenades. From its long history of making these nasty devices, your company knows that the mean time to explosion of their hand grenades is 10 seconds (after they are hand activated), with a standard deviation of σ. The hypotheses of interest when a new batch of hand grenades are produced are

$$\sigma < 1 \text{ second} \qquad \text{and} \qquad \sigma > 1 \text{ second}. \tag{10.9}$$

These two statements may be interpreted as follows. A soldier who may have to use a hand grenade would like the variability of the time to explosion to be low. If it is, the soldier can count on the fact that the grenade will not explode too early. A soldier would like to be quite certain that the time to explosion is fairly close to the mean time of 10 seconds. If σ is large, an early explosion cannot be excluded as a possibility. Given this concern, which of the two statements in (10.9) should play the role of the null hypothesis in testing these statements against each other? Let's consider the consequences of each of the two possible errors one might make here. Deciding that "$\sigma < 1$" when "$\sigma > 1$" is true can be thought of as reaching the conclusion that the hand grenades are "safe" when in fact they are "unsafe." On the other hand, deciding that "$\sigma > 1$" when "$\sigma < 1$" is true means that we have concluded that the hand grenades are "unsafe" when in fact they are "safe." The consequences of the first of these errors could be very serious indeed. A soldier who believes that the grenades are safe when they are not safe might be killed or seriously hurt by a grenade that explodes early. The consequences of the second type of error are clearly less serious. If a batch of grenades are classified as unsafe, even though they are in fact safe, they would be removed from service, resulting in some financial losses to the company. While no company likes this kind of outcome, it would readily prefer making the second error rather than the first, both because of the potentially serious physical harm that could befall users of their product and because malfunctions resulting from this type of error would also seriously damage their professional reputation. It is clear that you should set up the test of these two hypotheses so the first of the two errors above is a type I error, the error whose probability we plan to control quite tightly. If deciding that "$\sigma < 1$" when "$\sigma > 1$" is true is to be a type I error, then the statement "$\sigma > 1$" would play the role of the null hypothesis, that is, the test would be set up with

$$H_0 : \sigma \geq 1 \text{ and } H_1 : \sigma < 1. \tag{10.10}$$

This formulation of the test assumes that the hand grenades are unsafe (that is, H_0 is true) unless that data proves this assumption false. This seems like a completely appropriate setup—one of the many ways in which we should be protecting our troops. ■

I will close this section with a brief discussion of some extensions of the concepts that were introduced above. Let's suppose that the data available in a testing problem is a random sample from a distribution F_θ, where the unknown parameter might be a scalar, like the mean of a Poisson distribution, or a vector-valued parameter, like (μ, σ^2), the mean and variance of the normal distribution. Recall that a hypothesis that fully specifies the value of the parameter (for example, "$\theta = 10$" or "$(\mu, \sigma^2) = (0,1)$") is called a simple hypothesis, while a hypothesis that specifies more than one possible value of the parameter is referred to as a composite hypothesis. When H_0 is a simple hypothesis, the significance level of your test is defined quite unambiguously in (10.5). But when H_0 is a composite hypothesis, then the chances that one rejects H_0 when H_0 is true will vary, depending on what parameter value, among those specified in the null hypothesis, happens to be true. Since we wish to protect against incorrectly rejecting H_0 whenever the parameter takes a value satisfying H_0, the definition in (10.5) needs to be expanded. Indulging in a slight but harmless abuse of the notation above, let's for a moment think of H_0 as the set of parameter values designated by the statement H_0. For example, if H_0 is the statement "$\mu \leq 10$," then take the notation "$\mu \in H_0$" to mean that "the hypothesis $H_0 : \mu \leq 10$ is true."

Now, let's assume that we are testing a composite null hypothesis $H_0 : \theta \in \Theta_0$ against a composite alternative hypothesis $H_1 : \theta \in \Theta_1$, and suppose that the critical region of our test has been specified. In practice, this simply means that we have selected a test statistic on which to base the test and that the values of this statistic which lead to the rejection of H_0 have been determined. The significance level α of the test is then defined as the largest possible probability, for $\theta \in \Theta_0$, that H_0 will be rejected. In other words,

$$\alpha = \max_{\theta \in \Theta_0} P(\text{Reject } H_0 | \theta). \tag{10.11}$$

The probability α in (10.11) is called the *size* or *significance level* of the test. We will illustrate this definition in the context of a slightly expanded version of the automobile manufacturer example.

Example 10.1.5. Let's suppose that we have no prior intuition about how many miles per gallon the car of interest will get under normal driving conditions, and so we decide to test the auto manufacturer's claim that the mean mpg for the car is 40 by interpreting his claim as $H_0 : \mu \geq 40$ and we choose, as the natural (and typical) alternative $H_1 : \mu < 40$. Note that now the entire parameter space $\{\mu | -\infty \leq \mu \leq \infty\}$ is covered, so we needn't worry about the possibility that both the null and alternative hypotheses might be false, a possibility that naturally comes to mind when, for example, both H_0 and H_1 are simple hypotheses. Let's suppose that we decide, as in Example 10.1.2, to use the critical region "Reject H_0 if $\bar{X} < 38.36$." What is the significance level α of this test? Now the application of the definition in (10.11) can seem a bit imposing, requiring, as it does, the evaluation and comparison of the probability of landing in the critical region of the test at every parameter value μ that is consistent with $H_0 : \mu \geq 40$. As it happens, this task is not as daunting as it may seem, even though there are uncountably many parameter values to be considered. The following figure shows why. It is clear from this figure that the greatest chance of landing in the region $\{\bar{X} | \bar{X} < 38.36\}$ occurs when μ is the smallest possible value under H_0, that is, when $\mu = 40$.

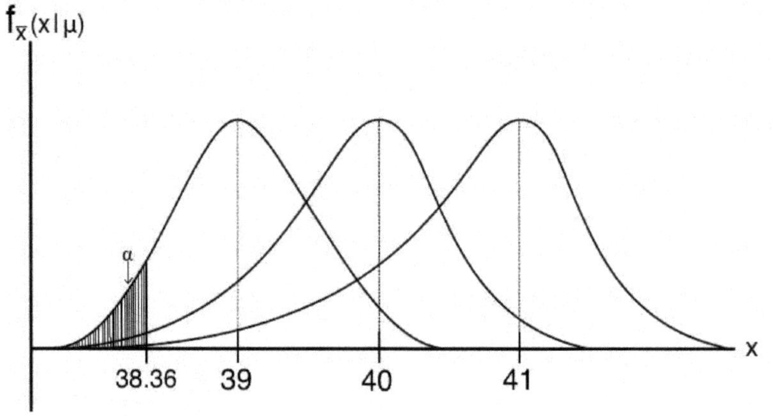

Figure 10.1.4. $P(\bar{X} < 38.36|\mu)$, under normality, for varied values of μ.

These considerations lead to the conclusion that, in this example,

$$\alpha = \max_{\mu \in H_0} P(\text{Reject } H_0|\mu) = P(\bar{X} < 38.36|\mu = 40) = .05. \quad \blacksquare$$

As you have no doubt foreseen, a similar expansion of our notion of the probability of type II error will be needed when H_1 is a composite hypothesis. We have used the symbol β to denote the probability of making a type II error when the parameter is a fixed value under the alternative hypothesis H_1. When H_1 is a composite hypothesis, a complete description of the probability of making an error of type II would require that the value of β be computed for all values of the parameter consistent with H_1. These computations turn out to be very important in comparing the performance of competing tests. Because of this, the calculation of type II errors for hypothesis tests that arise with some frequency has become something of a cottage industry in the field of Statistics, with tables and plots of many different types available in published books and at various Internet sites. When one or both of the hypotheses under consideration are composite, the notions of type I and type II errors tend to be summarized in a single object called the power function of the test. The latter characteristic of a hypothesis test is formally defined below.

Definition 10.1.3. Let $X_1, \ldots, X_n \overset{\text{iid}}{\sim} F_\theta$, where $\theta \in \Theta$, and consider testing the hypotheses $H_0 : \theta \in \Theta_0$ against $H_1 : \theta \in \Theta_1$, where $\Theta_0 \cup \Theta_1 = \Theta$, using the test statistic T. Denote the critical region of the test as "CR," that is, assume that H_0 will be rejected if and only if $T \in CR$. Then, the *power function* of the test is defined as

$$p(\theta) = P(\text{Rejecting } H_0|\theta) = P(T \in CR|\theta). \tag{10.12}$$

The power function of a test characterizes its behavior over the entire parameter space and is generally used as the sole measure of interest when assessing a test's performance or when comparing one test with another. One special feature of $p(\theta)$ is that it can be computed, and has a meaningful interpretation, at every parameter value. For $\theta_0 \in \Theta_0$, $p(\theta_0)$ represents the probability of making a type I error when the parameter takes the value θ_0, while for $\theta_1 \in \Theta_1$, $p(\theta_1)$ represents the probability of making the correct decision, that is, rejecting H_0 when the parameter takes the value θ_1, a value that is not covered by the null hypothesis. These interpretations help us recognize what the ideal power function would look like.

For concreteness, let us return to the setup in Example 10.1. Fear not, we are about to leave the automobile manufacturer behind forever! But, this one last time, let's consider testing the hypotheses in that example based on the observed sample mean $\bar{X}$. Recall that, in that example, $H_0 : \mu \geq 40$ and $H_1 : \mu < 40$. The test statistic $\bar{X}$ in the example has a $N(\mu, 1)$ distribution. The critical region "Reject H_0 if $\bar{X} < 38.36$" was decided upon. We will compute its power function, but first, we will plot the ideal power function in this testing problem. Ideally, we would like to be in the position of never making a mistake. If we could actually do it, we would use a testing procedure for which the probability of making a type I error was zero, as was the probability of making a type II error. The ideal, albeit unachievable, power function is displayed in the figure below.

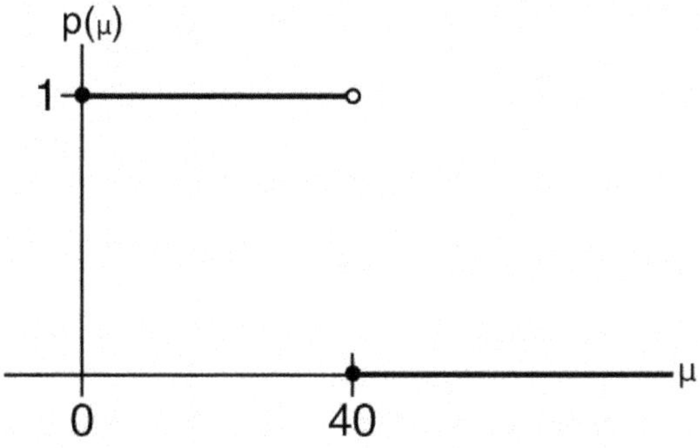

Figure 10.1.5. The ideal power function $p(\mu)$ for testing $H_0 : \mu \geq 40$ vs. $H_1 : \mu < 40$.

Let's now consider the power function for the test discussed in Example 10.1. Using the critical region "Reject H_0 if $\bar{X} < 38.36$," we noted above that the power function evaluated at $\mu = 40$ is

$$p(40) = P(\bar{X} < 38.36|\mu = 40) = P(Z < (38.36 - 40)/1) = P(Z < -1.64) = .05.$$

To fill out the power curve so that we can get a reasonable approximation by interpolation, we'll calculate the power function of our test at the values $\mu = 41, 39, 38, 37,$ and 36. The results from these calculations are:

$$\begin{aligned} p(41) &= P(\bar{X} < 38.36|\mu = 41) = P(Z < -2.64) = .0041 \\ p(39) &= P(\bar{X} < 38.36|\mu = 39) = P(Z < -0.64) = .2611 \\ p(38) &= P(\bar{X} < 38.36|\mu = 38) = P(Z < 0.36) = .6406 \\ p(37) &= P(\bar{X} < 38.36|\mu = 37) = P(Z < 1.36) = .9131 \\ p(36) &= P(\bar{X} < 38.36|\mu = 36) = P(Z < 2.36) = .9909. \end{aligned}$$

These calculations allow us to plot the following approximate power function.

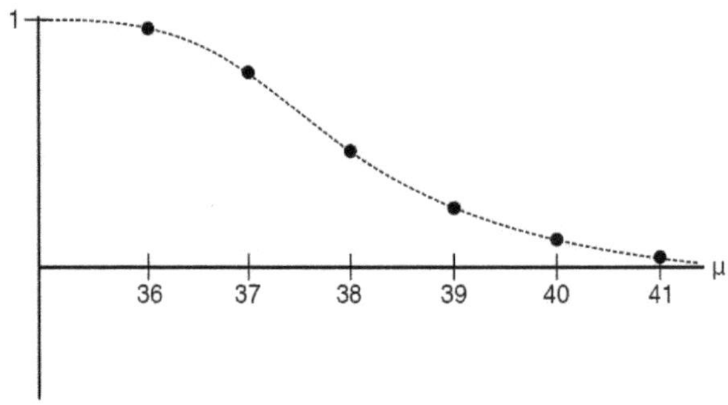

Figure 10.1.6. Approximate power function for the test in Example 10.1.5.

It is apparent that the real power function displayed in Figure 10.1.6 has the basic properties of the ideal power function. The power function is small under every μ associated with H_0 and the power function steadily increases as the value of μ grows more distant from H_0. If our test were based on a larger sample size, say $n = 100$ instead of $n = 9$, a test with significance level .05 would have a power function that even more closely resembles the ideal curve in Figure 10.1.5.

Let's now consider how two different tests of the same hypotheses H_0 and H_1 might be compared. You might wonder where two different tests would come from. By "different tests," I mean two tests having *different critical regions*. Suppose, for example, that we are interested in testing whether the center M (perhaps the mean or the median) of a population is above or below a certain threshold. Assume that a random sample $X_1, \dots, X_n \stackrel{\text{iid}}{\sim} F_M$ is available to test the hypothesis $H_0 : M \leq M_0$ against $H_1 : M > M_0$. Let $\bar{X}$ and m be the mean and median of the sample. Two critical regions that one might consider here are (1) "Reject H_0 if $\bar{X} > c$" for some constant c for which $P(\bar{X} > c|M = M_0) = .05$ and (2) "Reject H_0 if $m > d$" for some constant d for which $P(m > d|M = M_0) = .05$. If F_M has been specified (e.g., the precise form of the one-parameter model F_M is known), one could calculate the power functions of the two tests. Interestingly, test (1) will be better than test (2) for some models (e.g., if F_M is a normal distribution with mean M), and test (2) will be better than test (1) for other models (e.g., if F_M is a standard Cauchy distribution with median M). Thus, depending on the model in force in a particular application, the power function of either test may be as close as, or closer to, the ideal power curve than is the power function of the other test. Later in this chapter, we will pursue the comparison of power functions further and, in certain special circumstances, we'll give a prescription, based on the famous Neyman-Pearson Lemma, for finding a "uniformly most powerful (UMP)" test, that is, a test that is better than all other possible tests at every value of the model parameter. So, stay tuned!

We close this section by defining a few other terms that will be used in the sequel. First, there are two classes of testing problems that we want to distinguish from each other. The following pairs of hypotheses share a common feature:

(i) $H_0 : \theta = \theta_0$ vs. $H_1 : \theta = \theta_1$, where $\theta_1 \neq \theta_0$,

(ii) $H_0 : \theta \leq \theta_0$ vs. $H_1 : \theta > \theta_0$,

(iii) $H_0 : \theta \geq \theta_0$ vs. $H_1 : \theta < \theta_0$.

These types of testing problems are referred to as "one-sided" problems. The characteristic they share is that every parameter value in the alternative hypothesis is on one specific side (that is, to the left or to the right) of the parameter values in the null hypothesis. Usually, the best critical regions to use in such testing problems are so-called one-sided regions. If T is the test statistic used to resolve a one-sided testing problem, then the critical regions that naturally arise in most problems of this sort are of the form "Reject H_0 if $T > c$" or "Reject H_0 if $T < c$." Another class of testing problems of interest are "two-sided" problems like

(iv) $H_0 : \theta = \theta_0$ vs. $H_1 : \theta \neq \theta_0$,

(v) $H_0 : \theta \in [\theta_0 - a, \theta_0 + a]$ vs. $H_1 : \theta \notin [\theta_0 - a, \theta_0 + a]$.

These problems are characterized by the fact that there are parameter values in the alternative hypothesis that lie on either side of the parameter values in the null hypothesis. In two-sided problems, one will usually find that two-sided critical regions such as "Reject H_0 if $|T - \theta_0| > c$" will outperform most reasonable alternatives. You'll note that, both in one-sided and two-sided testing problems, I've left myself a little wiggle room regarding what a good test should look like. That's because so-called one-sided and two-sided tests are "best" under certain specific conditions, but they can be inferior to other tests under other conditions.

One final comment on the subject of one- and two-sided tests. When setting up a hypothesis test, one of the decisions you have to make is whether to carry out a one-sided or a two-sided test. In making this determination, it is helpful to think of the null hypothesis you're dealing with as a simple hypothesis, say $H_0 : \theta = \theta_0$, even though this value may in reality be a boundary value for the parameter values that will end up in your null hypothesis. You will make the choice between a one-sided or a two-sided alternative by asking yourself the following question: If $\theta \neq \theta_0$, do the alternative values of θ that seem most plausible lie to the right of θ_0 or to the left? Or do both right and left seem perfectly plausible? In the first instance, if, when θ_0 is rejected as a plausible value of θ, you consider larger values of θ the obvious alternative, then you should select the alternative hypothesis to be $H_1 : \theta > \theta_0$. We would then typically reformulate the null hypothesis as $H_0 : \theta \leq \theta_0$. Similar reasoning, in the other direction, would lead us to select $H_1 : \theta < \theta_0$ and $H_0 : \theta \geq \theta_0$. As mentioned previously, the equal sign (i.e., $\theta = \theta_0$) goes with the null hypothesis. In the remaining case where you really have no directional intuition about what θ might be if it is not equal to θ_0, you should take the alternative hypothesis to be $H_1 : \theta \neq \theta_0$ and retain $H_0 : \theta = \theta_0$ as the null hypothesis. A good example of this latter circumstance is a test about whether a coin you intend to flip a few times is a fair coin. The number of heads obtained in n tosses of this coin is the binomial variable $X \sim B(n, p)$. The hypothesis that the coin is fair is simply the statement "$p = 1/2$." If you ask the questions above, you would no doubt conclude that if the coin was not fair, we'd have no idea whether the coin was biased upward ($p > 1/2$) or downward ($p < 1/2$). Given that, we would be well-advised to let $H_0 : p = 1/2$ and $H_1 : p \neq 1/2$.

Exercises 10.1.

1. Jack and Jill disagree about how long it tends to take their friend Jojo to run up the hill. Both of them understand that the time is random, varying from day to day with the weather, with what Jojo ate for breakfast, and other factors. Jack believes that the time X that Jojo takes on a random day has distribution $X \sim N(30, 25)$, where X is measured in seconds, while Jill believes that $X \sim N(40, 25)$. (At least Jack and Jill agree on the variance!) Suppose they agree on the following test based on Jojo's next run up the hill: Reject $H_0 : \mu = 30$ in favor of $H_1 : \mu = 40$ if $X > 37$. (a) Find the significance level α of this test. (b) Find the

probability β of making a type II error. (c) Suppose Jojo runs up the hill in 39 seconds. What is the p-value of this observed value of X?

2. SlipperyRock Creamery sells one-pint (i.e., 16-ounce) containers of their hand-packed ice cream. Investigative reporter Vin D. Pulitzer thinks that SlipperyRock pints actually contain substantially less than 16 ounces of ice cream. Mr. Pulitzer plans to buy 25 pints of SlipperyRock hand-packed ice cream. He will then test $H_0 : \mu = 16$ against $H_1 : \mu = 15$. He will reject H_0 if and only if $\bar{X} < 15.25$, where $\bar{X}$ is the mean weight of his 25 pints. Assume that the weights are $N(\mu, \sigma^2)$ random variables with known standard deviation $\sigma = 1.5$. (a) Find the probability α of making a type I error with this test. (b) Find the probability β of making a type II error with this test.

3. *Consumer Reports* is investigating the performance characteristics of hand grenades made by the company Grenades R Us (GRU). It is generally acknowledged that the time to explosion (in seconds) of a random GRU grenade is $X \sim N(10, \sigma^2)$. GRU claims that $\sigma^2 = 1$, while critics claim that, in reality, $\sigma^2 = 4$, making GRU grenades more dangerous to the soldier who uses them. In testing the hypotheses $H_0 : \sigma^2 = 1$ vs. $H_1 : \sigma^2 = 4$, *Consumer Reports* has obtained a sample of 16 grenade explosion times, and has decided to use the critical region: Reject H_0 if $\bar{X} < 9.6$. (a) Find α, the probability of type I error for this test. (b) Find β, the probability of type II error for the test.

10.2 Standard Tests for Means and Proportions

The two problems that will be examined in this section are hypothesis tests involving the mean of a normal population and hypothesis tests involving a binomial proportion. By "standard tests," I mean the tests that are most widely used in applications. In Section 10.4, we will discuss the theoretical justification for using them.

Let's consider, first, tests involving a normal mean. Assume that the available data consists of a random sample of normally distributed observations, that is, $X_1, \ldots, X_n \overset{\text{iid}}{\sim} N(\mu, \sigma^2)$. Consider testing the hypothesis $H_0 : \mu \leq \mu_0$ against the alternative $H_1 : \mu > \mu_0$. We will treat three distinct cases: (1) σ^2 is a known number, (2) σ^2 is unknown, but n is large (which is generally interpreted as $n \geq 30$), and (3) σ^2 is unknown, but n is small (i.e., $n < 30$). The first of these problems is of course unrealistic—a "toy problem," if you will—but the solution to this case provides a simple and clear model for how to think about the other two cases. Actually, there are a few real applications in which one might be willing to assume that the variance σ^2 above is known. The prototypical example of such an application is an experiment that is performed repeatedly, perhaps once quarterly or annually, and it has been noticed that the variability of the data tends to be about the same each time the experiment is executed. An example would be national SAT tests, taken by a large number of high school students every time it is offered. While the mean score of these exams tend to trend up or trend down over certain periods of time, it's a known fact that the variance of scores changes very little from one year to another. But, of course, situations like this are not common, and in applications that you are likely to face in practice, the assumption of a known variance will rarely be justifiable. A colleague of mine likes to put it this way: "If Mother Nature is not willing to show you her μ, what makes you think that she'd be willing to show you her σ^2."

Case 1. $X_1, \ldots, X_n \overset{\text{iid}}{\sim} N(\mu, \sigma^2)$**, with** σ^2 **known.** We first note that the sample mean $\bar{X}$ is a sufficient statistic for μ and is thus the natural test statistic on which to base a test about μ. In testing $H_0 : \mu \le \mu_0$ vs. $H_1 : \mu > \mu_0$, it is clear that large values of $\bar{X}$ tend to favor H_1 while small values of $\bar{X}$ tend to favor H_0. It follows that the most sensible critical regions for this test have the form "Reject H_0 if $\bar{X} > c$," for some constant c. Suppose we want to test H_0 vs. H_1 at some fixed significance level α. We would then choose the constant c so that

$$\max_{\mu \le \mu_0} P(\bar{X} > c | \mu) = \alpha. \tag{10.13}$$

Since the probability in (10.13) is an increasing function of μ, it follows that (10.13) is equivalent to

$$P(\bar{X} > c | \mu = \mu_0) = \alpha. \tag{10.14}$$

Now if $\mu = \mu_0$, then $\bar{X} \sim N(\mu_0, \sigma^2/n)$ and thus

$$P\left(\frac{\bar{X} - \mu_0}{\sigma/\sqrt{n}} > Z_\alpha\right) = \alpha.$$

It then follows that

$$P(\bar{X} > \mu_0 + \frac{\sigma}{\sqrt{n}} Z_\alpha | \mu = \mu_0) = \alpha.$$

When σ^2 is known, the standard critical region of size α for testing $H_0 : \mu \le \mu_0$ vs. $H_1 : \mu > \mu_0$ is

$$\text{Reject } H_0 \text{ in favor of } H_1 : \mu > \mu_0 \text{ if } \bar{X} > \mu_0 + \frac{\sigma}{\sqrt{n}} Z_\alpha. \tag{10.15}$$

Case 2. $X_1, \ldots, X_n \overset{\text{iid}}{\sim} N(\mu, \sigma^2)$**, with** σ^2 **unknown,** n **large.** When a sample drawn from a normal population is "large," the sample variance s^2 is a reasonably precise estimate of the population variance σ^2, and the approximation of the standard normal variable $(\bar{X} - \mu)/(\sigma/\sqrt{n})$ by $(\bar{X} - \mu)/(s/\sqrt{n})$ may be considered reliable. Thus, under these circumstances, the test which simply mimics the test used in case 1, plugging in s in place of σ, provides satisfactory results. In essence, when n is sufficiently large, we treat the test statistic $(\bar{X} - \mu)/(s/\sqrt{n})$ as a standard normal variable, leading to the approximate probability statement

$$P(\bar{X} > \mu_0 + \frac{s}{\sqrt{n}} Z_\alpha | \mu = \mu_0) \approx \alpha.$$

Thus, when σ^2 is unknown but n is large, the standard critical region of size α for testing $H_0 : \mu \le \mu_0$ vs. $H_1 : \mu > \mu_0$ is

$$\text{Reject } H_0 \text{ in favor of } H_1 : \mu > \mu_0 \text{ if } \bar{X} > \mu_0 + \frac{s}{\sqrt{n}} Z_\alpha. \tag{10.16}$$

Remark 10.2.1. Some guidance should be given about how large the sample size n needs to be to justify the use of the approximate test in case 2. The threshold value $n = 30$ was mentioned earlier in this regard. If the assumptions of case 2 were strictly obeyed, then we know that the t statistic $(\bar{X} - \mu)/(s/\sqrt{n})$ is well approximated by a standard normal variable when n is at least 30. Evidence of this is provided in Table A.5 in which it can be seen that the cutoff points $t_{k,\alpha}$ and z_α get very close as the degrees of freedom k approach 30. But there is an interesting additional layer of complexity to consider. Case 2 is often used in practice even when it is known or suspected that the available data is not normally distributed. If you look over Case 2 carefully, you'll see that what we actually rely upon in the development of the test in that case is the normality of the sample mean $\bar{X}$. This suggests that the Case 2 test might be well justified whenever the Central Limit Theorem is applicable, that is, when the parent population has a finite variance and the sample size is sufficiently large. This approximate test does indeed work well in such settings. The main difference between these applications and those in which the normality assumption actually holds is that, in non-normal cases, the sample size n needed to ensure the approximate normality of $\bar{X}$ is not necessarily 30, and in fact may be much larger. It remains true that the test in Case 2 will be applicable in the latter scenarios. However, when the normality of the data is in doubt, it behooves the experimenter to examine his/her data carefully to determine the extent of non-normality present. If a histogram of the data has no resemblance to a bell-shaped curve, it is quite likely that a sample of size 30 will not be sufficient to provide reliable inferences about the mean μ using the Case 2 test. Available alternatives include the possibility of carrying out the Case 2 test with a larger sample or of using a nonparametric test which doesn't depend on an assumed model. Nonparametric tests are treated in Chapter 12.

Case 3. $X_1,\dots,X_n \overset{\text{iid}}{\sim} N(\mu,\sigma^2)$**, with** σ^2 **unknown,** n **small.** When the sample size is small, the idea of mimicking the Case 1 test with σ replaced by s does not work well. Even when the normality assumption is satisfied exactly, the distribution of the t statistic $(\bar{X}-\mu)/(s/\sqrt{n})$ is not well approximated by a standard normal distribution. However, under the normality assumption, the exact distribution of the t statistic is known. Indeed, for a normal random sample with $\mu=\mu_0$, we have that

$$\frac{\bar{X}-\mu_0}{s/\sqrt{n}} \sim t_{n-1}. \tag{10.17}$$

Thus, if $\mu=\mu_0$, we may write

$$P\left(\frac{\bar{X}-\mu_0}{s/\sqrt{n}} > t_{n-1,\alpha}\right) = \alpha.$$

It follows that

$$P(\bar{X} > \mu_0 + \frac{s}{\sqrt{n}} t_{n-1,\alpha} | \mu=\mu_0) = \alpha.$$

Thus, when σ^2 is unknown but n is small, the standard critical region of size α for testing $H_0: \mu \le \mu_0$ vs. $H_1: \mu > \mu_0$ is

$$\text{Reject } H_0: \mu \le \mu_0 \text{ in favor of } H_1: \mu > \mu_0 \text{ if } \bar{X} > \mu_0 + \frac{s}{\sqrt{n}} t_{n-1,\alpha}. \tag{10.18}$$

Technically, the critical region in (10.18) is exact and appropriate for use for any sample size n, provided the normality assumption is satisfied, but it is rarely used for $n \ge 30$ since, under the assumed normality, the Case 2 test gives an excellent approximation in such cases. As a practical matter, most published t tables don't provide α-cutoff points for degrees of freedom over 30, although these values aren't hard to find on the Internet.

The standard tests of other hypotheses involving a normal mean μ can be obtained using similar reasoning. Since the reasoning is straightforward, I will skip any further discussion of such developments, but will simply display the results in the table below. For economy of presentation in the table, critical regions are given in terms of the statistics Z and t, respectively, where

$$Z = \frac{\bar{X} - \mu_0}{\sigma/\sqrt{n}}$$

and

$$t = \frac{\bar{X} - \mu_0}{s/\sqrt{n}},$$

with both Z and t based on a random sample $X_1, \ldots, X_n \stackrel{\text{iid}}{\sim} N(\mu, \sigma^2)$.

H_0	H_1	Case	Critical Region of size α
$\mu \leq \mu_0$	$\mu > \mu_0$	σ^2 known	Reject H_0 if $Z > Z_\alpha$
$\mu \leq \mu_0$	$\mu > \mu_0$	σ^2 unknown, n large	Reject H_0 if $t > Z_\alpha$
$\mu \leq \mu_0$	$\mu > \mu_0$	σ^2 unknown, n small	Reject H_0 if $t > t_{n-1,\alpha}$
$\mu \geq \mu_0$	$\mu < \mu_0$	σ^2 known	Reject H_0 if $Z < -Z_\alpha$
$\mu \geq \mu_0$	$\mu < \mu_0$	σ^2 unknown, n large	Reject H_0 if $t < -Z_\alpha$
$\mu \geq \mu_0$	$\mu < \mu_0$	σ^2 unknown, n small	Reject H_0 if $t < -t_{n-1,\alpha}$
$\mu = \mu_0$	$\mu \neq \mu_0$	σ^2 known	Reject H_0 if $\|Z\| > Z_{\alpha/2}$
$\mu = \mu_0$	$\mu \neq \mu_0$	σ^2 unknown, n large	Reject H_0 if $\|t\| > Z_{\alpha/2}$
$\mu = \mu_0$	$\mu \neq \mu_0$	σ^2 unknown, n small	Reject H_0 if $\|t\| > t_{n-1,\alpha/2}$

Table 10.2.1. Standard tests for a normal mean μ.

While I won't treat extensions of the ideas above to the comparison of two normal means in great detail, I will treat one particular extension in order to give you a feeling for what a standard two-sample procedure looks like. The usual framework in which the comparison of two normal means is pursued entails the assumption that *independent* samples are available from two populations, with $X_1, \ldots, X_m \stackrel{\text{iid}}{\sim} N(\mu_x, \sigma_x^2)$ and $Y_1, \ldots, Y_n \stackrel{\text{iid}}{\sim} N(\mu_y, \sigma_y^2)$. The following distribution theory is essential in the comparison of interest: Given the Xs and the Ys above, the difference of the sample means $\bar{X}$ and $\bar{Y}$ is normally distributed, with

$$\bar{X} - \bar{Y} \sim N(\mu_x - \mu_y, \sigma_x^2/m + \sigma_y^2/n). \qquad (10.19)$$

The claim in (10.19) is easily proven using moment-generating functions (see Exercise 10.2.3). Now suppose we wish to test the hypothesis H_0: $\mu_x = \mu_y$ against the alternative H_1: $\mu_x \neq \mu_y$. As in the one-parameter testing problems treated above, there are several cases to consider, and there's a different testing procedure in each case. Let's consider the extension of the one-sample treatment of Case 1 to the two-sample problem in the case in which both population variances σ_x^2 and σ_y^2 are known. Under the null hypothesis of equal means, we have that

$$Z = \frac{\bar{X} - \bar{Y}}{\sqrt{\sigma_x^2/m + \sigma_y^2/n}} \sim N(0, 1). \qquad (10.20)$$

If H_0 is true, this statistic is expected to be reasonably close to zero, since under H_0, its distribution is centered at zero and its expected behavior is that of a standard normal variable. On the other hand, if H_0 is false, the distribution of $\bar{X} - \bar{Y}$ is centered on the difference $\mu_x - \mu_y$, so that the distribution of the statistic in (10.20) is not centered at zero and is thus expected to be farther from zero than it would be if H_0 were true. It thus seems reasonable to reject H_0 if the statistic Z in (10.20) is too far away from zero. Now this Z wouldn't be a "statistic" in the usual sense were it not for the fact that we assumed above that the parameters σ_x^2 and σ_y^2 are known. We would call this scenario Case 1 in the two-sample problem under consideration. Given that the statistic in (10.20) can in fact be computed from data under this latter assumption, the standard critical region of size α for testing H_0: $\mu_x = \mu_y$ vs. H_1: $\mu_x \neq \mu_y$ is

$$\text{Reject } H_0 : \mu_x = \mu_y \text{ in favor of } H_1 : \mu_x \neq \mu_y \text{ if } \left| \frac{\bar{X} - \bar{Y}}{\sqrt{\sigma_x^2/m + \sigma_y^2/n}} \right| > Z_{\alpha/2}.$$

Standard tests in the remaining cases are presented in most elementary statistics textbooks, and I leave it to you to check them out if you are interested. In Case 2, where σ_x^2 and σ_y^2 are unknown but m and n are large, the standard procedure resembles what we did in the one-sample problem, namely, it replaces the population variances in the Case 1 test by the corresponding sample variances and then uses the same critical region as above. When the population variances are unknown and not assumed equal, the problem of constructing an exact test of H_0: $\mu_x = \mu_y$ vs. H_1: $\mu_x \neq \mu_y$ is, perhaps surprisingly, an unsolved problem, a rather famous testing scenario known as the *Behrens-Fisher Problem.* Approximate tests by Welsh and by Satterthwaite for the case in which σ_x^2 and σ_y^2 are unknown and at least one of the sample sizes is small are presented in many elementary and intermediate Statistics textbooks. An exact test (the two-sample t test) does exist when one makes the additional assumption that the population variances, while taking unknown values, are known to be equal. (Mother Nature might well be consulted about this assumption before proceeding.) There is, in this instance, a statistical test for investigating the assumption, but as you know, testing a null hypothesis like H_0: $\sigma_x^2 = \sigma_y^2$ can't lead to airtight confirmation that it is true.) The following example illustrates the so called two-sample t test.

Example 10.2.1. The two-sample t test is valid under the following assumptions: Two independent samples $X_1, \dots, X_n \overset{\text{iid}}{\sim} N(\mu_x, \sigma_x^2)$ and $Y_1, \dots, Y_m \overset{\text{iid}}{\sim} N(\mu_y, \sigma_y^2)$, with $\sigma_x^2 = \sigma_y^2 = \sigma^2$, where σ^2 is unknown. Under these assumptions, and under H_0: $\mu_x = \mu_y$, it can be shown that the t statistic shown below has the indicated t distribution:

$$t = \frac{\bar{X} - \bar{Y}}{\sqrt{\frac{s_p^2}{n} + \frac{s_p^2}{m}}} \sim t_{n+m-2},$$

where s_p, the "pooled estimate" of the common variance σ^2, is given by

$$s_p = \frac{(n-1)s_x^2 + (m-1)s_y^2}{n+m-2}.$$

The critical region of size α of the two-sample t test of H_0: $\mu_x = \mu_y$ vs. H_0: $\mu_x \neq \mu_y$ is: Reject H_0 if

$$\left| \frac{\bar{X} - \bar{Y}}{\sqrt{\frac{s_p^2}{n} + \frac{s_p^2}{m}}} \right| > t_{n+m-2,\alpha/2}.$$

Suppose that the math SAT scores have been recorded for 8 randomly chosen psychology majors and 8 randomly chosen sociology majors, yielding these data:

Psych. Majors (X)	560	590	620	700	640	600	540	550
Soc. Majors (Y)	540	580	590	610	550	610	510	650

From these data, we find that $\bar{X} = 600$, $s_x^2 = 2,828.57$, $\bar{Y} = 580$, $s_y^2 = 1,885.71$ and $s_p^2 = 2,377.14$. (Incidentally, were we to test the hypothesis that $\sigma_x^2 = \sigma_y^2$ based on these data, this hypothesis would be "accepted" at any of the standard significance levels, i.e., at $\alpha = .01$ or higher.) In executing the t test, we have that

$$|t| = .7674 < 2.1448 = t_{14,.025}.$$

Thus, the hypothesis H_0: $\mu_x = \mu_y$ cannot be rejected at the .05 significance level. ■

Let us now turn our attention to another commonly occurring inference problem—tests involving a population proportion. The treatment given here is based on the assumption that the population is very large (effectively, infinite) so that our sample of subjects from whom a binary response (e.g., success/failure or yes/no) is obtained may be modeled as a sequence of independent Bernoulli trials. The basic data available is $X_1, \dots, X_n \stackrel{\text{iid}}{\sim} B(1, p)$, and any inference about p will be based on the sufficient statistic $X = \sum_{i=1}^n X_i \sim B(n, p)$, and, more specifically, on the sample proportion $\hat{p} = X/n$. We will also assume that the sample size n is sufficiently large to justify the normal approximation of $\hat{p}$. In testing hypotheses about p, one might be interested in either one-sided or two-sided hypotheses. We will develop the standard test for one of these three problems of interest and display the tests for the two other options without additional discussion.

Under the tacit assumptions made above about a dichotomous population and the sample drawn from it, we will treat the sample proportion as a normal random variable, that is, we will assume that

$$\hat{p} \sim N(p, pq/n), \text{ approximately.}$$

Let us consider testing $H_0 : p \leq p_0$ against the alternative $H_1 : p > p_0$. It is intuitively clear that small values of the statistic $\hat{p}$ favor the null hypothesis, while large values of $\hat{p}$ favor the alternative hypothesis. This suggests that we should reject H_0 when $\hat{p}$ is "too large. Since $P((\hat{p} - p_0)/\sqrt{p_0 q_0/n} > Z_\alpha | \hat{p} \sim N(p_0, p_0 q_0/n)) \approx \alpha$, we may conclude that we should reject H_0 if $\hat{p} > p_0 + \sqrt{p_0 q_0/n} Z_\alpha$. This is the standard critical region used in testing $H_0 : p \leq p_0$ vs. $H_1 : p > p_0$ based on a binomial sample. The three versions of the test for one- and two-sided testing problems are shown in the following table. For simplicity, the test statistic $(\hat{p} - p_0)/\sqrt{p_0 q_0/n}$ is denoted by Z.

H_0	H_1	Critical Region of size α
$p \leq p_0$	$p > p_0$	Reject H_0 if $Z > Z_\alpha$
$p \geq p_0$	$p < p_0$	Reject H_0 if $Z < -Z_\alpha$
$p = p_0$	$p \neq p_0$	Reject H_0 if $\|Z\| > Z_{\alpha/2}$

Table 10.2.2. Standard tests for a population proportion p.

In a two-sample test, the data consists of $X \sim B(m, p_1)$, and $Y \sim B(n, p_2)$, where X and Y are assumed independent. Let $\hat{p}_1 = X/m$ and $\hat{p}_2 = Y/n$. The standard critical region of size α for testing $H_0 : p_1 = p_2$ vs. $H_1 : p_1 \neq p_2$ is

$$\text{Reject } H_0 : p_1 = p_2 \text{ in favor of } H_1 : p_1 \neq p_2 \text{ if } \left| \frac{\hat{p}_1 - \hat{p}_2}{\sqrt{\hat{p}\hat{q}(1/m + 1/n)}} \right| > Z_{\alpha/2}.,$$

where $\hat{p} = (X+Y)/(m+n)$ is the so-called "pooled estimate of p" under the assumption that $p_1 = p_2 = p$ (which is implicit in H_0).

Exercises 10.2.

1. Nutritionist Lon B. U. Lyphe has developed an herbal supplement, and claims that 60-year-olds who begin taking the supplement regularly will live, on average, beyond age 80. A recent FDA study found that the lifetime X of 8 random participants in the Lyphe program were as follows: 72, 75, 75, 76, 78, 81, 83, 84. Let μ be the mean lifetime of practitioners of the Lyphe method. Test H_0: $\mu \geq 80$ against H_1: $\mu < 80$ at $\alpha = .05$.

2. Only 30 out of 80 high school students polled favored a year-around school calendar. If p is the proportion of all high school students in favor of year-around school, test the hypothesis H_0: $p \geq .5$ against $H_1 : p < .5$ at $\alpha = .05$ based on these data.

3. Suppose that independent samples are available from two populations, with
$$X_1, \dots, X_m \overset{\text{iid}}{\sim} N(\mu_x, \sigma_x^2) \quad \text{and} \quad Y_1, \dots, Y_n \overset{\text{iid}}{\sim} N(\mu_y, \sigma_y^2).$$
Use the known properties of moment-generating functions to show that $\bar{X} \sim N(\mu_x, \sigma_x^2/m)$, $\bar{Y} \sim N(\mu_y, \sigma_y^2/n)$, and $\bar{X} - \bar{Y} \sim N(\mu_x - \mu_y, \sigma_x^2/m + \sigma_y^2/n)$.

4. Suppose that independent samples are available from two populations, with
$$X_1, \dots, X_m \overset{\text{iid}}{\sim} N(\mu_x, \sigma_x^2) \quad \text{and} \quad Y_1, \dots, Y_n \overset{\text{iid}}{\sim} N(\mu_y, \sigma_y^2).$$
If $\mu_x \neq \mu_y$, what is the exact distribution of $(\bar{X} - \bar{Y}) / \sqrt{\sigma_x^2/m + \sigma_y^2/n}$?

5. Who's more "liked" by college students today, singer Elvis Presley or actor James Dean? Sociologist May B. Guessing tried to find out through two brief, entertaining biographical Facebook postings (each to a different set of random college "friends"). Suppose that 100 random college students responded to each posting and that the proportion of "likes" were $\hat{p}_1 = .37$ for Presley and $\hat{p}_2 = .50$ for Dean. Test $H_0 : p_1 = p_2$, that is, the population proportions of likes for Presley and Dean are equal, against a two-sided alternative, at $\alpha = .05$.

10.3 Sample Size Requirements for Achieving Pre-specified Power

Just as one might wish to determine the sample size needed to achieve a desired level of precision in a confidence interval for a parameter of interest, one might have a particular goal in a hypothesis test of a given H_0 regarding the power one would wish to achieve at a particular parameter value in the set of alternatives specified in H_1. In this section, I will illustrate the process of finding the needed sample size. For the sake of clarity and concreteness, I will do this in an example involving a normal random sample in a Case 1 problem, that is, a problem in which the population variance is assumed known.

Example 10.3.1. Let us suppose that a soup company produces tens of thousands of cans of soup every day, and that the label on each can states that the contents of the can weighs 12 ounces. The company undergoes periodic inspections by a state agency, and its products are inspected for a variety of issues ranging from the presence of contaminants to the true weight of their cans' contents. On the weight issue, the company must pay a hefty fine if it is found that the content of the sampled cans is, on average, .02 ounces or more below the advertized weight. In short, it is very costly for the company if the state inspector finds reason to believe that the mean net weight of the company's cans of soup is 11.98 ounces or less. To avoid this from happening, the company tests n cans each week with the intention of detecting the fact that $\mu \leq 11.98$. Their data may be represented as $X_1, \ldots, X_n \overset{\text{iid}}{\sim} N(\mu, \sigma^2)$. For simplicity, we will assume that the standard deviation σ is known to be 0.1 ounces. Each week, they test $H_0 : \mu \geq 12$ vs. $H_1 : \mu < 12$ at $\alpha = .01$. They use this small significance level because they don't wish to declare that a batch of soup cans are underweight when in fact their weight is acceptable (that is, their cans' weights meet or exceed their claim, on the average). Having to dispose of a large batch of good cans would be expensive for them. The question that remains to be answered is: How many cans should be tested? The calculation of the sample size n to use in this experiment will be determined by the power function of the associated test. ■

In the abstract, the problem above may be represented as the one-sided testing problem seeking to decide between $H_0 : \mu \geq \mu_0$ and $H_1 : \mu < \mu_0$, based on the measured weights from a random sample of n cans. The company will use the standard test, that is, H_0 will be rejected if $\bar{X} < \mu_0 - (\sigma\sqrt{n})Z_\alpha$, where α is the chosen significance level of the test. The further constraint that must be dealt with is the value of the power function at some important alternative value of μ which, in the present case, is some specific value $\mu_1 < \mu_0$. Let us suppose that we would like the power of the test to be equal to γ if $\mu = \mu_1$. Typically, $0 < \alpha << \gamma < 1$, that is, α would be chosen to be quite small and γ would be chosen to be quite large. The following figure illustrates the relationship between α and γ when H_0 is rejected if $\bar{X} < c$.

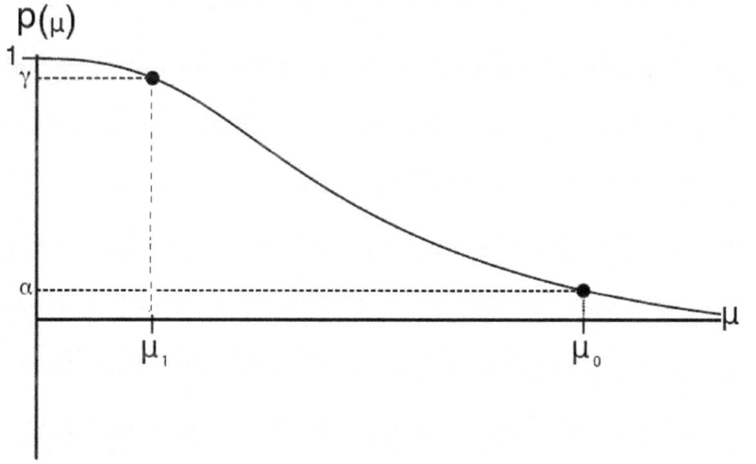

Figure 10.3.1. The power of a test of a normal mean at values $\mu = \mu_0$ and $\mu = \mu_1$.

The selection of the values α and γ give rise to the following equation, where we have written the test statistic above in the convenient standard normal form:

$$P\left(\frac{\bar{X} - \mu_0}{\sigma/\sqrt{n}} < -Z_\alpha \,|\, \mu = \mu_1\right) = \gamma. \tag{10.21}$$

From (10.21), we will determine the necessary sample size n for attaining the desired power at $\mu = \mu_1$. We may rewrite (10.21) as

$$P\left(\frac{\bar{X} - \mu_1 + \mu_1 - \mu_0}{\sigma/\sqrt{n}} < Z_\alpha | \mu = \mu_1\right) = \gamma \quad \text{or}$$

$$P\left(\frac{\bar{X} - \mu_1}{\sigma/\sqrt{n}} < \frac{\mu_0 - \mu_1}{\sigma/\sqrt{n}} - Z_\alpha | \mu = \mu_1\right) = \gamma, \tag{10.22}$$

or, finally, as

$$P\left(Z < \frac{\mu_0 - \mu_1}{\sigma/\sqrt{n}} - Z_\alpha\right) = \gamma, \tag{10.23}$$

where $Z \sim N(0,1)$. Now, since $P(Z < c) = \gamma$ means that $c = Z_{1-\gamma}$, we have that

$$\frac{\mu_0 - \mu_1}{\sigma/\sqrt{n}} - Z_\alpha = Z_{1-\gamma}. \tag{10.24}$$

Solving for n in (10.24) results in the desired equation for the sample size needed for a test of size α for testing $H_0 : \mu \geq \mu_0$ vs. $H_1 : \mu < \mu_0$ will have power γ at the alternative $\mu = \mu_1$:

$$n = \left(\frac{Z_{1-\gamma} + Z_\alpha}{\mu_0 - \mu_1}\right)^2 \sigma^2. \tag{10.25}$$

Example 10.3.1 (continued). In this example, the company would like the power of their test to be low (and has actually set it to be .01) when $\mu = 12$, and it is thus even lower when $\mu > 12$. Let's suppose that the company decides that they would like to be 99% sure of detecting a bad batch of cans. Utilizing the sample size equation in (10.25) with $\alpha = .01$, $\gamma = .99$, $\mu_0 = 12$, $\mu_1 = 11.98$, and $\sigma = .1$, we may compute the required sample size as

$$n = \left(\frac{Z_{.01} + Z_{.01}}{12.00 - 11.98}\right)^2 \sigma^2 = \left(\frac{2.33 + 2.33}{.02}\right)^2 (.1)^2 = 542.89. \tag{10.26}$$

We thus conclude that a test based on a sample of 543 cans from a given batch of soup cans would have probability .99 of detecting that the cans are underweight if the mean weight μ of the cans was actually equal to 11.98. By the way, while "destructive testing" in unavoidable in some testing scenarios, here, the company would simply weigh an empty can and subtract that weight from the sampled cans to determine the weights of the soup that these cans contain.

■

While the problem treated above is a simple one, it outlines the basic principles and tools that would be used in more complex situations. The extension to the normal problem in which the variance σ^2 is unknown is simple enough. One would simply estimate σ^2 by the sample variance s^2 from a preliminary sample and obtain an approximation of the sample size needed for the desired power at a specific value of μ. In other problems, one may follow the same agenda as in the example above. The elements of the analysis require that you know three things: (i) the distribution of the test statistic under the null hypothesis (so you can implement your choice of significance level), (ii) the distribution of the test statistic under the alternative hypothesis, and of particular importance, its distribution under the value of the parameter at which one wishes to set the value of the power function, and (iii) the probabilities α and γ that one wishes to set for the power function under particular parameter values in the null and alternative hypotheses. Of these, item (ii) is the most troublesome, as the distribution of the test statistic when H_0 fails is often complex and may be difficult to use in practice. Fortunately, there is a considerable amount of software available which can help in making the necessary calculations. You will find a host of such software if you search the Internet for "statistical calculators," and most standard statistical packages have sections dedicated to power analysis.

Exercises 10.3.

1. Let $X_1,\dots,X_n \stackrel{\text{iid}}{\sim} N(\mu,\sigma^2)$, where σ^2 is known to be equal to 100. In testing $H_0 : \mu \leq 25$ vs. $H_1 : \mu > 25$, the critical region of size $\alpha = .05$ is: Reject H_0 if $\bar{X} > 25 + (10/\sqrt{n})(1.645)$. What sample size n would be necessary if one wishes to reject H_0 with probability at least .95 if $\mu = 26$?

2. Suppose that a coin is to be tossed n times, and you wish to test the hypothesis $H_0 : p = 1/2$ vs. $H_1 : p > 1/2$ at $\alpha = .05$. What sample size n would be necessary if you want to reject H_0 with probability .90 if $p = .55$? Since the sample size will be reasonably large in this problem, the approximate normality of the sample proportion $\hat{p}$ may be assumed.

10.4 Optimal Tests: The Neyman-Pearson Lemma

Let's begin with the simplest possible testing problem. Suppose we wish to test the null hypothesis H_0: $\theta = \theta_0$ against the alternative H_1: $\theta = \theta_1$ based on a random sample whose distribution depends on the parameter θ. If we had two different critical regions in mind, how would we go about comparing the two tests and making a choice between them? In this simple circumstance, we would revert to the basic characteristics of tests examined in Section 10.1; we would compare the probabilities for type I and type II errors for the two tests. If one test had a smaller α and a smaller β, we would most certainly choose that test, as it would yield a more reliable result, whichever hypothesis happened to be true. Outcomes such as this are quite rare; what typically happens in practice is that the test with the smaller type I error α will have a larger type II error β (since, as you would expect, the increased level of protection against making a type I error comes at a cost). This type of trade-off is a natural consequence of the fact that we are not able, in a testing problem with some fixed sample size, to control both α and β simultaneously. Thus, what we would typically do is to set the significance level of the two competing tests equal to the same value α. Then, we would compare the power of the two tests at θ_1 (recall that the power of a test at θ_1 is equal to $1-\beta$, with β calculated at $\theta = \theta_1$) and choose the test with the larger power. This would be the standard procedure for deciding between two tests. But, in this section, we wish to go well beyond such comparisons. Here, we'll ask a very ambitious question. In the testing problem above, is there one test that's actually better than all others? The fact that the answer to this question is "yes" is fascinating. Even in this simple problem, there tend to be countless tests that could be considered, and one's initial intuition might be that having an airtight argument that shows one particular test is better than all others is too much to ask. This intuition isn't far off, actually; we'll see that the circumstances in which a single best test exists are quite special. But the result that proved the existence of a best test and its exact form will show us an appealing and effective way of proceeding in a rather broad collection of problems. We begin with the famous theorem that laid the foundations for the mathematical theory of hypothesis testing. Note that, in the stated result, both H_0 and H_1 are simple hypotheses, that is, the distribution of the available data is completely specified under both the null and alternative hypotheses. We'll gradually expand the range of applicability of the result. The theorem below is valid for both the discrete and continuous cases. We will state and prove the result for the continuous case. A proof for the discrete case is left as an exercise. The statement below assumes that the proposed test is based on a single real-valued random variable X. In standard applications, X would be a one-dimensional sufficient statistic based on a random sample of size n.

Theorem 10.4.1. (The Neyman-Pearson Lemma) Let X be a continuous random variable with density f_0 under H_0 and f_1 under H_1. Let $k > 0$, and consider the test that rejects H_0 if the observed value $X = x$ satisfies the inequality

$$\frac{f_1(x)}{f_0(x)} > k. \tag{10.27}$$

This test is the best test of its size for testing H_0 vs. H_1. More specifically, if the test in (10.27) has size α, then among all tests with size $\leq \alpha$, this test maximizes the power under H_1.

Remark 10.4.1. The fraction $f_1(x)/f_0(x)$ in (10.27) is well defined for any x for which either $f_0(x) > 0$ or $f_1(x) > 0$. If $f_0(x) = 0$ while $f_1(x) > 0$, the fraction is defined as $+\infty$, and the inequality in (10.27) holds for any positive number k; if $f_1(x) = 0$ while $f_0(x) > 0$, the fraction is equal to 0, and the inequality in (10.27) fails to hold for any positive number k. In the former case, the test will reject H_0 while in the latter case, H_0 will be accepted.

Remark 10.4.2. In the proof below, we'll make use of the following notation. For sets A for which the integral exists,

$$\int_A f(x)dx = \int_{-\infty}^{\infty} I_A(x)f(x)dx.$$

If a test has critical region CR, we may write:

$$\alpha = P(\text{reject } H_0|H_0 \text{ true}) = \int_{CR} f_0(x)dx = \int_{-\infty}^{\infty} I_{CR}(x)f_0(x)dx$$

and

$$1-\beta = P(\text{reject } H_0|H_1 \text{true}) = \int_{CR} f_1(x)dx = \int_{-\infty}^{\infty} I_{CR}(x)f_1(x)dx.$$

Proof of Theorem 10.4.1. Let $CR_{(1)}$ be the critical region specified in (10.27), and let us rewrite this set as

$$CR_{(1)} = \{x|f_1(x) > kf_0(x)\}.$$

Let $CR_{(2)}$ be the critical region of an arbitrary competing test. Assume that the first test has size α and that the second test has size $\leq \alpha$. It is thus assumed that

$$\int_{-\infty}^{\infty} I_{CR_{(2)}}(x)f_0(x)dx = P(CR_{(2)}|H_0) \leq \alpha = P(CR_{(1)}|H_0) = \int_{-\infty}^{\infty} I_{CR_{(1)}}(x)f_0(x)dx. \tag{10.28}$$

The inequality in (10.28) may be rewritten as

$$\int_{-\infty}^{\infty} \left(I_{CR_{(1)}}(x) - I_{CR_{(2)}}(x)\right) f_0(x)dx \geq 0. \tag{10.29}$$

We want to show that the first test has better power than the second test under H_1, that is, we want to show that

$$\int_{-\infty}^{\infty} I_{CR_{(1)}}(x)f_1(x)dx = P(CR_{(1)}|H_1) \geq P(CR_{(2)}|H_1) = \int_{-\infty}^{\infty} I_{CR_{(2)}}(x)f_1(x)dx. \tag{10.30}$$

We may rewrite the inequality in (10.30) as

$$\int_{-\infty}^{\infty} \left(I_{CR_{(1)}}(x) - I_{CR_{(2)}}(x)\right) f_1(x)dx \geq 0. \tag{10.31}$$

We will prove that, for any $k > 0$,

$$\int_{-\infty}^{\infty} \left(I_{CR_{(1)}}(x) - I_{CR_{(2)}}(x)\right)\left(f_1(x) - kf_0(x)\right) dx \geq 0, \tag{10.32}$$

from which the inequality in (10.31) will be shown to follow. The integral in (10.32) is nonnegative because the integrand is nonnegative for all values of x. To see this, consider the two cases below:

(i) If $x \in CR_{(1)}$, then $I_{CR_{(1)}}(x) = 1$, so that

$$I_{CR_{(1)}}(x) - I_{CR_{(2)}}(x) = 1 - I_{CR_{(2)}}(x) \geq 0;$$

moreover, by the definition of $CR_{(1)}$,

$$\frac{f_1(x)}{f_0(x)} > k,$$

so that

$$f_1(x) - kf_0(x) > 0.$$

We thus see that, for all $x \in CR_{(1)}$ (including x values for which $f_0(x) = 0$ while $f_1(x) > 0$), the integrand in (10.32) is the product of two nonnegative functions. Thus, the integrand is nonnegative for such x.

(ii) If $x \notin CR_{(1)}$, then $I_{CR_{(1)}}(x) = 0$ and

$$I_{CR_{(1)}}(x) - I_{CR_{(2)}}(x) = 0 - I_{CR_{(2)}}(x) \leq 0;$$

moreover, by the definition of $CR_{(1)}$,

$$\frac{f_1(x)}{f_0(x)} \leq k,$$

so that

$$f_1(x) - kf_0(x) \leq 0.$$

Thus, for all $x \notin CR_{(1)}$ (including x values for which $f_1(x) = 0$ while $f_0(x) > 0$), the integrand in (10.32) is the product of two nonpositive functions so, again, the integrand is nonnegative for such x.

Together, (i) and (ii) above imply that the inequality in (10.32) holds. Now, let us rewrite (10.32) as

$$\int_{-\infty}^{\infty} \left(I_{CR_{(1)}}(x) - I_{CR_{(2)}}(x)\right) f_1(x)dx - k\int_{-\infty}^{\infty} \left(I_{CR_{(1)}}(x) - I_{CR_{(2)}}(x)\right) f_0(x)dx \geq 0. \tag{10.33}$$

From this, we may establish the inequality in (10.31) and the desired conclusion:

$$\int_{-\infty}^{\infty} \left(I_{CR_{(1)}}(x) - I_{CR_{(2)}}(x)\right) f_1(x)dx \geq k\int_{-\infty}^{\infty} \left(I_{CR_{(1)}}(x) - I_{CR_{(2)}}(x)\right) f_0(x)dx \geq 0, \tag{10.34}$$

the final inequality following from the assumption in (10.29) that the competing test has size less than or equal to the size of the test with critical region $CR_{(1)}$. ■

The Neyman-Pearson Lemma is important for historical reasons, of course, as it was the first successful attempt to introduce the notion of optimality in hypothesis testing; it showed that some tests are better than others. It also provided a constructive proof which demonstrated precisely how "best tests" could be found. The lemma itself, and its extensions, can be usefully applied well beyond the tests of simple hypotheses treated in Theorem 10.4.1. But, as we shall also see, these generalizations do have their limitations.

The intuition behind the Neyman-Pearson test should not be overlooked, both because it is very reasonable, informative, and revealing and because it will form the basis for intuitive extensions which make useful inroads into multiparameter problems in which the lemma is not directly applicable. The Neyman-Pearson Lemma says that the values of the observed random variable X which lead to rejection of the hypothesis $H_0 : f(x) = f_0(x)$ in favor of $H_1 : f(x) = f_1(x)$ are the values x for which $f_1(x)/f_0(x)$ is sufficiently large. It is instructive to graph the ratio $f_1(x)/f_0(x)$ for a given f_0 and f_1. For example, if the constant k in (10.27) were set equal to 1, the N-P test would reject H_0 if and only if $f_1(x) > f_0(x)$, that is, precisely when $X = x$ has a greater likelihood under H_1 than under H_0. If $k = 1/2$, the criterion for rejection is broadened, so that one rejects H_0 only if the likelihood of $X = x$ is at least half as large under H_1 as it is under H_0. On the other hand, when $k = 2$, the requirement for rejecting H_0 is more stringent: H_0 will be rejected if and only if the likelihood of $X = x$ is at least twice as large under H_1 as it is under H_0. Neyman and Pearson's key idea was that one is attracted to the alternative hypothesis when it appears to provide a better explanation of the experimental outcome than does H_0 in terms of relative likelihoods. As is the case with many mathematical results, Neyman and Pearson no doubt thought of this intuition first and then went about finding a way to prove that acting upon this intuition resulted in the best possible testing procedure.

Let's now examine some applications of the Neyman-Pearson Lemma. We begin with an elementary application, a test of two simple hypotheses involving the parameter p of a binomial distribution.

Example 10.4.1. Suppose that you observe a sequence of n iid Bernoulli trials for which the probability of success p is unknown. Let X be the total number of successes obtained in the n trials, which is, of course, a sufficient statistic for p. We know that $X \sim B(n, p)$. Consider testing the simple hypothesis $H_0 : p = 1/2$ against the alternative $H_1 : p = 3/4$. Rejecting the null hypothesis when the observed X is "too big" seems like a reasonable criterion for rejecting H_0. Setting the threshold for rejection for such a test would depend on the significance level we would like to have. To confirm that this type of test is in fact the best thing to do, we will derive the form of a test that is best of its size for testing H_0 vs. H_1. The Neyman-Pearson (N-P) Lemma says that best tests are of the form: Reject H_0 if

$$\frac{f_1(x)}{f_0(x)} = \frac{\binom{n}{x}\left(\frac{3}{4}\right)^x\left(\frac{1}{4}\right)^{n-x}}{\binom{n}{x}\left(\frac{1}{2}\right)^x\left(\frac{1}{2}\right)^{n-x}} = \frac{3^x/4^n}{1/2^n} > k, \tag{10.35}$$

where the choice of the constant $k > 0$ is fixed but arbitrary. We thus may write the critical region in (10.35) as: Reject H_0 if

$$3^x > k', \tag{10.36}$$

where k' is a new constant (defined as $k' = k(2)^n$). Taking the logarithms of both sides of (10.36), we see a Neyman-Pearson test must have a critical region of the form: Reject H_0

if

$$x > k'', \tag{10.37}$$

where k'' is a newer constant (defined as $k'' = \ln k' / \ln 3$). We conclude from (10.37) that best tests have critical regions of the form: Reject H_0 if $X > c$ for some fixed constant c. If $n = 20$ and $X \sim B(20, p)$, then the test that rejects H_0 when $X > 13$ is the best test of size

$$\alpha = P(X \geq 14 | p = 1/2) = .0577 \tag{10.38}$$

for testing $H_0 : p = 1/2$ vs. $H_1 : p = 3/4$. The power of this test under H_1, that is, when $p = 3/4$, is

$$1 - \beta = P(X \geq 14 | p = 3/4) = .7858.$$

■

Remark 10.4.3. The odd-looking value of α in (10.38) is due to the discreteness of the variable we are working with. Rejecting H_0 if $X \geq 13$ yields $\alpha = .1316$, while rejecting H_0 if $X \geq 15$ yields $\alpha = .0207$. Our choices are limited. (See Problems 10.8.58 and 10.8.59 for a discussion about how this issue is usually addressed.)

The example above can be generalized to show that, in testing a small value of p, say p_0, vs. a large value of p, say p_1, based on a binomial variable X, a one-sided test that rejects when X is "too large" will always be the best test of its size. Interestingly, the very same test that is best of size $\alpha = .0577$ for testing $H_0 : p = 1/2$ vs. $H_1 : p = 3/4$ is also best of size $\alpha = .0577$ for testing $H_0 : p = 1/2$ vs. $H_1 : p = 4/5$. The verification of this claim is left as Exercise 10.4.1.

The form of a best test can vary substantially, depending on the model that is assumed to hold for the test statistic one employs. The following example derives the best test of a particular size for the median of a Cauchy variable. As you'll see, a best test can take a surprising and rather odd form.

Example 10.4.2. Let X be a single observation from a Cauchy distribution with median m and scale parameter 1. The density of X is given by

$$f(x) = \frac{1}{\pi} \frac{1}{1 + (x - m)^2} I_{(-\infty,\infty)}(x).$$

Consider testing the simple null hypothesis $H_0 : m = 0$ against the simple alternative $H_1 : m = 1$. We will show that the test which rejects H_0 when the observed value of X is in the interval $(1, 3)$ is best of its size in testing H_0 vs. H_1. Let's determine what form a Neyman-Pearson test will take in this problem. The Neyman-Pearson Lemma states that we should reject H_0 if

$$\frac{f_1(x)}{f_0(x)} = \frac{1/[1 + (x - 1)^2]}{1/[1 + x^2]} > k,$$

where $k > 0$ is fixed but arbitrary. This inequality is equivalent to

$$\frac{x^2 + 1}{x^2 - 2x + 2} > k,$$

which, in turn, may be written as

$$(1 - k)x^2 + 2kx + (1 - 2k) > 0. \tag{10.39}$$

For $k = 1$, (10.39) says that H_0 should be rejected if $x > 1/2$. For $k \neq 1$, the LHS of (10.39) is a quadratic function of x, and the x values at which this function takes the value zero may be obtained from the quadratic formula as

$$x = \frac{-2k \pm \sqrt{4k^2 - 4(1-k)(1-2k)}}{2(1-k)}. \tag{10.40}$$

For $k > 1$, the quadratic function in (10.39) is a parabola that opens downward, and if the function has two real zeros a and b, then the inequality in (10.39) is satisfied if and only if $x \in (a,b)$. For $k = 2$, the zeros identified through (10.40) are 2 ± 1, so that the test which rejects H_0 if $X \in (1,3)$ is best of its size for testing H_0 vs. H_1. See Exercise 10.4.2 for a quite different critical region when $k < 1$. ■

We use the following example, which involves one-sided testing problems about the mean of a normal population, as a vehicle for extending the range of the Neyman-Pearson Lemma from tests of simple hypotheses to tests in which both H_0 and H_1 are composite hypotheses. First, we define the following important concept.

Definition 10.4.1. Let $X_1, \dots, X_n \overset{\text{iid}}{\sim} F_\theta$, and consider a testing problem involving the two composite hypotheses $H_0 : \theta \in \Theta_0$ and $H_1 : \theta \in \Theta_1$, where $\Theta_0 \cap \Theta_1 = \emptyset$ and $\Theta_0 \cup \Theta_1 = \Theta$ and where the parameter space Θ is assumed to contain the true value of θ. Test A (or critical region A) based on $X_1, \dots, X_n$ is said to be the *Uniformly Most Powerful* (UMP) test of size α if test A has size α and, for any alternative test with size $\leq \alpha$, test A has the same or greater power than the alternative test at every $\theta \in \Theta_1$.

Example 10.4.3. Let $X_1, \dots, X_n \overset{\text{iid}}{\sim} N(\mu, \sigma^2)$, where σ^2 is assumed known. Since the sample mean $\bar{X}$ is a sufficient statistic for μ, tests for the hypothesis testing to be treated here will be based on $\bar{X}$. Consider first the problem of testing the simple hypotheses $H_0 : \mu = \mu_0$ vs. $H_1 : \mu = \mu_1$, where for concreteness, we take $\mu_0 < \mu_1$. Neyman-Pearson tests of H_0 vs. H_1 will reject H_0 if

$$\frac{\left(\frac{1}{\sqrt{2\pi}\sigma/\sqrt{n}} \cdot e^{-(\bar{x}-\mu_1)^2/(2\sigma^2/n)}\right)}{\left(\frac{1}{\sqrt{2\pi}\sigma/\sqrt{n}} \cdot e^{-(\bar{x}-\mu_0)^2/(2\sigma^2/n)}\right)} > k \tag{10.41}$$

for some constant $k > 0$. The inequality in (10.41) in each of the successively simpler forms is displayed below. Using the notation "$\Leftrightarrow$" for "is equivalent to," we have that (10.41) $\Leftrightarrow$

$$\begin{aligned}
&e^{[(\bar{x}-\mu_0)^2 - (\bar{x}-\mu_1)^2]/(2\sigma^2/n)} > k \\
&\Leftrightarrow [(\bar{x}-\mu_0)^2 - (\bar{x}-\mu_1)^2]/(2\sigma^2/n) > \ln k = k' \\
&\Leftrightarrow \bar{x}^2 - 2\bar{x}\mu_0 + \mu_0^2 - \bar{x}^2 + 2\bar{x}\mu_1 - \mu_1^2 > 2\sigma^2 k'/n = k'' \\
&\Leftrightarrow 2\bar{x}(\mu_1 - \mu_0) > k'' + \mu_1^2 - \mu_0^2 = k''' \\
&\Leftrightarrow \bar{x} > k'''/2(\mu_1 - \mu_0) = k''''(\text{since } (\mu_1 - \mu_0) > 0).
\end{aligned} \tag{10.42}$$

From (10.42), it is evident that any test of the form: Reject H_0 if $\bar{X} > c$ for some fixed c is the best test of its size for testing $H_0 : \mu = \mu_0$ vs. $H_1 : \mu = \mu_1$, where $\mu_0 < \mu_1$. Since, under H_0, $\bar{X} \sim N(\mu_0, \sigma^2)$, we have that the critical region of the best test of size α for testing H_0 vs. H_1 is:

$$\text{Reject } H_0 \text{ if } \bar{X} > \mu_0 + (\sigma/\sqrt{n})Z_\alpha. \tag{10.43}$$

■

Example 10.4.3 (extended). Now, suppose that we were interested in testing $H_0 : \mu = \mu_0$ vs. $H_1 : \mu = \mu_2$, where μ_2 differs from μ_1 but is also greater that μ_0, that is, $\mu_0 < \mu_2$. If we were to repeat the algebra above for deriving the form of best tests of H_0 vs. this new H_1, we would find that the critical region of a best test is again of the form "Reject H_0 if $\bar{X} > c$ for some fixed c". Since the null hypothesis is still $H_0 : \mu = \mu_0$, the best test of size α will again have the critical region displayed in (10.43). It should be clear that the best test of size α is the same, regardless of the alternative to μ_0 that is considered, provided that it is a value larger than μ_0. Since the test with critical region in (10.42) has size α and has maximal power at every alternative value of μ that is larger than μ_0, it follows that this test is the UMP test of size α for testing $H_0 : \mu = \mu_0$ vs. $H_1 : \mu > \mu_0$.

■

Example 10.4.3 (extended some more). Let us now examine how the test with critical region given in (10.43) behaves for values of μ that are smaller than μ_0. Keep in mind that $\bar{X} \sim N(\mu, \sigma^2/n)$, with σ^2 known, so that under the assumption that the mean μ takes a particular value μ^*, we may compute $P(\bar{X} > c|\mu = \mu^*)$ for any fixed constant c. Note that for any constant c, $P(\bar{X} > c|\mu = \mu^*)$ is an increasing function of μ^*. This follows from the simple observation that the area to the right of c under the $N(\mu, \sigma^2)$ density increases as μ moves from left to right on the real line. Of particular interest to us here is the probability $P(\bar{X} > \mu_0 + (\sigma/\sqrt{n})Z_\alpha|\mu = \mu^*)$ when $\mu^* < \mu_0$. Since this probability is an increasing function of μ^* and takes the value α when $\mu^* = \mu_0$, it follows that $P(\bar{X} > \mu_0 + (\sigma/\sqrt{n})Z_\alpha|\mu = \mu^*) < \alpha$ for all $\mu^* < \mu_0$. We thus see that

$$\max_{\mu \leq \mu_0} P(\bar{X} > \mu_0 + (\sigma/\sqrt{n})Z_\alpha|\mu) = \alpha.$$

Thus, the test with critical region given in (10.43) has size α in testing $H_0 : \mu \leq \mu_0$ vs. $H_1 : \mu > \mu_0$. We have already determined that this test maximizes the power at every $\mu > \mu_0$ among all tests with size (that is, power) no greater than α at $\mu = \mu_0$. Since any test with power $\leq \alpha$ at all $\mu \leq \mu_0$ necessarily has power $\leq \alpha$ at $\mu = \mu_0$, the test specified in (10.43) has maximal power at every $\mu > \mu_0$ among all tests with size $\leq \alpha$ in testing $H_0 : \mu \leq \mu_0$ vs. $H_1 : \mu > \mu_0$. We thus have that the "standard test" for testing $H_0 : \mu \leq \mu_0$ vs. $H_1 : \mu > \mu_0$ when σ^2 is known (the one-sided testing problem for a normal mean classified in Section 10.2 as Case 1) is the uniformly most powerful test of H_0 vs. H_1.

■

We conclude this section with some comments about some related problems. First, it should be mentioned that there is a sizable class of one-sided testing problems in which any one-sided test based on a sufficient statistic is the uniformly most powerful test of its size. The standard test for $H_0 : p \leq p_0$ vs. $H_1 : p > p_0$ (see Table 10.2.2) is an example of a UMP test, even though the size of the test only approximates the true size because of the normal approximation used. Some other examples are examined in the problems at the end of this chapter. It is worth mentioning that the Neyman-Pearson Lemma provides useful insights in other testing problems. For example, the Lemma can be used to show that the standard two-sided test concerning a normal mean is not a UMP test of H_0: $\mu = \mu_0$ vs. H_1: $\mu \neq \mu_0$ in the Case 1 version of that testing problem. This follows from the fact that the one-sided test with critical region in (10.43) maximizes the power for every $\mu > \mu_0$ among all tests with power α at $\mu = \mu_0$, and so it necessarily has better power than the standard two-sided test of size α when $\mu > \mu_0$. However, the one-sided test with CR in

(10.43) has terrible power (i.e., always less than α) for $\mu < \mu_0$. The one-sided test that rejects H_0: $\mu = \mu_0$ if $\bar{X} < \mu_0 - (\sigma/\sqrt{n})Z_\alpha$ is the UMP for testing $H_0 : \mu = \mu_0$ vs. $H_1 : \mu < \mu_0$ and achieves maximal power at all $\mu < \mu_0$. Since two different tests maximize the power above and below μ_0 when testing the hypothesis $H_0 : \mu = \mu_0$, it follows that a UMP test does not exist in the two-sided problem. When no universally best test exists, a common next step is to restrict the class of tests considered and then to seek a best test in the restricted class. An intuitively attractive restriction is the following.

Definition 10.4.2. A test is called *unbiased* if its power function $p(\theta)$ satisfies the inequality

$$\max_{\theta \in H_0}\{p(\theta)\} \leq \min_{\theta \in H_1}\{p(\theta)\}, \tag{10.44}$$

that is, if the probability of rejecting the null hypothesis is never higher when H_0 is true than it is when H_0 is false.

It can be shown that the two-sided test for a normal mean is the uniformly most powerful unbiased (UMPU) test (that is, the best possible test in the class of unbiased tests) in the Case 1 version of the problem of testing $H_0 : \mu = \mu_0$ vs. $H_1 : \mu \neq \mu_0$. Also, when σ^2 is unknown parameter, the one-sided t-test can be shown to be the UMPU test for testing $H_0 : \mu \leq \mu_0$ vs. $H_1 : \mu > \mu_0$. The Neyman-Pearson theory of hypothesis testing may be extended further (including, for example, locally best tests that maximize power at alternative parameter values "in a neighborhood" of the null hypothesis and uniformly most powerful invariant (UMPI) tests that place certain structural restrictions on the test statistic used). A detailed discussion of such extensions is beyond the scope of this book. In the next section, we turn our attention to procedures called likelihood ratio tests, which are very much in the spirit of the Neyman-Pearson approach to testing but are justified by their large sample behavior rather than by any type of fixed-sample size optimality.

Exercises 10.4.

1. Let $X \sim B(20, p)$. Consider testing $H_0 : p = 1/2$ in favor of $H_1 : p = 4/5$ based on the observed value of X. Show that the test which rejects H_0 if $X > 13$ is the best test of size .0577 for testing H_0 vs. H_1.

2. As in Example 10.4.2, let X be a single observation from a Cauchy distribution with median m and scale parameter 1. Show that the test which rejects $H_0 : m = 0$ in favor of $H_1 : m = 1$ if either $X > 0$ or $X < -2$ is best of its size for testing H_0 vs. H_1.

3. Let $X_1, \dots, X_n \overset{\text{iid}}{\sim} P(\theta)$, the Poisson distribution with mean θ. Consider testing the null hypothesis $H_0 : \theta = 1$ against $H_1 : \theta = 2$. Show that the test that rejects H_0 when $\bar{X} > c$, where c is an arbitrary positive constant, is best of its size for testing H_0 vs. H_1. Suppose that $n = 8$. Calculate the size α of the test that rejects H_0 if $\bar{X} > 1.5$. Calculate the power of this test when $\theta = 2$.

4. Let $X_1, \dots, X_n \overset{\text{iid}}{\sim} Exp(\lambda)$, the exponential distribution with failure rate λ. Show that the test which rejects the null hypothesis $H_0 : \lambda = 1$ in favor of the alternative $H_1 : \lambda = 2$ when $\bar{X} < k$ for a fixed constant $k > 0$ is best of its size.

10.5 Likelihood Ratio Tests

While the Neyman-Pearson Lemma will lead to the identification of optimal tests when testing two simple hypotheses, and it admits to extensions to certain testing scenarios involving composite hypotheses, it has its limitations. For example, in the last section, we noted that no uniformly most powerful test exists in the problem of testing the "pinpoint hypothesis" H_0: $\mu = \mu_0$ vs. the alternative H_1: $\mu \neq \mu_0$, where μ is a normal mean. However, under the additional restriction of unbiasedness, both the Z test and the t test can be shown to be UMPU tests in the separate "cases" in which they apply. But there are many testing problems which are too complex to be treated fruitfully using the Neyman-Pearson Lemma. Even then, however, the basic intuition of the lemma proves useful in constructing tests.

In testing the hypotheses H_0: $\theta \in \Theta_0$ vs. $H_1 : \theta \in \Theta_1$, where H_0 and H_1 are complex composite hypotheses, tests which reject the null hypothesis when the ratio of the likelihoods $L(x_1, \dots, x_n|\theta \in \Theta_0)/L(x_1, \dots, x_n|\theta \in \Theta_1)$ is small are intuitively reasonable. While this is a good first thought about such problems, there are some practical difficulties in dealing with such a statistic, not the least of which is the fact that the ratio will vary as θ varies within Θ_0 in the numerator and θ varies within Θ_1 in the denominator. The likelihood ratio test we now present constitutes a useful refinement of the initial intuition above.

Let us suppose that a random sample is available from a distribution F_θ, where θ may be a scalar- or vector-valued parameter. It is thus assumed that we will observe $X_1, \dots, X_n \stackrel{\text{iid}}{\sim} F_\theta$ for $\theta \in \Theta$, where Θ is the entire parameter space. When F_θ is the exponential distribution with mean $\theta, \Theta = (0, \infty)$, while when F_θ is the $N(\mu, \sigma^2)$ distribution, $\Theta = \{(\mu, \sigma^2) \in (-\infty, \infty) \times (0, \infty)\}$. Now, consider testing $H_0 : \theta \in \Theta_0$ vs. $H_1 : \theta \in \Theta_1$, where $\Theta_0 \cup \Theta_1 = \Theta$. Let $L(x_1, \dots, x_n|\theta)$ be the likelihood function of the observed data. We view the likelihood as a function of θ alone, with the observed sample being fixed constants. Then the likelihood ratio statistic λ is defined as

$$\lambda(\mathbf{x}) = \frac{\max_{\theta \in \Theta_0} L(\mathbf{x}|\theta)}{\max_{\theta \in \Theta} L(\mathbf{x}|\theta)}. \tag{10.45}$$

The statistic in (10.45) merits some comments. First, note that the ratio does not compare H_0 with H_1 directly, but rather, it compares how large the likelihood can be made under H_0 with how large it can be made when θ is free to take any value in the entire parameter space Θ. Since the denominator involves a maximization over a larger collection of values of θ, we must have that $0 \leq \lambda(\mathbf{x}) \leq 1$ for all $\mathbf{x}$. Secondly, note that the evaluation of the likelihood ratio statistic λ involves the derivation of maximum likelihood estimators under two sets of assumptions. If $\hat{\theta}_0$ is the MLE of θ among $\theta \in \Theta_0$ and $\hat{\theta}$ is the MLE of θ among $\theta \in \Theta$, then λ may be written as

$$\lambda(\mathbf{x}) = \frac{L(\mathbf{x}|\hat{\theta}_0)}{L(\mathbf{x}|\hat{\theta})}. \tag{10.46}$$

The numerator of (10.46) represents the best explanation of the data that can be provided by a parameter value $\theta \in \Theta_0$, while the denominator of (10.46) is the best explanation of the data that can be provided by the model F_θ for $\theta \in \Theta$. Thus, when $\hat{\theta} \in \Theta_0, \lambda(\mathbf{x}) = 1$, and $\lambda(\mathbf{x})$ being close to 1 may be interpreted as the saying that the null hypothesis provides a good explanation of the data, relative to what is possible under the general model F_θ. It is thus natural to reject H_0 when the likelihood ratio statistic $\lambda(\mathbf{x})$ is too small. This is in fact the usual critical region of the likelihood ratio test. However, it is typical that a slightly different form of this test is used in practice. As it turns out, it is a simple function of $\lambda(\mathbf{x})$ that has a familiar asymptotic distribution. A simple version of this problem, where θ is a parameter pair (θ_1, θ_2), is stated in the following theorem. A more general version of this theorem is stated later in this section.

Theorem 10.5.1. Let $\Theta = \{(\theta_1, \theta_2), \theta \in A \subseteq R^2\}$, where R represents the real line and R^2 represents the Euclidean plane. Suppose we wish to test $H_0 : \theta_1 = \theta_1^*$ (a fixed constant) vs. $H_1 : \theta_1 \neq \theta_1^*$ based on $X_1, \ldots, X_n \overset{\text{iid}}{\sim} F_\theta$. Then under suitable regularity conditions (i.e., conditions needed for the asymptotic theory of MLEs to hold), and under H_0,

$$-2\ln\lambda(\mathbf{x}) \xrightarrow{D} X \sim \chi_1^2 \text{ as } n \to \infty, \tag{10.47}$$

where $\lambda(\mathbf{x})$ is the likelihood ratio statistic in (10.45).

A prototypical application of the theorem above is the problem of testing the null hypothesis $H_0 : \mu = \mu_0$ vs. $H_1 : \mu \neq \mu_0$ based on a random sample from $N(\mu, \sigma^2)$ with both parameters unknown. This is the subject of the following:

Example 10.5.1. Let $X_1, \ldots, X_n \overset{\text{iid}}{\sim} N(\mu, \sigma^2)$. The likelihood ratio test for $H_0 : \mu = \mu_0$ vs. $H_1 : \mu \neq \mu_0$ is based on the statistic

$$\lambda(\mathbf{x}) = \frac{\max_{\sigma^2} \left(\frac{1}{2\pi\sigma^2}\right)^{n/2} e^{-\sum_{i=1}^n (x_i-\mu_0)^2/(2\sigma^2)}}{\max_{\mu,\sigma^2} \left(\frac{1}{2\pi\sigma^2}\right)^{n/2} e^{-\sum_{i=1}^n (x_i-\mu)^2/(2\sigma^2)}}. \tag{10.48}$$

The numerator of (10.48) is obtained by maximizing the log likelihood with respect to σ^2, holding μ_0 fixed, a process which yields

$$\ln L = -\frac{n}{2}\ln(2\pi\sigma^2) - \frac{\sum_{i=1}^n (x_i - \mu_0)^2}{2\sigma^2}$$

$$\frac{\partial}{\partial\sigma^2}\ln L = -\frac{n}{2\sigma^2} + \frac{\sum_{i=1}^n (x_i - \mu_0)^2}{2(\sigma^2)^2} = 0,$$

an equation with the unique solution

$$\hat{\sigma}^2 = \frac{\sum_{i=1}^n (x_i - \mu_0)^2}{n}.$$

The second derivative test shows that $\ln L$ is indeed maximized by $\hat{\sigma}^2$. The denominator of (10.48) is obtained by the usual MLE derivation. The results are the MLEs

$$\hat{\mu} = \bar{X} \text{ and } \hat{\sigma}^2 = \frac{\sum_{i=1}^n (x_i - \bar{x})^2}{n}.$$

Plugging is the respective MLEs for the parameters μ and σ^2 in $\lambda(\mathbf{x})$ yields, after some cancelations,

$$\lambda(\mathbf{x}) = \left(\frac{\sum_{i=1}^n (x_i - \bar{x})^2}{\sum_{i=1}^n (x_i - \mu_0)^2}\right)^{n/2}.$$

The asymptotic theory of the likelihood ratio test leads to rejecting $H_0 : \mu = \mu_0$ at significance level α if $-2\ln\lambda(\mathbf{x})$ is too large, or, more specifically, if

$$-2\ln\lambda(\mathbf{x}) = -n\ln\frac{\sum_{i=1}^n (x_i - \bar{x})^2}{\sum_{i=1}^n (x_i - \mu_0)^2} > \chi^2_{1,\alpha}.$$

The t test for testing $H_0 : \mu = \mu_0$ vs. $H_1 : \mu \neq \mu_0$ based on a normal sample of size n rejects H_0 if

$$|t| = \left| \frac{\bar{X} - \mu_0}{s/\sqrt{n}} \right|$$

is too large. One can show algebraically that

$$[\lambda(\mathbf{x})]^{2/n} = \frac{1}{1 + t^2/(n-1)}, \tag{10.49}$$

so that the t statistic is large when the likelihood ratio statistic $\lambda(\mathbf{x})$ is small. Thus, the likelihood ratio test in this problem is, in fact, the t test. ■

Likelihood ratio tests are often of interest in problems in which your stochastic model has r unknown parameters, where r is relatively large, and you wish to test the null hypothesis that some of these parameters have fixed, known values. The technique lends itself to the treatment of such problems, since the basic tool for implementing these tests is the derivation of maximum likelihood estimators under two different conditions, plus the execution of some algebra which reduces the test statistic to some reasonably manageable form. In such cases, Theorem 10.5.1 generalizes to the following.

Theorem 10.5.2. Let $\Theta = \{(\theta_1, \dots, \theta_k), \theta \in A \subseteq R^k\}$, where R^k represents k-dimensional Euclidean space. Suppose we wish to test $H_0 : \theta_{i_j} = \theta^*_{i_j}$ for $j = 1, \dots, s$ vs. $H_1 : \theta_{i_j} \neq \theta^*_{i_j}$ for at least one $j \in \{1, \dots, s\}$ based on $X_1, \dots, X_n \overset{\text{iid}}{\sim} F_\theta$, where $\{\theta^*_{i_1}, \dots, \theta^*_{i_s}\}$ are fixed, known values. Then under suitable regularity conditions (i.e., conditions needed for the asymptotic theory of MLEs to hold), and under H_0,

$$-2\ln\lambda(\mathbf{x}) \xrightarrow{D} X \sim \chi^2_s \text{ as } n \to \infty, \tag{10.50}$$

where $\lambda(\mathbf{x})$ is the likelihood ratio statistic in (10.45). In other words, if the parameter of the model satisfying the null hypothesis is $k - s$ dimensional and the parameter of the general, unconstrained model is k dimensional, then the convergence in distribution in (10.50) holds.

The linear models studied in the next chapter are typically indexed by multiple parameters, and likelihood ratio tests arise naturally there in testing hypotheses of interest regarding the model parameters. We will, for example, show that the standard F test in one-way analysis of variance (ANOVA) may be derived as a likelihood ratio test.

There is an extensive literature on the asymptotic performance of likelihood ratio tests. For alternative parameter values θ_1 that are close to H_0, it is possible to obtain a large-sample approximation for the power of the likelihood ratio test at θ_1. When such a θ_1 is treated as the true value of the parameter, the likelihood ratio statistic has an asymptotic distribution known as the "non-central" chi-square distribution. Details on this approximation and on power calculations using it may be found in Ferguson (1996). Asymptotic properties of likelihood ratio tests, under suitable regularity conditions, include an asymptotic "local UMP" property, meaning that as $n \to \infty$, the test achieves the best possible power at alternatives sufficiently close to H_0. An asymptotic property of tests that is also of interest is that of consistency. A test is said to be *consistent* if its power function $p(\theta)$ tends to zero at all parameter values in the interior of H_0 (written $\theta \in \Theta_0^I$) and tends to 1 at all parameter values in the interior of H_1 (written $\theta \in \Theta_1^I$). For tests, consistency just means that the probability that H_0 is rejected tends to 1 if H_0 is false and tends to 0 if H_0 is true (with boundary values of H_0 and H_1 excluded, since the power function remains constant at the predetermined value α for such values). Under mild regularity conditions, the likelihood ratio test is consistent. (See Stuart and Ord (2010) for details.)

Exercises 10.5.

1. Let $X_1,\dots,X_n \stackrel{\text{iid}}{\sim} N(\mu,\sigma_0^2)$, where σ_0^2 is known. Consider testing $H_0:\mu=\mu_0$ vs. $H_1:\mu\neq\mu_0$. Show that the test which rejects H_0 if $|Z|=|(\bar{X}-\mu_0)/(\sigma_0/\sqrt{n})|$ is too large, that is, the standard two-sided test for $H_0:\mu=\mu_0$ in Case 1, is the likelihood ratio test of H_0 vs. H_1.
2. In Example 10.5.1, verify the identity in (10.49) which shows that the two-sided t test for $H_0:\mu=\mu_0$ vs. $H_1:\mu\neq\mu_0$ based on a normal random sample is equivalent to the likelihood ratio test of H_0 vs. H_1.
3. Let $X_1,\dots,X_n \stackrel{\text{iid}}{\sim} B(1,p_1)$ and let $Y_1,\dots,Y_m \stackrel{\text{iid}}{\sim} B(1,p_2)$, where the Xs and Ys are assumed to be independent. Derive the likelihood ratio statistic for testing $H_0:p_1=p_2$ vs. $H_1:p_1\neq p_2$ and identify its asymptotic distribution.
4. Consider independent samples from k potentially distinct Poisson distributions, where $X_{i,1},\dots,X_{i,n_i} \stackrel{\text{iid}}{\sim} P(\lambda_i)$ for $i=1,\dots,k$. Let λ be a fixed positive constant. Derive the likelihood ratio statistic for testing $H_0:\lambda_i=\lambda$ for all i vs. $H_1:\lambda_i\neq\lambda_j$ for some $i\neq j$, and identify its asymptotic distribution.

10.6 Testing the Goodness of Fit of a Probability Model

We typically begin the process of statistical inference by postulating a model for the experimental data we will observe. We might, for example, model count data as Poisson variables, physical measurements as $N(\mu,\sigma^2)$ variables, or lifetime data as $\Gamma(\alpha,\beta)$ variables. Although we have referred to the modeling of data as the first stage in the inference process, there's a natural question that arises at this stage which we have postponed discussing until now. Before the model is assumed to hold, it is often necessary to ask the question: "Is the model correct?" The question is often phrased in a different but equivalent way, namely, "Does the model fit the data well?" The question is usually answered by determining whether or not the experimental data would be considered "rare" when the model is assumed to hold. The goal of this section is to help you understand how you might test the goodness of fit of a parametric model. We will conclude this discussion with an example of how one can test the hypothesis that the available data is normally distributed. But I will begin by treating a somewhat simpler problem.

Let us suppose we want to test the proposition that a particular six-sided die is balanced. It should be apparent that this proposition is equivalent to specifying the equi-probable model for the six possible outcomes of a given roll. In other words, if X is the digit facing upwards when the die is rolled and if $p_i=P(X=i)$ for $i=1,2,3,4,5,6$, the proposition that the die is balanced simply stipulates that $p_i=1/6$ for each i. To test this hypothesis, we would roll the die repeatedly to see if the outcomes tended to support this hypothesis. The data we would collect satisfies the requirements of a multinomial experiment, and the vector $\mathbf{X}=(X_1,\dots,X_6)$ representing the number of occurrences of the outcomes $1,\dots,6$ in n rolls of the die has the $M_6(n,\mathbf{p})$ distribution. We can define the null hypothesis of interest as

$$H_0 : p_i = 1/6 \text{ for } i = 1, \ldots, 6,$$

while the alternative hypothesis is the obvious one,

$$H_1 : p_i \neq 1/6 \text{ for some i.}$$

The motivation for the test we will treat here—the chi-square goodness-of fit-test—is the fact that if the null hypothesis is true, then the frequency of occurrence of each of the possible values of the variable X should be close to the expected frequency predicted by the model. The expected frequencies derived from the multinomial model are easily obtained. Since the marginal distributions of the number X_i of occurrences of the digit i in n trials is the $B(n, p_i)$ distribution, we have that $EX_i = np_i$. To test any specific multinomial model with a fully specified parameter **p**, we would compare these expected frequencies of the possible outcomes, which we will denote by e_i, with the observed frequencies, denoted by o_i, in the n trials. This leads to a table of expected and observed frequencies which represents the basic information about the model and the experiment that will be required in testing the goodness of fit of the model. If, for example, we rolled a die 60 times, we could record expected and observed frequencies under the hypothesis H_0 above as in the following table.

Outcome→	1	2	3	4	5	6
e_i	10	10	10	10	10	10
o_i	11	16	8	7	13	5

Table 10.6.1. Data for testing whether or not a die is balanced.

There are quite a few ways you could go about quantifying the extent to which the expected and observed frequencies in Table 10.6.1 are "close"; we will discuss just one. The chi-square goodness-of-fit test has a number of properties that recommend it for use. It is often referred to as an "omnibus test," meaning that it is widely applicable, applying to virtually any parametric model. It also has a well-established large-sample distribution theory. Finally, the test is intuitively sensible and is easy to apply. It is true that there are tests that are more powerful when testing the goodness of fit of a particular model; in fact, the literature on such matters is vast, but the topic will be left to the interested reader.

The chi-square goodness-of-fit test is based on the following statistic, referred to as the chi-square statistic:

$$\chi^2\text{stat} = \sum_{i=1}^{k} \frac{(o_i - e_i)^2}{e_i}. \tag{10.51}$$

A few comments justifying the use of this particular test statistic are in order. First, notice that the chi-square statistic is indeed a measure of the distance between the o's and the e's. The statistic is necessarily non-negative. Now, if $o_i = e_i$ for every i, the statistic will take the value 0. A small value of the statistic indicates that the model fits the data quite well, and such a result would generally be taken as evidence that the model is reasonable (if not exactly correct). You might wonder why a simple alternative statistic like the sum of squared differences of the o's and e's, that is $\sum(o_i - e_i)^2$, is not used instead of the statistic in (10.51). The reason is that some compensation for the scale of the o's and e's seems wise, since without some adjustment for scale, the terms $(100 - 105)^2$ and $(10 - 5)^2$ would be considered as equal in importance, an inappropriate weighting since, in many applications, the first difference would be considered small while the second difference would probably be seen as sizeable. Dividing such terms by the corresponding expected frequency helps us recognize when one difference is more important than another.

Executing the chi-square test will require some discussion of the behavior of the chi-square test statistic in (10.51). The exact distribution of the statistic is not easy to derive and is in general, not available in closed form. It is known, however, that the statistic converges in distribution to a particular chi-square distribution as the sample size n tends to ∞. This result is stated without proof in the theorem below.

Theorem 10.6.1. Consider a multinomial experiment with k "cells" in which the random vector $\mathbf{X} \sim M_k(n, \mathbf{p})$ is observed. For $i = 1, \dots, k$, let $e_i = np_i$ and denote the observed value of X_i as o_i. Then, as $n \to \infty$,

$$\chi^2\text{stat} = \sum_{i=1}^{k} \frac{(o_i - e_i)^2}{e_i} \xrightarrow{D} X \sim \chi^2_{k-1}. \tag{10.52}$$

Naturally, to apply this result, we would need to have some way of judging whether our sample size can be considered to be sufficiently large for the limiting distribution in Theorem 10.6.1 to be useable. A commonly used rule of thumb in this regard is to consider n as sufficiently large if the condition $np_i \geq 5$ holds for all i. This rule is known to be quite conservative, as it has been shown that the asymptotic approximation implicit in Theorem 10.6.1 is generally reliable even when there are a limited number of violations of these conditions. The conditions are nonetheless a useful guide in applied work, and it is a good practice to plan your experiments so that the n is large enough to satisfy the lower bound of 5 for each np_i. After all, the larger n is, the better is the chi-square approximation of the distribution of the chi-square statistic.

We will use the usual notation for the cutoff points of the chi-square distributions. Our notation is formally defined in the following:

Definition 10.6.1. The α-cutoff point of the χ^2_k distribution is denoted by $\chi^2_{k,\alpha}$ and represents the number such that

$$P(X > \chi^2_{k,\alpha}) = \alpha, \tag{10.53}$$

where $X \sim \chi^2_k$.

The cutoff points for the chi-square distribution with k degrees of freedom are given in Table A.3 in the Appendix. Because chi-square distributions are not symmetric, their cutoff points are shown for both small and large values of α.

We are now in a position to complete our analysis of the null hypothesis that a particular die is balanced. Suppose we wish to test this hypothesis at the 5% significance level. The test requires the calculation of the chi-square statistic from the data in Table 10.6.1 and the comparison of the value obtained with the 5% cutoff point of the approximating chi-square distribution. If the $\chi^2\text{stat} > \chi^2_{5,.05} = 11.07$, the hypothesis "$H_0$: the equiprobable model is correct" would be rejected at the 5% significance level. The test statistic is computed as

$$\begin{aligned}\chi^2\text{ stat} &= \frac{(11-10)^2}{10} + \frac{(16-10)^2}{10} + \frac{(8-10)^2}{10} + \frac{(7-10)^2}{10} + \frac{(13-10)^2}{10} + \frac{(5-10)^2}{10} \\ &= 1/10 + 36/10 + 4/10 + 9/10 + 9/10 + 25/10 \\ &= 8.4.\end{aligned}$$

We note that the χ^2stat is 8.4 and does not exceed the threshold for rejection of H_0. Thus, even though the various departures from the expected frequency of 10 look a bit suspicious, they are not, collectively, sufficiently large to constitute convincing evidence that the die is not balanced.

Most often, the chi-square goodness-of-fit test is used to investigate the fit of a parametric model to a random sample of experimental data. The null hypothesis in such applications is that some member of a specific family of distributions is the correct model for the data. For instance, it is common to encounter the need to investigate the hypothesis H_0: the data is normally distributed. The hypothesis simply means that there are particular values μ_0 and σ_0^2 such that the random sample is drawn from the $N(\mu_0, \sigma_0^2)$ model. This type of application differs from the example above in that, once the data is separated into cells, the exact expected frequency of each cell is not immediately known. The expected frequencies are obtained by first estimating the normal parameters μ and σ^2 from the data and then using the curve with the estimated parameters to calculate the probabilities $\{p_i\}$ associated with each cell. The expected frequencies are then obtained as $\{np_i\}$, where n is the sample size.

The distribution of the chi-square statistic when testing a model with unknown parameters involves some complications. The asymptotic distribution of the statistic is known when the unknown parameters are estimated by the "minimum chi-square (MCS)" method, a method used to describe the process of minimizing the criterion function

$$\chi^2\text{stat} = \sum_{i=1}^{k} \frac{(o_i - e_i(\theta))^2}{e_i(\theta)}, \tag{10.54}$$

where θ is the unknown scalar or the vector-valued parameter of the model. The process of obtaining the MCS estimator is tedious, and, in practice, the unknown parameters are usually estimated by the MLE $\hat{\theta}$ and the expected frequencies $e_i(\theta)$ in (10.54) are replaced by $e_i(\hat{\theta})$ in computing the chi-square statistic. This replacement results in a statistic whose asymptotic distribution is known only approximately. More specifically, it is known that the tail probabilities of the version of the chi-square statistic based on the MLE of θ lie somewhere between the tail probabilities of the χ^2_{k-1} and the χ^2_{k-1-r} distributions, where r is the number of estimated parameters. This approximation is due Chernoff and Lehmann (1954) and often provides enough information to carry out a useful approximate test in an application of interest. What we know is that when the χ^2stat is calculated using the MLE of θ based on the entire (ungrouped) sample, we have

$$P(X_{k-1-r} > t) \le P(\chi^2\text{stat} > t) \le P(X_{k-1} > t), \tag{10.55}$$

where $X_s \sim \chi^2_s$. It follows from (10.55) that when a chi-square statistic is obtained from grouped data with expected frequencies based on the fitted r-parameter model with its parameters estimated by MLEs, it is valid to reject (at significance level α) the null hypothesis that the parametric model is correct if the chi-square statistic exceeds $\chi^2_{k-1,\alpha}$, as this guarantees that the p-value of the goodness-of-fit statistic is less than α. On the other hand, if the chi-square statistic is smaller than $\chi^2_{k-1-r,\alpha}$, this null hypothesis may be "accepted" (with the usual interpretation), since in this case the p-value of the goodness-of-fit statistic is necessarily larger than α. Finally, if

$$\chi^2_{k-1-r,\alpha} \le \chi^2 stat \le \chi^2_{k-1,\alpha},$$

the test is inconclusive, and an alternative goodness-of-fit procedure should be executed. The chi-square goodness-of-fit test of a normal model with both parameters unknown is illustrated in the following:

Example 10.6.1. Standardized tests like the Scholastic Aptitude Test, or SAT, (now called the SAT Reasoning Test) are taken each year by several million high school seniors. Consider the following hypothetical experiment. A random sample of 200 students was drawn from among those who took last year's SAT, and their scores (from 0 to 800) from each of two sections, the quantitative (formerly "mathematics") and the critical reading (formerly "verbal") sections, are summed. The overall performance of these 200 students in the math and verbal sections is shown below, grouped into nine ranges.

Score (in hundreds)	<6	6–7	7–8	8–9	9–10	10–11	11–12	12–13	>13
Observed frequency	11	20	24	38	48	36	14	6	3

Table 10.6.2. Math plus verbal SAT scores for 200 students, with scores shown in hundreds.

Treating the 200 SAT scores as independent $N(\mu,\sigma^2)$ variables, suppose that the MLEs of the population mean and standard deviation were found to be 950 and 200, respectively. The probabilities of the nine ranges of scores in Table 10.6.2 under the $N(950,(200)^2)$ density is shown in the following table.

Score (in hundreds)	< 6	6–7	7–8	8–9	9–10	10–11	11–12	12–13	>13
Probability	.04	.07	.12	.17	.20	.17	.12	.07	.04

Table 10.6.3. Probabilities under the $N(950,(200)^2)$ curve for 9 ranges of scores, in hundreds.

The two tables above yield the following observed and expected frequencies:

Score (in hundreds)	<6	6–7	7–8	8–9	9–10	10–11	11–12	12–13	>13
Obs. Freq. (o_i)	11	20	24	38	48	36	14	6	3
Exp. Freq. (e_i)	8	14	24	34	40	34	24	14	8

Table 10.6.4. Observed and expected frequencies for 200 grouped SAT scores, in hundreds.

The chi-square GoF statistic in (10.51) may thus be computed as

$$\begin{aligned}\chi^2 \text{ stat} &= \frac{9}{8}+\frac{36}{14}+\frac{0}{24}+\frac{16}{34}+\frac{64}{40}+\frac{4}{34}+\frac{100}{24}+\frac{64}{14}+\frac{25}{8}\\ &= 17.7478.\end{aligned}$$

From Table A.3, we find that $\chi^2_{6,.05} = 12.59$ and $\chi^2_{8,.05} = 15.51$. Since the chi-square statistic exceeds 15.51, we would reject H_0: "Math plus verbal SAT scores are normally distributed" at the .05 significance level. Since $\chi^2_{6,.01} = 16.81$ and $\chi^2_{8,.01} = 20.09$, we note that the chi-square GoF test is inconclusive at the .01 significance level. ■

Exercises 10.6.

1. When operating properly, a casino roulette wheel will land on the outcomes red, black, and green with probabilities $9/19$, $9/19$, and $1/19$, respectively. Suppose that you witnessed 95 consecutive spins of the roulette wheel at Shakey Jake's Casino in Las Vegas and you observed that these three possible outcomes occurred with the frequencies shown below.

Outcome	Red	Black	Green
Obs. Freq.	38	47	10

Using the significance level $\alpha = .05$, test the hypothesis that Shakey Jake's roulette wheel is operating properly.

2. Mendelian inheritance theory indicates that seed types A, B, C, and D will occur in the ratio $9:3:3:1$ when two varieties of peas are crossed. Pea Wee Harmon performed a pea-crossing experiment in which 160 seeds were planted and the following frequencies were observed.

Seed Type	A	B	C	D
Obs. Freq.	80	25	35	20

Using the Mendelian theory as a model, test the null hypothesis that the model is correct at $\alpha = .01$.

3. The lifetimes (in hours) of 60-watt filament light bulbs are thought to be exponentially distributed. One hundred such light bulbs were tested by the manufacturer, resulting in the following grouped lifetime data.

Lifetime	0–300	300–600	600–900	900–1200	> 1200
Observed Frequency	25	15	10	5	5

Test, at $\alpha = .05$, the hypothesis that light bulbs of this type are exponentially distributed. Assume that, under the assumption of exponentiality, the MLE of the mean lifetime calculated from the sample of 100 bulbs was found to be 500 hours.

4. When he was in the fourth grade, Jack B. attempted 5 regular free throws in his backyard sports court after school every day. (He follows with ten reverse layups, just for fun, ten alley-oop free throws in honor of his grandpa's hero, Rick Barry, and ten sky hooks in honor of his grandpa's other hero, Kareem Abdul Jabbar. Then he goes in and does his homework, in honor of his mom and dad.) Assume that each day's regular free throws are independent of any other day's. In 180 school days, the frequency with which he made k free throws, for $k = 0, 1, 2, 3, 4, 5$, is shown in the table below.

Free Throws Made	0–2	3	4	5
Observed Frequency	10	30	70	70

Test, at $\alpha = .05$, the hypothesis that the number of free throws X_i that Jack makes each day is a binomial (i.e., $B(5, p)$) variable. (Note: The MLE of p based on the ungrouped data is simply the total number of free throws made divided by 900. For these data, take as given that $\hat{p} = 736/900$.)

10.7 Fatherly Advice about the Perils of Hypothesis Testing (Optional)

In this section, I will discuss the most common objections to and reservations about hypothesis testing. You might wonder why I would take the time, at this point, to take pot shots, rather gingerly, at the theory and approach that we been discussing so assiduously in the first six sections of this chapter. The reason is that there are misuses and misinterpretations of hypothesis testing methodology that are quite common in practice, and it is important that you know about them and try to avoid them in your own applications of the methodology. Following the "critique" of hypothesis testing in the next five paragraphs, I will make some comments on sensible ways to avoid some of the pitfalls mentioned and on alternative approaches which can assist a researcher in accomplishing his/her goals without resorting to a formal hypothesis test.

(1) It must be acknowledged that the development of a hypothesis test is a process that is vulnerable to various forms of abuse. Whenever data is available, one could easily identify some null hypothesis that would be rejected at significance level .05 (or any other level you wish to operate at, for that matter). The proper use of testing methodology requires that H_0 and H_1 be selected prior to having access to the experimental data on which the test will be based. Unfortunately, when test results are reported, it is often difficult or impossible to tell how H_0 and H_1 were chosen. Unless the reasons for testing a particular pair of hypotheses H_0 and H_1 are clearly discussed, the confidence one should have in the implications of the study will be unclear. The greatest threat to the validity of a test is the practice of "data-peeking," that is, looking at the data before you formulate your null and alternative hypotheses. You can't control what other investigators do, but I suggest that you make yourself a promise that, in any formal hypothesis tests you carry out in the future, you'll follow the simple "no-peeking" rule. In addition, for the benefit of the readers of a report on a test you have carried out, I'd suggest that you explain why your null hypothesis and its alternative are of interest in the problem you are studying. If they are the traditional or natural hypotheses to use in your study, you should say so.

(2) To make my second point extra-clear, I'll restrict attention to the testing of a pinpoint hypothesis, for example, the hypothesis that a particular freshly minted coin is fair. In Section 10.2, we discussed the standard test of this very hypothesis (namely, $H_0 : p = 1/2$) based on the observed variable $X \sim B(n,p)$. Let's now take note of the fact that this null hypothesis is known *a priori* to be false. In the real world, there is no such thing as a fair coin. For a coin to be fair, the probability of heads would have to be equal to the infinite decimal $.500000000\ldots$ Since the number of coins that have been minted in human history is finite, the chance that one of them has this exact probability of heads is nil. Thus, when setting up a test of the null hypothesis $H_0 : p = 1/2$, it must be recognized that we already know that H_0 is false. It's clear that what we really mean to do is to ask the question: Is p close enough to .5 for us to consider the coin to be "essentially fair." But then the adjusted null hypothesis would vary from application to application, and might be $p \in (.49, .51)$ for one person and $p \in (.45, .55)$ for someone else. Going back to the original pinpoint hypothesis, we next note that the power function of the standard test (or any reasonable test) will tend to 1 at any alternative value of p that differs from $.500000000\ldots$ This means that if we have a sufficiently large sample size, we are sure to

reject H_0. This is OK, of course, since we already know that H_0 is false. But the obvious question remains: why did we bother with this pointless exercise in the first place?

(3) A criticism similar to that put forward in (2) above applies to virtually all hypothesis tests. No matter what H_0 is selected, there are alternative values of the parameter of interest that are so close to a boundary point θ_0 of the null hypothesis that they are, in a practical sense, indistinguishable from θ_0. For example, in testing the hypothesis $H_0 : p \leq .5$ against $H_1 : p > .5$, where p is a binomial parameter, we would have to consider the value $p = .5001$ to be so close to satisfying the null hypothesis that we might well be reluctant to call H_0 false if the true value of p was in fact .5001. Still, if this was the true value of p, the standard test of H_0 vs. H_1, which rejects H_0 when $\hat{p}$ is "too large," would lead to a rejection of H_0 with limiting probability 1 as the sample size $n \to \infty$. One way to think about such a rejection of H_0 is to recognize it as a departure from the null hypothesis that is *statistically significant* but not *practically significant*. Concern about rejecting H_0 when the true parameter is very close to H_0 often motivates a "sample size calculation" on the part of the investigator. This calculation would begin with the process of identifying a parameter value θ_1 in H_1 that is far enough away from H_0 for the distance to be considered "practically important." The investigator would then calculate, using the methods discussed in Section 10.6, the sample size required to achieve suitably high power (say .95) at the parameter value θ_1. Doing this does two things: it makes it likely that the test will detect a departure from the null hypothesis that is considered important and it also keeps the power of the test from growing too large if the departure from H_0 is less striking. This may sound like a perfectly acceptable way of handling the problem of practical vs. statistical significance. The conundrum introduced by this strategy is this: the apparent resolution suggests that a particular sample size would be good, but that a larger sample size would cause difficulties and that a huge sample size would cause huge difficulties. However, our intuition tells us that we should be wary about any statistical methodology that indicates that increasing the sample size would be unwise. Since this dilemma lies at the very core of hypothesis testing, it can be resolved only by using a different approach to the problem. We'll discuss this possibility later in this section.

(4) As we have seen, when testing two hypotheses H_0 and H_1, the standard protocol controls the probability α of making a type I error at the expense of allowing the probability β of making a type II error to be potentially quite large. This setup allows us to have a reasonable amount of confidence in rejecting H_0 when the test statistic falls in the critical region of the test, as the probability of doing so when H_0 is true is, by design, quite small. However, when H_0 is not rejected, we may say, for convenience, that we "accept H_0," but because the probability of failing to reject H_0 when H_1 is true can be large, we cannot have much confidence that H_0 is true when we have failed to reject it. All this notwithstanding, it is not uncommon for an investigator to interpret the failure to reject H_0 as evidence that H_0 is true. A good example of this is the test for goodness of fit treated in Section 10.6. Suppose that 50 randomly selected type A light bulbs are tested until failure. An investigator might consider a Weibull model as an appropriate description of the lifetime behavior of light bulbs of this type. If a chi-square goodness-of-fit test is carried out on these data, it might well lead to the acceptance of the null hypothesis that the Weibull model is correct. Should the investigator feel confident in analyzing the data as if it was drawn from a Weibull distribution? What one often finds is that there are one

or more alternative models that will also fit the data quite well. The same data might lead to accepting a gamma model for the data, and a lognormal model might also pass the goodness-of-fit test. If that turned out to be the case, what could we conclude from our testing? First, we would need to consider the possibility that the goodness-of-fit test used, based on a sample of size 50, does not have very high power and that the probability of detecting departures from any of these models is small unless the departure itself is quite dramatic. Thus, one way to address this dilemma is to increase the sample size and rerun the test. It is likely that one or more of the models you are entertaining will drop out of the competition. It is not unreasonable to have greater confidence in the Weibull model if it is "the last one standing" in a process such as this. The basic reservation we have about treating an "accepted null hypothesis" as true remains, of course, as there is no way to prove statistically that a particular model is correct in describing the sampled data in a given experiment. Moreover, if one takes a large enough sample, any fixed parametric model will be rejected as a null hypothesis simply because no hypothesized model in a random experiment will be exactly correct.

(5) The final reservation that I will discuss about accepting the outcome of a hypothesis test at face value is sometimes referred to as the "file drawer" problem. Imagine 20 researchers doing the same experiment concerning some cutting-edge problem in their field. Suppose that each of them is testing the hypothesis that the mean measurement μ is at least equal to 12. Suppose that the true value of μ is actually 12 so that, in reality, the null hypothesis is true. If all twenty researchers test this null hypothesis at significance level .05, then it is quite likely (with actual probability .6415) that at least one of these researchers will reject H_0 at $\alpha = .05$. Now, it is a standard practice for subject-matter journals to publish a paper describing the outcome of a statistical study only if the paper reports a statistically significant result. In this instance, the researcher(s) who obtained a significant result (at $\alpha = .05$) would typically submit a paper announcing a significant outcome of the hypothesis test, while the other researchers would simply toss their notes on the experiment in a file drawer, labeling the test as "inconclusive." The journal in this field would publish the "significant outcome," and it would be treated as a scientific finding to be believed and built upon. Later, it is possible that rumors would emerge that other researchers have been unable to replicate the result, and this might lead to a careful reconsideration of the original null hypothesis. However, such reexaminations of testing results are, at best, haphazard and do not appear to happen with great frequency.

The criticisms and/or reservations about hypothesis testing raised above are real and, in my opinion, quite well taken. There are, however, a number of good reasons why you should know about hypothesis testing methods, about the construction of optimal tests, and about some of the tests that arise most frequently. First, in spite of the criticisms of hypothesis testing above, and others that I have not mentioned, the methodology is widely used in scientific applications. Because of this, anyone who needs to stay apprised of advances in a particular area of applied science should be able to assess the validity of published test results and to understand the elements of the construction and properties of the tests used. Secondly, there are most certainly situations in which hypothesis tests are valid and useful. Clearly, when one needs to determine if one procedure (like a new treatment for a disease) is better that an alternative procedure (like the currently favored treatment), a one-sided test which sets "current is as good as new" as H_0 and "new is better than current" as H_1 is a natural vehicle for making the determination. Thirdly, while tests in which the truth of the null hypothesis is of special interest would appear to run contrary to the way in which hypothesis tests are designed to work, they are nonetheless informative, both because one can sometimes discover that a contemplated model provides, perhaps unexpectedly, a poor fit of the available data, but also because the simultaneous testing of a collection of competing models may result in some clarity with regard to which model is the most appropriate.

Among the guidelines that the above critique of testing suggest are (i) specify your choice of null and alternative hypotheses before your data becomes available, and defend the choice as the natural and/or traditional specification when this is possible, (ii) don't test simple null hypotheses, (iii) avoid taking the acceptance of a null hypothesis as evidence that H_0 is true, but follow up such a test outcome with further tests aimed at investigating whether there are alternative hypotheses that are also supported by the data. The consequences of accepting a particular null hypothesis as true should be carefully studied. In particular, when several models remain competitive after multiple goodness-of-fit tests, the consequences of each of the potential modeling assumptions on the estimation of parameters of interest (such as the population mean and variance, for example) should be investigated and compared.

Criticisms like (3) and (5) above are not easily addressed. I will not say more about the file drawer problem, as there is no simple solution to this problem, and it is mentioned here as food for thought rather than as a call to action. Regarding (3), it should be noted that the development of a confidence interval for a parameter of interest can often shed light on the questions of practical and statistical significance. Before elaborating on this point, let me mention a simple and direct connection between hypothesis tests and confidence intervals. (This connection was first alluded to in Example 8.1.5.) The connection is easiest to understand in the context of pinpoint hypothesis testing, so I will discuss that particular case in detail. The pertinent ideas can be easily extended to more general tests. Let's suppose that we are interested in testing hypotheses of the form $H_0 : \theta = \theta_0$ vs. $H_1 : \theta \neq \theta_0$ at $\alpha = .05$ based on the test statistic T, a function of a random sample from a one-parameter family of distributions $\{F_\theta\}$. Now, if (θ_L, θ_U) is a 95% confidence interval for θ, we may think of the interval as the collection of all values of the parameter θ that are consistent with the observed value of the statistic T at the probability level .05. In reverse, if one were to test, for various values of θ_0, the hypotheses $H_0 : \theta = \theta_0$ vs. $H_1 : \theta \neq \theta_0$ at $\alpha = .05$, the values of θ_0 for which the observed value of the statistic T falls in the acceptance region of the test are precisely the values of the parameter θ which are compatible with the observed T. Thus, one may formulate a "confidence statement" about θ on the basis of these values. Typically, these values will be an interval of real numbers, and the interval (θ_L, θ_U) constitutes a 95% confidence interval for θ. This latter process is usually referred to as "inverting a hypothesis test" to construct the associated confidence interval. Whether or not the confidence interval derived by this type of inversion yields the confidence interval that is most commonly in use depends on precisely how the test was constructed, but it is often the case that the standard tests for a parameter θ will yield, by inversion, the standard confidence interval.

As an example of the ideas above, suppose you are interested in hypothesis testing problems involving a bent coin with an unknown probability of heads p. Suppose that you flip the coin a bunch of times and the observed sample proportion is $\hat{p}$. Then, assuming n is suitably large, an approximate 95% confidence interval for p has the confidence limits

$$\hat{p} \pm \sqrt{\hat{p}\hat{q}/n}(1.96). \tag{10.56}$$

The interval bounded by these limits can be thought of as all values of the parameter p that are consistent with the data or, equivalently, with the sample proportion $\hat{p}$, at the probability level .05. This statement is exactly the same as the statement that the observed value of $\hat{p}$ would be in the acceptance region of the standard test for $H_0 : p = p_0$ vs $H_1 : p \neq p_0$ for any value of p_0 in the confidence interval in (10.56). Recall that the acceptance region for testing a particular value of p is precisely the collection of values of the statistic $\hat{p}$ is for which $|\hat{p} - p| \leq \sqrt{\hat{p}\hat{q}}(1.96)$. Clearly, the values of p for which this will happen are precisely the values in the interval $(\hat{p} - \sqrt{\hat{p}\hat{q}/n}(1.96), \hat{p} + \sqrt{\hat{p}\hat{q}/n}(1.96))$, the standard large-sample 95% confidence interval for p.

If, as discussed above, hypothesis tests and confidence intervals are so intimately related, why and when would a confidence interval be a more informative statistical procedure than the corresponding test. Consider the following example. Suppose you are interested in testing $H_0 : p \leq 1/2$ against $H_1 : p > 1/2$, and suppose that $\hat{p} = .51$. If the sample size n is sufficiently large, the null hypothesis would be rejected at the chosen level α on the basis of this observation. The corresponding confidence interval might be something like $(.503, .517)$. If you were simply using the confidence interval as another way to carry out the hypothesis test, you would reject the null hypothesis, as suggested by the fact that the $100(1 - \alpha)\%$ confidence interval contains only parameter values that belong to the alternative hypothesis. But the confidence interval allows one to make a more nuanced judgment. In this example, we see that while the confidence interval and test indicate that the observed outcome is statistically significant, we also see that the true value of p appears to be very close to the boundary value $p = .5$ of the null hypothesis. We are then able to make a judgment about whether the departure from H_0 is of practical significance. Depending on the purpose of the study, one might decide to proceed under the assumption that p is greater than $1/2$, but you might decide, alternatively, to treat the departure as of minor importance. For example, if I was prepared to take some cumbersome action, like refinancing my home provided that the probability of particular financial benefits was well over $1/2$, I would probably consider an outcome like the one above to be insignificant in the practical sense, and it would probably lead to a decision not to refinance.

My final comment on this general topic has to do with goodness-of-fit testing. My main intent here is to call your attention to the many graphical techniques that exist for investigating the goodness of fit of a probability model. The simplest of all these methods is the widely used process of drawing a histogram of your data. This involves creating "cells" or "bins" based on intervals of real numbers, counting the frequency of occurrence of each cell for the experimental data obtained, and drawing successive rectangles over each cell with heights equal to these frequencies. A histogram is usually most informative if each of the intervals used to classify the data is of the same length. The shape of the histogram can tell you a great deal about whether a model you are considering fits the data. For example, if the data was drawn from a normal population, the histogram for your data should look reasonably bell-shaped, that is, should look like a rough approximation of a normal density. Of course, one needs more than unimodality and approximate symmetry of the histogram to confirm the normality of the data, but these two properties are, by themselves, a good start toward that end. In addition, it is helpful to check that the relative frequency counts of the intervals bounded by $\bar{X} \pm k(s/\sqrt{n})$ are close to the probabilities of these intervals under the $N(\bar{X}, s^2)$ density. Besides comparing the histogram of your grouped data to the density of the model you are considering, there are many plots of transformed data that are known to take particular shapes if the data belongs to a certain family of distributions.

For example, it is known that, if data is drawn from an exponential distribution with some unspecified failure rate λ, then $-\ln \bar{F}(x) = \lambda x$, that is, the negative of the logarithm of the survival function $\bar{F}(x) = 1 - F(x)$ is a straight line through the origin. A plot of $-\ln \bar{F}_n(x)$, where $\bar{F}_n(x)$ is the empirical survival function given by

$$\bar{F}_n(x) = \frac{1}{n} \sum_{i=1}^{n} I_{(x,\infty)}(X_i),$$

is often used to check whether the assumption of exponentiality is reasonable. Many similar graphical vehicles are available for checking other distributional assumptions. While formal hypothesis tests are not typically involved in these procedures, they are rather widely recognized as useful evidence for or against the assumption that a particular family of parametric distributions is appropriate as a model in a given application.

Exercises 10.7.

1. Reread Section 10.7 and write a brief (200-word) essay on the pros and cons of hypothesis testing.

10.8 Chapter Problems

1. Local TV weather forecaster Seymour Storms tends to overestimate the chances of rain. In investigating this rumor, the station manager decides to keep track of X, the number of rainy days out of the next four days for which Seymour predicts a $2/3$ chance of rain. (Note that $X \sim B(4, p)$, where $p =$ the probability of rain.) The manager plans to reject $H_0 : p = 2/3$ in favor of the alternative $H_1 : p = 1/2$ if $X \leq 1$. (a) Compute the probability α of type I error using this test. (b) Compute the probability β of type II error using this test.
2. The distance that a randomly chosen state worker commutes to work is thought to be normally distributed. Two legislative analysts disagree on the exact distribution of this random variable. One believes that it is $N(10,4)$, while the other thinks it's $N(15,9)$. The analysts have agreed on the following way of testing the hypotheses $H_0 : X \sim N(10,4)$ vs. $H_1 : X \sim N(15,9)$. One state worker will be chosen randomly, and H_0 will be rejected if $X > 12$. (a) Find the probability of type I error associated with this test. (b) Find the probability of type II error associated with this test.
3. The registrar of a college in California's central valley was overheard bragging at a cocktail party that the average high school GPA of his college's freshman class was $\mu = 3.5$, with a standard deviation of $\sigma = .2$. A reporter who happened to overhear the remark decided to test the registrar's claim about μ. Assuming that the GPAs are normally distributed with $\sigma = .2$, the reporter interviewed a random sample of 16 freshmen at the college and found that their GPAs had an average of $\bar{X} = 3.32$. In testing $H_0 : \mu = 3.5$ vs. $H_1 : \mu < 3.5$, what p-value would be associated with the reporter's findings?
4. Last year, School Superintendent Frida B. Mee changed the curriculum in her school district. Ms. Mee's students took three standardized tests per day. Last year, 46% of the students in her district scored above grade level on the statewide test. This year, in a random sample of 256 students from Ms. Mee's district (the sample size required by the state's "bonus" program), 128 students scored above grade level. Let p be the proportion of Ms.

Mee's current students who score above their grade level this year. In order for Ms. Mee's school district to qualify for a \$100,000 bonus, the district must show a statistically significant improvement in students scoring above grade level. Test the hypothesis $H_0 : p \leq .46$ against the alternative $H_1 : p > .46$ at $\alpha = .05$. Will Ms. Mee's district get the bonus?

5. In a random sample of 100 college freshmen, 63 opined that an introductory course in statistics should be required of every undergraduate (with the possible exception of the students in their particular major). Last year, 70 randomly selected college freshmen held the same view. Test the hypothesis that the proportions of freshman from these two classes who have this opinion are equal, against a two-sided alternative, at $\alpha = .05$.
6. You observe a single X. Two different rejection regions are under consideration for testing the hypotheses $H_0 : X \sim N(0,1)$ vs. $H_1 : X \sim N(1,4)$: (a) Reject H_0 if $|X| > 1.96$ and (b) Reject H_0 if $X > 1.64$. Both tests have significance level $\alpha = .05$. Which is the better test, that is, which test has a smaller probability β of type II error?
7. Assume that the data $X_1, X_2, \ldots, X_{16}$ are independently drawn from a normal population with known standard deviation $\sigma = 12$. The mean of this sample was found to be 64.2. Find the p-value of $\bar{X} = 64.2$ when testing $H_0 : \mu = 60$ against $H_1 : \mu > 60$.
8. Horticulturist I. C. Bloom claims that the average germination time for the tulip bulbs I bought from him is 40 days. My 16 tulip bulbs took an average of $\bar{X} = 44$ days to germinate and had a standard deviation of $s = 9$ days. Assuming that germination times are normally distributed, use my data to test $H_0 : \mu = 40$ vs. $H_1 : \mu > 40$ at $\alpha = .05$.
9. Suppose you toss a bent coin 200 times and obtain 116 heads. Test the null hypothesis $H_0 : p = 1/2$, that is, that the coin is "fair," against the alternative $H_1 : p \neq 1/2$ at significance level $\alpha = .02$.
10. McDougals' Restaurant claims that their McDougalburger contains 14 ounces of meat, after cooking. Consumer advocate Will B. Acrate accepts McDougals' claim that the population standard deviation of these weights is $\sigma = 2$ oz., but plans to test the claim that $\mu = 14$ oz. Suppose that Will takes a sample of 16 McDougalburgers and that his test rejects $H_0 : \mu = 14$ in favor of $H_1 : \mu = 12$ if and only if the sample mean is less that 12.75 ounces. (a) Compute the probability α of making a type I error with this test. (b) Compute the probability β of making a type I error with this test.
11. Let X be the number facing up when a die is rolled. For a balanced die, each integer $1, 2, 3, 4, 5, 6$ occurs with probability $1/6$. For a loaded die, the probability distribution of X is given below.

X	1	2	3	4	5	6
$P(X)$	.05	.05	.4	.4	.05	.05

Suppose you plan to roll a particular die once and will reject H_0: "the die is balanced" in favor of H_1: "the die is loaded" if $X = 3$ or 4. (a) Calculate α, the probability of type I error for this test. (b) Calculate β, the probability of type II error for this test.

12. The IRS has taken a random sample of the number of shares of the various stocks held personally by financier Rob deBank. They believe that the population of all his holdings is normal with an unknown mean μ and a standard deviation $\sigma = 100$. Based on a sample of size 25 of deBank's stocks, they will test the hypothesis $H_0 : \mu = 400$ (he's honest) against the rumored alternative $H_1 : \mu = 450$ (he's been embezzling his clients) using the critical region: Reject H_0 if $\bar{X} > 430$. (a) Calculate α, the probability of type I error for this test. (b) Calculate β, the probability of type II error for this test.

13. Arizona State University and UC Santa Barbara are widely considered to be the top party schools in the country. A recent poll asked random juniors at each school whether or not they were going to a wild and crazy party this weekend. Thirty of the 50 UCSB students said yes, while 45 of the 60 ASU students said yes. Let p_A and p_{SB} be the corresponding proportions for all ASU and UCSB juniors who plan to attend a wild and crazy party this weekend. Using the data above, test $H_0 : p_A = p_{SB}$ against $H_1 : p_A \neq p_{SB}$ at $\alpha = .05$.
14. A coin is tossed 1000 times, and the outcome heads occurred 560 times. Is this outcome sufficient evidence to declare that the coin is biased in favor of heads?
15. Dewey Disk likes playing Frisbee Golf while riding a skate board in New York's Central Park. The course has exactly 5 targets. Dewey plays the game every day that he isn't in bed with game-related injuries. The probability p that Dewey gets a "yahoo" on a given target (i.e., hits it on his first try) is unknown. His dad believes $H_0 : p = .6$, while his mom favors $H_1 : p = .8$. They plan to settle the issue based on Dewey's first game tomorrow. If X is number of yahoos Dewey gets in the game, then $X \sim B(5, p)$. They've agreed to reject H_0 in favor of H_1 if $X \geq 4$. (a) Calculate α, the probability of type I error of this test. (b) Calculate β, the probability of type II error of this test.
16. Find the p-value of the observation $X = 7$, when $X \sim B(10, p)$ and you are testing (a) $H_0 : p = 1/2$ vs. $H_1 : p > 1/2$ or (b) $H_0 : p = 1/2$ vs. $H_1 : p \neq 1/2$. (Hint: The second p-value is larger.)
17. Presidential contenders Morgan Madison and Madison Morgan agree on a lot of things, e.g., foreign policy, the economy, the best color for the Oval Office. But they disagree about the proportion p of super delegates (SDs) who will vote for a nominee for president based on the majority vote in their state primaries. Morgan Madison believes $H_0 : p = 0.5$, while Madison Morgan strongly believes in the independence of SDs and thus espouses the hypothesis $H_1 : p = 0.3$. Morgan and Madison have jointly hired a polling firm to resolve the issue. The firm will randomly sample 100 SDs and will reject H_0 if $\hat{p} < 0.4$, where $\hat{p}$ is the proportion of these SDs who will vote in accordance with their state's majority. Treat $\hat{p}$ as approximately normal. (a) Calculate α, the probability of type I error for this test. (b) Calculate β, the probability of type II error for this test. (c) What is the p-value associated with the outcome $\hat{p} = .416$?
18. The lifetime in years of a newly developed laptop is well modeled by the "Gnarly distribution" F_θ with density function

$$f_\theta(x) = \frac{\theta}{(1+x)^{\theta+1}} \text{ for } x > 0.$$

Deal Computers makes two versions of its laptop—a cheap one (with $\theta = 2$) and a super-cheap one (with $\theta = 4$). A friend got you one for your birthday, but you don't know which one it is. You plan to test $H_0 : \theta = 2$ vs. $H_1 : \theta = 4$ based on the lifetime X of your new laptop. You've chosen the critical region: reject H_0 if $X < 0.1$ years. Show that this test is best of size 0.173 for testing H_0 vs. H_1.
19. Consider testing the hypotheses $H_0 : X \sim Be(3, 1)$ against $H_1 : X \sim Be(2, 1)$ based on the single observation X. (a) Find the best test of size .05. (b) Calculate the power of this test when H_1 is true.
20. Consider testing the hypotheses $H_0 : X \sim U(0, 1)$ against $H_1 : X \sim U(0, 2)$ based on the single observation X. (a) Describe the collection of all tests of size $\alpha = 0$. (b) Identify the test of size 0 that achieves the largest possible power when H_1 is true.
21. Find the best test of size $\alpha = .01$ for testing $H_0 : X \sim U(0, 1)$ vs. $H_1 : X \sim N(0, 1)$. Calculate the power of this test when H_1 is true.

22. Suppose that $X \sim \Gamma(\alpha,\beta)$. Consider testing the hypotheses $H_0 : X \sim \Gamma(3,1)$ against $H_1 : X \sim \Gamma(2,1)$ based on the single observation X. Find the best test of size .05. (Hint: Recall the relationship between that gamma and the Poisson distributions.)

23. Suppose that the single observation X is drawn from a distribution F_θ with pdf

$$f_\theta(x) = 2(1-\theta)x + \theta \text{ for } 0 < x < 1,$$

where θ can take any value in the interval $[0,2]$. (a) Find the best test of size α for testing $H_0 : \theta = 1/2$ vs. $H_1 : \theta = 3/2$. (b) Compute the probability β of making a type II error.

24. Wes Knyle, an entomologist whose integrity is questionable, is studying properties of mosquitoes that are known to carry the West Nile virus. He is interested in testing the pinpoint hypothesis $H_0 : \theta = \theta_0$ against $H_1 : \theta \neq \theta_0$. He has determined to test the hypothesis repeatedly at significance level α until H_0 is finally rejected. When it is rejected, Knyle will write a research paper on the experiment and will submit the very interesting result for publication in a highly regarded Entomology journal. (a) What is the overall size of Knyle's testing procedure? (b) If H_0 is true, what is the expected number of experiments that Knyle will have to perform to obtain a result that is significant at level α?

25. Suppose that a random sample of size $n = 25$ was drawn from a normal population with unknown mean μ and known variance 100. The critical region for testing $H_0 : \mu \leq 5$ vs. $H_1 : \mu > 5$ was: Reject H_0 if $\bar{X} > 8.28$. Let $p(\mu)$ be the power function of this test. Compute $p(\mu)$ at the values $\mu = 3,5,7,9$, and 11, and sketch the increasing curve $p(\mu)$ passing through these points.

26. Repeat the power calculation in Problem 10.8.25, using the critical region: Reject H_0 if $\bar{X} > 5 + \frac{10}{\sqrt{n}}(1.64)$, for the sample sizes $n = 9, 36$, and 100 and plot the four power functions on the same graph.

27. Suppose that $X \overset{\text{iid}}{\sim} B(20,p)$ and that you test $H_0 : p \leq 1/2$ vs. $H_1 : p > 1/2$ using the critical region: Reject H_0 if $X \geq 15$. Calculate the power function of your test at the values $p = .3, .4., .5, .6, .7, .8$, and graph the function by interpolating between the computed values.

28. Suppose that $X_1, \dots, X_n \overset{\text{iid}}{\sim} N(\mu, 16)$. How large a sample size would be required if you wish to test the hypothesis $H_0 : \mu = 70$ vs. $H_1 : \mu \neq 70$ at significance level $\alpha = .05$ and you would like your test to have power .9 when either $\mu = 72$ or $\mu = 68$?

29. How large a sample size n would be required to obtain power .9 at $\mu = 18$ if you test $H_0 : \mu \leq 16$ vs. $H_1 : \mu > 16$ at $\alpha = .05$ based on a random sample from an exponential distribution with mean μ.

30. Suppose that $X_1, \dots, X_n \overset{\text{iid}}{\sim} U(0,\theta)$. Consider the test of $H_0 : \theta \geq 2$ vs. $H_1 : \theta < 2$ which rejects H_0 if the maximum order statistic $X_{(n)}$ is smaller than 1.5. (a) Determine the size α of this test. (b) Plot the power function of this test.

31. Suppose that the single observation X is drawn from a distribution F_θ with pdf

$$F_\theta(x) = 2(1-\theta)x + \theta \text{ for } 0 < x < 1,$$

where θ can take any value in the interval $[0,2]$. Consider testing $H_0 : \theta \geq 1$ vs. $H_1 : \theta < 1$ by rejecting H_0 if $X > .9$. (a) Calculate the size of this test. (b) Sketch the power function of this test.

32. The company Prize Pickles, Inc. makes dill pickles reputed to be the best in the business. They test 25 pickles from each batch and determine, by a chemical reading, whether the pickles are up to their standards. Let p be the proportion of defective pickles. Prize Pickles

are pretty fussy, and they will test an entire batch (an expensive process) unless their preliminary test rejects $H_0 : p \geq .1$ in favor of $H_1 : p < .1$. In a given test, PPI rejects H_0 only if $X = 0$, where $X \sim B(25, p)$. (a) Determine the size α of this test. (b) Plot the power function of this test.

33. Derive the most powerful test for testing $H_0 : X \sim \Gamma(2,3)$ vs. $H_1 : X \sim \Gamma(3,2)$ at significance level $\alpha = .05$.

34. Let X be a single observation from the one-parameter beta distribution with density
$$f(x) = \theta x^{\theta-1} \text{ for } 0 < x < 1,$$
where θ is a positive parameter. (a) Find the most powerful test of size $\alpha = .05$ for testing $H_0 : \theta = 1$ vs. $H_1 : \theta = 2$. (Hint: Use the Neyman-Pearson Lemma to find the form of the best test. Note that when $\theta = 1, X \sim U[0,1]$.) (b) Argue that the test in part (a) is the uniformly most powerful test of size .05 for testing $H_0 : \theta \leq 1$ vs. $H_1 : \theta > 1$.

35. Suppose you are sure that the mean of the count data you will be sampling is 2, but you're not sure about how to model the experiment. Find the best test of size $\alpha \cong .05$ for testing $H_0 : X_1, \ldots, X_{10} \stackrel{\text{iid}}{\sim} P(2)$ vs. $H_1 : X_1, \ldots, X_{10} \stackrel{\text{iid}}{\sim} G(1/2)$ based on the ten observations you have available.

36. Derive the most powerful test for testing $H_0 : X \sim N(0,4)$ vs. $H_1 : X \sim N(1,1)$ at significance level $\alpha = .05$.

37. Suppose a single observation X is drawn from a distribution F with density f. Derive the best test of size α of the hypotheses $H_0 : f = f_0$ vs. $H_1 : f = f_1$, where
$$f_0(x) = 1 \text{ for } 0 < x < 1 \text{ and } f_1(x) = 2x \text{ for } 0 < x < 1.$$

38. Suppose that $X \sim \Gamma(\alpha, \beta)$, the gamma distribution with shape parameter α and scale parameter β. In testing $H_0 : \alpha = 1, \beta = 1$ vs. $H_1 : \alpha = 2, \beta = 2$, show that the test which rejects H_0 if $X > 4$ is the best test of size .0183 for testing H_0 vs. H_1. (Hint: You'll find the following fact useful: for any constant $A > 0$, the function $g(x) = xe^{Ax}$ is strictly increasing for $x \in (0, \infty)$).

39. Suppose that $X_1, \ldots, X_n \stackrel{\text{iid}}{\sim} N(0, \sigma^2)$. Show that the most powerful test for testing the hypotheses $H_0 : \sigma^2 = \sigma_0^2$ vs. $H_1 : \sigma^2 = \sigma_1^2$, where $\sigma_1^2 > \sigma_0^2$, is the test that rejects H_0 when
$$\sum_{i=1}^{n} X_i^2 > c,$$
where c is a positive constant.

40. Let X_1 and X_2 be independent $U(\theta, \theta + 1)$ random variables. Consider testing $H_0 : \theta = 0$ vs. $H_1 : \theta > 0$ using one of the following tests:
$$\text{Test A: Reject } H_0 \text{ if } X_1 > .95.$$
$$\text{Test B: Reject } H_0 \text{ if } X_1 + X_2 > 2 - \sqrt{.1}.$$
(a) Verify that both of these tests have size .05. (b) Calculate and graph the power functions of both tests. (c) Is one test uniformly better than the other?

41. A single observation $Y = X^\theta$ is obtained, where $X \sim Exp(1)$. (a) Find the form of the best test for testing $H_0 : \theta = 1$ vs. $H_1 : \theta = 2$. Find the best test of size $\alpha = .05$ in this testing problem, and calculate the power of this test at $\theta = 2$.

42. Suppose that $X_1, \ldots, X_n \stackrel{\text{iid}}{\sim} U(\theta, \theta+1)$ and you are interested in testing $H_0 : \theta = 0$ vs. $H_1 : \theta > 0$. Consider the test with critical region: Reject H_0 if
$$X_{(n)} > 1 \text{ or } X_{(1)} > c,$$
where c is a positive constant. Find the constant c for which this test has size α.

43. Let X be a single observation having the $Be(1,\theta)$ distribution with density $f_\theta(x) = \theta(1-x)^{\theta-1} I_{(0,1)}(x)$, where $\theta > 0$. Suppose you wish to test the hypotheses $H_0 : \theta = 2$ vs. $H_1 : \theta = 3$. Show that the test which rejects H_0 if $X < .05$ is the best test of size $\alpha = .0975$.

44. Let $X_1, \ldots, X_n \stackrel{\text{iid}}{\sim} F_{\lambda,\theta}$ a two-parameter distribution with density function
$$f(x) = \lambda e^{-\lambda(x-\theta)} I_{(\theta,\infty)}(x).$$
Derive the form of the Neyman-Pearson test for testing the two simple hypotheses $H_0 : \lambda = \lambda_0, \theta = \theta_0$ vs. $H_1 : \lambda = \lambda_1, \theta = \theta_1$, where (λ_0, θ_0) and (λ_1, θ_1) are two distinct points in the plane, with λ_0 and λ_1 positive.

45. The existence of uniformly most powerful one-sided tests for the parameters of a one-parameter model F_θ can be guaranteed when the family $\{F_\theta\}$ of distributions has the "monotone likelihood ratio (MLR)" property. Assume the family $\{F_\theta\}$ has corresponding densities $\{f_\theta\}$. Then $\{F_\theta\}$ has the MLR property if, when $\theta_1 < \theta_2$, the likelihood ratio $f(x|\theta_2)/f(x|\theta_1)$ is a nondecreasing function of x. Show that the following discrete families of distributions have the MLR property. (a) the $B(n, p)$ family, (b) the $P(\lambda)$ family, and (c) the $HG(N, r, n)$ family, where N and n are assumed known.

46. (See Problem 10.8.45.) Show that the following continuous families of distributions have the MLR property. (a) The $U(\theta, \theta+1)$ family, (b) the $N(\mu, \sigma_0^2)$ family (where σ_0^2 is known), and (c) the logistic family $L(\mu, 1)$.

47. (See Problem 10.8.45.) Let $X_1, \ldots, X_n \stackrel{\text{iid}}{\sim} \Gamma(\alpha, \beta)$. (a) Assume that α is a known. Show that the joint pdf of $X_1, \ldots, X_n$ has monotone likelihood ratio in the statistic $-\bar{X}_n$. (b) Assume that β is known. Show that the joint distribution of $X_1, \ldots, X_n$ has monotone likelihood ratio in the statistic $\prod_{i=1}^n X_i$.

48. (See Problem 10.8.45.) Let $X_1, \ldots, X_n \stackrel{\text{iid}}{\sim} F_\theta$, where F_θ is a member of a one-parameter exponential family with density given in (6.36). Assume that the function $A(\theta)$ in that density is a strictly increasing function of θ. Show that the joint pdf of $X_1, \ldots, X_n$ has monotone likelihood ratio in the sufficient statistic $\sum_{i=1}^n B(X_i)$.

49. (See Problem 10.8.45.) Show that the one-parameter Cauchy family $\{C(m, 1)\}$ does not have the monotone likelihood ratio property.

50. (See Problem 10.8.45.) Prove the following results. If the distribution F_θ of X has the MLR property, then (a) any one-sided test, that is, any test with a critical region of the form: Reject H_0 if $X > x_0$ has a non-decreasing power function and (b) any one-sided test is uniformly most powerful of its size for testing $H_0 : \theta \le \theta_0$ vs. $H_1 : \theta \ge \theta_0$.

51. (See Problem 10.8.50.) Suppose that $X_1, \ldots, X_n \stackrel{\text{iid}}{\sim} \Gamma(\alpha, \beta)$. (a) Assume that α is known. Show that a UMP test of size $p \in (0,1)$ exists for testing $H_0 : \beta \le \beta_0$ vs. $H_1 : \beta > \beta_0$. (b) Assume that β is known. Show that a UMP test of size $p \in (0,1)$ exists for testing $H_0 : \alpha \le \alpha_0$ vs. $H_1 : \alpha > \alpha_0$.

52. (See Problem 10.8.50.) Suppose that $X_1, \ldots, X_n \stackrel{\text{iid}}{\sim} IG(\mu, \tau)$, the inverse Gaussian distribution with positive parameters μ and τ and with density
$$f(x) = \sqrt{\frac{\tau}{2\pi x^3}} e^{-\tau(x-\mu)^2/2x\mu^2} \text{ for } x > 0.$$

(a) Assume that τ is known. Show that a UMP test of size $p \in (0,1)$ exists for testing $H_0 : \mu \leq \mu_0$ vs. $H_1 : \mu > \mu_0$. (b) Assume that μ is known. Show that a UMP test of size $p \in (0,1)$ exists for testing $H_0 : \tau \leq \tau_0$ vs. $H_1 : \tau > \tau_0$.

53. (See Problem 10.8.50.) Let $X_1, \ldots, X_n \stackrel{\text{iid}}{\sim} W(\alpha, \beta)$, the two-parameter Weibull distribution. (a) Assume that the shape parameter α is known. Show that a UMP test of size $p \in (0,1)$ exists for testing $H_0 : \beta \leq \beta_0$ vs. $H_1 : \beta > \beta_0$.

54. (See Problem 10.8.50.) Suppose $X_1, \ldots, X_n \stackrel{\text{iid}}{\sim} N(0, \sigma^2)$. Show that, for any significance level $\alpha \in (0,1)$, there exists a UMP test of size α for testing $H_0 : \sigma^2 \leq \sigma_0^2$ vs. $H_1 : \sigma^2 > \sigma_0^2$.

55. (See Problem 10.8.50.) Let $X \sim C(m, 1)$, the one-parameter Cauchy distribution with unknown median m. Show that, for any fixed significance level α, there does not exist a UMP test of $H_0 : m = 0$ vs. $H_1 : m > 0$.

56. (See Problem 10.8.50.) Suppose that the single observation X is drawn from a distribution F_θ with pdf

$$F_\theta(x) = 2(1-\theta)x + \theta \text{ for } 0 < x < 1,$$

where θ can take any value in the interval $[0,2]$. (a) Show that $F_\theta(x)$ has monotone likelihood ratio in the statistic $T = -X$. (b) Find the UMP test of size α for testing $H_0 : \theta \leq 1/2$ vs. $H_1 : \theta > 1/2$.

57. (See Problem 10.8.50.) Let $X_1, X_2, \ldots, X_n$ be a random sample from a chi-square distribution with the degrees of freedom k unknown. Show that the test that rejects $H_0 : k \leq k_0$ in favor of $H_1 : k > k_0$ if $\sum_{i=1}^n \ln X_i > c$ is the UMP test of its size.

58. You may recall that in Example 10.4.1, the significance level of the test described there, based on the variable $X \sim B(20, p)$, was a somewhat odd, irregular decimal .0577. There's a reason for this, of course. When you are operating with a discrete distribution, your choices regarding the cutoff points which define your rejection region will be limited. See Remark 10.4.3.) It is possible to use "artificial randomization" to ameliorate the situation. The idea is to use a randomization device to adjust the probability of rejecting H_0 to a desired level. Let's consider a simple example of the process. Suppose you wish to test the hypothesis $H_0 : p = 1/2$ vs. $H_1 : p > 1/2$, where p represents the probability that a certain coin is heads. The coin will be tossed 5 times, and the hypotheses will be tested based on the data $X \sim B(5, p)$. Among the choices we have for reasonable critical regions (and, in fact, best tests) are Reject if $X \geq 4$ and Reject if $X = 5$. Now the first CR has significance level $\alpha = 6/32 = .1875$, which of course seems too large. The second has significance level $\alpha = 1/32 = .03125$, which seems a bit small. But we can devise a test with the significance level we probably wanted all along, $\alpha = .05$, by randomizing in the following way: Using a random number generator, draw a random variable Y that has probability $\pi = 01875/.15625 \cong .1200$ of being 1 and has probability .8800 of being 0. Then, reject H_0 according to the following randomized procedure:

$$\text{If } X = 4 \text{ and } Y = 1, \text{ reject } H_0; \text{ if } X = 5, \text{ reject } H_0. \tag{10.57}$$

(a) Calculate the probability that $H_0 : p = 1/2$ is rejected based on $X \sim B(5, p)$ when the critical region in (10.57) is used. (b) Reread Example 10.4.1 and devise a randomized test that achieves significance level $\alpha = .05$ in that testing problem.

59. For discussion. (See Problem 10.8.58.) A middle-aged man with whom your family have been long time friends recently had a biopsy on a tumor in one of his internal organs. Today, he's going to see the surgeon who performed the biopsy. The surgeon has run a test which yields an outcome X that he uses to help him classify the tumor as malignant or

benign. The variable X takes values on the scale from 1 to 10. When your friend arrives, the doctor has just received the report. He looks at the value of X and says "hmmmm"—not what your friend really wanted to hear. Then, the doctor tosses a coin, and your friend sees that it comes up heads. The doctor turns to your friend and says "Whew, the cancer is not malignant!" The doctor is just reporting on the outcome of his randomized testing procedure. What should your friend think about the diagnosis he's just received?

60. Your evil cousin Eddy is coming over in 5 minutes. You bought some loaded dice last month and plan to trick Eddy into making a bad bet. Serves him right! But you've forgotten a key fact: Is the probability p of rolling a "7" equal to $1/3$ or is it equal to $2/3$? You distinctly recall that it is one or the other. You decide to carry out a quick test of the hypothesis $H_0 : p = 1/3$ vs. $H_1 : p = 2/3$ before Eddy arrives. Your test will be based on rolling the dice until you get your third "7." The number of trials X required to complete your experiment has a negative binomial distribution $NB(r, p)$, with $r = 3$. Show that the test that rejects H_0 if $X < 5$ is the best test of size $1/9$ for testing H_0 vs. H_1.

61. John (the Mole) Mollay does a fair amount of consulting for the CIA. Last week, they came to him with a very delicate question. They wouldn't tell John how they got their data, but simply described it as the result of "a probe." They were quite sure that the measured X had one of two discrete distributions: either X has a Poisson distribution with parameter $\lambda = 1/2$ or X had the probability mass function

$$p(x) = (1/2)(1/2)^x \text{ for } x = 0, 1, 2, \ldots.$$

Let's call the two distributions $P(1/2)$ and $P^*(1/2)$. They asked John for his advice regarding a test for $H_0 : X \sim P(1/2)$ vs. $H_1 : X \sim P^*(1/2)$. After some careful thought, and a quick call to his old statistics instructor, John allowed as to how rejecting H_0 when $X > 2$ was a pretty good test. (a) What is the significance level α of the test John recommends? (b) What is β, the probability of type II error of John's test? (c) Use the Neyman-Pearson Lemma to show that John's test is the best of its size for testing H_0 vs. H_1.

62. An observation X (a sample of size $n = 1$) is taken from a distribution on the positive real line. Using X, you wish to test the hypothesis $H_0 : X \sim$ half normal (also called the folded normal) against $H_1 : X \sim \Gamma(1, 1)$ (i.e., the exponential distribution with mean 1). The density of the half-normal distribution is

$$f(x) = \sqrt{\frac{2}{\pi}} exp\{-x^2/2\} \text{ for } x > 0,$$

where $exp\{A\} = e^A$. The density $f(x)$ is equal to exactly twice the $N(0, 1)$ density for values of $x \in (0, \infty)$. Note also that under $H_0, P(X > Z_\alpha) = 2\alpha$. (a) Consider testing H_0 vs. H_1 using the critical region: Reject H_0 if $X > 2$. Calculate α and β, the probabilities of type I error and of type II error for this test. (b) Since H_0 and H_1 are simple hypotheses, the Neyman-Pearson Lemma may be used to find the best test of any given size. Show that the test which rejects H_0 when $(X - 1)^2 > k$, where k is a fixed constant, is the best test of its size. (c) Is the test that rejects H_0 if $X > 2$ the best test of its size for testing H_0 vs. H_1, that is, is it a Neyman-Pearson test for some choice of the constant k in part (b)?

63. Suppose that $X_1, X_2, \ldots, X_n \sim N(\mu_x, \sigma^2)$ and $Y_1, Y_2, \ldots, Y_m \sim N(\mu_y, \sigma^2)$ are independent random samples from normal populations with a common variance σ^2. For a fixed constant $c \neq 0$, construct a t test for testing hypotheses $H_0 : \mu_x - \mu_y = c$ vs. $H_1 : \mu_x - \mu_y \neq c$ at significance level α. (Hint: Think about testing $\mu_u = \mu_y$ after replacing each X_i by $U_i = X_i - c$.)

64. Let $X_1, \ldots, X_n \overset{\text{iid}}{\sim} F_\theta$. Consider testing the simple hypotheses $H_0 : \theta = \theta_0$ against the simple

alternative $H_1 : \theta = \theta_1$. Show that the likelihood ratio test which rejects H_0 when the likelihood ratio statistic $\lambda(\mathbf{x})$ is small can be rewritten in the form of a "best test" identified in the Neyman-Pearson Lemma.

65. Suppose that $X_1, \dots, X_n \stackrel{\text{iid}}{\sim} B(1, p)$, the Bernoulli distribution with probability of success p. Show that the critical rejoin of the likelihood ratio test of $H_0 : p \leq p_0$ vs. $H_1 : p > p_0$ has the form: Reject H_0 if $\sum_{i=1}^{n} X_i > c$, where c is a positive constant.

66. Let $X_1, \dots, X_n \stackrel{\text{iid}}{\sim} P(\alpha, \theta)$, the two-parameter Pareto distribution. Show that the likelihood ratio test of the hypothesis $H_0 : \alpha = 1$ vs. $H_1 : \alpha \neq 1$ (with θ unknown) has the critical region: Reject H_0 if $T < c_1$ or $T > c_2$, where $0 < c_1 < c_2 < \infty$, and T is given by

$$T = \left[\frac{\prod_{i=1}^{n} X_i}{X_{(1)}^n} \right],$$

where $X_{(1)} = \min\{X_1, \dots, X_n\}$.

67. Let $X_1, \dots, X_n \stackrel{\text{iid}}{\sim} W(\alpha, \beta)$, the Weibull distribution with shape parameter α and scale parameter β. The Weibull model is widely used in engineering reliability as a model for the lifetime of an engineered system. When this model is used, testing the hypotheses $H_0 : \alpha = 1$ vs. $H_1 : \alpha \neq 1$ is often of special interest, as the null hypothesis asserts that the data is exponentially distributed. Derive the likelihood ratio test statistic $\lambda(\mathbf{x})$ for testing H_0 vs. H_1. (Note: Since the MLE of the parameter pair (α, β) cannot be obtained analytically, show that the MLE $\left(\hat{\alpha}, \hat{\beta}\right)$ of (α, β) may be expressed as $\left(\hat{\alpha}, (1/n)\sum_{i=1}^{n} X_i^{\hat{\alpha}}\right)$. Use the latter expression in the denominator of the likelihood ratio statistic.)

68. Let $X_1, \dots, X_n \stackrel{\text{iid}}{\sim} \Gamma(1, \mu, \theta)$, the three-parameter gamma distribution with density

$$f(x) = (1/\mu) e^{-(x-\theta)/\mu} I_{(\theta, \infty)}(x),$$

where both μ and θ are unknown. Derive the likelihood ratio test of the hypotheses $H_0 : \theta = 0$ vs. $H_1 : \theta \neq 0$.

69. Suppose that $X_1, \dots, X_n \stackrel{\text{iid}}{\sim} N(\theta, a\theta)$, where both θ and a are unknown positive constants. Derive the likelihood ratio test of the hypotheses H_0: a = 1 vs. H_1: a $\neq$ 1.

70. Suppose that $X_1, \dots, X_n \stackrel{\text{iid}}{\sim} N(\mu, \sigma^2)$, with both μ and σ^2 unknown. Show that the likelihood ratio test for $H_0 : \sigma^2 = \sigma_0^2$ vs. $H_1 : \sigma^2 \neq \sigma_0^2$ may be simplified to the inequality $\sum_{i=1}^{n} (x_i - \bar{x})^2 < c_1$ or $\sum_{i=1}^{n} (x_i - \bar{x})^2 > c_2$, where $0 < c_1 < c_2$.

71. Suppose that $X_1, \dots, X_n \stackrel{\text{iid}}{\sim} N(\theta, a\theta^2)$, where both θ and a are unknown, with $a > 0$. Derive the likelihood ratio test of the hypotheses $H_0 : a = 1$ vs. $H_1 : a \neq 1$.

72. Two giants in the meat preservative business offer quite different additives to the major U.S. meat processors. Chemist Chui Chorizo offers a product that gives processed meat a $\Gamma(1, \mu)$ lifetime, while Biochemist Mal Practice's product gives processed meat a $\Gamma(1, \nu)$ lifetime. *Consumer Reports* plans to test H_0: $\mu = \nu$ vs. H_1: $\mu \neq \nu$ using a likelihood ratio test. If $X_1, X_2, \dots, X_n \sim \Gamma(1, \mu)$ and $Y_1, Y_2, \dots, Y_m \sim \Gamma(1, \nu)$ are independent random samples of the lifetimes of similar meat products using the competing additives, the likelihood function of the joint samples is

$$L(\mu, \nu | \mathbf{x}, \mathbf{y}) = (1/\mu)^n exp\{-(1/\mu) \sum_{i=1}^{n} x_i\} \times (1/\nu)^m exp\{-(1/\nu) \sum_{j=1}^{m} y_j\}.$$

(a) Assume that $H_0 : \mu = \nu$ is true, and label the common mean as "θ." Show that the MLE of θ is $\hat{\theta} = (n\bar{X} + m\bar{Y})/(n+m)$. (b) If μ and ν can be arbitrary positive numbers, their respective MLEs are $\bar{X}$ and $\bar{Y}$. Obtain an exact expression for the likelihood ratio statistic $\lambda(\mathbf{x},\mathbf{y})$. (A helping hand: Your answer should be equivalent to $\lambda = (\bar{X}/\hat{\theta})^n(\bar{Y}/\hat{\theta})^m$.) (c) If $\sum_{i=1}^n x_i = 300$ and $\sum_{j=1}^m y_j = 75$, where $n = 50$ and $m = 25$, carry out the LR test for H_0 vs. H_1 at significance level .05, and state your conclusion.

73. FDA agents at two Salinas-area farms are inspecting the leaves of randomly selected heads of Romaine lettuce in search of a certain bacteria. The number of contaminated leaves on each head is recorded. The first inspector tests n randomly selected heads of lettuce and finds a total of X contaminated heads. The second tests m randomly selected heads of lettuce and finds a total of Y contaminated heads. Assume that the two random samples are independent of each other and that each is drawn from Poisson distributions with respective parameters θ_1 and θ_2. It then follows that $X \sim P(n\theta_1)$ and that $Y \sim P(m\theta_2)$. Because these two farms are situated quite differently relative to a large dairy farm outside of Salinas (thought to be the source of the bacteria), it is of interest to test the hypothesis $H_0 : \theta_1 = \theta_2 \equiv \theta$ against $H_1 : \theta_1 \neq \theta_2$. Both inspectors have doubts, a priori, that H_0 is true for these particular farms. (a) Under the hypothesis $H_1 : \theta_1 \neq \theta_2$, we know that the MLEs of the population parameters are $\hat{\theta}_1 = X/n$ and $\hat{\theta}_2 = Y/m$. Find the MLE of the parameter θ under the hypothesis $H_0 : \theta_1 = \theta_2 \equiv \theta$. (b) Obtain the likelihood ratio statistic λ for testing H_0 vs. H_1. (c) Suppose that $n = m = 50$, and that $X = 10$ and $Y = 2$. Carry out an approximate likelihood ratio test for H_0 vs. H_1 at level $\alpha = .1$.

74. Suppose independent samples are drawn from each of two normal populations. Denote the data as $X_1, X_2, \ldots, X_n \sim^{iid} N(\mu_1, \sigma_1^2)$ and $Y_1, Y_2, \ldots, Y_m \overset{\text{iid}}{\sim} N(\mu_2, \sigma_2^2)$. All four parameters are unknown. Suppose we wish to derive the likelihood ratio statistic λ for testing $H_0 : \sigma_1^2 = \sigma_2^2$ vs. $H_1 : \sigma_1^2 \neq \sigma_2^2$ at some pre-specified significance level α. (a) Show that the MLEs of the parameters μ_1, μ_2, and σ^2 under the null hypothesis of equal variances, that is, under the assumption that $\sigma_1^2 \equiv \sigma^2 \equiv \sigma_2^2$, are $\hat{\mu}_1 = \bar{X}, \hat{\mu}_2 = \bar{Y}$, and

$$\hat{\sigma}^2 = \frac{\sum_{i=1}^n (X_i - \bar{X})^2 + \sum_{i=1}^m (Y_i - \bar{Y})^2}{n+m}.$$

(b) Show that the MLEs of models parameters in the unconstrained parameter space are given by

$$\hat{\mu}_1 = \bar{X}, \hat{\mu}_2 = \bar{Y}, \hat{\sigma}_1^2 = \frac{\sum_{i=1}^n (X_i - \bar{X})^2}{n} \text{ and } \hat{\sigma}_2^2 = \frac{\sum_{i=1}^n (Y_i - \bar{Y})^2}{m}.$$

(c) Using the MLEs in (a) and (b), obtain an expression for the likelihood ratio statistic λ. (d) Show that the LR statistic λ may be written as

$$\lambda = \left(\frac{n+m}{n}\right)^{n/2}\left(\frac{n+m}{m}\right)^{m/2}\frac{((n-1)s_1^2)^{n/2}((m-1)s_2^2)^{m/2}}{((n-1)s_1^2 + (m-1)s_2^2)^{(n+m)/2}},$$

where s_i^2 is the usual sample variance, that is, $s_i^2 = [n/(n-1)]\hat{\sigma}_i^2$. (e) From (c), it follows that the LR statistic λ is a constant multiple of the ratio

$$R = \frac{\left(\frac{(n-1)s_1^2}{(m-1)s_2^2}\right)^{n/2}}{\left(\frac{(n-1)s_1^2}{(m-1)s_2^2} + 1\right)^{(n+m)/2}}.$$

Use the latter expression to argue that rejecting H_0 when λ (or R) is small is equivalent to rejecting H_0 when the ratio of sample variances s_1^2/s_2^2 is either too small or too large. (Hint: Note that $R \geq 0$, and that it approaches its minimum when s_1^2/s_2^2 approaches either 0 or ∞.)

75. As an amateur gardener, Hulk Hooligan likes to plant pansies in the spring. (That's right, real men like pansies too!) Anyway, Hulk bought a pack of 100 mixed pansies that claimed to contain the color distribution: 20%, 20%, 10%, 20%, 30% for the colors yellow, red, white, blue, and mauve (Hulk's personal favorite). The actual color distribution of his planted flowers is shown below.

Color	Y	R	W	B	M
Observed frequency	25	25	8	22	20

Needless to say, Hulk is disappointed! (It's the mauve thing.) Test the hypothesis that the probabilities mentioned above $(.2, .2, .1, .2, .3)$ are valid for flower packs of the type that Hulk bought. Use $\alpha = .05$.

76. The Hardy-Weinberg principle states that, under random mating, allele frequencies in a population remain constant (i.e., "in equilibrium") from generation to generation. For the simple case of one locus with two alleles, let the relative frequency of the dominant allele **A** be p and that of the recessive allele **a** be q, where $q = 1 - p$. If the population is in equilibrium, the proportion of **AA** and **aa** "homozygotes" in the population will be p^2 and q^2, respectively, and the proportion of **Aa** "heterozygotes" will be $2pq$. Each individual in the population has two alleles, one from each "parent." In other words, the odds of finding *AA*, *aa*, and *Aa* individuals in such a population should be $p^2 : q^2 : 2pq$ (for some value of p). A population of simple organisms is to be tested for Hardy-Weinberg equilibrium. The data collected on 100 organisms is shown below:

AA	aa	Aa
20	20	60

From these observed frequencies, the standard (maximum likelihood) estimate of p is .5. Test H_0: "This population is in Hardy Weinberg equilibrium" at $\alpha = .05$.

77. U.S. fraternities are sometimes thought of as a bit snobbish, but they don't hold a candle, in this regard, to British fraternities. For example, the Iota Beta Upsilon Fraternity at Oxford (usually called IBU and translated as "I'm better than you") boasts that only bluebloods are admitted. A sociologist at Oxford thinks this claim is overstated, and has decided to test the distribution of blood types of IBU members. (All were found to have red blood, by the way.) Suppose that the distribution of blood types among British males is 25% A, 15% B, 5% AB, and 50% O. The distribution of blood types of the 100 IBU members sampled was 35% A, 15% B, 10% AB, and 40% O. Test, at $\alpha = .01$, that the overall distribution of blood types among British males applies to IBU members.

78. The randomness of the California Lotto Game was challenged in court in 1990. At that time, the state used a form of the game that chose six numbers "at random" from the numbers 1–49. An independent firm was retained to test the hypothesis of randomness. The firm chose a single Lotto number at random from 140 consecutive games and obtained the following distribution of outcomes:

Outcome	1–7	8–14	15–21	22–28	29–35	36–42	43–49
Observed frequency	24	16	21	26	23	20	10

Given that the expected frequency of each class is 20, test the hypothesis H_0: "all outcomes are equally likely" at $\alpha = .05$.

79. Geneticist Seymour Blooms has been performing a plant breeding experiment in which the four possible types of plants that may bloom will occur, according to Blooms' model, with probabilities shown in the table below.

Plant Type i	1	2	3	4
Probability p_i	θ	$\theta/2$	$\theta/2$	$1-2\theta$

The value taken by θ is a decimal between 0 and $1/2$, which depends on how much sunlight the plants are exposed to. Dr. Blooms bred $n = 80$ plants and observed the following frequencies for the four plant types.

Plant Type i	1	2	3	4
Frequency o_i	28	7	5	40

Show that, based on these data, the maximum likelihood estimator of θ is $\hat{\theta} = 1/4$. Obtain the expected frequencies of the four plant types with θ estimated by $\hat{\theta}$, and test the null hypothesis that Dr. Blooms' model fits the data he obtained. Use $\alpha = .05$.

80. Suppose you flip a coin 100 times and get 58 heads. (a) Test that heads and tails are equally likely at $\alpha = .05$ using a chi-square goodness-of-fit test. (b) Test the hypothesis $H_0 : p = P(\text{Heads}) = 1/2$ using the standard test for proportions in Section 10.2. (c) Do the tests lead to the same conclusion? Are the tests equivalent, that is, can the test statistic in one test be written as a function of the test statistic of the other in such a way that they both always lead to the same conclusion regarding H_0, at the same significance level, given the same data?

11

Estimation and Testing for Linear Models

This chapter requires a brief preamble. The subject of linear statistical models is very broad. There are quite a few good books on the subject, and it is often the case that universities offer separate courses on particular aspects of this subfield of Statistics. Most often, the subject of linear models is separated into two parts, regression analysis and the analysis of variance (ANOVA), and each of these parts is taught in a course exclusively devoted to it. In a single chapter of a book like the present one, the scope of coverage of linear model theory is necessarily limited. I will cover only the simplest models in both of these areas: simple linear (or straight-line) regression and one-way (or single-factor) analysis of variance. I will discuss some of the mathematical aspects of both topics, my aim being to illustrate how the concepts we have studied so far shed some light on the behavior of, and the justification for, the standard procedures in these two areas. For example, I will show that, even without the usual assumption of normality, the standard estimators of the coefficients in a simple linear regression model are the best linear unbiased estimators of these parameters, and that under the assumption of normality, the standard estimators of the coefficients in a simple linear regression model are the maximum likelihood estimators of these parameters. I will also show that the standard F test in one-way ANOVA may be derived as a likelihood ratio test.

My aim in treating the specific developments in this section is to provide some insights into the models and analyses treated here that are complementary to what a student might see in a "methods course" on these topics. In the discussion here, I will emphasize the "why" rather than the "how" that is emphasized in more applied courses. It is fair to say that any student interested in the use of linear models in real applications will be well served by having had some exposure to both the "how" and the "why" behind these very useful and widely utilized statistical methods.

11.1 Simple Linear Regression

Let us begin with a discussion of the type of problem in which regression models typically arise. Regression analysis is about the possible relationships between variables. For example, one might be interested in the relationship between a random variable Y and a collection of random variables $X_1, X_2, \ldots, X_k$ that, at least intuitively, might well be associated with Y. The usual goal in a regression problem is to estimate how Y is related to these Xs so that the estimated relationship can help us better understand the behavior of Y (the descriptive function of regression) and can also help us in predicting future values of Y when we happen to know the X values associated with it (regression's predictive function). It is often the case that Y is a variable that is harder to obtain than are the Xs, so that the ability to predict the value of Y from the observed value of the Xs can be very useful in practical applications. A simple example of this is a problem in which Y is a variable that isn't available presently (like what a high school senior's performance (say GPA) will be in his/her first year of college) and the Xs are a collection of variables that are immediately available (like a current high school senior's grades, SAT scores or the amount of his/her participation is extra-curricular activities). A second example would be a problem in which Y is the measured outcome of an invasive medical procedure (like a biopsy) and the Xs are the results obtained from a straightforward physical examination and/or a blood or urine sample. In cases such as these, a reliable representation of the relationship between Y and some subset of the Xs should be quite helpful. The fact that many colleges use regression models as an aid in making their admissions decisions and many medical practitioners use regression models to assist in the diagnosis of various illnesses represents, in itself, a testament to the utility of the approach.

Let us now assume that we are interested in studying the relationship between a variable Y and a single variable X. We will refer to X as the *independent variable* (though it is often called the "predictor" or the "concomitant variable") and we will refer to Y as the *dependent variable*. The model we will now discuss can be thought of as a way of quantifying the idea that X and Y are "linearly related." If X and Y were actually linearly related, we could write the equation $Y = b_0 + b_1 X$, and a couple of observations of pairs (X, Y) would suffice in identifying the values of the constants b_0, the "y-intercept" and b_2, the "slope" which completely define the relationship. If the (X, Y) pairs always lie on the straight line $Y = b_1 + b_2 X$, we would refer to the relationship between X and Y as "deterministic," and the phenomenon would be considered simple and not very interesting. But if X and Y are both random variables, the deterministic relationship above would virtually never occur, and we would be interested, instead, in whether or not X and Y have a *linear statistical relationship*. There is more than one way to make this notion concrete, but the most frequently encountered one involves the idea that Y is a linear function of X "on the average." We'll now discuss the modeling assumptions through which this idea is typically quantified. We will take the view that whenever we observe a value x of the variable X, there is a population of possible responses Y to that x. Suppose that we are interested in an experiment in which n (X, Y) pairs will be observed. We will assume that, for any $i \in \{1, \ldots, n\}$, the conditional distribution of Y_i, given $X_i = x_i$, may be represented as

$$Y_i = \beta_0 + \beta_1 x_i + \varepsilon_i, \tag{11.1}$$

where

(1) Y_i is the random response obtained when $X_i = x_i$ is observed in the ith trial,

(2) β_0 and β_1 are unknown parameters representing the y-intercept and slope of the linear relationship between X and Y, on the average, that is, $E(Y_i | X_i = x_i) = \beta_0 + \beta_1 x_i$, and

(3) ε_i is the random error in the ith trial, where the random errors $\{\varepsilon_i, i = 1, \ldots, n\}$ each have mean 0 and a common variance σ^2, and are, in addition, uncorrelated, that is, for any $i \neq j$, $Cov(\varepsilon_i, \varepsilon_j) = 0$.

The three assumptions above are the minimum necessary for any statistical work with the model in (11.1). However, in typical applications of regression analysis, a stronger assumption is made on the random errors. We shall see that assumption (3) above will often be replaced by

(3*) $\varepsilon_1, \ldots, \varepsilon_n \overset{iid}{\sim} N(0, \sigma^2)$.

The assumption (3*) retains the zero-mean and constant variance assumptions in (3), but, in addition, stipulates that these errors are *normally distributed*. This additional assumption, together with zero-correlation assumption made in (3), will imply that the errors $\{\varepsilon_i, i = 1, \ldots, n\}$ are independent. Both the independence of the errors, and the common normal distribution they enjoy, are conveniently embodied in the statement of condition (3*). The modeling assumptions made, be they (1), (2), and (3) or (1), (2), and (3*), need to be confirmed in applications of regression analysis, and there are a number of formal and graphical methods for checking such assumptions.

When the assumptions (1), (2), and (3*) are made, there is a particularly simple way of representing the model. In short, these three assumptions are equivalent to the assumption that the collection of pairs $\{(X_i, Y_i), i = 1, \ldots, n\}$ are independent and

$$Y_i | X_i = x_i \sim N(\beta_0 + \beta_1 x_i, \sigma^2) \text{ for } i = 1, \ldots, n. \tag{11.2}$$

The model description in (11.2) is particularly easy to interpret when viewed graphically. The assumptions made simply indicate that for every observed value x of X, the response Y is a normally distributed random variable with mean on the straight line $Y = \beta_0 + \beta_1 x$, and common variance σ^2. This interpretation of (11.2) is pictured in Figure 11.1.1.

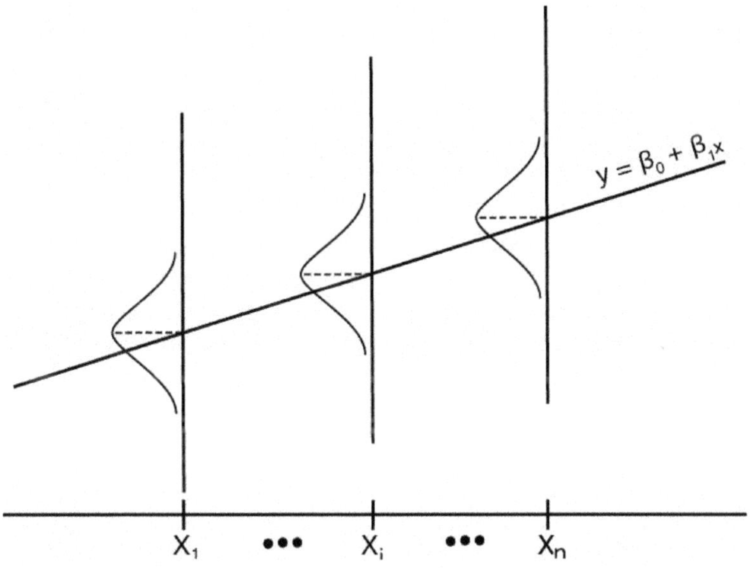

Figure 11.1.1. The distribution of Y as a function of the observation $X_i = x_i$.

Linear regression models are defined as models which are *linear in the parameters*. In the model displayed in (11.1) or (11.2), it is clear that the mean of Y is modeled as a linear combination (or a linear function) of the parameters β_0 and β_1. More complex linear models might posit that the mean of Y, given $\mathbf{X} = \mathbf{x}$, has the form $\beta_0 + \sum_{i=1}^{k} \beta_i x_i$ (when the relationship between Y and k X variables is under study) or perhaps the form $\beta_0 + \sum_{i=1}^{k} \beta_i x^i$ (when it is thought that Y is well represented, on the average, by a kth degree polynomial in x). In both of these cases, we would be dealing with linear statistical models, as both models are based on the assumption that the mean value of Y, given $\mathbf{X} = \mathbf{x}$, is a linear function of the parameters $\beta_0, \beta_1, \ldots, \beta_k$. The model in (11.2) is linear in another way. It assumes that the mean value of Y, given $X = x$, is a linear function of x. This makes the linear regression model especially simple and easy to work with, and it is the reason that problems which use the model in (11.1) or (11.2) are referred to as *simple linear regression* (SLR). As simple as this particular regression model is, it is very useful, as there are many applications in which the SLR model fits the available (X,Y) data well and provides a means of predicting future values of Y quite reliably. We now turn to the practical problem of fitting such models to data.

The fitting of a straight line to data involves two important steps. Logically, the first question one should settle is whether or not there is convincing evidence that there actually is a linear relationship between X and Y. This question is investigated by a hypothesis test typically referred to as a "test for the significance of regression." The null hypothesis of such a test is that there is no linear relationship between X and Y. We will find ways of stating that hypothesis in terms of model parameters. When this hypothesis is rejected, it is natural to try to identify the linear relationship that has been detected by our test. This, we will do through the estimation of the parameters β_0 and β_1, a process that results in an estimated regression line

$$\hat{Y} = \hat{\beta}_0 + \hat{\beta}_1 X, \tag{11.3}$$

a line from which the behavior of Y as a function of X may be studied and from which predictions of future Y values associated with future observed X values may be made. We treat both elements—testing for the significance of regression and estimating the linear relationship, assuming there is one—but we will do so in the opposite order, simply as a matter of convenience stemming from the notation that we are about to build up.

To introduce some regression jargon, the data set that we work with when examining a regression problem is generally referred to as a *training sample*. In simple linear regression, the sample can be thought of as the observed values of n (X,Y) pairs, that is, a set of points $\{(x_1,y_1),\dots,(x_n,y_n)\}$. Since the model in (11.2) only stipulates the *conditional behavior* of Y, given X, the "observed" X values in our data set may be a set of fixed, non-random values chosen by the experimenter (like the dose levels at which a medication will be administered), although they may also be treated as outcomes of a random experiment when the X values are obtained in that way. In the latter case, the regression analysis that follows would be referred to as a "conditional" analysis, based on the observed values of the variable X.

We will acknowledge, in passing, that the Ys in these data pairs are often harder to record than the Xs. Be that as it may, unless we are able to observe a collection of (X,Y) pairs, we will have no basis to ask and answer the two questions posed above. So we will assume that, by hook or by crook, the collection of (X,Y) pairs alluded to above have been collected. As you can imagine, the process of developing an equation for predicting a first year college GPA from a high school senior's SAT score will require that we keep track of a group of high school seniors from the time they take their SAT test to the time they complete their freshman year of college so that we can record an X and a Y for each of them.

Let's consider how we would go about fitting a straight line to data. We would probably begin by staring at the data. Staring at the data is facilitated by what is called a *scatter plot*, which is nothing more than a graph of the set of observed (x,y) points. A typical scatter plot is shown in Figure 11.1.2.

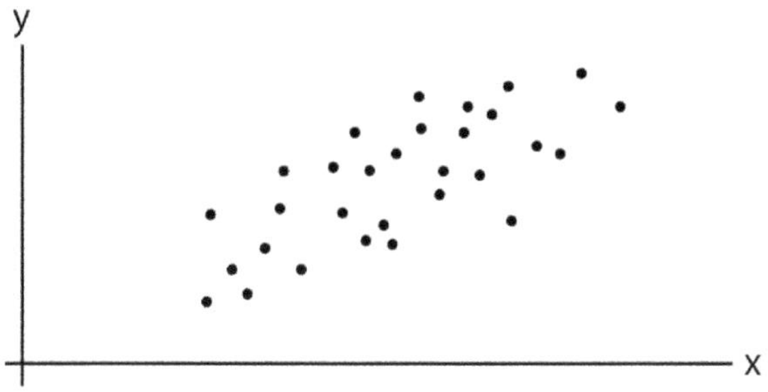

Figure 11.1.2. A scatter plot for observed (x, y) pairs.

Having graphed the data, and assuming that some type of linear trend is discernible from the graph or confirmed via a formal test, our next challenge would be to determine which straight line would most appropriately represent the linear trend in Y as a function of X. The standard approach to this task utilizes the so-called "least-squares criterion" for assessing the fit of a candidate line to the data. Consider the following measure, which we will denote as "SS" for "sum of squares":

$$SS = SS(\beta_0, \beta_1) = \sum_{i=1}^{n} (y_i - \beta_0 - \beta_1 x_i)^2. \tag{11.4}$$

The sum of squares in (11.4) measures, for each point (x_i, y_i) in our scatter plot, the vertical distance between that point and the point on the line $Y = \beta_0 + \beta_1 x_i$ at the x-coordinate x_i. The set of vertical distances $d_i = |y_i - \beta_0 - \beta_1 x_i|\}$ between each point (x_i, y_i) and the line $Y = \beta_0 + \beta_1 x_i$ is pictured in Figure 11.1.3.

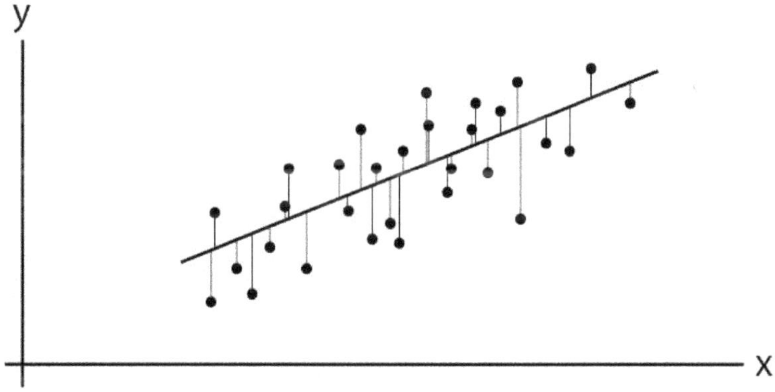

Figure 11.1.3. Vertical distances between points (x_i, y_i) and the line $Y = \beta_0 + \beta_1 x$.

In $SS(\beta_0, \beta_1)$, the vertical distances in Figure 11.1.3 are squared and then added together. It is thus clear that SS really does measure how well a particular line fits the data $\{(x_i, y_i)\}$. If every point in the set $\{(x_i, y_i), i = 1, \ldots, n\}$ were on the line $Y = \beta_0 + \beta_1 x$, each vertical distance from the point to the line would be 0, so that $SS = 0$ in that case. This is, of course, the smallest possible value of SS, and that value is achieved only when the line fits the data perfectly. When real data is involved, SS will take different positive values for the different choices of (β_0, β_1) that we might make. The *least-squares line* is the line which makes the sum of squares SS as small as possible, given the data $\{(x_i, y_i), i = 1, \ldots, n\}$

with which we are working. A minimization problem such as this, that is, the problem of minimizing $SS(\beta_0, \beta_1)$ over all possible real numbers β_0, and β_1, is amenable to the application of the calculus. Indeed, we can find the coefficients of the least-squares line by simultaneously solving the two equations,

$$\frac{\partial}{\partial \beta_0} SS(\beta_0, \beta_1) = 0 \tag{11.5}$$

and

$$\frac{\partial}{\partial \beta_1} SS(\beta_0, \beta_1) = 0. \tag{11.6}$$

These two equations have traditionally been called "the normal equations"; this is perhaps an unfortunate name, as the word "normal" has already been preempted by a widely used continuous probability model, but we will stick with tradition and use the word in these two distinct contexts. When the differentiation in (11.5) and (11.6) is done, we obtain the more informative version of the normal equations below:

$$\sum_{i=1}^{n} y_i - n\beta_0 - \beta_1 \sum_{i=1}^{n} x_i = 0 \tag{11.7}$$

and

$$\sum_{i=1}^{n} x_i y_i - \beta_0 \sum_{i=1}^{n} x_i - \beta_1 \sum_{i=1}^{n} x_i^2 = 0. \tag{11.8}$$

You'll recognize the equations in (11.7) and (11.8) as being two linear equations in two unknowns, a problem you learned to solve in high school, if not earlier. This is the same problem as that of finding the point at which two straight lines in the plane intersect. To obtain the solution $(\hat{\beta}_0, \hat{\beta}_1)$, we might solve for β_0 in the first equation and substitute the expression we got for the variable β_0 in the second equation. Doing this leads to one equation in the one unknown β_1, from which we would obtain the solution for the variable β_1 in terms of the Xs and Ys. This solution may be written as

$$\hat{\beta}_1 = \frac{\sum_{i=1}^{n} x_i y_i - n \cdot \bar{x} \cdot \bar{y}}{\sum_{i=1}^{n} x_i^2 - n \cdot \bar{x}^2}. \tag{11.9}$$

Having $\hat{\beta}_1$ in hand, we can obtain the solution for the variable β_0 as

$$\hat{\beta}_0 = \bar{y} - \hat{\beta}_1 \bar{x}. \tag{11.10}$$

The uniqueness of the solution of (11.7) and (11.8) guarantees that $SS(\beta_0, \beta_1)$ is minimized at this point. The line

$$y = \hat{\beta}_0 + \hat{\beta}_1 x \tag{11.11}$$

is referred to as the least-squares line, and the regression equation

$$\hat{Y} = \hat{\beta}_0 + \hat{\beta}_1 x \tag{11.12}$$

is used to predict the value of Y at the value $X = x$ obtained at some future time.

The process of fitting a straight line to data is generally executed using a statistical computing package like Minitab, R, SAS, or SPSS. For large data sets, one would not wish to do the computations required to obtain $\hat{\beta}_0$ and $\hat{\beta}_1$, not to mention other quantities of interest, manually—with or without a hand-held calculator. I will, however, illustrate these calculations for the following small data set.

Example 11.1.1. Dr. E. Nee-Meany, Dean of the College of Engineering at Upancomin University wants to develop a regression equation for predicting the number of research papers Y a college faculty member will publish in a given year as a function of the number X of extramural research grants he or she received during the previous year. Dean Nee-Meany has collected the following data for the last two academic years for 5 college faculty members selected at random:

x	0	0	1	2	2
y	1	2	3	4	5

A scatter plot of the five points above shows a definite linear trend with a positive slope. Let's obtain the least-squares line from these data. The following summary table is helpful in making the needed computations.

x	0	0	1	2	2	$\Sigma x = 5$
y	1	2	3	4	5	$\Sigma y = 15$
xy	0	0	3	8	10	$\Sigma xy = 21$
x^2	0	0	1	4	4	$\Sigma x^2 = 9$

From the table above, we note that $\bar{x} = 1$ and $\bar{y} = 3$, and we thus find that the slope of the least-squares line is

$$\hat{\beta}_1 = \frac{21 - 5 \cdot 1 \cdot 3}{9 - 5(1)^2} = 1.5,$$

and its y-intercept is

$$\hat{\beta}_0 = 3 - (1.5)(1) = 1.5.$$

The least-squares line is given by $\hat{Y} = 1.5 + 1.5x$. If Dean Nee-Meany were to use the least-squares line for prediction, he would, for example, predict that a college faculty member with 2 extramural research grants this year would be expected to publish, on average, $\hat{Y} = 1.5 + 1.5(2) = 4.5$ papers next year. ■

Let me add a word about the use of a fitted regression line for prediction purposes. When it is determined that the linear relationship between variables X and Y is significant, the relationship has been confirmed for x-values in a certain given range. Predictions based on the fitted line would be considered valid only for X values within this same range of x's. For X values outside of that range, we have no information about the relationship between X and Y, and it would be inappropriate, and perhaps quite misleading, to predict the response Y for such Xs. This is a rule which is quite understandable and is also quite easy to apply in simple linear regression. The word of caution about extrapolation from a fitted regression equation applies as well to more complex regression models, but the implementation of the rule is somewhat more challenging.

In developments later in this chapter, we will study the properties of the estimated coefficients of the SLR model when the Ys are treated as random responses to a set of observed X values. For that purpose, we will need to treat $\hat{\beta}_0$ and $\hat{\beta}_1$ as random variables, as is reflected in the following notation for $\hat{\beta}_0$ and $\hat{\beta}_1$:

$$\hat{\beta}_1 = \frac{\sum_{i=1}^{n} x_i Y_i - n \cdot \bar{x} \cdot \bar{Y}}{\sum_{i=1}^{n} x_i^2 - n \cdot \bar{x}^2} \tag{11.13}$$

and

$$\hat{\beta}_0 = \bar{Y} - \hat{\beta}_1 \bar{x}. \tag{11.14}$$

In what follows, the two equivalent expressions given below for the slope $\hat{\beta}_1$ of the least-squares line will prove useful. Demonstrations of their equivalence to $\hat{\beta}_1$ in (11.13) are left as exercises at the end of this section. Note that

$$\hat{\beta}_1 = \frac{\sum_{i=1}^{n}(x_i - \bar{x})(Y_i - \bar{Y})}{\sum_{i=1}^{n}(x_i - \bar{x})^2} \tag{11.15}$$

and also that

$$\hat{\beta}_1 = \frac{\sum_{i=1}^{n}(x_i - \bar{x})Y_i}{\sum_{i=1}^{n}(x_i - \bar{x})^2}. \tag{11.16}$$

We now turn our attention to a particular version of the test for the significance of regression. Recall the correlation $\rho_{X,Y}$ between variables X and Y from Chapter 4. This parameter is defined as

$$\rho_{X,Y} = \frac{Cov(X,Y)}{\sigma_X \cdot \sigma_Y} = \frac{EXY - (EX)(EY)}{\sqrt{[EX^2 - (EX)^2][EY^2 - (EY)^2]}}. \tag{11.17}$$

The correlation coefficient $\rho_{X,Y}$ is a measure of the linearity between X and Y. Indeed, if the linear relationship between X and Y is exact, that is, if X is a random variable and $Y = a + bX$ for fixed constants a and b, then $\rho_{X,Y}$ will take on an extreme value of either -1 or 1 (recall that $|\rho_{X,Y}| \leq 1$), while when the linear relationship between X and Y is weak or nonexistent, $\rho_{X,Y}$ takes a value near or equal to 0. The standard test for significance of regression may be expressed in a variety of forms. We will examine a procedure for testing the hypotheses $H_0 : \rho = 0$ vs. $H_1 : \rho \neq 0$, where $\rho = \rho_{X,Y}$. The test is based on the sample correlation coefficient $r = r_{X,Y}$ given by

$$r_n = \frac{\sum_{i=1}^{n} x_i y_i - n \cdot \bar{x} \cdot \bar{y}}{\sqrt{\left(\sum_{i=1}^{n} x_i^2 - n \cdot \bar{x}^2\right)\left(\sum_{i=1}^{n} y_i^2 - n \cdot \bar{y}^2\right)}}, \tag{11.18}$$

where the subscript on r in (11.18) simply reflects the size of the sample from which r is calculated. When viewed as a function of the n random responses $\{Y_i\}$ to the n observed values $\{x_i\}$, the correlation coefficient is written as

$$r_n = \frac{\sum_{i=1}^{n} x_i Y_i - n \cdot \bar{x} \cdot \bar{Y}}{\sqrt{\left(\sum_{i=1}^{n} x_i^2 - n \cdot \bar{x}^2\right)\left(\sum_{i=1}^{n} Y_i^2 - n \cdot \bar{Y}^2\right)}}. \tag{11.19}$$

Now, as noted in Problem 5.5.41, the general asymptotic distribution of r_n is given by

$$\sqrt{n}(r_n - \rho) \xrightarrow{D} Y \sim N(0, (1 - \rho^2)^2). \tag{11.20}$$

Under the null hypothesis $H_0 : \rho = 0$ above, the random variable r_n converges in distribution to a standard normal variable Z. Thus, the test which rejects H_0 when $|r_n| > (1/\sqrt{n})Z_{\alpha/2}$ is an approximate test of H_0 of size α if n is sufficiently large. But this approximate test is generally found unsatisfactory because the convergence in (11.20) is notoriously slow. Fortunately, an approximate test is not needed, as it is possible to transform r_n to a variable $T = g(r_n)$ which has an exact t distribution under the assumption that $\rho = 0$. It can be shown that, under $H_0 : \rho = 0$, the statistic

$$T = g(r_n) = \frac{r_n \sqrt{n-2}}{\sqrt{1 - r_n^2}} \tag{11.21}$$

has a t distribution with $(n-2)$ degrees of freedom. Thus, an exact test of $H_0 : \rho = 0$ vs. $H_1 : \rho \neq 0$ of size α may be formulated in terms of the statistic T in (11.21). The critical region of this test is: Reject $H_0 : \rho = 0$ in favor $H_1 : \rho \neq 0$ if

$$\left| \frac{r_n\sqrt{n-2}}{\sqrt{1-r_n^2}} \right| > t_{n-2,\alpha/2}. \tag{11.22}$$

A one-sided test is of course similar, rejecting if the statistic T exceeds $t_{n-2,\alpha}$ in one instance or if $T < -t_{n-2,\alpha}$ in the other. But the two-sided test above is by far the most common. When $n > 30$, t cutoff points are replaced by Z cutoff points. We return to the example above to illustrate the t-test for the significance of regression in that situation.

Example 11.1.1 (continued). Let's consider again the data in Example 11.1.1. What was left undone in that example was the execution of a test for the significance of regression. We expand the table of computations in that example to include what we need to compute the sample correlation r:

x	0	0	1	2	2	$\Sigma x = 5$
y	1	2	3	4	5	$\Sigma y = 15$
xy	0	0	3	8	10	$\Sigma xy = 21$
x^2	0	0	1	4	4	$\Sigma x^2 = 9$
y^2	1	4	9	16	25	$\Sigma y^2 = 55$

Using the expression in (11.18), we obtain the value of r_5 as

$$r_n = \frac{21 - 5 \cdot 1 \cdot 3}{\sqrt{(9-5(1)^2)(55-5(3)^2)}} = \frac{6}{\sqrt{4 \cdot 10}} = .9487.$$

The value of our test statistic for testing $H_0 : \rho = 0$ vs. $H_1 : \rho \neq 0$ is

$$T = \frac{(.9487)\sqrt{3}}{\sqrt{1-(.9487)^2}} = 5.197.$$

The critical region of size .05 for testing $H_0 : \rho = 0$ vs. $H_1 : \rho \neq 0$ is: Reject $H_0 : \rho = 0$ if $|T| > t_{3,.025} = 3.182$. Here $|T| = 5.197 > 3.182$, so the null hypothesis is rejected at the 5% significance level. This result serves as justification for fitting the least-square line to these data. ■

Remark 11.1.1. A brief historical note regarding the so-called "regression fallacy" is in order. Sir Francis Galton, a leading scientist in the late nineteenth century, was one of the principle architects of statistical regression analysis. His work was celebrated in his day, seen by many as path-breaking but by others as daring and controversial. Even today, he is admired for the breadth of his contributions to psychology, anthropology, and human genetics. He is also credited with important contributions to the field of statistics, notwithstanding an embarrassing slip in his attempts to interpret the regression analysis he was applying to some anthropological data that he was earnestly investigating. One of his applications involved the heights (X, Y) of fathers and their first-born sons. Galton noticed that, with amazing regularity, the sons of tall fathers tended to be shorter than their fathers, while the sons of short fathers tended to be taller than their fathers. Galton surmised that height was an attribute which enjoyed the property of regression toward the mean. (The name "regression" actually came from this observation.) If tall men had shorter sons and short men had taller sons, Galton deduced that, in the long run, all men would be of the same height, that is, the variability around the mean height would eventually shrink to zero. This turned out to be quite an embarrassing mistake for a certified genius to make. What happens, actually, is that some tall men have taller sons, some short men have shorter sons, and men of average height sometimes have very tall sons and sometimes have very short sons. As a consequence, the distribution of men's heights tends to vary rather little from generation to generation, and the variation that does occur tends to depend on things like changing diets and the heights of the mothers. They are not driven by some natural tendency for heights to regress toward the mean. Still, Galton's overall contributions to the advancement of science continue to be appreciated as a notable and cherished legacy.

Remark 11.1.2. It is important to understand that the existence of a strong relationship between the variables X and Y need not imply that there is a *causal relationship* between X and Y. If two variables X and Y are strongly correlated, we know only that they vary together and that a future value of Y can be reliably predicted from a future observed X. If the correlation is positive, we know that high values of Y tend to occur when the value of X is high. But it may well be that neither variable directly causes the other to have a similar (high or low) value. For example, there may be a third variable that causes both X and Y to be high or low together. A simple example will suffice in making this point clear. As you might imagine, the cost Y of the damages associated with an urban fire is likely to be high precisely when the number X of firemen working on the fire was large. Indeed, data on urban fires would no doubt show that X and Y are strongly correlated. But it is also clear that the damages are not high *because* there are many firemen working on extinguishing the fire. Rather, the size of the fire itself (measured, say, by the dollar value or the size of the property involved) is the primary determining factor in both the amount of damages sustained and in the number of firemen called in to work on the fire. The term *spurious correlation* is often used in reference to variables that vary together due to an extraneous cause.

Exercises 11.1.

1. Horticulturist Flora Plentie has been studying the relationship between the age X of a papaya tree and its height Y. Her latest data is shown below:

X (age)	1	2	3	4	5
Y (ht.)	5	7	9	10	14

   (a) Test the linearity of the relationship between X and Y, that is, test the hypothesis $H_0 : \rho = 0$ against $H_1 : \rho \neq 0$ at significance level .05. (b) Assume that the hypothesis $H_0 : \rho = 0$ above has been rejected. Obtain the least-squares regression line for predicting the height Y of a papaya tree on the basis of its age X. (c) What is your prediction for the height of a 3.5-year-old papaya tree?

2. Show that the expressions for the slope of the least-squares line given in (11.13) and (11.15) are equivalent.

3. Show that the expressions for the slope of the least-squares line given in (11.13) and (11.16) are equivalent.

4. In an effort to predict which companies that offer lightning-fast Internet connections have real staying power, economist C. Moore-DeMande gathered profitability data on five companies for which last year was their first full year of operation. Her data is shown below.

Salesmen-to-Exec. ratio (X)	1	2	4	8	10
No. of Profitable months (Y)	0	1	4	9	8

(a) Test the hypothesis $H_0 : \rho = 0$ against $H_1 : \rho \neq 0$ at significance level .05. (b) Assume that the hypothesis $H_0 : \rho = 0$ above has been rejected. Obtain the least-squares regression line for predicting the profitability of a similar company having its first full year of operation this year on the basis of its salesmen–executive ratio.

5. Show that $\hat{\beta}_1 = (S_x/S_y)r$, where $s_x = \sqrt{\sum_{i=1}^n (x_i - \bar{x})^2}$ and $s_y = \sqrt{\sum_{i=1}^n (y_i - \bar{y})^2}$. This shows, in particular, that $\hat{\beta}_1 = 0$ when $r = 0$, and thus leads to the suspicion that one might be able to test the significance of regression by testing the null hypothesis $H_0 : \beta_1 = 0$ based on the value of $\hat{\beta}_1$. This suspicion is confirmed as correct in Section 11.3.

11.2 Some Distribution Theory for Simple Linear Regression

In this section, I have gathered a number of useful distributional results for various statistics that arise in fitting the SLR model to data. I will provide some comments or hints about the derivation of these results, but I will not be concerned here with rigorous proofs. This section is meant to provide a quick review for you if you've taken a course in regression analysis and to give you a quick reference to these results if you haven't. Let's assume that the data $(X_1, Y_1), \ldots, (X_n, Y_n)$ conform to the SLR model, that is, assume that for $i = 1, \ldots, n$,

$$Y_i | X_i = x_i \sim N(\beta_0 + \beta_1 x_i, \sigma^2) \text{ for } i = 1, \ldots, n. \tag{11.23}$$

Since we treat the "observed" x's as fixed constants, we can see that the intercept $\hat{\beta}_0$ and the slope $\hat{\beta}_1$ of the least-squares line are simply linear combinations of the independent normally distributed random variables $\{Y_i\}$. Specifically, using (11.14), we have

$$\hat{\beta}_1 = \frac{\sum_{i=1}^n (x_i - \bar{x}) Y_i}{\sum_{i=1}^n (x_i - \bar{x})^2} = \sum_{i=1}^n \frac{(x_i - \bar{x})}{\sum_{i=1}^n (x_i - \bar{x})^2} \cdot Y_i. \tag{11.24}$$

and

$$\hat{\beta}_0 = \bar{Y} - \hat{\beta}_1 \bar{x} = \frac{1}{n} \sum_{i=1}^n Y_i - \bar{x} \sum_{i=1}^n \frac{(x_i - \bar{x})}{\sum_{i=1}^n (x_i - \bar{x})^2} \cdot Y_i = \sum_{i=1}^n \left[\frac{1}{n} - \bar{x} \cdot \frac{(x_i - \bar{x})}{\sum_{i=1}^n (x_i - \bar{x})^2} \right] Y_i. \tag{11.25}$$

A simple argument using moment-generating functions will confirm that a linear combination of independent normal variables is also normal. Further, applying Corollary 4.2.1 and Corollary 4.3.1 regarding the mean and variance of a linear combination of independent random variables, we may identify the exact distributions of $\hat{\beta}_0$ and $\hat{\beta}_1$ as

$$\hat{\beta}_0 \sim N\left(\beta_0, \frac{\sigma^2 \sum_{i=1}^n x_i^2}{n \sum_{i=1}^n (x_i - \bar{x})^2}\right) \tag{11.26}$$

and

$$\hat{\beta}_1 \sim N\left(\beta_1, \frac{\sigma^2}{\sum_{i=1}^n (x_i - \bar{x})^2}\right). \tag{11.27}$$

These distributional results imply that $\hat{Y}$, the predicted value of Y at a future observed value of X, is also normally distributed. Since, given $X = x$, where x is within the range of Xs on which the regression relationship was established, $\hat{Y} = \hat{\beta}_0 + \hat{\beta}_1 x$ is a simple linear combination of two normal random variables $\hat{\beta}_0$ and $\hat{\beta}_1$, the mean and variance of $\hat{Y}$ are easily obtained:

$$E(\hat{Y}) = E(\hat{\beta}_0 + \hat{\beta}_1 x) = \beta_0 + \beta_1 x$$

and

$$\begin{aligned} V(\hat{Y}) &= V(\hat{\beta}_0 + \hat{\beta}_1 x) \\ &= V(\bar{Y} - \hat{\beta}_1 \bar{x} + \hat{\beta}_1 x) \\ &= V(\bar{Y} + \hat{\beta}_1 (x - \bar{x})) \\ &= V(\bar{Y}) + (x - \bar{x})^2 V(\hat{\beta}_1), \end{aligned} \tag{11.28}$$

where (11.28) follows from the independence of $\bar{Y}$ and $\hat{\beta}_1$ (see Exercise 11.2.1, where a proof of their independence is examined). We thus have that

$$V(\hat{Y}) = \frac{\sigma^2}{n} + \frac{(x - \bar{x})^2 \sigma^2}{\sum_{i=1}^n (x_i - \bar{x})^2}. \tag{11.29}$$

The predicted value $\hat{Y}$ of Y at a future observed value $X = x$ also serves as the natural estimator of the conditional mean $\mu_{Y|X=x}$. Letting $\hat{\mu}_{Y|X=x} = \hat{\beta}_0 + \hat{\beta}_1 x$, we have

$$\mu_{Y|\hat{X}=x} = \hat{\beta}_0 + \hat{\beta}_1 x \sim N\left(\beta_0 + \beta_1 x, \frac{\sigma^2}{n} + \frac{(x - \bar{x})^2 \sigma^2}{\sum_{i=1}^n (x_i - \bar{x})^2}\right). \tag{11.30}$$

It is often of interest to obtain confidence intervals for the parameters β_0, β_1, and $\mu_{Y|X=x} = \beta_0 + \beta_1 x$. This will require having an estimator of the variance σ^2 in the SLR model. It can be shown that the sum of squared "residuals" $\{Y_i - \hat{Y}_i\}$, usually called the "sum of squares for error" and denoted by SSE, is independent of each of the estimators $\hat{\beta}_0$ and $\hat{\beta}_1$, and has a known distribution as indicated below:

$$\frac{SSE}{\sigma^2} \sim \chi^2_{n-2}. \tag{11.31}$$

From (11.31), we see that the so-called mean squared error (MSE), given by

$$\hat{\sigma}^2 = MSE = \frac{SSE}{n-2} \tag{11.32}$$

is an unbiased estimator of σ^2.

It's really rather satisfying that the distribution theory here works out as nicely as it does. (This is, in fact, one of the reasons for the "popularity" and wide spread use of linear models, as the exact distributions of the statistics of interest are readily available.) Notice that we have normal variables $\hat{\beta}_0$ and $\hat{\beta}_1$ that are independent of the chi-square variable SSE/σ^2, suggesting that we should be able to use Theorem 4.6.2 to construct a ratio having a t distribution. The details of such constructions are left as Exercises 11.2.2 and 11.2.3. These developments lead to the following distributions for $\hat{\beta}_0$, $\hat{\beta}_1$ and $\hat{\mu}_{Y|X=x} = \hat{\beta}_0 + \hat{\beta}_1 x$ and confidence intervals for β_0, β_1 and $\mu_{Y|X=x} = \beta_0 + \beta_1 x$. For $\hat{\beta}_0$, we have

$$\frac{\hat{\beta}_0 - \beta_0}{\hat{\sigma}\sqrt{\sum_{i=1}^n x_i^2 / n \sum_{i=1}^n (x_i - \bar{x})^2}} \sim t_{n-2}, \tag{11.33}$$

and $100(1-\alpha)\%$ confidence limits for β_0 are given by

$$\hat{\beta}_0 \pm \hat{\sigma}\sqrt{\frac{\sum_{i=1}^{n} x_i^2}{n\sum_{i=1}^{n}(x_i-\bar{x})^2}}\, t_{n-2,\alpha/2}; \tag{11.34}$$

for $\hat{\beta}_1$, we have

$$\frac{\hat{\beta}_1-\beta_1}{\hat{\sigma}\sqrt{1/\sum_{i=1}^{n}(x_i-\bar{x})^2}} \sim t_{n-2}, \tag{11.35}$$

and $100(1-\alpha)\%$ confidence limits for β_1 are given by

$$\hat{\beta}_1 \pm \hat{\sigma}\sqrt{\frac{1}{\sum_{i=1}^{n}(x_i-\bar{x})^2}}\, t_{n-2,\alpha/2}; \tag{11.36}$$

and for $\hat{\mu}_{Y|X=x}$, we have

$$\frac{\hat{\mu}_{Y|x}-\mu_{Y|x}}{\hat{\sigma}\sqrt{1/n+(x-\bar{x})^2/\sum_{i=1}^{n}(x_i-\bar{x})^2}} \sim t_{n-2}, \tag{11.37}$$

and $100(1-\alpha)\%$ confidence limits for $\mu_{Y|X=x}$ are given by

$$\hat{\mu}_{Y|x} \pm \hat{\sigma}\sqrt{\frac{1}{n}+\frac{(x-\bar{x})^2}{\sum_{i=1}^{n}(x_i-\bar{x})^2}}\, t_{n-2,\alpha/2}. \tag{11.38}$$

The final inferential result we will discuss in this section is the development of a $100(1-\alpha)\%$ prediction interval for a new (yet to be observed) Y value at a future observed value $X=x$. The structure of the prediction interval we will develop is essentially the same as that of the intervals above. The interval of interest will be centered at $\hat{\mu}_{Y|X=x}$, but it is wider than the confidence interval for $\mu_{Y|X=x}$, the difference stemming from the fact that the difference $Y-\hat{\mu}_{Y|X=x}$ has a larger variance than $\hat{\mu}_{Y|X=x}$ itself, as is evident from the representation

$$Y-\hat{\mu}_{Y|X=x} = (Y-\mu_{Y|X=x})+(\mu_{Y|X=x}-\hat{\mu}_{Y|X=x}). \tag{11.39}$$

Since the new Y is independent of $Y_1,\ldots,Y_n$, it follows from (11.39) that

$$\begin{aligned} V(Y-\hat{\mu}_{Y|X=x}|X=x) &= V(Y|X=x)+V(\hat{\mu}_{Y|X=x}) \\ &= \sigma^2+\frac{\sigma^2}{n}+\frac{(x-\bar{x})^2\sigma^2}{\sum_{i=1}^{n}(x_i-\bar{x})^2} \\ &= \sigma^2\left(1+\frac{1}{n}+\frac{(x-\bar{x})^2}{\sum_{i=1}^{n}(x_i-\bar{x})^2}\right). \end{aligned}$$

We may, again, construct a t statistic using the SSE in (11.31) to obtain

$$\frac{Y-\hat{\mu}_{Y|x}}{\hat{\sigma}\sqrt{1+1/n+(x-\bar{x})^2/\sum_{i=1}^{n}(x_i-\bar{x})^2}} \sim t_{n-2}.$$

It follows that $100(1-\alpha)\%$ prediction limits for a new Y value at a future observed value $X=x$ are given by

$$\hat{\mu}_{Y|x} \pm \hat{\sigma}\sqrt{1+\frac{1}{n}+\frac{(x-\bar{x})^2}{\sum_{i=1}^{n}(x_i-\bar{x})^2}}\, t_{n-2,\alpha/2}.$$

Exercises 11.2.

1. Show that, under the assumptions on the SLR model made in (11.23), two linear combinations of the independent variables $Y_1,\ldots,Y_n$ given by
$$\bar{Y}=\sum_{i=1}^{n}\frac{1}{n}\cdot Y_i \quad\text{and}\quad \hat{\beta}_1=\sum_{i=1}^{n}\frac{(x_i-\bar{x})}{\sum_{i=1}^{n}(x_i-\bar{x})^2}\cdot Y_i,$$
are uncorrelated and thus, since they are normally distributed, are independent. (Hint: Use the expression in Problem 4.8.26 for the covariance of two linear combinations of random variables.)
2. Use the known normal distribution of $\hat{\beta}_0$, the chi-square distribution of SSE/σ^2, and the independence of the two variables to derive the t distribution of the statistic displayed in (11.33).
3. Use the known normal distribution of $\hat{\beta}_1$, the chi-square distribution of SSE/σ^2, and the independence of the two variables to derive the t distribution of the statistic displayed in (11.35).
4. Consider the data in Example 11.1.1. Find 95% confidence intervals for the parameters β_0, β_1, and $\mu_{Y|X=1}$, and obtain a 95% prediction interval for a new value of Y to be observed at a future value $X=2$.

11.3 Theoretical Properties of Estimators and Tests under the SLR Model

We'll begin with an examination of the least-squares estimators of β_0 and β_1 in simple linear regression. There is a very general theorem in linear model theory which states that, under nominal assumptions, the least-squares estimators (LSEs) of the mean parameters of a linear model are the best linear unbiased estimators of these parameters. A rigorous treatment of the famous Gauss-Markov Theorem is beyond the scope of this book. However, we will prove the theorem in the special case we are studying in this chapter, the case of simple linear regression. This has the advantages of allowing us to use the theoretical results we developed in earlier chapters in an important application and of helping us to appreciate why the general result might be true. In the theorem below, we will prove that the LSE $\hat{\beta}_1$ of β_1 is the BLUE of β_1. We will need the following tool.

Lemma 11.3.1. Let $\hat{\beta}_1$ be the least-squares estimator of the slope parameter β_1 in the SLR model, that is, let
$$\hat{\beta}_1=\frac{\sum_{i=1}^{n}(x_i-\bar{x})Y_i}{\sum_{i=1}^{n}(x_i-\bar{x})^2}=\sum_{i=1}^{n}c_iY_i, \tag{11.40}$$
where the constants $\{c_i\}$ are given by
$$c_i=\frac{(x_i-\bar{x})}{\sum_{i=1}^{n}(x_i-\bar{x})^2}\ \text{for}\ i=1,\ldots,n. \tag{11.41}$$
Then, (a) $\sum_{i=1}^{n}c_i=0$, (b) $\sum_{i=1}^{n}c_ix_i=1$ and (c) $\sum_{i=1}^{n}c_i^2=1/\sum_{i=1}^{n}(x_i-\bar{x})^2$.

Proof. (a) This property of the c's is transparent:
$$\sum_{i=1}^{n}c_i=\sum_{i=1}^{n}\frac{(x_i-\bar{x})}{\sum_{j=1}^{n}(x_j-\bar{x})^2}=\frac{\sum_{i=1}^{n}(x_i-\bar{x})}{\sum_{j=1}^{n}(x_j-\bar{x})^2}=\frac{0}{\sum_{i=1}^{n}(x_i-\bar{x})^2}=0.$$

(b) Here, we may write

$$\sum_{i=1}^{n} c_i x_i = \sum_{i=1}^{n} \frac{(x_i-\bar{x})x_i}{\sum_{i=1}^{n}(x_i-\bar{x})^2}$$
$$= \frac{\sum_{i=1}^{n}(x_i-\bar{x})(x_i-\bar{x})}{\sum_{i=1}^{n}(x_i-\bar{x})^2} \tag{11.42}$$
$$= \frac{\sum_{i=1}^{n}(x_i-\bar{x})^2}{\sum_{i=1}^{n}(x_i-\bar{x})^2} = 1,$$

where (11.42) holds because $\sum_{i=1}^{n}(x_i-\bar{x})\bar{x} = \bar{x}\sum_{i=1}^{n}(x_i-\bar{x}) = 0$.
(c) Finally, we have that

$$\sum_{i=1}^{n} c_i^2 = \sum_{i=1}^{n}\left(\frac{(x_i-\bar{x})}{\sum_{i=1}^{n}(x_i-\bar{x})^2}\right)^2$$
$$= \frac{\sum_{i=1}^{n}(x_i-\bar{x})^2}{(\sum_{i=1}^{n}(x_i-\bar{x})^2)^2}$$
$$= 1/\sum_{i=1}^{n}(x_i-\bar{x})^2.$$

■

We are now in a position to prove the following special case of the Gauss-Markov Theorem.

Theorem 11.3.1. Consider the simple linear regression model in (11.1), where the errors $\{\varepsilon_i, i=1,\ldots,n\}$ are assumed to be uncorrelated (but not necessarily normal). The least-squares estimator $\hat{\beta}_1$, given in (11.14), of the slope parameter β_1 in the SLR model is the best linear unbiased estimator of β_1.

Proof. Note, first, that we may write $\hat{\beta}_1$ as $\sum_{i=1}^{n} c_i Y_i$, where the constants $\{c_i\}$ are defined in (11.41). This confirms that $\hat{\beta}_1$ is indeed a linear estimator of β_1. The fact that it is unbiased follows from Lemma 11.3.1, parts (a) and (b):

$$E(\hat{\beta}_1) = E\left(\sum_{i=1}^{n} c_i Y_i\right) = \sum_{i=1}^{n} c_i EY_i = \beta_0 \sum_{i=1}^{n} c_i + \beta_1 \sum_{i=1}^{n} c_i x_i = \beta_0(0) + \beta_1(1) = \beta_1.$$

Now, the variance of β_1 may be identified, using Lemma 11.3.1, part (c), as

$$V(\hat{\beta}_1) = V\left(\sum_{i=1}^{n} c_i Y_i\right) = \sum_{i=1}^{n} c_i^2 V(Y_i) = \sigma^2 \sum_{i=1}^{n} c_i^2 = \frac{\sigma^2}{\sum_{i=1}^{n}(x_i-\bar{x})^2}.$$

We will now show that, among linear unbiased estimators of β_1, $\hat{\beta}_1$ has the smallest possible variance. Let us suppose that $\tilde{\beta}_1 = \sum_{i=1}^{n} d_i Y$ is a linear unbiased estimator of β_1 that differs from $\hat{\beta}_1$. The unbiasedness of $\tilde{\beta}_1$ imposes requirements on the coefficients $\{d_i\}$ similar to those satisfied by $\{c_i\}$, namely

$$\sum_{i=1}^{n} d_i = 0 \text{ and } \sum_{i=1}^{n} d_i x_i = 1. \tag{11.43}$$

The difference between $\hat{\beta}_1$ and $\tilde{\beta}_1$ is reflected in the fact that the coefficients $\{d_i\}$ of the alternative estimator $\tilde{\beta}_1$ are not uniformly equal to the coefficients $\{c_i\}$ of the estimator $\hat{\beta}_1$. We may thus express the relationship between the two sets of coefficients as

$$d_i = c_i + \Delta_i, \tag{11.44}$$

where at least some of the Δ_i are non-zero. We may thus compute the variance of the competing estimator as

$$\begin{aligned} V(\tilde{\beta}_1) &= V\left(\sum_{i=1}^{n} d_i Y_i\right) \\ &= \sum_{i=1}^{n} (c_i + \Delta_i)^2 V(Y_i) \\ &= \sigma^2 \left[\sum_{i=1}^{n} c_i^2 + \sum_{i=1}^{n} \Delta_i^2 + 2\sum_{i=1}^{n} c_i \Delta_i\right]. \end{aligned} \tag{11.45}$$

Since $V(\hat{\beta}_1) = \sigma^2 \sum_{i=1}^{n} c_i^2$ and $\sum_{i=1}^{n} \Delta_i^2 > 0$, the conclusion that $V(\hat{\beta}_1) < V(\tilde{\beta}_1)$ will follow from (11.45) if we can show that $\sum_{i=1}^{n} c_i \Delta_i = 0$. But, using (11.43), this latter identity follows from the calculation below:

$$\begin{aligned} \sum_{i=1}^{n} c_i \Delta_i &= \sum_{i=1}^{n} c_i (d_i - c_i) \\ &= \sum_{i=1}^{n} c_i d_i - \sum_{i=1}^{n} c_i^2 \\ &= \sum_{i=1}^{n} \frac{d_i (x_i - \bar{x})}{\sum_{i=1}^{n} (x_i - \bar{x})^2} - \frac{1}{\sum_{i=1}^{n} (x_i - \bar{x})^2} \\ &= \frac{\sum_{i=1}^{n} d_i x_i - \bar{x} \sum_{i=1}^{n} d_i)}{\sum_{i=1}^{n} (x_i - \bar{x})^2} - \frac{1}{\sum_{i=1}^{n} (x_i - \bar{x})^2} \\ &= \frac{1 - 0}{\sum_{i=1}^{n} (x_i - \bar{x})^2} - \frac{1}{\sum_{i=1}^{n} (x_i - \bar{x})^2} \quad \text{from (11.43)} \\ &= 0. \end{aligned}$$

This completes the proof of the theorem. ■

Remark 11.3.1. The method of proof utilized in Theorem 11.3.1 is of interest in its own right, as it takes an approach that can often be useful in optimization problems. I will refer to it as a "variational argument," as what it actually does is (1) it examines the effect of small perturbations (in any direction) from a *candidate point* at which a minimum (or, more generally, an extreme value) of a function might be obtained and (2) it determines whether the value of the function is diminished (or, changed in a uniform direction) by such permutations. If the value of a function at the point cannot be decreased by arbitrary variations of its coordinates (subject to the restrictions on the points under consideration), the function is in fact minimized at the candidate point. The following theorem, stated without proof, can be established by a variational argument similar to the one above. The proof of this theorem is left as Exercise 11.3.1.

Theorem 11.3.2. Consider the simple linear regression model in (11.1), where the errors

$\{\varepsilon_i, i = 1, \ldots, n\}$ are assumed to be uncorrelated (but not necessarily normal). The least-squares estimator $\hat{\beta}_0$, given in (11.12), of the intercept parameter β_0 in the SLR model is the best linear unbiased estimator of β_0.

Under the additional assumption that the errors in the SLR model are normally distributed, we can say quite a bit more about the least-squares estimators $\hat{\beta}_0$ and $\hat{\beta}_1$. Let's examine the likelihood function $L(\beta_0, \beta_1, \sigma^2 | \mathbf{x}, \mathbf{y})$ associated with the model in (11.2). We have

$$L(\beta_0, \beta_1, \sigma^2 | \mathbf{x}, \mathbf{y}) = \prod_{i=1}^{n} \left(\frac{1}{\sqrt{2\pi}\sigma} \right) \exp\left(-\frac{(y_i - \beta_0 - \beta_1 x_i)^2}{2\sigma^2} \right)$$
$$= \left(\frac{1}{2\pi\sigma^2} \right)^{n/2} \exp\left(-\frac{\sum_{i=1}^{n} (y_i - \beta_0 - \beta_1 x_i)^2}{2\sigma^2} \right).$$

More conveniently, the log likelihood may be written as

$$\ln L = -(n/2)\ln(2\pi\sigma^2) - \frac{\sum_{i=1}^{n} (y_i - \beta_0 - \beta_1 x_i)^2}{2\sigma^2}. \tag{11.46}$$

The resulting likelihood equations obtained by taking partial derivatives of $\ln L$ are

$$\frac{\partial}{\partial \beta_0} \ln L = \frac{\sum_{i=1}^{n} (y_i - \beta_0 - \beta_1 x_i)}{\sigma^2} = 0 \tag{11.47}$$

$$\frac{\partial}{\partial \beta_1} \ln L = \frac{\sum_{i=1}^{n} x_i (y_i - \beta_0 - \beta_1 x_i)}{\sigma^2} = 0 \tag{11.48}$$

and

$$\frac{\partial}{\partial \sigma^2} \ln L = -\frac{n}{2\sigma^2} + \frac{\sum_{i=1}^{n} (y_i - \beta_0 - \beta_1 x_i)^2}{2\sigma^4} = 0. \tag{11.49}$$

You'll notice that the likelihood equations in (11.47) and (11.48) are equivalent to the normal equations given in (11.7) and (11.8). Thus, the least-squares estimators of β_0 and β_1 are, under the assumption of normality, also the unique maximum likelihood estimators of β_0 and β_1. The MLE of σ^2 is a biased estimator of σ^2 that may be written as

$$\tilde{\sigma}^2 = \frac{\sum_{i=1}^{n} (Y_i - \hat{Y}_i)^2}{n}, \tag{11.50}$$

where $\hat{Y}_i = \hat{\beta}_0 + \hat{\beta}_1 X_i$. In Exercise 11.3.2, you are asked to verify that the estimators $(\hat{\beta}_0, \hat{\beta}_1, \tilde{\sigma}^2)$ actually maximize the log-likelihood in (11.46).

Under the normality assumption, there is a good deal more that we can claim about the character and performance of the least-squares estimators $\hat{\beta}_0$ and $\hat{\beta}_1$ of the parameters β_0 and β_1. For starters, it is easy to confirm (see (and do) Exercise 11.3.3) that the vector ($\hat{\beta}_0$, $\hat{\beta}_1$, $\tilde{\sigma}^2$) is a jointly sufficient statistic for the parameter vector $(\beta_0, \beta_1, \sigma^2)$. Further, the fact that $\hat{\beta}_0$ and $\hat{\beta}_1$ are MLEs under a regular model implies that the least-squares estimators are consistent, asymptotically normal, and asymptotically efficient estimators of β_0 and β_1. Finally, under the normality assumption, it is possible to show that both $\hat{\beta}_0$ and $\hat{\beta}_1$ are more than the BLUEs of β_0 and β_1 based on a fixed sample of size n, but they are, in fact, the MVUEs of the parameters β_0 and β_1 (see Exercise 11.3.4). That's quite a bit of mileage from a simple little assumption! Of course the question of whether the normality assumption is appropriate should always be asked. There are a good many graphical and formal approaches to investigating that assumption.

A fair percentage of the procedures that are used to check on the assumptions made in the SLR model deal with the residuals $\{(Y_i - \hat{Y}_i), i = 1, \ldots, n\}$. Where they get their name should be apparent; $Y_i - \hat{Y}_i$, being the difference between the observed and fitted Ys, give us some idea of what the model has failed to explain, what is left over after the model has been fit to the data, the "residual," if you will. We will take a moment here to point out some properties of these residuals. Let us adopt the simplifying notation $e_i = Y_i - \hat{Y}_i$. The easiest possible graphical procedure that, at the same time, can be very informative is a plot of residuals in the plane, that is, the plot of the n points (i, e_i). If the model has fit the data well, this plot will look something like the plot in Figure 11.3.1.

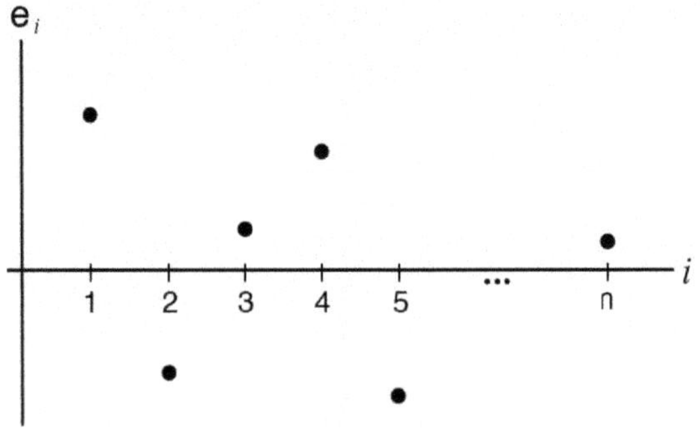

Figure 11.3.1. The residual plot for the set $\{(i, e_i)\}$.

If the plot above shows any kind of trend or has an odd shape, it is usually an indication that the model may be inadequate. A variety of other residual plots are used to detect departures from the modeling assumptions. The plot of e_i vs. X_i can often help detect departures from the constant variance assumption, and the plot of e_i vs. a new independent variable X^* is often used to determine whether the new variable should be added to the regression equation. Of course, there are formal methods like *stepwise regression* for approaching the latter question.

The final topic of this section amounts to a reformulation of the test for significance of regression. This too requires a closer look at residuals, this time from a mathematical perspective. We will see, shortly, that the statistic defined as the sum of squared residuals in useful in regression analysis. We will later refer to the fact that this sum has $(n-2)$ "degrees of freedom." The following result provides the justification for that statement. The theorem essentially says that the residuals $Y_1 - \hat{Y}_1, \ldots, Y_n - \hat{Y}_n$ satisfy exactly two linear constraints, so that if you know that values of $n-2$ of these residuals, the remaining two are completely determined.

Theorem 11.3.3. Suppose that, given the observed X values $x_1, \ldots, x_n$, the variables $Y_1, \ldots, Y_n$ satisfy (11.1), that is, $Y_i = \beta_0 + \beta_1 x_i + \varepsilon_i$ for $i = 1, \ldots, n$, where the errors $\{\varepsilon_i\}$ are uncorrelated. Let $\hat{Y} = \hat{\beta}_0 + \hat{\beta}_1 x$, where $\hat{\beta}_0$ and $\hat{\beta}_1$ are the least squares estimators of β_0 and β_1. Finally, let $Y_i - \hat{Y}_i$, the ith residual of the fitted model, be denoted by e_i. Then (a) $\sum_{i=1}^n e_i = 0$, (b) $\sum_{i=1}^n x_i e_i = 0$, and (c) $\sum_{i=1}^n \hat{Y}_i e_i = 0$.

Proof. The identities $\sum_{i=1}^n e_i = 0$ and $\sum_{i=1}^n x_i e_i = 0$ may be rewritten as the first and second normal equations, and thus hold by the virtue of the fact that the least-squares estimators $\hat{\beta}_0$ and $\hat{\beta}_1$ are the unique solutions of these equations. The identity in (c) follows from (a) and (b) since

$$\sum_{i=1}^{n} \hat{Y}_i e_i = \sum_{i=1}^{n} (\hat{\beta}_0 - \hat{\beta}_1 x_i) e_i = \hat{\beta}_0 \sum_{i=1}^{n} e_i - \hat{\beta}_1 \sum_{i=1}^{n} x_i e_i = 0.$$

■

The identities established in Theorem 11.3.3 allow us to prove the important representation below. The theorem below is restricted to the SLR framework, but is valid for the general linear model.

Theorem 11.3.4. (The Fundamental Decomposition of Variability in Simple Linear Regression). Suppose that, given the observed X values $x_1, \ldots, x_n$, the variables $Y_1, \ldots, Y_n$ satisfy (11.1), that is, $Y_i = \beta_0 + \beta_1 x_i + \varepsilon_i$ for $i = 1, \ldots, n$, where the errors $\{\varepsilon_i\}$ are uncorrelated. Let $\hat{Y} = \hat{\beta}_0 + \hat{\beta}_1 x$, where $\hat{\beta}_0$ and $\hat{\beta}_1$ are the least-squares estimators of β_0 and β_1. The variability of $Y_1, \ldots, Y_n$ may be decomposed as

$$SST = SSR + SSE, \tag{11.51}$$

where the total sum of squares SST is given by $\sum_{i=1}^{n} (Y_i - \bar{Y})^2$, the sum of squares for regression SSR is given by $\sum_{i=1}^{n} (\hat{Y}_i - \bar{Y})^2$ and the error sum of squares SSE is given by $\sum_{i=1}^{n} (Y_i - \hat{Y}_i)^2$.

Proof. We may expand SST as follows:

$$\begin{aligned} SST &= \sum_{i=1}^{n} (Y_i - \bar{Y})^2 \\ &= \sum_{i=1}^{n} (Y_i - \hat{Y}_i + \hat{Y}_i - \bar{Y})^2 \qquad (11.52) \\ &= \sum_{i=1}^{n} (Y_i - \hat{Y}_i)^2 + \sum_{i=1}^{n} (\hat{Y}_i - \bar{Y})^2 + 2\sum_{i=1}^{n} (Y_i - \hat{Y}_i)(\hat{Y}_i - \bar{Y}). \qquad (11.53) \end{aligned}$$

The first two terms on the RHS of (11.53) are SSE and SSR, respectively. Thus, the proof will be complete if we show that

$$\sum_{i=1}^{n} (Y_i - \hat{Y}_i)(\hat{Y}_i - \bar{Y}) = 0. \tag{11.54}$$

But the equation in (11.54) is a direct consequence of parts (a) and (c) of Theorem 11.3.3:

$$\sum_{i=1}^{n} (Y_i - \hat{Y}_i)(\hat{Y}_i - \bar{Y}) = \sum_{i=1}^{n} e_i(\hat{Y}_i - \bar{Y}) = \sum_{i=1}^{n} \hat{Y}_i e_i - \bar{Y} \sum_{i=1}^{n} e_i = 0.$$

■

Remark 11.3.2. It is worth mentioning that Theorems 11.3.3 and 11.3.4 hold without any modeling assumptions whatever. The theorems apply to any arbitrary collection of pairs (x_i, Y_i) and the approximations of the Ys by the best-fitting (least-squares) line $\hat{Y} = \hat{\beta}_0 + \hat{\beta}_1 x$. In statistical applications, such as in the test for the significance of regression developed below, we will use the results above in a context in which the modeling assumptions are essential.

Theorem 11.3.4 has a simple and natural interpretation. The total variability in a set of n Ys is measured in the usual way, that is, by SST, the sum of squared deviations of the Ys from their mean $\bar{Y}$. The sum of squared differences SSR between the Ys at the various x values and the value $\hat{Y} = \hat{\beta}_0 + \hat{\beta}_1 x$ on the least-squares line measures the quality of the fit of the line to the data. It accounts for the amount of the variability in the Ys that the linear trend in the data can explain. When the fit is perfect, that is, every pair (x, Y) is on the line, the SSR is equal to the SST. If SSR is small relative to SST, the regression line explains only a small part of the variability of Y, an indication that the linear relationship between X and Y is weak. The ratio SSR/SST of the two measures is known as R^2, the *coefficient of determination*, and is a standard measure of the quality of fit of a straight line to the data. It is often described as the percentage of the variability of Y explained by the least-squares line. Its importance as a measure of linearity is no surprise, given the fact that R^2 is algebraically equal to the square of the sample correlation coefficient r (see Problem 11.6.10).

The test that I will discuss below will be our first official encounter with a probability model called the F distribution. I will take a brief digression here to discuss this distribution. The F distribution was the subject of Problem 4.8.66. As mentioned there, the F distribution arises from the following construction: If U and V are independent chi-square variables, with $U \sim \chi^2_k$ and $V \sim \chi^2_r$, then the variable $X = (U/k)/(V/r)$ has an F distribution with k and r degrees of freedom. The density of the $F_{k,r}$ distribution is given by

$$f_X(x) = \frac{\Gamma((k+r)/2)\,(k/r)^{k/2}}{\Gamma(k/2)\Gamma(r/2)} \frac{x^{k/2-1}}{(1+kx/r)^{(k+r)/2}} I_{(0,\infty)}(x).$$

Because the F distribution arises as a ratio of two positive random variables, an F variable is necessarily positive. It can be shown that if $X_{k,r} \sim F_{k,r}$, then the mean and variance of $X_{k,r}$ are given by

$$EX_{k,r} = \frac{r}{r-2} \text{ for } r > 2$$

and

$$V(X_{k,r}) = 2\left(\frac{r}{r-2}\right)^2 \frac{(k+r-2)}{k(r-4)} \text{ for } r > 4.$$

The density of a typical $F_{k,r}$ distribution is displayed in Figure 11.3.1. Also shown in that figure is the α-cutoff point $F_{k,r,\alpha}$. If $X \sim F_{k,r}$, then $P(X > F_{k,r,\alpha}) = \alpha$.

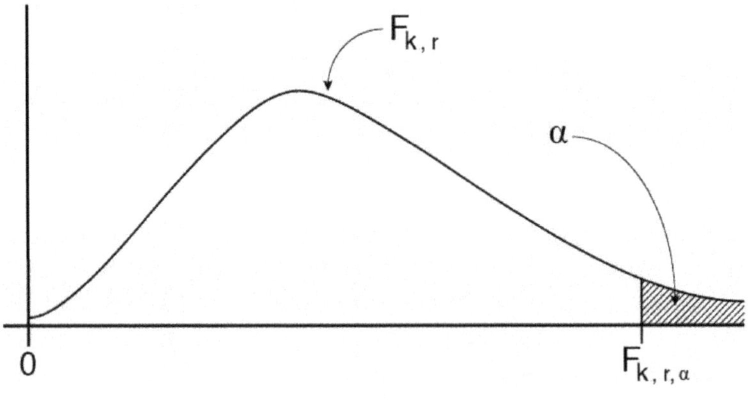

Figure 11.3.2. An $F_{k,r}$ density with α-cutoff point $F_{k,r,\alpha}$.

The test for significance of regression is often presented through what is called an analysis of variance table. Such a table accounts for the various sources of variability of a random variable Y and records the amount of that variability that is due to each of these sources. For simple linear regression, the ANOVA table is shown below. The table identifies two sources of variation, sums of squares for each, their attendant degrees of freedom (df), the mean squares (SS/df) for each source, and their expected values (EMS).

Source of Variation	SS	df	MS	EMS
Regression	SSR	1	$MSR = SSR/1$	$\sigma^2 + \beta_1^2 \sum (x_i - \bar{x})^2$
Error	SSE	$n-2$	$MSE = SSE/(n-2)$	σ^2
Total	SST	$n-1$		

Table 11.3.1. ANOVA table for simple linear regression.

The test for significance of regression is most often represented as a test of the hypotheses $H_0 : \beta_1 = 0$ vs. $H_1 : \beta_1 \neq 0$. If H_0 is true, then the true regression line is a horizontal line and the value of X has no effect on the mean of the dependent variable Y. From Table 11.3.1, you can see that, under the null hypothesis, the expected mean square for regression is equal to σ^2, the same as the expected mean square for error (which is equal to σ^2 under either hypothesis). This suggests that the ratio MSR/MSE will be reasonably close to the value 1 when the null hypothesis is true. Under the alternative hypothesis, we expect a larger value of MSR. It follows that it is a large value of the ratio MSR/MSE that might lead us to consider rejecting H_0.

Under the hypothesis $H_0 : \beta_1 = 0$, the exact distribution of the ratio MSR/MSE is known and extensively tabled. The construction of this distribution utilizes the following facts, all of which hold under the null hypothesis: SSR and SSE are independent random variables, with $SSR/\sigma^2 \sim \chi_1^2$ and $SSE/\sigma^2 \sim \chi_{n-2}^2$ and the ratio $F = MSR/MSE$ has an F distribution with 1 and $(n-2)$ degrees of freedom, that is, $F \sim F_{1,n-2}$. Since it is large values of F that tend to discredit H_0, the standard test of size α for testing $H_0 : \beta_1 = 0$ vs. $H_1 : \beta_1 \neq 0$ has critical region:

$$\text{Reject } H_0 \text{ if } F > F_{1,n-2,\alpha}. \tag{11.55}$$

The F test for the significance of regression is equivalent to the test treated in Section 11.1, namely, the test based on the correlation coefficient r_n. That this is that case is obvious from the fact that $(r_n)^2 = R^2 = F/[(n-2)+F]$, so that rejecting H_0 when F is large is equivalent to rejecting H_0 when $|r_n|$ is large which, in turn, is equivalent to rejecting H_0 when $|r_n\sqrt{n-2}/\sqrt{1-r_n^2}|$ is large.

We end this section with an example of the F test applied to Example 11.1.

Example 11.3.1. Recall that Example 11.1.1 deals with the variables X and Y, where X is the number of extramural research grants a faculty member received during the previous year and Y is the number of research papers the faculty member publishes the following year. The data obtained on a sample of size 5 is shown below, as is the predicted value $\hat{y}$ for each x drawn from the least-squares line $\hat{Y} = 1.5 + 1.5x$.

x	0	0	1	2	2
y	1	2	3	4	5
$\hat{y}$	1.5	1.5	3	4.5	4.5

We may thus calculate SSR, SSE, and SST as follows:

$$SSR = \sum_{i=1}^{n} (\hat{Y}_i - \bar{Y})^2 = (-1.5)^2 + (-1.5)^2 + (0)^2 + (1.5)^2 + (1.5)^2 = 9$$

$$SSE = \sum_{i=1}^{n} (Y_i - \hat{Y}_i)^2 = (-.5)^2 + (.5)^2 + (0)^2 + (-.5)^2 + (.5)^2 = 1$$

and

$$SST = \sum_{i=1}^{n} (Y_i - \bar{Y})^2 = (-2)^2 + (-1)^2 + (0)^2 + (1)^2 + (2)^2 = 10.$$

Of course, SST may be obtained as $SSR + SSE$, but there's no harm in checking one's arithmetic. The resulting ANOVA table is

Source of Variation	SS	df	MS	EMS
Regression	9	1	9	$\sigma^2 + \beta_1^2 \sum (x_i - \bar{x})^2$
Error	1	3	.3333	σ^2
Total	10	4		

The F statistic for these data is $F = 27 > F_{1,3,.05} = 10.125$, so that $H_0 : \beta_1 = 0$ would be rejected at the 5% significance level. ■

Exercises 11.3.

1. Prove Theorem 11.3.2.
2. Verify that the estimators $(\hat{\beta}_0, \hat{\beta}_1, \tilde{\sigma}^2)$ actually maximize the log-likelihood in (11.45).
3. Suppose that the available data of size n satisfies the form of the SLR model as stated in (11.23). Show that the vector $(\hat{\beta}_0, \hat{\beta}_1, \tilde{\sigma}^2)$ of MLEs of the parameters of the model is a jointly sufficient statistic for the parameter vector $(\beta_0, \beta_1, \sigma^2)$.
4. Suppose that the available data of size n satisfies the form of the SLR model as stated in (11.23). Show that the least-squares estimators $\hat{\beta}_0$ and $\hat{\beta}_1$ of the parameters β_0 and β_1 are the MVUEs of these parameters.
5. Employing the definitions of e_i, $\hat{Y}_i$, and $\hat{\beta}_1$, give a direct algebraic proof of part (a) of Theorem 11.3.3.
6. People are always asking for concrete examples. Well, here's one for you. A construction engineer wants to establish a formula for the final (28-day) strength Y of concrete in terms of the strength X of concrete after two days of hardening. The data he has available, measured in 100 pounds of pressure per unit volume, is as follows:

x	3	3	4	4	6
y	4	5	6	7	8

(a) From a scatter plot, it appears evident that the X and Y variables above are linearly related. Obtain the least-squares regression line for predicting Y from X. (b) Develop an ANOVA table and carry out a formal F test for the significance of regression.

7. In applications of the Simple Linear Regression (SLR) model to processes measuring output as a function of input, the intercept parameter β_0 is generally presumed to be zero. In this case, the SLR model simplifies to

$$Y_i = \beta x_i + \varepsilon_i, \qquad \text{for } i = 1, 2, \ldots, n,$$

where the x_is are known and $\varepsilon_i \stackrel{\text{iid}}{\sim} N(0, \sigma^2)$. (a) Show that $\hat{\beta} = \sum_{i=1}^n x_i Y_i / \sum_{i=1}^n x_i^2$ is the least-squares estimator of β. It is known that $\hat{\beta}$ is the best linear unbiased estimator (BLUE) of β when the errors are assumed to be uncorrelated. Under the assumption that $\varepsilon_i \stackrel{\text{iid}}{\sim} N(0, \sigma^2)$, one can show that $\hat{\beta}$ is the MVUE of β as follows. (b) Using that $Y_i | X_i = x_i \sim N(\beta x_i, \sigma^2)$, compute the Fisher Information $I_{Y_i}(\beta)$ in the single observation Y_i. (c) Obtain the Cramér-Rao Lower Bound (CRLB) for the variance of an unbiased estimator of β. (Hint: Although the Ys above are not identically distributed, the independence assumption implies that

$$I_{\mathbf{Y}}(\beta) = I_{(Y_1, Y_2, \ldots, Y_n)}(\beta) = \sum_{i=1}^n I_{Y_i}(\beta).$$

This expression is the right alternative to the Fisher Information $nI_Y(\theta)$ from an iid sample of Ys.) (d) Show that $\hat{\beta}$ is the minimum variance unbiased estimator of β, that is, confirm that $V(\hat{\beta}) = \text{CRLB}$.

11.4 One-Way Analysis of Variance

The analysis of variance is often thought of, and is referred to as, the "other half" of linear model theory. It seems appropriate to begin our discussion of ANOVA by indicating exactly what makes it different from regression analysis. The primary difference is the character of the independent variable X. In linear regression, the X variables are typically continuous, and our goal is to describe how one or more X variables affect the value that a dependent variable Y will take. The analysis of variance deals, instead, with a qualitative variable X that might keep track of particular conditions under which experiments are performed. A typical example would be the examination of the effects of three types of fertilizer on crop growth by measuring the growth of a certain crop in three separate sections of an experimental field in which all factors other than the fertilizers used are held constant. In experiments such as this, observations of the dependent variable are taken under a finite collection of experimental conditions and, often, these conditions cannot be expressed in terms of particular values of some concomitant variable X. The conditions under which experiments are run in the analysis of variance are typically called "factors," with each factor having a finite set of "levels." A "one-way ANOVA" is the phrase used to describe experiments in which there is only one factor of interest; in a one-way ANOVA, every experimental condition is viewed as a different level of that single factor. The fertilizer example above would represent a one-way ANOVA with the factor "fertilizer" available at three levels, that is, in three distinct types. If one wishes to study the effect that the octane level of gasoline has on mileage, the factor octane would probably be divided into low, medium, and high, as those divisions aptly describe the way that gasoline is marketed.

The data that one typically faces in a one-way ANOVA problem will look like the data in the table below. The standard assumptions that are made about these data are (1) independent random samples are drawn from the k populations corresponding to the k

levels of the factor under study, (2) sample sizes may vary among from factor levels, and (3) if $Y_{i,j}$ is the jth observation at factor level i, then

$$Y_{i,j} \stackrel{\text{iid}}{\sim} N(\mu_i, \sigma^2) \text{ for } j = 1, \ldots, n_i \text{ and for } i = 1, \ldots, k. \tag{11.56}$$

An ANOVA data set in a "one-way layout" may be displayed as shown below.

Factor Level	Independent samples at each factor level
1	$Y_{1,1} Y_{1,2} \cdots\cdots\cdots\cdots\cdots Y_{1,n_3}$
2	$Y_{2,1} Y_{2,2} \cdots\cdots Y_{2,n_2}$
3	$Y_{3,1} Y_{3,2} \cdots\cdots\cdots\cdots Y_{3,n_3}$
...	......
...	......
$k-1$	$Y_{k-1,1} Y_{k-1,2} \cdots\cdots\cdots Y_{k-1,n_{k-1}}$
k	$Y_{k,1} Y_{k,2} \cdots\cdots\cdots\cdots\cdots\cdots Y_{k,n_k}$

Table 11.4.1. Data in a one-way ANOVA.

The analysis of data such as that shown in Table 11.4.1 will benefit from the introduction of some special language and notation. It is common to refer to the different levels of the factor of interest as *treatments*. As is clear from the table and also from (11.56), the size of the sample drawn from treatment i is n_i where the $n_i s$ may vary. We will denote the total sample size as N, that is, we define

$$N = \sum_{i=1}^{k} n_i; \tag{11.57}$$

we will use the standard "dot notation" for averages of interest, that is, we denote the means of interest by

$$\bar{Y}_{i\cdot} = \frac{1}{n_i} \sum_{j=1}^{n_i} Y_{i,j} \tag{11.58}$$

and

$$\bar{Y}_{\cdot\cdot} = \frac{1}{N} \sum_{i=1}^{k} \sum_{j=1}^{n_i} Y_{i,j}. \tag{11.59}$$

The averages $\bar{Y}_{i\cdot}$ and $\bar{Y}_{\cdot\cdot}$ are called, respectively, the treatment means and the grand mean. It is easy to verify (see Exercise 11.4.1) that the sample means $\{\bar{Y}_{i\cdot}\}$ of the k treatments are the least-squares estimators of the theoretical treatment means $\{\mu_i\}$. The analysis of variance got its name from the fact that the variability of the data set $\{Y_{i,j}\}$ can be dissected to determine precisely what proportion of that variability could be attributed to possible differences among the treatments. The intuitive content of ANOVA is derived from the simple observation that the variability *within* a given treatment, say the ith, as reflected in the sum of squares $\sum_{j=1}^{n_i} (y_{i,j} - y_{i\cdot})^2$, sheds light on the amount of random variation in an iid sample, while the variability we see in the treatment means $\{\bar{Y}_{i\cdot}\}$ sheds light on whether or not the treatment means $\{\mu_i\}$ are the same or different. The latter type of variability is usually referred to as the variability *between* treatments. The relative size of the estimated variability due to each of these sources (within and between treatments) is the basis for testing the standard null hypothesis is one-way ANOVA, namely

$$H_0 : \mu_1 = \mu_2 = \cdots = \mu_k.$$

The alternative hypothesis H_1 is taken to be the complementary statement, namely, that the means are not all equal. The logic of the standard F test comes from the following representation of the total variability in an ANOVA data set.

Theorem 11.4.1. (The Fundamental Decomposition of Variability in One-Way ANOVA). Suppose that data is available from k groups, and denote the jth observation from the ith group as $Y_{i,j}$. Denote the mean of the ith group as $\bar{Y}_{i\cdot}$ and the overall mean of the data as $\bar{Y}_{\cdot\cdot}$. Then, the variability in data $\{Y_{i,j}\}$ may be decomposed as

$$SST = SSB + SSW, \tag{11.60}$$

where the total sum of squares SST is given by $\sum_{i=1}^{k}\sum_{j=1}^{n_i}(Y_{i,j} - \bar{Y}_{\cdot\cdot})^2$, the sum of squares SSB between treatments is given by $\sum_{i=1}^{k} n_i(\bar{Y}_{i\cdot} - \bar{Y}_{\cdot\cdot})^2$, and the error sum of squares SSW within treatments is given by $\sum_{i=1}^{k}\sum_{j=1}^{n_i}(Y_{i,j} - \bar{Y}_{i\cdot})^2$.

Proof. As the proof of this result is very similar to that of Theorem 11.3.4, we will simply outline the main elements of the proof and leave certain statements to be verified in the exercises. First, we note that, by writing $Y_{i,j} - \bar{Y}_{\cdot\cdot} = Y_{i,j} - \bar{Y}_{i\cdot} + \bar{Y}_{i\cdot} - \bar{Y}_{\cdot\cdot}$, SST may be expanded as

$$SST = \sum_{i=1}^{k}\sum_{j=1}^{n_i}(Y_{i,j} - \bar{Y}_{i\cdot} + \bar{Y}_{i\cdot} - \bar{Y}_{\cdot\cdot})^2 \tag{11.61}$$

$$= \sum_{i=1}^{k}\sum_{j=1}^{n_i}(Y_{i,j} - \bar{Y}_{i\cdot})^2 + \sum_{i=1}^{k}\sum_{j=1}^{n_i}(\bar{Y}_{i\cdot} - \bar{Y}_{\cdot\cdot})^2 + 2\sum_{i=1}^{k}\sum_{j=1}^{n_i}(Y_{i,j} - \bar{Y}_{i\cdot})(\bar{Y}_{i\cdot} - \bar{Y}_{\cdot\cdot}) \tag{11.62}$$

$$= \sum_{i=1}^{k}\sum_{j=1}^{n_i}(Y_{i,j} - \bar{Y}_{i\cdot})^2 + \sum_{i=1}^{k} n_i(\bar{Y}_{i\cdot} - \bar{Y}_{\cdot\cdot})^2 + 2\sum_{i=1}^{k}\sum_{j=1}^{n_i}(Y_{i,j} - \bar{Y}_{i\cdot})(\bar{Y}_{i\cdot} - \bar{Y}_{\cdot\cdot}), \tag{11.63}$$

since the summand in the middle term of (11.61) doesn't depend on j. Note that the first two terms in (11.62) are precisely equal to SSW and SSB. Thus, the proof of (11.60) may be completed by showing that

$$\sum_{i=1}^{k}\sum_{j=1}^{n_i}(Y_{i,j} - \bar{Y}_{i\cdot})(\bar{Y}_{i\cdot} - \bar{Y}_{\cdot\cdot}) = 0. \tag{11.64}$$

This final component of the proof is left as Exercise 11.4.2. ■

The elements of the standard test of the hypothesis $H_0 : \mu_1 = \mu_2 = \cdots = \mu_k$ may be presented through the following analysis of variance table. The degrees of freedom in the table are easily explained. Clearly, the degrees of freedom associated with SST is n − 1, as the differences $Y_{i,j} - \bar{Y}_{\cdot\cdot}$ involved in that sum of squares are subject to a single linear constraint, namely,

$$\sum_{i=1}^{k}\sum_{j=1}^{n_i}(Y_{i,j} - \bar{Y}_{\cdot\cdot}) = 0.$$

For similar reasons, the sum of squares $\sum_{j=1}^{n_i}(y_{i,j} - y_{i\cdot})^2$ has $n_i - 1$ degrees of freedom, a fact that implies that $\sum_{i=1}^{k}\sum_{j=1}^{n_i}(y_{i,j} - y_{i\cdot})^2$ has $\sum_{i=1}^{k}(n_i - 1) = N - k$ degrees of freedom. Finally, the ANOVA table refers to the "grand mean" μ, which is yet to be defined. The parameter μ represents the expected value of the average $\bar{Y}_{..}$, which, under the general assumption in (11.56), is equal to

$$\mu = \frac{\sum_{i=1}^{k} n_i \mu_i}{N}. \tag{11.65}$$

Under $H_0 : \mu_1 = \cdots = \mu_k$, the parameter μ in (11.65) is the common value of the μ_i.

Source of Variation	SS	df	MS	EMS
Between Treatments	SSB	$k-1$	$MSB = SSB/(k-1)$	$\sigma^2 + \frac{1}{k-1}\sum_i n_i(\mu_i - \mu)^2$
Within Treatments	SSW	$N-k$	$MSW = SSW/(N-k)$	σ^2
Total	SST	$N-1$		

Table 11.4.2. ANOVA table for a one-way layout.

The standard test for the hypothesis $H_0 : \mu_1 = \mu_2 = \cdots = \mu_k$ is based on the ratio MSR/MSE. From Table 11.4.2, you can see that, under the null hypothesis, the expected mean square between treatments is equal to σ^2, the same as the expected mean square within treatments (which is equal to σ^2 under both H_0 and H_1). Thus, we expect that MSR/MSE will be reasonably close to the value 1 when the null hypothesis is true. Under the alternative hypothesis, we expect a larger value of MSR. It follows that it is a large value of the ratio MSR/MSE that might lead us to consider rejecting H_0.

Under the hypothesis $H_0 : \mu_1 = \mu_2 = \cdots = \mu_k$, the exact distribution of the ratio MSR/MSE is well known and is tabled in all textbooks that treat ANOVA. The construction of this distribution utilizes the following facts, all of which hold under the null hypothesis: SSB and SSW are independent random variables with distributions $SSB/\sigma^2 \sim \chi^2_{k-1}$ and $SSW/\sigma^2 \sim \chi^2_{N-k}$ and the ratio $F = MSB/MSW$ has an F distribution with $k-1$ and $(N-k)$ degrees of freedom, that is, $F \sim F_{k-1,N-k}$. Since it is large values of F that lead us to suspect that H_0 may be false, the standard test of size α for testing $H_0 : \mu_1 = \mu_2 = \cdots = \mu_k$ has critical region: Reject H_0 if $F > F_{k-1,N-k,\alpha}$.

Example 11.4.1. (Adapted from Johnson (2005)) Three brands of drying compounds were applied to ordinary library paste, and the time in minutes for the paste to dry after application was recorded. The data obtained are shown below.

Brand	Dry Is Us	Best Dry Buy	Dry Your Tears
Data	13, 10, 8, 11, 8	13, 11, 14, 14	4, 1, 3, 4, 2, 4

The three treatment means are easily seen to be $\bar{X}_{1\cdot} = 10$, $\bar{X}_{2\cdot} = 13$, and $\bar{X}_{3\cdot} = 5$. The table below summarizes the necessary calculations for carrying out the usual F test.

Source of Variation	SS	df	MS	F
Between Treatments	270	2	135	50.6
Within Treatments	32	12	2.667	
Total	302	14		

Since $F = 50.6 > F_{2,12,.05} = 3.89$, the null hypothesis of equal means would be readily rejected. ■

When the null hypothesis of equal means is rejected, it is natural to ask the question: Can specific treatment means be identified that are significantly larger or smaller than other means? There are various methods for making *multiple comparisons* among the treatment means. The following method, due to Henry Scheffé (1953), handles a somewhat larger collection of questions involving linear combinations of treatment means. The terminology involved is introduced in the following:

Definition 11.4.1. Consider a one-way ANOVA layout as modeled in (11.56). A *contrast* L is a linear combination of means, that is,

$$L = \sum_{i=1}^{k} c_i \mu_i, \tag{11.66}$$

such that the constants $\{c_i, i = 1, \ldots, k\}$ satisfy the constraint

$$\sum_{i=1}^{k} c_i = 0. \tag{11.67}$$

Typical contrasts of interest would be differences in means, like $\mu_1 - \mu_2$, and differences between averages of means, like $(\mu_1 + \mu_2)/2 - (\mu_3 + \mu_4 + \mu_6)/3$, but other contrasts, such as $\mu_1 - 3\mu_2 + 2\mu_3$, occasionally arise. Now, the natural estimator of a contrast is its sample analog, that is, one would typically estimate $L = \sum_{i=1}^{k} c_i \mu_i$ by the unbiased estimator

$$\hat{L} = \sum_{i=1}^{k} c_i \bar{X}_{i\cdot}. \tag{11.68}$$

As asserted in Problems 11.6.44 and 11.6.46, $\hat{L}$ is, in fact, the best linear unbiased estimate (i.e., the BLUE) of L, even without the assumption of normality in (11.56), and $\hat{L}$ is the minimum variance unbiased estimator of L when the normality assumption in (11.56) holds. It is easy to confirm that, under the model in (11.56), $\hat{L}$ is normally distributed with variance given by

$$V(\hat{L}) = MSW \sum_{i=1}^{k} \frac{c_i^2}{n_i}. \tag{11.69}$$

Scheffé proved that interesting and highly useful result that the probability is $1 - \alpha$ that all contrasts L simultaneously satisfy the bounds

$$\hat{L} - S \cdot s(\hat{L}) \leq L \leq \hat{L} + S \cdot s(\hat{L}), \tag{11.70}$$

where $S = \sqrt{(k-1)F(k-1, N-k, \alpha)}$ and $s(\hat{L}) = \sqrt{V(\hat{L})}$. The formal probability statement justifying the Scheffé method of simultaneous confidence intervals for all contrasts is

$$P\left(\hat{L} - S \cdot s(\hat{L}) \leq L \leq \hat{L} + S \cdot s(\hat{L}) \quad \forall L \in A, \hat{L} \in B, \mathbf{c} \in C\right) = (1 - \alpha), \tag{11.71}$$

where $A = \{L = \sum_{i=1}^{k} c_i \mu_i\}$, $B = \{\hat{L} = \sum_{i=1}^{k} c_i \bar{X}_{i\cdot}\}$ and $C = \{c_1, \ldots, c_k | \sum_{i=1}^{k} c_i = 0\}$. One could, by invoking (11.71), construct simultaneous confidence intervals for all $\binom{k}{2}$ pairwise contrasts $(\mu_r - \mu_s)$ of treatment means for $1 \leq r < s \leq k$ with an overall confidence coefficient bounded below by $1 - \alpha$. One could, in fact, include simultaneous confidence intervals for complex contrasts as well while maintaining the same overall confidence level. The Scheffé method of simultaneous inference is illustrated in the following continuation of our earlier example.

Example 11.4.1 (cont'd). In this example, $n_{1\cdot} = 5$, $\bar{X}_{1\cdot} = 10$, $n_{2\cdot} = 4$, $\bar{X}_{2\cdot} = 13$, $n_{1\cdot} = 6$, $\bar{X}_{3\cdot} = 5$, and $MSW = 2.667$. Since, for pairwise contrasts, all the constants c_i are equal to $-1, 0$, or $+1$ and only two are non-zero, we find that

$$\hat{L}_{1,2} = \bar{X}_{1\cdot} - \bar{X}_{2\cdot} = -3, s(\hat{L}_{1,2}) = \sqrt{(2.667)(9/20)} = 1.0955,$$

$$\hat{L}_{1,3} = \bar{X}_{1\cdot} - \bar{X}_{3\cdot} = 5, s(\hat{L}_{1,3}) = \sqrt{(2.667)(11/30)} = .9889,$$

and

$$\hat{L}_{2,3} = \bar{X}_{2\cdot} - \bar{X}_{3\cdot} = 8, s(\hat{L}_{2,3}) = \sqrt{(2.667)(10/24)} = 1.0542.$$

Further, we find that $S = \sqrt{2F_{2,12,.05}} = \sqrt{2(3.89)} = 2.7893$. Thus, at a confidence level no smaller than 95%, we may compute the simultaneous confidence limits for the three contrasts of interest to be: -3 ± 3.0557 for $\mu_1 - \mu_2$, 5 ± 2.7583 for $\mu_1 - \mu_3$, and 8 ± 2.9405 for $\mu_2 - \mu_3$. Since the confidence interval for $\mu_1 - \mu_2$ contains 0, we cannot claim a significant difference between μ_1 and μ_2. On the other hand, we may assert with 95% confidence that $\mu_1 > \mu_3$ and that $\mu_2 > \mu_3$. ■

Exercises 11.4.

1. Consider the sum of squares associated with a one-way analysis of variance: $SS = \sum_{i=1}^{k} \sum_{j=1}^{n_i} (Y_{i,j} - \mu_i)^2$ Show that the treatment means $\{\hat{\mu}_i = \bar{Y}_{i\cdot}, i = 1, \ldots, k\}$ minimize SS and are thus least-squares estimators of the population means $\{\mu_i\}$.

2. Prove the identity in (11.64).

3. Prove the following element of the distribution of the F statistic in the one-way analysis of variance: $SSW/\sigma^2 \sim \chi^2_{N-k}$. (Hint: Use what you know about the behavior of sample variances of iid normal samples, and then utilize the overall assumptions of the one-way ANOVA model.)

4. A trivia question. Who named the standard test in the analysis of variance the "F test" in honor of Ronald Fisher?

5. Twenty-five patients between the ages of 45 and 75 underwent physical therapy after knee-replacement surgery. Therapy sessions continued until patients reached a specific performance standard. Patients are classified below by age (i), and by the number of days (j) of therapy required to reach the prescribed performance goal.

$i\downarrow / j\rightarrow$	1	2	3	4	5	6	7	8	9	10
45–55	21	30	18	25	16	22	20			
55–65	34	38	27	31	33	26	38	25	23	36
65–75	40	29	43	37	40	32	42	41		

Assume that the model in (11.56) fits the data above. Test the hypothesis of equal means at $\alpha = .05$. (a) Show the ANOVA table, calculate the standard F statistic, and state your conclusion. (b) Denoting the three age groups above as young (y), middle-aged (m), and old (o), use the Scheffé method to determine which of the means μ_y, μ_m, and μ_o can be considered different from others at an overall 95% confidence level.

11.5 The Likelihood Ratio Test in One-Way ANOVA

The F test presented in Section 11.4 was motivated by two things—the fact that it makes some intuitive sense (given the general expressions for the expected mean squares) and the fact that the exact distribution theory of the test statistic is known under the usual ANOVA assumptions and the null hypothesis of equal means. The first pleases our senses, and the second is of great practical importance, since it enables us to set up exact tests at the significance level of choice. For many statistical procedures, motivation like this is deemed to be enough, especially when questions about the optimality of a procedure are difficult to address and connections with tried-and-true theoretical developments are not particularly obvious. In the case of the F test we have just examined, there is a bit more that can be said. In this section, we will motivate the F test using quite a different approach to the one-way ANOVA problem.

Let us suppose that we have, as in Section 11.4, a k-sample problem. For simplicity of exposition, I will assume that the sample is balanced, that is, the sample sizes $n_1, n_2, \ldots, n_k$ are all equal to n. (The general version of the problem is left as Exercise 11.5.1.) Thus, I will assume in what follows that for every integer i such that $1 \le i \le k$,

$$Y_{i,j} \stackrel{\text{iid}}{\sim} N(\mu_i, \sigma^2) \text{ for } j = 1, \ldots, n. \tag{11.72}$$

We are interested in testing the hypothesis $H_0 : \mu_1 = \mu_2 = \cdots = \mu_k \equiv \mu$ against the alternative $H_1 : \mu_h \neq \mu_j$ for some $h \neq j$. For the general model, the parameter space may be represented as

$$\Theta = \{(\mu_1, \ldots, \mu_k, \sigma^2) \in (-\infty, \infty)^k \times (0, \infty)\},$$

while the subspace of Θ that contains the parameter under the null hypothesis is

$$\Theta_0 = \{(\mu, \sigma^2) \in (-\infty, \infty) \times (0, \infty)\}.$$

The likelihood functions under H_0 and in general are given by

$$L(\mu, \sigma^2) = \left(\frac{1}{2\pi\sigma^2}\right)^{kn/2} \exp\left[-\sum_i \sum_j (y_{i,j} - \mu)^2 / 2\sigma^2\right] \tag{11.73}$$

and

$$L(\mu_1, \ldots, \mu_k, \sigma^2) = \left(\frac{1}{2\pi\sigma^2}\right)^{kn/2} \exp\left[-\sum_i \sum_j (y_{i,j} - \mu_i)^2 / 2\sigma^2\right]. \tag{11.74}$$

To derive the likelihood ratio statistic λ, note that under the null hypothesis we have that $Y_{i,j} \stackrel{\text{iid}}{\sim} N(\mu, \sigma^2)$ for all i and j, so that the MLE's for μ and σ^2 are

$$\hat{\mu} = \frac{\Sigma_i \Sigma_j Y_{i,j}}{kn} \text{ and } \hat{\sigma}^2 = \frac{\Sigma_i \Sigma_j (Y_{i,j} - \bar{Y}_{\cdot\cdot})^2}{kn}. \tag{11.75}$$

Thus,

$$\max_{(\mu,\sigma^2)} L(\mu, \sigma^2) = \left[\frac{kn}{2\pi \Sigma_i \Sigma_j (y_{i,j} - \bar{y}_{\cdot\cdot})^2}\right]^{kn/2} \exp\left[-\frac{kn}{2}\right]. \tag{11.76}$$

To find the MLE within Θ, note that the $k+1$ likelihood equations are given by

$$\frac{\partial}{\partial \mu_i} \ln \mathrm{L}\left(\mu_1, \ldots, \mu_k, \sigma^2\right) = \sum_j (y_{i,j} - \mu_i)/\sigma^2 = 0 \text{ for } i = 1, \ldots, k \tag{11.77}$$

and

$$\frac{\partial}{\partial\sigma^2}\ln L\left(\mu_1,\ldots,\mu_k,\sigma^2\right) = -\frac{kn}{2\sigma^2} + \sum_i\sum_j (y_{i,j}-\mu_i)^2/2\sigma^4 = 0. \tag{11.78}$$

The unique solutions of the equations in (11.77) are

$$\hat{\mu}_i = \frac{\sum_j Y_{i,j}}{n} = \bar{Y}_{i\cdot\cdot} \tag{11.79}$$

Substituting these solutions into (11.78) yields the unique solution

$$\hat{\sigma}^2 = \frac{\sum_i\sum_j (Y_{i,j}-\bar{Y}_{i\cdot})^2}{kn}. \tag{11.80}$$

I could examine the Hessian matrix in this problem and show, formally, that the estimators in (11.79) and (11.80) do indeed maximize the likelihood in (11.74). But I haven't shown you a new trick in quite a while, so here's a particularly useful one. I know that the estimators above are the MLEs because they constitute a *unique* critical point of the likelihood function, and the likelihood function, which is everywhere positive, tends to the value 0 as any of the $(k+1)$ parameters tends to infinity. This means that outside of a sufficiently large ball containing this critical point, the likelihood function can be made arbitrarily small. So the critical point couldn't possibly correspond to a minimum. It's pretty easy to determine that it couldn't be a saddle point either. So what's left? Hey, it must be the MLE! We thus have that

$$\max_{(\mu_1,\ldots,\mu_k,\sigma^2)} L\left(\mu_1,\ldots,\mu_k,\sigma^2\right) = \left[\frac{kn}{2\pi\sum_i\sum_j (Y_{i,j}-\bar{Y}_{i\cdot})^2}\right]^{kn/2} \exp\left[-\frac{kn}{2}\right].$$

We may therefore calculate the likelihood ratio statistic as

$$\begin{aligned}\lambda &= \frac{\max_{(\mu_1,\ldots,\mu_k,\sigma^2)} L\left(\mu,\sigma^2\right)}{\max_{(\mu,\sigma^2)} L\left(\mu_1,\ldots,\mu_k,\sigma^2\right)}\\ &= \left[\frac{\sum_i\sum_j (Y_{i,j}-\bar{Y}_{i\cdot})^2}{\sum_i\sum_j (Y_{i,j}-\bar{Y}_{\cdot\cdot})^2}\right]^{kn/2}\\ &= \left[\frac{SSW}{SST}\right]^{kn/2}\\ &= \left[\frac{1}{1+\frac{k-1}{k(n-1)}F}\right]^{kn/2},\end{aligned}$$

where SSW and SST are the sum of squares within treatments and the total sum of squares, respectively, and F is the standard F statistic defined in Section 11.4. Since the statistic λ is a decreasing function of F, λ will be small (and $-2ln\lambda$ is large) when F is large. This implies that the F test and the test based on λ are equivalent, that is, the F test as a likelihood ratio test. What this adds to our intuition about the F test is the knowledge that the critical region of the F test consists of precisely those data configurations which have relatively low probability of occurrence under H_0 when compared to their probability under some models which fall under the alternative hypothesis. The test will also enjoy the general asymptotic optimality of likelihood ration tests (see Section 10.5).

Exercises 11.5.

1. Show that the F test is the likelihood ratio test of the hypothesis $H_0 : \mu_1 = \cdots = \mu_k$, against the alternative $H_1 : \mu_h \neq \mu_j$ for some $h \neq j$, in the unbalanced case of a one-way ANOVA, that is, where, for every $i \in \{1, \ldots, k\}$,

$$Y_{i,j} \overset{\text{iid}}{\sim} N(\mu_i, \sigma^2) \text{ for } j = 1, \ldots, n_i,$$

where the n_is are not necessarily equal.

2. In a balanced one-way ANOVA with k treatments and n observations per treatment, verify the algebraic identity $SSW/SST = \{1 + [(k-1)/k(n-1)]F\}^{-1}$.

11.6 Chapter Problems

1. A high-ranking government economist has suggested that the U.S. Common Stock Index (Y) is related to the price of tea in China (X). The changes in both of these variables in the last six months is given below:

X	10	20	20	30	40	60
Y	20	10	−10	−20	−30	−30

(a) Test the hypothesis for the significance of regression at $\alpha = .05$. (b) Obtain the least-squares regression line for these data.

2. Consider the data below representing the dose level of a prescription drug and the change in blood pressure of the patient.

Dose	1	2	3	4	5
BP Change	5	8	12	14	17

(a) Display the scatter plot for these data treating the dose levels as the Xs and the changes in blood pressure as the Ys. (b) Derive the least-squares regression line when the dose levels are treated as the Xs and the changes in blood pressure are treated as the Ys, and draw the line on our scatter plot. (c) Derive the least-squares regression line when the changes in blood pressure are treated as the Xs and the dose levels are treated as the Ys, and draw the line on your scatter plot. Note that the lines are not the same. Why not?

3. A certain Statistics professor claims that the grade a student gets in his course is well predicted by the number of book problems that the student worked out for practice. At the beginning of last term, he selected six students at random and asked them to keep track of the number of book problems they worked for practice. The results of this experiment are shown below.

X (No. of worked problems)	40	65	90	100	120	185
Y (Grade on a 4 point scale)	1.3	2.0	2.7	2.3	3.0	3.7

(a) Test the significance of regression at $\alpha = .05$. (b) Obtain the least-squares regression line for these data.

4. Let X = the number of courses a student takes per term, and let Y = the number of aspirin the student consumes per week during the term. The following bivariate data was obtained from a random sample of 7 students.

X	2	3	3	4	5	5	6
Y	2	2	3	4	6	8	10

(a) Test the significance of regression at $\alpha = .05$. (b) Obtain the least-squares regression line for these data.

5. An educational researcher has compiled a list of non-standard indices of high-school performance from which academic success in college might be predicted. On this list are things like (1) writing for a high school newspaper, (2) singing in a madrigal choir, (3) holding a student body office, and (4) posting a record score on some popular video game. For six randomly chosen students, the number X of activities on the list and the variable Y, the student's GPA in his/her freshman year of college, were recorded and are shown below.

X (No. of activities on the list)	0	1	1	4	4	5
Y (Freshman year GPA)	2.0	2.5	2.0	3.0	4.0	3.5

(a) Test the hypothesis for the significance of regression at $\alpha = .05$. (b) Obtain the least-squares regression line for these data.

6. The dean of students at Straight Arrow University is studying the effects of the frequently held dorm parties on her campus. Ten freshmen dorm residents were chosen at random from among those dormies enrolled last term in both calculus and statistics classes that assigned homework following every class meeting. The following information was recorded:

$$X = \text{the number of dorm parties a student attended per week}$$

$$Y = \text{the number of homework assignments (out of six) a student turned in per week}$$

The data collected is shown below.

X_i	0	0	1	2	2	2	3	4	4	5
Y_i	6	4	4	3	4	2	4	0	1	0

(a) Test the hypothesis for the significance of regression at $\alpha = .05$. (b) Obtain the least-squares regression line for these data.

7. In a simple linear regression problem, show that the least-squares regression line must pass through the point $(\bar{x}, \bar{y})$.

8. In a simple linear regression problem with n iid normally distributed errors, define the random variable $U = \hat{\beta}_0 + \hat{\beta}_1 x$. Show that the random variables U and $\hat{\beta}_1$ are uncorrelated if and only if $x = \bar{x}$.

9. Consider an SLR problem in which a $100(1-\alpha)\%$ confidence interval for $\mu_{Y|X=x}$ for all x values in the interval $(x_{(1)}, x_{(n)})$. Show that the widths of these intervals take on their minimal value when $X = \bar{x}$.

10. Let $(X_1, Y_1), \ldots, (X_n, Y_n)$ be a arbitrary collection of continuous bivariate data. Show that R^2, the *coefficient of determination* defined as SSR/SST, is algebraically equal to the square of the sample correlation coefficient r.

11. Two researchers are trying to find prediction equations for a variable Y, one using the independent variable X_1 as a predictor, and the other using the independent variable X_2 as a predictor. The experiments were similar, using randomly observed Ys at five fixed values of the predictor variables. The experimental outcomes are shown below.

X_1	1	3	5	7	9
Y	5	8	6	9	12

and

X_2	1	3	5	7	9
Y	4	6	7	9	10

(a) Obtain the least-squares prediction equations for each of these two data sets, that is, obtain the lines

$$\hat{Y} = \hat{\beta}_0 + \hat{\beta}_1 X_1$$

and

$$\hat{Y} = \hat{\alpha}_0 + \hat{\alpha}_1 X_2.$$

(b) Which of the two researchers can claim to have a better prediction equation for the variable Y. Justify your answer.

12. Suppose a least-squares regression line is developed for a set of bivariate data. Explain the difference between the "mean response at the value $X = x$" and "the mean of m new observations at $X = x$".

13. In conducting statistical tests concerning the parameter β_1 in the SLR model, justify the claim that the t test is more versatile than the F test.

14. Suppose that three researchers study the relationship between jogging at a specific speed X, in miles per hour, and blood pressure Y after a 10-minute exercise. Each researcher designed their own experiment and took measurements at the following x values:

Researcher 1: $x = 3, 3, 4, 4, 5, 5, 6, 6, 7, 7,$
Researcher 2: $x = 3, 3, 3, 3, 3, 7, 7, 7, 7, 7,$
Researcher 3: $x = 3, 4, 4, 5, 5, 5, 5, 6, 6, 7.$

If all three researchers had the same exact MSE for their experiments, which of the three researchers had the narrowest confidence intervals for the coefficients β_0 and β_1 of their prediction equations?

15. Let $t \sim t_k$. Show that $t^2 \sim F_{1,k}$. (Hint: You may assume that the t variable is equivalent to a t_k variable constructed as in Theorem 4.6.2.)

16. (See Problem 11.6.15 above.) Consider a bivariate data set satisfying the SLR model given in (11.2). Show that the t test for testing the hypothesis $H_0 : \beta_1 = 0$ vs. $H_1 : \beta_1 \neq 0$ is algebraically equivalent to the F test for significance of regression.

17. (See Problem 11.6.15 above.) Consider a bivariate data set satisfying the SLR model given in (11.2). Show that the t test for testing the hypothesis $H_0 : \rho = 0$ against $H_1 : \rho \neq 0$ based on the statistic $r\sqrt{n-2}/\sqrt{1-r_n^2}$ is algebraically equivalent to the F test for significance of regression.

18. Prove Theorem 11.3.2.

19. (See Problem 11.6.2.) Suppose a bivariate data set $(x_1, y_1), \ldots, (x_n, y_n)$ is available and you calculate the least-squares lines for predicting Y from X and for predicting X from Y, that is, you obtain the prediction equations

$$\hat{Y} = \hat{\beta}_0 + \hat{\beta}_1 X \quad \text{and} \quad \hat{X} = \hat{\alpha}_0 + \hat{\alpha}_1 Y. \tag{11.81}$$

(a) State a set of conditions on your data that will imply that these two lines are one and the same. (b) Prove that the conditions stated in your response to part (a) guarantee that the two lines in (11.81) are the same.

20. There are certain applications of the SLR model that require that the y-intercept of the simple linear regression model is equal to 0. For example, a salesperson's monthly commission as a function of monthly sales would require such a model. Suppose that the appropriate model for such an experiment, given $X_i = x_i$ is

$$Y_i = \beta x_i + \varepsilon_i, \text{ for } i = 1, \ldots, n, \tag{11.82}$$

where $\varepsilon_1, \ldots, \varepsilon_n$ are assumed to be uncorrelated errors with zero means and common variance σ^2. (a) Show that the least-squares estimator $\hat{\beta} = \sum_{i=1}^{n} c_i Y_i$, of β, where the constants $\{c_i\}$ are given by

$$c_i = \frac{x_i}{\sum_{i=1}^{n} x_i^2}, \text{ for } i = 1, \ldots, n, \tag{11.83}$$

is the best linear unbiased estimator of β. (Hint: Use a variational argument as in the proof of Theorem 11.3.1.)

For Problems 11.6.21–11.6.25. The director of admissions of a small college has devised, with the help from a faculty advisory board, a diagnostic test whose score X is thought to be an effective predictor of Y, a freshman student's first year GPA at her college. The test was administered to a random sample of 10 entering freshman during fall semester orientation. The data obtained is shown below.

X_i	12	20	14	18	15	10	19	20	11	16
Y_i	1.8	3.7	2.5	3.5	2.6	1.6	3.3	4.0	1.6	2.7

Table 11.6.1. Data from pre-test vs. Frosh GPA experiment.

21. For the data in Table 11.6.1, construct an ANOVA table and test the significance of regression for these data, first using at test for testing H_0: $\rho = 0$ and then using an F test for testing H_0: $\beta_1 = 0$. Use $\alpha = .01$.

22. Derive the least-squares regression line for the data in Table 11.6.1. Use this line to predict the freshman's first year GPA if this student had a score of 17 on the pre-test. What is your estimate of the change in the mean response when the pre-test score changes by 1.

23. For the data in Table 11.6.1, derive a 95% confidence interval for the slope parameter β_1.

24. For the data in Table 11.6.1, derive a 95% confidence interval for $\mu_{Y|X=15}$.

25. For the data in Table 11.6.1, derive a 95% prediction interval for the GPA of a future freshman at this college who scored 17 on the pretest.

For Problems 11.6.26–11.6.30. In 1857, Scottish scientist J. D. Forbes published data from which he obtained a formula for relating the boiling point X of water (in degrees Fahrenheit) to atmospheric (barometric) pressure Y. His experiment involved boiling water at a variety of altitudes and recording the atmospheric pressure at that altitude as a dependent variable. Forbes' ultimate goal was to estimate altitude from a measurement on the boiling point of water at that altitude (as a barometer would only rarely be available at high altitude). The data shown below is taken from Forbes' published paper.

X_i	194.5	197.9	199.4	200.9	201.4	203.6	209.5	210.7	212.2
Y_i	20.79	22.40	23.15	23.89	24.02	25.14	28.49	29.04	30.06

Table 11.6.2. J. D. Forbes data on water boiling point and atmospheric pressure.

26. For the data in Table 11.6.2, construct an ANOVA table and test the significance of regression for these data, first using at test for testing H_0: $\rho = 0$ and then using an F test for testing H_0: $\beta_1 = 0$.
27. Derive the least-squares regression line for the data in Table 11.6.2. Use this line to predict the barometric pressure Y if water is boiled at 205°F. What is your estimate of the change in the mean response when the boiling point of water changes by 1 degree?
28. For the data in Table 11.6.2, derive a 95% confidence interval for the slope parameter β_1.
29. For the data in Table 11.6.2, derive a 95% confidence interval for $\mu_{Y|X=200}$.
30. For the data in Table 11.6.2, derive a 95% prediction interval for the barometric pressure Y at a given altitude if water is boiled at 205°F at that altitude.

For Problems 11.6.31–11.6.35. What's the relationship between a car's gas consumption and the speed at which it travels? To answer this question, the Department of Energy funded a study which tested a collection of 100 cars, 10 cars each of 10 specific models, including some foreign and some domestic cars. Each car was driven through a ten-mile course at an assigned "miles per hour" (X). The fraction (Y) of a gallon of gas used in the process was recorded for each car. The data obtained from one particular model is shown below.

X_i	50	50	55	55	60	60	65	65	70	70
Y_i	.25	.20	.27	.28	.32	.32	.38	.34	.38	.42

Table 11.6.3. Data on speed and gas consumption.

31. For the data in Table 11.6.3, construct an ANOVA table and test the significance of regression for these data, first using a t test for testing H_0: $\rho = 0$ and then using an F test for testing H_0: $\beta_1 = 0$.
32. Derive the least-squares regression line for the data in Table 11.6.3. Use this line to predict the gas consumption Y if a car of this particular model is driven at 58 miles per hour. What is your estimate of the change in the mean response when the miles per hour changes by 10?
33. For the data in Table 11.6.3, derive a 95% confidence interval for the slope parameter β_1.
34. For the data in Table 11.6.3, derive a 95% confidence interval for $\mu_{Y|X=65}$.
35. For the data in Table 11.6.3, derive a 95% prediction interval for the gas consumption of a car of this particular model that will be driven at 62 miles per hour.

For Problems 11.6.36–11.6.39. The term "ordinary least squares" (OLS) is sometimes used to describe the estimation of regression parameters by minimizing the sum SS of squared errors displayed in (11.4). This approach is considered appropriate under the constant variance assumption of the SLR model. In circumstances in which that assumption is in question, and the variance of Y differs at different values of x, an alternative approach called "weighted least squares" (WLS) is often used. Given the weights $w_1, \ldots, w_n$ on the bivariate data $(X_1, Y_1), \ldots, (X_n, Y_n)$, the WLS criterion function is given by

$$SS_w = \sum_{i=1}^{n} w_i(y_i - \beta_0 - \beta_1 x_i)^2. \tag{11.84}$$

36. Derive the normal equations corresponding to weighted least-squares regression, that is, obtain expressions for the equations

$$\frac{\partial}{\partial\beta_0} SS_w = 0 \text{ and } \frac{\partial}{\partial\beta_1} SS_w = 0. \tag{11.85}$$

Note that, if $w_i = c$ for $i = 1, \ldots, n$, the WLS normal equations are the same as the OLS normal equations. Solve the WLS normal equations corresponding to (11.85) and obtain expressions for the WLS estimators of β_0 and β_1.

37. In principle, data points with a smaller variance should receive greater weight in weighted least-squares regression, as the dependent variable in such a data point has been measured with greater precision. If the variances $\{V(Y_i|X_i = x_i) = \sigma_i^2\}$ were known, then the weights in WLS regression would be chosen as $w_i = 1/\sigma_i^2$. It is not uncommon to encounter problems in which the variability of Y, given $X = x$, grows as x itself grows. Suppose, in a given application, it is assumed that

$$V(Y|X = x) = kx^2. \tag{11.86}$$

Obtain expressions for the WLS estimators of β_0 and β_1 when the weights are taken to be $w_i = 1/x_i^2$.

38. Assume that Equation (11.86) is appropriate for fitting a WLS regression model, with weights $w_i = 1/x_i^2$, to the data shown below.

X_i	1	1	2	3	3	4	6	6	7	8	10	11
Y_i	10	11	15	15	23	20	25	35	28	48	42	62
w_i	1	1	1/4	1/9	1/9	1/16	1/36	1/36	1/49	1/64	1/100	1/121

(a) Display a scatter plot of the twelve data points above. Note that the variability of the response variable Y does indeed increase with the value of x. (b) Obtain the WLS regression line and plot it in your scatter plot.

39. Fit an OLS regression line to the 12 data points in Problem 11.6.38. Examine the fit of these two lines to the points in your scatter plot. Does the WLS line give a better fit to data points with small x (say, $x \leq 4$)? Compare the residuals

$$\{e_i = Y_i - \hat{Y}_i, i = 1, \ldots, 12\}$$

from the two models and comment on the difference.

40. Data on the amount of radioactivity in eggs was obtained from three major egg farms in or around Egg Harbor, Wisconsin. The concentration of the radioactive isotope Strontium-90 was measured for the eggs of ten different hens chosen at random from each farm. The data obtained is shown below:

Farm	Strontium-90 measurement
Eggs a Plenty, Ltd.	5.7, 6.3, 6.0, 7.1, 6.2, 5.1, 5.3, 5.8, 6.7, 4.8
Mess O' Eggs Partnership	8.3, 10.7, 9.6, 11.4, 9.9, 9.1, 8.6, 10.5, 89.0, 10.9
Eggstordinary, Inc.	9.5, 9.2, 10.7, 9.3, 10.9, 10.0, 9.5. 10.2, 9.9, 9.8

(a) Under the usual assumptions of one-way ANOVA, test the hypothesis, at $\alpha = .05$, that the mean levels of Stronium-90 at the three egg farms are equal. (b) Use the Scheffé method to determine which, if any, of the three pairwise contrasts of population means can be regarded as different from 0.

41. Class action lawsuits have been filed against three entertainment promoters for the unhealthy decibel levels of the rock concerts they promote at indoor venues. The data collected over the last year is shown below.

Promoter	Decibel (dB) Level Measurement
Hard Rock, Inc.	115, 124, 116, 110, 107, 125, 118, 115, 120, 110
Harder Rock, Inc.	122, 126, 129, 134, 118, 120, 122, 121
Hardest Rock, Inc.	125, 130, 134, 126, 139, 133, 129, 140, 132

(a) Under the usual assumptions of one-way ANOVA, test the hypothesis, at $\alpha = .05$, that the mean decibel levels at concerts sponsored by these three promoters are equal. (b) Use the Scheffé method to determine which, if any, of the three pairwise contrasts of population means can be regarded as different from 0.

42. The following data on the calories in randomly sampled hot dogs of four specific types was published in *Consumer Reports* in June 1986. Soon thereafter, consumer activist Ralph Nader declared that the hot dog was "America's deadliest missile." (I suspect that it was the foul hot dogs that got to him!)

Variety	Calorie Count
Beef	186, 181, 179, 149, 184, 190, 158, 139, 175, 148, 152, 111, 141, 153, 190, 157, 131 149, 135, 132
Pork	173, 191, 182, 190, 172, 147, 146, 139, 175, 136, 179, 153, 107, 195, 135, 140, 138
Foul	129, 132, 102, 106, 94, 102, 87, 99, 107, 113, 135, 142, 86, 143, 152, 146, 144
Mixed	155, 170, 114, 191, 162, 146, 140, 187, 180

(a) Under the usual assumptions of one-way ANOVA, test the hypothesis, at $\alpha = .05$, that the mean calorie counts of the four populations from which these hot dogs were sampled are equal. (b) Use the Scheffé method to determine which, if any, of the six pairwise contrasts of population means can be regarded as different from 0.

43. Consider a one-way ANOVA problem satisfying the assumptions of (11.56) with the exception of the assumption that data are normally distributed. Let $L = \sum_{i=1}^{k} c_i \mu_i$ and $\hat{L} = \sum_{i=1}^{k} c_i \bar{X}_{i\cdot}$. Show that $\hat{L}$ is the BLUE of L.

44. Suppose that you have a one-way ANOVA problem in which the entire data set has been summarized by the values of the sizes of the random samples taken under each of the k treatments, the k sample means, and the k sample variances, that is, you are given

$$n_1, \ldots, n_k, \bar{X}_{1\cdot}, \ldots, \bar{X}_{k\cdot} \text{ and } s_1^2, \ldots, s_k^2.$$

Show that this information is sufficient to construct a complete ANOVA table.

45. Consider a one-way ANOVA problem satisfying the assumptions of (11.56), including the assumption that the observations are normally distributed. Define the linear contrast L and the unbiased estimator $\hat{L}$ of L as

$$L = \sum_{i=1}^{k} c_i \mu_i \qquad \text{and} \qquad \hat{L} = \sum_{i=1}^{k} c_i \bar{X}_{i\cdot}.$$

Show that $\hat{L}$ is the MVUE of L. (Hint: Use what you know about data drawn from a distribution in a $(k+1)$ parameter exponential family.)

12

Nonparametric Statistical Methods

In this chapter, we will discuss approaches to statistical inference in which minimal assumptions are made about the underlying distribution F from which data is drawn. We will continue to assume that our inferences are based on one or more random samples from the distributions of interest. In one-sample problems, we will assume that the data may be represented as $X_1, \dots, X_n \stackrel{\text{iid}}{\sim} F$, where F is a distribution that, in some applications, will be completely general (that is, can be any distribution, continuous or discrete) and in other settings, might be assumed to obey some modest restrictions like continuity and/or symmetry. In any case, the functional form of F is not assumed to be known. In particular, the distribution F is not indexed by a finite number of parameters as are the discrete and continuous models we discussed in Chapters 2 and 3. The term "nonparametric" is thus seen as a natural way to describe the models and methods utilized in this quite general framework.

Before embarking on our tour of "nonparametric statistics," it seems worthwhile discussing, albeit briefly, why this subfield of statistics is important and, in some instances, essential. The process of parametric modeling and inference is appropriate and useful in many problems to which Statistics is applied, but it does have its difficulties and limitations. While the physical circumstances of an experiment, or logical reasoning about the experimental process, can often suggest a particular parametric model as appropriate, there are many situations in which the modeling of experimental data is not easily accomplished. Since goodness-of-fit tests are more reliable in discrediting a proposed model than in confirming that a model holds, deciding on the appropriate parametric model can be a challenging proposition. Using a nonparametric approach to modeling and inference represents an obvious and potentially quite useful alternative.

Two important questions that arise when a nonparametric procedure is contemplated are: "What are the risks?" and "What are the potential benefits?" The fact that the answers to both of these questions tend to support the nonparametric approach is perhaps the main reason for the substantial growth of interest in and usage of nonparametric techniques in modern statistical work. Here's a simple example of how these questions would be answered in a specific instance. In Section 12.5, we will study a nonparametric test called the Wilcoxon Rank Sum test for testing the hypothesis that two distributions F and G are equal. Typically, the available data consist of two independent samples $X_1, \dots, X_n \stackrel{\text{iid}}{\sim} F$ and $Y_1, \dots, Y_m \stackrel{\text{iid}}{\sim} G$. The Wilcoxon test is based solely on the respective ranks of one sample within the combined sample. This test is the leading nonparametric alternative to the two-sample t test, a test that is developed under the assumption that F and G are normal distributions with a common (though unknown) variance. The two-sample t test is described in detail in Example 10.2.1. One standard measure of the performance of one test relative to another is "Asymptotic Relative Efficiency (ARE)" which, roughly speaking, is the limiting ratio of sample sizes needed for both tests to have the same power at alternative values that are suitably close to the null hypothesis. The following interesting facts are known about the two-sample t test: The ARE of the Wilcoxon test relative to the t test is .955 when the assumptions of the t test (independence, normality, equal variances) are met exactly. When these assumptions do not hold in the typical two-sample location parameter problem, (where $f_Y(t) = f_X(t - \Delta)$ for some constant Δ and it is assumed that $V(X) = V(Y)$ is finite), the ARE of the Wilcoxon test relative to the t test is never smaller than .864 (a value which actually occurs if X has a density of the form $f(x) = A(B^2 - x^2)I_{(-B,B)}(x)$). For most non-normal densities f, the ARE of the Wilcoxon test relative to the t test is larger than 1, and it can take arbitrarily large values depending on the actual distributions of the Xs and Ys. The practical implications of these facts are quite profound. Even when the standard conditions for using the t test hold, the Wilcoxon test is 95.5% as efficient as the t test (that is, would have the same performance, near the null hypothesis, with just a slightly larger a sample size). In situations in which these conditions fail, the Wilcoxon test will typically outperform the t test, sometimes quite dramatically. This suggests that a decision to always use the Wilcoxon test instead of the two-sample t test would be entirely sensible.

In the following sections, I will treat both nonparametric estimation and nonparametric tests. In treating the first of these topics, I will present the standard methods for estimating the distribution F (or the survival function $\bar{F} = 1 - F$) nonparametrically, first by the empirical distribution (or survival) function based on "complete data" and secondly by the widely used Kaplan-Meier estimator based on data subject to random censoring. I will also introduce the nonparametric bootstrap, a very useful tool for estimating characteristics of a statistic of interest or generating a valid confidence interval for a particular population characteristic. Following our discussion of nonparametric estimation, I will present three nonparametric procedures for testing hypotheses concerning the distributions of two random observables X and Y. Our coverage will include the simple but useful "sign test" for testing the null hypothesis that $P(X < Y) = P(X > Y)$ based on paired data (X,Y), and the Wald-Wolfowitz "runs test" and the Wilcoxon rank-sum test, both based on independent random samples drawn from two distributions F and G. The latter procedures are both aimed at testing the hypothesis $H_0 : F = G$ against a one-sided or two-sided alternative.

12.1 Nonparametric Estimation

Proceeding without any assumptions about the distribution that describes the stochastic behavior of our experimental data may seem quite daunting, at least at first. Our minimalistic approach will assume, simply, that $X_1, \dots, X_n \overset{\text{iid}}{\sim} F$, where F is completely unspecified. What can the data tell us about F? Well, if we were to construct a histogram using these data, we would consider it to be a rough approximation of the density or probability mass function of F. From that histogram, we would have a rough idea of the mean μ and variance σ^2 of F, and, from the data itself, we would have unbiased estimators $\bar{X}$ and s^2 of μ and σ^2. The traditional estimator of F itself is the *empirical distribution function* (edf) given by

$$\hat{F}_n(x) = \frac{1}{n} \sum_{i=1}^{n} I_{(-\infty,x)}(x_i), \tag{12.1}$$

where x_i represents the observed value of X_i for $i = 1, \ldots, n$. The edf has a simple interpretation; $\hat{F}_n(x)$ is the proportion of the sample values $\{x_i, i = 1, \ldots, n\}$ that are less than or equal to x. Because random samples tend to resemble the population from which they are drawn, it is reasonable to expect that $\hat{F}_n(x)$ would be a decent estimator of $F(x)$, the proportion of the population that is less than or equal to x. Indeed, under our sampling assumptions $I_{(-\infty,x)}(X_1), \ldots, I_{(-\infty,x)}(X_n) \stackrel{\text{iid}}{\sim} B(1, F(x))$, so that the statistic $\hat{F}_n(x) = (1/n)\sum_{i=1}^{n} I_{(-\infty,x)}(X_i)$ is an unbiased estimator of $F(x)$. Further, from what we know about binomial sampling, we may assert that the statistic $\hat{F}_n(x)$ is the MVUE of $F(x)$ with variance equal to $F(x)[1-F(x)]/n$. Since $\hat{F}_n(x)$ is an average, it also follows from the WLLN and the Central Limit Theorem, that, for all x,

$$\hat{F}_n(x) \xrightarrow{p} F(x) \text{ as } n \to \infty \tag{12.2}$$

and

$$\sqrt{n}(\hat{F}_n(x) - F(x)) \xrightarrow{D} Y \sim N(0, F(x)[1-F(x)]) \text{ as } n \to \infty. \tag{12.3}$$

Thus, as it turns out, we can learn quite a bit about F from $\hat{F}_n$. The convergence stated in (12.2) holds for each fixed x, but an even stronger claim can be made. We state without proof an important result that asserts that the convergence in (12.2) is uniform in x. In common language, this means that, no matter how close we want $\hat{F}_n$ to be to F, there exists an integer N such that if $n > N$, then $\hat{F}_n(x)$ is that close to $F(x)$ for all $x \in (-\infty, \infty)$.

Theorem 12.1.1. (Glivenko-Cantelli) Let $\hat{F}_n$ be the empirical distribution function based on the data $X_1, \ldots, X_n \stackrel{\text{iid}}{\sim} F$. Then,

$$\sup_{-\infty<x<\infty} |\hat{F}_n(x) - F(x)| \xrightarrow{p} 0 \text{ as } n \to \infty, \tag{12.4}$$

where $\sup_{x \in A} h(x)$ represents the least upper bound of the set $\{h(x), x \in A\}$ or its maximum, if it exists. While $\hat{F}_n$ is a discrete distribution which assigns probability $1/n$ to each of the n observed data points $\{x_i\}$, and its graph is a step function taking jumps at these x_is, the edf still provides a very good approximation of the unknown cdf F when the sample size is at least moderate. The figure below is a typical simulation of the edf plotted against the true distribution F.

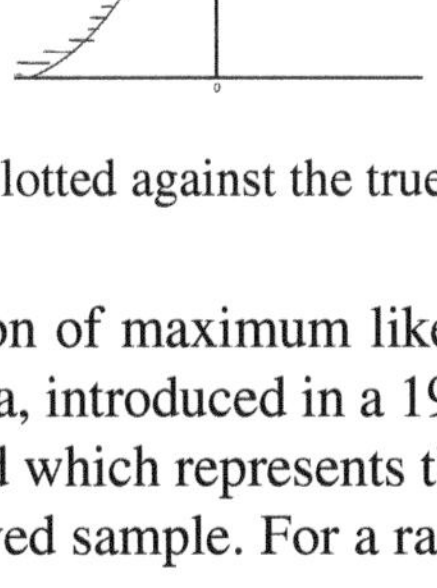

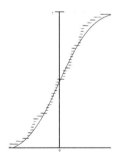

Figure 12.1.1. The edf plotted against the true $F = N(0,1)$ for $n = 50$.

There is a nonparametric version of maximum likelihood estimation that we will examine in this section. The basic idea, introduced in a 1956 paper by Kiefer and Wolfowitz, involves a nonparametric likelihood which represents the weight that a general probability model F would place on the observed sample. For a random sample from F, that is, given

$X_1,\ldots,X_n \stackrel{\text{iid}}{\sim} F$, the nonparametric likelihood is given by

$$L(F) = \prod_{i=1}^{n} [F(x_i) - F(x_i^-)] = \prod_{i=1}^{n} p_i, \tag{12.5}$$

where $p_i = F(x_i) - F(x_i^-)$ is simply the probability that the model F places on the observed point x_i. The nonparametric maximum likelihood estimator (NPMLE) of F is the model $\hat{F}$ that maximizes $L(F)$, that is, the NPMLE is defined as

$$\hat{F}(x) = \arg\min_{F} L(F(x)). \tag{12.6}$$

Three things about the likelihood in (12.5) are immediately evident. First, you can see that $L(F) = 0$ if the distribution F is continuous, since for continuous F, every p_i in $L(F)$ is equal to 0. Secondly, among discrete distributions, only distributions that place positive mass on each observed x_i will yield a positive value of $L(F)$. Finally, the NPMLE will be a discrete distribution that places all of its mass on the observed data $\mathbf{D} = \{x_i, i = 1,\ldots,n\}$, since if F^* is a discrete distribution that places positive mass on each x_i but places a total mass on the n observations that is less than 1, then $L(F)$ can be increased simply by shifting the mass that F^* places on $\mathbf{D}^c$ onto the set $\mathbf{D}$. In the following theorem, the edf $\hat{F}_n$ is shown to be the unique NPMLE of F in the problem of current interest. This result adds the natural heuristics associated with the maximum likelihood approach to the somewhat ad hoc defense of the edf with which we started. It also establishes the fact, using MLE theory for proportions, that for each fixed x, the edf $\hat{F}_n(x)$ is the best asymptotically normal (BAN) estimator of the cdf $F(x)$ at any fixed value of x.

Theorem 12.1.2. Let $\hat{F}_n$ be the empirical distribution function based on the data $X_1,\ldots,X_n \stackrel{\text{iid}}{\sim} F$. Then, $\hat{F}_n$ is the NPMLE of F, that is, $\hat{F}_n$ uniquely maximizes the likelihood function $L(F) = \prod_{i=1}^{n} p_i$ in (12.5).

Proof. We will use the method of Lagrange multipliers (see Remark 6.5.1) to maximize $\ln L(F)$. Define the criterion function $G(\mathbf{p},\lambda)$ as

$$G(\mathbf{p},\lambda) = \sum_{i=1}^{n} \ln(p_i) + \lambda(\sum_{i=1}^{n} p_i - 1). \tag{12.7}$$

Setting the partial derivatives of G with respect to $p_1,\ldots,p_n$ and λ equal to 0, we obtain the following $(n+1)$ "estimating equations":

$$\frac{1}{p_i} + \lambda = 0 \text{ for } i = 1,\ldots,n \tag{12.8}$$

and

$$\sum_{i=1}^{n} p_i = 1. \tag{12.9}$$

The equations in (12.8) indicate that $p_i = -1/\lambda$ for all i, that is, all the p_is take on the same value. In view of (12.9), that common value must be equal to $1/n$. Thus, the unique maximizer of $\ln L(\mathbf{p})$, and therefore also of $L(\mathbf{p})$, is the vector $\mathbf{p}$ for which

$$p_i = F(x_i) - F(x_i^-) = 1/n,$$

In other words, the edf $\hat{F}_n(x)$ is the NPMLE of F. ■

We now turn to a more complex problem, one that occurs with some frequency in applications in the fields of medicine and public health, and occurs in a variety of other fields as well. Let us suppose that a given study provides us with data obtained from a random sample of n subjects. We are typically interested in estimating the survival function $\bar{F}_X(t) = P(X > t)$ of the random variable X, the failure time of a randomly chosen member of the population of interest. While our ideal sample would consist of n failure times, it is often the case that some subjects are "lost to the study" before their failure time is observed. What is then observed is the time L at which the loss occurred. Such observations are referred to as *censored data*. These data are not as informative about $\bar{F}_X$ as an observed failure time, but they do carry some useful information. What an observed loss L tells us is that this particular subject's failure time X is larger than the observed value L. We assume that, for every subject involved in a given experiment, we will observe either a failure time X or a loss to the study at time L. The specific reason that a subject is lost to the study is not specified, but we will assume that L and X are independent random variables. Since applications of the scenario above typically involve "time until failure," we will assume that $P(X_i > 0) = 1$ and $P(L_i > 0) = 1$. For a given subject, the observation made may be represented as

$$T = \min(X, L). \tag{12.10}$$

Further, it will be assumed that it is known whether an observation T is a failure time or a loss. This assumption may be executed by defining the indicator function

$$I_i = I_i(X_i, L_i) = I_{(0,L_i)}(X_i). \tag{12.11}$$

The observed data will consist of the "identified minima" $(T_1, I_1), \ldots, (T_n, I_n)$.

The following simple example was utilized by Kaplan and Meier (1958) both to illustrate the concept of censored data and to motivate the need for using all the available information the data provides.

Example 12.1.1. Suppose that in year 1 of the study of a particularly deadly disease, 100 patients were treated and 70 of them died within that year. Of the 30 that survived the first year of the study, 15 of them died during the second year. At the beginning of the second year of the study, 1000 new patients were recruited to the study, receiving the same treatment; 750 of them died by the end of that year. Suppose that we wish to estimate $\bar{F}(2) = P(X > 2)$, the probability of surviving the disease for at least two years. One simple and immediate estimate of $\bar{F}(2)$ is 15/100, the proportion of the patients who participated in the entire two-year study and survived past the two-year mark. Under the assumption that these 100 patients were randomly selected from the target population, this estimate of $\bar{F}(2)$ is perfectly valid, being the sample proportion under binomial sampling. It may seem at first view that the second sample can't be useful here because none of those patients were observed over a two-year period. Noting that this initial impression is not correct is quite instructive, as the logic of using this additional data is an essential ingredient in the Kaplan-Meier approach to the analysis of censored data. Note that the entire second sample was censored after one year in the study. It can thus be used indirectly in estimating $\bar{F}(2) = P(X > 2)$ by writing

$$P(X > 2) = P(X > 1) \cdot P(X > 2 | X > 1).$$

Clearly, the second sample is useful in the problem of estimating $P(X > 1)$. Using our standard approach to estimating proportions, we have

$$\hat{P}(X > 1) = \frac{30 + 250}{100 + 1000} = .255$$

and

$$\hat{P}(X > 2|X > 1) = \frac{15}{30} = .5.$$

It is thus apparent that a better estimate of $\bar{F}(2)$ is available, one that uses the quite substantial information (and, thus, additional precision) that the data provides for estimating $\bar{F}(1)$. That better estimate is $\hat{\bar{F}}(2) = (.255) \cdot (.5) = .1275$. ■

Let's now consider the derivation of the NPMLE of the underlying failure-time survival function $\bar{F}$ (that is, the survival function of the Xs) based on a random sample of identified minima. Typically, $\bar{F}$ will be taken to be a continuous distribution, so that ties among the Xs have probability 0 of occurrence. We will use the following notation. Suppose that k of the n observations are known to be failure times. For notational simplicity, let us take $X_1 < X_2 < \cdots < X_k$ as the k distinct ordered observed failure times (writing X_i instead of the more cumbersome $X_{(i)}$). In addition, we will use the following notation:

$$n = \text{ the total sample size,} \tag{12.12}$$

$$X_0 = 0 \quad \text{and} \quad X_{k+1} = \infty, \tag{12.13}$$

$$\delta_j = \text{ the number of failures observed at time } t = X_j, \quad j = 1, \ldots, k \tag{12.14}$$

$$\lambda_j = \text{ the number of losses in the half-open interval } [X_j, X_{j+1}), \quad j = 0, \ldots, k, \tag{12.15}$$

$$L_i^{(j)} = \text{ the time of the } i^{th} \text{ loss in } [X_j, X_{j+1}) \text{ for } i = 1, \ldots, \lambda_j, \quad j = 0, \ldots, k, \tag{12.16}$$

$$n_j = \text{ the number of subjects at risk at time } X_j - 0 = n - \sum_{i=1}^{j-1} \delta_i - \sum_{i=0}^{j-1} \lambda_i. \tag{12.17}$$

Note that the definition of δ_j above permits multiple failures at the observed failure time $t = X_j$. If failure times were recorded with sufficient precision, there would be no ties among the observed failure times, and we would have $\delta_j = 1$ for $j = 1, \ldots, k$. In practice, ties might occur due to rounding. From the notation above, it should be clear that

$$n = \sum_{j=1}^{k} \delta_j + \sum_{j=0}^{k} \lambda_j, \tag{12.18}$$

that is, the sample size n is simply the sum of the number $\sum_{j=1}^{k} \delta_j$ of observed failure times and the number $\sum_{j=0}^{k} \lambda_j$ observed losses among the n observed values $T_1, \ldots, T_n$.

For a sample that includes censored failure times, the likelihood of the observed data is more complex than the likelihood in (12.5) since here, we must account for both censored and uncensored data, that is, both observed Ls and observed Xs. The likelihood element corresponding to the observed failure time $X_j = x_j$ is given by

$$L(x_j) = [\bar{F}(x_j - 0) - \bar{F}(x_j)]^{\delta_j}, \tag{12.19}$$

a value that represents the probability mass that would be placed at this observed failure time. An observed loss, that is, a censored failure time, does carry information about F, but it is information of a different type. If a given T is an observed loss, say $T = L^*$, then what has been observed is simply the fact that the failure time for that particular subject is in the interval (L^*, ∞). The likelihood element corresponding to an observed loss $L^* = l^*$ is thus given by

$$L(l^*) = [\bar{F}(l^*)]. \tag{12.20}$$

Using the notation established in (12.12)–(12.17), we may write the full likelihood of the observed data as

$$L(\bar{F}) = \prod_{i=0}^{\lambda_0} \bar{F}(l_i^{(0)}) \cdot \prod_{j=1}^{k} \left([\bar{F}(x_j - 0) - \bar{F}(x_j)]^{\delta_j} \prod_{i=1}^{\lambda_j} \bar{F}(l_i^{(j)}) \right). \tag{12.21}$$

We will now consider maximizing $L(\bar{F})$ in (12.21) with respect to $\bar{F}$. This may seem like quite a challenging task. The fact that one can identify the unique maximizer of $L(\bar{F})$ in closed form is perhaps the sweetest part of the Kaplan-Meier paper. To obtain an expression for this estimator, we first note that the function $\bar{F}$ that maximizes $L(\bar{F})$ has some properties that substantially simplify the problem. From (12.21), we see that $L(\bar{F})$ may be maximized by making, for all i and j,

$$\bar{F}(l_i^{(j)}) \text{ as large as possible} \tag{12.22}$$

$$\bar{F}(x_j - 0) \text{ as large as possible, and} \tag{12.23}$$

$$\bar{F}(x_j) \text{ as small as possible,} \tag{12.24}$$

subject to condition (a) that $\bar{F}$ is a non-increasing function and to the assumption (b) that $x_j \leq l_i^{(j)} < x_{j+1}$. An immediate consequence of (12.22) is the fact that

$$\bar{F}(l_i^{(0)}) = 1, \quad \text{for } i = 1, \ldots, \lambda_0, \tag{12.25}$$

as these losses occur before the first observed failure time, so that each $\bar{F}(l_i^{(0)})$ is constrained only by the upper bound 1. Next, we note that the goals in (12.22)–(12.24) can be achieved by setting

$$\bar{F}(x_j) = \bar{F}(l_i^{(j)}) = \bar{F}(x_{j+1} - 0). \tag{12.26}$$

Checking that the NPMLE must satisfy (12.26) is left as Exercise 12.1.4. The identities in (12.26) imply that the NPMLE is a step function that may only take jumps at the observed failure times $[x_j, j = 1, \ldots, k]$. Letting P_j be the common value of the probabilities in (12.26), with $P_0 \equiv 1$, we may then rewrite $L(\bar{F})$ as

$$L(\bar{F}) = \prod_{j=1}^{k} [P_{j-1} - P_j]^{\delta_j} P_j^{\lambda_j}. \tag{12.27}$$

We'll make one further substitution that, very pleasantly, transforms the problem of maximizing $L(\bar{F})$ into a simple problem whose solution is obvious by inspection. For $j = 1, \ldots, k$, let $p_j = P_j / P_{j-1}$. Then, denoting $1 - p_j$ as q_j, we have that

$$P_j = \prod_{i=1}^{j} p_i \text{ and } [P_{j-1} - P_j] = \left(\prod_{i=1}^{j-1} p_i \right) q_j. \tag{12.28}$$

We may thus express the likelihood $L(\bar{F})$ as a function of these p's and q's:

$$L(\bar{F}) = \prod_{j=1}^{k} p_j^{\Sigma_{i=j+1}^{k} \delta_i + \Sigma_{i=j}^{k} \lambda_i} q_j^{\delta_j} = \prod_{j=1}^{k} p_j^{n_j - \delta_j} q_j^{\delta_j}, \tag{12.29}$$

an identity that results from making appropriate substitutions in (12.27). It is evident that the likelihood in (12.29) factors into n partial likelihoods, that is,

$$L(\bar{F}) = \prod_{j=1}^{k} L(p_j), \tag{12.30}$$

where

$$L(p_j) = p_j^{n_j-\delta_j} q_j^{\delta_j} \text{ for } j = 1, \ldots, k. \tag{12.31}$$

The partial likelihoods in (12.31) are of the familiar binomial form. It is thus evident by inspection that $L(\bar{F})$ in (12.30) is maximized by the estimator

$$\hat{\bar{F}}(t) = \prod_{\{j:X_j \le t\}} \hat{p}_j,$$

where

$$\hat{p}_j = \frac{n_j - \delta_j}{n_j}, j = 1, \ldots, k.$$

We summarize the developments above in the following:

Theorem 12.1.3. Let $X_1 < \cdots < X_k$ be the distinct observed failure times obtained from an iid sample of the failure times of n subjects, subject to random censoring but assumed to have a common survival function $\bar{F}$. Let δ_j be the number of failures observed at time $t = X_j$. Each of the remaining $n - \sum_{i=1}^{k} \delta_i$ failure times are censored (or "lost to the study") at a random time L, where L is assumed to be drawn independently and at random from a censoring distribution G. The data may be represented as a set of n observed minima $[(T,I)]$, where $T = \min(X,L)$ and $I = I_{(0,L)}(X)$, where $X \sim F$ and $L \sim G$, with X and L independent. Assuming that the number of losses λ_j in the interval $\{X_j, X_{j+1}\}$ is known for $i = 1, \ldots, \lambda_k$ and $j = 1, \ldots, k$, with $X_0 \equiv 0$ and $X_{k+1} \equiv \infty$, and given the data $\{X_j, \delta_j, \lambda_j, L_i^{(j)}, n_j \text{ for } j = 1, \ldots, k\}$ as defined above and in (12.14)–(12.17), the NPMLE $\hat{\bar{F}}(t)$ of $\bar{F}(t)$ at a fixed point $t > 0$ is given by

$$\hat{\bar{F}}(t) = \prod_{\{j:X_j \le t\}} \frac{n_j - \delta_j}{n_j}. \tag{12.32}$$

Remark 12.1.1. Kaplan and Meier called their new estimator of $\bar{F}(t)$ "the product limit estimator (PLE)," and some authors continue to refer to the estimator by that name. Most researchers and practitioners who work with or use the estimator refer to it as the Kaplan-Meier estimator (KME). It may interest you to know that the paper that introduced this estimator, that is, Kaplan and Meier (1958), has been, for some time now, the most frequently cited paper in the field of Statistics. I heard Paul Meier speak at a conference in the 1980s about this estimator and its extensions. He mentioned that he and Kaplan had to fight vigorously with the referees regarding the merits of the paper, and they got it accepted for publication only after many arguments and painful revisions. He recalled that one particular referee objected to the paper on the grounds that the proposed estimator provided only a modest improvement over the so-called actuarial estimator. With 20-20 hindsight, it is easy to see the paper now as a truly path-breaking contribution to the field.

Remark 12.1.2. When the data is uncensored, the estimator in (12.32) reduces to the empirical survival function. In the presence of censoring, the Kaplan-Meier estimator $\hat{\bar{F}}(t)$ has a "peculiarity" that sets it apart from the NPMLE of $\bar{F}(t)$ based on uncensored data. Depending on the way in which that data is configured, the KME may not actually be a valid survival function. If the largest observed T_i happens to be a loss rather than a failure time, then $\hat{\bar{F}}(x_k) > 0$; more specifically, in this circumstance, the estimator of the survival function $\bar{F}(t)$ takes on the constant, positive value $\hat{\bar{F}}(x_k)$ for $t > x_k$, where x_k is the largest observed failure time. Thus, in this case, the KME does not approach 0 as $t \to \infty$; thus, the estimator $\hat{\bar{F}}(t)$ provides a valid estimate of $\bar{F}(t)$ only for $t \leq x_k$. Since, in this case, the mean of $\hat{F}$ cannot be obtained from the usual formula

$$\hat{\mu} = \int_0^\infty \hat{\bar{F}}(t)dt,$$

it is common to estimate central tendency by the median of the KME rather than by its mean.

Remark 12.1.3. The Kaplan-Meier estimator is known to have a small bias that tends to 0 as $n \to \infty$. It is known that, for a sufficiently large sample, the variance of Kaplan-Meier estimator is well approximated by

$$V\left(\hat{\bar{F}}(t)\right) \cong \bar{F}^2(t) \sum_{\{j:T_j \leq t\}} \frac{q_j}{n_j p_j}. \tag{12.33}$$

The formula in (12.33) is generally known as "Greenwood's formula." Further, under suitable assumptions about censoring (the most common of which is the "random censorship model" which postulates that the losses $\{Li, i = 1, \ldots, n\}$ are iid variables with distribution G and that censoring times and failure times are independent), the Kaplan-Meier estimator is known to be "root n consistent" and is asymptotically normal for any fixed t. Kaplan and Meier's derivation (described above) of the estimator in (12.33) as the nonparametric maximum likelihood estimator of $\bar{F}$ based on censored data is considered to be a heuristic demonstration rather than a rigorous argument. For example, Lawless (1982, p. 7) writes, "This derivation ignores technical difficulties that arise in attempting to precisely specify $L(\bar{F})$ and its parameter space. These can be circumvented by discretizing the time axis and then passing to a limit by letting the number of intervals go to infinity." Johansen (1978) is generally credited with providing a formal proof that the KME is the NPMLE of $\bar{F}$.

I close this section with an example illustrating the computation of the KME.

Example 12.1.2. Suppose that in a small empirical study of stage 4 lung cancer, a sample of eight patients provided the following data: deaths at .8, 3.1, 5.4, and 9.2 months and losses at 1.0, 2.7, 7.0, and 12.1 months. These may be ordered in terms of failures T and losses $L+$ as .8, 1.0+, 2.7+, 3.1, 5.4, 7.0+, 9.2, 12.1+. The Kaplan-Meier estimator of $\bar{F}$ is a step function with jumps at the four observed failure times. The calculation of the estimator is illustrated in Table 12.1.1.

j	X_j or $\max(L)$	n_j	δ_j	λ_j	$F(t_j)$
1	.8	8	1	2	$7/8$
2	3.1	5	1	0	$7/10$
3	5.4	4	1	1	$21/40$
4	9.2	2	1	0	$21/80$
$\max(L)$	12.1+	1	0	1	$21/80$

Table 12.1.1. Computation of the Kaplan-Meier estimator for a small data set.

The Kaplan-Meier estimator obtained above is graphed in the following figure.

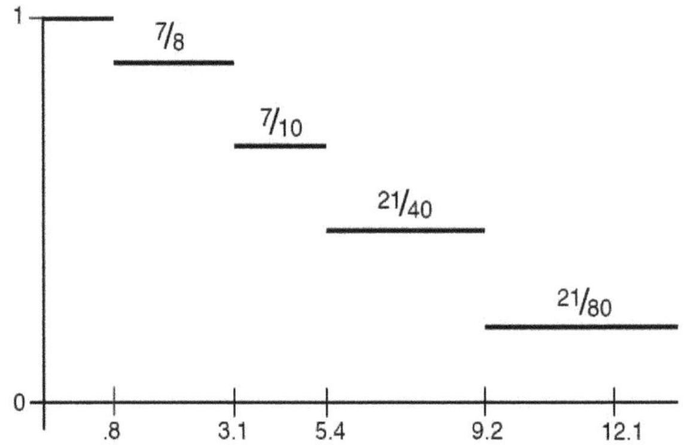

Figure 12.1.2. Graph of the KME for the data in Example 12.1.2.

■

Exercises 12.1.

1. Use the formula for the empirical distribution function in (12.1) to calculate the edf for the following discrete data: 8, 4, 3, 4, 7, 1, 6, 6, 8, 10.
2. Show that the identity in (12.26) actually maximizes the likelihood in (12.21).
3. Show that the likelihood in (12.27) may be written as the expression in (12.29) by virtue of the transformation in (12.28).
4. Nine random participants in the Weight Watcher's weight loss program with the goal of losing at least 20 pounds were followed until they lost twenty or more pounds. The number of weeks it took them to achieve this plateau was recorded. Three of the participants dropped out of the program before the plateau was reached. The following data was obtained: 4+, 6, 8, 9+, 10, 10, 12+, 15, 20. Calculate the Kaplan-Meier estimator of the theoretical survival function $\bar{F}$ of the time required to lose 20 pounds under this weight loss program.

12.2 The Nonparametric Bootstrap

Let us suppose that $X_1, \ldots, X_n \stackrel{\text{iid}}{\sim} F$, where F is a probability model of unknown form, and suppose that $T_n = T(X_1, \ldots, X_n)$ is a statistic computed from this sample and utilized to estimate some population characteristic θ. Suppose further, as would be expected in this general framework, that the exact distribution of T is unknown. Let us consider how one might construct a confidence interval for θ. If we knew that $\sqrt{n}(T_n - \theta)$ was asymptotically normal with mean 0 and asymptotic variance AV and n was sufficiently large, we could use the normal approximation of the distribution of T_n to obtain values

$$T_n \pm \frac{\sqrt{AV}}{\sqrt{n}} Z_{\alpha/2} \tag{12.34}$$

that may serve as approximate $100(1-\alpha)\%$ confidence limits for θ. If AV above was a function of unknown parameters, then it would also need to be estimated in order to make use of (12.34).

There are some potential difficulties in carrying out the tentative program above. It may well be that the statistic T_n is not asymptotically normal. Even if it is, it may be difficult to derive the asymptotic variance of T_n analytically. Even if the asymptotic variance AV can be obtained in closed form, it may be difficult to identify a suitably precise estimator of AV. Even if all of these difficulties can be overcome, the sample size n may be too small to justify an asymptotic approximation. Let's suppose that the tentative program above breaks down for one or more of the reasons I've mentioned. To make this scenario concrete, think of F as unspecified, think of θ as the population mean μ, and let T_n be a "trimmed mean," that is, the average of the middle observations where, say, the smallest 10% of the Xs and the largest 10% of the Xs are dropped from the sample before the data is averaged, and suppose that n is small, say $n = 25$. (Trimmed means are considered to be safe and reliable (a.k.a., robust) estimators of the mean in problems in which F is taken to be a symmetric distribution about its mean and some small fraction of the data are expected to be "outliers," that is, observations that are fairly distant from the center due to contamination of the data set or, perhaps, due to random errors in recording the data.) An asymptotic approximation of the distribution of T_{25} may be quite unreliable. What else could one do?

For simplicity, let us make the (not uncommon) assumption that the statistic T is an unbiased estimator of θ and that its distribution is symmetric about θ. In an idealized world in which we could observe as many Xs as we would like, we could generate a large collection (of size N, say) of independent and identically distributed Ts (each based on a random sample of n Xs). We could then view the empirical distribution $\hat{F}_{T,N}$ of the collection

$$\{T_{n,i}, i = 1, \ldots, N\} \tag{12.35}$$

as an estimator of the distribution of T. Further, from this sample, we could do two things that we are unable to do in the real world. First, we could estimate the standard error of the statistic T with considerable precision. Secondly, we could develop a confidence interval for θ as follows. Since the $\alpha/2$ and $1-\alpha/2$ quantiles of $\hat{F}_{T,N}$, denoted by $Q_{N,\alpha/2}$ and $Q_{N,1-\alpha/2}$ respectively, have approximate probability $1-\alpha$ of capturing the random variable T between them, we may write

$$\begin{aligned} 1-\alpha &\cong P\{Q_{N,\alpha/2} < T < Q_{N,1-\alpha/2}\} \\ &= P\{Q_{N,\alpha/2} - \theta < T - \theta < Q_{N,1-\alpha/2} - \theta\} \\ &= P\{T - (Q_{N,1-\alpha/2} - \theta)\} < \theta < T + (\theta - Q_{N,\alpha/2})\}. \end{aligned} \tag{12.36}$$

But, under the assumption that F_T is symmetric about θ, we have that

$$-(Q_{N,\alpha/2} - \theta) \cong Q_{N,1-\alpha/2} - \theta \cong (Q_{N,1-\alpha/2} - Q_{N,\alpha/2})/2.$$

It thus follows from (12.36) that

$$1-\alpha \cong P\{T - (Q_{N,1-\alpha/2} - Q_{N,\alpha/2})/2 < \theta < T + (Q_{N,1-\alpha/2} - Q_{N,\alpha/2})/2\},$$

implying that the values

$$T \pm (Q_{N,1-\alpha/2} - Q_{N,\alpha/2})/2 \tag{12.37}$$

are approximate $100(1-\alpha)\%$ confidence limits for θ. Unfortunately, since we don't live in an idealized world, such confidence intervals are unavailable to us.

This brings us, finally, to the nonparametric bootstrap. An intriguing alternative to repeated sampling from F is repeated sampling from an estimator of F. But it is by no means obvious that doing this will be helpful. After all, any estimator $\hat{F}$ of F is based on the available sample, so it would seem that sampling from $\hat{F}$ can't give us more information than the original sample contains. Although the notion of "resampling" had been studied earlier (for example, by Quenouille (1949) and by Hartigan (1969)), Bradley Efron's 1979 paper showed its flexibility, power, and promise. Efron showed convincingly that what resampling could do was to provide insights beyond the value of a particular statistic of interest; indeed, it could shed light on the variability of that statistic, could serve as a vehicle for approximating the distribution of that statistic, regardless of the distribution F from which the sample was drawn, and could be used to develop confidence statements about unknown population parameters. Of course, there are situations in which the approach doesn't work as intended, but it has been shown to apply in a wide range of problems, including some scenarios that are extremely complex and quite resistant to analytical treatment. Besides introducing the bootstrap, Efron's paper is rightly considered to be a seminal contribution to the area of computer-intensive statistical methods. In the paragraphs that follow, we will examine how the nonparametric bootstrap is implemented, how it may be used to estimate the variability of a statistic T and to obtain an approximate confidence interval for the statistic's expected value θ. We close the section with an example of its use in a recent biomedical study. As I am sure you have noticed, I use the adjective "nonparametric" in describing the version of the bootstrap to be treated here. We will forgo discussion of the so-called "parametric bootstrap."

Suppose that we have a random sample $X_1,\dots,X_n$ of size n drawn from the distribution F. Let $\hat{F}_n$ be the empirical distribution based on the observed sample values $x_1,\dots,x_n$. A bootstrap sample is obtained by drawing a sample of size n, with replacement, from the population of values $\{x_1,\dots,x_n\}$. This sample can also be described as a sample, drawn with replacement, from the edf $\hat{F}_n$. The fact that this sampling is *with replacement* is crucial here, since if we were sampling without replacement, we would always recover the original sample after n draws, and nothing would be learned from the exercise. When we sample with replacement, any particular sample value x_i might be sampled more than once, and some of the sample values may be absent altogether from the bootstrap sample. Most importantly, since this sampling can be automated through a simple computer program, it can be repeated many times. Typically, hundreds or even thousands of bootstrap samples are obtained, and the statistic $T_n = T(X_1,\dots,X_n)$, which we think of as an estimator of the parameter θ, can be computed for each bootstrap sample. We will denote the B bootstrap samples as $\mathbf{x}_1^*,\dots,\mathbf{x}_B^*$, where $\mathbf{x}_i^* = (x_{i,1}^*,\dots,x_{i,n}^*)$ is the vector of x values drawn in the ith bootstrap sample. Since there is variability from one bootstrap sample to another, the value of the statistic T will tend to vary when evaluated from different bootstrap samples. The variability of T is precisely what the bootstrap is intended to reveal. Let $T_i^* = T(\mathbf{x}_i^*)$ for $i = 1,\dots,B$. We turn now to a discussion of how the variability in $\{\mathbf{T}_i^*, i = 1,\dots,B\}$ can be exploited.

Consider first the estimation of the standard error of T, that is, the standard deviation of the distribution of T. The bootstrap estimator $\hat{\sigma}_T$ of σ_T is remarkably simple:

$$\hat{\sigma}_{T_n} = \left(\frac{1}{B-1}\sum_{i=1}^{B}(T_i^* - \bar{T}^*)^2\right)^{1/2}, \tag{12.38}$$

where $\bar{T}^* = (1/B)\sum_{i=1}^{B} T_i^*$; that is, we estimate σ_T by the standard deviation of the T values calculated from the B bootstrap samples. Let us examine why one might think that such an estimation strategy could work. Intuitively, one could argue as follows: the edf $\hat{F}_n$ of the original sample is a pretty good estimator of F. So taking an iid sample from $\hat{F}_n$ should resemble the process of taking an iid sample from F, especially if n is reasonably large. So at least for large n, taking many (namely, B) random samples from $\hat{F}_n$ and evaluating the statistic T each time should resemble the process of obtaining many T values based on random samples from F. And the latter process clearly gives us a solid, reliable estimator of σ_T. It makes sense, then, that the bootstrap samples might indeed provide a good estimate of σ_T. It has been shown, both empirically and theoretically, that this intuition is correct. Surprisingly, the bootstrap often provides a good estimator of σ_T even when the sample size is small. Efron and Tibshirani (1993) use the bootstrap in an example involving the survival times of 16 mice, which underwent a surgical procedure, with 7 receiving a postsurgical treatment.

You might wonder what good the standard error of T is if you don't know the distribution of T. Now if T was thought to be approximately normal, we could do quite a bit with the estimated standard error. An approximate confidence interval for the parameter $\theta = ET$ would then be possible. As it happens, many statistics of interest are approximately normally distributed when n is large, so there is a sizable class of problems which can be handled by such an approximation. Without the assumption of approximate normality, you cannot do as much, but you should keep in mind that the application of Chebyshev's inequality can often resolve questions of interest.

Another extremely useful purpose served by bootstrapping is the development of confidence intervals for certain population parameters. I will describe the method which is used to construct the "percentile interval." This is a quick and easy method, both to understand and to apply. There are various "adjustments" one can make to this method to yield better results (that is, more accurate confidence limits) in selected circumstances. Among them, Efron and Tibshirani (1993) recommend the BC_a method (that is, the "bias-corrected and accelerated" bootstrap); a detailed discussion of the BC_a method is beyond the scope of this book (an author's favorite cop-out, I know, but in this case, a remark that is totally fitting). By the way, the BC_a method isn't needed under the assumption made above that the target of our estimation is the parameter $\theta = ET$ as, in this case, there is no bias to correct. Applications of this framework include such statistics T as the sample mean and sample variance, in general, and trimmed means and the median when the distribution is assumed to be symmetric with finite variance σ^2.

So how does the percentile method work? Again, its simplicity, along with its utility, are rather amazing. Assume that we have the B bootstrap samples $\mathbf{x}_1^*, \dots, \mathbf{x}_B^*$ described above, and that the statistic T has been computed for each such sample, yielding the collection $\{\mathbf{T}_i^*, i = 1, \dots, B\}$. Suppose we wish to a have a $100(1-\alpha)\%$ confidence interval for $\theta = ET$. Recall that we are not generally able to use the confidence limits $T \pm (Q_{N,1-\alpha/2} - Q_{N,\alpha/2})/2$ for θ, as a confidence interval with these limits assumes that we have access to many independently sampled values of the statistic T, something that is available only in our ideal-world scenario. But with our bootstrap sample, we can approximate $Q_{N,1-\alpha/2} - Q_{N,\alpha/2}$ by $T^*_{(1-\alpha/2)} - T^*_{(\alpha/2)}$, where $T^*_{(p)}$ represents the pth quantile of the collection $\{T_i^*\}$, and using a little mathematics, we can justify the use of $(T^*_{(1-\alpha/2)}, T^*_{(\alpha/2)})$ as an approximate $100(1-\alpha)\%$ confidence interval for θ. For concreteness, let's take B to be 1000, a commonly chosen number of bootstrap replications, and let $\alpha = .05$. If $T^*_{(.025)}$ represents the 25th smallest T^* value and $T^*_{(.975)}$ represents the 25th largest T value among the 1000 T^*s computed from our bootstrap samples, then the interval $(T^*_{(.025)}, T_{(.975)})^*$ is

Treatment group (X):	94,	197,	16,	38,	99,	141,	23		
Control group (Y):	52,	104,	146,	10,	51,	30,	40,	27,	46

Table 12.2.1. Survival lines for 16 mice.

an approximate 95% confidence interval for the parameter θ. The following example illustrates the process and utility of obtaining bootstrap confidence intervals for a parameter of interest.

Example 12.2.1. In Table 12.2.1 data is given of the survival time, in days, of sixteen mice which underwent a particular surgical procedure, appears in Efron and Tibshirani (1993). Seven of the mice received a particular post-surgical treatment, while the other mice received no treatment and served as controls in the study.

Code to compute a 95% bootstrap interval in the statistical package R might look like the following:

```
Treatment <- c(94, 197, 16, 38, 99, 141, 23)
Control <- c(52, 104, 146, 10, 51, 30, 40, 27, 46)
n1 <- length(Treatment)
n2 <- length(Control)
B <- 1000

BS.Treat <- sapply(1:B, function(i) sample(Treatment, size=n1,
    replace=TRUE))
BS.Control <- sapply(1:B, function(i) sample(Control, size=n2,
    replace=TRUE))

BS.Mean.Treat <- colMeans(BS.Treat)
BS.Mean.Control <- colMeans(BS.Control)

BS.Mean.Diffs <- BS.Mean.Treat - BS.Mean.Control
c(quantile(BS.Mean.Diffs, .025), quantile(BS.Mean.Diffs, .975))
```

A 95% confidence interval for the mean difference is thus $(-26.14603, 83.24365)$, which includes zero. ■

Some practical advice is in order. It is generally understood that the number of bootstrap samples B required for reasonable accuracy will vary with the goals for its use. For estimating the standard error of an estimator, B in the range of 100–200 will generally suffice, while the generation of confidence intervals for a parameter of interest requires larger B, with 1000 being considered acceptable in most problems. It is known that the bootstrap can fail to provide accurate results when the underlying distribution F is heavy tailed. It is also known that the accuracy of the bootstrap is affected by the form of the statistic T on which the bootstrap is based. Bickel and Freedman (1981) were among the first to identify conditions under which bootstrap estimates will be consistent as $n \to \infty$. It is known that special adjustments of the bootstrap (generally some form of "smoothing" of the data) are needed to ensure that bootstrap estimates are consistent for various classes of "awkward" statistics T (see Davison and Hinkley (1997, p. 60)). Peter Hall (1988, 1992) showed that a special form of the bootstrap—the studentized bootstrap—generally leads to confidence intervals and tests with better performance characteristics than other bootstrapping methods. The literature on the bootstrap is vast, and the introduction provided here will hopefully whet your appetite for exploring some of its trajectories.

I close with a substantive example involving recent breast cancer research.

Example 12.2.2. It is thought that mutated versions of BRCA1 and BRCA2 genes increase the risk of breast cancer. A Stanford University research group led by Allison Kurian and Alice Whittemore has studied subjects with breast cancer, each referred to as a "proband," and their close (first-degree) relatives, referred to as FDRs. They use an approach called segregation analysis. Survival data is collected on all participants. Data is collected on the genotypes of each proband, and when available, genotype data is collected from the FDRs as well. The probability model for the data collected involves the probability of the family genotype multiplied by the probability of family survival data given the family genotype. Since genotypes are not obtained for every participant, the probability of the observed data involves summing over the unobserved genotypes. Finally, the contribution of a given family to the overall likelihood is modeled as the ratio of the probability of the observed family data divided by the probability of the proband's survival data. The goal of the study is to construct a confidence interval for the relative risk of disease associated with being a non-carrier FDR of a BRCA1- and BRCA2-carrying proband. Maximizing the likelihood function is analytically intractable. Even numerical methods pose major challenges, as numerical instabilities have been encountered in attempting to obtain the second derivative of the log likelihood at the maximum. The Stanford group used the bootstrap to obtain an approximate 95% confidence interval $(.04, 3.81)$ for the relative risk. Note that this confidence interval includes the value 1 that is associated with equal risks. The research team's conclusion: The study does not provide statistically significant evidence of increased breast cancer risk among non-carrier FDRs of carrier probands as compared to FDRs from families without the BRCA1 and BRCA2 mutations. ■

Exercises 12.2.

1. Ten UCLA undergraduates who graduated last year were chosen at random. The number of units taken in their undergraduate careers were recorded as: 172, 188, 177, 190, 210, 200, 184, 199, 178, 192. Suppose you are interested in the mean μ of the population from which these students were drawn. Let your estimator T of μ be the sample mean. Generate 1000 bootstrap samples and use them to (a) estimate the standard error of T and (b) obtain a 95% confidence interval for μ.

2. Consider the data in Exercise 12.2.1. Suppose you are interested in the median m of the population from which these students were drawn. Let your estimator T of m be the sample median. Generate 1000 bootstrap samples and use them to (a) estimate the standard error of T and (b) obtain a 95% confidence interval for m.

3. Prove the "transformation-respecting property" of the percentile interval based on bootstrapping: if (a,b) is the $100(1-\alpha)$ percentile interval for the parameter θ and g is a one-to-one transformation, then $(\min\{g(a),g(b)\},\max\{g(a),g(b)\})$ is the $100(1-\alpha)$ percentile interval for $g(\theta)$.

4. (See Exercise 12.2.3.) Use the transformation-respecting property of the percentile interval to prove the "percentile interval lemma": Consider a statistic T for which

$\theta = ET$ and a transformation g such that $S = g(T) \sim N(\lambda, \sigma^2)$, where $\lambda = g(\theta)$. Then $(g^{-1}(S - \sigma Z_{\alpha/2}), g^{-1}(S + \sigma Z_{\alpha/2}))$ is the $100(1-\alpha)\%$ percentile interval for θ based on T.

12.3 The Sign Test

We'll begin our relatively brief tour of nonparametric testing with a treatment of a very simple procedure that is thought to be the oldest test of this kind. The origins of the sign test date back at least to Arbuthnot (1710), who used it to test the equality of birth rates for boys and girls over an 82-year period. Technically, this isn't our first encounter with nonparametric testing, as the chi-square goodness-of-fit test treated in Section 10.6 is a fully nonparametric test and is treated as such in most textbooks on this subfield of statistics. But the sign test will be the first of three tests that we will treat which are concerned with two populations and the question of whether they are the "same" in some sense or they are different. The simplicity of the test comes at a cost—it is not a particularly powerful test, and it is not difficult to construct tests that have a stronger statistical performance. You might ask why we are spending time on a test that is known to be inefficient. There are several good reasons, including its important status in the history of the subject and the fact that this test applies more widely than many others, as it does not require the independence of the available samples of Xs and Ys (as do, for example, the tests discussed in the next two sections). Finally, you should recognize that simple and somewhat inefficient tests (or as statistician John Tukey liked to call them, "quick and dirty tests") can still be quite useful. If you reject the hypothesis of equality using the sign test, then a more powerful test will most likely also lead to rejection, but since the difference between the populations has already been noted, we needn't bother with more complex tests. It's only when the sign test fails to reject the hypothesis of equality that we might wish to explore whether a more sophisticated test will detect a difference.

The sign test applies to bivariate data which obeys an iid framework. Let F be the bivariate distribution for the random pair (X,Y), and suppose that

$$(X_1,Y_1),\ldots,(X_{n^*},Y_{n^*}) \stackrel{\text{iid}}{\sim} F. \tag{12.39}$$

Note that the Xs and Ys need not be independent; that assumption is not needed when investigating the hypotheses of interest here. The sign test is used to test the hypothesis $H_0 : P(X < Y) = P(X > Y)$; the usual alternative is the two-sided hypothesis $H_1 : P(X < Y) \neq P(X > Y)$. When F is an absolutely continuous distribution with density function f, then $P(X = Y) = 0$ and the null hypothesis may be written as $P(X \leq Y) = 1/2$. The probability $P(X \leq Y)$ is an important quantity in many statistical studies. For example, in stress–strength testing in engineering reliability, this probability represents the reliability of the material being tested (like a welded steel bar used in building roads and bridges), as it is readily seen to be the probability that the strength Y of a bar exceeds the stress X that is placed upon it. To test $H_0 : P(X < Y) = P(X > Y)$, the data is divided into three groups:

Data (X_i,Y_i) for which $X_i = Y_i$ are set aside

Data (X_i,Y_i) for which $X_i < Y_i$ are labeled as "+"s

Data (X_i,Y_i) for which $X_i > Y_i$ are labeled as "−"s.

While the original sample size was denoted by n^*, we will denote the total number of "+"s and "−"s as n. The power of the sign test depends on n rather than n^*, so in performing an experiment in which the sign test will be used, one would typically perform enough trials to ensure that the value of n is large enough to provide the desired power at a particular alternative of interest. Now, define the statistic T to be the number of "+"s observed among our n signed observations. If we let

$$p = \frac{P(X < Y)}{P(X \neq Y)}, \tag{12.40}$$

then $T \sim B(n, p)$. Suppose that we wish to test the null hypotheses $H_0 : p = 1/2$ against the alternative $H_1 : p \neq 1/2$. For small n, say $n \leq 20$, we would typically use the binomial table to determine the p-value of the observed test statistic. Given the observed value of the test statistic is $T = t$, we would reject H_0 at the significance level α if $P(T \leq t) < \alpha/2$ or $P(T \geq t) < \alpha/2$, where $T \sim B(n, 1/2)$. For a one-sided test of size α, the null hypothesis would be rejected when $P(T \leq t) < \alpha$ when $H_0 : p \geq 1/2$ and would be rejected when $P(T \geq t) < \alpha$ when $H_0 : p \leq 1/2$. A one-sided sign test is illustrated in the example below.

Example 12.3.1. In many large cities, the title of "Chess Master–X" is given to the students in Grade X who rank among the city's top 15 chess players in that grade. The 8th-grade Chess Masters from Los Angeles and New York played in the U. S. Championship last month. The wins (1), losses (−1) and ties (0) from 15 games between randomly paired players appear below.

Pair	1	2	3	4	5	6	7	8	9	10	11	12	13	14	15
LA	0	1	1	1	0	1	0	1	1	0	0	−1	1	1	1
NY	0	−1	−1	−1	0	−1	0	−1	−1	0	0	1	−1	−1	−1

In using the sign test for testing the null hypothesis that the teams are evenly matched, against a two-sided alternative, note that, among the 10 pairs that are not tied, there are $X = 9$ plusses (LA is better than NY). The p-value of the observed X is equal to $2P(X \geq 9)$, where $X \sim B(10, p)$, with $p = 1/2$, under H_0. From Table A.1, we have $P(X \geq 9) = .0107$, so the p-value of the observed X is .0214. It follows that $H_0 : p = 1/2$ would be rejected at significance level .05. ■

If we wish to test the hypotheses $H_0 : p = 1/2$ vs. $H_1 : p \neq 1/2$ and the sample size n is sufficiently large (say $n > 20$), then the standard critical region of size α is: Reject H_0 if

$$\left| \frac{\hat{p} - 1/2}{1/\sqrt{4n}} \right| > Z_{\alpha/2}, \tag{12.41}$$

where $\hat{p} = T/n$ is the proportion of "+"s obtained in the sample of n signed observations. The obvious adjustments are made of one-sided tests: reject $H_0 : p \leq 1/2$ in favor of $H_1 : p > 1/2$ if $(\hat{p} - 1/2)/(1/\sqrt{4n}) > Z_\alpha$ and reject $H_0 : p \geq 1/2$ in favor of $H_1 : p < 1/2$ if $(\hat{p} - 1/2)/(1/\sqrt{4n}) < -Z_\alpha$. The following example illustrates the use of the sign test based on a large sample.

Example 12.3.2. In a carefully designed experiment, a new drug for hypoglycemia is given to a random sample of patients known to have high blood sugar. There were a total of 100 of these patients with an initial blood sugar measurement X_i that was different than their post-treatment blood sugar level Y_i. The experimenter would like to determine if the drug is helpful. To test the hypothesis $H_0 : p \geq 1/2$ vs. $H_1 : p < 1/2$ at significance level $\alpha = .01$, where $p = P(X < Y)/P(X \neq Y)$, the critical region would be: Reject H_0 if $(\hat{p} - 1/2)/(1/\sqrt{400}) < -Z_{.01}$. In the actual experiment, it was found that 63 of the 100 patients had lower blood sugar after the treatment. Thus, for these data, $\hat{p} = .37$ and $(.37 - .5)/(1/20) = -2.6 < -2.33 = -Z_{.01}$. It follows that the null hypothesis is rejected at the .01 significance level and that the drug may be described as efficacious. ■

The sign test is often used for other purposes. For example, it can be used in a univariate setting to test the hypothesis that the median m of an unspecified distribution F is equal to the value m_0. If F were known to be continuous, so that $P(X = m_0) = 0$ when $X \sim F$, a sample of size n would yield $X = x$ values below m_0 and $n - x$ values that were larger than m_0. Labeling values below m_0 as "+"s, we have that under $H_0 : m = m_0$, $X \sim B(n, 1/2)$. The null hypothesis would be rejected at significance level α if

$$\left| \frac{\hat{p} - 1/2}{1/\sqrt{4n}} \right| > Z_{\alpha/2},$$

where $\hat{p} = X/n$. This test is sometimes referred to as the median test; it is easily generalized to test the pth quantile ξ_p of a distribution F is equal to a specified value, where ξ_p is defined as the number for which $P(X \leq \xi_p) = p$ when $X \sim F$. The pth quantile is associated with the $(100)p^{th}$ percentile, so that a student in the 95th percentile of her class has a score or ranking at or above the quantile $\xi_{.95}$.

Finally, because the sign test is a very simple test that treats matched-pairs data $\{(X_i, Y_i)\}$ using a very basic summary statistic that discards much of the information that the data contains, it is not a very powerful test. If the normality assumption of the paired t test is satisfied, the latter test is substantially more powerful in testing the parametric hypothesis $\mu_x = \mu_y$. If the normality assumption is questionable, there are several non-parametric tests that are known to be more powerful than the sign test. Among these, the Wilcoxon signed ranks test (SRT) is usually recommended. While this alternative for handling matched pairs data nonparametrically is not treated in this book, any one of the three nonparametric texts mentioned at the end of Section 12.5 may be consulted for a careful presentation of the SRT.

Exercises 12.3.

1. Suppose that the variables X and Y are independent, with $X \sim B(3, .5)$ and $Y \sim B(3, .6)$. If n^* (X, Y) pairs are sampled, what is the expected number of ties, that is, pairs for which $X = Y$? How large would the sample size n^* need to be if an experimenter might expect to have $n = 50$ pairs for which $X \neq Y$?

2. Two high school tennis teams, each with twelve players, were matched from first to last seed on the team (a measure of their past performance in competitive tennis). Twelve separate singles matches were played, and Team A players beat Team B players in 10 of the matches. Does this result justify rejecting the null hypothesis that the teams are evenly matched against a two-sided alternative, at $\alpha = .05$?

3. A random sample of 50 pairs of boy-girl twin births in Los Angeles hospitals last year was obtained. It was noted that, in 31 of these births, the girl was delivered first. Use the sign test, at $\alpha = .1$, to test, against a two-sided alternative, the null hypothesis that the probability of being born first is the same for boy and girl twins.

4. Consider performing the sign test for the hypothesis $H_0 : P(+) = P(-)$ based on a sample of size n (the total number of plusses and minuses). Let $T = t$ be the number of observed plusses. (a) What is the two-sided p-value of the test statistic T if $n = 15$ and $T = 2$? (b) What is the two-sided p-value of the test statistic T if $n = 100$ and $T = 37$?

12.4 The Runs Test

One of the very early nonparametric tests for the equality of two distributions was the "runs test" proposed by Wald and Wolfowitz in 1940. The nonparametric approach to such problems relies on being able to find a test statistic that effectively detects differences between populations while not depending on the exact distributions involved. A test based on such a statistic could then be described as "distribution free" and would be applicable to the comparison of any two distributions, regardless of their exact form. In our treatment of the nonparametric tests presented in this section and the next, we will make the simplifying assumption that the distributions from which samples are drawn are continuous, so that ties among the observations have probability 0 of occurrence. Both of these tests can be extended to arbitrary distributions, but the treatment of continuous distributions makes the properties of the nonparametric procedures more easily understood.

Suppose that two independent samples are obtained, namely $X_1, \ldots, X_n \stackrel{\text{iid}}{\sim} F_x$ and $Y_1, Y_2, \ldots, Y_m \stackrel{\text{iid}}{\sim} F_y$, and we are interested in testing the hypothesis $H_0 : F_x = F_y$. While various possible statements might come to mind as possible formulations of the alternative hypothesis (with $F_x \neq F_y$ being the most obvious), we are usually interested in the alternative that one of the distribution functions is actually smaller than the other. It is useful to give this type of ordering its own name, and this is done in the following:

Definition 12.4.1. Let X and Y be independent random variables, with $X \sim F_x$ and $Y \sim F_y$. Then X is said to be *stochastically smaller* than Y, denoted by $X \leq_{st} Y$, if

$$F_x(a) \geq F_y(a) \text{ for all } a \in (-\infty, \infty), \tag{12.42}$$

with $F_x(a_0) > F_y(a_0)$ for at least one real number a_0; X is *strictly stochastically smaller* than Y, written $X <_{st} Y$, if

$$F_x(a) > F_y(a) \text{ for all } a \in (-\infty, \infty). \tag{12.43}$$

The interpretation of the "stochastic order" $\leq_{st}$ as defined in (12.42) is quite simple. If $F_x(a)$ is larger than $F_y(a)$ for all a, the X values are, as a rule, farther to the left than the Y values. You can think of the density of the Xs as being situated to the left of the density of the Ys. Because the inequalities $X \leq_{st} Y$ and $F_x(a) \geq F_y(a)$ seem backwards, some authors prefer to define the relationship in terms of the survival functions of X and Y, that is, in terms of $\bar{F} = 1 - F$. In particular, we may write

$$X \leq_{st} Y \text{ if and only if } \bar{F}_x(a) \leq \bar{F}_y(a) \text{ for all } a \in (-\infty, \infty). \tag{12.44}$$

Given the definition above, we may write the "usual" alternative hypothesis to $H_0 : F_x = F_y$ as

$$H_1 : X \leq_{st} Y \quad \text{or} \quad Y \leq_{st} X. \tag{12.45}$$

In spite of the appearance of H_1 in (12.45), the runs test is a one-sided test, as when either of the alternatives in (12.45) is true, one expects to see a small number of runs. Thus, H_0 is rejected when the number of runs R is determined to be too small.

The runs test is extremely simple to describe and to implement. Suppose you have the samples of Xs and Ys above in hand. The combined data can then be ordered from smallest to largest, and the number R of "runs" of like elements (that is, Xs in a row and Ys in a row) are counted. For example, the sequence

$$\text{XXX Y XXX YYYY XX YYY XX} \tag{12.46}$$

has a total of 7 runs. The question one would need to resolve in testing $H_0 : F_x = F_y$ is whether having $R = 7$ runs in a sequence of 18 observations (with 10 Xs and 8 Ys) is consistent with this null hypothesis. We will address this type of question shortly. To see why counting runs has some discriminatory power, consider what one might expect if the Xs are stochastically smaller than the Ys. Since, in this case, the Xs would tend to fall to the left of the Ys, we're more likely to see a sequence with somewhat fewer runs than that in (12.46), perhaps something like

$$\text{XXXXXXXXX YYYY X YYYY.} \tag{12.47}$$

When one distribution is smaller than the other (in the "st" sense) one expects more clustering of the Xs among themselves and the Ys among themselves, a phenomenon that would result in fewer runs. In short, we expect more runs when $F_x = F_y$ than when one F is larger than the other (that is, when $X \le_{st} Y$ or $Y \le_{st} X$). The runs test qualifies as a nonparametric test because the distribution of the test statistic R, the number of observed runs, is the same for any pair of continuous distributions F_x and F_y, that is, the behavior of the test statistic is distribution-free. The test is especially useful because, under the null hypothesis, the distribution of the test statistic R can be obtained exactly using elementary combinatorics. This allows us to set the significance level of the test quite easily. We turn now to the derivation of the null distribution of R.

Recall that we have assumed that $X_1,\dots,X_n \overset{\text{iid}}{\sim} F_x$ and $Y_1,Y_2,\dots,Y_m \overset{\text{iid}}{\sim} F_y$, and that the Xs are independent of the Ys. Under the null hypothesis $H_0 : F_x = F_y$, we have that the Xs and Ys are simply a random sample of size $n+m$ from common continuous distribution F. To obtain the distribution of R, we first recognize the fact that every permutation of the combined Xs and Ys has the same chance of occurrence. Among the $n+m$ "slots" into which the n Xs can be placed, we can choose the slots that will contain an X in exactly

$$\binom{n+m}{n}$$

equally likely ways. We will now identify the number of different ways in which the runs statistic R can take the value r, where r is an integer that may range from a low of 2 to a high of $2\min\{n,m\}$ or $2\min\{n,m\}+1$, the upper bound being the number of runs you get when you alternate the Xs and Ys until you run out of one or the other. The difference of upper bounds results from whether or not $n = m$. The formulae we develop below for $P(R = r)$ will be different when r is even and when r is odd.

Let k be a positive integer such that $1 \le k \le \min\{n,m\}$. We will first calculate $P(R = 2k)$. Consider the n Xs lined up in a row. Now, choose $k-1$ spaces among the $n-1$ spaces between the Xs. These choices will define the k runs of Xs that the sequence of $2k$ runs will contain. We can choose these spaces in

$$\binom{n-1}{k-1}$$

ways. Shifting focus to the Ys, choose $k-1$ spaces among the $m-1$ spaces between the Ys. This can be done in $\binom{m-1}{k-1}$ ways. These choices will define the k runs of Ys that the sequence of $2k$ runs will contain. From the basic rule of counting, the process of defining the groupings of the Xs and the Ys, each into k clusters, can be done in

$$\binom{n-1}{k-1}\binom{m-1}{k-1}$$

ways. Now, we combine (or "perfectly shuffle") the X clusters with the Y clusters, yielding a sequence of either X and Y clusters in succession or Y and X clusters in succession, these 2 types of sequences resulting from whether the sequence starts with Xs or with Ys. From this, we may surmise that

$$P(R=2k)=\frac{2\binom{n-1}{k-1}\binom{m-1}{k-1}}{\binom{n+m}{n}} \text{ for } k=1,\ldots,\min\{n,m\}. \tag{12.48}$$

When R is odd, we may obtain by a similar argument (which accounts for the fact that when $R=2k+1$, one can have either k clusters of Xs and $(k+1)$ clusters of Ys or k clusters of Ys and $(k+1)$ clusters of Xs) the following formula:

$$P(R=2k+1)=\frac{\binom{n-1}{k}\binom{m-1}{k-1}+\binom{n-1}{k-1}\binom{m-1}{k}}{\binom{n+m}{n}} \text{ for } k=1,\ldots,\min\{n,m\}. \tag{12.49}$$

In addition to having the exact distribution of R under the null hypothesis, it is useful to have formulae for the mean and variance of R as well. In fact, when n and m are sufficiently large, these two values arise in the normal approximation for the distribution of R. The mean and variance of R can be calculated from the distribution above, but we will forgo the calculation and just provide the answers. (This time, as a change of pace, I won't leave the calculations as exercises for you to do for homework. You might take this problem on though, as an interesting challenge or, alternatively, you might just check these answers for suitably small choices of n and m.) Here, we simply record the following facts. For fixed n and m, let $R_{n,m}$ be the number of "runs" of like elements among the ordered combined sample of $n+m$ observations. Then

$$E(R_{n,m})=\mu_{n,m}=\frac{2nm}{n+m}+1 \quad \text{and} \quad V(R_{n,m})=V_{n,m}=\frac{(\mu_{n,m}-1)(\mu_{n,m}-2)}{n+m-1}. \tag{12.50}$$

Now, consider testing the hypothesis $H_0: F_x=F_y$ vs. the two-sided alternative $H_1: X\le_{st} Y$ or $Y\le_{st} X$. The probabilities in (12.48) and (12.49) put us in position to calculate the p-value of the observed R statistic. Let us denote the p-value of the R statistic by p_R. The critical region of size α for this test may then be written as: Reject H_0 if $p_R\le\alpha$, as this outcome would indicate that the R statistic is in an extreme tail of the distribution of R. This testing process is illustrated in the following.

Example 12.4.1. Let us suppose that independent random samples of 10 Xs and 10 Ys were drawn and that the Xs and Ys are clustered as

3X, 1Y, 5X, 2Y, 2X, 7Y,

so that the observed number of runs is $R=6$. From (12.50), we know that the expected number of runs for these data is $\mu_{10,10}=11$, so the number of runs here seems small. To determine whether the hypothesis $H_0: F_x=F_y$ should be rejected in favor of a two-sided alternative at significance level $\alpha = .05$, we will calculate the p-value of the observation $R = 6$. We have

$$P(R=2) = \frac{2\binom{9}{0}\binom{9}{0}}{\binom{20}{10}} = \frac{2}{184,756}$$

$$P(R=3) = \frac{\binom{9}{1}\binom{9}{0}+\binom{9}{0}\binom{9}{1}}{\binom{20}{10}} = \frac{18}{184,756}$$

$$P(R=4) = \frac{2\binom{9}{1}\binom{9}{1}}{\binom{20}{10}} = \frac{162}{184,756}$$

$$P(R=5) = \frac{\binom{9}{2}\binom{9}{1}+\binom{9}{1}\binom{9}{2}}{\binom{20}{10}} = \frac{648}{184,756}$$

$$P(R=6) = \frac{2\binom{9}{2}\binom{9}{2}}{\binom{20}{10}} = \frac{2592}{184,756}$$

From these calculations, we find that

$$P(R \leq 6) = \frac{3,422}{184,756} = .0185.$$

Since the p-value of the observed runs R for these data is equal to $.0185 < .05$, we would reject the hypothesis that $F_X = F_Y$, against the alternative in (12.45), at the significance level $\alpha = .05$. ■

As mentioned above, the runs test statistic R is approximately normally distributed when n and m are sufficiently large. The literature on this approximation suggests that the normal approximation is usable when n and m are both at least 10. The asymptotic distribution of the standardized statistic R under the null hypothesis $H_0 : F_x = F_y$ is given in the following:

Theorem 12.4.1. Consider two independent samples $X_1, \dots, X_n \overset{\text{iid}}{\sim} F_x$ and $Y_1, \dots, Y_m \overset{\text{iid}}{\sim} F_y$, and let $R_{n,m}$ be the number of runs of like elements among the ordered combined sample of $n+m$ observations. If $n \to \infty$ and $m \to \infty$ in such a way that $n/(n+m) \to p \in (0,1)$, then

$$\frac{R_{n,m} - \mu_{n,m}}{\sqrt{V(R_{n,m})}} \xrightarrow{D} Z \sim N(0,1), \tag{12.51}$$

where $\mu_{n,m}$ and $V_{n,m}(R)$ are given in (12.50).

The result above is illustrated in the following.

Example 12.4.2. Suppose the Math Graduate Record Exam (GRE) scores are obtained for two independent random samples of 50 graduating seniors from Harvard (the Xs) and Yale (the Ys), respectively. Suppose that 40 runs are observed in the ordered combined sample of these 100 seniors. To test the null hypothesis $H_0 : F_x = F_y$ vs. the two-sided alternative in (12.45) at $\alpha = .05$, we would reject H_0 if the observed number of runs R is too small. Since the expected number of runs is

$$\mu_{50,50} = \frac{2(50)(50)}{100} + 1 = 51,$$

and the variance of

$$V_{50,50} = \frac{(50)(49)}{99} = 24.7475,$$

we may compute the p-value as follows:

$$P(R_{50,50} \leq 40) \simeq P(Z \leq (40.5 - 51)/4.9747) = -2.11) = .0174. \qquad (12.52)$$

The p-value of the observed R is seen to be smaller than .05. Based on this calculation, we would reject, at the 5% significance level, the hypothesis $H_0 : F_x = F_y$ in favor of the alternative hypothesis in (12.45). Note that our calculation in (12.52) uses the continuity correction $(+.5)$ in approximating a probability for the discrete variable R by the probability associated with the continuous variable Z. ■

The runs test covered in this section is often referred to as a "test for randomness." In this case, the null hypothesis is framed as:

$$H_0 : X_1, \ldots, X_n, Y_1, \ldots, Y_m \overset{\text{iid}}{\sim} F, \qquad (12.53)$$

and the alternative hypothesis is simply the complementary statement, that is, the statement that H_0 is not true. Now, the null hypothesis $H_0 : F_x = F_y$ put forth above, together with the tacit assumption of independence of the Xs and the Ys, is identical to (12.53). When the null hypothesis is rejected, it is correct to conclude that the Xs and Ys are not consistent with the iid assumption. One difficulty with thinking of this test as a test for randomness is the fact that it has very poor power against some alternatives to randomness. For example, the test would have a rather poor chance of detecting non-randomness if the Xs are standard normal variables and the Ys are standard Cauchy variables, as such as alternative would typically give rise to a moderate number of runs and would thus rarely reject H_0. There are alternative tests of randomness (for example, the Kolomogorov-Smirnoff test) that are known to be considerably more powerful. In spite of this, the runs test can still be useful as a test for randomness, as it is quite simple to execute, and it will often detect departures from the randomness assumption. The runs test is known to be especially effective against alternatives to H_0 that stipulate that the true distribution of the Ys has the same shape as that of the Xs but has simply been shifted by a fixed amount c, that is, $F_Y(t) = F_X(t - c)$.

Exercises 12.4.

1. Provide a combinatorial argument which proves (12.49).
2. Animal physiologist Whitey Rodent has been studying the level of serotonin in field mice (the Xs) and in lab-grown mice (the Ys). In a recent experiment, Dr. Rodent measured the serotonin level in 17 mice and recorded the following measurements (in microliters per milliliter of cerebral fluid):
   X: 1.57, 2.34, 2.56, 1.25, 1.53. 1.48, 2.31, 1.02, 2.01
   Y: 1.79, 2.88, 1.94, 2.59, 1.80, 3.12, 3.42, 1.59
   Use the runs test to test the hypothesis that $F_x = F_y$, against H_1 in (12.45), at $\alpha = .05$.
3. Fifty workers in a toy company's assembly plant each assembled 400 toy cars, a task that takes about a minute per car. Twenty-five of the workers did their tasks while soft classical music was played in the background, and the number of defective cars (the Xs) was noted for each worker. The remaining twenty-five workers assembled the toy cars in silence, and the number of defective cars (the Ys) was noted for each of them. When the Xs and Ys

were combined and put in order, it was found that there were 18 runs of like observations. Does this outcome suggest that the hypothesis $H_0 : F_x = F_y$ should be rejected at the 5% level of significance? Use the normal approximation in Theorem 12.4.1.

12.5 The Rank Sum Test

Nonparametric testing took a major leap forward in the 1940s with the introduction of tests based on "ranks." Initially, the targeted application involved a two-sample problem in which the question of interest was whether or not the populations from which the samples were drawn were the same or different. If the available samples are $X_1, X_2, \dots, X_n \overset{\text{iid}}{\sim} F_x$ and $Y_1, Y_2, \dots, Y_m \overset{\text{iid}}{\sim} F_y$, then a "rank test" concerning properties of F_1 and F_2 would be based on the ranks associated with one of the samples (say the Xs) among the combined sample of Xs and Ys. Wilcoxon (1945) proposed the use of the "rank sum" statistic $S = \sum_{i=1}^{n} R(X_i)$, where $R(X_i)$ represents the rank of the ith observed X among the combined sample of size $n+m$. For instance, your sample, when ranked from smallest to largest, looked like XXYXYYYXXYYX, the ranks of the X sample would be 1, 2, 4, 8, 9, and 12. Wilcoxon argued that either an unusually small or an unusually large value of S would provide strong evidence against the hypothesis that $F_x = F_y$. Mann and Whitney (1947) proposed a seemingly different approach to the same problem, but then showed that their now famous statistic U was in fact equivalent to S (a fact that you are asked to establish for yourself in Problem 12.6.40 at the end of the chapter). For continuous data (or, more generally, for data with no ties), Mann and Whitney's U was defined as

$$U = \sum_{i=1}^{n} \sum_{j=1}^{m} I_{(-\infty, X_i)}(Y_j).$$

In words, U counts the total number of times an X is preceded by a Y. Operationally, you rank the combined data from smallest to largest and record the number n_i of Y values that are smaller than $X_{(i)}$, the ith smallest X; then, $U = \sum_{i=1}^{n} n_i$. It can be shown (heh, heh) that U and S are related as follows:

$$U = S - \frac{n(n+1)}{2}. \tag{12.54}$$

Both the U and the S statistics can be adapted to account for ties in the data. For the rank sum statistic S, you simply apply the average rank to all tied observations and calculate S with those assigned ranks. If, for example, the 8th and 9th smallest observations are equal, they would both be assigned the rank 8.5. Similarly, for the Mann-Whitney statistic U, any Y value that is tied with one or more X values is counted as $1/2$ instead of 1 in terms of preceding each of these X values. Tied Ys may be treated as if they were untied but adjacent. The main difficulty with ties is that the exact distribution of these statistics becomes difficult to identify and the asymptotic distribution of S, as n and m grow, is somewhat more complex. However, the inference is still manageable, and virtually all textbooks in nonparametric statistics cover the treatment of data with ties. Later in this section, we will provide guidance on the treatment of ties in your data, but our main exposition will assume that the data is continuous and there are no ties in the observed data. Under this assumption, the exact null distributions of the statistics S and U are known (and are tabled in textbooks on nonparametric testing) and their asymptotic distributions may be rigorously established. For simplicity, we will work exclusively, from here on out, with the rank sum test, as the equivalence of the two tests makes their parallel treatment unnecessary.

The asymptotics we will mention are due to Mann and Whitney, who also were the first to provide extensive tables for the test for small or moderate sample sizes. Frank Wilcoxon is generally credited with the idea of the test, and for demonstrating its utility in experiments in chemistry (his native field, with a Ph.D. from Cornell in 1924) and in other applications. H. B. Mann was a highly regarded mathematician with strong interests in number theory and algebra and varied statistical interests including experimental design, an area in which he wrote the first dedicated text. D. Ransom Whitney's dissertation, under Mann's direction, focused on the study of the U statistic. Much of the theoretical underpinnings of the Mann-Whitney test were established through their collaboration at The Ohio State University in the mid-forties, and they both served on the OSU faculty until their respective retirements. I had the good fortune of taking a course from Ransom Whitney during the 1966–67 academic year, and I a learned a good deal from him. He had an uncanny intuition about good approaches to data analysis. I admired Ransom Whitney as a teacher, but I have to admit that it was the person who he was that I liked best. He had the habit of inviting groups of graduate students over to enjoy his well-equipped game room in the basement of his home. For a graduate student who was a couple thousand miles from home for the first time in his life, that was an invitation that I especially enjoyed and appreciated.

Let us begin our treatment of the statistic S with an exact calculation of its mean and variance. These formulae follow directly from the facts established in the theorem below. We restate, as a lemma, two mathematical identities that were stated earlier in Theorem 2.4.1 and Exercise 2.4.2. Both are easily established using mathematical induction.

Lemma 12.5.1. Consider the positive integers $\{1,2,\ldots,n-1,n\}$. Then

$$\sum_{i=1}^{n} i = \frac{n(n+1)}{2} \quad \text{and} \quad \sum_{i=1}^{n} i^2 = \frac{n(n+1)(2n+1)}{6}. \tag{12.55}$$

We put this lemma to immediate use in the following:

Theorem 12.5.1. Let X be the sum of n integers drawn randomly, w/o replacement, from the N integers $1,2,\ldots,N$. Then

$$E(X) = \frac{n(N+1)}{2} \text{ and } V(X) = \frac{n(N+1)(N-n)}{12}. \tag{12.56}$$

Proof. Let X_i be the number selected on the ith draw. Since draws are made without replacement, $X_1, X_2, \ldots, X_n$ are dependent variables, but each has the same marginal distribution, that is, for every i,

$$P(X_i = k) = \frac{1}{N} \text{ for } k = 1,2,\ldots,N. \tag{12.57}$$

The easiest way to see this is to note that, when the values of the other Xs are unknown, any particular value that X_i might take on is neither more likely nor less likely than any other value. Thus, the values $1,2,\ldots,N$ are equally likely. For example,

$$\begin{aligned} P(X_2 = 1) &= P(X_1 = 1)P(X_2 = 1|X_1 = 1) + P(X_1 \neq 1)P(X_2 = 1|X_1 \neq 1) \\ &= (1/N)(0) + ((N-1)/N)(1/(N-1) = 1/N. \end{aligned}$$

From (12.57), we may verify the expectation claimed in (12.56), as follows:

$$E(X_i) = \sum_{k=1}^{N} k\left(\frac{1}{N}\right) = \frac{1}{N}\sum_{k=1}^{N} k = \frac{1}{N}\left(\frac{N(N+1)}{2}\right) = \frac{N+1}{2}$$

and

$$E(X) = E\left(\sum_{i=1}^{n} X_i\right) = \sum_{i=1}^{n} EX_i = \frac{n(N+1)}{2}.$$

Further,

$$\begin{aligned} V(X_i) &= EX_i^2 - (EX_i)^2 = \sum_{k=1}^{N} k^2\left(\frac{1}{N}\right) - \left(\frac{N+1}{2}\right)^2 \\ &= \frac{1}{N}\left(\frac{N(N+1)(2N+1)}{6}\right) - \frac{(N+1)^2}{4} = \frac{(N+1)(N-1)}{12}, \end{aligned} \tag{12.58}$$

and, for $i \neq j$,

$$\begin{aligned} Cov(X_i, X_j) &= E(X_i - \frac{N+1}{2})(X_j - \frac{N+1}{2}) \\ &= \sum_{k,s=1,\ldots,N;k\neq s} \left(k - \frac{N+1}{2}\right)\left(s - \frac{N+1}{2}\right)\frac{1}{N}\frac{1}{N-1}. \end{aligned} \tag{12.59}$$

Adding and subtracting the terms in which $k = s$ to the summand in (12.59), we find that

$$\begin{aligned} Cov(X_i, X_j) &= \frac{1}{N(N-1)}\sum_{k=1}^{N}\left(k - \frac{N+1}{2}\right)\sum_{s=1}^{N}\left(s - \frac{N+1}{2}\right) - \sum_{k=1}^{N}\left(k - \frac{N+1}{2}\right)^2\frac{1}{N}\frac{1}{N-1} \\ &= \frac{0\cdot 0}{N(N-1)} - \frac{1}{N-1}V(X_i) \\ &= -\frac{N+1}{12}. \end{aligned} \tag{12.60}$$

Finally, we verify $V(X)$ by computing

$$\begin{aligned} V(X) &= V\left(\sum_{i=1}^{n} X_i\right) = \sum_{i=1}^{n} V(X_i) + \sum_{i,j=1,\ldots,n;i\neq j} Cov(X_i, X_j) \\ &= n\left(\frac{(N+1)(N-1)}{12}\right) + n(n-1)\left(-\frac{N+1}{12}\right) \\ &= \frac{n(N+1)}{12}((N-1) - (n-1)) \\ &= \frac{n(N+1)(N-n)}{12}, \end{aligned} \tag{12.61}$$

completing the proof. ■

Theorem 12.5.1 can be mapped directly onto the problem of determining the mean and variance of the rank sum statistic S under the null hypothesis of interest. Assume, as we have above, that the available data consist of two independent random samples with potentially different distributions, that is, suppose that $X_1, X_2, \ldots, X_n \stackrel{\text{iid}}{\sim} F_x$ and $Y_1, Y_2, \ldots, Y_m \stackrel{\text{iid}}{\sim} F_y$. Consider using the rank sum statistic $S = \sum_{i=1}^{n} R(X_i)$ for testing $H_0 : F_x = F_y$ against a stochastic ordering alternative, that is, either $F_x <_{st} F_y$ or $F_x >_{st} F_y$. Under the null hypothesis, all possible permutations of the Xs and Ys are equally likely, each having a common probability $1/(n+m)!$. In particular, the ranks $R(X_1),\ldots,R(X_n)$ of the X sample are equally likely to be any subset of integers of the superset $\{1,2,\ldots,n+m\}$. It follows that the process of choosing the values of these ranks is identical to the process of selecting n integers, without replacement, from the set $\{1,2,\ldots,n+m\}$. Theorem 12.5.1 provides us with the mean and variance of the sum of these numbers. We may thus apply the theorem directly to the statistic S. The desired result is:

Theorem 12.5.2. Let $X_1,X_2,\ldots,X_n \overset{\text{iid}}{\sim} F_x$ and $Y_1,Y_2,\ldots,Y_m \overset{\text{iid}}{\sim} F_y$ be two independent samples from the respective distributions F_x and F_y. Under the hypothesis $H_0 : F_x = F_y$, the rank sum statistic $S = \sum_{i=1}^{n} R(X_i)$ has a discrete distribution on the integers $\{(n(n+1)/2,\ldots,n(n+2m+1)/2\}$ with mean and variance given by

$$E(S) = \frac{n(n+m+1)}{2} \quad \text{and} \quad V(S) = \frac{nm(n+m+1)}{12}. \tag{12.62}$$

Proof. The lower bound of the rank sum statistic S follows immediately from the fact that the smallest possible values the ranks of the X sample can take are the integers from 1 to n, and the sum of these integers, as noted in Lemma 12.5.1, is $n(n+1)/2$. On the other hand, the largest possible values that the ranks of the X sample can take on are $m+1,\ldots,m+n$. Again by Lemma 12.5.1, the sum of these ranks is $nm+n(n+1)/2 = n(n+2m+1)/2$. Now, suppose that the ranks of the Xs are selected at random, without replacement from the set $\{1,\ldots,n+m\}$. The n ranks so selected are simply a sample of size n selected without replacement from the integers $1,\ldots,N$, where $N = n+m$. It follows from Theorem 12.5.1 that the sum S of the ranks of the Xs has mean

$$E(S) = \frac{n(N+1)}{2} = \frac{n(n+m+1)}{2}$$

and variance

$$V(S) = \frac{n(N+1)(N-n)}{12} = \frac{n(n+m+1)(m)}{12} = \frac{nm(n+m+1)}{12}$$

as claimed. ∎

Let us now turn to a discussion of the probability distribution of the rank sum statistic S under the sampling assumptions of Theorem 12.5.2 and the additional assumption that $H_0 : F_x = F_y = F$ is true, that is, that the Xs and the Ys have the same distribution. We will thus assume that the combined sample of Xs and Ys are an iid sample from a common continuous distribution F. It is easy to confirm that, under these conditions, the distribution of the statistic S is a discrete distribution that is symmetric about its center $n(n+m+1)/2$. (See Problem 12.6.39.) The distribution is not known in closed form, but a table of cutoff points for the exact distribution of S is easily constructed for small to moderate n and m. We will use the letter W, for Wilcoxon, to denote the distribution of the Wilcoxon rank sum statistic, and the notation $W_{n,m}$ for the distribution of S based on n Xs and m Ys. For $n \leq 20$ and $m \leq 20$, the entry $W_{n,m,\alpha}$ in Table A.7 is the smallest integer x for which $P(S \leq x) \leq \alpha$ when $S \sim W_{n,m}$. To accommodate both one-tailed and two-tailed tests, the α-cutoff points for the distribution of S are tabled for $\alpha = .05, .025, .01$, and $.005$. Since, for example, $W_{6,8,.025} = 30$, we have that if $S \sim W_{6,8}$, then $P(S \leq 30) \leq .025$ and $P(S \leq 31) > .025$. Thus, a rank sum as small as or smaller than $S = 30$ would be considered a rare event under the null hypothesis $H_0 : F_x = F_y$ when $n = 6$ and $m = 8$. Since the standard test of this null hypothesis is a two-sided test, we would reject H_0 if S was either too small or too large. By the symmetry of the distribution of S, we would reject H_0 if either $S \leq 30 (= \mu - 15)$ or $S \geq 60 (= \mu + 15)$. If $S = 30$ were the observed value of S from a sample with $n = 6$ and $m = 8$, the (two-sided) p-value of the observation would be $2P(S \leq 30) \leq 2(.025) = .05$, a result that would lead us to reject the null hypothesis that $F_x = F_y$.

It would not be difficult to provide a larger table than Table A.7, but as it happens, the distribution of the Wilcoxon rank-sum statistic S can be reliably approximated by the normal distribution, provided n and m are not too small. Lehman (1975) indicates that the approximation is acceptable when both n and m are at least 10. Since the distribution of S is discrete, a continuity correction is generally used in approximating probabilities for S. The details of the normal approximation of the distribution of S are given in the following:

Theorem 12.5.3. (Mann and Whitney) Assume the conditions of Theorem 12.5.2 hold. If n and m tend to ∞ in such a way that $n/m \to c \in (0, \infty)$, the suitably standardized Wilcoxon statistic converges to a standard normal variable Z, that is, for n and m sufficiently large,

$$\frac{S - \mu_S}{\sqrt{V(S)}} \sim N(0, 1), \text{ approximately} \tag{12.63}$$

where

$$\mu_S = \frac{n(n+m+1)}{2} \quad \text{and} \quad V(S) = \frac{nm(n+m+1)}{12}. \tag{12.64}$$

The accuracy of the normal approximation of the rank sum statistic is apparent from the example below.

Example 12.5.1. Suppose that $n = m = 10$, so that the rank sum statistic has mean $(10)(21)/2 = 105$ and variance $(10)(10)(21)/12 = 175$. From tables available in Lehman (1975) and elsewhere, the exact probability $P(S \leq 75)$ may be identified as .0116. The normal approximation to this probability is obtained as

$$P(S \leq 75.5) = P\left(\frac{S - 105}{\sqrt{175}} \leq \frac{75.5 - 105}{\sqrt{175}}\right) = P(Z \leq -2.23) = .0129.$$

This is a pretty impressive performance with such small samples. For moderate or large sample sizes, it is clear that tables for exact cutoff points for the distribution of the rank sum statistic will not be necessary. ■

The data analysis required in applying the Wilcoxon rank sum test is illustrated in the following:

Example 12.5.2. Ginkgo Biloba (GB) is thought to stimulate brain activity, and so is Korean Gensing Root (KGR). Thirty (blind) mice were involved in a double-blind randomized study in which 15 ingested GB and 15 ingested KGR. Each of the thirty mice then ran through an identical maze, and their time, in seconds, to completion was recorded. The data collected is shown below

GB Group (X)			KGR Group (Y)		
25.3	26.2	26.3	28.6	28.7	30.7
30.1	24.8	28.5	29.9	29.3	29.6
26.7	29.1	26.6	32.0	29.7	31.0
25.9	25.5	25.2	30.2	26.4	30.5
28.2	29.0	26.9	28.3	29.5	27.4

When we order the data from smallest to largest, with the GB Group times bolded, we may easily determine the ranks of the Xs and the Ys.

24.8, 25.2, 25.3, 25.5, 25.9, 26.2, 26.3, 26.4, **26.6,**
26.7, 26.9, 27.4, **28.2,** 28.3, **28.5,** 28.6, 28.7, **29.0, 29.1**,
29.3, 29.5, 29.7, 29.9, **30.1,** 30.2, 30.5, 30.7, 31.0, 32.0.

Letting $S = \sum_{i=1}^{n} R(X_i)$, we see that the value of S for these data is $S = 147$. Using the asymptotic approximation for the distribution of S, we have that

$$S \sim N(232.5, 581.25) \text{ approximately.}$$

Therefore, we may approximate the p-value of the observed value of S to be

$$p-\text{value} = P(S \leq 147) = P(Z \leq (147.5 - 232.5)/\sqrt{581.25}) = P(Z \leq -3.53) < .0002.$$

It is thus clear that the hypothesis that $F_x = F_y$ should be rejected in favor of the one-sided alternative $H_1 : F_x > F_y$ (that is, the Xs tend to be smaller than the Ys) at any significance level $\alpha > .0002$. So at the standard levels of α used in practice (that is, $\alpha = .05$ or .01), we would reject the hypothesis that the Xs and the Ys are generated from the same probability distribution. ■

So far, the treatment of the Wilcoxon rank sum test has been nice and tidy. But, as perhaps you have noticed, there's an elephant in the room. The difficulty that we must deal with in practical applications is that fact that "ties" often occur in real data, and the results discussed thus far assume that $n+m$ distinct values have occurred among the n Xs and m Ys. In reality, ties often occur in the data we'd like to handle via the Wilcoxon rank sum test. It's often the case that the data we are dealing with is discrete, in which case the probability of ties is positive and possibly non-negligible. But even if the data at hand is continuous, it is often the case that data is rounded to a certain level of accuracy. For example, 10.12 and 10.09 aren't really the same numbers, but when our data is rounded to the first decimal point, both of these observations would be recorded as 10.1. So the question naturally arises: how should ties in the data be treated in an analysis involving the Wilcoxon rank sum test? The definition of the rank of an observation that is tied with other values in the sample is the easy issue to settle. As mentioned earlier in this section, if two or more observations are recorded as equal, then the ranks assigned to each is the average of the ranks they would have enjoyed if they had been adjacent but not equal. For example, if the value 10.7 occurred as the 14th and 15th smallest data values in the sample, then each would be assigned the so-called mid-rank of the potential ranks of the three data points, that is, the rank $(14+15)/2 = 14.5$. Given this procedure for assigning ranks to the data, one can again calculate the sum of "ranks" for the X sample or for the Y sample. When the data has ties, we will denote the Wilcoxon statistic by S^*, and we will refer to it as the Wilcoxon mid-rank sum statistic (even though it would typically contains some true ranks along with some mid-ranks). The question that remains concerns the issue of how to treat the distribution of the mid-rank sum statistic S^*.

The exact distribution of the mid-rank sum statistic $S^* = \sum_{i=1}^{n} R(X_i)$ is difficult to obtain, though in principle, it involves simply finding the number of permutations of the combined data, when classified as n Xs and m Ys, that can give rise to a particular value of S^*. However, we do not include a table of the exact cutoff points for S^* for small n and m, as ties are less frequently encountered in small problems than they are in large problems. Fortunately, a normal approximation of the distribution of S^* is quite reliable for moderate or large values of n and m, although some adjustments are needed in the normal approximation that is used for the distribution of S. The approximate normal distribution that is used for S^* is described in the following paragraph.

It can be shown theoretically that the statistic S^* has the same expected value as S itself, that is

$$E(S^*) = \frac{n(n+m+1)}{2}. \tag{12.65}$$

This result is not surprising given that the ranks of tied observations are assigned in a balanced and unbiased manner. (The proof of the claim in (12.65) is left as Problem 12.6.36.) However, the variance of S^* is not the same as that of S. This too is expected, as the replacement of a set of ranks by their average clearly reduces the variability of the collection. The approximate distribution of S^* may be obtained as follows. Suppose that in the available sample of n Xs and m Ys, with some ties among them, there is a total of k distinct values. Some of these will be untied observations and will be assigned their true ranks and others will be assigned mid-ranks. Let's denote the number of ties at the ith of these k distinct values by d_i. For the data 2, 7, 7, 7, 8, 8, we have $k = 3$ and $d_1 = 1, d_2 = 3$ and $d_3 = 2$. With this notation, we are able to state the following result.

Theorem 12.5.4. Assume the conditions of Theorem 12.5.2 hold. Assume that for a given n and m, there are a total of $k = k(n,m)$ distinct values in the combined data, and let d_i be the number of data taking the ith value, where $i = 1, \ldots, k$. If n and m tend to ∞ in such a way that $n/m \to c \in (0, \infty)$, then the standardized Wilcoxon mid-rank sum statistic S^* converges to a standard normal variable Z, that is, for n and m sufficiently large,

$$\frac{S^* - \mu_{S^*}}{\sqrt{V(S^*)}} \sim N(0,1), \text{ approximately} \tag{12.66}$$

where

$$\mu_{S^*} = \frac{n(n+m+1)}{2} \text{ and } V(S^*) = \frac{nm(n+m+1)}{12} - \frac{nm\sum_{i=1}^{k}(d_i^3 - d_i)}{12(n+m)(n+m+1)}. \tag{12.67}$$

OK, so there is a bit more arithmetic to do when you have ties in your data, but basically, you have a reliable approximation for the distribution of the test statistic S^*, and you can feel confident in carrying out the Wilcoxon test with ties in your data when the sample sizes are moderate to large. In practice, the second term in the standard error formula is negligibly small unless there are quite a substantial number of ties in the data. When there are few ties, the asymptotic distribution of the mid-rank sum statistic S^* is actually indistinguishable for that of the rank sum statistic S. The Wilcoxon test is applied to a data set with ties in the final example of the section (and the chapter, and the book.)

Example 12.5.3. Two calculus classes were taught using quite different teaching methods. Twenty students were assigned at random to each course from an initial random sample of forty participants. The first method was quite traditional, with regular lectures, graded homework, and biweekly quizzes. The second method was a self-paced Internet-based course with multiple choice quizzes after each major topic and a tutor available via e-mail.

Both classes took a common problem-solving final exam. The scores of the final exam are shown below.

Self-paced (X)	66, 77, 83, 64, 68, 79, 90, 84, 35, 50 81, 86, 73, 68, 74, 87, 44, 79, 65, 70
Traditional (Y)	77, 89, 99, 47, 65, 80, 85, 95, 91, 77, 64, 78, 85, 81, 97, 98, 72, 82, 90, 75

The artifact of a stem and leaf plot is useful in ordering one's data. Such a plot is pictured below, with the final exam scores from the self-paced course bolded.

3	**5**
4	**4** 7
5	**0**
6	**6 4 8 8 5** 5 4
7	**7 9 3 4 9 0** 7 7 8 2 5
8	**3 4 1 6 7** 9 0 5 5 1 2
9	**0** 9 5 1 7 8 0

The ties within the combined data set are easily identified: 64, 64, 65, 65, 68, 68, 77, 77, 77, 79, 79, 81, 81, 85, 85, 90, 90. This indicates that there were 31 distinct test scores among these forty students. In the notation above, we see that there are 7 d_is that take the value 2 and 1 d_i that takes the value 3, with the remaining 23 d_is taking the value 1. (Notice that identifying the exact subscripts of the $d_i > 1$ will not be needed in our computation.) From this, we have that

$$\mu_{S^*} = \frac{20(41)}{2} = 410$$

and

$$V(S^*) = \frac{20^2(41)}{12} - \frac{20^2[7(2^3-2)+(3^3-3)]}{12(40)(41)} = 1366.67 - 1.34 = 1365.33.$$

We may calculate S^* as

$$\begin{aligned} S^* &= 1+2+4+5.5+7.5+9+10.5+10.5+12+14+15 \\ &\quad +18+21.5+21.5+24.5+27+28+31+32+34.5 \\ &= 329. \end{aligned}$$

The standardized value of S^* is thus equal to

$$\frac{S^* - \mu_{S^*}}{\sqrt{V(S^*)}} = \frac{329-410}{36.95} = -2.19.$$

Since $P(Z \leq -2.19) = .0146$ when $Z \sim N(0,1)$, we see that S^* has a one sided p-value of .0146 and a two-sided p-value of .0292, so that $H_0 : F_x = F_y$ would be rejected at the significance level .05 against either a one-sided or a two-sided alternative. ∎

There is a great deal more to know about nonparametric testing, and the treatment given in this and the preceding two sections is intended only as a brief introduction to the subject. When taken together with the nonparametric goodness-of-fit test treated in Chapter 10 and the introduction to nonparametric estimation provided in Sections 12.1 and 12.2, the treatment here constitutes, I believe, a reasonable entrée into the nonparametric approach to statistics. However, appreciating the range of the subject and understanding the details of its theoretical underpinnings is achievable only with further study. I can recommend three books on the subject, each with a different aim. Conover's (1999) *Practical Nonparametric Statistics* treats the subject comprehensively, but at a pre-calculus level and from an applied viewpoint, without many mathematical details. Lehman's (1975) *Nonparametrics* focuses exclusively on tests based on ranks and provides an outstanding exposition of both the theory and applications of such tests. Finally, Kvam and Vidakovic's (2007) *Nonparametric Statistics with Applications to Science and Engineering* treats both estimation and testing, including traditional nonparametric topics and a selection of modern nonparametric tools (such as bootstrapping, wavelet shrinkage, neural networks, and Bayesian nonparametric estimation).

I'd like to close this section with a true story from years ago. A student who was taking an introductory statistics course from me came to one of my office hours. There were other students already there when he arrived, so he took a seat and, with a little time on his hands, started looking over my bookshelves. There was rather little in them that interested him, as they were filled with statistics textbooks, reference books, and multiple issues of the several statistical journals I subscribed to. But suddenly, he burst out laughing and couldn't stop. All of us in the room were highly intrigued. He pointed to a book on my shelf and blurted out the title .... "Rank Tests." He said "That's the coolest title for a book!" You see, at the time, the word "rank" was a popular pejorative, used like you might use the word "lame" today. To him, the title read like the title "Lame Tests" would to you. He thought it was fascinating that some statisticians would have the incredible foresight to compile all the lame tests they could think of in a single volume. This would serve to alert the otherwise unprotected public about a bunch of tests that they shouldn't waste their time on.

Exercises 12.5.

1. Let S be the Wilcoxon rank sum statistic and let U be the Mann-Whitney precedence statistic. Prove that the relationship in (12.54) holds between them.
2. Suppose that U^* is defined as the total number of Xs that precede the Ys, that is,
$$U^* = \sum_{i=1}^{m}\sum_{j=1}^{n} I_{(-\infty,Y_i)}(X_j).$$
Show that $U + U^* = nm$, where U is the Mann-Whitney statistic.
3. Obtain the exact distribution of the Wilcoxon rank sum test under $H_0 : F_X = F_Y$ for each of the following cases in which n is the number of Xs and m is the number of Ys: (a) $n = 2, m = 4$, (b) $n = 2, m = 5$, and (c) $n = 3 = m$.
4. Use the calculations in Exercise 12.5.3 to find all possible significance levels for one- and two-sided tests of $H_0 : F_x = F_y$ for the three cases considered above. In each case, find the attainable significance level for a one- or a two-sided test under $H_1 : F_X > F_Y$ (X is stochastically smaller than Y) or $H_1 : F_X \neq F_Y$, respectively, that is closest to $\alpha = .1$.
5. Use the Wilcoxon rank sum test to test the hypothesis $H_0 : F_x = F_y$ against a two-sided alternative, at $\alpha = .05$, based on the following independent random samples:
X: 48, 77, 62, 74, 75, 90, 82, 89, 98, 57
Y: 72, 40, 52, 86, 59, 66, 70, 78, 93, 45.

12.6 Chapter Problems

1. Ten random customers of Safeway markets were selected from customer checkout files, and the amount of money they spent at Safeway in the last 30 days was recorded as follows: 32, 314, 260, 451, 255, 72, 182, 377, 314, 275. (a) Plot the empirical distribution function $\hat{F}_{10}(x)$ of the data. Using the formula (see Problem 3.7.7)
$$EX = \int_0^\infty \bar{F}(x)dx$$
for computing the mean of a random variable X, where $\bar{F}(x) = 1 - F(x)$, verify that the mean $\bar{X}_{10}$ of this sample is the mean of the random variable having the distribution function $\hat{F}_{10}(x)$.

2. Write the likelihood function in (12.5) as
$$L(F) = \prod_{i=1}^{n-1} p_i \times \left(1 - \sum_{1=1}^{n-1} p_i\right),$$
where, for $i \leq n$, $p_i = F(x_i) - F(x_i^-)$ is the probability that the model F places on the observed point x_i and $p_n = 1 - \sum_{i=1}^{n-1} p_i$. This formulation implies that $\{p_i, i = 1, \ldots, n-1\}$ must satisfy the constraints $p_i \geq 0$ for $i = 1, \ldots, n-1$ and $\sum_{i=1}^{n-1} p_i \leq 1$. Show, using ordinary calculus, that $L(F)$ is maximized by the edf in (12.1), that is, by the values $p_i = 1/n$ for $i = 1, \ldots, n-1$.

3. Suppose that a random sample of fifth grade boys was taken and their heights, in inches, were recorded as: 48, 52, 55, 56, 57, 59, 62, 64, 65, 67. Obtain the empirical distribution function $F_{10}(x)$ from these data, graph it in the X-Y plane, and derive a 95% confidence interval for $F(60)$, the probability that a fifth grade boy is no greater than five feet tall.

4. Consider two independent samples taking the values shown below:

   X:10, 14, 22, 27, 18, 21, 24, 20, 16, 30, 18, 22, 25, 28, 14, 21, 15, 32, 22, 15
   Y:12, 18, 14, 18, 21, 25, 11, 16, 19, 24, 28, 14, 15, 28, 21, 16, 19, 24, 18, 12

   Construct $\bar{F}_X(x)$ and $\bar{F}_Y(y)$, the edf's of the X and the Y samples, and graph them together in the plane. Test that hypothesis $H_0 : F_X(20) = F_Y(20)$ against $H_1 : F_x(20) \neq F_Y(20)$ at $\alpha = .05$. (Hint: See Section 10.2).

5. Consider the following sample of "full" recovery times from hip replacement surgery (after appropriate physical therapy) for 14 patients who had the surgery in a particular New York hospital 12 months ago. Four of the patients left the study before the experiment was completed. Their censored observed recovery times (in months) are marked with a "+". The data collected was: 2.0. 2.2., 2.7+, 3.3., 3.3+, 4.0, 4.2, 4.5+, 5.0+, 5.2, 5.6, 6.0, 6.2, 6.6. Obtain the Kaplan-Meier for the survival function for these data. Even though n is small here, use Greenwood's formula to calculate the approximate variance of the estimate $\hat{\bar{F}}(5)$ of $\bar{F}(5)$, the probability that a random patient would take more than 5 months to make a full recovery.

6. Show that when the data is uncensored, that is, when observed lifetimes consist exclusively of failure times, the Kaplan-Meier estimator reduces to the empirical survival function.

7. The "reduced sample" method for estimating a survival function $\bar{F}$ simply uses the empirical distribution of the complete data (obtained by removing the censored data from consideration). Calculate the reduced sample estimator $\hat{\bar{F}}_{RS}(x)$ of $\bar{F}(x)$ for the data in Problem

12.5.5. Calculate the approximate variance for $\hat{\bar{F}}_{RS}(5)$, and compare it to the approximate variance of the Kaplan-Meier estimator of $\bar{F}(5)$.

8. In a study of some historical importance, Boag (1949) reported on 121 breast cancer patients that received the standard treatment of the era. The data below shows the survival times of these patients over the first three years of the study: .3, .3+, 4+, 5.0 ,5.6, 6.2, 6.3, 6.6, 6.8, 7.4+, 7.5, 8.4, 8.4, 10.3, 11.0. 11.8, 12.2, 13.5, 14.4, 14.4, 14.8, 15.5+, 15.7, 16.2, 16.3, 16.5, 16.8, 17.2, 17.3, 17.5, 17.9, 19.8, 20.4, 20.9, 21.0, 21.0, 21.1, 23.0, 23.4+. 23.6, 24.0, 27.9, 28.2, 29.1, 30.0, 31.0, 31.0, 32.0, 35.0. 35.0, 36+, 36+, ..., 36+. From these data, use the Kaplan-Meier estimator to estimate the 1-year, 2-year, and 3-year survival probability for patients from the sampled population.

9. The following data on the survival times from vaginal cancer in 19 female rats was published in a public health journal. The data obtained is as follows: 188, 206, 230, 216+, 190, 246, 304, 213, 143, 265, 244+, 164, 192, 216, 234, 220, 227, 188, 192. Derive and graph the Kaplan-Meier estimator of the survival function of the population from which this data was drawn. Estimate the mean and variance of this population.

10. The following algorithm for constructing the Kaplan-Meier estimator was suggested by Efron (1967). For right censored failure time data, we observe $X_i = \min(T_i, L_i)$ and $\delta_i = I\{T_i \leq L_i\}$ for $i = 1, \ldots, n$, where T_i and L_i are the failure time and censoring time, respectively. The "Redistribute to the Right" algorithm can be summarized as follows. For a sample of size n, each item has an initial mass of $1/n$. Starting with the first censored item, redistribute its mass equally to all items to its right. Then redistribute the mass of the second censored item in the same way. Repeat the process until you have done so for the last censored item. Show that this process results in the Kaplan-Meier estimator in Equation (12.31).

11. Suppose that your data consists of a random sample of size n from an unknown distribution F. Show that the number of possible (distinct) bootstrap samples is

$$\binom{2n-1}{n}.$$

(Note: This problem is a special case of Problem 1.8.72. See the hint for the latter problem.)

12. Suppose that your data consists of a random sample of size n from an unknown distribution F. Suppose that the statistic $T_n = T(X_1, \ldots, X_n)$ is the estimator of choice of the population parameter θ, and let B be the number of bootstrap samples drawn in the process of estimating the standard error of T_n. If $\hat{\sigma}_{T_n}^{(B)}$ (as given in (12.35)) represents the bootstrap estimator of the standard error σ_{T_n} of T_n based on B bootstrap samples, then the ideal bootstrap estimate $\hat{\sigma}_{T_n}^{(\infty)}$ of σ_{T_n} is defined as

$$\lim_{B\to\infty} \hat{\sigma}_{T_n}^{(B)} = \hat{\sigma}_{T_n}^{(\infty)}.$$

Show that, for any B, $E[(\hat{\sigma}_{T_n}^{(B)})^2] = [\hat{\sigma}_{T_n}^{(\infty)}]^2$.

13. Consider the data in Exercise 12.1.1. Let $T(\mathbf{X})$ be the 40% trimmed mean which averages the middle 6 data points. Calculate the bootstrap estimator of the standard error of T for bootstrap replications of size $B = 500, 1000, 1500$, and 2000. From these results, estimate the ideal bootstrap estimate $\hat{\sigma}_{T_n}^{(\infty)}$ of σ_{T_n}.

14. (A failed bootstrap.) Generate a random sample $X_1, \ldots, X_{50} \stackrel{\text{iid}}{\sim} U(0, \theta)$; for convenience,

let the θ in your simulation take the value 1. For the general model, the maximum order statistic $X_{(50)}$ is the MLE of θ. Obtain 1000 bootstrap samples and calculate the statistic $T = X_{(50)}$ for each, and summarize the empirical distribution of these 1000 replicates by a histogram with 50 bins of size .001 ranging from .950 to 1.000. What's wrong with this picture? (Explain the problem by focusing of the edf $\hat{F}_T(\theta)$ as an estimator of the underlying F. Substantiate the claim that $\hat{F}_T(\theta)$ of the bootstrap sample of Ts is a poor estimate of F in the extreme right-hand tail of the distribution.)

15. In the scenario described in Problem 12.6.14, show that, for a random sample of size n from $U(0,\theta)$,

$$P(T_n^* = \hat{\theta}) = 1 - (1 - 1/n)^n \to 1 - e^{-1} \cong .632 \text{ as } n \to \infty,$$

where T_n^* is the value of T in a bootstrap replication and $\hat{\theta}$ is the MLE based on the original sample.

16. There is a tradition among Members of the Scottish Parliament (MSPs) that several units (drams) of whiskey (scotch, of course) are consumed at a celebration after the last session of parliament preceding the national holidays of Christmas and Hogmanay. Eighteen MSPs were chosen at random, and the number of drams consumed at each of these two celebrations were recorded. The data for a random sample of 18 MSPs are shown below:

C Drams	2	3	3	2	4	0	3	6	2	2	5	4	3	6	0	3	3	0
H Drams	5	1	5	6	4	7	5	9	0	4	15	6	8	9	0	6	5	12

Use the sign test to examine the hypothesis that the MSPs will consume the same amount of whiskey at each holiday event against the two-sided alternative that MSPs drink more whiskey at one particular holiday event than at the other. Use $\alpha = .05$.

17. A random sample of 75 UC Davis students was taken. Each student was asked whether they were for or against the proposal being considered by the UC Davis Student Council regarding the elimination of funding for Club Sports. The students expressed their opinions on a scale from 1 to 5, as shown below:

    1. strongly opposed (8 students)
    2. mildly opposed (12 students)
    3. neutral (45 students)
    4. mildly in favor (8 students)
    5. strongly in favor (2 students)

    (a) Carry out a one-sided sign test, with $\alpha = .05$, to determine whether one may conclude that the number of UCD students (mildly or strongly) opposed to the proposal is significantly greater than the number of students (mildly or strongly) in favor. (b) Compute the one-sided p-value of the observed data above.

18. Psychologist Earnest Ree-Sirch would like to publish his recent study on the differences in academic performance between first-born and second-born identical twins. His data on math SAT scores (X,Y) of 10 identical twins are shown below:

Twin pair:	1	2	3	4	5	6	7	8	9	10
Twin 1's score (X):	600	550	720	500	680	560	800	710	640	520
Twin 2's score (Y):	580	600	700	500	650	550	800	690	600	480

Determine whether the Sign Test leads to rejection of $H_0 : P(X > Y) = P(X < Y)$ in favor of $H_1 : P(X > Y) > P(X < Y)$ at $\alpha = .05$.

19. The popular weight loss program So Long Posterior claims that they specialize in helping their clients maintain their weight loss (or lose more weight) once they have reached their weight loss goal. A random sample of 10 clients were tracked after having achieved their weight loss goal, and their weight at the beginning and end of the six-month follow-up period was recorded. The data collected is shown below.

Client	1	2	3	4	5	6	7	8	9	10
Start Wt	125	165	210	185	120	175	120	190	240	200
End Wt	132	168	240	182	135	190	125	235	250	215

Use a one-sided sign test to determine if the SLP's claim should be rejected at the 5% significance level.

20. Suppose a random sample of 80 high school students was drawn from among the population of students who took the SAT exam in the spring of their junior year, took an intensive SAT preparation course the following summer, and retook the SAT exam in the fall of their senior year. Of these 80 students, 42 improved their scores in the fall, 6 got exactly the same score, and 32 got lower scores. Use the sign test, with "+" representing an improved score, to test the hypothesis $H_0 : P(+) = 1/2 = P(-)$, against $H_1 : P(+) > P(-)$, at $\alpha = .05$. Use the test based on the normal approximation and the critical region in (12.41), appropriately adapted for a one-sided testing problem.

21. (McNemar's test) A popular variation of the sign test is often used when only nominal data is available, that is, data is available for which the exact values of underlying data (the Xs and Ys) are not observed, but the fact that each X or Y is in a specific category, say, small or large, has been recorded. Using "0" for small and "1" for large, we can consider the available data to be summarized in a table such as

	$\#Y = 0$	$\#Y = 1$
$\#X = 0$	n_{00}	n_{01}
$\#X = 1$	n_{10}	n_{11}

McNemar proposed the test statistic

$$T = \frac{(n_{10} - n_{01})^2}{n_{10} + n_{01}}$$

for testing the hypothesis $H_0 : P(X = 0, Y = 1) = P(X = 1, Y = 0)$ against a two-sided alternative. It can be shown that, if $n = n_{10} + n_{01}$ is sufficiently large, then

$$Z = \frac{n_{01} - [(n_{10} + n_{01})/2]}{(1/2)\sqrt{n_{10} + n_{01}}} \sim N(0, 1) \text{ approximately.}$$

If we think of n_{01} as the number of plusses and n_{10} as the number of minuses in a sample of size $n_{01} + n_{10}$, then a test based on Z is precisely the sign test for the hypothesis that $P(+) = P(-)$. Show that $T = Z^2$, that is, McNamar's test is in fact equivalent to the sign test.

22. (See Problem 12.6.21) Carry out McNamar's test at $\alpha = .05$ (using the statistic T) for the data in the following table in which Y records the number of low or high scores for a random sample of college math majors on the quantitative section of the law school

aptitude test (LSAT) and X records the number of low or high scores for these students on the verbal part of the LSTA.

	$\#Y = 0$	$\#Y = 1$
$\#X = 0$	20	50
$\#X = 1$	30	20

23. True story. Sir Ronald Fisher (the guy who invented the analysis of variance, put maximum likelihood estimation on the map and, basically, made statistics a household word) happened to have a colleague who claimed that she could tell, when served a cup of tea with milk, whether the tea was poured first or the milk was poured first. One afternoon, Fisher prepared a little experiment. He had 8 cups of tea and milk prepared, and he asked his colleague to take a sip from each and determine whether the tea or the milk was poured first. She was told that four cups were poured one way and four were poured the other way. This gave rise to a 2×2 table in which the rows represented the truth (MF—milk first or TF—tea first) and the columns represented Fisher's colleague's guess (again, MF or TF). His colleague got six cups right, i.e., the contingency table for these data looked like:

	Guess MF	Guess TF
Actual MF	3	1
Actual TF	1	3

Fisher noted that, since the marginals in the table above are fixed and known (each being equal to 4), the entries in the table have a single degree of freedom, and the hypothesis of "random guessing" can be tested treating the correct guesses in one given row. If X is the number of correct guesses of MF, then $X \sim HG(8,4,4)$. Use "Fisher's exact test" to test the hypotheses that Fisher's colleague was just guessing. Use $\alpha = .05$. (Hint: Simply calculate the p-value of the observed X, that is, calculate $P(X \geq 3)$.) (Note: This is how Fisher conceived of the test that bears his name, or so the story goes.)

24. (See Problem 12.6.23.) The LAPD plans to hire a psychic to assist them with their homicide investigations. They will test all applicants using a collection of photographs of 12 LA citizens, 6 of whom are convicted murderers, while the other 6 have no criminal records. Applicants are shown the back (blank) side of each picture and asked to classify each as a murderer (M) or a non-murderer (N). Applicant Claire Voyant's results seem quite impressive:

	Claire's guess: M	Claire's guess: N	Total
True status: M	5	1	6
True Status: N	1	5	6
Total:	6	6	12

Use Fisher's exact test to examine H_0: "Claire's result was obtained by random guessing." Calculate the p-value associated with the outcome $X = 5$, where X is the number of murderers she correctly classified. Should H_0 be rejected against a one-sided alternative at $\alpha = .05$?

25. (See See Problem 12.6.23.) Proud parents Mr. and Mrs. I. B. Smart claim that their one-year old daughter Varrie has an uncanny ability to distinguish between vowels (V) and consonants (C). To test this claim, the host of NBC's *Today* show picked out 16 letters

from his own scrabble game—8 vowels and 8 consonants—and presented them to Varrie. The Smarts explained to Varrie (in their own special way) that there were exactly 8 vowels among the 16 letters, and they stated emphatically that she understood them. When Varrie was asked to pick out the 8 vowels from the 16 letters, the outcome was as shown in the table below.

	Varrie's guess: V	Varrie's guess: C	Total
True status: V	6	2	8
True Status: C	2	6	8
Total:	8	8	16

Using Fisher's Exact Test, test H_0: "Varrie's outcome resulted from random guessing." Compute the p-value $P(X \geq 6)$, where X is the number of vowels she correctly classified. Should one reject H_0 in favor of the one-sided alternative H_1: "Varrie has talent" at $\alpha = .05$?

26. Two nationally ranked college basketball teams (say A and B) are thought to be very well matched. However, before posting a prediction on their upcoming game, sports prognosticator Bud Goode (a real person) decided to compare the shooting percentages of each team's 12 players and test the hypothesis H_0: the two distributions are equal. Suppose that when the shooting percentages of these 24 players are ordered from smallest to largest, the data reveals precisely four runs: 7A, 4B, 5A, 8B. (a) If Bud carries out the Wald-Wolfowitz runs test on these data, would H_0 be rejected at $\alpha = .05$? (b) Calculate the p-value associated with the outcome $R = 4$, where R is the number of runs observed.

27. Suppose that random samples of 15 UCD graduates and 15 U. Betcha graduates from the year 2000 were asked about the highest annual salary they had earned since graduation. It was found that when these salaries were listed from lowest to highest, there were only four runs R in the data: 8 UB, 3 UCD, 7 UB, 12 UCD. Suppose the standard Wald-Wolfowitz runs test is applied to these data. Calculate the p-value of the observation $R = 4$, that is, compute $P(R \leq 4)$.

28. Five random Google employees (the Xs) were chosen to compete with five random Facebook employees (the Ys) in a *San Francisco Chronicle* trivia contest. When their scores were ranked from smallest to largest, the outcomes looked like YYY X YY XXXX. Compute the probability (or p-value) $P(R \leq 4)$, where R is the number of runs observed, and use it to test H_0: the Xs and the Ys have the same distribution, against the alternative in (12.45), at $\alpha = .05$.

29. Under the null hypothesis $H_0 : F_X = F_Y$, where X and Y are continuous variables, and assuming that the data consists of two independent samples with $n = 2$ Xs and $m = 3$ Ys, identify the exact distribution of the Wilcoxon rank sum statistic S.

30. Two weight-loss regimens were compared. Ten adult males who were each 50 pounds overweight participated in the experiment. Five participants were chosen at random and assigned to regimen A and the rest assigned were assigned to regimen B. The loss of weight over a 30-day period was recorded for each participant. When ranked from smallest to largest, the rank ordering of weight losses was AAABABABBB. Calculate the exact two-sided p-value of the observed rank sum $T = 18$ for the A sample. On the basis of the p-value, should the $H_0 : F_A = F_B$ be rejected in favor of $H_0 : F_A \neq F_B$ at $\alpha = .05$, where F_A and F_B are the respective weight-loss distributions under the two regimens?

31. In an article published in 1974 in the British medical journal *Lancet*, Bolander and his coauthors reported on the effect of vitamin D as an anticonvulsant in the treatment of epilepsy. Fifteen patients with severe epilepsy were randomly separated into two groups. The treatment group (X) received daily megadoses of vitamin D, while the control group (Y) received a placebo. Data on the number of seizures experienced over a fixed period was recorded as: X: 4, 1, 3, 12, 19, 8, 7 and Y: 2, 6, 21, 17, 34, 30, 53, 5. Use the rank sum statistic to test $H_0 : F_X = F_y$ against $F_X > F_Y$ (the X's tend to be smaller) at $\alpha = .05$.
32. Fifteen random members of the Harvard band's wind section were tested for lung capacity, yielding the data $X_1, \ldots, X_{15}$. Fifteen random members of the Stanford band's wind section were also tested for lung capacity, yielding the data $Y_1, \ldots, Y_{15}$. The Wilcoxon rank sum statistic was computed as $T = \sum_{i=1}^{15} R(X_i) = 265$. Assume there were no ties among the Xs and Ys. Test H_0: "the Xs and Ys have the same distribution" in two different ways. (a) (The exact method.) Use the exact quantiles of the distribution of the rank sum statistic T in Table A7 to test $H_0 : F_H = F_S$ vs. $H_1 : F_S > F_H$ (i.e., $X \geq_{st} Y$) at $\alpha = .05$. (b) (The approximate method.) Use the approximate normal distribution of T to test $H_0 : F_H = F_S$ vs. $H_1 : F_S > F_H$ (again, $X \geq_{st} Y$) at $\alpha = .05$.
33. Suppose that Aesop was dissatisfied with his classic experiment in which one tortoise was found to beat one hare in a race, and he decided to carry out a significance test to discover whether the results could be extended to tortoises and hares in general. He collected random samples of 6 tortoises and 6 hares, and made them all run his race at once. The order in which they reached the finish line (their rank order, from first to last crossing the finish line) is as follows, writing T for a tortoise and H for a hare: T H H H H H T T T T T H
Calculate the value of S, the sum of the ranks for tortoises, and use it to test the hypothesis $H_0 : F_t = F_h$, against $H_1 : F_t > F_h$ (tortoises tend to be slower) at $\alpha = .05$.
34. Take as given: the X data: 6, 9, 4, 12, 7; the Y data: 8, 10, 12, 18, 13, 15. Notice that there is a tie between two ranked values. Calculate the Wilcoxon rank sum statistic, and using Table A.7 for the necessary quantile (OK since the number of ties is clearly small), test $H_0 : F_X = F_Y$ vs. $H_1 : F_X > F_Y$ (the Xs tend to be smaller than the Ys) at $\alpha = .05$.
35. Suppose samples of size $n = m = 5$ are obtained from two populations (the Xs and the Ys) and that $S = \sum_{i=1}^{5} R(X_i)$ is the Wilcoxon rank sum statistic based on these data. Verify the calculations tabled below.

$S = s$	15	16	17	18	19	...
$p(s)$	1/252	1/252	2/252	3/252	5/252	...

Extend these calculations to determine the values of $P(S = 20)$ and $P(S = 21)$.
36. Let S^* be the Wilcoxon mid-rank sum statistic. Prove that S^* has the same expected value as the Wilcoxon rank sum statistic, that is, that
$$E(S^*) = \frac{n(n+m+1)}{2}.$$
37. Suppose that two independent random samples of size 10 are drawn. One group received a treatment consisting of an oral briefing and a hands-on demonstration. The other group received a basic introduction to the task to be performed. The performance of each subject, classified on a five-point scale, is shown below.

Score	Very Poor	Poor	Average	Good	Very Good
Control (X)	3	2	4	1	0
Treatment (Y)	0	2	1	3	4

Assign mid-ranks to all the tied observations. Calculate the statistic S^*, the sum of the mid-ranks of the X sample, and test the hypothesis $H_0 : F_x = F_y$ against a one-sided hypothesis at $\alpha = .01$. (Use the large sample approximation for the distribution of the statistic S^*.)

38. Fifty Cal Tech students entered a national competition in the engineering sciences, as did 25 students from MIT. Through the "Freedom of Only Slightly Interesting Information" Act, the scores of these 75 students were released to the *Pasadena News and Review*. In the article on the results published in the *PNR*, it was noted that the rank sum S (they actually called it W) of the scores from the Cal Tech students was 2050, and the conclusion that Cal Tech students are superior to MIT students in this particular subject matter was proudly annunciated. Carry out a one-sided Wilcoxon rank sum test at $\alpha = .05$ and determine whether or not the outcome above justifies such a conclusion. (Assume the data contained no ties.)

39. Under the sampling assumptions of Theorem 12.5.2 and the assumption that the null hypothesis $F_x = F_y$ is true, show that the distribution of the rank sum statistic S is symmetric about its mean $\mu = n(n+m+1)/2$, that is, show that for arbitrary $k > 0$, $P(S \leq \mu - k) = P(S \geq \mu + k)$.

40. Consider testing $H_0 : F_x = F_y$ against a two-sided alternative based on the following artificial data of 5 Xs and 5 Ys:

$$1,2,3,4,5,6,7,8,9,10.$$

Suppose that the Xs are 1, 2, 3, 5, and 7. (a) Carry out the runs test for $H_0; F_x = F_y$ by calculating the p-value of the runs statistic R. Note that this value is the smallest significance level at which H_0 can be rejected using this test. (b) Carry out the two-sided rank sum test. What is the smallest significance level (approximately) at which H_0 can be rejected using this test. (c) Which test seems the most appropriate?

Appendix

Tables

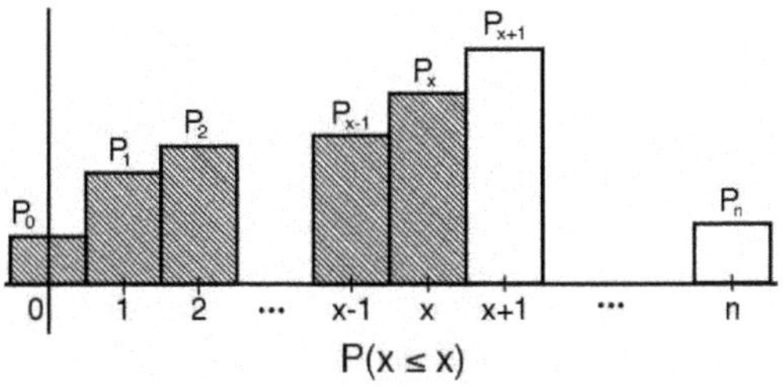

Table A.1. The Binomial Distribution
Tabled value is $P(X \leq x)$, where $X \sim B(n, p)$

		p									
n	x	0.05	0.1	0.15	0.2	0.25	0.3	0.35	0.4	0.45	0.5
2	0	0.9025	0.8100	0.7225	0.6400	0.5625	0.4900	0.4225	0.3600	0.3025	0.2500
	1	0.9975	0.9900	0.9775	0.9600	0.9375	0.9100	0.8775	0.8400	0.7975	0.7500
	2	1.0000	1.0000	1.0000	1.0000	1.0000	1.0000	1.0000	1.0000	1.0000	1.0000
3	0	0.8574	0.7290	0.6141	0.5120	0.4219	0.3430	0.2746	0.2160	0.1664	0.1250
	1	0.9928	0.9720	0.9392	0.8960	0.8438	0.7840	0.7182	0.6480	0.5747	0.5000
	2	0.9999	0.9990	0.9966	0.9920	0.9844	0.9730	0.9571	0.9360	0.9089	0.8750
	3	1.0000	1.0000	1.0000	1.0000	1.0000	1.0000	1.0000	1.0000	1.0000	1.0000
4	0	0.8145	0.6561	0.5220	0.4096	0.3164	0.2401	0.1785	0.1296	0.0915	0.0625
	1	0.9860	0.9477	0.8905	0.8192	0.7383	0.6517	0.5630	0.4752	0.3910	0.3125
	2	0.9995	0.9963	0.9880	0.9728	0.9492	0.9163	0.8735	0.8208	0.7585	0.6875
	3	1.0000	0.9999	0.9995	0.9984	0.9961	0.9919	0.9850	0.9744	0.9590	0.9375
	4	1.0000	1.0000	1.0000	1.0000	1.0000	1.0000	1.0000	1.0000	1.0000	1.0000
5	0	0.7738	0.5905	0.4437	0.3277	0.2373	0.1681	0.1160	0.0778	0.0503	0.0312
	1	0.9774	0.9185	0.8352	0.7373	0.6328	0.5282	0.4284	0.3370	0.2562	0.1875
	2	0.9988	0.9914	0.9734	0.9421	0.8965	0.8369	0.7648	0.6826	0.5931	0.5000
	3	1.0000	0.9995	0.9978	0.9933	0.9844	0.9692	0.9460	0.9130	0.8688	0.8125
	4	1.0000	1.0000	0.9999	0.9997	0.9990	0.9976	0.9947	0.9898	0.9815	0.9688
	5	1.0000	1.0000	1.0000	1.0000	1.0000	1.0000	1.0000	1.0000	1.0000	1.0000
6	0	0.7351	0.5314	0.3771	0.2621	0.1780	0.1176	0.0754	0.0467	0.0277	0.0156
	1	0.9672	0.8857	0.7765	0.6554	0.5339	0.4202	0.3191	0.2333	0.1636	0.1094
	2	0.9978	0.9842	0.9527	0.9011	0.8306	0.7443	0.6471	0.5443	0.4415	0.3437
	3	0.9999	0.9987	0.9941	0.9830	0.9624	0.9295	0.8826	0.8208	0.7447	0.6562
	4	1.0000	0.9999	0.9996	0.9984	0.9954	0.9891	0.9777	0.9590	0.9308	0.8906
	5	1.0000	1.0000	1.0000	0.9999	0.9998	0.9993	0.9982	0.9959	0.9917	0.9844
	6	1.0000	1.0000	1.0000	1.0000	1.0000	1.0000	1.0000	1.0000	1.0000	1.0000

		p									
n	x	0.05	0.1	0.15	0.2	0.25	0.3	0.35	0.4	0.45	0.5
7	0	0.6983	0.4783	0.3206	0.2097	0.1335	0.0824	0.0490	0.0280	0.0152	0.0078
	1	0.9556	0.8503	0.7166	0.5767	0.4449	0.3294	0.2338	0.1586	0.1024	0.0625
	2	0.9962	0.9743	0.9262	0.8520	0.7564	0.6471	0.5323	0.4199	0.3164	0.2266
	3	0.9998	0.9973	0.9879	0.9667	0.9294	0.8740	0.8002	0.7102	0.6083	0.5000
	4	1.0000	0.9998	0.9988	0.9953	0.9871	0.9712	0.9444	0.9037	0.8471	0.7734
	5	1.0000	1.0000	0.9999	0.9996	0.9987	0.9962	0.9910	0.9812	0.9643	0.9375
	6	1.0000	1.0000	1.0000	1.0000	0.9999	0.9998	0.9994	0.9984	0.9963	0.9922
	7	1.0000	1.0000	1.0000	1.0000	1.0000	1.0000	1.0000	1.0000	1.0000	1.0000
8	0	0.6634	0.4305	0.2725	0.1678	0.1001	0.0576	0.0319	0.0168	0.0084	0.0039
	1	0.9428	0.8131	0.6572	0.5033	0.3671	0.2553	0.1691	0.1064	0.0632	0.0352
	2	0.9942	0.9619	0.8948	0.7969	0.6785	0.5518	0.4278	0.3154	0.2201	0.1445
	3	0.9996	0.9950	0.9786	0.9437	0.8862	0.8059	0.7064	0.5941	0.4770	0.3633
	4	1.0000	0.9996	0.9971	0.9896	0.9727	0.9420	0.8939	0.8263	0.7396	0.6367
	5	1.0000	1.0000	0.9998	0.9988	0.9958	0.9887	0.9747	0.9502	0.9115	0.8555
	6	1.0000	1.0000	1.0000	0.9999	0.9996	0.9987	0.9964	0.9915	0.9819	0.9648
	7	1.0000	1.0000	1.0000	1.0000	1.0000	0.9999	0.9998	0.9993	0.9983	0.9961
	8	1.0000	1.0000	1.0000	1.0000	1.0000	1.0000	1.0000	1.0000	1.0000	1.0000
9	0	0.6302	0.3874	0.2316	0.1342	0.0751	0.0404	0.0207	0.0101	0.0046	0.0020
	1	0.9288	0.7748	0.5995	0.4362	0.3003	0.1960	0.1211	0.0705	0.0385	0.0195
	2	0.9916	0.9470	0.8591	0.7382	0.6007	0.4628	0.3373	0.2318	0.1495	0.0898
	3	0.9994	0.9917	0.9661	0.9144	0.8343	0.7297	0.6089	0.4826	0.3614	0.2539
	4	1.0000	0.9991	0.9944	0.9804	0.9511	0.9012	0.8283	0.7334	0.6214	0.5000
	5	1.0000	0.9999	0.9994	0.9969	0.9900	0.9747	0.9464	0.9006	0.8342	0.7461
	6	1.0000	1.0000	1.0000	0.9997	0.9987	0.9957	0.9888	0.9750	0.9502	0.9102
	7	1.0000	1.0000	1.0000	1.0000	0.9999	0.9996	0.9986	0.9962	0.9909	0.9805
	8	1.0000	1.0000	1.0000	1.0000	1.0000	1.0000	0.9999	0.9997	0.9992	0.9980
	9	1.0000	1.0000	1.0000	1.0000	1.0000	1.0000	1.0000	1.0000	1.0000	1.0000
10	0	0.5987	0.3487	0.1969	0.1074	0.0563	0.0282	0.0135	0.0060	0.0025	0.0010
	1	0.9139	0.7361	0.5443	0.3758	0.2440	0.1493	0.0860	0.0464	0.0233	0.0107
	2	0.9885	0.9298	0.8202	0.6778	0.5256	0.3828	0.2616	0.1673	0.0996	0.0547
	3	0.9990	0.9872	0.9500	0.8791	0.7759	0.6496	0.5138	0.3823	0.2660	0.1719
	4	0.9999	0.9984	0.9901	0.9672	0.9219	0.8497	0.7515	0.6331	0.5044	0.3770
	5	1.0000	0.9999	0.9986	0.9936	0.9803	0.9527	0.9051	0.8338	0.7384	0.6230
	6	1.0000	1.0000	0.9999	0.9991	0.9965	0.9894	0.9740	0.9452	0.8980	0.8281
	7	1.0000	1.0000	1.0000	0.9999	0.9996	0.9984	0.9952	0.9877	0.9726	0.9453
	8	1.0000	1.0000	1.0000	1.0000	1.0000	0.9999	0.9995	0.9983	0.9955	0.9893
	9	1.0000	1.0000	1.0000	1.0000	1.0000	1.0000	1.0000	0.9999	0.9997	0.9990
	10	1.0000	1.0000	1.0000	1.0000	1.0000	1.0000	1.0000	1.0000	1.0000	1.0000
11	0	0.5688	0.3138	0.1673	0.0859	0.0422	0.0198	0.0088	0.0036	0.0014	0.0005
	1	0.8981	0.6974	0.4922	0.3221	0.1971	0.1130	0.0606	0.0302	0.0139	0.0059
	2	0.9848	0.9104	0.7788	0.6174	0.4552	0.3127	0.2001	0.1189	0.0652	0.0327
	3	0.9984	0.9815	0.9306	0.8389	0.7133	0.5696	0.4256	0.2963	0.1911	0.1133
	4	0.9999	0.9972	0.9841	0.9496	0.8854	0.7897	0.6683	0.5328	0.3971	0.2744
	5	1.0000	0.9997	0.9973	0.9883	0.9657	0.9218	0.8513	0.7535	0.6331	0.5000
	6	1.0000	1.0000	0.9997	0.9980	0.9924	0.9784	0.9499	0.9006	0.8262	0.7256
	7	1.0000	1.0000	1.0000	0.9998	0.9988	0.9957	0.9878	0.9707	0.9390	0.8867
	8	1.0000	1.0000	1.0000	1.0000	0.9999	0.9994	0.9980	0.9941	0.9852	0.9673
	9	1.0000	1.0000	1.0000	1.0000	1.0000	1.0000	0.9998	0.9993	0.9978	0.9941

		p									
n	x	0.05	0.1	0.15	0.2	0.25	0.3	0.35	0.4	0.45	0.5
	10	1.0000	1.0000	1.0000	1.0000	1.0000	1.0000	1.0000	1.0000	0.9998	0.9995
	11	1.0000	1.0000	1.0000	1.0000	1.0000	1.0000	1.0000	1.0000	1.0000	1.0000
12	0	0.5404	0.2824	0.1422	0.0687	0.0317	0.0138	0.0057	0.0022	0.0008	0.0002
	1	0.8816	0.6590	0.4435	0.2749	0.1584	0.0850	0.0424	0.0196	0.0083	0.0032
	2	0.9804	0.8891	0.7358	0.5583	0.3907	0.2528	0.1513	0.0834	0.0421	0.0193
	3	0.9978	0.9744	0.9078	0.7946	0.6488	0.4925	0.3467	0.2253	0.1345	0.0730
	4	0.9998	0.9957	0.9761	0.9274	0.8424	0.7237	0.5833	0.4382	0.3044	0.1938
	5	1.0000	0.9995	0.9954	0.9806	0.9456	0.8822	0.7873	0.6652	0.5269	0.3872
	6	1.0000	0.9999	0.9993	0.9961	0.9857	0.9614	0.9154	0.8418	0.7393	0.6128
	7	1.0000	1.0000	0.9999	0.9994	0.9972	0.9905	0.9745	0.9427	0.8883	0.8062
	8	1.0000	1.0000	1.0000	0.9999	0.9996	0.9983	0.9944	0.9847	0.9644	0.9270
	9	1.0000	1.0000	1.0000	1.0000	1.0000	0.9998	0.9992	0.9972	0.9921	0.9807
	10	1.0000	1.0000	1.0000	1.0000	1.0000	1.0000	0.9999	0.9997	0.9989	0.9968
	11	1.0000	1.0000	1.0000	1.0000	1.0000	1.0000	1.0000	1.0000	0.9999	0.9998
	12	1.0000	1.0000	1.0000	1.0000	1.0000	1.0000	1.0000	1.0000	1.0000	1.0000
13	0	0.5133	0.2542	0.1209	0.0550	0.0238	0.0097	0.0037	0.0013	0.0004	0.0001
	1	0.8646	0.6213	0.3983	0.2336	0.1267	0.0637	0.0296	0.0126	0.0049	0.0017
	2	0.9755	0.8661	0.6920	0.5017	0.3326	0.2025	0.1132	0.0579	0.0269	0.0112
	3	0.9969	0.9658	0.8820	0.7473	0.5843	0.4206	0.2783	0.1686	0.0929	0.0461
	4	0.9997	0.9935	0.9658	0.9009	0.7940	0.6543	0.5005	0.3530	0.2279	0.1334
	5	1.0000	0.9991	0.9925	0.9700	0.9198	0.8346	0.7159	0.5744	0.4268	0.2905
	6	1.0000	0.9999	0.9987	0.9930	0.9757	0.9376	0.8705	0.7712	0.6437	0.5000
	7	1.0000	1.0000	0.9998	0.9988	0.9944	0.9818	0.9538	0.9023	0.8212	0.7095
	8	1.0000	1.0000	1.0000	0.9998	0.9990	0.9960	0.9874	0.9679	0.9302	0.8666
	9	1.0000	1.0000	1.0000	1.0000	0.9999	0.9993	0.9975	0.9922	0.9797	0.9539
	10	1.0000	1.0000	1.0000	1.0000	1.0000	0.9999	0.9997	0.9987	0.9959	0.9888
	11	1.0000	1.0000	1.0000	1.0000	1.0000	1.0000	1.0000	0.9999	0.9995	0.9983
	12	1.0000	1.0000	1.0000	1.0000	1.0000	1.0000	1.0000	1.0000	1.0000	0.9999
	13	1.0000	1.0000	1.0000	1.0000	1.0000	1.0000	1.0000	1.0000	1.0000	1.0000
14	0	0.4877	0.2288	0.1028	0.0440	0.0178	0.0068	0.0024	0.0008	0.0002	0.0001
	1	0.8470	0.5846	0.3567	0.1979	0.1010	0.0475	0.0205	0.0081	0.0029	0.0009
	2	0.9699	0.8416	0.6479	0.4481	0.2811	0.1608	0.0839	0.0398	0.0170	0.0065
	3	0.9958	0.9559	0.8535	0.6982	0.5213	0.3552	0.2205	0.1243	0.0632	0.0287
	4	0.9996	0.9908	0.9533	0.8702	0.7415	0.5842	0.4227	0.2793	0.1672	0.0898
	5	1.0000	0.9985	0.9885	0.9561	0.8883	0.7805	0.6405	0.4859	0.3373	0.2120
	6	1.0000	0.9998	0.9978	0.9884	0.9617	0.9067	0.8164	0.6925	0.5461	0.3953
	7	1.0000	1.0000	0.9997	0.9976	0.9897	0.9685	0.9247	0.8499	0.7414	0.6047
	8	1.0000	1.0000	1.0000	0.9996	0.9978	0.9917	0.9757	0.9417	0.8811	0.7880
	9	1.0000	1.0000	1.0000	1.0000	0.9997	0.9983	0.9940	0.9825	0.9574	0.9102
	10	1.0000	1.0000	1.0000	1.0000	1.0000	0.9998	0.9989	0.9961	0.9886	0.9713
	11	1.0000	1.0000	1.0000	1.0000	1.0000	1.0000	0.9999	0.9994	0.9978	0.9935
	12	1.0000	1.0000	1.0000	1.0000	1.0000	1.0000	1.0000	0.9999	0.9997	0.9991
	13	1.0000	1.0000	1.0000	1.0000	1.0000	1.0000	1.0000	1.0000	1.0000	0.9999
	14	1.0000	1.0000	1.0000	1.0000	1.0000	1.0000	1.0000	1.0000	1.0000	1.0000
15	0	0.4633	0.2059	0.0874	0.0352	0.0134	0.0047	0.0016	0.0005	0.0001	0.0000
	1	0.8290	0.5490	0.3186	0.1671	0.0802	0.0353	0.0142	0.0052	0.0017	0.0005
	2	0.9638	0.8159	0.6042	0.3980	0.2361	0.1268	0.0617	0.0271	0.0107	0.0037
	3	0.9945	0.9444	0.8227	0.6482	0.4613	0.2969	0.1727	0.0905	0.0424	0.0176

		p									
n	x	0.05	0.1	0.15	0.2	0.25	0.3	0.35	0.4	0.45	0.5
	4	0.9994	0.9873	0.9383	0.8358	0.6865	0.5155	0.3519	0.2173	0.1204	0.0592
	5	0.9999	0.9978	0.9832	0.9389	0.8516	0.7216	0.5643	0.4032	0.2608	0.1509
	6	1.0000	0.9997	0.9964	0.9819	0.9434	0.8689	0.7548	0.6098	0.4522	0.3036
	7	1.0000	1.0000	0.9994	0.9958	0.9827	0.9500	0.8868	0.7869	0.6535	0.5000
	8	1.0000	1.0000	0.9999	0.9992	0.9958	0.9848	0.9578	0.9050	0.8182	0.6964
	9	1.0000	1.0000	1.0000	0.9999	0.9992	0.9963	0.9876	0.9662	0.9231	0.8491
	10	1.0000	1.0000	1.0000	1.0000	0.9999	0.9993	0.9972	0.9907	0.9745	0.9408
	11	1.0000	1.0000	1.0000	1.0000	1.0000	0.9999	0.9995	0.9981	0.9937	0.9824
	12	1.0000	1.0000	1.0000	1.0000	1.0000	1.0000	0.9999	0.9997	0.9989	0.9963
	13	1.0000	1.0000	1.0000	1.0000	1.0000	1.0000	1.0000	1.0000	0.9999	0.9995
	14	1.0000	1.0000	1.0000	1.0000	1.0000	1.0000	1.0000	1.0000	1.0000	1.0000
	15	1.0000	1.0000	1.0000	1.0000	1.0000	1.0000	1.0000	1.0000	1.0000	1.0000
16	0	0.4401	0.1853	0.0743	0.0281	0.0100	0.0033	0.0010	0.0003	0.0001	0.0000
	1	0.8108	0.5147	0.2839	0.1407	0.0635	0.0261	0.0098	0.0033	0.0010	0.0003
	2	0.9571	0.7892	0.5614	0.3518	0.1971	0.0994	0.0451	0.0183	0.0066	0.0021
	3	0.9930	0.9316	0.7899	0.5981	0.4050	0.2459	0.1339	0.0651	0.0281	0.0106
	4	0.9991	0.9830	0.9209	0.7982	0.6302	0.4499	0.2892	0.1666	0.0853	0.0384
	5	0.9999	0.9967	0.9765	0.9183	0.8103	0.6598	0.4900	0.3288	0.1976	0.1051
	6	1.0000	0.9995	0.9944	0.9733	0.9204	0.8247	0.6881	0.5272	0.3660	0.2272
	7	1.0000	0.9999	0.9989	0.9930	0.9729	0.9256	0.8406	0.7161	0.5629	0.4018
	8	1.0000	1.0000	0.9998	0.9985	0.9925	0.9743	0.9329	0.8577	0.7441	0.5982
	9	1.0000	1.0000	1.0000	0.9998	0.9984	0.9929	0.9771	0.9417	0.8759	0.7728
	10	1.0000	1.0000	1.0000	1.0000	0.9997	0.9984	0.9938	0.9809	0.9514	0.8949
	11	1.0000	1.0000	1.0000	1.0000	1.0000	0.9997	0.9987	0.9951	0.9851	0.9616
	12	1.0000	1.0000	1.0000	1.0000	1.0000	1.0000	0.9998	0.9991	0.9965	0.9894
	13	1.0000	1.0000	1.0000	1.0000	1.0000	1.0000	1.0000	0.9999	0.9994	0.9979
	14	1.0000	1.0000	1.0000	1.0000	1.0000	1.0000	1.0000	1.0000	0.9999	0.9997
	15	1.0000	1.0000	1.0000	1.0000	1.0000	1.0000	1.0000	1.0000	1.0000	1.0000
	16	1.0000	1.0000	1.0000	1.0000	1.0000	1.0000	1.0000	1.0000	1.0000	1.0000
20	0	0.3585	0.1216	0.0388	0.0115	0.0032	0.0008	0.0002	0.0000	0.0000	0.0000
	1	0.7358	0.3917	0.1756	0.0692	0.0243	0.0076	0.0021	0.0005	0.0001	0.0000
	2	0.9245	0.6769	0.4049	0.2061	0.0913	0.0355	0.0121	0.0036	0.0009	0.0002
	3	0.9841	0.8670	0.6477	0.4114	0.2252	0.1071	0.0444	0.0160	0.0049	0.0013
	4	0.9974	0.9568	0.8298	0.6296	0.4148	0.2375	0.1182	0.0510	0.0189	0.0059
	5	0.9997	0.9887	0.9327	0.8042	0.6172	0.4164	0.2454	0.1256	0.0553	0.0207
	6	1.0000	0.9976	0.9781	0.9133	0.7858	0.6080	0.4166	0.2500	0.1299	0.0577
	7	1.0000	0.9996	0.9941	0.9679	0.8982	0.7723	0.6010	0.4159	0.2520	0.1316
	8	1.0000	0.9999	0.9987	0.9900	0.9591	0.8867	0.7624	0.5956	0.4143	0.2517
	9	1.0000	1.0000	0.9998	0.9974	0.9861	0.9520	0.8782	0.7553	0.5914	0.4119
	10	1.0000	1.0000	1.0000	0.9994	0.9961	0.9829	0.9468	0.8725	0.7507	0.5881
	11	1.0000	1.0000	1.0000	0.9999	0.9991	0.9949	0.9804	0.9435	0.8692	0.7483
	12	1.0000	1.0000	1.0000	1.0000	0.9998	0.9987	0.9940	0.9790	0.9420	0.8684
	13	1.0000	1.0000	1.0000	1.0000	1.0000	0.9997	0.9985	0.9935	0.9786	0.9423
	14	1.0000	1.0000	1.0000	1.0000	1.0000	1.0000	0.9997	0.9984	0.9936	0.9793
	15	1.0000	1.0000	1.0000	1.0000	1.0000	1.0000	1.0000	0.9997	0.9985	0.9941
	16	1.0000	1.0000	1.0000	1.0000	1.0000	1.0000	1.0000	1.0000	0.9997	0.9987
	17	1.0000	1.0000	1.0000	1.0000	1.0000	1.0000	1.0000	1.0000	1.0000	0.9998
	18	1.0000	1.0000	1.0000	1.0000	1.0000	1.0000	1.0000	1.0000	1.0000	1.0000
	19	1.0000	1.0000	1.0000	1.0000	1.0000	1.0000	1.0000	1.0000	1.0000	1.0000
	20	1.0000	1.0000	1.0000	1.0000	1.0000	1.0000	1.0000	1.0000	1.0000	1.0000

n	x	p: 0.05	0.1	0.15	0.2	0.25	0.3	0.35	0.4	0.45	0.5
25	0	0.2774	0.0718	0.0172	0.0038	0.0008	0.0001	0.0000	0.0000	0.0000	0.0000
	1	0.6424	0.2712	0.0931	0.0274	0.0070	0.0016	0.0003	0.0001	0.0000	0.0000
	2	0.8729	0.5371	0.2537	0.0982	0.0321	0.0090	0.0021	0.0004	0.0001	0.0000
	3	0.9659	0.7636	0.4711	0.2340	0.0962	0.0332	0.0097	0.0024	0.0005	0.0001
	4	0.9928	0.9020	0.6821	0.4207	0.2137	0.0905	0.0320	0.0095	0.0023	0.0005
	5	0.9988	0.9666	0.8385	0.6167	0.3783	0.1935	0.0826	0.0294	0.0086	0.0020
	6	0.9998	0.9905	0.9305	0.7800	0.5611	0.3407	0.1734	0.0736	0.0258	0.0073
	7	1.0000	0.9977	0.9745	0.8909	0.7265	0.5118	0.3061	0.1536	0.0639	0.0216
	8	1.0000	0.9995	0.9920	0.9532	0.8506	0.6769	0.4668	0.2735	0.1340	0.0539
	9	1.0000	0.9999	0.9979	0.9827	0.9287	0.8106	0.6303	0.4246	0.2424	0.1148
	10	1.0000	1.0000	0.9995	0.9944	0.9703	0.9022	0.7712	0.5858	0.3843	0.2122
	11	1.0000	1.0000	0.9999	0.9985	0.9893	0.9558	0.8746	0.7323	0.5426	0.3450
	12	1.0000	1.0000	1.0000	0.9996	0.9966	0.9825	0.9396	0.8462	0.6937	0.5000
	13	1.0000	1.0000	1.0000	0.9999	0.9991	0.9940	0.9745	0.9222	0.8173	0.6550
	14	1.0000	1.0000	1.0000	1.0000	0.9998	0.9982	0.9907	0.9656	0.9040	0.7878
	15	1.0000	1.0000	1.0000	1.0000	1.0000	0.9995	0.9971	0.9868	0.9560	0.8852
	16	1.0000	1.0000	1.0000	1.0000	1.0000	0.9999	0.9992	0.9957	0.9826	0.9461
	17	1.0000	1.0000	1.0000	1.0000	1.0000	1.0000	0.9998	0.9988	0.9942	0.9784
	18	1.0000	1.0000	1.0000	1.0000	1.0000	1.0000	1.0000	0.9997	0.9984	0.9927
	19	1.0000	1.0000	1.0000	1.0000	1.0000	1.0000	1.0000	0.9999	0.9996	0.9980
	20	1.0000	1.0000	1.0000	1.0000	1.0000	1.0000	1.0000	1.0000	0.9999	0.9995
	21	1.0000	1.0000	1.0000	1.0000	1.0000	1.0000	1.0000	1.0000	1.0000	0.9999
	22	1.0000	1.0000	1.0000	1.0000	1.0000	1.0000	1.0000	1.0000	1.0000	1.0000
	23	1.0000	1.0000	1.0000	1.0000	1.0000	1.0000	1.0000	1.0000	1.0000	1.0000
	24	1.0000	1.0000	1.0000	1.0000	1.0000	1.0000	1.0000	1.0000	1.0000	1.0000
	25	1.0000	1.0000	1.0000	1.0000	1.0000	1.0000	1.0000	1.0000	1.0000	1.0000

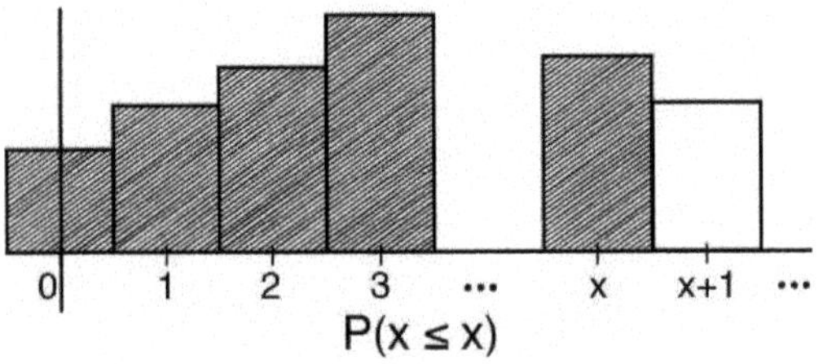

Table A.2. The Poisson Distribution
Tabled value is $P(X \leq x)$ where $X \sim P(\lambda)$.

	$\lambda = E(X)$									
x	0.1	0.2	0.3	0.4	0.5	0.6	0.7	0.8	0.9	1
0	0.9048	0.8187	0.7408	0.6703	0.6065	0.5488	0.4966	0.4493	0.4066	0.3679
1	0.9953	0.9825	0.9631	0.9384	0.9098	0.8781	0.8442	0.8088	0.7725	0.7358
2	0.9998	0.9989	0.9964	0.9921	0.9856	0.9769	0.9659	0.9526	0.9371	0.9197
3	1.0000	0.9999	0.9997	0.9992	0.9982	0.9966	0.9942	0.9909	0.9865	0.9810
4	1.0000	1.0000	1.0000	0.9999	0.9998	0.9996	0.9992	0.9986	0.9977	0.9963
5	1.0000	1.0000	1.0000	1.0000	1.0000	1.0000	0.9999	0.9998	0.9997	0.9994
6	1.0000	1.0000	1.0000	1.0000	1.0000	1.0000	1.0000	1.0000	1.0000	0.9999

	$\lambda = E(X)$									
x	1.1	1.2	1.3	1.4	1.5	1.6	1.7	1.8	1.9	2
0	0.3329	0.3012	0.2725	0.2466	0.2231	0.2019	0.1827	0.1653	0.1496	0.1353
1	0.6990	0.6626	0.6268	0.5918	0.5578	0.5249	0.4932	0.4628	0.4337	0.4060
2	0.9004	0.8795	0.8571	0.8335	0.8088	0.7834	0.7572	0.7306	0.7037	0.6767
3	0.9743	0.9662	0.9569	0.9463	0.9344	0.9212	0.9068	0.8913	0.8747	0.8571
4	0.9946	0.9923	0.9893	0.9857	0.9814	0.9763	0.9704	0.9636	0.9559	0.9473
5	0.9990	0.9985	0.9978	0.9968	0.9955	0.9940	0.9920	0.9896	0.9868	0.9834
6	0.9999	0.9997	0.9996	0.9994	0.9991	0.9987	0.9981	0.9974	0.9966	0.9955
7	1.0000	1.0000	0.9999	0.9999	0.9998	0.9997	0.9996	0.9994	0.9992	0.9989
8	1.0000	1.0000	1.0000	1.0000	1.0000	1.0000	0.9999	0.9999	0.9998	0.9998

	$\lambda = E(X)$									
x	2.2	2.4	2.6	2.8	3	3.2	3.4	3.6	3.8	4
0	0.1108	0.0907	0.0743	0.0608	0.0498	0.0408	0.0334	0.0273	0.0224	0.0183
1	0.3546	0.3084	0.2674	0.2311	0.1991	0.1712	0.1468	0.1257	0.1074	0.0916
2	0.6227	0.5697	0.5184	0.4695	0.4232	0.3799	0.3397	0.3027	0.2689	0.2381
3	0.8194	0.7787	0.7360	0.6919	0.6472	0.6025	0.5584	0.5152	0.4735	0.4335
4	0.9275	0.9041	0.8774	0.8477	0.8153	0.7806	0.7442	0.7064	0.6678	0.6288
5	0.9751	0.9643	0.9510	0.9349	0.9161	0.8946	0.8705	0.8441	0.8156	0.7851
6	0.9925	0.9884	0.9828	0.9756	0.9665	0.9554	0.9421	0.9267	0.9091	0.8893
7	0.9980	0.9967	0.9947	0.9919	0.9881	0.9832	0.9769	0.9692	0.9599	0.9489
8	0.9995	0.9991	0.9985	0.9976	0.9962	0.9943	0.9917	0.9883	0.9840	0.9786
9	0.9999	0.9998	0.9996	0.9993	0.9989	0.9982	0.9973	0.9960	0.9942	0.9919
10	1.0000	1.0000	0.9999	0.9998	0.9997	0.9995	0.9992	0.9987	0.9981	0.9972
11	1.0000	1.0000	1.0000	1.0000	0.9999	0.9999	0.9998	0.9996	0.9994	0.9991
12	1.0000	1.0000	1.0000	1.0000	1.0000	1.0000	0.9999	0.9999	0.9998	0.9997

	$\lambda = E(X)$									
x	4.2	4.4	4.6	4.8	5	5.2	5.4	5.6	5.8	6
0	0.0150	0.0123	0.0101	0.0082	0.0067	0.0055	0.0045	0.0037	0.0030	0.0025
1	0.0780	0.0663	0.0563	0.0477	0.0404	0.0342	0.0289	0.0244	0.0206	0.0174
2	0.2102	0.1851	0.1626	0.1425	0.1247	0.1088	0.0948	0.0824	0.0715	0.0620
3	0.3954	0.3594	0.3257	0.2942	0.2650	0.2381	0.2133	0.1906	0.1700	0.1512
4	0.5898	0.5512	0.5132	0.4763	0.4405	0.4061	0.3733	0.3422	0.3127	0.2851
5	0.7531	0.7199	0.6858	0.6510	0.6160	0.5809	0.5461	0.5119	0.4783	0.4457
6	0.8675	0.8436	0.8180	0.7908	0.7622	0.7324	0.7017	0.6703	0.6384	0.6063

x	4.2	4.4	4.6	4.8	5	5.2	5.4	5.6	5.8	6
	$\lambda = E(X)$									
7	0.9361	0.9214	0.9049	0.8867	0.8666	0.8449	0.8217	0.7970	0.7710	0.7440
8	0.9721	0.9642	0.9549	0.9442	0.9319	0.9181	0.9027	0.8857	0.8672	0.8472
9	0.9889	0.9851	0.9805	0.9749	0.9682	0.9603	0.9512	0.9409	0.9292	0.9161
10	0.9959	0.9943	0.9922	0.9896	0.9863	0.9823	0.9775	0.9718	0.9651	0.9574
11	0.9986	0.9980	0.9971	0.9960	0.9945	0.9927	0.9904	0.9875	0.9841	0.9799
12	0.9996	0.9993	0.9990	0.9986	0.9980	0.9972	0.9962	0.9949	0.9932	0.9912
13	0.9999	0.9998	0.9997	0.9995	0.9993	0.9990	0.9986	0.9980	0.9973	0.9964
14	1.0000	0.9999	0.9999	0.9999	0.9998	0.9997	0.9995	0.9993	0.9990	0.9986
15	1.0000	1.0000	1.0000	1.0000	0.9999	0.9999	0.9998	0.9998	0.9996	0.9995
16	1.0000	1.0000	1.0000	1.0000	1.0000	1.0000	0.9999	0.9999	0.9999	0.9998

x	6.5	7	7.5	8	8.5	9	9.5	10	10.5	11
	$\lambda = E(X)$									
0	0.0015	0.0009	0.0006	0.0003	0.0002	0.0001	0.0001	0.0000	0.0000	0.0000
1	0.0113	0.0073	0.0047	0.0030	0.0019	0.0012	0.0008	0.0005	0.0003	0.0002
2	0.0430	0.0296	0.0203	0.0138	0.0093	0.0062	0.0042	0.0028	0.0018	0.0012
3	0.1118	0.0818	0.0591	0.0424	0.0301	0.0212	0.0149	0.0103	0.0071	0.0049
4	0.2237	0.1730	0.1321	0.0996	0.0744	0.0550	0.0403	0.0293	0.0211	0.0151
5	0.3690	0.3007	0.2414	0.1912	0.1496	0.1157	0.0885	0.0671	0.0504	0.0375
6	0.5265	0.4497	0.3782	0.3134	0.2562	0.2068	0.1649	0.1301	0.1016	0.0786
7	0.6728	0.5987	0.5246	0.4530	0.3856	0.3239	0.2687	0.2202	0.1785	0.1432
8	0.7916	0.7291	0.6620	0.5925	0.5231	0.4557	0.3918	0.3328	0.2794	0.2320
9	0.8774	0.8305	0.7764	0.7166	0.6530	0.5874	0.5218	0.4579	0.3971	0.3405
10	0.9332	0.9015	0.8622	0.8159	0.7634	0.7060	0.6453	0.5830	0.5207	0.4599
11	0.9661	0.9467	0.9208	0.8881	0.8487	0.8030	0.7520	0.6968	0.6387	0.5793
12	0.9840	0.9730	0.9573	0.9362	0.9091	0.8758	0.8364	0.7916	0.7420	0.6887
13	0.9929	0.9872	0.9784	0.9658	0.9486	0.9261	0.8981	0.8645	0.8253	0.7813
14	0.9970	0.9943	0.9897	0.9827	0.9726	0.9585	0.9400	0.9165	0.8879	0.8540
15	0.9988	0.9976	0.9954	0.9918	0.9862	0.9780	0.9665	0.9513	0.9317	0.9074
16	0.9996	0.9990	0.9980	0.9963	0.9934	0.9889	0.9823	0.9730	0.9604	0.9441
17	0.9998	0.9996	0.9992	0.9984	0.9970	0.9947	0.9911	0.9857	0.9781	0.9678
18	0.9999	0.9999	0.9997	0.9993	0.9987	0.9976	0.9957	0.9928	0.9885	0.9823
19	1.0000	1.0000	0.9999	0.9997	0.9995	0.9989	0.9980	0.9965	0.9942	0.9907
20	1.0000	1.0000	1.0000	0.9999	0.9998	0.9996	0.9991	0.9984	0.9972	0.9953
21	1.0000	1.0000	1.0000	1.0000	0.9999	0.9998	0.9996	0.9993	0.9987	0.9977
22	1.0000	1.0000	1.0000	1.0000	1.0000	0.9999	0.9999	0.9997	0.9994	0.9990
23	1.0000	1.0000	1.0000	1.0000	1.0000	1.0000	0.9999	0.9999	0.9998	0.9995

x	11.5	12	12.5	13	13.5	14	14.5	15	15.5	16
	$\lambda = E(X)$									
0	0.0000	0.0000	0.0000	0.0000	0.0000	0.0000	0.0000	0.0000	0.0000	0.0000
1	0.0001	0.0001	0.0001	0.0000	0.0000	0.0000	0.0000	0.0000	0.0000	0.0000
2	0.0008	0.0005	0.0003	0.0002	0.0001	0.0001	0.0001	0.0000	0.0000	0.0000
3	0.0034	0.0023	0.0016	0.0011	0.0007	0.0005	0.0003	0.0002	0.0001	0.0001
4	0.0107	0.0076	0.0053	0.0037	0.0026	0.0018	0.0012	0.0009	0.0006	0.0004
5	0.0277	0.0203	0.0148	0.0107	0.0077	0.0055	0.0039	0.0028	0.0020	0.0014
6	0.0603	0.0458	0.0346	0.0259	0.0193	0.0142	0.0105	0.0076	0.0055	0.0040
7	0.1137	0.0895	0.0698	0.0540	0.0415	0.0316	0.0239	0.0180	0.0135	0.0100
8	0.1906	0.1550	0.1249	0.0998	0.0790	0.0621	0.0484	0.0374	0.0288	0.0220
9	0.2888	0.2424	0.2014	0.1658	0.1353	0.1094	0.0878	0.0699	0.0552	0.0433
10	0.4017	0.3472	0.2971	0.2517	0.2112	0.1757	0.1449	0.1185	0.0961	0.0774
11	0.5198	0.4616	0.4058	0.3532	0.3045	0.2600	0.2201	0.1848	0.1538	0.1270
12	0.6329	0.5760	0.5190	0.4631	0.4093	0.3585	0.3111	0.2676	0.2283	0.1931
13	0.7330	0.6815	0.6278	0.5730	0.5182	0.4644	0.4125	0.3632	0.3171	0.2745
14	0.8153	0.7720	0.7250	0.6751	0.6233	0.5704	0.5176	0.4657	0.4154	0.3675

$\lambda = E(X)$

x	11.5	12	12.5	13	13.5	14	14.5	15	15.5	16
15	0.8783	0.8444	0.8060	0.7636	0.7178	0.6694	0.6192	0.5681	0.5170	0.4667
16	0.9236	0.8987	0.8693	0.8355	0.7975	0.7559	0.7112	0.6641	0.6154	0.5660
17	0.9542	0.9370	0.9158	0.8905	0.8609	0.8272	0.7897	0.7489	0.7052	0.6593
18	0.9738	0.9626	0.9481	0.9302	0.9084	0.8826	0.8530	0.8195	0.7825	0.7423
19	0.9857	0.9787	0.9694	0.9573	0.9421	0.9235	0.9012	0.8752	0.8455	0.8122
20	0.9925	0.9884	0.9827	0.9750	0.9649	0.9521	0.9362	0.9170	0.8944	0.8682
21	0.9962	0.9939	0.9906	0.9859	0.9796	0.9712	0.9604	0.9469	0.9304	0.9108
22	0.9982	0.9970	0.9951	0.9924	0.9885	0.9833	0.9763	0.9673	0.9558	0.9418
23	0.9992	0.9985	0.9975	0.9960	0.9938	0.9907	0.9863	0.9805	0.9730	0.9633
24	0.9996	0.9993	0.9988	0.9980	0.9968	0.9950	0.9924	0.9888	0.9840	0.9777
25	0.9998	0.9997	0.9994	0.9990	0.9984	0.9974	0.9959	0.9938	0.9909	0.9869
26	0.9999	0.9999	0.9997	0.9995	0.9992	0.9987	0.9979	0.9967	0.9950	0.9925
27	1.0000	0.9999	0.9999	0.9998	0.9996	0.9994	0.9989	0.9983	0.9973	0.9959
28	1.0000	1.0000	1.0000	0.9999	0.9998	0.9997	0.9995	0.9991	0.9986	0.9978
29	1.0000	1.0000	1.0000	1.0000	0.9999	0.9999	0.9998	0.9996	0.9993	0.9989
30	1.0000	1.0000	1.0000	1.0000	1.0000	0.9999	0.9999	0.9998	0.9997	0.9994
31	1.0000	1.0000	1.0000	1.0000	1.0000	1.0000	1.0000	0.9999	0.9998	0.9997
32	1.0000	1.0000	1.0000	1.0000	1.0000	1.0000	1.0000	1.0000	0.9999	0.9999
33	1.0000	1.0000	1.0000	1.0000	1.0000	1.0000	1.0000	1.0000	1.0000	0.9999
34	1.0000	1.0000	1.0000	1.0000	1.0000	1.0000	1.0000	1.0000	1.0000	1.0000
35	1.0000	1.0000	1.0000	1.0000	1.0000	1.0000	1.0000	1.0000	1.0000	1.0000

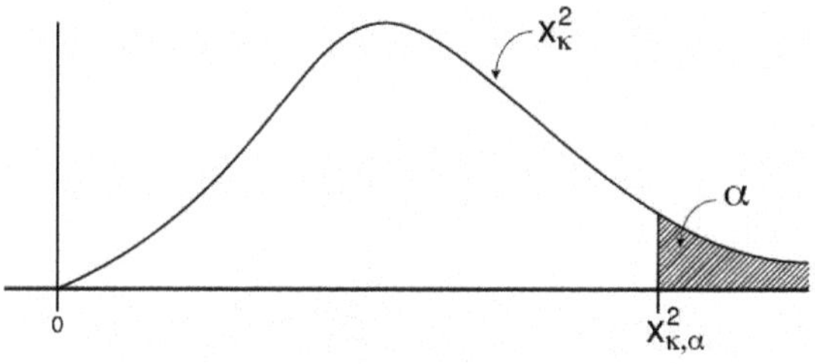

Table A.3. The Chi-Square Distribution

	$P(X \geq \chi^2_{k,\alpha})$, where $X \sim \chi^2_k$							
k	$\chi^2_{k,0.99}$	$\chi^2_{k,0.975}$	$\chi^2_{k,0.95}$	$\chi^2_{k,0.90}$	$\chi^2_{k,0.10}$	$\chi^2_{k,0.05}$	$\chi^2_{k,0.025}$	$\chi^2_{k,0.01}$
1	0.000	0.001	0.004	0.016	2.706	3.841	5.024	6.635
2	0.020	0.051	0.103	0.211	4.605	5.991	7.378	9.210
3	0.115	0.216	0.352	0.584	6.251	7.815	9.348	11.345
4	0.297	0.484	0.711	1.064	7.779	9.488	11.143	13.277
5	0.554	0.831	1.145	1.610	9.236	11.070	12.833	15.086
6	0.872	1.237	1.635	2.204	10.645	12.592	14.449	16.812
7	1.239	1.690	2.167	2.833	12.017	14.067	16.013	18.475
8	1.646	2.180	2.733	3.490	13.362	15.507	17.535	20.090
9	2.088	2.700	3.325	4.168	14.684	16.919	19.023	21.666
10	2.558	3.247	3.940	4.865	15.987	18.307	20.483	23.209
11	3.053	3.816	4.575	5.578	17.275	19.675	21.920	24.725
12	3.571	4.404	5.226	6.304	18.549	21.026	23.337	26.217
13	4.107	5.009	5.892	7.042	19.812	22.362	24.736	27.688
14	4.660	5.629	6.571	7.790	21.064	23.685	26.119	29.141
15	5.229	6.262	7.261	8.547	22.307	24.996	27.488	30.578
16	5.812	6.908	7.962	9.312	23.542	26.296	28.845	32.000
17	6.408	7.564	8.672	10.085	24.769	27.587	30.191	33.409
18	7.015	8.231	9.390	10.865	25.989	28.869	31.526	34.805
19	7.633	8.907	10.117	11.651	27.204	30.144	32.852	36.191
20	8.260	9.591	10.851	12.443	28.412	31.410	34.170	37.566
21	8.897	10.283	11.591	13.240	29.615	32.671	35.479	38.932
22	9.542	10.982	12.338	14.041	30.813	33.924	36.781	40.289
23	10.196	11.689	13.091	14.848	32.007	35.172	38.076	41.638
24	10.856	12.401	13.848	15.659	33.196	36.415	39.364	42.980
25	11.524	13.120	14.611	16.473	34.382	37.652	40.646	44.314
26	12.198	13.844	15.379	17.292	35.563	38.885	41.923	45.642
27	12.879	14.573	16.151	18.114	36.741	40.113	43.195	46.963
28	13.565	15.308	16.928	18.939	37.916	41.337	44.461	48.278
29	14.256	16.047	17.708	19.768	39.087	42.557	45.722	49.588
30	14.953	16.791	18.493	20.599	40.256	43.773	46.979	50.892
40	22.164	24.433	26.509	29.051	51.805	55.758	59.342	63.691
50	29.707	32.357	34.764	37.689	63.167	67.505	71.420	76.154
60	37.485	40.482	43.188	46.459	74.397	79.082	83.298	88.379
70	45.442	48.758	51.739	55.329	85.527	90.531	95.023	100.425
80	53.540	57.153	60.391	64.278	96.578	101.879	106.629	112.329

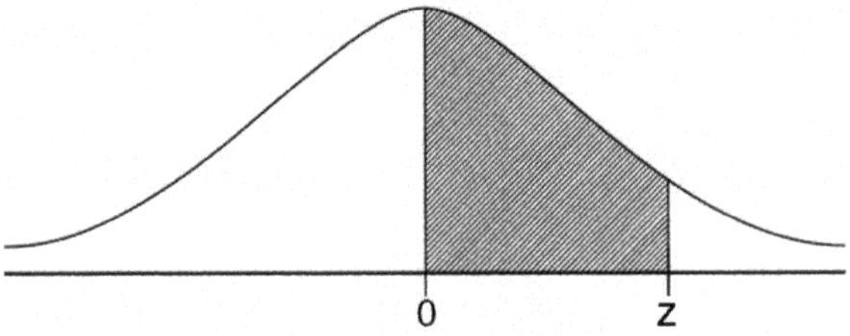

Table A.4. The Normal Distribution
Tabled value is $P(0 \leq Z \leq z)$ where $Z \sim N(0,1)$.

z	0.00	0.01	0.02	0.03	0.04	0.05	0.06	0.07	0.08	0.09
0.0	0.0000	0.0040	0.0080	0.0120	0.0160	0.0199	0.0239	0.0279	0.0319	0.0359
0.1	0.0398	0.0438	0.0478	0.0517	0.0557	0.0596	0.0636	0.0675	0.0714	0.0753
0.2	0.0793	0.0832	0.0871	0.0910	0.0948	0.0987	0.1026	0.1064	0.1103	0.1141
0.3	0.1179	0.1217	0.1255	0.1293	0.1331	0.1368	0.1406	0.1443	0.1480	0.1517
0.4	0.1554	0.1591	0.1628	0.1664	0.1700	0.1736	0.1772	0.1808	0.1844	0.1879
0.5	0.1915	0.1950	0.1985	0.2019	0.2054	0.2088	0.2123	0.2157	0.2190	0.2224
0.6	0.2257	0.2291	0.2324	0.2357	0.2389	0.2422	0.2454	0.2486	0.2517	0.2549
0.7	0.2580	0.2611	0.2642	0.2673	0.2704	0.2734	0.2764	0.2794	0.2823	0.2852
0.8	0.2881	0.2910	0.2939	0.2967	0.2995	0.3023	0.3051	0.3078	0.3106	0.3133
0.9	0.3159	0.3186	0.3212	0.3238	0.3264	0.3289	0.3315	0.3340	0.3365	0.3389
1.0	0.3413	0.3438	0.3461	0.3485	0.3508	0.3531	0.3554	0.3577	0.3599	0.3621
1.1	0.3643	0.3665	0.3686	0.3708	0.3729	0.3749	0.3770	0.3790	0.3810	0.3830
1.2	0.3849	0.3869	0.3888	0.3907	0.3925	0.3944	0.3962	0.3980	0.3997	0.4015
1.3	0.4032	0.4049	0.4066	0.4082	0.4099	0.4115	0.4131	0.4147	0.4162	0.4177
1.4	0.4192	0.4207	0.4222	0.4236	0.4251	0.4265	0.4279	0.4292	0.4306	0.4319
1.5	0.4332	0.4345	0.4357	0.4370	0.4382	0.4394	0.4406	0.4418	0.4429	0.4441
1.6	0.4452	0.4463	0.4474	0.4484	0.4495	0.4505	0.4515	0.4525	0.4535	0.4545
1.7	0.4554	0.4564	0.4573	0.4582	0.4591	0.4599	0.4608	0.4616	0.4625	0.4633
1.8	0.4641	0.4649	0.4656	0.4664	0.4671	0.4678	0.4686	0.4693	0.4699	0.4706
1.9	0.4713	0.4719	0.4726	0.4732	0.4738	0.4744	0.4750	0.4756	0.4761	0.4767
2.0	0.4772	0.4778	0.4783	0.4788	0.4793	0.4798	0.4803	0.4808	0.4812	0.4817
2.1	0.4821	0.4826	0.4830	0.4834	0.4838	0.4842	0.4846	0.4850	0.4854	0.4857
2.2	0.4861	0.4864	0.4868	0.4871	0.4875	0.4878	0.4881	0.4884	0.4887	0.4890
2.3	0.4893	0.4896	0.4898	0.4901	0.4904	0.4906	0.4909	0.4911	0.4913	0.4916
2.4	0.4918	0.4920	0.4922	0.4925	0.4927	0.4929	0.4931	0.4932	0.4934	0.4936
2.5	0.4938	0.4940	0.4941	0.4943	0.4945	0.4946	0.4948	0.4949	0.4951	0.4952
2.6	0.4953	0.4955	0.4956	0.4957	0.4959	0.4960	0.4961	0.4962	0.4963	0.4964
2.7	0.4965	0.4966	0.4967	0.4968	0.4969	0.4970	0.4971	0.4972	0.4973	0.4974
2.8	0.4974	0.4975	0.4976	0.4977	0.4977	0.4978	0.4979	0.4979	0.4980	0.4981
2.9	0.4981	0.4982	0.4982	0.4983	0.4984	0.4984	0.4985	0.4985	0.4986	0.4986
3.0	0.4987	0.4987	0.4987	0.4988	0.4988	0.4989	0.4989	0.4989	0.4990	0.4990
3.1	0.4990	0.4991	0.4991	0.4991	0.4992	0.4992	0.4992	0.4992	0.4993	0.4993
3.2	0.4993	0.4993	0.4994	0.4994	0.4994	0.4994	0.4994	0.4995	0.4995	0.4995
3.3	0.4995	0.4995	0.4995	0.4996	0.4996	0.4996	0.4996	0.4996	0.4996	0.4997
3.4	0.4997	0.4997	0.4997	0.4997	0.4997	0.4997	0.4997	0.4997	0.4997	0.4998

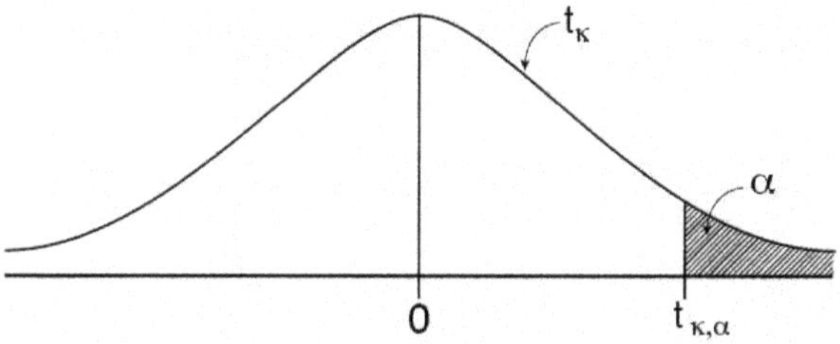

Table A.5. The t Distribution

$P(T \geq t_{k,\alpha}) = \alpha$, where $T \sim t_k$				
k	$t_{k,0.05}$	$t_{k,0.025}$	$t_{k,0.01}$	$t_{k,0.005}$
1	6.314	12.706	31.821	63.657
2	2.920	4.303	6.965	9.925
3	2.353	3.182	4.541	5.841
4	2.132	2.776	3.747	4.604
5	2.015	2.571	3.365	4.032
6	1.943	2.447	3.143	3.707
7	1.895	2.365	2.998	3.499
8	1.860	2.306	2.896	3.355
9	1.833	2.262	2.821	3.250
10	1.812	2.228	2.764	3.169
11	1.796	2.201	2.718	3.106
12	1.782	2.179	2.681	3.055
13	1.771	2.160	2.650	3.012
14	1.761	2.145	2.624	2.977
15	1.753	2.131	2.602	2.947
16	1.746	2.120	2.583	2.921
17	1.740	2.110	2.567	2.898
18	1.734	2.101	2.552	2.878
19	1.729	2.093	2.539	2.861
20	1.725	2.086	2.528	2.845
21	1.721	2.080	2.518	2.831
22	1.717	2.074	2.508	2.819
23	1.714	2.069	2.500	2.807
24	1.711	2.064	2.492	2.797
25	1.708	2.060	2.485	2.787
26	1.706	2.056	2.479	2.779
27	1.703	2.052	2.473	2.771
28	1.701	2.048	2.467	2.763
29	1.699	2.045	2.462	2.756
30	1.697	2.042	2.457	2.750
∞	1.645	1.960	2.326	2.576

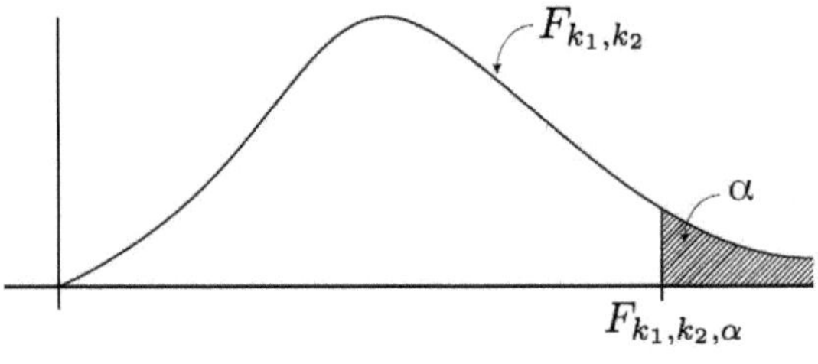

Table A.6. The F Distribution.

$F_{k_1,k_2,.10}$

	$k_1=1$	2	3	4	5	6	7	8	9	10
$k_2=1$	39.86	49.50	53.59	55.83	57.24	58.20	58.91	59.44	59.86	60.19
2	8.53	9.00	9.16	9.24	9.29	9.33	9.35	9.37	9.38	9.39
3	5.54	5.46	5.39	5.34	5.31	5.28	5.27	5.25	5.24	5.23
4	4.54	4.32	4.19	4.11	4.05	4.01	3.98	3.95	3.94	3.92
5	4.06	3.78	3.62	3.52	3.45	3.40	3.37	3.34	3.32	3.30
6	3.78	3.46	3.29	3.18	3.11	3.05	3.01	2.98	2.96	2.94
7	3.59	3.26	3.07	2.96	2.88	2.83	2.78	2.75	2.72	2.70
8	3.46	3.11	2.92	2.81	2.73	2.67	2.62	2.59	2.56	2.54
9	3.36	3.01	2.81	2.69	2.61	2.55	2.51	2.47	2.44	2.42
10	3.29	2.92	2.73	2.61	2.52	2.46	2.41	2.38	2.35	2.32
11	3.23	2.86	2.66	2.54	2.45	2.39	2.34	2.30	2.27	2.25
12	3.18	2.81	2.61	2.48	2.39	2.33	2.28	2.24	2.21	2.19
13	3.14	2.76	2.56	2.43	2.35	2.28	2.23	2.20	2.16	2.14
14	3.10	2.73	2.52	2.39	2.31	2.24	2.19	2.15	2.12	2.10
15	3.07	2.70	2.49	2.36	2.27	2.21	2.16	2.12	2.09	2.06
16	3.05	2.67	2.46	2.33	2.24	2.18	2.13	2.09	2.06	2.03
17	3.03	2.64	2.44	2.31	2.22	2.15	2.10	2.06	2.03	2.00
18	3.01	2.62	2.42	2.29	2.20	2.13	2.08	2.04	2.00	1.98
19	2.99	2.61	2.40	2.27	2.18	2.11	2.06	2.02	1.98	1.96
20	2.97	2.59	2.38	2.25	2.16	2.09	2.04	2.00	1.96	1.94
21	2.96	2.57	2.36	2.23	2.14	2.08	2.02	1.98	1.95	1.92
22	2.95	2.56	2.35	2.22	2.13	2.06	2.01	1.97	1.93	1.90
23	2.94	2.55	2.34	2.21	2.11	2.05	1.99	1.95	1.92	1.89
24	2.93	2.54	2.33	2.19	2.10	2.04	1.98	1.94	1.91	1.88
25	2.92	2.53	2.32	2.18	2.09	2.02	1.97	1.93	1.89	1.87
26	2.91	2.52	2.31	2.17	2.08	2.01	1.96	1.92	1.88	1.86
27	2.90	2.51	2.30	2.17	2.07	2.00	1.95	1.91	1.87	1.85
28	2.89	2.50	2.29	2.16	2.06	2.00	1.94	1.90	1.87	1.84
29	2.89	2.50	2.28	2.15	2.06	1.99	1.93	1.89	1.86	1.83
30	2.88	2.49	2.28	2.14	2.05	1.98	1.93	1.88	1.85	1.82
40	2.84	2.44	2.23	2.09	2.00	1.93	1.87	1.83	1.79	1.76
50	2.81	2.41	2.20	2.06	1.97	1.90	1.84	1.80	1.76	1.73
60	2.79	2.39	2.18	2.04	1.95	1.87	1.82	1.77	1.74	1.71
100	2.76	2.36	2.14	2.00	1.91	1.83	1.78	1.73	1.69	1.66
200	2.73	2.33	2.11	1.97	1.88	1.80	1.75	1.70	1.66	1.63
1000	2.71	2.31	2.09	1.95	1.85	1.78	1.72	1.68	1.64	1.61

$F_{k_1,k_2,.05}$

	$k_1 = 1$	2	3	4	5	6	7	8	9	10
$k_2 = 1$	161.45	199.50	215.71	224.58	230.16	233.99	236.77	238.88	240.54	241.88
2	18.51	19.00	19.16	19.25	19.30	19.33	19.35	19.37	19.38	19.40
3	10.13	9.55	9.28	9.12	9.01	8.94	8.89	8.85	8.81	8.79
4	7.71	6.94	6.59	6.39	6.26	6.16	6.09	6.04	6.00	5.96
5	6.61	5.79	5.41	5.19	5.05	4.95	4.88	4.82	4.77	4.74
6	5.99	5.14	4.76	4.53	4.39	4.28	4.21	4.15	4.10	4.06
7	5.59	4.74	4.35	4.12	3.97	3.87	3.79	3.73	3.68	3.64
8	5.32	4.46	4.07	3.84	3.69	3.58	3.50	3.44	3.39	3.35
9	5.12	4.26	3.86	3.63	3.48	3.37	3.29	3.23	3.18	3.14
10	4.96	4.10	3.71	3.48	3.33	3.22	3.14	3.07	3.02	2.98
11	4.84	3.98	3.59	3.36	3.20	3.09	3.01	2.95	2.90	2.85
12	4.75	3.89	3.49	3.26	3.11	3.00	2.91	2.85	2.80	2.75
13	4.67	3.81	3.41	3.18	3.03	2.92	2.83	2.77	2.71	2.67
14	4.60	3.74	3.34	3.11	2.96	2.85	2.76	2.70	2.65	2.60
15	4.54	3.68	3.29	3.06	2.90	2.79	2.71	2.64	2.59	2.54
16	4.49	3.63	3.24	3.01	2.85	2.74	2.66	2.59	2.54	2.49
17	4.45	3.59	3.20	2.96	2.81	2.70	2.61	2.55	2.49	2.45
18	4.41	3.55	3.16	2.93	2.77	2.66	2.58	2.51	2.46	2.41
19	4.38	3.52	3.13	2.90	2.74	2.63	2.54	2.48	2.42	2.38
20	4.35	3.49	3.10	2.87	2.71	2.60	2.51	2.45	2.39	2.35
21	4.32	3.47	3.07	2.84	2.68	2.57	2.49	2.42	2.37	2.32
22	4.30	3.44	3.05	2.82	2.66	2.55	2.46	2.40	2.34	2.30
23	4.28	3.42	3.03	2.80	2.64	2.53	2.44	2.37	2.32	2.27
24	4.26	3.40	3.01	2.78	2.62	2.51	2.42	2.36	2.30	2.25
25	4.24	3.39	2.99	2.76	2.60	2.49	2.40	2.34	2.28	2.24
26	4.23	3.37	2.98	2.74	2.59	2.47	2.39	2.32	2.27	2.22
27	4.21	3.35	2.96	2.73	2.57	2.46	2.37	2.31	2.25	2.20
28	4.20	3.34	2.95	2.71	2.56	2.45	2.36	2.29	2.24	2.19
29	4.18	3.33	2.93	2.70	2.55	2.43	2.35	2.28	2.22	2.18
30	4.17	3.32	2.92	2.69	2.53	2.42	2.33	2.27	2.21	2.16
40	4.08	3.23	2.84	2.61	2.45	2.34	2.25	2.18	2.12	2.08
50	4.03	3.18	2.79	2.56	2.40	2.29	2.20	2.13	2.07	2.03
60	4.00	3.15	2.76	2.53	2.37	2.25	2.17	2.10	2.04	1.99
100	3.94	3.09	2.70	2.46	2.31	2.19	2.10	2.03	1.97	1.93
200	3.89	3.04	2.65	2.42	2.26	2.14	2.06	1.98	1.93	1.88
1000	3.85	3.00	2.61	2.38	2.22	2.11	2.02	1.95	1.89	1.84

$F_{k_1,k_2,.01}$

	$k_1 = 1$	2	3	4	5	6	7	8	9	10
$k_2 = 1$	4052.18	4999.50	5403.35	5624.58	5763.65	5858.99	5928.36	5981.07	6022.47	6055.85
2	98.50	99.00	99.17	99.25	99.30	99.33	99.36	99.37	99.39	99.40
3	34.12	30.82	29.46	28.71	28.24	27.91	27.67	27.49	27.35	27.23
4	21.20	18.00	16.69	15.98	15.52	15.21	14.98	14.80	14.66	14.55
5	16.26	13.27	12.06	11.39	10.97	10.67	10.46	10.29	10.16	10.05
6	13.75	10.92	9.78	9.15	8.75	8.47	8.26	8.10	7.98	7.87
7	12.25	9.55	8.45	7.85	7.46	7.19	6.99	6.84	6.72	6.62
8	11.26	8.65	7.59	7.01	6.63	6.37	6.18	6.03	5.91	5.81
9	10.56	8.02	6.99	6.42	6.06	5.80	5.61	5.47	5.35	5.26
10	10.04	7.56	6.55	5.99	5.64	5.39	5.20	5.06	4.94	4.85
11	9.65	7.21	6.22	5.67	5.32	5.07	4.89	4.74	4.63	4.54
12	9.33	6.93	5.95	5.41	5.06	4.82	4.64	4.50	4.39	4.30
13	9.07	6.70	5.74	5.21	4.86	4.62	4.44	4.30	4.19	4.10
14	8.86	6.51	5.56	5.04	4.69	4.46	4.28	4.14	4.03	3.94
15	8.68	6.36	5.42	4.89	4.56	4.32	4.14	4.00	3.89	3.80
16	8.53	6.23	5.29	4.77	4.44	4.20	4.03	3.89	3.78	3.69
17	8.40	6.11	5.18	4.67	4.34	4.10	3.93	3.79	3.68	3.59
18	8.29	6.01	5.09	4.58	4.25	4.01	3.84	3.71	3.60	3.51
19	8.18	5.93	5.01	4.50	4.17	3.94	3.77	3.63	3.52	3.43
20	8.10	5.85	4.94	4.43	4.10	3.87	3.70	3.56	3.46	3.37
21	8.02	5.78	4.87	4.37	4.04	3.81	3.64	3.51	3.40	3.31
22	7.95	5.72	4.82	4.31	3.99	3.76	3.59	3.45	3.35	3.26
23	7.88	5.66	4.76	4.26	3.94	3.71	3.54	3.41	3.30	3.21
24	7.82	5.61	4.72	4.22	3.90	3.67	3.50	3.36	3.26	3.17
25	7.77	5.57	4.68	4.18	3.85	3.63	3.46	3.32	3.22	3.13
26	7.72	5.53	4.64	4.14	3.82	3.59	3.42	3.29	3.18	3.09
27	7.68	5.49	4.60	4.11	3.78	3.56	3.39	3.26	3.15	3.06
28	7.64	5.45	4.57	4.07	3.75	3.53	3.36	3.23	3.12	3.03
29	7.60	5.42	4.54	4.04	3.73	3.50	3.33	3.20	3.09	3.00
30	7.56	5.39	4.51	4.02	3.70	3.47	3.30	3.17	3.07	2.98
40	7.31	5.18	4.31	3.83	3.51	3.29	3.12	2.99	2.89	2.80
50	7.17	5.06	4.20	3.72	3.41	3.19	3.02	2.89	2.78	2.70
60	7.08	4.98	4.13	3.65	3.34	3.12	2.95	2.82	2.72	2.63
100	6.90	4.82	3.98	3.51	3.21	2.99	2.82	2.69	2.59	2.50
200	6.76	4.71	3.88	3.41	3.11	2.89	2.73	2.60	2.50	2.41
1000	6.66	4.63	3.80	3.34	3.04	2.82	2.66	2.53	2.43	2.34

Table A.7*. Quantiles of the Wilcoxon Rank-Sum Statistic.

n	α	$m=2$	3	4	5	6	7	8	9	10	11	12	13	14	15	16	17	18	19	20
2	0.005	3	3	3	3	3	3	3	3	3	3	3	3	3	3	3	3	3	4	4
	0.01	3	3	3	3	3	3	3	3	3	3	3	4	4	4	4	4	4	5	5
	0.025	3	3	3	3	3	3	4	4	4	5	5	5	5	5	5	6	6	6	6
	0.05	3	3	3	4	4	4	5	5	5	5	6	6	7	7	7	7	8	8	8
3	0.005	6	6	6	6	6	6	6	7	7	7	8	8	8	9	9	9	9	10	10
	0.01	6	6	6	6	6	7	7	8	8	8	9	9	9	10	10	11	11	11	12
	0.025	6	6	6	7	8	8	9	9	10	10	11	11	12	12	13	13	14	14	15
	0.05	6	7	7	8	9	9	10	11	11	12	12	13	14	14	15	16	16	17	18
4	0.005	10	10	10	10	11	11	12	12	13	13	14	14	15	16	16	17	17	18	19
	0.01	10	10	10	11	12	12	13	14	14	15	16	16	17	18	18	19	20	20	21
	0.025	10	10	11	12	13	14	15	15	16	17	18	19	10	11	21	21	23	14	25
	0.05	10	11	11	13	14	15	16	17	18	19	10	21	21	23	25	16	27	28	29
5	0.005	15	15	15	16	17	17	18	19	20	21	21	23	23	24	25	16	27	28	29
	0.01	15	15	16	17	18	19	20	21	21	23	14	25	26	27	28	19	30	31	32
	0.025	15	16	17	18	19	11	21	23	24	25	27	28	29	30	31	33	34	35	36
	0.05	16	17	18	20	21	21	24	25	27	28	19	31	32	34	35	36	38	39	41
6	0.005	21	21	21	23	24	25	26	27	28	29	31	32	33	34	35	37	38	39	40
	0.01	21	21	23	24	25	26	28	29	30	31	33	34	35	37	38	40	41	41	44
	0.025	21	23	24	25	27	28	30	32	33	35	36	38	39	41	43	44	46	47	49
	0.05	21	24	25	27	29	30	32	34	36	38	39	41	43	45	47	48	50	52	54
7	0.005	28	28	29	30	32	33	35	36	38	39	41	42	44	45	47	48	50	51	53
	0.01	28	19	30	32	33	35	36	38	40	41	43	45	46	48	50	52	53	55	57
	0.025	28	30	32	34	35	37	39	41	43	45	47	49	51	53	55	57	59	61	63
	0.05	29	31	33	35	37	40	42	44	46	48	50	53	55	57	59	62	64	66	68
8	0.005	36	36	38	39	41	43	44	46	48	50	52	54	55	57	59	61	63	65	67
	0.01	36	37	39	41	43	44	46	48	50	52	54	56	59	61	63	65	67	69	71
	0.025	37	39	41	43	45	47	50	52	54	56	59	61	63	66	68	71	73	75	78
	0.15	38	40	42	45	47	50	52	55	57	60	63	65	68	70	73	76	78	81	84
9	0.005	45	46	47	49	51	53	55	57	59	62	64	66	68	70	73	75	77	79	82
	0.01	45	47	49	51	53	55	57	60	61	64	67	69	72	74	77	79	82	84	86
	0.025	46	48	50	53	56	58	61	63	66	69	72	74	77	80	83	15	88	91	94
	0.05	47	50	52	55	58	61	64	67	70	73	76	79	82	85	88	91	94	97	100
10	0.005	55	56	58	60	62	65	67	69	73	74	77	80	82	85	87	90	93	95	98
	0.01	55	57	59	62	64	67	69	73	75	78	80	83	86	89	92	94	97	100	103
	0.025	56	59	61	64	67	70	73	76	79	82	85	89	92	95	98	101	104	108	111
	0.05	57	60	63	67	70	73	76	80	83	87	90	93	97	100	104	107	111	114	118
11	0.005	66	67	69	73	74	77	80	83	15	88	91	94	97	100	103	106	109	112	115
	0.01	66	68	71	74	76	79	82	85	89	92	95	98	101	104	108	111	114	117	120
	0.025	67	70	73	76	80	83	86	90	93	97	100	104	107	111	114	118	122	125	129
	0.05	68	73	75	79	83	86	90	94	98	101	105	109	113	117	121	124	128	132	136
12	0.005	78	80	82	85	88	91	94	97	100	103	106	110	113	116	120	123	126	130	133
	0.01	78	81	84	87	90	93	96	100	103	107	110	114	117	121	125	128	132	135	139
	0.025	80	83	86	90	93	97	101	105	108	112	116	120	124	128	132	136	140	144	148
	0.05	81	84	88	92	96	100	105	109	111	117	121	126	130	134	139	143	147	151	156
13	0.005	91	93	95	99	102	105	109	112	116	119	123	126	130	134	137	141	145	149	152
	0.01	92	94	97	101	104	108	112	115	119	123	127	131	135	139	143	147	151	155	159
	0.025	93	96	100	104	108	112	116	120	125	129	133	137	142	146	151	155	159	164	168
	0.05	94	98	102	107	111	116	120	125	129	134	139	143	148	153	157	162	167	172	176
14	0.005	105	107	110	113	117	121	124	128	132	136	140	144	148	152	156	160	164	169	173
	0.01	106	108	112	116	119	123	128	132	136	140	144	149	151	157	162	166	171	175	179
	0.025	107	111	115	119	123	128	132	137	142	146	151	156	161	165	170	175	180	184	189
	0.05	109	111	117	122	127	132	137	142	147	152	157	162	167	172	177	183	188	193	198
15	0.005	120	123	126	129	133	137	141	145	150	154	158	163	167	172	176	111	185	190	194
	0.01	121	124	128	132	136	140	145	149	154	158	163	168	172	172	182	187	191	196	201
	0.025	122	126	131	135	140	145	150	155	160	165	170	175	180	115	191	196	201	206	211
	0.05	124	128	133	139	144	149	154	160	165	171	176	182	187	193	198	204	209	215	221

n	α	$m=2$	3	4	5	6	7	8	9	10	11	12	13	14	15	16	17	18	19	20
16	0.005	136	139	142	146	150	155	159	164	168	173	178	182	187	192	197	202	207	211	216
	0.01	137	140	144	149	153	158	163	168	172	178	183	188	193	198	203	208	213	219	224
	0.025	138	143	148	152	158	163	168	174	179	184	190	196	201	207	212	218	223	229	235
	0.05	140	145	151	156	162	167	173	179	185	191	197	202	208	214	220	226	232	238	244
17	0.005	153	156	160	164	169	173	178	183	188	193	198	203	208	214	219	224	229	235	240
	0.01	154	158	162	167	172	177	182	187	192	198	203	209	214	220	225	231	236	242	247
	0.025	156	160	165	171	176	182	188	193	199	205	211	217	223	229	235	241	247	253	259
	0.05	157	163	169	174	180	187	193	199	205	211	218	224	231	237	243	250	256	263	269
18	0.005	171	174	178	183	188	193	198	203	209	214	219	225	230	236	242	247	253	259	264
	0.01	172	176	181	186	191	196	202	208	213	219	225	231	237	242	248	254	260	266	272
	0.025	174	179	184	190	196	202	208	214	220	227	233	239	246	252	258	265	271	278	284
	0.05	176	181	188	194	200	207	213	220	227	233	240	247	254	260	267	274	281	288	295
19	0.005	191	194	198	203	208	213	219	224	230	236	242	248	254	260	265	272	278	284	290
	0.01	192	195	200	206	211	217	223	229	235	241	247	254	260	266	273	279	285	292	298
	0.025	193	198	204	210	216	223	229	236	243	249	256	263	269	276	283	290	297	304	310
	0.05	195	201	208	214	221	228	235	242	249	256	263	271	278	285	292	300	307	314	321
20	0.005	211	214	219	224	229	235	241	247	253	259	265	271	278	284	290	297	303	310	316
	0.01	212	216	221	227	233	239	245	251	258	264	271	278	284	291	298	304	311	318	325
	0.025	213	219	225	231	238	245	251	259	266	273	280	287	294	301	309	316	323	330	338
	0.05	215	222	229	236	243	250	258	265	273	280	288	295	303	311	318	326	334	341	349

$*$ Adapted, with permission, from Table 7, Conover, W.J. (1999) *Practical Nonparametric Statistics*, 3rd Edition, New York, John Wiley and Sons.

Table A.8. Probability Models

I. Discrete Models

1.1. **Hypergeometric**: $X \sim HG(N,r,n)$

pmf: For non-negative integers N, r, and n satisfying $\max(r,n) \leq N$,

$$p(x) = \frac{\binom{r}{x}\binom{N-r}{n-x}}{\binom{N}{n}} \text{ for } \max(0,n+r-N) \leq x \leq \min(r,n).$$

mean and variance: $E(X) = nr/N$ and $V(X) = nr(N-r)(N-n)/N^2(N-1)$.

mgf: not available in closed form

Notes: This model is appropriate when drawing a random sample of size n, without replacement, from a finite dichotomous population of size N. The population is viewed as consisting of r individuals of "type R" and $N-r$ individuals that are not of type R. The random variable X represents the number of type R individuals drawn in a sample of size n under the "simple random sampling" assumption that all possible samples of size n have the same probability of being chosen.

1.2. **Bernoulli**: $X \sim B(1,p)$

pmf: For $p \in (0,1)$, with $q = 1-p$,

$$p(x) = p^x q^{1-x} \text{ for } x = 0,1.$$

mean and variance: $E(X) = p$ and $V(X) = pq$.

mgf: $m_X(t) = q + pe^t$ for $-\infty < t < \infty$.

Notes: The model applies to a single random draw from a finite or infinite dichotomous population separated into an "S" group (associated with the word "success") and an "F" group (associated with the word "failure"). It is assumed that the proportion of the population belonging to the S group is p, that is, $p = P(S)$. A single experiment yielding a Bernoulli random variable is called a Bernoulli trial. A sequence of independent, identically distributed Bernoulli trials may be obtained either from repeated draws, with replacement, from a finite population for which the proportion of successes is p or from repeated draws from an infinite population for which the probability of success in a single draw is p.

1.3. **Binomial**: $X \sim B(n,p)$

pmf: For $p \in (0,1)$, with $q = 1-p$, and for any fixed positive integer n,

$$p(x) = \binom{n}{x} p^x q^{n-x}, \text{ for } x = 0,1,2,\ldots,n.$$

mean and variance: $E(X) = np$ and $V(X) = npq$.

mgf: $m_X(t) = (q + pe^t)^n$ for $-\infty < t < \infty$.

Notes: If $X_1,\ldots,X_n$ are a sequence of independent Bernoulli trials with probability p of success, then $X = \sum_{i=1}^n X_i$ represents the number of successes in n trials and has the $B(n,p)$ distribution. It can be shown that when N is large and the parameter p is set equal to r/N, the hypergeometric distribution $HG(N,rn)$ is well approximated by the $B(n,p)$

distribution. (See Theorem 2.5.4.)

1.4. **Geometric**: $X \sim G(p)$

pmf: For $p \in (0,1)$, with $q = 1-p$,

$$p(x) = p \cdot q^{x-1} \text{ for } x = 1,2,3,\ldots$$

mean and variance: $E(X) = 1/p$ and $V(X) = q/p^2$.

mgf: $m_X(t) = pe^t/(1-qe^t)$ for $t < \ln(1/q)$

Notes: If $X_1,\ldots,X_n,\ldots$ are a sequence of independent Bernoulli trials with probability p of success, and X is defined as the number of trials that are required to obtain the first success, then $X \sim G(p)$. An alternative form of the geometric model may be obtained as the distribution of Y, the number of failures that occur before the first success. Since $Y = X-1$, Y has pmf $p(y) = p \cdot q^y$ for $y = 0,1,2,\ldots$. The mean and variance of Y are $E(Y) = q/p$, $V(Y) = q/p^2$ and the mgf of Y is $m_y(t) = p/(1-qe^t)$ for $t < \ln(1/q)$. The geometric distribution is the unique discrete distribution with the memoryless property: if $X \sim G(p)$, then $P(X > x+k|X > k) = P(X > x)$ for arbitrary positive integers x and k.

1.5. **Negative Binomial**: $X \sim NB(r,p)$

pmf: For $p \in (0,1)$, with $q = 1-p$, and for any fixed positive integer r,

$$p(x) = \binom{x-1}{r-1} p^r q^{x-r} \text{ for } x = r, r+1, r+2, \ldots.$$

mean and variance: $E(x) = r/p$ and $V(X) = r \cdot q/p^2$.

mgf: $m_X(t) = [pe^t/(1-qe^t)]^r$ for $t < \ln(1/q)$.

Notes: Consider an experiment yielding a sequence of independent Bernoulli trials, and suppose that the experiment is terminated when the rth success is obtained. If X is the number of trials required to obtain the rth success, then X has the $NB(r,p)$ distribution. If $X_1,\ldots,X_r \overset{\text{iid}}{\sim} G(p)$ and $X = \sum_{i=1}^r X_i$, then $X \sim NB(r,p)$. This result follows from the fact that when X_i is viewed as the number of trials required, after the $(i-1)^{st}$ success is obtained, to obtain the ith success in a sequence of independent Bernoulli trials with probability p of success, then $X = \sum_{i=1}^r X_i$ is clearly the total number of trials required to obtain the rth success. An alternative form of the negative binomial model is obtained as the distribution of $Y_{r,p} = X_{r,p} - r$, where $X_{r,p} \sim NB(r,p)$, and $Y_{r,p}$ represents the number of failures that occur before the rth success. It can be shown (see Theorem 5.2.3) that if $r \to \infty, q \to 0$ and $rq \to \lambda > 0$, then $Y_{r,p} \xrightarrow{D} P(\lambda)$, the Poisson distribution with mean λ. (See 1.6 below.)

1.6. **Poisson**: $X \sim P(\lambda)$

pmf: For $\lambda > 0, p(x) = \frac{\lambda^x e^{-\lambda}}{x!}$ for $x = 0, 1, 2, \ldots$.

mean and variance: $E(X) = \lambda$ and $V(X) = \lambda$.

mgf: $m_x(t) = e^{\lambda(e^t - 1)}$ for $-\infty < t < \infty$.

Notes: The wide range of applications which are well modeled by the Poisson distribution is due, in large part, to the Law of Rare Events (see Theorem 2.7.3) which states that if $X_{n,p} \sim B(n,p)$ and if $n \to \infty$, $p \to 0$ and $np \to \lambda > 0$, then $X_{n,p} \xrightarrow{D} P(\lambda)$. In other words, if "success" is a rare event (with a small probability p of occurrence in a single trial) and if the number n of independent trials of the experiment is sufficiently large, then the number of successes in n trials is approximately distributed as a Poisson random variable with mean $\lambda = np$.

1.7. **Multinomial**: $(X_1, \ldots, X_k) \sim M_k(N, (p_{1,\ldots,}p_k)$

pmf: For $\mathbf{p} = (p_1, p_2, \ldots, p_k)$, with $p_i \geq 0 \ \forall$ i, for which $\sum_{i=1}^k p_i = 1$, and for any fixed positive integer n,

$$p(x_1, \ldots, x_k) = \frac{n!}{\prod_{i=1}^k x_i!} \prod_{i=1}^k p_i^{x_i} \text{ for } \mathbf{x} \text{ such that } x_i \text{ is a positive integer for all i and} \sum_{i=1}^k x_i = n,$$

marginal means, variances, and covariances: For $i = 1, \ldots, k$, $E(X_i) = np_i$ and $V(X_i) = np_i q_i$ and, for $i \neq j, Cov(X_i, X_j) = -np_i p_j$.

mgf: $m_{X_1,\ldots,X_{k-1}}(t_1, \ldots, t_{k-1}) = E\left(e^{\sum_{i=1}^{k-1} t_i X_i}\right) = \left(1 - \sum_{i=1}^{k-1} p_i + \sum_{i=1}^{k-1} p_i e^{t_i}\right)^n$.

Notes: The multinomial model is the natural extension of the binomial distribution to experiments with k possible outcomes, where $k > 2$. If is evident that $B(n,p) = M_2(n, (p, 1-p))$. The chi-square goodness-of-fit tests for multinomial models is treated in Section 10.6.

II. Continuous models.

2.1. **Uniform**: $X \sim U(a,b)$

pdf: $f(x) = 1/(b-a) I_{(a,b)}(x)$.

cdf: $F(x) = (x-a)/(b-a)$ for $x \in (a,b)$.

mean and variance: $E(X) = (a+b)/2$ and $V(X) = (b-a)^2/12$.

mgf: $m_X(t) = (e^{bt} - e^{at})/(bt - at)$.

Notes: The uniform distribution is sometimes referred to as the rectangular distribution due to its density's shape. It plays a key role in computer simulations of data from a continuous, strictly increasing distribution function F. If the random variable X is drawn from such an F, that is, if $X \sim F$, then $F(X) \sim U(0,1)$. If U is a computer-generated random decimal, that is, $U \sim U(0,1)$, then $F^{-1}(U) \sim F$.

2.2. **Beta**: $X \sim Be(\alpha,\beta)$

pdf: For $\alpha > 0$ and $\beta > 0$, $f(x) = \frac{\Gamma(\alpha+\beta)}{\Gamma(\alpha)\Gamma(\beta)} x^{\alpha-1}(1-x)^{\beta-1} I_{(0,1)}(x)$.

cdf: For general α and β, $F(x)$ has no closed form expression. However, if $X \sim Be(k, n-k+1)$ for positive integers $k \le n$, then $F(x) = \sum_{i=k}^{n} \binom{n}{i} x^i (1-x)^{n-i}$ for x $\in (0,1)$.

mean and variance: $E(X) = \alpha/(\alpha+\beta)$ and $V(X) = \alpha\beta/(\alpha+\beta)^2(\alpha+\beta+1)$.

mgf: Can only be expressed as an infinite series.

Notes: The Beta distribution on the interval $(0,1)$ is a flexible model that can take a variety of shapes—bell-shaped, U-shaped, increasing, decreasing and flat. The latter distribution is the $U(0,1)$ model and occurs when $\alpha = \beta = 1$. If $X_1, \ldots, X_n$ is a random sample from $U(0,1)$, then the order statistics $X_{(1)}, \ldots, X_{(n)}$ have Beta marginal distributions, that is, $X_{(i)} \sim Be(i, n-i+1)$ for $i = 1, \ldots, n$. The beta distribution is the standard conjugate prior distribution for the binomial model. If $p \sim Be(\alpha,\beta)$ and $X|p \sim B(n,p)$, then $p|X = x \sim Be(\alpha+x, \beta+n-x)$.

2.3. **Exponential**: $X \sim Exp(\lambda)$

pdf: For $\lambda > 0$, $f(x) = \lambda e^{-\lambda x} I_{(0,\infty)}(x)$.

cdf: $F(x) = 1 - e^{-\lambda x}$ for $x > 0$.

mean and variance: $E(X) = 1/\lambda$ and $V(X) = 1/\lambda^2$.

mgf: $m_X(t) = (1 - t/\lambda)^{-1}$.

Notes: When parameterized as above, the parameter λ is called the failure rate of the Exp(λ) model. It is easily verified that $f(x)/(1-F(x)) = \lambda$. The exponential distribution is the unique continuous distribution with the memoryless property: if $X \sim Exp(\lambda)$, then for arbitrary $x > 0$ and $t > 0$, $P(X > x+t | X > t) = P(X > x)$. If $X \sim Exp(\lambda)$, then for $\theta \in (-\infty, \infty)$, the variable $Y = X + \theta$ has the translated exponential distribution with density function $f(y) = \lambda e^{-\lambda(x-\theta)} I_{(\theta,\infty)}(y)$.

2.4. **Gamma**: $X \sim \Gamma(\alpha,\beta)$

pdf: For $\alpha > 0$ and $\beta > 0$, $f(x) = \frac{1}{\Gamma(\alpha)\beta^\alpha} x^{\alpha-1} e^{-\frac{x}{\beta}} I_{(0,\infty)}(x)$.

cdf: In general, the cdf $F(x)$ of the $\Gamma(\alpha,\beta)$ distribution cannot be obtained in closed form. However, if the shape parameter $\alpha = k$, a positive integer, and $X \sim \Gamma(k,\beta)$, then (see Section 3.4),

$$F_X(x) = 1 - \sum_{i=0}^{k-1} \frac{(x/\beta)^i e^{-x/\beta}}{i!}.$$

mean and variance: $E(X) = \alpha\beta$ and $V(X) = \alpha\beta^2$.

mgf: $m_X(t) = (1 - \beta t)^{-\alpha}$.

Notes: The exponential distribution Exp(λ) is a special case of the gamma model, that is, $Exp(\lambda) = \Gamma(1, 1/\lambda)$. The general gamma model is often used as a model for lifetimes. It can be shown that the failure rate $f(x)/(1-F(x))$ of the $\Gamma(\alpha,\beta)$ model is decreasing for $\alpha \le 1$ and is increasing for $\alpha \ge 1$. The three-parameter or translated gamma model is denoted by $\Gamma(\alpha,\beta,\theta)$, with shape parameter α, scale parameter β, location parameter θ. Its density function is given by

$$f(x) = \frac{1}{\Gamma(\alpha)\beta^\alpha}(x-\theta)^{\alpha-1} e^{-\frac{(x-\theta)}{\beta}} I_{(\theta,\infty)}(x).$$

If $\{X(t), t > 0\}$ is a Poisson process with rate λ, then the waiting time X_k for the occurrence of the kth event, that is, the time required to first observe $X(t) = k$ has the $\Gamma(k, 1/\lambda)$ distribution (see Section 3.4). The gamma distribution is the standard conjugate prior distribution for the Poisson model. If $\lambda \sim \Gamma(\alpha, \beta)$ and $X|\lambda \sim P(\lambda)$, then $\lambda|X = x \sim \Gamma(\alpha + x, \beta/(\beta + 1))$. The most widely used form of the gamma model is the chi-square distribution displayed below.

2.5. **Chi-Square with** k **Degrees of Freedom**: $X \sim \chi^2_k$ (or $X \sim \Gamma(k/2, 2)$).
pdf: $f(x) = \frac{1}{\Gamma(k/2) \cdot 2^{k/2}} x^{(k/2)-1} e^{-\frac{x}{2}} I_{(0,\infty)}(x)$.
cdf: See notes above for the gamma model.
mean and variance: $E(X) = k$ and $V(X) = 2k$.
mgf: $m_X(t) = (1 - 2t)^{-k/2}$.

Notes: The chi-square distribution often arises as the distribution of a sum of squared random variables. For example, if $X_1, \ldots, X_n \overset{\text{iid}}{\sim} N(0, 1)$, then $\sum_{i=1}^n X_i^2 \sim \chi^2_n$. It is also well known that if $X_1, \ldots, X_n \overset{\text{iid}}{\sim} N(\mu, \sigma^2)$, then $\sum_{i=1}^n (X_i - \bar{X})^2/\sigma^2 = (n-1)s^2/\sigma^2 \sim \chi^2_{n-1}$, where s^2 is the sample variance. The chi-square distribution arises prominently in linear model theory as the distribution of certain sums of squares under suitable assumptions.

2.6. **Normal**: $X \sim N(\mu, \sigma^2)$

pdf: $f(x) = \frac{1}{\sqrt{2\pi}\sigma} e^{-\frac{(x-\mu)^2}{2\sigma^2}} I_{(-\infty,\infty)}(x)$.

cdf: The cdf of the normal distribution cannot be obtained in closed form, but the values of $F_Z(z) = P(Z \leq z)$ are given in Table A.4 for $0 \leq z \leq 3.49$, where $Z \sim N(0, 1)$, the standard normal distribution, with z rounded to 2 decimal places. For $z < 0, P(Z \leq z) = P(Z \geq -z)$. If $X \sim N(\mu, \sigma^2)$, then $(X - \mu)/\sigma \sim N(0, 1)$; thus, $F_X(x) = F_Z((x - \mu)/\sigma)$ may also be obtained from Table A.4.

mean and variance: $E(X) = \mu$ and $V(X) = \sigma^2$.

mgf: $m_X(t) = e^{\mu t + \frac{\sigma^2 t^2}{2}}$.

Notes: Often called the Gaussian distribution, the normal model is perhaps the most widely used parametric model in statistical modeling and inference. Its frequent occurrence in nature is often attributed to the Central Limit Theorem which states that, under mild conditions, sample averages are approximately normally distributed. When observed data may be considered to result as averages of a large numbers of small perturbations, their approximate normality may be anticipated. In the statistical subfield of Linear Model Theory, random errors are often modeled as independent, identically distributed normal variables. The normal distribution is the standard conjugate prior distribution for the normal distribution. If $\mu \sim N(v, \tau^2)$ and $X|\mu \sim N(\mu, \sigma_0^2)$, where σ_0^2 is known, then $\mu|X = x \sim N\left(\frac{v\sigma_0^2 + \tau^2 \sum_{i=1}^n x_i}{\sigma_0^2 + n\tau^2}, \frac{\sigma_0^2 + n\tau^2}{\sigma_0^2 \tau^2}\right)$.

2.7. **Double Exponential**: $X \sim DE(\lambda, \mu)$
pdf: For $\mu \in (-\infty, \infty)$ and $\lambda > 0$, $f(x) = \frac{\lambda}{2} e^{-\lambda|x-\mu|} I_{(-\infty,\infty)}(x)$.

cdf: $F(x) = 1 - \frac{1}{2}e^{-\lambda(x-\mu)}$ if $x \geq \mu$ and $F(x) = \frac{1}{2}e^{\lambda(x-\mu)}$ if $x < \mu$.

mean and variance: $E(X) = \mu$ and $V(X) = 2/\lambda^2$.

mgf: $m_X(t) = e^{\mu t}/(1-(t/\lambda)^2)$ for $|t| < \lambda$.

Notes: The double exponential model is sometimes used in place of the normal model when an application calls for a symmetric distribution with heavier tails than the normal.

2.8. **Lognormal**: $Y \sim LN(\mu, \sigma^2)$

pdf: $f(y) = (1/\sqrt{2\pi\tau})(1/y)exp\{-(1/2\tau)(lny - \mu)^2\}I_{(0,\infty)}(y)$,

cdf: Does not exist in closed form.

moments: For any positive integer k, $E(Y^k) = exp\{k\mu + k^2\sigma^2/2\}$.

mgf: Does not exist.

Notes: If $X \sim N(\mu, \sigma^2)$, then $Y = e^X \sim LN(\mu, \sigma^2)$. This explains why moments of Y are so easy to obtain: If $X \sim N(\mu, \sigma^2)$, then $EY^k = E(e^{kX}) = m_X(t)|_{t=k}$. The lognormal distribution gets its name from the fact that if $Y \sim LN(\mu, \sigma^2)$, then $\ln Y \sim N(\mu, \sigma^2)$. If $Y_1, \ldots Y_n \sim LN(\mu, \sigma^2)$, then the Central Limit Theorem implies that the sample geometric mean $GM(Y_1, \ldots, Y_n) = \left(\prod_{i=1}^{n} Y_i\right)^{1/n}$ has a lognormal distribution. This motivates the use of the lognormal distribution for random processes that are multiplicative rather than additive as is, for example, compound interest.

2.9. **Pareto**: $X \sim P(\alpha, \theta)$

pdf: For $\alpha > 0$ and $\theta > 0$, $f(x) = \alpha\theta^\alpha/x^{\alpha+1}I_{(\theta,\infty)}(x)$.

cdf: $F(x) = 1 - (\theta/x)^\alpha$ for $x > \theta$.

mean and variance: If $\alpha > 1$, $E(X) = \alpha\theta/(\alpha - 1)$; if $\alpha > 2, V(X) = \alpha\theta^2/(\alpha-1)^2(\alpha-2)$.

mgf: Does not exist.

Notes: The Pareto distribution has proven useful in modeling economic variables like income. It has also used as a lifetime model in applications in engineering and biostatistics, especially when a "heavy-tailed distribution" is called for. An alternative form of the Pareto model is sometimes referred to as the Lomax distribution. If $X \sim P(\alpha, \theta_0)$, with θ_0 known and $Y = X - \theta_0$, then Y has the Lomax distribution with density $f(y) = \left[\alpha/(1+y)^{\alpha+1}\right] I_{(0,\infty)}(y)$. If $X_1, X_2, \ldots, X_n \overset{\text{iid}}{\sim} P(\alpha, \theta)$, then the minimum order statistic $X_{(1)}$ has the $P(n\alpha, \theta)$ distribution.

2.10. **Weibull**: $X \sim W(\alpha, \beta)$

pdf: $f(x) = \frac{\alpha x^{\alpha-1}}{\beta^\alpha} e^{-(x/\beta)^\alpha} I_{(0,\infty)}(x)$.

cdf: $F(x) = 1 - e^{-(x/\beta)^\alpha}$ for $x > 0$.

mean and variance: $EX = \beta\Gamma(\frac{\alpha+1}{\alpha})$ and $V(X) = \beta^2[\Gamma(1+\frac{2}{\alpha}) - \Gamma^2(1+\frac{1}{\alpha})]$.

mgf: Can be obtained in closed form only in special cases.

Notes: The Weibull distribution is widely used as a lifetime distribution in engineering reliability and biostatistics. The model has the distinction of being one of the three "extreme value" distributions. This accounts for the motivation that is often given of the Weibull model as the appropriate lifetime distribution of a system that tends to fail due to the failure of its weakest component. The failure rate of a $W(\alpha, \beta)$ variable is given by $r(t) = f(t)/1 - F(t) = \alpha t^{\alpha-1}/\beta^\alpha$. The Weibull distribution has a decreasing failure rate (DFR) if $\alpha \leq 1$ and has an increasing failure rate (IFR) if $\alpha \geq 1$. When $\alpha = 1$, the Weibull variable has the $\Gamma(1, \beta)$ distribution. If $X \sim W(\alpha, \beta)$, then $X^\alpha \sim \Gamma(1, \beta^\alpha)$. The $W(2, \beta)$ model is referred to as the Rayleigh distribution.

2.11. **Cauchy**: $X \sim C(m, \theta)$

pdf: $f(x) = \frac{1}{\pi\theta\left[1+\left(\frac{x-m}{\theta}\right)^2\right]} I_{(-\infty,\infty)}(x)$.

cdf: $F(x) = \frac{1}{\pi}\left[\arctan(x-m)/\theta + \frac{\pi}{2}\right]$, where $\arctan y$ is the angle α, measured in radians, in the interval $(-\pi/2, \pi/2)$ for which $\tan\alpha = y$.

mean and variance: The mean and variance do not exist.

mgf: The mgf of the Cauchy distribution does not exist.

Notes: The Cauchy distribution is perhaps the most famous example of a distribution which does not satisfy the conditions of the Central Limit Theorem. Indeed, it can be shown that if $X_1, \ldots, X_n \overset{\text{iid}}{\sim} C(m, \theta)$, then the sample mean $\bar{X}_n$ has a Cauchy distribution for any integer n

2.12. **Logistic**: $X \sim L(\mu, \beta)$

pdf: $f(x) = \frac{1}{\beta} \frac{e^{-(x-\mu)/\beta}}{[1+e^{-(x-\mu)/\beta}]^2} I_{(-\infty,\infty)}(x)$.

cdf: $F(x) = \frac{1}{1+e^{-(x-\mu)/\beta}}$ for $x \in (-\infty, \infty)$.

mean and variance: $E(X) = \mu$ and $V(X) = \frac{\pi^2\beta^2}{3}$.

mgf: $m_X(t) = e^{\mu t}\Gamma(1-\beta t)\Gamma(1+\beta t)$ for $|t| < \frac{1}{\beta}$.

Notes: The logistic distribution cdf is often used to model growth, and is used as the basis for modeling the relationship between a dichotomous variable Y and a continuous variable X by positing that $P(Y = 1|X = x) = 1/(1+e^{-(x-\mu)/\beta})$ for $x \in (-\infty, \infty)$.

2.13. t **with** k **degrees of freedom**: $X \sim t_k$

pdf: $f(t) = \frac{\Gamma((k+1)/2)}{\Gamma(k/2)\sqrt{\pi k}(1+t^2/k)^{k+1/2}}$ for $-\infty < t < \infty$.

cdf: Can not be obtained in closed form.

mean and variance: $EX = 0$ for $k > 1$, $V(X) = k/(k+2)$ for $k > 2$.

mgf: Does not exist.

Notes: If U and V are independent random variables, with $U \sim N(0,1)$ and $V \sim \chi^2_k$, then the variable $t = U/\sqrt{V/k} \sim t_k$. If $X_1,\ldots,X_n \sim^{iid} N(\mu,\sigma^2)$, then $(\bar{X}-\mu)/(s/\sqrt{n}) \sim t_{n-1}$.

2.14. F **with** k **and** r **degrees of freedom**: $X \sim F_{k,r}$

pdf: $f_X(x) = \frac{\Gamma((k+r)/2)(k/r)^{k/2}}{\Gamma(k/2)\Gamma(r/2)} \frac{x^{k/2-1}}{\left(1+\frac{kx}{r}\right)^{(k+r)/2}} I_{(0,\infty)}(x)$.

cdf: Cannot be obtained in closed form.

mean and variance: If $X_{k,r} \sim F_{k,r}$, then

$$EX_{k,r} = \frac{r}{r-2} \text{ for } r > 2 \text{ and } V(X_{k,r}) = 2\left(\frac{r}{r-2}\right)^2 \frac{(k+r-2)}{k(r-4)} \text{ for } r > 4.$$

mgf: Does not exist.

Notes: If U and V are independent chi-square variables, with $U \sim \chi^2_k$ and $V \sim \chi^2_r$, then the variable $X = (U/k)/(V/r) \sim F_{k,r}$.

2.15. **Bivariate Normal:** $\binom{X}{Y} \sim N_2(\mu,\Sigma)$, where the mean vector μ and covariance matrix Σ are given by $\mu = \binom{\mu_x}{\mu_y}$ and $\Sigma = \begin{bmatrix} \sigma_x^2 & \rho\sigma_x\sigma_y \\ \rho\sigma_x\sigma_y & \sigma_y^2 \end{bmatrix}$.

pdf: $f_{X,Y}(x,y) = \frac{1}{2\pi\sigma_x\sigma_y\sqrt{1-\rho^2}} e^{-\left(\frac{(x-\mu_x)^2}{2\sigma_x^2}+\rho\left(\frac{x-\mu_x}{\sigma_x}\right)\left(\frac{y-\mu_y}{\sigma_y}\right)+\frac{(y-\mu_y)^2}{2\sigma_y^2}\right)/(1-\rho^2)} I_{(-\infty,\infty)\times(-\infty,\infty)}(x,y)$.

cdf: Not available in closed form.

means, variances, and covariance: $E(X) = \mu_x$, $V(X) = \sigma_x^2$, $E(Y) = \mu_y$, $V(Y) = \sigma_y^2$, and $Cov(X,Y) = \rho\sigma_x\sigma_y$.

mgf: $m_{(X,Y)} = E(e^{tX+sY}) = e^{\mu_X t + \mu_Y s + (\sigma_X^2 t^2 + 2\rho\sigma_X\sigma_Y st + \sigma_Y^2 s^2)/2}$.

Notes: If $\binom{X}{Y} \sim N_2(\mu,\Sigma)$, then X and Y are independent if and only if $\rho = 0$. Further,

$$X \sim N(\mu_x,\sigma_x^2) \text{ and } X|Y=y \sim N\left(\mu_x + \rho\frac{\sigma_x}{\sigma_y}(y-\mu_y), \sigma_x^2(1-\rho^2)\right),$$

and similarly,

$$Y \sim N(\mu_y,\sigma_y^2) \text{ and } Y|X=x \sim N\left(\mu_y + \rho\frac{\sigma_y}{\sigma_x}(x-\mu_x), \sigma_y^2(1-\rho^2)\right).$$

Table A.9. Answers to Selected Exercises

Chapter 1.

Exercises 1.1

1. (b) $\{2,6,10\}$; (c) $\{4,8\}$; (e) $\{6\}$; (f) $\{5,7\}$; (g) $\{6\}$; (h) $\{1,2,4,8,9,10\}$; (i) $\{5,7\}$.
2. Draw three overlapping circles. Notice that you could color seven distinct areas with seven different colored crayons. Label the areas 1–7, with the set of $A-(B\cup C)$, as number 1, etc.
4. To prove $A\cap B^c \subseteq A-B$, let $x\in A\cap B^c$. Then $x\in A$ and $x\notin B$, so that $x\in A-B$. To prove $A-B\subseteq A\cap B^c$, reverse these steps.
5. To prove $A\cup B\subseteq A\cup(B-A)$, let $x\in A\cup B$. Then $x\in A$ or $x\in B$, or both. If $x\in A$, then clearly $x\in A\cup(B-A)$. If $x\notin A$, then since $x\in A\cup B$, x must be in B, so that $x\in B-A$. Thus, $x\in A\cup(B-A)$, completing the proof. To prove $A\cup(B-A)\subseteq A\cup B$, reverse the steps.
6. (i) $B\subseteq \bigcup_{i=1}^{n}(B\cap A_i)$: let $x\in B$. Since $\{A_i, i=1,\ldots,n\}$ is a partition of the universe U, and $B\subseteq U$, $x\in A_j$ for some $j\Rightarrow x\in B\cap A_j \Rightarrow x\in \bigcup_{i=1}^{n}(B\cap A_i)$. (ii) $\bigcup_{i=1}^{n}(B\cap A_i)\subseteq B$: reverse these steps.

Exercises 1.2

1. The sample space is $\{HHH\}, \{HHT\}, \{HTH\}, \{THH\}, \{HTT\}, \{THT\}, \{TTH\}$ and $\{TTT\}$. Model these simple events as equally likely. (a) $3/8$, (b) $3/8$, (c) $1/4$.
2. (a) 1/4, (b) 1/2, (c) 6/36, (d) 5/18, (e) 5/9.
3. (a) 1/4, (b) 3/13, (c) 11/26.
4. For Hamilton, $P(X<3)=11/24$. For Madison, $P(X>2)=0$. The evidence favors Hamilton.

Exercises 1.3

1. Assume $A\subseteq B$. Then, write $B=A\cup(B-A)$. Since $A\cap(B-A)=\emptyset$, it follows that $P(B)=P(A)+P(B-A)$. Thus, $P(B-A)=P(B)-P(A)$ holds when $A\subseteq B$.
2. $P(A\cup B\cup C)=P(A\cup B)+P(C)-P((A\cup B)\cap C)$ (by Thm. 1.3.5),

$$P(A)+P(B)-P(A\cap B)+P(C)-P((A\cup B)\cap C)(\text{by } Thm.\ 1.3.5),$$
$$=P(A)+P(B)+P(C)-P(A\cap B)-P((A\cap C)\cup(B\cap C))$$
$$(\text{by distributive law of }\cap\text{ with respect to }\cup)$$
$$=P(A)+P(B)+P(C)-P(A\cap B)-P(A\cap C)-P(B\cap C)+P(A\cap B\cap C)$$
$$(\text{by } Thm.\ 1.3.5).$$

3. To show that $\bigcup_{i=1}^{k}B_i\subseteq\bigcup_{i=1}^{k}A_i$, note that the definition of B_i implies that $B_i\subseteq A_i$. It follows that $\bigcup_{i=1}^{k}B_i\subseteq\bigcup_{i=1}^{k}A_i$. To show that $\bigcup_{i=1}^{k}A_i\subseteq\bigcup_{i=1}^{k}B_i$, let $x\in\bigcup_{i=1}^{k}A_i$. Then $x\in A_i$ for some i. Let j be the smallest index for which $x\in A_j$. Since $x\in A_j-\bigcup_{i=1}^{j-1}A_i\Rightarrow x\in B_j$, $x\in\bigcup_{i=1}^{k}B_i$. The same argument holds if $k=\infty$.
4. (a) 7/8, (b) 1/16, (c) 1/7.
5. (a) 1/17, (b) 8/17.
6. 0.60.

Exercises 1.4

1. $P(A^c|B) = \frac{P(A^c \cap B)}{P(B)} = \frac{P(B) - P(A \cap B)}{P(B)} = 1 - \frac{P(A \cap B)}{P(B)} = 1 - P(A|B)$.

2. $P(A \cup B|C) = \frac{P((A \cup B) \cap C)}{P(C)} = \frac{P((A \cap C) \cup (B \cap C))}{P(C)} = \frac{P(A \cap C) + P(B \cap C) - P(A \cap B \cap C)}{P(C)}$
$= P(A|C) + P(B|C) - P(A \cap B|C)$.

3. $P(W_2) = P(W_1 \cap W_2) + P(R_1 \cap W_2) = 0 + 1/5 = 1/5$, while $= P(W_2|R_1)$
$= P((R_1 \cap W_2)/P(R_1) = (1/5)/(4/5) = 1/4$.

4. (a) P(Spade on first draw) $= 1/4$.
(b) P(spade on second draw) $= (1/4)/(12/51) + (3/4)(13/51) = 1/4$.
Intuition: With no information about the outcomes of any earlier draws, the probability of a spade on the kth draw is neither larger nor smaller that the probability of drawing any other suit on the kth draw.

5. (a) 5/18, (b) 13/18, (c) 3/10.

6. 0.7772.

7. (a) 23/360, (b) 12/23.

Exercises 1.5

1. Draw a tree with three paths $M \cap D$, $M \cap D^c$, and $M^c \cap D^c$ and respective probabilities .72, .08, and .2. Then, compute $P(M^c|D^c) = (.2)/(.28) = 5/7$.

2. Draw a tree with three paths $K \cap R$, $K^c \cap R$, and $K^c \cap R^c$ and respective probabilities .8, .04, and .16. Then, compute $P(K|R) = (.8)/(.84) = 20/21$.

3. Let $R =$ "Rifle used". $P(T|R) = 8/13$, $P(J|R) = 4/13$, $P(S|R) = 1/13$.

4. If $I =$ "inebriated" and $L =$ "light on," then $P(I|L) = (.42)/(.48) = 7/8$.

5. 15/23.

6. 1/13.

Exercises 1.6

1. If $P(A|B) = P(A) > 0$, then $P(B|A) = \frac{P(B \cap A)}{P(A)} = \frac{P(B)P(A|B)}{P(A)} = \frac{P(B)P(A)}{P(A)} = P(B)$.

2. 8/27.

3. The events are independent.

4. 43/64.

5. The event "W wins" is independent of the number of initial consecutive ties. Thus, the problem reduces to the calculation $P(W \text{ wins} \mid \text{either } W \text{ or } B \text{ wins})$ $= p_2/(p_2 + p_3)$.

Exercises 1.7

1. $2^{n+1} - 2$.

2. There are 6 possible committees with T and B serving, and 6 with U and P. There are $\binom{8}{3} = 56$ possible committees, so there are $56 - 12 = 44$ committees with people who will work together.

3. (a) 7.06×10^{-4}; (b) 1.228×10^{-5}; (c) 4.356×10^{-8}.

4. .02596.

Chapter 2.

Exercises 2.1

1. $p(x) = (2x-1)/36$ for $x = 1,2,3,4,5,6$.
2. $P(x = x) = (1/2)^{x+1}$ for $x = 0,1,2,3,\ldots$; $P(Y = y) = (1/2)^y$ for $y = 1,2,3,\ldots$; $X = Y - 1$.
3. $p(0) = 1/6$, $p(1) = 1/2$, $p(2) = 3/10$, $p(3) = 1/30$.
4. $p(0) = 1/16 = p(4)$, $p(1) = 1/4 = p(3)$, $p(2) = 3/8$.

Exercises 2.2

1. $EZ = 4.472$, $V(Z) = 1.971$.
2. $p(2) = 15/28$, $p(21) = 12/28$, $p(40) = 1/28$; $E(W) = 11.5$.
3. $EX = 3.33$, $V(X) = 2.22$.
4. \$4.00.
5. Let $W =$ winnings. $EW = 3(.391) - 2(.609) = -.045$. Game is unfavorable to you.

Exercises 2.3

1. $p(0) = 14/99$, $p(1) = 224/495$, $p(2) = 168/495$, $p(3) = 32/495$, $p(4) = 1/495$; $EX = 4/3$.
2. $p(0) = 35/120$, $p(1) = 63/120$, $p(2) = 21/120$, $p(3) = 1/120$; $EX = .9$, $V(X) = .49$.
3. $P(X \leq 2) = .98136$, $p(X = 3) = .01765$, $P(X = 6) = 7.15 \times 10^{-8}$.
5. Use the trick to evaluate $EX(X-1) = A$. Then, get $V(X)$ as $A + nr/N + (nr/N)^2$.

Exercises 2.4

1. $P(1)$ is obvious. Assume $P(n)$. Then $\sum_{i=1}^{n+1}(2i-1) = n^2 + [2(n+1)-1] = n^2 + 2n + 1 = (n+1)^2$, that is, $P(n+1)$ is true. Thus, $P(n)$ is true for all positive integers n.
3. $\sum_{i=0}^{n}\binom{n}{k} = \sum_{i=0}^{n}\binom{n}{k}(1)^k(1)^{n-k} = (1+1)^n = 2^n$.
4. $\sum_{i=0}^{n}\binom{n}{k}(-1)^k = \sum_{i=0}^{n}\binom{n}{k}(-1)^k(1)^{n-k} = (1-1)^n = 0$.

Exercises 2.5

1. (a) .032, (b) .264.
2. $P(X \leq 18) = .9264$, $P(X = 12) = .0760$, $P(15 \leq X \leq 20) = .5763$.
3. $P(X \leq 2) = .6856$, $P(Y \leq 2) = .678$.
4. This expectation can be written as

$$\sum_{x=k}^{n}\frac{(n-k)!}{(x-k)!(n-x)!}p^x q^{n-x} = p^k \sum_{x=k}^{n}\binom{n-k}{x-k}p^{x-k}q^{n-x} = p^k \sum_{y=0}^{n-k}\binom{n-k}{y}p^y q^{n-k-y} = p^k.$$

Exercises 2.6

2. If $S_\infty = \sum_{i=0}^{\infty}A^i$, then $1 + AS_\infty = 1 + \left(\sum_{i=1}^{\infty}A^i\right) = \sum_{i=0}^{\infty}A^i = S_\infty$.
3. (a) 1/24, (b) 2/33.
4. 10/19 or .5263.

5. .4275.

Exercises 2.7

1. Use the trick to show $EX(X-1)=\lambda^2$. Then, $V(X)=EX(X-1)+EX-(EX)^2$ $=\lambda^2+\lambda-\lambda^2$.

2. (a) .1512, (b) .08392, (c) From Table A.2 with $\lambda=6$,

$$P(0)+P(2)+\cdots+P(18)+.0001=.4999.$$

3. .5578.

4. .8649.

Exercises 2.8

1. $e^t(1-e^{nt})/n(1-e^t)$

2. $m_Y(t)=E(e^{tY})=E(e^{t(aX+b)})=e^{bt}E(e^{atX})=e^{bt}m_x(at)$.

3. Let $m(t)$ be the mgf of X, and let $R(t)=\ln m(t)$. Then $R'(t)=m'(t)/m(t)$, so that $R'(0)=m'(0)/m(0)=EX/1=EX$. Further, $R''(t)=m''(t)m(t)-(m'(t))^2/m^2(t)$ $=m''(t)/m(t)-(m'(t))^2/m^2(t)$ and $R''(0)=m''(0)m(0)-(m'(0))^2/m^2(0)$ $=EX^2-(EX)^2=V(X)$.

4. $EX=6$, $V(x)=12$.

5. $p(1)=.3$, $p(3)=.4$, $P(5)=.3$.

Chapter 3.

Exercises 3.1

1. $k=1/4$.

3. $k=.1875$.

4. $k=2$.

5. $k=\alpha\theta^\alpha$.

Exercises 3.2

1. $Ee^{tx}=\int_0^\theta e^{tx}\frac{1}{\theta}dx=\left[\frac{e^{tx}}{t\theta}\right]_{x=0}^{\theta}=\frac{e^{t\theta}-1}{t\theta}$.

2. $EX=(a+b)/2$, $V(X)=(b-a)^2/12$.

3. $m_Y(t)=E(e^{tY})=E(e^{t(aX+b)})=e^{bt}E(e^{atX})=m_x(at)$.

4. $I=1/20$.

5. (a) $k=\alpha+1$; (b) $EX=(\alpha+1)/(\alpha+2)$; (c) $m=[1/2]^{1/(\alpha+1)}$.

6. $EX=\frac{1}{\lambda}-\frac{te^{-\lambda t}}{1-e^{-\lambda t}}$.

Exercises 3.3

1. $f(x)=(\alpha+1)x^\alpha I_{(0,1)}(x)$ and $F(x)=x^{\alpha+1}$ for $0<x<1$.

2. $F(x)=5x^4-4x^5$ for $0<x<1$.

3. $F(x)=1-1/x^\theta$ for $x\geq 1$.

4. (a) $F(x)$ is increasing, $F(0)=0$ and $F(1)=1$; (b) $f(x)=1/\pi(1+x^2)I_{(-\infty,\infty)}(X)$; (c) $x_0=\tan(\pi/4)$.

5. $\int_{-\infty}^{\infty} h(x)dx = c\int_{-\infty}^{\infty} f(x)dx + (1-c)\int_{-\infty}^{\infty} g(x)dx = c+(1-c)=1$.

Exercises 3.4

1. $N \sim G(p)$, with $p=1=e^{-\lambda}$.

2. $EX^{1/2} = \frac{\Gamma(\alpha+1/2)\theta^{1/2}}{\Gamma(\alpha)}$.

3. $P(X \leq 10) = .4405$; Note that $X \sim \chi^2_{10}$, so that $x_0 = \chi^2_{10,.05} = 18.307$.

4. The mode of χ^2_k is $k-2$.

Exercises 3.5

1. .99; .896; .0548; .3632; .00964; .95.

2. .877; .7939; .0668; .3085; .1539; .7392.

3. 1.28; 1.65; 1.96; 2.17; 2.33; 2.58.

4. Pee Wee weights 444 pounds.

5. If $u=z$ and $dv=ze^{-z^2/2}$, then

$$\frac{1}{\sqrt{2\pi}}\int_{-\infty}^{\infty} z^2 e^{-z^2/2}dz = \frac{1}{\sqrt{2\pi}}\left[-ze^{-z^2/2}\right]_{-\infty}^{\infty} + \frac{1}{\sqrt{2\pi}}\int_{-\infty}^{\infty} e^{-z^2/2}dz = 0 + \frac{1}{\sqrt{2\pi}}\cdot\sqrt{2\pi} = 1.$$

6. $EX=m'(0)=\mu$ and $EX^2=m''(0)=\sigma^2+\mu^2$. Thus, $V(X)=(\sigma^2+\mu^2)-\mu^2=\sigma^2$.

Exercises 3.6

1. $EX = \frac{\Gamma(\alpha+\beta)\Gamma(\alpha+1)}{\Gamma(\alpha+\beta+1)\Gamma(\alpha)}\cdot 1 = \frac{\alpha}{\alpha+\beta}$; $EX^2 = \frac{\Gamma(\alpha+\beta)\Gamma(\alpha+2)}{\Gamma(\alpha+\beta+2)\Gamma(\alpha)}\cdot 1 = \frac{(\alpha+1)\alpha}{(\alpha+\beta+1)(\alpha+\beta)}$. It follows that $V(X) = \frac{\alpha\beta}{(\alpha+\beta)^2(\alpha+\beta+1)}$.

2. $EX=m'(0)=\mu$ and $EX^2=m''(0)=2/\lambda^2+\mu^2$. Thus, $V(X)=2/\lambda^2$.

3. $EY=E(e^{tX})=m_X(t)$, where $X\sim N(\mu,\sigma^2)$. Thus, $EY=e^{\mu+\sigma^2/2}$.

4. $F(x) = \int_\theta^\infty(\alpha\theta^\alpha/x^{\alpha+1})dx = 1-\theta^\alpha/x^\alpha$; for $\alpha>1$, $EX=\int_\theta^\infty(\alpha\theta^\alpha/x^\alpha)dx$ $=\alpha\theta/(\alpha-1)$; for $\alpha>2$, $EX^2=\frac{\alpha\theta^2}{\alpha-2}$; thus, for $\alpha>2$, $V(X)=\frac{\alpha\theta^2}{(\alpha-1)^2(\alpha-2)}$.

5. $X_{(1)} \sim P(n\alpha,\theta)$.

Chapter 4.

Exercises 4.1

1. $p_X(x) = .4, .4, .2$ for $x=1,2,3$; $p_Y(y) = .1, .2, .4, .3$ for $y=1,2,4,5$; $p_{X|Y=4}(x|4) = .5, .25, .25$ for $x=1,2,3$; $EXY=7$, $EX=1.8$, $EY=3.6$, $E(X|Y=4)=1.75$.

3. (a) $\int_0^1\int_0^y xyf_{X,Y}(x,y)dxdy$, $\int_0^1\int_1^x xyf_{X,Y}(x,y)dydx$; (b) $\int_0^1\int_{y^2}^y xyf_{X,Y}(x,y)dxdy$, $\int_0^1\int_x^{\sqrt{x}} xyf_{X,Y}(x,y)dydx$; (c) $\int_0^\infty\int_0^{1/y} xyf_{X,Y}(x,y)dxdy$, $\int_0^\infty\int_0^{1/x} xyf_{X,Y}(x,y)dydx$; (d) $\int_0^1\int_0^{1-y} xyf_{X,Y}(x,y)dxdy$, $\int_0^1\int_0^{1-x} xyf_{X,Y}(x,y)dydx$; (e) $\int_0^4\int_y^{y+2} xyf_{X,Y}(x,y)dxdy$, $\int_0^2\int_0^x xyf_{X,Y}(x,y)dydx + \int_2^4\int_{x-2}^x xyf_{X,Y}(x,y)dydx + \int_4^6\int_{x-2}^4 xyf_{X,Y}(x,y)dydx$; Note that in part (e) above, the $dxdy$ integration is much simpler than the $dydx$ integration.

4. (a) $f_Y(y) = 5y^4 I_{(0,1)}(y)$, (b) $f_{X|Y=y}(x|y) = 2x/y^2$ for $x \in (0,y)$, (c) $E(X|Y=y) = 2y/3$.

5. $P(X > Y) = 1/3$.

6. (a) $k = 6$; (b) $P(X < Y) = 2/5$.

Exercises 4.2

1. (a) $p_{X,Y}(x,y) = 1/8$ for $(x,y) = (0,0)$, $(0,1)$, $(1,0)$, $(2,1)$, $(1,2)$ or $(2,2)$, $p_{X,Y}(1,1) = 1/4$; (b) $p_X(x) = 1/4,\ 1/2,\ 1/4$ for $x = 0,1,2$ and $p_Y(y) = 1/4,\ 1/2,\ 1/4$ for $y = 0,1,2$; (d) $\rho_{X,Y} = 1/2$.

2. $EXY = 1/2$, $EX = 1/2 = EY$, $cov(X,Y) = 1/4 - (1/2)(1/2) = 0$

3. $V(2X_1) = \theta^2/3$, $V(4X_2/3) = 7\theta^2/81$, $V(-X_1/4 + 3X_2/2) = \theta^2/12$.

4. (a) $X \sim \Gamma(2,3)$, (b) $Y/x|X = x \sim Be(3,1)$, (c) $E(Y|X = x) = 3x/4$, $EY = E(3X/4) = 4.5$, (d) $V(Y) = 12.15$.

Exercises 4.3

1. (a) yes, (b) no, (c) no, (d) yes

2. $m_{\sum_{i=1}^n X_i}(t) = E\left(e^{t\sum_{i=1}^n X_i}\right) = \prod_{i=1}^n Ee^{tX_i} = \prod_{i=1}^n m_{X_i}(t)$.

3. If $X_1,\dots,X_n \overset{\text{iid}}{\sim} F$ with mgf $m(t)$, then $m_{\sum_{i=1}^n X_i}(t) = \prod_{i=1}^n m_{X_i}(t) = (m(t))^n$.

4. If $X_1,\dots,X_r \overset{\text{iid}}{\sim} G(p)$ and $Y = \sum_{i=1}^r X_i$, then $Y \sim NB(r,p)$ and $m_Y(t) = \left(\frac{pe^t}{1-qe^t}\right)^n$.

Exercises 4.4

1. If $(X_1,X_2,X_3,\dots,X_k) \sim M_k(n,(p_1,p_2,\dots,p_k))$, and if classes 1 and 2 are combined, with $Y = X_1 + X_2$ and $p_Y(1) = p_1 + p_2$, then $(Y,X_3,\dots,X_k)$ has the multinomial distribution $M_{k-1}(n,(p_1+p_2,p_3,\dots,p_k))$.

2. For $i \neq j$, $V(X_i + X_j) = n(p_i)(1-p_i) + n(p_j)(1-p_j) - np_ip_j$.

3. 2.4797×10^{-5}.

Exercises 4.5

2. $E(XY) = E(YE(X|Y)) = E[Y\{\mu_x + \rho(\sigma_x/\sigma_y)(Y-\mu_y)\}] = \mu_x\mu_y + \rho(\sigma_x/\sigma_y)[E(Y^2) - \mu_y^2]$ $= \mu_x\mu_y + \rho\sigma_x\sigma_y$. Thus, $Cov(X,Y) = E(XY) - \mu_x\mu_y = \rho\sigma_x\sigma_y$.

4. Since the range of X, given $Y = y$, depends on y, X, and Y are not independent. Since marginal distributions of X and Y are symmetric about the origin, $EX = 0 = EY$. Also, $E(XY) = \int_{-1}^{1}\int_{-\sqrt{1-y^2}}^{\sqrt{1-y^2}} xy\left(\frac{1}{\pi}\right)dxdy = \left[\frac{x^2}{2}\right]_{-\sqrt{1-y^2}}^{\sqrt{1-y^2}} = 0$. Thus, $Cov(X,Y) = E(XY) - (EX)(EY) = 0$.

Exercises 4.6.1

1. $m_Y(t) = m_{cX}(t) = m_X(ct) = (1 - c\beta t)^{-\alpha}$. Thus, $Y \sim \Gamma(\alpha, c\beta)$.

2. $m_X(t) = Ee^{tX} = E(E(e^{tX}|Y)) = E(e^{tY+t^2/2}) = e^{t^2/2}E(e^{tY}) = e^{t^2/2}\cdot e^{t^2/2} = e^{t^2}$. Thus, $X \sim N(0,2)$.

3. $X + Y \sim N(2\mu, 2\sigma^2)$, $X - Y \sim N(0, 2\sigma^2)$;

$$\begin{aligned} m_{X+Y,X-Y}(t_1,t_2) &= E(e^{(t_1+t_2)X+(t_1-t_2)Y}) = Ee^{(t_1+t_2)X}Ee^{(t_1-t_2)Y} \\ &= e^{(t_1+t_2)\mu+\sigma^2(t_1+t_2)^2/2}\cdot e^{(t_1-t_2)\mu+\sigma^2(t_1-t_2)^2/2} \\ &= e^{t_1(2\mu)+2\sigma^2t_1^2/2}\cdot e^{2\sigma^2t_2^2/2} = m_{X+Y}(t_1)\cdot m_{X-Y}(t_2). \end{aligned}$$

4. (a) $Ee^{tX} = E(E(e^{tX}|p) = E(q+pe^t)^n$
$= E_p\left(\sum_{k=0}^{n}\binom{n}{k}q^k(pe^t)^{n-k}\right) = \sum_{k=0}^{n}\binom{n}{k}\frac{\Gamma(k+1)\Gamma(n-k+1)}{\Gamma(n+2)}e^{(n-k)t}$; the last expression reduces to $\sum_{i=0}^{n}\frac{1}{n+1}e^{it}$, the mgf of the uniform distribution on $\{0,1,\ldots,n\}$. (b) Since $f(p|x) \propto p^x(1-p)^{n-x}$, it follows that $f(p|x) = \frac{\Gamma(n+2)}{\Gamma(x+1)\Gamma(n-x+1)}p^x(1-p)^{n-x}I_{(0,1)}(x)$.

Exercises 4.6.2

1. $P(Y \leq y) = P\left(\frac{X}{1-X} \leq y\right) = P\left(X \leq \frac{y}{1+y}\right) = \int_0^{y/(1+y)} 6x(1-x)dx = 3\left(\frac{y}{1+y}\right)^2 - 2\left(\frac{y}{1+y}\right)^3$, for $0 < y < \infty$. From this, one may obtain $f_Y(y) = 6y/(1+y)^4 I_{(0,\infty)}(y)$.
2. $F_Y(y) = P(Y \leq y) = P(X^\alpha \leq y) = P(X \leq y^{1/\alpha}) = F_X(y^{1/\alpha}) = 1 - e^{-y/\beta^\alpha}$. Thus, $Y \sim \Gamma(1, \beta^\alpha)$.
3. Solve $y = 1 - e^{-(x/\beta)^\alpha}$ for x; $x = \beta[-\ln(1-y)]^{1/\alpha}$. Thus, $X = \beta[-\ln(1-Y)]^{1/\alpha} \sim W(\alpha,\beta)$.
4. If $Y \sim U(0,1)$, then $X = \theta/(1-Y)^{1/\alpha} \sim P(\alpha,\theta)$.
5. If $X \sim Be(\theta,1)$, then $Y = X^\theta \sim U(0,1)$.
6. $P(U \leq u) = P(Y \geq X/u) = \int_0^\infty \int_{x/u}^\infty e^{-x-y}dydx = \int_0^\infty e^{-x(1+1/u)}dx = \frac{u}{1+u}$; thus, $f_U(u) = \frac{1}{(1+u)^2}I_{(0,\infty)}(u)$.

Exercises 4.6.3

1. $x = y/(1+y)$, $|J| = 1/(1+y)^2$, $y \in (0,\infty)$; $f_Y(y) = \frac{6y}{(1+y)^4}I_{(0,\infty)}(y)$.
2. $x = y^{1/\alpha}$, $|J| = (1/\alpha)y^{(1/\alpha)-1}$, $y \in (0,\infty)$; $f_Y(y) = (1/\beta^\alpha)e^{-y/\beta^\alpha}I_{(0,\infty)}(y)$.
3. $x = 1/y$, $|J| = 1/y^2$, $y \in (-\infty,\infty)$; $f_Y(y) = 1/\pi(1+y^2)$.
4. $x = uv$, $y = v$, $|J| = v$, $(u,v) \in (0,\infty)^2$; $f(u,v) = ve^{-uv}e^{-v}I_{(0,\infty)^2}(u,v)$; $f_U(u) = 1/(1+u)^2 I_{(0,\infty)}(u)$.
5. $x = u/v$, $y = v$, $|J| = 1/v$, $0 < u < v < 1$, $f(u,v) = 8uv$ for $0 < u < v < 1$; $f_U(u) = 4(u-u^3)I_{(0,1)}(u)$.
6. $x = (u+v)/2$, $y = (u-v)/2$, $|J| = 1/2$, $(u,v) \in (-\infty,\infty)^2$; $f(u,v) \propto e^{-u^2/4} \cdot e^{-v^2/4}$. Thus, U and V are independent, with $U \sim N(0,2)$ and $V \sim N(0,2)$.
7. $x = u$, $y = v - u$, $|J| = 1$, $0 < u < v < 1$, $f(u,v) = 6u$ for $0 < u < v < 1$, $f_U(u) = 6u(1-u)I_{(0,1)}(u)$.

Exercises 4.7

2. $X_{(1)} \sim P(n\alpha, \theta)$.

3. $X_{(1)} \sim Exp(n\lambda)$.

4. $f_{X_{(1)},X_{(n)}}(x_{(1)},x_{(n)}) = \frac{n(n-1)}{2^n}\left(x_{(n)} - x_{(1)}\right)^{n-2}$ for $\theta - 1 < x_{(1)} < x_{(n)} < \theta + 1$; $P(X_{(1)} < \theta < X_{(n)})$ may be obtained by integrating this joint density over the set $\{(x_{(1)},x_{(n)})|x_{(1)} < \theta < x_{(n)}),$ but can also be calculated as $P(X_{(1)} < \theta < X_{(n)}) = 1 - P(X_{(1)} \geq \theta) - P(X_{(n)} \leq \theta) = 1 - (1/2)^n - (1/2)^n = 1 - (1/2)^{n-1}$.

Chapter 5.

Exercises 5.1

1. If $A = \{x : |x - \mu| \geq \varepsilon\}$, then, $\sigma^2 \geq \sum_{x \in A}(x-\mu)^2 p(x) \geq \varepsilon^2 P(|X - \mu| \geq \varepsilon)$. Theorem 5.1.1. follows.

2. $P(|X - \mu| \geq a) = P((X - \mu)^2 \geq a^2) \leq \sigma^2/a^2$ (by (5.2)); setting $a = \varepsilon$, one obtains $P(|X - \mu| \geq \varepsilon) \leq \sigma^2/\varepsilon^2$.

3. By (5.2) and the assumption that $T_n \xrightarrow{qm} c$, $P((T_n - c)^2 \geq \varepsilon^2) \leq E(T_n - c)^2/\varepsilon^2 \to 0$ as $n \to \infty$. Thus, if $T_n \xrightarrow{qm} c$, then for any $\varepsilon > 0$, $P(|T_n - c| \geq \varepsilon) \to 0$ as $n \to \infty$, that is, $T_n \xrightarrow{p} c$.

4. If $X \sim P(\theta, \alpha)$, with $\alpha > 2$, then $P(|X - \mu| > 2\sigma) = P(X < \mu - 2\sigma) + P(X > \mu + 2\sigma)$ $= 0 + \left(\frac{\alpha - 1}{\alpha + 2\sqrt{\alpha/(\alpha-2)}}\right)^{\alpha}$. It can be seen by plotting the function in R that this is way smaller than Chebyshev's bound 0.25.

Exercises 5.2

1. (a) Choose an arbitrary $\varepsilon > 0$. Then $P(X_n \leq t) = P(X_n \leq t, X \leq t + \varepsilon) + P(X_n \leq t, X > t + \varepsilon)$. Thus sum is $\leq P(X \leq t + \varepsilon) + P(|X_n - X| > \varepsilon)$. Similarly, $P(X \leq t - \varepsilon) \leq P(X_n \leq t) + P(|X_n - X| > \varepsilon)$. Combining these, and using the fact that $X_n \xrightarrow{D} X$, it follows that $P(|X_n - X| > \varepsilon) \to 0$. This implies that $P(X \leq t - \varepsilon) \leq \lim\inf P(X_n \leq t) \leq \lim\sup P(X_n \leq t) \leq P(X \leq t + \varepsilon)$. If t is a continuity point of the limiting distribution, then both $P(X \leq t - \varepsilon)$ and $P(X \leq t + \varepsilon)$ converge to $P(X \leq t)$ as $\varepsilon \to 0$. Hence, $P(X_n \leq t) \to P(X \leq t)$. Part (b) follows from part (a).

2. Let t be a continuity point of $X + c$. Then $(t - c)$ is a continuity point of X. Choose an arbitrary $\varepsilon > 0$. Then, we have

$$\begin{aligned} P(X_n + Y_n \leq t) &= P(X_n \leq t - Y_n, Y_n \leq c - \varepsilon) + P(X_n \leq t - Y_n, Y_n > c - \varepsilon) \\ &\leq P(X_n \leq t - c + \varepsilon) + P(|Y_n - c| > \varepsilon) \to P(X + c \leq t + \varepsilon). \end{aligned}$$

Since ε is arbitrary, this limit tends to $P(X + c \leq t)$ as $\varepsilon \to 0$.

4. By the weak law of large numbers, if $X_1, X_2, X_3, \ldots, X_n$ are iid with mean μ and finite variance, then $\bar{X} \xrightarrow{p} \mu$. Here, the population mean is λ. Hence, $\bar{X} \xrightarrow{p} \lambda$, or $\bar{X}/\lambda \xrightarrow{p} 1$. From the Central Limit Theorem (Thm. 5.3.1), we have that $\sqrt{n}(\bar{X} - \lambda)/\sqrt{\lambda} \xrightarrow{D} U \sim N(0,1)$. It follows from Theorem 5.2.1 (iv) that $\sqrt{n}(\bar{X} - \lambda)/\sqrt{\bar{X}} \xrightarrow{D} U$.

Exercises 5.3

1. $P(\bar{X} > 1020) = P(Z > 1.6) = 0.0548$.
2. Let $X \sim B(250, .9)$ and $Y \sim N(225, 22.5)$. Then, $P(X \leq 225) \cong P(Y \leq 225.5)$ $= P(Z \leq .11) = .5438$.
3. $\mu = 1$ and $\sigma^2 = 1.4$. If X_i is the number of students that drop by on day i, then for the entire 140 days, $P\left(\sum_{i=1}^{140} X_i > 120\right) = P(\bar{X} > 6/7) = P\left(Z \geq \frac{.8571-1}{1/10}\right)$ $= P(Z \geq -1.43) = .9236$.
4. The mean and standard deviation are $\mu = 0.8$ and $\sigma = .98$. Therefore, if X_i is the number of suits he sells on day i, then $P\left(\sum_{i-1}^{30} X_i \leq 21\right) = P(\bar{X} \leq .7)$ $\cong P(Z \leq -.56) = .2877$. Note that the continuity correction has been ignored.
5. The mean and standard deviation are $\mu = 0$ and $\sigma = 1.5811$. Then $P(\bar{X} > 1/3)$ $= P(Z \geq 2) = .0228$.

Exercises 5.4

1. If $g(x) = \sqrt{x}$, then $g'(x) = 1/(2\sqrt{x})$. By the δ-method,
$$\sqrt{n}\left(\sqrt{T_n} - \sqrt{\theta}\right) \xrightarrow{D} Y \sim N(0, \theta/4).$$
2. If $g(x) = \ln x$, then $g'(x) = 1/x$. By the δ-method, $\sqrt{n}(\ln T_n - \ln\theta) \xrightarrow{D} Y \sim N(0,1)$.
3. If $g(x) = (x+1)/x$, then $g'(x) = -1/x^2$ and
$$\sqrt{n}((T_n+1)/T_n - (\theta+1)/\theta) \xrightarrow{D} Y \sim N(0, 1/\theta^2).$$
4. If $g(x) = x^3$, then $g'(x) = 3x^2$. By the δ-method,
$$\sqrt{n}\left((\bar{X}_n)^3 - \mu^3\right) \xrightarrow{D} Y \sim N(0, 9\mu^4\sigma^2).$$
5. If $g(x) = \arcsin\sqrt{x}$, then $g'(x) = 1/\sqrt{x(1-x)}$. The δ-method theorem shows that $g(\hat{p})$ is a VST:
$$\sqrt{n}\left(\arcsin\sqrt{\hat{p}} - \arcsin\sqrt{p}\right) \xrightarrow{D} Y \sim N\left(0, \left(\frac{1}{\sqrt{p(1-p)}}\right)^2 p(1-p))\right) = N(0,1).$$

Chapter 6.

Exercises 6.1

1. $E(\hat{\lambda}_1) = EX/2 + EY/6 = \lambda/2 + 3\lambda/6 = \lambda$; $E(\hat{\lambda}_2) = EX/10 + 3EY/10 = \lambda/10 + 9\lambda/10 = \lambda$. $V(\hat{\lambda}_1) = V(X)/4 + V(Y)/36$ $= \lambda/3; V(\hat{\lambda}_2) = V(X)/100 + 9V(Y)/100 = 7\lambda/25$; $\hat{\lambda}_2$ is better.
2. $V(\bar{X}) = 1/10; V(\bar{Y}/2) = 1/20$. Sample (b) provides a better unbiased estimator of μ.
3. Clearly, $\bar{X}$ is unbiased. Using $Y_i = X_i - \theta + 1/2 \sim U(0,1)$, we have that $Y_{(i)} \sim Be(i, 3-i)$. It follows that $E[(X_{(1)} + X_{(3)})/2] = E[(Y_{(1)} + Y_{(3)})/2] + \theta - 1/2$ $= 1/2 + \theta - 1/2 = \theta$. But $V(\bar{X}) = 1/36$, while $V[(X_{(1)} + X_{(3)})/2]$ $= V[(Y_{(1)} + Y_{(3)})/2] = (1/4)[V(Y_{(1)}) + V(Y_{(3)}) + 2Cov(Y_{(1)}, Y_{(3)})]$, which reduces to $(1/4)[3/80 + 3/80 + 2/80] = 1/40$; thus, $(X_{(1)} + X_{(3)})/2$ is the better estimator of θ.

4. $E(T_1) = 0(1-\theta)+1(\theta/2+\theta/2) = \theta$. $E(T_2) = (2/3)E(X) = (2/3)[0+\theta/2+\theta] = \theta$. $V(T_1) = \theta-\theta^2$, while $V(T_2) = 10\theta/9-4\theta^2/9$. T_1 is the better estimator.

Exercises 6.2

1. $E\left(\sum_{i=1}^k c_iT_i\right) = \sum_{i=1}^k c_iET_i = \theta\sum_{i=1}^k c_i = \theta$.

2. If $T_1 = I_{\{1\}}(X)$, then $E(T_1) = 1(p)+0(1-p) = p$; if $T_2 = 1-T_1$, then $E(T_2) = q$; $EX = 1/p$; Since $q/p = 1/p-1$, $E(X-1) = q/p$. If $T_3 = I_{\{6,7,\ldots\}}(X)$, then $E(T_3) = P(X>5) = \sum_{i=6}^\infty pq^{i-1} = q^5\sum_{i=1}^\infty pq^{i-1} = q^5$.

3. $EX = \lambda$ and $EX^2 = \lambda+\lambda^2$. Therefore, $E(X^2-X) = EX^2-EX = \lambda^2$.

4. Since $m_X(t) = E(e^{tX}) = e^{\lambda(e^t-1)}$, solve $e^t-1 = 2$, which yields $t = \ln(3)$. Therefore, $E(e^{X\cdot\ln(3)}) = e^{2\lambda}$.

5. Since $X \sim B(n,1/4+p/2)$, $EX = n/4+np/2$. Thus, $E[(X-n/4)/(n/2)] = p$, that is, $2X/n-1/2$ is an unbiased estimator of p.

Exercises 6.3

1. (a) $I_X(\theta) = 1/\theta^2(1-\theta)$ and the $CRLB = \theta^2(1-\theta)/n$; (b) $I_X(\theta) = 1/\theta^2$ and the $CRLB = \theta^2/n$.

2. (a) $I_X(\theta) = (1/2\theta^2)$, $CRLB = 2\theta^2/n$; (b) Since $(1/\theta)\sum_{i=1}^n X_i^2 \sim \chi^2_n$, $V[(1/n)\sum_{i=1}^n X_i^2] = 2\theta^2/n$, which is the $CRLB$. Thus, $(1/n)\sum_{i=1}^n X_i^2$ is the MVUE of θ.

3. Under regularity conditions, $E\left(\frac{\partial^2}{\partial\theta^2}\ln f(X|\theta)\right) = E\left(\frac{f''(X|\theta)}{f(X|\theta)}\right) - E\left(\frac{f'(X|\theta)}{f(X|\theta)}\right)^2 = 0-I_X(\theta)$.

4. $I_X(p) = 1/4(1/4+p/2)(3/4-p/2)$, $V(2X/n-1/2) = 4(1/4+p/2)(3/4-p/2)/n = CRLB$; therefore, $\tilde{p} = 2X/n-1/2$ is the MVUE of p.

5. (a) $F(Y \le y) = P(X \le e^y) = 1-e^{-y/\theta}$ and $E(lnX) = \theta$. (b) $I_X(\theta) = 1/\theta^2$. Let $T = (1/n)\sum_{i=1}^n \ln X_i$. Then $ET = \theta$ and $V(T) = \theta^2/n = CRLB$. Thus, T is the MVUE of θ.

Exercises 6.4

1. (a) $E(S) = (1/2)\lambda+(1/4)2\lambda = \lambda$. $V(S) = 3\lambda/8$; $V(S) < V(Y/2) < V(X)$. (b) $P(X=x) = e^{-2\lambda}\lambda^{x+y}(1/x!y!)$, showing that $T = X+Y$ is a sufficient statistic for λ. (c) $V(T/3) = \lambda/3 < V(S)$.

2. (a) $I_X(\beta^{\alpha_0}) = 1/\beta^{2\alpha_0}$, and $E\left((1/n)\sum_{i=1}^n X_i^{\alpha_0}\right) = \beta^{\alpha_0}$, (b) $\sum_{i=1}^n X_i^{\alpha_0}$ is a complete sufficient statistic for β^{α_0}. (c) By the Lehmann-Scheffé Theorem, $(1/n)\sum_{i=1}^n X_i^{\alpha_0}$ is the MVUE of the parameter β^{α_0}.

3. From (6.36), one can identify the $Be(\theta,1)$ distribution as a member of an exponential family with $A(\theta) = \theta$, $B(x) = \ln(x)$, $C(x) = \ln(x)$ and $D(\theta) = 0$. Thus, $\sum_{i=1}^n \ln(X_i)$ is a complete sufficient statistic for θ and $-(1/n)\sum_{i=1}^n \ln(X_i)$ is the unique MVUE of $1/\theta$ (the mean of $-\ln(X)$).

4. Let $g(x)$ be a function for which $E_\theta(g(X_{(n)})) = 0$ for all $\theta > 0$. Then, for $0 < a < b < \infty$, $\int_a^b g(y)ny^{n-1}dy = 0$. If a and b are such that $g(y) > (<)0$ for $a < y < b$, then the aforementioned integral must be positive (negative). This shows that g must be the zero function. Since $E[nX_{(n)}/(n+1) = \theta, nX_{(n)}/(n+1)]$ is the MVUE of θ by Theorem 6.4.4.

5. (a) Sufficiency follows from the factorization theorem; (b) See Theorem 3.6.5. That $(1/n)\sum_{i=1}^n X_i^{\alpha_0}$ is the MVUE of θ follows from the Cramér-Rao or Lehmann-Scheffé Theorems.

6. (a) Sufficiency of $(X_{(n)}, X_{(1)})$ follows from the factorization theorem.
(b) Since $E[X_{(n)} - X_{(1)} - (n-1)/(n+1)] = 0$, it follows that the sufficient statistic $(X_{(n)}, X_{(1)})$ is not complete.

7. The pair $(\prod X_i, \sum X_i)$ is jointly sufficient for (α, β).

Exercises 6.5

1. The BLUE of μ is $\hat{\mu} = X$.

2. $E\left(\sum_{i=1}^k a_i X_i\right) = \mu \Leftrightarrow \sum_{i=1}^n a_i = 1$. For such vectors $\mathbf{a}, V\left(\sum_{i=1}^k a_i X_i\right)$ is minimized when $a_i \equiv 1/n$.

3. $E\left(\sum_{i=1}^k a_i X_i\right) = \lambda \Leftrightarrow \sum_{i=1}^k a_i k_i = 1$. BLUE has $a_i = 1/\sum_{i=1}^n k_i$.

4. The BLUE estimator of θ is $\hat{\theta} = X + Y/2$.

Exercises 6.6

1. The estimator of the form $c\bar{X}$ with the smallest MSE is $[n/(n+1)]\bar{X}$.

2. $ET = 1(p) + (-1)(p)(1-p) = p^2$. However, T is inadmissible, since it will estimate p^2 by a negative number when $X = 2$ is observed. The estimator $\max(T, 0)$ has a smaller MSE.

3. The estimator of the form cs^2 with the smallest MSE is $[(n-1)/(n+1)]s^2$.

4. (a) $c = 2/3$; (b) the estimator of the form cX with the smallest MSE is $(9/14)X$.

5. $E(\bar{X}^2) = \mu^2 + 1/n$. Thus, $E(\bar{X}^2 - 1/n) = \mu^2$. But $\bar{X}^2 - 1/n$ is inadmissible; $\max(\bar{X}^2 - 1/n, 0)$ has a smaller MSE.

Chapter 7.

Exercises 7.1

1. Use Adam's rule to evaluate $E(X_{(1)}X_{(n)})$ and to get $Cov(X_{(1)}, X_{(n)})$, and then $V([X_{(1)} + X_{(n)}]/2)$. Both estimators converge to θ at the rate $1/\sqrt{n}$.

2. $\bar{X}_n \xrightarrow{p} 1/p$. Let $g(x) = 1/x$. By Theorem 5.2.1 (i), $1/\bar{X}_n = g(\bar{X}_n) \xrightarrow{p} g(1/p) = p$.

3. The proof is embedded in the hint.

Exercises 7.2

1. Let m_1 and m_2 be the first two sample moments. Then the estimators $\hat{\alpha} = m_1 - \sqrt{3(m_2 - m_1^2)}$ and $\hat{\beta} = m_1 + \sqrt{3(m_2 - m_1^2)}$ are MMEs of the parameters α and β.

2. The estimators $\hat{\mu} = \bar{X}$ and $\hat{\sigma}^2 = (1/n)\sum_{i=1}^n (X_i - \bar{X})^2$ are MMEs of μ and σ^2.

3. The estimators $\hat{\mu} = \bar{X}$ and $\hat{\beta} = \sqrt{(3/\pi^2)[(1/n)\sum_{i=1}^n X_i^2 - \bar{X}^2]}$ are MMEs of μ and σ^2.

4. $\hat{p} = 1/\bar{X}$ is an MME of p; $\sqrt{n}(1/\bar{X} - p) \xrightarrow{D} Y \sim N(0, (1-p)p^2)$.

5. $\hat{\alpha} = \bar{X}/(\bar{X} - 1)$ is an MME of p;

$$\sqrt{n}(\bar{X}/(\bar{X} - 1) - \alpha) \xrightarrow{D} Y \sim N(0, \alpha(\alpha - 1)^2/(\alpha - 2)).$$

Exercises 7.3

1. The MLE of p is $\hat{p} = 1/\bar{X}$; $\sqrt{n}(\hat{p}-p) \xrightarrow{D} Y \sim N(0, p^2(1-p))$.
2. $\hat{\sigma}^2 = (1/n)\sum_{i=1}^n X_i^2$; $\sqrt{n}(\hat{\sigma}^2 - \sigma^2) \xrightarrow{D} Y \sim N(0, 2\sigma^4)$.
3. The MLEs of θ and α are $\hat{\theta} = X_{(1)}$ and $\hat{\alpha} = n/[\sum_{i=1}^n \ln(X_i) - n\ln(X_{(1)})]$.
4. $\hat{\theta} = n/\sum_{i=1}^n \ln X_i$; $\sqrt{n}(\hat{\theta}-\theta) \xrightarrow{D} Y \sim N(0, \theta^2)$. The AV of the MME in Example 7.2.5 is $\theta(\theta+1)^2/(\theta+2)$. The inequality $\theta^2 < \theta(\theta+1)^2/(\theta+2)$ holds for all $\theta > 0$, so the MLE is superior.

Exercises 7.4

1. If $p_1 = P(D|E)$ and $p_2 = P(D|E^c)$, then $P(D|E)/P(D^c|E) = p_1/(1-p_1)$ and $P(D|E^c)/P(D^c|E^c) = p_2/(1-p_2)$. Since the function $g(x) = x/(1-x)$ is a strictly increasing function of x, we have that $g(p_1) = g(p_2)$ if and only if $p_1 = p_2$, which is the desired conclusion.
2. If $g(x) = x/(1-x)$, then $\omega > 1 \Leftrightarrow g(P(D|E)) > g(P(D|E^c)) \Leftrightarrow P(D|E)) > P(D|E^c)$.
3. (a) $\hat{\omega} = 7$ and $s_{\hat{\omega}} = .111$. A 95% CI for ω is given by (6.7824, 7.2170).
4. $r_E = .485$.

Exercises 7.5

2. The NR procedure yields $x_{i+1} = x_i - (x_i^3 - 2x_i + 2)/(3x_i - 2)$ for $i = 0, 1, 2, \ldots$. If $x_0 = 0$, then $x_1 = 1$, $x_2 = 0$, $x_3 = 1$, $x_4 = 0$, etc.

Exercises 7.6

2. If $\mathbf{Y}$ is the incomplete data and $\mathbf{X}$ is the complete data, $Y_1 = X_1 + X_2$, $Y_2 = X_3 + X_4$ and $Y_3 = X_5$, where the "complete data" model is $\mathbf{X} \sim M_5(100, \mathbf{p})$, with $p_1 = p = p_3$, $p_2 = .2$, $p_4 = .4$, and $p_5 = .4 - 2p$. Then, given complete data, the maximization step yields the MLE $\hat{p} = .2(X_1 + X_3)/(X_1 + X_3 + X_5)$.

Chapter 8.

Exercises 8.1

1. Using the pivotal statistic $\sum_{i=1}^n X_i^2/\sigma^2$, a $100(1-\alpha)\%$ CI for σ^2 is $\left(\frac{\sum_{i=1}^n X_i^2}{\chi^2_{n,\alpha/2}}, \frac{\sum_{i=1}^n X_i^2}{\chi^2_{n,1-\alpha/2}}\right)$.
2. Using the pivotal statistic $(n-1)s^2/\sigma^2$, a $100(1-\alpha)\%$ CI for σ^2 is $\left(\frac{(n-1)s^2}{\chi^2_{n-1,\alpha/2}}, \frac{(n-1)s^2}{\chi^2_{n-1,1-\alpha/2}}\right)$.
3. Using the pivotal statistic $2\sum_{i=1}^n X_i/\beta$, a $100(1-\alpha)\%$ CI for β is $\left(\frac{2\sum_{i=1}^n X_i}{\chi^2_{2n,\alpha/2}}, \frac{(n-1)s^2}{\chi^2_{2n,1-\alpha/2}}\right)$.
4. $S = \sum_{i=1}^{10} X_i \sim P(10\lambda)$; $P(S \geq 14|\lambda = .75) = .0216$; $P(S \leq 14|\lambda = 2.35) = .0249$. Thus, $(\lambda_L, \lambda_U) = (.75, 2.35)$ is an approximate 95% confidence interval for λ.

Exercises 8.2

1. If $\hat{p} = 1/\bar{X}$, then $\sqrt{n}(\hat{p}-p) \xrightarrow{D} Y \sim N(0, p^2 q)$. Thus, a $100(1-\alpha)\%$ CI for p is given by $\left(\hat{p} - (\hat{p}\sqrt{\hat{q}}/\sqrt{n})z_{\alpha/2}, \hat{p} + (\hat{p}\sqrt{\hat{q}}/\sqrt{n})z_{\alpha/2}\right)$.

2. $\sqrt{n}(r_n - \rho) \xrightarrow{D} Y \sim N(0,(1-\rho^2)^2)$. Let $g(x) = (1/2)\ln[(1+x)/(1-x)]$. Then $g'(x) = 1/(1-x^2)$. Thus, $\sqrt{n}(g(r_n) - g(\rho)) \xrightarrow{D} Y \sim N(0,1)$. It is evident that g is a VST.

Exercises 8.3

1. (a) A 95% confidence interval for p is $(14.02, 15.98)$ (b) Set $p = .5$, $n = 385$; set $p = .43$, $n = 377$.

2. (a) If one ignores the continuity correction (since the sample size $n = 100$ is "large"), $P(X \leq 43) = P(\hat{p} \leq .43) = P(Z \leq -1.4) = .0749$; (b) An approximate 95% CI for p is $(.333, .527)$; (c) $n = 385$.

3. (a) A 95% confidence interval for p is $(.483, .677)$. (b) $n = 1040$.

Exercises 8.4

1. $\sum_{i=0}^{99}\binom{100}{i}(.9705)^i(.0295)^{100-i} = .95$; $(X_{(1)}, X_{(100)})$ is a 95% tolerance interval for 97.05% of the population.

Chapter 9.

Exercises 9.1

1. Let $X \sim B(n,p)$. The risk function of $\hat{p} = X/n$ is $R(p,\hat{p}) = E_{X|p}(\hat{p} - p)^2$. Since $\hat{p}$ is unbiased for p, its risk function is equal to its variance. Thus, $R(p,\hat{p}) = p(1-p)/n$.

2. Let $X_1, \ldots, X_n \overset{\text{iid}}{\sim} N(\mu, \sigma^2)$. The risk function of $\hat{\mu} = \bar{X}$ is $R(\mu, \bar{X}) = E_{\bar{X}|\mu}(\bar{X} - \mu)^2$. Since $\bar{X}$ is unbiased for μ, its risk function is equal to its variance. Thus, $R(\mu, \bar{X}) = \sigma^2/n$.

3. Suppose $\hat{\theta}$ is a Bayes rule with respect to the prior G. Suppose, further, that $\hat{\theta}$ is an equalizer rule, that is, $R(\theta, \hat{\theta}) = c$, where c is a fixed constant. If $\hat{\theta}$ is not a minimax rule, then there exists an estimator $\tilde{\theta}$ for which $\sup_\theta R(\theta, \tilde{\theta}) < c$. But then $r(G, \tilde{\theta}) < r(G, \hat{\theta})$, a contradiction.

4. Under this loss function and the $U(0,1)$ prior, $\hat{p} = X/n$ is easily shown to be the Bayes estimator of p. Further, $R(p,\hat{p}) = E\left(\frac{(\hat{p}-p)^2}{p(1-p)}\right) = 1/n$. Thus, $\hat{p}$ is an equalizer rule. It is minimax by Ex. 9.1.3.

Exercises 9.2

1. The joint probability function of X and λ is

$$f(x,\lambda) = \frac{\lambda^x e^{-\lambda}}{x!} \cdot \frac{1}{\Gamma(\alpha)\beta^\alpha}\lambda^{\alpha-1}e^{-\lambda/\beta} \quad \text{for } x = 0,1,2,\ldots \quad \text{and } 0 < \lambda < \infty.$$

The posterior distribution of λ is: $\lambda | X = x \sim \Gamma[(\alpha + x), \beta/(\beta+1)]$. The Bayes estimator of λ is $E(\lambda|x) = (\alpha + x)\beta/(\beta+1)$.

2. The posterior distribution of θ, given $X = x$, is the Pareto distribution $P(\alpha + n, \theta_0^*)$, where the parameter $\theta_0^* = \max\{\theta_0, X_1, \ldots, X_n\}$. The Bayes estimator of θ is $\hat{\theta} = [(\alpha+n)\theta_0^*/(\alpha+n-1)]$.

4. Let m be a median of the distribution of X, and suppose that $a = m + c > m$. Then, note that the variable $U = |X - a| - |X - m|$ is such that $U = c$ if $X \leq m$, $-c \leq U \leq c$ if $m < X < m + c$ and $U = -c$ if $X \geq m + c$. Letting $Y = c$ if $X \leq m$ and $Y = -c$ if $X > m$, we have that $Y \leq |X - a| - |X - m|$. Since $E(Y) = cP(Y = c) - cP(Y = -c) = cP(X \leq m) - cP(X > m)$, the fact that $P(X \leq m) \geq 1/2$ implies that $E(Y) \geq 0$. Since $Y \leq |X - a| - |X - m|$, we have that $0 \leq E(Y) \leq E|X - a| - E|X - m|$. We conclude that for any real number a, $E|X - m| \leq E|X - a|$, so that $a = m$ minimizes $E|X - a|$ among all possible values of a.

5. $\hat{\theta} = x + [\ln(1 + c)]/c$.

Exercises 9.4

1. From Exercise 9.2.1, we have that $\hat{\lambda} = (\alpha + x)\beta/(\beta + 1)$ is the Bayes estimator of λ. We may write $\hat{\lambda}$ as $\hat{\lambda} = \alpha\beta[1/(\beta + 1)] + x[\beta/(\beta + 1)]$, that is, $\hat{\lambda}$ is a convex combination of the prior mean $\alpha\beta$ and X, the MVUE of λ.

3. Note that when $\alpha = n/(n + \omega)$, the multiplier $(1 + \eta)/(1 - \eta)$ in (9.52) reduces to $(\omega + 2n)/\omega$. The Cramér-Rao inequality yields $r(G_0, \hat{\theta}) = E_{G_0} I^{-1}(\theta)/n$ when $\hat{\theta}$ is efficient. Appropriate substitutions in (9.52) yield the desired result.

4. Under the stated conditions, the fraction $(1 + \eta)/(1 - \eta)$ in (9.44) may be written as $(\omega + 2n)/\omega$, and the expression $r(G_0, \hat{\theta}) = \frac{1}{n} E_{G_0} I^{-1}(\theta)$ follows from the Cramér-Rao inequality. Substitution in (9.44) gives the desired result.

Exercises 9.5

1. Note that $P(\max(X_1, X_2) < \theta + 1) = 1/4)$ and that $P(\min(X_1, X_2) > \theta + 1) = 1/4)$. It follows that $P(X_{(1)} < \theta + 1 < X_{(2)}) = 1/2$. Thus, $(X_{(1)} - 1, X_{(2)} - 1)$ is a 50% confidence interval for θ. However, if $x_{(2)} - x_{(1)} > 1$, we know with 100% certainty that the interval $(x_{(1)} - 1, x_{(2)} - 1)$ contains θ. The Bayesian view would be that the 50% claim is incoherent.

2. If $X = 1$, the interval $(.5407, .9958)$ is a central 95% credibility interval for p. If $X = 0$, the upper limit of the 95% credibility interval is the solution of the equation $6p^5 - 5p^6 = .975$ and the lower limit of the 95% credibility interval is the solution of the equation $6p^5 - 5p^6 = .025$.

3. A central 95% credibility interval for μ is given by $(.199983, 1.80017)$. A 95% confidence interval for μ is given by $(.32346, 2.0654)$. The confidence interval is a bit wider, and it does not use the prior intuition that μ is reasonably close to 0.

Chapter 10.

Exercises 10.1

1. $\alpha = .0808$, $\beta = .2743$, and the p-value of $X = 39$ is $.0359$.

2. $\alpha = .0082$, $\beta = .2033$.

3. $\alpha = .0548$, $\beta = .7881$.

Exercises 10.2

1. $n = 8$, $\bar{X} = 78$, $s = 4.276$; CR of size .05: Reject H_0 if $t = (\bar{X} - 80)/(s/\sqrt{n}) < -t_{7,.05} = -1.895$. Here, $t = -1.32737$. Accept H_0.

2. CR of size .05: Reject H_0 if $Z = \frac{\hat{p}-.5}{\sqrt{(.5)(.5)/80}} < -Z_{.05} = -1.64$. Here, $Z = -2.24 < -1.64$. The null hypothesis should be rejected.

3. $m_{\bar{X}}(t) = e^{\mu_x t + \sigma_x^2 t^2/2m}$, $m_{\bar{Y}}(t) = e^{\mu_y t + \sigma_y^2 t^2/2n}$.
Therefore, $m_{\bar{X}-\bar{Y}}(t) = e^{(\mu_x - \mu_y)t + (\sigma_x^2/m + \sigma_y^2/n)t^2/2}$. From this, it follows that $\bar{X} - \bar{Y} \sim N\left(\mu_x - \mu_y, \frac{\sigma_x^2}{m} + \frac{\sigma_y^2}{n}\right)$.

4. $\frac{\bar{X}-\bar{Y}}{\sqrt{\sigma_x^2/m + \sigma_y^2/n}} \sim N\left(\frac{\mu_x - \mu_y}{\sqrt{\sigma_x^2/m + \sigma_y^2/n}}, 1\right)$.

5. Pooled estimate of p is $\hat{p} = .435$. CR of size .05: Reject H_0 if $|Z| = \left|\frac{\hat{p}_1 - \hat{p}_2}{\sqrt{\hat{p}\hat{q}(1/100 + 1/100)}}\right| > Z_{.025} = 1.96$. Here, $|Z| = 1.854 < 1.96$. Accept H_0.

Exercises 10.3

1. $n = 1079$.

2. $n = 841$.

Exercises 10.4

1. By the N-P Lemma, any test with a critical region of the following form is best of its size: "Reject H_0 if $\binom{20}{x}(4/5)^x(1/5)^{20-x}/\binom{20}{x}(1/2)^x(1/2)^{20-x} > k$ for some fixed $k > 0$." This inequality is equivalent to the inequality $x > c$, where c is a function of k. Thus, best tests reject H_0 if X is sufficiently large. The CR: Reject if $X > 13$ has size $\alpha = P(X > 13|n = 20, p = 1/2) = 1 - .9433 = .0577$. Thus, this test is the best test of size .0577 of $H_0 : p = 1/2$ vs. $H_1 : p = 4/5$.

2. Best tests reject H_0 when $f_1(x)/f_0(x) > k$ for some fixed $k > 0$. Here, this inequality is equivalent to the inequality $g(x) \equiv (1-k)x^2 + 2kx + (1-2k) > 0$. If $k < 1$, then the quadratic function $g(x)$ is a parabola that opens upward. When $k = 1/2$, the two roots of the quadratic function are -2 and 0, and thus for this choice of k, $g(x) > 0$ if and only if either $x < -2$ or $x > 0$. Thus, the test which rejects H_0 for such x is the best test of its size.

3. Let $T = \sum_{i=1}^n X_i$. T is sufficient for θ, and $T \sim P(n\theta)$. By the NP-Lemma, best tests of $H_0 : \theta = 1$ vs. $H_1 : \theta = 2$ reject H_0 when $f(t|\theta = 2)/f(t|\theta = 1) > k$ for some fixed $k > 0$. This inequality is equivalent to the inequality $x > [\ln(k) + n]/\ln(2)$. The test which rejects H_0 if $\bar{X} > c$ (or, equivalently, if $T > 8c$) is best of its size for testing H_0 vs. H_1. If $n = 8$, the test that rejects H_0 when $T > 12$ is best of size $\alpha = .0638$. The power of this test at $\theta = 2$ is $P(T > 12|\theta = 2) = .8069$.

4. By the N-P Lemma, a test with critical region of the form: Reject H_0 if $f(t|\lambda = 2)/f(t|\lambda = 1) > k$, where $t = \sum_{i=1}^n x_i$ and $k > 0$ is arbitrary, is best of its size for testing $H_0 : \lambda = 1$ vs. $H_1 : \lambda = 2$. This inequality reduces to $2^n e^{-2t}/e^{-t} > k$ and is equivalent to $-t > k' = \ln(k/2^n) \Leftrightarrow t < k = -k'$. Thus, best tests reject H_0 if $T = \sum_{i=1}^n X_i$ is sufficiently small, or equivalently, when $\bar{X}$ is sufficiently small.

Exercises 10.5

1. The numerator of $\lambda(\mathbf{x})$ is simply the likelihood with $\mu = \mu_0$ and $\sigma^2 = \sigma_0^2$. The denominator of $\lambda(\mathbf{x})$ is the likelihood with $\mu = \bar{x}$ and $\sigma^2 = \sigma_0^2$. A little algebra shows that the inequality $\lambda(\mathbf{x}) < k$ is equivalent to $\sum_{i=1}^{n}(x_i - \mu_0)^2 - \sum_{i=1}^{n}(x_i - \bar{x})^2 > k' \Leftrightarrow n(\bar{x} - \mu_0)^2 > k' \Leftrightarrow \left|\frac{\bar{x}-\mu_0}{\sigma_0/\sqrt{n}}\right| > k''$, which is the critical region of the standard test.

3. The numerator of $\lambda(\mathbf{x},\mathbf{y})$ is the likelihood with p_1 and p_2 replaced by

$$\hat{p} = \frac{\sum_{i=1}^{n} x_i + \sum_{j=1}^{m} y_j}{n+m}.$$

The denominator of $\lambda(\mathbf{x},\mathbf{y})$ is the likelihood with p_1 and p_2 replaced by $\hat{p}_1 = \frac{\sum_{i=1}^{n} x_i}{n}$ and $\hat{p}_2 = \frac{\sum_{j=1}^{m} y_j}{m}$. $\lambda(\mathbf{x},\mathbf{y}) = \hat{p}^{A+B}(1-\hat{p})^{n+m-A-B}/\hat{p}_1^A(1-\hat{p}_1)^{n-A}\ \hat{p}_2^B(1-\hat{p}_2)^{m-B}$, where $A = \sum_{i=1}^{n} x_i$ and $B = \sum_{j=1}^{m} y_j$. The LR statistic $-2\ln\lambda(\mathbf{x},\mathbf{y}) \sim \chi_1^2$.

4. The numerator of $\lambda(\mathbf{x_1},\ldots,\mathbf{x_k})$ is the likelihood with each λ_i replaced by $\hat{\lambda} = \frac{\sum_{i=1}^{k}\sum_{j=1}^{n_i} x_{i,j}}{\sum_{i=1}^{k} n_i}$. The denominator of $\lambda(\mathbf{x_1},\ldots,\mathbf{x_k})$ is the likelihood with λ_i replaced by $(1/n_i)\sum_{j=1}^{n_i} x_{i,j} \forall i$. Moreover, $-2\ln\lambda \sim \chi_{k-1}^2$.

Exercises 10.6

1. Chi-Square Stat $= 6.18 > 5.991 = \chi_{2,.05}^2$. Reject H_0.

2. Chi-Square Stat $= 7.78 < 9.38 = \chi_{3,.01}^2$. Accept H_0.

3. The estimated probabilities for intervals $i = 1,\ldots,5$ are $p(i) = e^{-(3/5)(i-1)} - e^{-(3/5)(i)}$. Chi-Square Stat $= .6755 < \chi_{3,.05}^2 = 7.815 < 9.488 = \chi_{4,.05}^2$. Accept H_0.

4. The estimated probabilities for intervals $i = 1,\ldots,4$ are obtained from the $B(5,.8178)$ distribution. Chi-Square Stat $= 1.078 < \chi_{2,.05}^2 = 5.991 < 7.815 = \chi_{3,.05}^2$. Accept H_0.

Chapter 11.

Exercises 11.1

1. $r = .979$, $|t| = \left|.979 \times \sqrt{3}/\sqrt{1-(.979)^2}\right| = 8.344 > 3.182 = t_{3,.025}$. Thus, $H_0: \rho = 0$ is rejected; the least-squares line is $\hat{Y} = 2.7 + 2.1x$; if $x = 3.5$, $\hat{Y} = 10.05$.

2. First, note that $\sum_{i=1}^{n}(x_i-\bar{x})^2 = \sum_{i=1}^{n} x_i^2 - 2\bar{x}\sum_{i=1}^{n} x_i + n\bar{x}^2 = \sum_{i=1}^{n} x_i^2 - n\bar{x}^2$. Also, note that $= \sum_{i=1}^{n}(x_i-\bar{x})Y_i - \bar{Y}\sum_{i=1}^{n}(x_i-\bar{x}) = \sum_{i=1}^{n}(x_i-\bar{x})Y_i$. Thus, (11.15) is equivalent to (11.13).

3. First, note that $\sum_{i=1}^{n}(x_i-\bar{x})^2 = \sum_{i=1}^{n} x_i^2 - 2\bar{x}\sum_{i=1}^{n} x_i + n\bar{x}^2 = \sum_{i=1}^{n} x_i^2 - n\bar{x}^2$. Then, note that $\sum_{i=1}^{n}(x_i-\bar{x})Y_i = \sum_{i=1}^{n} x_iY_i - \bar{x}\sum_{i=1}^{n} Y_i = \sum_{i=1}^{n} x_iY_i - n\bar{x}\bar{Y}$. Thus, (11.16) is equivalent to (11.13).

4. $r = .9593$, $|t| = \left|.9593 \times \sqrt{3}/\sqrt{1-(.9593)^2}\right| = 5.8839 > 3.182 = t_{3,.025}$. Thus, $H_0: \rho = 0$ is rejected. The least-squares line is $\hat{Y} = -.6 + x$.

5. $r\frac{s_y}{s_x} = \frac{\sum_{i=1}^{n}(x_i-\bar{x})(y_i-\bar{y})}{\sqrt{\sum_{i=1}^{n}(x_i-\bar{x})^2\sum_{i=1}^{n}(y_i-\bar{y})^2}}\frac{\sqrt{\sum_{i=1}^{n}(y_i-\bar{y})^2}}{\sqrt{\sum_{i=1}^{n}(x_i-\bar{x})^2}} = \frac{\sum_{i=1}^{n}(x_i-\bar{x})(y_i-\bar{y})}{\sum_{i=1}^{n}(x_i-\bar{x})^2} = \hat{\beta}_1$.

Exercises 11.2

1. $Cov(\bar{Y},\hat{\beta}_1) = \sum_{i=1}^n \sum_{j=1}^n \frac{1}{n} \frac{(x_j-\bar{x})}{\sum_{k=1}^n (x_k-\bar{x})^2} Cov(Y_i,Y_j) = \frac{\sigma^2}{n\sum_{k=1}^n (x_k-\bar{x})^2} \sum_{j=1}^n (x_j-\bar{x}) = 0.$

2. Since $\hat{\beta}_0 \sim N\left(\beta_0, \frac{\sigma^2 \sum_{i=1}^n x_i^2}{n\sum_{i=1}^n (x_i-\bar{x})^2}\right)$, $SSE/\sigma^2 \sim \chi^2_{n-2}$ and $\hat{\beta}_0$ and SSE are independent, we have $\frac{(\hat{\beta}_0-\beta_0)/\sqrt{\frac{\sigma^2\sum_{i=1}^n x_i^2}{n\sum_{i=1}^n (x_i-\bar{x})^2}}}{\sqrt{SSE/(n-2)\sigma^2}} = \frac{\hat{\beta}_0-\beta_0}{\hat{\sigma}\sqrt{\sum_{i=1}^n x_i^2/n\sum_{i=1}^n (x_i-\bar{x})^2}} \sim t_{n-2}$, where $\hat{\sigma} = \sqrt{SSE/(n-2)}$.

3. See solution to 11.2.2.

Exercises 11.3

1. Note that $\hat{\beta}_0 = \sum_{i=1}^n h_i Y_i$, where $h_i = 1/n + c_i\bar{X}$, with the constants c_i defined as in Lemma 11.3. It is easily verified that $\sum_{i=1}^n h_i = 1$ and $\sum_{i=1}^n h_i X_i = 0$. This will imply that $E(\hat{\beta}_0) = \beta_0$. Using known properties of c_i, one can show that $V(\hat{\beta}_0) = \left(\frac{1}{n} + \frac{\bar{x}}{\sum_{i=1}^n (x_i-\bar{x})^2}\right)\sigma^2$. Note that $\sum_{i=1}^n h_i^2 = \frac{1}{n} + \frac{\bar{x}}{\sum_{i=1}^n (x_i-\bar{x})^2}$. Now if $\tilde{\beta}_0 = \sum_{i=1}^n k_i Y_i$ is another unbiased estimator of β_0, with $k_i = h_i + \Delta_i$, with $\Delta_i > 0$ for at least one i, then $V(\tilde{\beta}_0) = \sigma^2\left(\sum h_i^2 + \sum \Delta_i^2 + 2\sum h_i\Delta_i\right)$. Using properties of h_i, we can obtain that $\sum h_i\Delta_i = 0$. Thus, $V(\tilde{\beta}_0) = \sigma^2\left(\sum h_i^2 + \sum \Delta_i^2\right) > \sigma^2\left(\sum h_i^2\right) = V(\hat{\beta}_0)$.

3. This is an easy consequence of the Neyman-Fisher Factorization Theorem.

5. $\sum e_i = \sum(Y_i - \hat{Y}) = \sum(Y_i - \hat{\beta}_0 - x_i\hat{\beta}_1) = \sum(Y_i - \bar{Y}) - \hat{\beta}_1\sum(x_i-\bar{x}) = 0 - \hat{\beta}_1(0) = 0.$

6. The least-squares line is $\hat{Y} = 1.333 + 1.167x$. $SSR = 8.1667$. $SSE = 1.8333$, $F = 13.364 > 10.13 = F_{1,3,.05}$. The hypothesis $H_0 : \beta_1 = 0$ is rejected.

Exercises 11.4

1. First, note that for arbitrary numbers $x_1,\ldots,x_n$, $\frac{\partial}{\partial a}\sum_{i=1}^n (x_i-a)^2 = -2\sum_{i=1}^n x_i^2 + 2na = 0$ has the unique solution $a = \bar{x}$, and $a = \bar{x}$ minimizes $\sum_{i=1}^n (x_i-a)^2$. Thus, for every i, $\sum_{j=1}^{n_i} (Y_{i,j}-\mu_i)^2$ is minimized by $\mu_i = \bar{Y}_i$. It follows that $\sum_{i=1}^k \sum_{j=1}^{n_i} (Y_{i,j}-\mu_i)^2$ is minimized by $\mu_i = \bar{Y}_i$ for $i = 1,\ldots,k$, that is, $\bar{Y}_i$ is the least-squares estimator of μ_i for $i = 1,\ldots,k$.

2. $\sum_{i=1}^k \sum_{j=1}^{n_i} (Y_{i,j}-\bar{Y}_{i\cdot})(\bar{Y}_{i\cdot}-\bar{Y}_{\cdot\cdot}) = \sum_{i=1}^k (\bar{Y}_{i\cdot}-\bar{Y}_{\cdot\cdot})\sum_{j=1}^{n_i}(Y_{i,j}-\bar{Y}_{i\cdot}) = \sum_{i=1}^k \bar{Y}_{i\cdot} = 0.$

3. For each fixed i, $\frac{\sum_{j=1}^{n_i}(Y_{i,j}-\bar{Y}_{i\cdot})}{\sigma^2} \sim \chi^2_{n_i-1}$. Since the Ys are independent, it follows that $(1/\sigma)\sum_{i=1}^k \sum_{j=1}^{n_i}(Y_{i,j}-\bar{Y}_{i\cdot})$ can be viewed as the sum of k independent chi-square variables, and thus, by a moment-generating function argument, $(1/\sigma)\sum_{i=1}^k \sum_{j=1}^{n_i}(Y_{i,j}-\bar{Y}_{i\cdot}) \sim \chi^2_{\Sigma n_i-k} = \chi^2_{N-k}$.

4. George Snedecor.

5. $SSB = 993.11$, $SSW = 582.33$, $F = 18.76 > 3.44 = F_{2,22,.05}$. Thus, $H_0 : \mu_y = \mu_m = \mu_o$ is rejected. Simultaneous Scheffé confidence limits with overall 95% confidence: for $\mu_2 - \mu_1$, 9.3857 ± 6.6504, for $\mu_3 - \mu_1$, 16.2857 ± 6.9955, for $\mu_3 - \mu_2$, 6.9 ± 6.4012. Therefore, we conclude that $\mu_1 < \mu_2$, $\mu_1 < \mu_3$ and $\mu_2 < \mu_3$.

Exercises 11.5

2. Noting that in a balanced one-way ANOVA, $k(n-1) = N-k$, we have that $F = \frac{SSB/(k-1)}{SSW/(N-k)}$, so that $\frac{k-1}{N-k}F = \frac{SSB}{SSW}$. It follows that $\left(1+\frac{k-1}{N-k}F\right)^{-1} = \left(1+\frac{SSB}{SSW}\right)^{-1} = \left(\frac{SST}{SSW}\right)^{-1} = \frac{SSW}{SST}$.

Chapter 12.

Exercises 12.1

1. 1. $F_{10}(x)$ is the step function tabled below with jumps at the points $x = 1,3,4,6,7,8$ and 10.

$X = x$	1-	1	3	4	6	7	8	10
$F_{10}(x)$	0.0	0.1	0.2	0.4	0.6	0.7	0.9	1.0

2. From (11.21), it is clear that making the distance between $\bar{F}(t_j - 0)$ and $\bar{F}(t_j)$ larger increases the likelihood L. Since the differences $\bar{F}(t_j - 0) - \bar{F}(t_j)$ and $\bar{F}(t_{j+1} - 0) - \bar{F}(t_{j+1})$ are adjacent terms in L, and $\bar{F}(t_{j+1} - 0) \leq \bar{F}(t_j)$, it follows that setting $\bar{F}(t_{j+1} - 0) = \bar{F}(t_j)$ maximizes the likelihood. Since $t_j \leq l_i^{(j)} \leq t_{j+1} - 0$, the monotonicity of $\bar{F}$ requires that $\bar{F}\left(l_i^{(j)}\right) = \bar{F}(t_j)$ for all i and j.

3. Writing out the terms of L in (11.27) for $j = 1,2,3,\ldots$ and multiplying shows, using the definitions in (12.14)–(12.17), that the resulting expression is a product of powers of p_j with the exponents as claimed.

4. The KME $\hat{\bar{F}}$ of the survival function $\bar{F}$ based on these data is the step function tabled below.

$X = x$	6-	6	8	10	15	20
$\hat{\bar{F}}(x)$	1	7/8	3/4	9/20	9/40	0

Exercises 12.2

1. Answers will vary.

2. Answers will vary.

3. The transformation g is necessarily monotonic. Assume that g is increasing. If a and b are the $\alpha/2$ and $1 - \alpha/2$ quantiles of the statistic T^* based on B bootstrap samples, then $g(a)$ and $g(b)$ are the $\alpha/2$ and $1 - \alpha/2$ quantiles of the statistic $g(T^*)$ and thus $(g(a), g(b))$ is the $100(1 - \alpha)$ percentile interval for $g(T^*)$. If g is decreasing, the same argument establishes that $(g(b), g(a))$ is the $100(1 - \alpha)$ percentile interval for $g(T^*)$.

4. See Efron and Tibshirani (1998, p. 173).

Exercises 12.3

1. $P(X = Y) = .305$. If n^* observations are taken, the expected number of pairs for which $X \neq Y$ is $n^* \times .695$. Thus, $n* = 59/.695 = 72$.

2. $H_0 : T \sim B(12, 0.5)$, $\alpha = 0.05$, $N = 12$. $P(T \geq 10 | T \sim B(12, .5)) = 0.0193$. The two-sided p-value of the observed T is $0.0193 \cdot 2 = 0.0386 < \alpha$. Therefore, one should reject H_0 in favor of H_1 at the 5% significant level. The two teams are not evenly matched.

3. Assume $T \sim B(50, 0.5)$, $H_0 : p = 0.5$, $H_1 : p \neq 0.5$. By (12.41), the standard test of H_0 vs. H_1 rejects H_0 at $\alpha = .1$ if $|(\hat{p} - 1/2)/(1/\sqrt{800}| > Z_{0.05} \Leftrightarrow \hat{p} > 0.6163$. Since $\hat{p} = 31/50 = 0.62 > 0.6163$, we reject H_0 at $\alpha = .1$.

4. (a) $H_0 : P(+) = P(-)$, $H_1 : P(+) \neq P(-)$, $P(T \leq 2) == 0.003$. Therefore, the two-sided p-value of the observed T is 0.006. (b) Let $Y \sim N(50, 25)$. Then $P(T \leq 37) \cong P(Y \leq 37.5) = P(Z \leq -2.5) = .0062$. The two-sided p value of the observed T is .0124.

Exercises 12.4

1. Suppose $R = 2k+1$ for some positive integer k. Then either there are $k+1$ runs of Xs and k runs of Ys, with Xs as the first and last runs, or there are $k+1$ runs of Ys and k runs of Xs, with Ys as the first and last runs. In the first instance, think about the $n-1$ "slots" between the n consecutive ordered Xs. They can be divided into $k+1$ runs by selecting k slots, thus separating the Xs into k subgroups or "runs." The number of ways one can do this is $\binom{n-1}{k}$. Treating the m Ys similarly, there are $\binom{m-1}{k-1}$ possible ways of separating the m ordered Ys into k runs. Once these runs are identified, they are simply shuffled perfectly to obtain the $k+1$ runs of Xs and the k runs of Ys. This leads to the probability of obtaining such a configuration: $\binom{n-1}{k}\binom{m-1}{k-1}/\binom{n+m}{n}$. The denominator of this fraction is the number of locations, among $n+m$ locations, in which the n Xs can be placed, each resulting in a specified number of runs ranging from 2 to $\max\{2n+1, 2m+1\}$. A similar argument applies if there are $k+1$ runs of Ys and k runs of Xs. Putting the two parts together results in (11.50).

2. $n_x = 9$, $n_y = 8$, $N = 17$, $H_0 : F_x = F_y$, $H_1 : F_x \neq F_y$, $\alpha = 0.05$; Xs and Ys are clustered as $5X, 4Y, 4X, 4Y$, so that the observed number of runs is $R = 4$. Since $P(R=2) = 0.000082$, $P(R=3) = 0.000617$, $P(R=4) = 0.004607$, we find that $P(R \leq 4) = 0.005306 < 0.025$. we reject H_0 and in favor of H_1 and conclude that the level of serotonin in field mice and in lab-grown mice is different.

3. Let $H_0 : F_x = F_y$, $H_1 : F_x \neq F_y$ and $\alpha = 0.05$. Given that $n_x = n_y = 25$, we have that $\mu_{25,25} = 26$ and that $V_{25,25} = 12.2449$. The p-value may be approximated as $P(R_{25,25} \leq 18) = P(Z \leq -2.1433) = 0.016044$. The p-value of the observed R is smaller than .05, so that the null hypothesis is rejected at significance level $\alpha = .05$.

Exercises 12.5

1. For fixed i, $\sum_{j=1}^{m} I(-\infty, X_i)(Y_j) =$ the number of $Y_j < X_i$. If we know k_i, the number of Xs which are less than or equal to X_i for $i = 1, \ldots, n$, then $R(X_i)$, the rank of X_i, is equal to $\sum_{j=1}^{m} I(-\infty, X_i)(Y_j) + k_i$. Therefore,

$$U = \sum_{i=1}^{n} (R(X_i) - k_i) = \sum_{i=1}^{n} R(X_i) - \sum_{i=1}^{n} k_i = S - \frac{n(n+1)}{2}.$$

2. From Exercise 12.5.1, $U = S(X) - \frac{n(n+1)}{2}$ and $U^* = S(Y) - \frac{m(m+1)}{2}$; therefore,

$$U + U^* = S(X) - n(n+1)/2 + S(Y) - m(m+1)/2 = S(X) + S(Y) - (n^2 + n + m^2 + m)/2$$
$$= (m+n)(m+n+1)/2 - (n^2 + n + m^2 + m)/2 = 2mn/2 = mn.$$

3. (a) $n = 2$, $m = 4$.

R	3	4	5	6	7	8	9	10	11
$P(R)$	1/15	1/15	2/15	2/15	3/15	2/15	2/15	1/15	1/15

(b) $n = 2, m = 5$.

R	3	4	5	6	7	8	9	10	11	12	13
$P(R)$	1/21	1/21	2/21	2/21	3/21	3/21	3/21	2/21	2/21	1/21	1/21

(c) $n = m = 3$.

R	6	7	8	9	10	11	12	13	14	15
$P(R)$	1/20	1/20	2/20	3/20	3/20	3/20	3/20	2/20	1/20	1/20

4. (a) $H_1 : F_x > F_y$, $P(R \geq 10|H_0) = .1333$; $H_1 : F_x < F_y$, $P(R \leq 4|H_0) = .1333$; $H_1 : F_x \neq F_y$, $P(R \leq 3 \cup R \geq 10|H_0) = .2666$.
(b) $H_1 : F_x > F_y$, $P(R \geq 12|H_0) = .095238$; $H_1 : F_x < F_y$, $P(R \leq 4|H_0) = .095238$; $H_1 : F_x \neq F_y$, $P(R \leq 4 \cup R \geq 13|H_0) = .190476$. (c) $H_1 : F_x > F_y$, $P(R \geq 14|H_0) = .1$; $H_1 : F_x < F_y$, $P(R \leq 7|H_0) = .1$; $H_1 : F_x \neq F_y$, $P(R \leq 7 \cup R \geq 14|H_0) = .2$.

5. $W = 66$, p-value $= 0.2475$. Accept H_0.

Bibliography

* The books prefixed by an asterisk are textbooks written at a level comparable to the present book and are recommended for collateral reading and as a source of worked examples and additional problems.

Adams, W. J. (2009). *The Life and Times of the Central Limit Theorem*, 2nd Edition, Providence, RI: The American Mathematical Society.

Bayes, T. (1763). An essay towards solving a problem in the doctrine of chances, *Philosophical Transactions of the Royal Society of London*, **53**, 370–418.

Berger, J. O. and Wolpert, R. L. (1984). *The Likelihood Principle*, Hayward, CA: Institute of Mathematical Statistics.

Berkson, J. (1955). Maximum likelihood and minimum χ^2 estimates of the logistic function, *Journal of the American Statistical Association*, **50**, 130–162.

Bickel, P. J. and Doksum, K. (2001). *Mathematical Statistics: Basic Ideas and Selected Topics, Volume I*, 2nd Edition, New Jersey: Prentice Hall.

Bickel, P. J. and Freedman, D. A. (1981). Some asymptotic theory for the bootstrap, *Annals of Statistics*, **9**, 1196–1217.

Boyles, R. A. (1983). On the convergence of the EM algorithm, *Journal of the Royal Statistical Society*, **B**, **45**, 47–50.

Breslow, N. E. and Day, N. E. (1981). *Statistical Methods in Cancer Research, Volume 1, The Analysis of Case Control Studies*, Lyon: International Agency for Research on Cancer.

*Casella, G. and Berger, R. L. (2002). *Statistical Inference*, 2nd Edition, Pacific Grove, CA: Brooks/Cole.

Chernoff, H. and Lehmann E. L. (1954). The use of maximum likelihood estimates in chi square tests for goodness-of-fit, *The Annals of Mathematical Statistics,* **25**, 579–586.

Chung, K. L. (1968). *A Course in Probability Theory*, New York: Harcourt, Brace and World.

Cochran, W. G. (1977). *Sampling Techniques*, New York: John Wiley and Sons.

Conover, W. J. (1999). *Practical Nonparametric Statistics,* 3rd Edition, New York: John Wiley and Sons.

Davison, A. C. and Hinkley, D. V. (1997). *Bootstrap Methods and their Application*, Cambridge: Cambridge University Press.

DeGroot, M. H. (1970). *Optimal Statistical Decisions*, New York: McGraw-Hill.

*DeGroot, M. H. and Schervish, M. J. (2002). *Probability and Statistics*, 3rd Edition, Menlo Park, CA: Addison-Wesley.

Dempster, A. P., Laird, N. M., and Rubin, D. B. (1977). Maximum likelihood from incomplete data via the EM algorithm (with discussion), *Journal of the Royal Statistical Society*, B, **39**, 1–38

Diaconis, P. and Ylvisaker, N. D. (1979). Conjugate priors for exponential families, *Annals of Statistics*, **7**, 269–281.

Efron, B. (1979). Bootstrap methods: Another look at the jackknife, *Annals of Statistics*, **7**, 1–26.

Efron, B. and Gong, G. (1983). A leisurely look at the bootstrap, the jackknife and cross validation, *The American Statistician*, **37**, 36–48.

Efron, B. and Tibshirani, R. J. (1993). *An Introduction to the Bootstrap*, New York: Chapman and Hall/CRC.

Ferguson, T. S. (1967). *Mathematical Statistics: A Decision Theoretic Approach*, Orlando, FL: Academic Press.

Ferguson, T. S. (1996). *A Course in Large Sample Theory*, New York: Chapman and Hall.

Fleiss, J. L. (1981). *Statistical Methods for Rates and Proportions*, New York: John Wiley and Sons.

Galton, F. (1889). *Natural Inheritance*, New York: McMillan.

Gamerman, D. and Lopes, H. F. (2006). *Markov Chain Monte Carlo: Stochastic Simulation for Bayesian Inference*, 2nd Edition, New York: Chapman & Hall/CRC.

Gong, G. and Samaniego, F. J. (1981). Pseudo maximum likelihood estimation: Theory and applications, *Annals of Statistics*, **9**, 861–69.

Hajek, J. and Sidak, Z. (1967). *Theory of Rank Tests*, New York: Academic Press.

Hall, P. (1992). *The Bootstrap and Edgeworth Expansion*, New York, Springer.

Hartigan, J. (1969). Using subsample values as typical values, *Journal of the American Statistical Association*, **64**, 1303–17.

*Hoel, P., Port, S., and Stone, C. (1971). *Introduction to Probability Theory*, Boston: Houghton Mifflin.

*Hoel, P., Port, S., and Stone, C. (1971). *Introduction to Statistical Theory*, Boston: Houghton Mifflin.

*Hogg, R. V., McKean, J., and Craig , A. T. (2012). *Introduction to Mathematical Statistics*, 7th Edition, New Jersey: Pearson College Division.

*Hogg, R. V. and Tanis, E. (2009). *Probability and Statistical Inference*, 8th Edition, New Jersey: Pearson Prentice Hall.

Johnson, N. L., Kotz, S., and Balakrishnan, N. (1995). *Continuous Univariate Distributions, Volumes 1 and 2*, 2nd Editions, New York: John Wiley and Sons.

*Johnson, R. A. (2005). *Miller and Freund's Probability and Statistics for Engineers*, New Jersey: Pearson Prentice Hall.

Kaplan, E. L. and Meier, P. (1958). Nonparametric estimation from incomplete observations, *Journal of the American Statistical Association*, **53**, 457–481.

Kiefer, J. and Wolfowitz, J. (1956). Consistency of the maximum likelihood estimator in the presence of infinitely many incidental parameters, *Annals of Mathematical Statistics*, **27**, 887–906.

Kvam, P. H. and Vidakovic, B. (2007). *Nonparametric Statistics with Applications to Science and Engineering*, New York: John Wiley and Sons.

*Larsen, R. J. and Marx, M. L. (2001). *An Introduction to Mathematical Statistics and its Applications*, New Jersey: Prentice Hall.

Lawless, J. F. (1982). *Statistical Models and Methods for Lifetime Data*, New York: John Wiley and Sons.

Lehmann, E. (1975). *Nonparametrics: Statistical Methods Based on Ranks*, San Francisco: Holden-Day.

Lehmann, E. (1986). *Testing Statistical Hypotheses*, 2nd Edition, New York: John Wiley and Sons.

Lehmann, E. and Casella, G. (1998). *Theory of Point Estimation*, 2nd Edition, New York: Springer.

Mann, H. B. and Whitney, D. R. (1947). On a test of whether one of two random variables is stochastically larger than the other, *Annals of Mathematical Statistics*, **18**, 50–60.

Mosteller, F. and Wallace, D. F. (1964). *Inference and Disputed Authorship: The Federalist*, Reading, MA: Addison-Wesley.

Neter, J. and Wasserman, W. (1974). *Applied Linear Statistical Models*, Homewood, IL: Irwin.

Quenouille, M. (1949). Approximate tests of correlation in time series, *Journal of the Royal Statistical Society*, B, **11**, 18–44.

Rao, C. R. (1973). *Linear Statistical Inference and Its Applications*, 2nd Edition, New York: John Wiley and Sons.

*Rice, J. A. (2007). *Mathematical Statistics and Data Analysis*, Belmont, CA: Thompson Brooks/Cole.

*Ross, S. (2009). *A First Course in Probability*, 8th Edition, New Jersey: Pearson Prentice Hall.

*Roussas, G. G. (1997). *A Course in Mathematical Statistical Inference*, San Diego, CA: Academic Press/Elsevier Science.

*Roussas, G. G. (2003). *Introduction to Probability and Statistical Inference*, San Diego, CA: Academic Press/Elsevier Science.

Samaniego, F. J. (1976). A characterization of convoluted Poisson distributions with applications to estimation, *Journal of the American Statistical Association*, **71**, 475–79.

Samaniego, F. J. (1982). Moment identities for nonnegative variables via integrated survival curves, *IEEE Transactions on Reliability*, **TR-31**, 455–57.

Samaniego, F. J. (1996). On estimating the "commingled rate" in recycling studies: A case study on alternatives to sampling by weight, *Journal of Agricultural, Biological and Environmental Statistics,* **1**, 1–16.

Samaniego, F. J. (2007). *System Signatures and Their Applications in Engineering Reliability*, New York: Springer.

Samaniego, F. J. (2010). *A Comparison of the Bayesian and Frequentist Approaches to Estimation*, New York: Springer.

Samaniego, F. J. and Neath, A. A. (1996). How to be a better Bayesian, *Journal of the American Statistical Association,* **91**, 733–42.

Samaniego, F. J. and Reneau, D. (1994). Towards a reconciliation of the Bayesian and frequentist approaches to estimation, *Journal of the American Statistical Association,* **89**, 947–59.

Samaniego, F. J. and Whitiker, L. R. (1986). On estimating population characteristics from record-breaking observations. I. Parametric results, *Naval Research Logistics Quarterly*, **33**, 531–544.

Scheffé, H. (1953). A method for judging all contrasts in the analysis of variance. *Biometrika*, **40**, 87–104.

Singh, S. (1997). *Fermat's Enigma*, New York: Doubleday.

Stuart, A. and Ord, K. (2010). *Kendall's Advanced Theory of Statistics*, Volume 1, New York: John Wiley and Sons.

Wald, A. (1950). *Statistical Decision Functions*, New York: John Wiley and Sons.

Wald, A. and Wolfowitz, J. (1940). On a test whether two samples are from the same population, *Annals of Mathematical Statistics,* **11**, 147–162.

*Walpole, R.E., Myers, R. H., Myers, S. L., and Ye, K. (2007). *Probability and Statistics for Engineers and Scientists*, 8th Edition, New Jersey: Pearson Prentice Hall.

*Weiss, N. A. (2005). *A Course in Probability*, New York: Pearson Addison Wesley.

Wilcoxon, F. (1945). Individual comparisons by ranking methods, *Biometric Bulletin*, **1**, 80–83.

*Woodroofe, M. (1975). *Probability with Applications*, New York: McGraw-Hill.

Wu, C. F. J. (1983). On the convergence properties of the EM algorithm, *Annals of Statistics*, **11**, 95–103.

Index

For Product Safety Concerns and Information please contact our EU
representative GPSR@taylorandfrancis.com
Taylor & Francis Verlag GmbH, Kaufingerstraße 24, 80331 München, Germany

www.ingramcontent.com/pod-product-compliance
Ingram Content Group UK Ltd.
Pitfield, Milton Keynes, MK11 3LW, UK
UKHW051829180425
457613UK00007B/262
* 9 7 8 1 4 6 6 5 6 0 4 6 8 *